Polymer Chemistry

Second Edition

Polymer Chemistry

Second Edition

Paul C. Hiemenz
Timothy P. Lodge

CRC Press
Taylor & Francis Group
Boca Raton London New York

CRC Press is an imprint of the
Taylor & Francis Group, an informa business

CRC Press
Taylor & Francis Group
6000 Broken Sound Parkway NW, Suite 300
Boca Raton, FL 33487-2742

International Standard Book Number-10: 1-57444-779-3 (Hardcover)
International Standard Book Number-13: 978-1-57444-779-8 (Hardcover)

Library of Congress Cataloging-in-Publication Data

Hiemenz, Paul C., 1936-
    Polymer chemistry / Paul C. Hiemenz and Tim Lodge. -- 2nd ed.
        p. cm.
    Includes bibliographical references and index.
    ISBN-13: 978-1-57444-779-8 (alk. paper)
    ISBN-10: 1-57444-779-3 (alk. paper)
    1. Polymers. 2. Polymerization. I. Lodge, Tim. II. Title.

QD381.H52 2007
547.7--dc22   2006103309

Visit the Taylor & Francis Web site at
http://www.taylorandfrancis.com

and the CRC Press Web site at
http://www.crcpress.com

Preface to the Second Edition

Polymer science is today a vibrant field. Its technological relevance is vast, yet fundamental scientific questions also abound. Polymeric materials exhibit a wealth of fascinating properties, many of which are observable just by manipulating a piece in your hands. Yet, these phenomena are all directly traceable to molecular behavior, and especially to the long chain nature of polymer molecules. The central goal of this book is to develop a molecular level understanding of the properties of polymers, beginning with the underlying chemical structures, and assuming no prior knowledge beyond undergraduate organic and physical chemistry. Although such an understanding should be firmly based in chemistry, polymer science is a highly interdisciplinary endeavor; concepts from physics, biology, materials science, chemical engineering, and statistics are all essential, and are introduced as needed.

The philosophy underlying the approach in this book is the same as that in the first edition, as laid out in the previous preface. Namely, we endeavor to develop the fundamental principles, rather than an encyclopedic knowledge of particular polymers and their applications; we seek to build a molecular understanding of polymer synthesis, characterization, and properties; we emphasize those phenomena (from the vast array of possibilities) that we judge to be the most interesting. The text has been extensively reorganized and expanded, largely to reflect the substantial advances that have occurred over the intervening years. For example, there is now an entire chapter (Chapter 4) dedicated to the topic of controlled polymerization, an area that has recently undergone a revolution. Another chapter (Chapter 11) delves into the viscoelastic properties of polymers, a topic where theoretical advances have brought deeper understanding. The book also serves as a bridge into the research literature. After working through the appropriate chapters, the student should be able to make sense of a large fraction of the articles published today in polymer science journals.

There is more than enough material in this book for a full-year graduate level course, but as with the first edition, the level is (almost) always accessible to senior level undergraduates. After an introductory chapter of broad scope, the bulk of the text may be grouped into three blocks of four chapters each. Chapter 2 through Chapter 5 describe the many ways in which polymers can be synthesized and how the synthetic route influences the resulting molecular structure. This material could serve as the basis for a single quarter or semester chemistry course that focuses on polymer synthesis. Chapter 6 through Chapter 9 emphasize the solution properties of polymers, including their conformations, thermodynamics, hydrodynamics, and light scattering properties. Much of this material is often found in a quarter or semester course introducing the physical chemistry of polymers. Chapter 10 through Chapter 13 address the solid state and bulk properties of polymers: rubber elasticity, viscoelasticity, the glass transition, and crystallization. These topics, while presented here from a physical chemical point of view, could equally well serve as the cornerstone of an introductory course in materials science or chemical engineering.

The style of the presentation, as with the previous edition, is chosen with the student in mind. To this end, we may point out the following features:

- There are over 60 worked example problems sprinkled throughout the book.
- There are 15 or more problems at the end of every chapter, to reinforce and develop further understanding; many of these are based on data from the literature.

- There are almost 200 figures, to illustrate concepts or to present experimental results from the literature.
- Studies chosen for the examples, problems, and figures range in vintage from very recent to over 50 years old; this feature serves to give the reader some sense of the historical progression of the field.
- Concise reviews of many topics (such as thermodynamics, kinetics, probability, and various experimental techniques) are given when the subject is first raised.
- A conscious effort has been made to cross-reference extensively between chapters and sections within chapters, in order to help tie the various topics together.
- Important equations and mathematical relations are almost always developed step by step. We have avoided, wherever possible, the temptation to pull equations out of a hat. Occasionally this leads to rather long stretches of algebra, which the reader is welcome to skip. However, at some point the curious student will want to know where the result comes from, and then this book should be a particularly valuable resource. Surprisingly, perhaps, the level of mathematical sophistication is only about the same as needed in undergraduate chemical thermodynamics. As a further help in this regard, an Appendix reviews many of the important mathematical tools and tricks.

An undertaking such as writing a textbook can never be completed without important contributions from many individuals. Large sections of manuscript were carefully typed by Becky Matsch and Lynne Johnsrud; Lynne also helped greatly with issues of copyright permissions and figure preparation. My colleagues past and present in the Polymer Group at Minnesota have been consistently encouraging and have provided both useful feedback and insightful examples: Frank Bates, Shura Grosberg, Marc Hillmyer, Chris Macosko, Wilmer Miller, David Morse, Steve Prager, and Matt Tirrell. In large measure the style adopted in this second edition has been inspired by the example set by my graduate instructors and mentors at the University of Wisconsin: R. Byron Bird, John Ferry, Arthur Lodge, John Schrag, and Hyuk Yu. In particular, it was in his graduate course Chemistry 664 that Hyuk Yu so ably demonstrated that no important equation need come out of thin air.

I would like extend a special thank you to all of the students enrolled in Chemistry/Chemical Engineering/Materials Science 8211 over the period 2002–2005, who worked through various drafts of Chapter 6 through Chapter 13, and provided many helpful suggestions: Sayeed Abbas, David Ackerman, Sachin Agarwal, Saurabh Agarwal, Julie Alkatout, Pedro Arrechea, Carlos Lopez-Barron, Soumendra Basu, Jeff Becker, Joel Bell, A.S. Bhalla, Michael Bluemle, Paul Boswell, Bryan Boudouris, Adam Buckalew, Xiuyu Cai, Neha Chandra, Joon Chatterjee, Liang Chen, Ying Chen, Juhee Cho, Seongho Choi, Jin-Hwa Chung, Kevin Davis, Michail Dolgovskij, Jingshan Dong, Will Edmonds, Sandra Fritz, Carolyn Gamble, Piotr Grzywacz, Jeong-Myeong Ha, Benjamin Hamilton, Amanda Haws, Nazish Hoda, Hao Hou, Deanna Huehn, Shengxiang Ji, Karan Jindal, Young Kang, Aaron Khieu, Byeong-Su Kim, BongSoo Kim, Hyunwoo Kim, Jin-Hong Kim, Seung Ha Kim, Chunze Lai, Castro Laicer, Qiang Lan, Sangwoo Lee, Zhibo Li, Elizabeth Lugert, Nate Lynd, Sudeep Maheshwari, Huiming Mao, Adam Meuler, Yoichiro Mori, Randy Mrozek, Siddharthya Mujumdar, Jaewook Nam, Dan O'Neal, Sahban Ozair, Matt Panzer, Alhad Phatak, William Philip, Jian Qin, Benjamin Richter, Scott Roberts, Josh Scheffel, Jessica Schommer, Kathleen Schreck, Peter Simone, Zach Thompson, Kristianto Tjiptowidjojo, Mehul Vora, Jaye Warner, Tomy Widya, Maybelle Wu, Jianyan Xu, Dan Yu, Ilan Zeroni, Jianbin Zhang, Ling Zhang, Yu Zhang, Ning Zhou, Zhengxi Zhu, and John Zupancich. Last but not least the love, support, and tolerance of my family, Susanna, Hannah, and Sam, has been a constant source of strength.

Tim Lodge

Preface to the First Edition

Physical chemistry has been defined as that branch of science that is fundamental, molecular, and interesting. I have tried to write a polymer textbook that could be described this way also. To the extent that one subscribes to the former definition and that I have succeeded in the latter objective, then the approach of this book is physical chemical. As a textbook, it is intended for students who have completed courses in physical and organic chemistry. These are the prerequisites that define the level of the book; no special background in physics or mathematics beyond what is required for physical chemistry is assumed. Since chemistry majors generally study physical chemistry in the third year of the undergraduate curriculum, this book can serve as the text for a senior-level undergraduate or a beginning graduate-level course. Although I use chemistry courses and chemistry curricula to describe the level of this book, students majoring in engineering, materials science, physics, and various specialties in the biological sciences will also find numerous topics of interest contained herein.

Terms like "fundamental," "molecular," and "interesting" have different meanings for different people. Let me explain how they apply to the presentation of polymer chemistry in this text.

The words "basic concepts" in the title define what I mean by "fundamental." This is the primary emphasis in this presentation. Practical applications of polymers are cited frequently–after all, it is these applications that make polymers such an important class of chemicals–but in overall content, the stress is on fundamental principles. "Foundational" might be another way to describe this. I have not attempted to cover all aspects of polymer science, but the topics that have been discussed lay the foundation–built on the bedrock of organic and physical chemistry–from which virtually all aspects of the subject are developed. There is an enormous literature in polymer science; this book is intended to bridge the gap between the typical undergraduate background in polymers–which frequently amounts to little more than occasional "relevant" examples in other courses–and the professional literature on the subject. Accordingly, the book assumes essentially no prior knowledge of polymers, and extends far enough to provide a usable level of understanding.

"Molecular" describes the perspective of the chemist, and it is this aspect of polymeric materials that I try to keep in view throughout the book. An engineering text might emphasize processing behavior; a physics text, continuum mechanics; a biochemistry text, physiological function. All of these are perfectly valid points of view, but they are not the approach of this book. It is polymer molecules–their structure, energetics, dynamics, and reactions–that are the primary emphasis throughout most of the book. Statistics is the type of mathematics that is natural to a discussion of molecules. Students are familiar with the statistical nature of, say, the kinetic molecular theory of gases. Similar methods are applied to other assemblies of molecules, or in the case of polymers, to the assembly of repeat units that comprise a single polymer molecule. Although we frequently use statistical arguments, these are developed quite thoroughly and do not assume any more background in this subject than is ordinarily found among students in a physical chemistry course.

The most subjective of the words which (I hope) describe this book is "interesting." The fascinating behavior of polymers themselves, the clever experiments of laboratory researchers, and the elegant work of the theoreticians add up to an interesting total. I have tried to tell about these topics with clarity and enthusiasm, and in such a way as to make them intelligible to students. I can only hope that the reader agrees with my assessment of what is interesting.

This book was written with the student in mind. Even though "student" encompasses persons with a wide range of backgrounds, interests, and objectives; these are different than the corresponding experiences and needs of researchers. The following features have been included to assist the student:

1. Over 50 solved example problems are sprinkled throughout the book.
2. Exercises are included at the end of each chapter which are based on data from the original literature.
3. Concise reviews of pertinent aspects of thermodynamics, kinetics, spectrophotometry, etc. are presented prior to developing applications of these topics to polymers.
4. Theoretical models and mathematical derivations are developed in enough detail to be comprehensible to the student reader. Only rarely do I "pull results out of a hat," and I scrupulously avoid saying "it is obvious that ..."
5. Generous cross-referencing and a judicious amount of repetition have been included to help unify a book which spans quite a wide range of topics.
6. SI units have been used fairly consistently throughout, and attention is paid to the matter of units whenever these become more than routine in complexity.

The book is divided into three parts of three chapters each, after an introductory chapter which contains information that is used throughout the book.

In principle, the three parts can be taken up in any order without too much interruption in continuity. Within each of the parts there is more carryover from chapter to chapter, so rearranging the sequence of topics within a given part is less convenient. The book contains more material than can be covered in an ordinary course. Chapter 1 plus two of the three parts contain about the right amount of material for one term. In classroom testing the material, I allowed the class to decide–while we worked on Chapter 1–which two of the other parts they wished to cover; this worked very well.

Material from Chapter 1 is cited throughout the book, particularly the discussion of statistics. In this connection, it might be noted that statistical arguments are developed in less detail further along in the book as written. This is one of the drawbacks of rearranging the order in which the topics are covered. Chapters 2 through 4 are concerned with the mechanical properties of bulk polymers, properties which are primarily responsible for the great practical importance of polymers. Engineering students are likely to have both a larger interest and a greater familiarity with these topics. Chapers 5 through 7 are concerned with the preparation and properties of several broad classes of polymers. These topics are closer to the interests of chemistry majors. Chapters 8 through 10 deal with the solution properties of polymers. Since many of the techniques described have been applied to biopolymers, these chapters will have more appeal to students of biochemistry and molecular biology.

Let me conclude by acknowledging the contributions of those who helped me with the preparation of this book. I wish to thank Marilyn Steinle for expertly typing the manuscript. My appreciation also goes to Carol Truett who skillfully transformed my (very) rough sketches into effective illustrations. Lastly, my thanks to Ron Manwill for preparing the index and helping me with the proofreading. Finally, let me acknowledge that some errors and/or obscurities will surely elude my efforts to eliminate them. I would appreciate reports about these from readers so that these mistakes can eventually be eliminated.

Paul C. Hiemenz

Contents

1

Introduction to Chain Molecules

1.1 Introduction

"I am inclined to think that the development of polymerization is perhaps the biggest thing chemistry has done, where it has had the biggest impact on everyday life" [1]. This assessment of the significance of polymer chemistry to modern society was offered 25 years ago by Lord Todd (President of the Royal Society and 1957 Nobel Laureate in Chemistry), and subsequent developments have only reinforced this sentiment. There is hardly an area of modern life in which polymer materials do not play an important role. Applications span the range from the mundane (e.g., packaging, toys, fabrics, diapers, nonstick cookware, pressure-sensitive adhesives, etc.) to demanding specialty uses (e.g., bulletproof vests, stealth aircraft, artificial hip joints, resorbable sutures, etc.). In many instances polymers are the main ingredients, and the ingredients whose characteristic properties are essential to the success of a particular technology: rubber tires, foam cushions and insulation, high-performance athletic shoes, clothing, and equipment are good examples. In other cases, polymers are used as additives at the level of a few percent by volume, but which nevertheless play a crucial role in the properties of the final material; illustrations of this can be found in asphalt (to suppress brittle fracture at low temperature and flow at high temperature), shampoo and other cosmetics (to impart "body"), automobile windshields (to prevent shattering), and motor oil (to reduce the dependence of viscosity on temperature, and to suppress crystallization).

For those polymer scientists "of a certain age," the 1967 movie "The Graduate" [2] provided an indelible moment that still resonates today. At his college graduation party, the hero Benjamin Braddock (played by Dustin Hoffman) is offered the following advice by Mr. McGuire (played by Walter Brooke):

> MR. MCGUIRE. I want to say one word to you. Just one word.
> BENJAMIN. Yes, sir.
> MR. MCGUIRE. Are you listening?
> BENJAMIN. Yes I am.
> MR. MCGUIRE. Plastics.

In that period, the term "plastic" was often accompanied by negative connotations, including "artificial," as opposed to "natural," and "cheap," as opposed to "valuable." Today, in what we might call the "post-graduate era," the situation has changed. To the extent that the advice offered to Benjamin was pointing him to a career in a particular segment of the chemical industry, it was probably very sound advice. The volume of polymer materials produced annually has grown rapidly over the intervening years, to the point where today several hundred pounds of polymer materials are produced each year for each person in the United States. More interesting than sheer volume, however, is the breadth of applications for polymers. Not only do they continue to encroach into the domains of "classical" materials such as metal, wood, and glass (note the inexorable transformation of polymers from minor to major components in automobiles), but they also play a central role in many emerging technologies. Examples include "plastic electronics," gene therapy, artificial prostheses, optical data storage, electric cars, and fuel cells. In short, a reasonable appreciation of the properties of chain molecules, and how these result in the many desirable attributes of polymer-containing materials, is a necessity for a well-trained chemist, materials scientist, or chemical engineer today.

Science tends to be plagued by clichés, which make invidious comparison of its efforts; "they can cure such and such a dreaded disease, but they cannot do anything about the common cold" or "we know more about the surface of the moon than the bottom of the sea." If such comparisons were popular in the 1920s, the saying might have been, "we know more about the structure of the atom than about those messy, sticky substances called polymers." Indeed, Millikan's determination of the charge of an electron, Rutherford's idea of the nuclear atom, and Bohr's model of the hydrogen atom were all well-known concepts before the notion of truly covalent macromolecules was accepted. This was the case in spite of the great importance of polymers to human life and activities. Our bodies, like all forms of life, depend on polymer molecules: carbohydrates, proteins, nucleic acids, and so on. From the earliest times, polymeric materials have been employed to satisfy human needs: wood and paper; hides; natural resins and gums; fibers such as cotton, wool, and silk.

Attempts to characterize polymeric substances had been made, of course, and high molecular weights were indicated, even if they were not too accurate. Early workers tended to be more suspicious of the interpretation of the colligative properties of polymeric solutions than to accept the possibility of high molecular weight compounds. Faraday had already arrived at C_5H_8 as the empirical formula of "rubber" in 1826, and isoprene was identified as the product resulting from the destructive distillation of rubber in 1860. The idea that a natural polymer such as rubber somehow "contained" isoprene emerged, but the nature of its involvement was more elusive.

During the early years of the 20th century, organic chemists were enjoying success in determining the structures of ordinary-sized organic molecules, and this probably contributed to their reluctance to look beyond structures of convenient size. Physical chemists were interested in intermolecular forces during this period, and the idea that polymers were the result of some sort of association between low molecular weight constituent molecules prevailed for a long while. Staudinger is generally credited as being the father of modern polymer chemistry, although a foreshadowing of his ideas can be traced through older literature. In 1920, Staudinger proposed the chain formulas we accept today, maintaining that structures are held together by covalent bonds, which are equivalent in every way to those in low molecular weight compounds. There was a decade of controversy before this "macromolecular hypothesis" began to experience widespread acceptance. Staudinger was awarded the Nobel Prize in 1953 for his work with polymers. By the 1930s, Carothers began synthesizing polymers using well-established reactions of organic chemistry such as esterification and amidation. His products were not limited to single ester or amide linkages, however, but contained many such groups: they were polyesters and polyamides. Physical chemists also got in on the act. Kuhn, Guth, Mark, and others were soon applying statistics and crystallography to describe the multitude of forms a long-chain molecule could assume [3].

Our purpose in this introduction is not to trace the history of polymer chemistry beyond the sketchy version above; interesting and extensive treatments are available [4,5]. Rather, the primary objective is to introduce the concept of chain molecules, which stands as the cornerstone of all polymer chemistry. In the next few sections we shall explore some of the categories of polymers, some of the reactions that produce them, and some aspects of isomerism which multiply the structural possibilities. A common feature of all synthetic polymerization reactions is the statistical nature of the individual polymerization steps. This leads inevitably to a distribution of molecular weights, which we would like to describe. As a consequence of these considerations, another important part of this chapter is an introduction to some of the statistical concepts that also play a central role in polymer chemistry.

1.2 How Big Is Big?

The term *polymer* is derived from the Greek words *poly* and *meros*, meaning many parts. We noted in the Section 1.1 that the existence of these parts was acknowledged before the nature of the interaction which held them together was known. Today we realize that ordinary covalent bonds are the intramolecular forces that keep the polymer molecule intact. In addition, the usual types of intermolecular forces—hydrogen bonds, dipole–dipole interactions, London forces, etc.—hold assemblies of these molecules together in the bulk state. The only thing that is remarkable about these molecules is their size, but that feature is remarkable indeed. Another useful term is *macromolecule*, which of course simply means "large (or long) molecule." Some practitioners draw a distinction between the two: all polymers are macromolecules, but not all macromolecules are polymers. For example, a protein is not made by repeating one or two chemical units many times, but involves a precise selection from among 20 different amino acids; thus it is a macromolecule, but not a polymer. In this text we will not be sticklers for formality, and will use the terms rather interchangeably, but the reader should be aware of the distinction.

1.2.1 Molecular Weight

One of the first things we must consider is what we mean when we talk about the "size" of a polymer molecule. There are two possibilities: one has to do with the number of repeat units and the other to the spatial extent. In the former case, the standard term is *molecular weight* (although again the reader must be aware that *molar mass* is often preferred). A closely related concept, the *degree of polymerization* is also commonly used in this context. A variety of experimental techniques are available for determining the molecular weight of a polymer. We shall discuss a few such methods in Section 1.8 and postpone others until the appropriate chapters. The expression molecular weight and molar mass should always be modified by the word *average*. This too is something we shall take up presently. For now, we assume that a polymer molecule has a molecular weight M, which can be anywhere in the range 10^3–10^7 or more. (We shall omit units when we write molecular weights in this book, but the student is advised to attach the units g/mol to these quantities when they appear in problem calculations.)

Since polymer molecules are made up of chains of repeat units, after the chain itself comes the repeat unit as a structural element of importance. Many polymer molecules are produced by covalently bonding together only one or two types of repeat units. These units are the parts from which chains are generated; as a class of compounds they are called monomers. Throughout this book, we shall designate the molecular weight of a repeat unit as M_0.

The degree of polymerization of a polymer is simply the number of repeat units in a molecule. The degree of polymerization N is given by the ratio of the molecular weight of the polymer to the molecular weight of the repeat unit:

$$N = \frac{M}{M_0} \tag{1.2.1}$$

One type of polymerization reaction is the addition reaction in which successive repeat units add on to the chain. No other product molecules are formed, so the molecular weight of the monomer and that of the repeat unit are identical in this case. A second category of polymerization reaction is the condensation reaction, in which one or two small molecules such as water or HCl are eliminated for each chain linkage formed. In this case the molecular weight of the monomer and the repeat unit are somewhat different. For example, suppose an acid (subscript A) reacts with an alcohol (subscript B) to produce an ester linkage and a water molecule. The molecular weight of the ester—the repeat unit if an entire chain is built up this way—differs from the combined weight of the reactants by twice the molecular weight of the water; therefore,

$$N = \frac{M}{M_0} = \frac{M}{M_A + M_B - 2M_{H_2O}} \tag{1.2.2}$$

The end units in a polymer chain are inevitably different from the units that are attached on both sides to other repeat units. We see this situation in the n-alkanes: each end of the chain is a methyl group and the middle parts are methylene groups. Of course, the terminal group does not have to be a hydrogen as in alkanes; indeed, it is often something else. Our interest in end groups is concerned with the question of what effect they introduce into the evaluation of N through Equation (1.2.2). The following example examines this through some numerical calculations.

Example 1.1

As a polymer prototype consider an n-alkane molecule consisting of $N-2$ methylenes and 2 methyl groups. How serious an error is made in M for different Ns if the difference in molecular weight between methyl and methylene groups is ignored?

Solution

The effect of different end groups on M can be seen by comparing the true molecular weight with an approximate molecular weight, calculated on the basis of a formula $(CH_2)_N$. These Ms and the percentage difference between them are listed here for several values of N

N	M	$M_{approx.}$	% Difference
3	44	42	4.5
7	100	98	2.0
12	170	168	1.2
52	730	728	0.3
102	1,430	1,428	0.14
502	7,030	7,028	0.028
1002	14,030	14,028	0.014

Although the difference is almost 5% for propane, it is closer to 0.1% for the case of $N \approx 100$, which is about the threshold for polymers. The precise values of these numbers will be different, depending on the specific repeat units and end groups present. For example, if $M_0 = 100$ and $M_{end} = 80$, the difference would be 0.39% in a calculation such as that above for $N \approx 100$.

The example shows that the contribution of the ends becomes progressively less important as the number of repeat units in a structure increases. By the time polymeric molecular sizes are reached, the error associated with failure to distinguish between segments at the end and those within the chain is generally less than experimental error. In Section 1.8.2 we shall consider a method for polymer molecular weight determination based on chemical analysis for the end groups in a polymer. A corollary of the present discussion is that the method of end group analysis is applicable only in the case of relatively low molecular weight polymers.

As suggested above, not all polymers are constructed by bonding together a single kind of repeat unit. For example, although protein molecules are polyamides in which N amino acid repeat units are bonded together, the degree of polymerization is a less useful concept, since an amino acid unit might be any one of the 20-odd molecules that are found in proteins. In this case the molecular weight itself, rather than the degree of polymerization, is generally used to describe the molecule. When the actual content of individual amino acids is known, it is their sequence that is of special interest to biochemists and molecular biologists.

1.2.2 Spatial Extent

We began this section with an inquiry into how to define the size of a polymer molecule. In addition to the molecular weight or the degree of polymerization, some linear dimension that characterizes the molecule could also be used for this purpose. As an example, consider a hydrocarbon molecule stretched out to its full length but without any bond distortion. There are several features to note about this situation:

1. The tetrahedral geometry at the carbon atoms gives bond angles of $109.5°$.
2. The equilibrium bond length of a carbon–carbon single bond is 0.154 nm or 1.54 Å.
3. Because of the possibility of rotation around carbon–carbon bonds, a molecule possessing many such bonds will undergo many twists and turns along the chain.
4. Fully extended molecular length is not representative of the spatial extension that a molecule actually displays. The latter is sensitive to environmental factors, however, so the extended length is convenient for our present purposes to provide an idea of the spatial size of polymer molecules.

A fully extended hydrocarbon molecule will have the familiar all-*trans* zigzag profile in which the hydrogens extend in front of and in back of the plane containing the carbons, with an angle of $109.5°$ between successive carbon–carbon bonds. The chain may be pictured as a row of triangles resting corner to corner. The length of the row equals the product of the number of triangles and the length of the base of each. Although it takes three carbons to define one of these triangles, one of these atoms is common to two triangles; therefore the number of triangles is the same as the number of pairs of carbon atoms, except where this breaks down at the ends of the molecule. If the chain is sufficiently long, this end effect is inconsequential. The law of cosines can be used to calculate the length of the base of each of these triangles: $[2(0.154)^2(1 - \cos 109.5°)]^{1/2} = 0.252$ nm. If the repeat unit of the molecule contributes two carbon atoms to the backbone of the polymer—as is the case for vinyl polymers—the fully extended chain length is given by $N(0.252)$ nm. For a polymer with $N = 10^4$, this corresponds to 2.52 μm. Objects which actually display linear dimensions of this magnitude can be seen in an ordinary microscope, provided that they have suitable optical properties to contrast with their surroundings; an example will be given in Figure 1.1a. Note that the distance between every other carbon atom we have used here is also the distance between the substituents on these carbons for the fully extended chains.

We shall see in Chapter 6 that, because of all the twists and turns a molecule undergoes, the actual average end-to-end distance of the jumbled molecules increases as $N^{1/2}$. With the same repeat distance calculated above, but the square root dependence on N, the actual end-to-end distance of the coiled chain with $N = 10^4$ is closer to $(10^4)^{1/2} \times 0.252$ nm ≈ 25 nm. If we picture one end of this jumbled chain at the origin of a coordinate system, the other end might be anywhere on the surface of a sphere whose radius is given by this end-to-end distance. This spherical geometry comes about because the random bends occurring along the chain length can take the end of the chain anywhere in a spherical domain whose radius depends on $N^{1/2}$.

The above discussion points out the difficulty associated with using the linear dimensions of a molecule as a measure of its size: it is not the molecule alone that determines its dimensions, but also the shape or *conformation* in which it exists. Fully extended, linear arrangements of the sort described above exist in polymer crystals, at least for some distance, although usually not over the full length of the chain. We shall take up the structure of polymer crystals in Chapter 13. In the solution and bulk states, many polymers exist in the coiled form we have also described. Still other structures are important, notably the rod or semiflexible chain, which we shall discuss in Chapter 6. The overall shape assumed by a polymer molecule can be greatly affected by the environment. The shape of a molecule in solution plays a key role in determining many properties of polymer solutions. From a study of these solutions, some conclusions can be drawn regarding the shape of

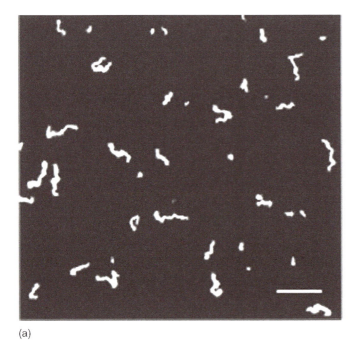

(a)

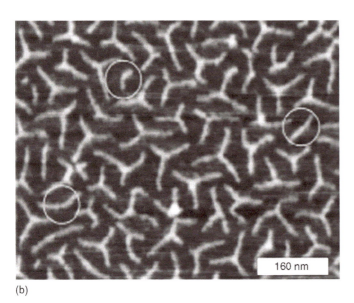

(b)

Figure 1.1   (a) Individual molecules of DNA of various sizes, spread on a fluid positively charged surface, imaged by fluorescence. The scale bar is 10 μm. (Reproduced from Maier, B. and Rädler, J.O. *Macromolecules* 33, 7185, 2000. With permission.) (b) Atomic force microscopy images of three-arm star polymers, where each arm is a heavily branched comb. The circles indicate linear molecules. (Reproduced from Matyjaszewski, K., Qin, S., Boyce, J.R., Shirvanyants, D., and Sheiko, S.S. *Macromolecules* 36, 1843, 2003. With permission.)

the molecule in the environment. Relevant aspects of polymer solutions are taken up in Chapter 6 through Chapter 9.

Figure 1.1a and Figure 1.1b are rather striking images of individual polymer molecules. Figure 1.1a shows single molecules of DNA that have been heavily labeled with fluorescent dyes; the dyes intercalate between the base pairs along the chain, without seriously altering the conformation of the molecule. Under illumination the resulting fluorescence provides a good representation of the molecules themselves. In this particular image, the DNA molecules are spread out in two dimensions, on a cationically charged imitation lipid membrane. DNA, as it turns out, is an excellent example of a semiflexible chain, which can actually be inferred from these images; the molecules are not straight rods, but neither are they heavily coiled around themselves. The scale bar corresponds to 10 μm, indicating that these molecules are of very high molecular weight indeed. In Figure 1.1b, the image is of a star-shaped polymer, but one in which each arm of the star is a heavily branched comb or "bottlebrush." The molecule is thus akin to a kind of starfish, with very hairy arms. This picture was obtained by atomic force microscopy (AFM), one of a series of surface-sensitive analysis techniques with exquisite spatial resolution. The molecules themselves were deposited from a Langmuir–Blodgett trough onto a mica substrate. Both situations depicted in Figure 1.1a and Figure 1.1b raise the question of the relationship between the conformation observed on the surface and that at equilibrium in solution. In Chapter 6 through Chapter 9 we will encounter several ways in which the solution conformation can be determined reliably, which can serve to confirm the impression derived from figures such as these.

We conclude this section by questioning whether there is a minimum molecular weight or linear dimension that must be met for a molecule to qualify as a polymer. Although a dimer is a molecule for which $N = 2$, no one would consider it a polymer. The term *oligomer* has been coined to designate molecules for which $N < 10$. If they require a special name, apparently the latter are not full-fledged polymers either. At least as a first approximation, we shall take the attitude that there is ordinarily no discontinuity in behavior with respect to observed properties as we progress through a homologous series of compounds with different N values. At one end of the series, we may be dealing with a simple low molecular weight compound, and at the other end with a material that is unquestionably polymeric. The molecular weight and chain length increase monotonically through this series, and a variety of other properties vary smoothly also. This point of view emphasizes continuity with familiar facts concerning the properties of low molecular weight compounds. There are some properties, on the other hand, which follow so closely from the chain structure of polymers that the property is simply not observed until a certain critical molecular size has been reached. This critical size is often designated by a threshold molecular weight. The elastic behavior of rubber and several other mechanical properties fall into this latter category. In theoretical developments, large values of N are often assumed to justify neglecting end effects, using certain statistical methods and other mathematical approximations. With these ideas in mind, $M = 1000$ seems to be a convenient round number for designating a compound to be a polymer, although it should be clear that this cutoff is arbitrary (and on the low side).

1.3 Linear and Branched Polymers, Homopolymers, and Copolymers

1.3.1 Branched Structures

Most of the preceding section was based on the implicit assumption that polymer chains are linear (with the striking exception of Figure 1.1b). In evaluating both the degree of polymerization and the extended chain length, we assumed that the chain had only two ends. While linear polymers are important, they are not the only type of molecules possible: branched and cross-linked molecules are also common. When we speak of a branched polymer, we refer to the presence of additional polymeric chains issuing from the backbone of a linear molecule. (Small substituent groups such as methyl or phenyl groups on the repeat units are generally not considered branches, or, if they are, they should be specified as "short-chain branches.") Branching can arise through several routes. One is to introduce into the polymerization reaction some monomer with the capability of serving as a branch. Consider the formation of a polyester.

The presence of difunctional acids and difunctional alcohols allows the polymer chain to grow. These difunctional molecules are incorporated into the chain with ester linkages at both ends of each. Trifunctional acids or alcohols, on the other hand, produce a linear molecule by the reaction of two of their functional groups. If the third reacts and the resulting chain continues to grow, a branch has been introduced into the original chain. A second route is through adventitious branching, for example, as a result of an atom being abstracted from the original linear molecule, with chain growth occurring from the resulting active site. This is quite a common occurrence in the free-radical polymerization of ethylene, for example. A third route is *grafting*, whereby pre-formed but still reactive polymer chains can be added to sites along an existing backbone (so-called "grafting to"), or where multiple initiation sites along a chain can be exposed to monomer (so-called "grafting from").

The amount of branching introduced into a polymer is an additional variable that must be specified for the molecule to be fully characterized. When only a slight degree of branching is present, the concentration of junction points is sufficiently low that these may be simply related to the number of chain ends. For example, two separate linear molecules have a total of four ends. If the end of one of these linear molecules attaches itself to the middle of the other to form a *T*, the resulting molecule has three ends. It is easy to generalize this result. If a molecule has ν branches, it has $\nu + 2$ chain ends if the branching is relatively low. Two limiting cases to consider, illustrated in Figure 1.2, are *combs* and *stars*. In the former, a series of relatively uniform branches emanate from along the length of a common backbone; in the latter, all branches radiate from a central junction. Figure 1.1b gave an example of both of these features.

If the concentration of junction points is high enough, even branches will contain branches. Eventually a point can be reached at which the amount of branching is so extensive that the polymer molecule becomes a giant three-dimensional network. When this condition is achieved, the molecule is said to be cross-linked. In this case, an entire macroscopic object may be considered to consist of essentially one molecule. The forces that give cohesiveness to such a body are covalent bonds, not intermolecular forces. Accordingly, the mechanical behavior of cross-linked bodies is much different from those without cross-linking. This will be discussed at length in Chapter 10. However, it is also possible to suppress cross-linking such that the highly branched molecules remain as discrete entities, known as *hyperbranched polymers* (see Figure 1.2). Another important class of highly branched polymers illustrated in Figure 1.2 are *dendrimers*, or treelike molecules. These are completely regular structures, with well-defined molecular weights, that are made by the successive condensation of branched monomers. For example, begin with a

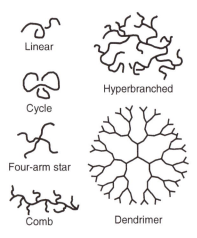

Figure 1.2   Illustration of various polymer architectures.

trifunctional monomer "B_3," or "generation 0." This is reacted with an excess of AB_2 monomers, leading to a generation 1 dendrimer with 6 B groups. A second reaction with AB_2 leads to generation 2 with 12 pendant B groups. Eventually, perhaps at generation 6 or 7, the surface of the molecule becomes so congested that addition of further complete generations is impossible. Note that the "B" part of the AB_2 monomer needs to be protected in some way so that only one generation can be added at one time.

A final class of nonlinear polymers to consider are *cycles* or *rings*, whereby the two ends of the molecule react to close the loop. Such polymers are currently more of academic interest than commercial importance, as they are tricky to prepare, but they can shed light on various aspects of polymer behavior. Interestingly, nature makes use of this architecture; the DNA of the Lambda bacteriophage reversibly cyclizes and uncyclizes during gene expression.

1.3.2 Copolymers

Just as it is not necessary for polymer chains to be linear, it is also not necessary for all repeat units to be the same. We have already mentioned proteins, where a wide variety of different repeat units are present. Among synthetic polymers, those with a single kind of repeat unit are called *homopolymers*, and those containing more than one kind of repeat unit are *copolymers*. Note that these definitions are based on the repeat unit, not the monomer. An ordinary polyester is not really a copolymer, even though two different monomers, acids and alcohols, are its monomers. By contrast, copolymers result when different monomers bond together in the same way to produce a chain in which each kind of monomer retains its respective substituents in the polymer molecule. The unmodified term copolymer is generally used to designate the case where two different repeat units are involved. Where three kinds of repeat units are present, the system is called a *terpolymer*; where there are more than three, the system is called a multicomponent copolymer. The copolymers we discuss in this book will be primarily two-component molecules. We shall explore aspects of the synthesis and characterization of copolymers in both Chapter 4 and Chapter 5.

The moment we admit the possibility of having more than one kind of repeat unit, we require additional variables to describe the polymer. First, we must know how many kinds of repeat units are present and what they are. To describe the copolymer quantitatively, the relative amounts of the different kinds of repeat units must be specified. Thus the empirical formula of a copolymer may be written $A_x B_y$, where A and B signify the individual repeat units and x and y indicate the relative number of each. From a knowledge of the molecular weight of the polymer, the molecular weights of A and B, and the values of x and y, it is possible to calculate the number of each kind of monomer unit in the copolymer. The sum of these values gives the degree of polymerization of the copolymer. The following example illustrates some of the ways of describing a copolymer.

Example 1.2

A terpolymer is prepared from vinyl monomers A, B, and C; the molecular weights of the repeat units are 104, 184, and 128, respectively. A particular polymerization procedure yields a product with the empirical formula $A_{3.55} B_{2.20} C_{1.00}$. The authors of this research state that the terpolymer has "an average unit weight of 134" and "the average molecular weight per angstrom of 53.5." Verify these values.[†]

Solution

The empirical formula gives the relative amounts of A, B, and C in the terpolymer. The total molecular weight of this empirical formula unit is given by adding the molecular weight contributions of A, B, and C: $3.44(104) + 2.20(184) + 1.00(128) = 902$ g/mol per empirical formula unit.

[†] A. Ravve and J.T. Khamis, *Addition and Condensation Polymerization Processes*, Advances in Chemistry Series, Vol. 91, American Chemical Society Publications, Washington, DC, 1969.

The total amount of chain repeat units possessing this total weight is $3.55 + 2.20 + 1.00 = 6.75$ repeat units per empirical formula unit. The ratio of the total molecular weight to the total number of repeat units gives the average molecular weight per repeat unit:

$$\frac{902}{6.75} = 134 \text{ g/mol per repeat unit}$$

Since the monomers are specified to be vinyl monomers, each contributes two carbon atoms to the polymer backbone, with the associated extended length of 0.252 nm per repeat unit. Therefore, the total extended length of the empirical formula unit is

$$6.75(0.252 \text{ nm}) = 1.79 \text{ nm} = 17.0 \text{ Å}$$

The ratio of the total weight to the total extended length of the empirical formula unit gives the average molecular weight per length of chain:

$$\frac{902}{17} = 53 \text{ g/mol per Å}$$

Note that the average weight per repeat unit could be used to evaluate the overall degree of polymerization of this terpolymer. For example, if the molecular weight were 43,000, the corresponding degree of polymerization would be

$$\frac{43,000}{134} = 321 \text{ repeat units per molecule}$$

With copolymers, it is far from sufficient merely to describe the empirical formula to characterize the molecule. Another question that must be asked concerns the location of the different kinds of repeat units within the molecule. Starting from monomers A and B, the following distribution patterns can be obtained in linear polymers:

1. *Random* (or *statistical*). The A–B sequence is governed strictly by chance, subject only to the relative abundances of repeat units. For equal proportions of A and B, we might have structures like

   –AAABABAABBABBB–

   Such a polymer could be called poly(A-*stat*-B) or poly(A-*ran*-B).
2. *Alternating*. A regular pattern of alternating repeat units in poly(A-*alt*-B):

   –ABABABABABAB–
3. *Block*. Long, uninterrupted sequence of each monomer is the pattern:

   –AAAAAAAAAAAAAABBBBBBBBBBBBBBBAAAAAAAAAA–

The above structure has three blocks, and is called poly(A-*block*-B-*block*-A), or an ABA triblock copolymer. If a copolymer is branched with different repeat units occurring in the branches and the backbone, we can have the following:

4. *Graft*. This segregation is often accomplished by first homopolymerizing the backbone. This is dissolved in the second monomer, with sites along the original chain becoming the origin of the comonomer side-chain growth:

                         BBBBBBBBBB–
                         |
   –AAAAAAAAAAAAAAAAAA–
                | |
   –BBBBBBBBBB   BBBBBBB–

In a cross-linked polymer, the junction units are different kinds of monomers than the chain repeat units, so these molecules might be considered to be still another comonomer. While the chemical reactions that yield such cross-linked substances are technically copolymerizations, the products are described as cross-linked rather than as copolymers. In this instance, the behavior due to cross-linking takes precedence over the presence of an additional type of monomer in the structure.

It is apparent from items 1–3 above that linear copolymers—even those with the same proportions of different kinds of repeat units—can be very different in structure and properties. In classifying a copolymer as random, alternating, or block, it should be realized that we are describing the average character of the molecule; accidental variations from the basic patterns may be present. Furthermore, in some circumstances, nominally "random" copolymers can have substantial sequences of one monomer or the other. In Chapter 5, we shall see how an experimental investigation of the sequence of repeat units in a copolymer is a valuable tool for understanding copolymerization reactions.

1.4 Addition, Condensation, and Natural Polymers

In the last section, we examined some of the categories into which polymers can be classified. Various aspects of molecular structure were used as the basis for classification in that section. Next we shall consider the chemical reactions that produce the molecules as a basis for classification. The objective of this discussion is simply to provide some orientation and to introduce some typical polymers. For this purpose, many polymers may be classified as being either addition or condensation polymers; both of these classes are discussed in detail in Chapter 2 and Chapter 3, respectively. Even though these categories are based on the reactions which produce the polymers, it should not be inferred that only two types of polymerization reactions exist. We have to start somewhere, and these two important categories are the usual places to begin.

1.4.1 Addition and Condensation Polymers

These two categories of polymers can be developed along several lines. For example, in addition-type polymers the following statements apply:

1. The repeat unit in the polymer and the monomer has the same composition, although, of course, the bonding is different in each.
2. The mechanism of these reactions places addition polymerizations in the kinetic category of chain reactions, with either free radicals or ionic groups responsible for propagating the chain reaction.
3. The product molecules often have an all-carbon chain backbone, with pendant substituent groups.

In contrast, for condensation polymers:

4. The polymer repeat unit arises from reacting together two different functional groups, which usually originate on different monomers. In this case, the repeat unit is different from either of the monomers. In addition, small molecules are often eliminated during the condensation reaction. Note the words *usual* and *often* in the previous statements; exceptions to both statements are easily found.
5. The mechanistic aspect of these reactions can be summarized by saying that the reactions occur in steps. Thus, the formation of an ester linkage between two small molecules is not essentially different from that between a polyester and a monomer.
6. The product molecules have the functional groups formed by the condensation reactions interspersed regularly along the backbone of the polymer molecule:

–C–C–Y–C–C–Y–

Next let us consider a few specific examples of these classes of polymers. The addition polymerization of a vinyl monomer $CH_2 = CHX$ involves three distinctly different steps. First, the reactive center must be initiated by a suitable reaction to produce a free radical, anionic, or cationic reaction site. Next, this reactive entity adds consecutive monomer units to propagate the polymer chain. Finally, the active site is capped off, terminating the polymer formation. If one assumes that the polymer produced is truly a high molecular weight substance, the lack of uniformity at the two ends of the chain—arising in one case from the initiation and in the other from the termination—can be neglected. Accordingly, the overall reaction can be written

(1.A)

Again, we emphasize that end effects are ignored in writing Reaction (1.A). These effects as well as the conditions of the reaction and other pertinent information will be discussed when these reactions are considered in Chapter 3 and Chapter 4. Table 1.1 lists several important addition polymers, showing each monomer and polymer structure in the manner of Reaction (1.A). Also included in Table 1.1 are the molecular weights of the repeat units and the common names of the polymers. The former will prove helpful in many of the problems in this book; the latter will be discussed in the next section. Poly(ethylene oxide) and poly(ε-caprolactam) have been included in this list as examples of the hazards associated with classification schemes. They resemble addition polymers because the molecular weight of the repeat unit and that of the monomer are the same; they resemble condensation polymers because of the heteroatom chain backbone. The reaction mechanism, which might serve as arbiter in this case, can be either of the chain or the step type, depending on the reaction conditions. These last reactions are examples of *ring-opening polymerizations*, yet another possible category of classification.

The requirements for formation of condensation polymers are twofold: the monomers must possess functional groups capable of reacting to form the linkage, and they ordinarily require more than one reactive group to generate a chain structure. The functional groups can be distributed such that two difunctional monomers with different functional groups react or a single monomer reacts, which is difunctional with one group of each kind. In the latter case especially, but also with condensation polymerization in general, the tendency to form cyclic products from intramolecular reactions may compete with the formation of polymers. Condensation polymerizations are especially sensitive to impurities. The presence of monofunctional reagents introduces the possibility of a reaction product forming which would not be capable of further growth. If the functionality is greater than 2, on the other hand, branching becomes possible. Both of these modifications dramatically alter the product compared to a high molecular weight linear product. When reagents of functionality less than or greater than 2 are added in carefully measured and controlled amounts, the size and geometry of product molecules can be manipulated. When such reactants enter as impurities, the undesired results can be disastrous. Marvel has remarked that more money has been wasted in polymer research by the use of impure monomers than in any other manner [6].

Table 1.2 lists several examples of condensation reactions and products. Since the reacting monomers can contain different numbers of carbon atoms between functional groups, there are quite a lot of variations possible among these basic reaction types. The inclusion of poly(dimethylsiloxane) in Table 1.2 serves as a reminder that polymers need not be organic compounds. The physical properties of inorganic polymers follow from the chain structure of these molecules, and the concepts developed in this volume apply to them and to organic polymers equally well. We shall not examine explicitly the classes and preparations of the various types of inorganic polymers in this text.

Table 1.1 Reactions by Which Several Important Addition Polymers Can Be Produced

Monomer	M(g/mol)	Repeat unit	Chemical name(s)
	28.0		Polyethylene
	104		Polystyrene
	62.5		Poly(vinyl chloride), "vinyl"
	53.0		Polyacrylonitrile, "acrylic"
	97.0		Poly(vinylidene chloride)
	100		Poly(methyl methacrylate), Plexiglas®, Lucite®
	56.0		Polyisobutylene
	100		Poly(tetrafluoroethylene), Teflon®
	44.0		Poly(ethylene oxide), poly(ethylene glycol)
	113		Poly(ε-caprolactam), Nylon-6

1.4.2 Natural Polymers

We conclude this section with a short discussion of naturally occurring polymers. Since these are of biological origin, they are also called biopolymers. Although our attention in this volume is primarily directed toward synthetic polymers, it should be recognized that biopolymers, like inorganic polymers, have physical properties which follow directly from the chain structure of

Table 1.2   Reactions by Which Several Important Condensation Polymers Can Be Produced

1. Polyester

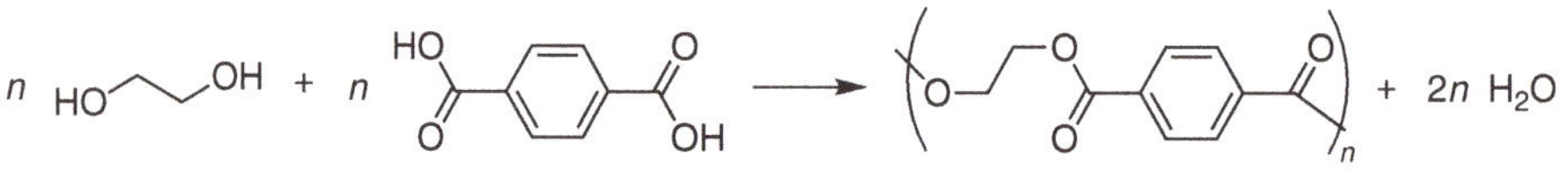

Poly(ethylene terephthalate), Terylene®,
Dacron®, Mylar®

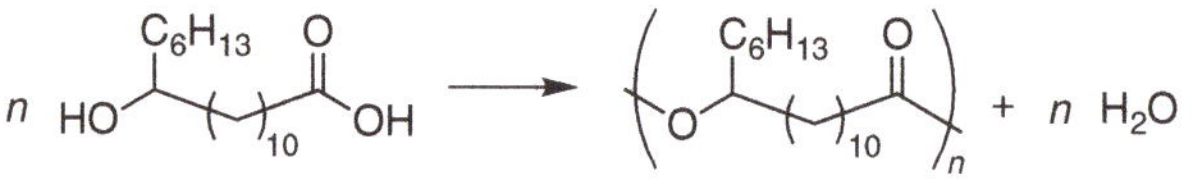

Poly(12-hydroxystearic acid)

2. Polyamide

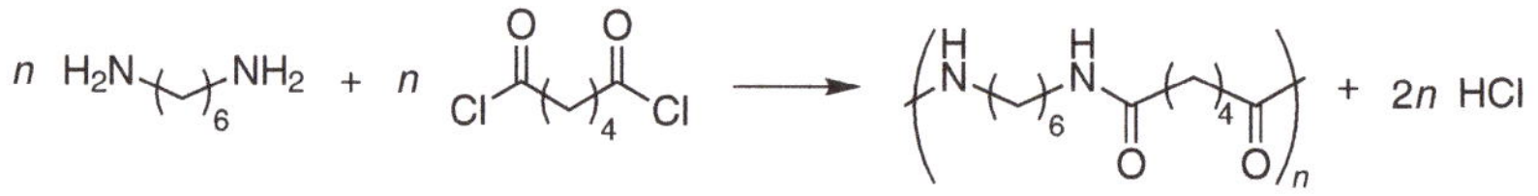

Poly(hexamethylene adipamide), Nylon-6,6

3. Polyurethane

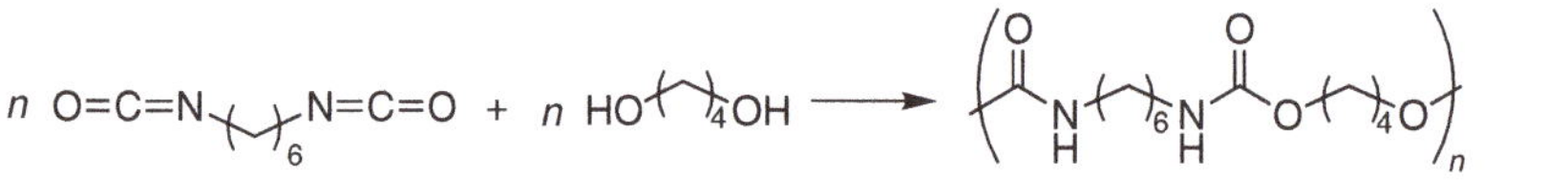

Poly(tetramethylenehexamethylene urethane),
Spandex®, Perlon®

4. Polycarbonate

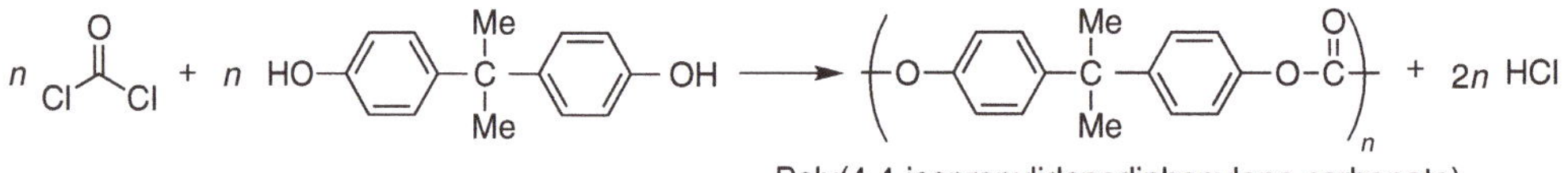

Poly(4,4-isopropylidenediphenylene carbonate)
bisphenol A polycarbonate, Lexan®

5. Inorganic

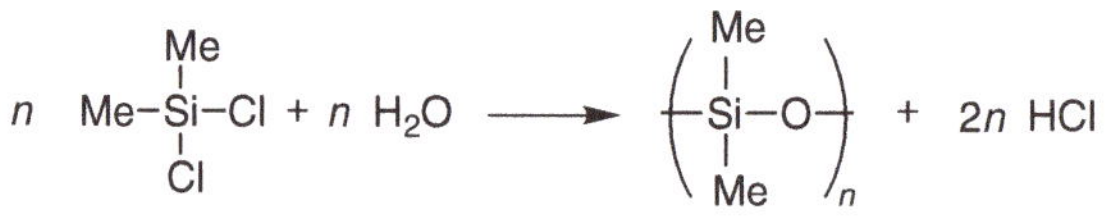

Poly(dimethylsiloxane)

their molecules. For example, the denaturation of a protein involves an overall conformation change from a "native" state, often a compact globule, to a random coil. As another example, the elasticity and integrity of a cell membrane is often the result of an underlying network of fibrillar proteins, with the origin of the elasticity residing in the same conformational entropy as in a rubber band. Consequently, although we will not discuss the synthesis by, and contribution to the function of, living organisms by such biopolymers, many of the principles we will develop in detail apply equally well to natural polymers.

As examples of natural polymers, we consider *polysaccharides*, *proteins*, and *nucleic acids*. Another important natural polymer, polyisoprene, will be considered in Section 1.6. Polysaccharides are macromolecules which make up a large part of the bulk of the vegetable kingdom. *Cellulose* and *starch* are, respectively, the first and second most abundant organic compounds in plants. The former is present in leaves and grasses; the latter in fruits, stems, and roots. Because of their abundance in nature and because of contemporary interest in renewable resources, there is a great deal of interest in these compounds. Both cellulose and starch are hydrolyzed by acids to D-glucose, the repeat unit in both polymer chains. The configuration of the glucoside linkage is different in the two, however. Structure (1.I) and Structure (1.II), respectively, illustrate that the linkage is a β-acetal— hydrolyzable to an equatorial hydroxide—in cellulose and an α-acetal—hydrolyzable to an axial hydroxide—in amylose, a starch:

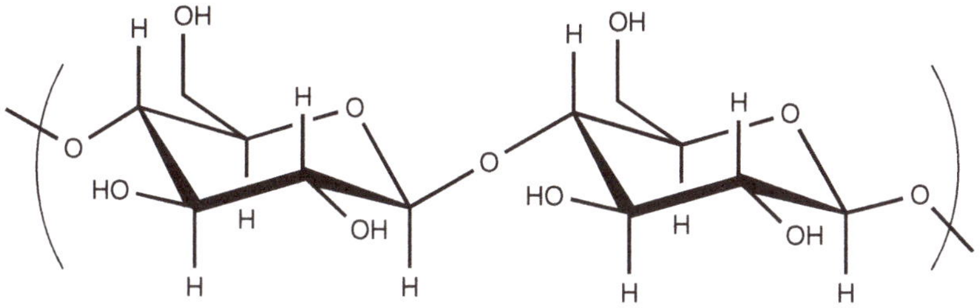

(1.I)

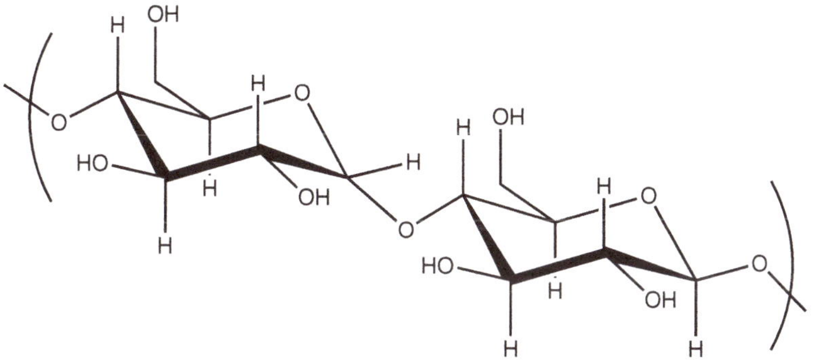

(1.II)

Amylopectin and glycogen are saccharides similar to amylose, except with branched chains.

The cellulose repeat unit contains three hydroxyl groups, which can react and leave the chain backbone intact. These alcohol groups can be esterified with acetic anhydride to form cellulose acetate; this polymer is spun into the fiber acetate rayon. Similarly, the alcohol groups in cellulose react with CS_2 in the presence of strong base to produce cellulose xanthates. When extruded into fibers, this material is called viscose rayon, and when extruded into sheets, cellophane. In both the acetate and xanthate formation, some chain degradation also occurs, so the resulting polymer chains are shorter than those in the starting cellulose. The hydroxyl groups are also commonly methylated, ethylated, and hydroxypropylated for a variety of aqueous applications, including food products. A closely related polysaccharide is chitin, the second most abundant polysaccharide in nature, which is found for example in the shells of crabs and beetles. Here one of the hydroxyls on each repeat unit of cellulose is replaced with an –NHCO–CH_3 amide group. This is converted to a primary amine –NH_2 in chitosan, a derivative of chitin finding increasing applications in a variety of fields.

As noted above, proteins are polyamides in which α-amino acids make up the repeat units, as shown by Structure (1.III):

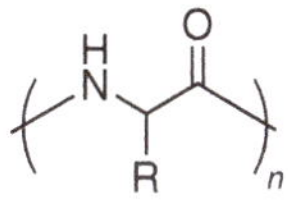

(1.III)

These molecules are also called *polypeptides*, especially when $M \leq 10,000$. The various amino acids differ in their R groups. The nature of R, the name, and the abbreviation used to represent

Table 1.3 Name, Abbreviations, and R Group for Common Amino Acids

Name	Abbreviation	Letter code	R Group
Alanine	Ala	A	Me
Arginine	Arg	R	
Asparagine	Asn	N	
Aspartic acid	Asp	D	
Cysteine	Cys	C	SH
Glutamic acid	Glu	E	
Glutamine	Gln	O	
Glycine	Gly	G	H
Histidine	His	H	
Isoleucine	Ile	I	
Leucine	Leu	L	
Lysine	Lys	K	

Table 1.3 (continued)

Name	Abbreviation	Letter Code	R Group	$\begin{array}{c}\text{HOOC} \quad \text{NH}_2 \\ \text{R} \end{array}$
Methionine	Met	M		
Phenylalanine	Phe	F		
Proline	Pro	P		
Serine	Ser	S		
Threonine	Thr	T		
Tryptophan	Trp	W		
Tyrosine	Tyr	Y		
Valine	Val	V		

some of the more common amino acids are listed in Table 1.3. In proline (Pro) the nitrogen and the α-carbon are part of a five-atom pyrrolidine ring. Since some of the amino acids carry substituent carboxyl or amino groups, protein molecules are charged in aqueous solutions, and hence can migrate in electric fields. This is the basis of electrophoresis as a means of separating and identifying proteins.

It is conventional to speak of three levels of structure in protein molecules:

1. *Primary structure* refers to the sequence of amino acids in the polyamide chain.
2. *Secondary structure* refers to the regions of the molecule that have particular spatial arrangements. Examples in proteins include the α-helix and the β-sheet.

3. *Tertiary structure* refers to the overall shape of the molecule, for example, a globule perhaps stabilized by disulfide bridges formed by the oxidation of cysteine mercapto groups. By extension the full tertiary structure implies knowledge of the relative spatial positions of all the residues.

Hydrogen bonding stabilizes some protein molecules in helical forms, and disulfide cross-links stabilize some protein molecules in globular forms. Both secondary and tertiary levels of structure are also influenced by the distribution of polar and nonpolar amino acid molecules relative to the aqueous environment of the protein molecules. In some cases, individual proteins associate in particular aggregates, which are referred to as quaternary structures.

Examples of the effects and modifications of the higher-order levels of structures in proteins are found in the following systems:

1. *Collagen* is the protein of connective tissues and skin. In living organisms, the molecules are wound around one another to form a three-stranded helix stabilized by hydrogen bonding. When boiled in water, the collagen dissolves and forms gelatin, thereby establishing a new hydrogen bond equilibrium with the solvent. This last solution sets up to form the familiar gel when cooled, a result of shifting the hydrogen bond equilibrium.
2. *Keratin* is the protein of hair and wool. These proteins are insoluble because of the disulfide cross-linking between cysteine units. Permanent waving of hair involves the rupture of these bonds, reshaping of the hair fibers, and the reformation of cross-links, which hold the chains in the new positions relative to each other. We shall see in Chapter 10 how such cross-linked networks are restored to their original shape when subjected to distorting forces.
3. The globular proteins *albumin* in eggs and *fibrinogen* in blood are converted to insoluble forms by modification of their higher-order structure. The process is called denaturation and occurs, in the systems mentioned, with the cooking of eggs and the clotting of blood.
4. *Actin* is a fascinating protein that exists in two forms: G-actin (globular) and F-actin (fibrillar). The globular form can polymerize (reversibly) into very long filaments under the influence of various triggers. These filaments play a crucial role in the cytoskeleton, i.e., in allowing cells to maintain their shape. In addition, the uniaxial sliding of actin filaments relative to filaments of a related protein, *myosin*, is responsible for the working of muscles.

Ribonucleic acid (RNA) and *deoxyribonucleic acid* (DNA) are polymers in which the repeat units are substituted esters. The esters are formed between the hydrogens of phosphoric acid and the hydroxyl groups of a sugar, D-ribose in the case of RNA and D-2-deoxyribose in the case of DNA. The sugar rings in DNA carry four different kinds of substituents: adenine (A) and guanine (G), which are purines, and thymine (T) and cytosine (C), which are pyramidines. The familiar double-helix structure of the DNA molecule is stabilized by hydrogen bonding between pairs of substituent base groups: G–C and A–T. In RNA, thymine is usually replaced by uracil (U). The replication of these molecules, the template model of their functioning, and their role in protein synthesis and the genetic code make the study of these polymers among the most exciting and actively researched areas in science. As with the biological function of proteins, we will not discuss these phenomena in this book. However, as indicated previously, DNA plays a very important role as a prototypical semiflexible polymer, as it is now readily obtainable in pure molecular fractions of varying lengths, and because it is readily dissolved in aqueous solution. It is also a charged polymer, or *polyelectrolyte*, and thus serves as a model system in this arena as well.

1.5 Polymer Nomenclature

Considering that a simple compound like C_2H_5OH is variously known as ethanol, ethyl alcohol, grain alcohol, or simply alcohol, it is not too surprising that the vastly more complicated polymer molecules are also often known by a variety of different names. The International Union of Pure and Applied Chemistry (IUPAC) has recommended a system of nomenclature based on the structure of the monomer or repeat unit [7]. A semisystematic set of trivial names is also in widespread usage; these latter names seem even more resistant to replacement than is the case with low molecular weight compounds. Synthetic polymers of commercial importance are often widely known by trade names that have more to do with marketing considerations than with scientific communication. Polymers of biological origin are often described in terms of some aspect of their function, preparation, or characterization.

If a polymer is formed from a single monomer, as in addition and ring-opening polymerizations, it is named by attaching the prefix *poly* to the name of the monomer. In the IUPAC system, the monomer is named according to the IUPAC recommendations for organic chemistry, and the name of the monomer is set off from the prefix by enclosing the former in parentheses. Variations of this basic system often substitute a common name for the IUPAC name in designating the monomer. Whether or not parentheses are used in the latter case is influenced by the complexity of the monomer name; they become more important as the number of words in the monomer name increases. Thus the polymer $(CH_2-CHCl)_n$ is called poly(1-chloroethylene) according to the IUPAC system; it is more commonly called poly(vinyl chloride) or polyvinyl chloride. Acronyms are not particularly helpful but are an almost irresistible aspect of polymer terminology, as evidenced by the initials PVC, which are widely used to describe the polymer just named. The trio of names poly(1-hydroxyethylene), poly(vinyl alcohol), and polyvinyl alcohol emphasizes that the polymer need not actually be formed from the reaction of the monomer named; this polymer is actually prepared by the hydrolysis of poly(1-acetoxyethylene), otherwise known as poly(vinyl acetate). These same alternatives are used in naming polymers formed by ring-opening reactions; for example, poly(6-aminohexanoic acid), poly(6-aminocaproic acid), and poly(ε-caprolactam) are all more or less acceptable names for the same polymer.

Those polymers which are the condensation products of two different monomers are named by applying the preceding rules to the repeat unit. For example, the polyester formed by the condensation of ethylene glycol and terephthalic acid is called poly(oxyethylene oxyterphthaloyl) according to the IUPAC system, but is more commonly referred to as poly(ethylene terephthalate) or polyethylene terephthalate. The polyamides poly(hexamethylene sebacamide) and poly(hexamethylene adipamide) are also widely known as nylon-6,10 and nylon-6,6, respectively. The numbers following the word nylon indicate the number of carbon atoms in the diamine and dicarboxylic acids, in that order. On the basis of this system, poly(ε-caprolactam) is also known as nylon-6.

Many of the polymers in Table 1.1 and Table 1.2 are listed with more than one name. Also listed are some of the registered trade names by which these substances—or materials which are mostly of the indicated structure—are sold commercially. Some commercially important cross-linked polymers go virtually without names. These are heavily and randomly cross-linked polymers which are insoluble and infusible and therefore widely used in the manufacture of such molded items as automobile and household appliance parts. These materials are called resins and, at best, are named by specifying the monomers that go into their production. Often even this information is sketchy. Examples of this situation are provided by phenol–formaldehyde and urea–formaldehyde resins, for which typical structures are given by Structure (1.IV) and Structure (1.V), respectively:

(1.IV)

(1.V)

1.6   Structural Isomerism

In this section, we shall consider three types of isomerism that are encountered in polymers. These are *positional isomerism, stereo isomerism*, and *geometrical isomerism*. We shall focus attention on synthetic polymers and shall, for the most part, be concerned with these types of isomerism occurring singly, rather than in combinations. Some synthetic and analytical aspects of stereo isomerism will be considered in Chapter 5. Our present concern is merely to introduce the possibilities of these isomers and some of the associated vocabulary.

1.6.1   Positional Isomerism

Positional isomerism is conveniently illustrated by considering the polymerization of a vinyl monomer. In such a reaction, the adding monomer may become attached to the growing chain end (designated by $*$) in either of two orientations:

Structure (1.VI) and Structure (1.VII), respectively, are said to arise from *head-to-tail* or *head-to-head* orientations. In this terminology, the substituted carbon is defined to be the head and the methylene is the tail. Tail-to-tail linking is also possible.

For most vinyl polymers, head-to-tail addition is the dominant mode of addition. Variations from this generalization become more common for polymerizations which are carried out at higher temperatures. Head-to-head addition is also somewhat more abundant in the case of halogenated monomers such as vinyl chloride. The preponderance of head-to-tail additions is understood to arise from a combination of resonance and steric effects. In many cases, the ionic or free-radical reaction center occurs at the substituted carbon due to the possibility of resonance stabilization or electron delocalization through the substituent group. Head-to-tail attachment is also sterically favored, since the substituent groups on successive repeat units are separated by a methylene carbon. At higher polymerization temperatures, larger amounts of available thermal energy make the less-favored states more accessible. In vinyl fluoride, no resonance stabilization is possible and steric effects are minimal. This monomer adds primarily in the head-to-tail orientation at low temperatures and tends toward a random combination of both at higher temperatures. The styrene radical, by contrast, enjoys a large amount of resonance stabilization in the bulky phenyl group and polymerizes almost exclusively in the head-to-tail mode. The following example illustrates how chemical methods can be used to measure the relative amounts of the two positional isomers in a polymer sample.

Example 1.3

1,2-Glycol bonds are cleaved by reaction with periodate; hence poly(vinyl alcohol) chains are broken at the site of head-to-head links in the polymer. The fraction of head-to-head linkages in poly(vinyl alcohol) may be determined by measuring the molecular weight before (subscript b) and after (subscript a) cleavage with periodate according to the following formula: Fraction $= 44(1/M_a - 1/M_b)$. Derive this expression and calculate the value for the fraction in the case of $M_b = 10^5$ and $M_a = 10^3$.

Solution

Begin by recognizing that a molecule containing x of the head-to-head links will be cleaved into $x + 1$ molecules upon reaction. Hence if n is the number of polymer molecules in a sample of mass w, the following relations apply before and after cleavage: $n_a = (x + 1) n_b$ or $w/M_a = (x + 1)$ (w/M_b). Solving for x and dividing the latter by the total number of linkages in the original polymer gives the desired ratio. The total number of links in the original polymer is M_b/M_0. Therefore the ratio is $xM_0/M_b = M_0(1/M_a - 1/M_b)$. For poly(vinyl alcohol) M_0 is 44, so the desired formula has been obtained. For the specific data given, $x/n_b = 44(10^{-3} - 10^{-5}) = 0.044$, or about 4% of the additions are in the less favorable orientation. We shall see presently that the molecular weight of a polymer is an average, which is different depending on the method used for its determination. The present example used molecular weights as a means for counting the number of molecules present. Hence the sort of average molecular weight used should also be one which is based on counting.

1.6.2 Stereo Isomerism

The second type of isomerism we discuss in this section is stereo isomerism. Again we consider the number of ways a singly substituted vinyl monomer can add to a growing polymer chain:

$$(1.\text{VIII})$$

$$(1.\text{C})$$

$$(1.\text{IX})$$

Structure (1.VIII) and Structure (1.IX) are not equivalent; they would not superimpose if the extended chains were overlaid. The difference has to do with the stereochemical configuration at the asymmetric carbon atoms. Note that the asymmetry is more accurately described as pseudoasymmetry, since two sections of chain are bonded to these centers. Except near chain ends, which we ignore for high polymers, these chains provide local symmetry in the neighborhood of the carbon under consideration. The designations of D and L or R and S are used to distinguish these structures, even though true asymmetry is absent.

We use the word *configuration* to describe the way the two isomers produced by Reaction (1.C) differ. It is only by breaking bonds, moving substituents, and reforming new bonds that the two structures can be interconverted. This state of affairs is most readily seen when the molecules are drawn as fully extended chains in one plane, and then examining the side of the chain on which substituents lie. The configurations are not altered if rotation is allowed to occur around the various bonds of the backbone to change the shape of the molecule to a jumbled coil. We shall use the term *conformation* to describe the latter possibilities for different molecular shapes. The configuration is not influenced by conformational changes, but the stability of different conformations may be affected by differences in configuration. We shall return to these effects in Chapter 6.

In the absence of any external influence, such as a catalyst that is biased in favor of one configuration over the other, we might expect Structure (1.VIII) and Structure (1.IX) to occur at random with equal probability as if the configuration at each successive addition were determined by the toss of a coin. Such indeed is the ordinary case. However, in the early 1950s, stereospecific catalysts were discovered; Ziegler and Natta received the Nobel Prize for this discovery in 1963. Following the advent of these catalysts, polymers with a remarkable degree of stereoregularity have been formed. These have such a striking impact on polymer science that a substantial part of

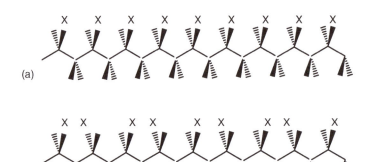

Figure 1.3 Sections of "polyvinyl X" chains of differing tacticity: (a) isotactic, (b) syndiotactic, and (c) atactic.

Chapter 5 is devoted to a discussion of their preparation and characterization. For now, only the terminology involved in their description concerns us. Three different situations can be distinguished along a chain containing pseudoasymmetric carbons:

1. *Isotactic.* All substituents lie on the same side of the extended chain. Alternatively, the stereoconfiguration at the asymmetric centers is the same, say, –DDDDDDDDD–.
2. *Syndiotactic.* Substituents on the fully extended chain lie on alternating sides of the backbone. This alternation of configuration can be represented as –DLDLDLDLDLDL–.
3. *Atactic.* Substituents are distributed at random along the chain, for example, DDLDLLLDLDLL–.

Figure 1.3 shows sections of polymer chains of these three types; the substituent X equals phenyl for polystyrene and methyl for polypropylene. The general term for this stereoregularity is *tacticity*, a term derived from the Greek word meaning "to put in order." Polymers of different tacticity have quite different properties, especially in the solid state. As we will see in Chapter 13, one of the requirements for polymer crystallinity is a high degree of microstructural regularity to enable the chains to pack in an orderly manner. Thus atactic polypropylene is a soft, tacky substance, whereas both isotactic and syndiotactic polypropylene are highly crystalline.

1.6.3 Geometrical Isomerism

The final type of isomerism we take up in this section is nicely illustrated by the various possible structures that result from the polymerization of 1,3-dienes. Three important monomers of this type are 1,3-butadiene, 1,3-isoprene, and 1,3-chloroprene, Structure (1.X) through Structure (1.XII), respectively:

$$
\text{H}_2\text{C}=\text{CH}-\text{CH}=\text{CH}_2
$$

(1.X)

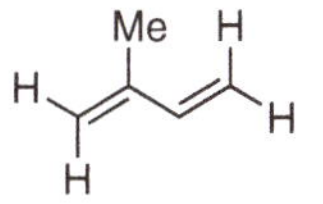

(1.XI)

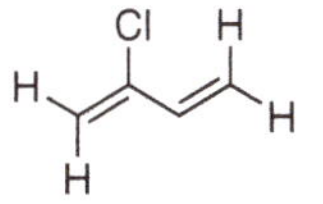

(1.XII)

To illustrate the possible modes of polymerization of these compounds, consider the following reactions of isoprene:

1. 1,2- and 3,4-Polymerizations. As far as the polymer chain backbone is concerned, these compounds could just as well be mono-olefins, since the second double bond is relegated to the status of a substituent group. Because of the reactivity of the latter, however, it might become involved in cross-linking reactions. For isoprene, 1,2- and 3,4-polymerizations yield different products:

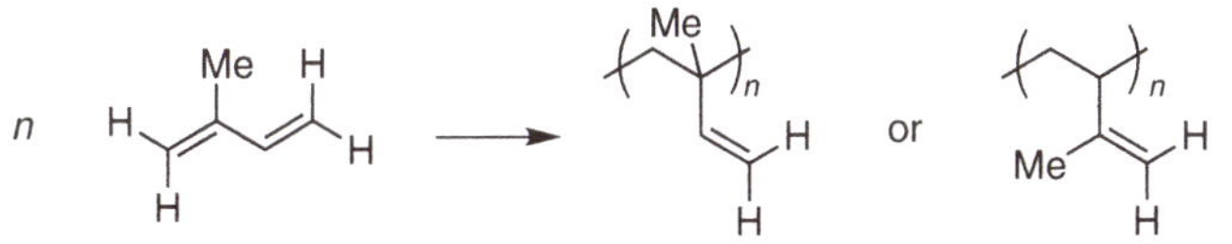

(1.D)

(1.XIII)   (1.XIV)

These differences do not arise from 1,2- or 3,4-polymerization of butadiene. Structure (1.XIII) and Structure (1.XIV) can each exhibit the three different types of tacticity, so a total of six structures can result from this monomer when only one of the olefin groups is involved in the backbone formation.

2. 1,4-Polymerization. This mode of polymerization gives a molecule with double bonds along the backbone of the chain. Again using isoprene as the example,

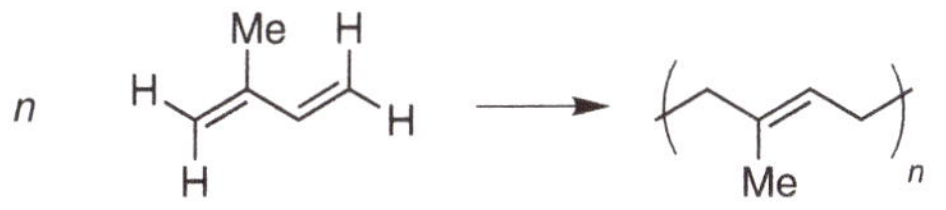

(1.E)

As in all double-bond situations, the adjacent chain sections can be either *cis* or *trans*— Structure (1.XV) and Structure (1.XVI), respectively—with respect to the double bond, producing the following geometrical isomers:

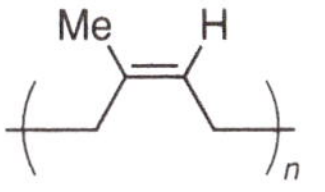

(1.XV)

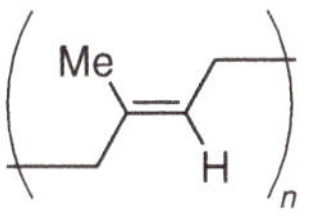

(1.XVI)

Figure 1.4 shows several repeat units of *cis*-1,4-polyisoprene and *trans*-1,4-polyisoprene. Natural rubber is the *cis* isomer of 1,4-polyisoprene and gutta-percha is the *trans* isomer.

3. Polymers of chloroprene (Structure (1.XII)) are called neoprene and copolymers of butadiene and styrene are called SBR, an acronym for styrene–butadiene rubber. Both are used for many of the same applications as natural rubber. Chloroprene displays the same assortment of possible isomers

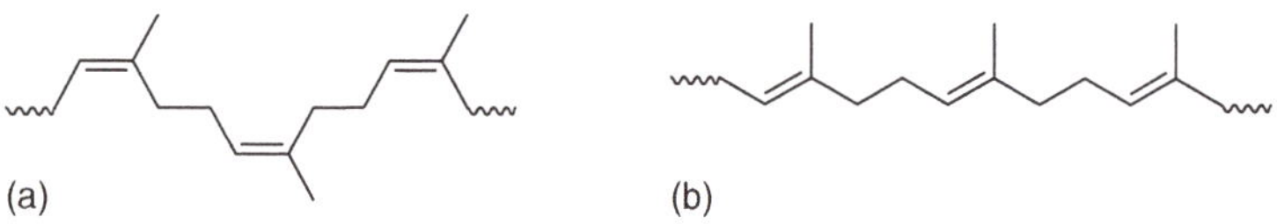

(a)   (b)

Figure 1.4   1,4-Polyisoprene (a) all-*cis* isomer (natural rubber) and (b) all-*trans* isomer (gutta-percha).

as isoprene; the extra combinations afforded by copolymer composition and structure in SBR offset the fact that Structure (1.XIII) and Structure (1.XIV) are identical for butadiene.

4. Although the conditions of the polymerization reactions may be chosen to optimize the formation of one specific isomer, it is typical in these systems to have at least some contribution of all possible isomers in the polymeric product, except in the case of polymers of biological origin, like natural rubber and gutta-percha.

Example 1.4

Suppose you have just ordered a tank car of polybutadiene from your friendly rubber company. By some miracle, all the polymers in the sample have $M = 54,000$. The question we would like to consider is this: what are the chances that any two molecules in this sample have exactly the same chemical structure?

Solution

We will not attempt to provide a precise answer to such an artificial question; what we really want to know is whether the probability is high (approximately 1), vanishing (approximately 0), or finite.

From the discussion above, we recognize three geometrical isomers: *trans*-1,4, *cis*-1,4, and 1,2. We will ignore the stereochemical possibilities associated with the 1,2 linkages. Assuming all three isomers occur with equal probability, the total number of possible structures is $3 \times 3 \times 3 \times \cdots \times 3 = 3^N$, where N is the degree of polymerization. (Recall that the combined probability of a sequence of events is equal to the products of the individual probabilities.) In this case $N = 54,000/54 = 1000$, and thus there are about $3^{1000} \approx 10^{500}$ possible structures. Now we need to count how many molecules we have. Assuming for simplicity that the tank car is 3.3 m $\times$ 3.3 m $\times$ 10 m = 100 m^3 = 10^8 cm^3, and the density of the polymer is 1 g/cm^3 (it is actually closer to 0.89 g/cm^3), we have 10^8 g of polymer. As $M = 54,000$ g/mol, we have about 2000 moles, or 2000 $\times$ 6 $\times 10^{23} \approx 10^{27}$ molecules. Clearly, therefore, there is essentially no chance that any two molecules have the identical structure, even without taking the molecular weight distribution into account.

This example, as simplistic as it is, actually underscores two important points. First, polymer chemists have to get used to the idea that while all carbon atoms are identical, and all 1,3-butadiene molecules are identical, polybutadiene actually refers to an effectively infinite number of distinct chemical structures. Second, almost all synthetic polymers are heterogeneous in more than one variable: molecular weight, certainly; isomer and tacticity distribution, probably; composition and sequence distribution, for copolymers; and branching structure, when applicable.

1.7   Molecular Weights and Molecular Weight Averages

Almost every synthetic polymer sample contains molecules of various degrees of polymerization. We describe this state of affairs by saying that the polymer shows *polydispersity* with respect to molecular weight or degree of polymerization. To see how this comes about, we only need to think of the reactions between monomers that lead to the formation of polymers in the first place. Random encounters between reactive species are responsible for chain growth, so statistical descriptions are appropriate for the resulting product. The situation is reminiscent of the distribution of molecular velocities in a sample of gas. In that case, also, random collisions impart extra energy to some molecules while reducing the energy of others. Therefore, when we talk about the molecular weight of a polymer, we mean some characteristic average molecular weight. It turns out there are several distinct averages that may be defined, and that may be measured experimentally; it is therefore appropriate to spend some time on this topic. Furthermore, one might well encounter two samples of a particular polymer that were equivalent in terms of one kind of average, but different in terms of another; this, in turn, can lead to the situation where the two polymers behave identically in terms of some important properties, but differently in terms of others.

In Chapter 2 through Chapter 4 we shall examine the expected distribution of molecular weights for condensation and addition polymerizations in some detail. For the present, our only concern is how such a distribution of molecular weights is described. We will define the most commonly encountered averages, and how they relate to the distribution as a whole. We will also relate them to the standard parameters used for characterizing a distribution: the mean and standard deviation. Although these are well-known quantities, many students are familiar with them only as results provided by a calculator, and so we will describe them in some detail.

1.7.1 Number-, Weight-, and z-Average Molecular Weights

Suppose we have a polymer sample containing many molecules with a variety of degrees of polymerization. We will call a molecule with degree of polymerization i an "i-mer", and the associated molecular weight $M_i = iM_0$, where M_0 is the molecular weight of the repeat unit. (Conversion between a discussion couched in terms of i or in terms of M_i is therefore straightforward, and we will switch back and forth when convenient.) The number of i-mers we will denote as n_i (we could also refer to n_i as the number of moles of i-mer, but again this just involves a factor of Avogadro's number). The first question we ask is this: if we choose a molecule at random from our sample, what is the probability of obtaining an i-mer? The answer is straightforward. The total number of molecules is $\sum_i n_i$, and thus this probability is given by

$$x_i = \frac{n_i}{\sum_i n_i} \tag{1.7.1}$$

The probability x_i is the number fraction or *mole fraction* of i-mer. We can use this quantity to define a particular average molecular weight, called the *number-average molecular weight*, M_n. We do this by multiplying the probability of finding an i-mer with its associated molecular weight, $x_i M_i$, and adding all these up:

$$M_n = \sum_i x_i M_i = \frac{\sum_i n_i M_i}{\sum_i n_i} = M_0 \frac{\sum_i i n_i}{\sum_i n_i} \tag{1.7.2}$$

The other expressions on the right-hand side of Equation 1.7.2 are equivalent, and will prove useful subsequently. You should convince yourself that this particular average is the one you are familiar with in everyday life: take the value of the property of interest, M_i in this case, add it up for all the (n_i) objects that possess that value of the property, and divide by the total number of objects.

So far, so good. We return for a moment to our hypothetical sample, but instead of choosing a molecule at random, we choose a repeat unit or monomer at random, and ask about the molecular weight of the molecule to which it belongs. We will get a different answer, as a simple argument illustrates. Suppose we had two molecules, one a 10-mer and another a 20-mer. If we choose molecules at random, we would choose each one 50% of the time. However, if we choose monomers at random, 2/3 of the monomers are in the 20-mer, so we would pick the larger molecule twice as often as the smaller. The total number of monomers in a sample is $\sum_i i n_i$, and the chance of picking a particular i-mer will be determined by the product in_i. The resulting ratio is, in fact, the *weight fraction* or mass fraction of i-mer in the sample, w_i:

$$w_i = \frac{in_i}{\sum_i in_i} \tag{1.7.3}$$

Accordingly, we define the *weight-average molecular weight* of the sample, M_w, by

$$M_w = \sum_i w_i M_i = \frac{\sum_i in_i M_i}{\sum_i in_i} = \frac{\sum_i n_i M_i^2}{\sum_i n_i M_i} = M_0 \frac{\sum_i i^2 n_i}{\sum_i in_i} \tag{1.7.4}$$

Of course, mass-average would be the preferred descriptor, but it is not in common usage. Qualitatively, we can say that M_n is the characteristic average molecular weight of the sample when the number of molecules is the crucial factor, whereas M_w is the characteristic average molecular weight when the size of each molecule is the important feature. Although knowledge of M_w and M_n is not sufficient to provide all the information about a polydisperse system, these two averages are by far the most important and most commonly encountered, as we shall see throughout the book.

Comparison of the last expressions in Equation 1.7.2 and Equation 1.7.4 suggests a trend; we can define a new average by multiplying the summation terms in the numerator and denominator by i. The so-called *z-average molecular weight*, M_z, is constructed in just such a way:

$$M_z = M_0 \frac{\sum_i i^3 n_i}{\sum_i i^2 n_i} \tag{1.7.5}$$

Although M_z is not directly related to a simple fraction like x_i or w_i, it does have some experimental relevance. We could continue this process indefinitely, just by incrementing the power of i by one in both numerator and denominator of Equation 1.7.5, but it will turn out that there is no real need to do so. However, there is a direct relationship between this process and something well-known in statistical probability, namely the moments of a distribution, as we will see in the next section.

1.7.2 Polydispersity Index and Standard Deviation

Although the values M_w or M_n tell us something useful about a polymer sample, individually they do not provide information about the breadth of the distribution. However, the ratio of the two turns out to be extremely useful in this regard, and it is given a special name, the *polydispersity index* (PDI) or just the polydispersity:

$$PDI = \frac{M_w}{M_n} \tag{1.7.6}$$

The PDI is always greater than 1, unless the sample consists of exactly one value of M, in which case the PDI $= 1$; such a sample is said to be *monodisperse*. We will see in Chapter 2 and Chapter 3 that typical polymerization schemes are expected to give PDIs near 2, at least in the absence of various side reactions; in industrial practice, such side reactions often lead to PDIs as large as 10 or more. In Chapter 4, in contrast, so-called living polymerizations give rise to PDIs of 1.1 or smaller. Thus, distributions for which the PDI < 1.5 are said to be "narrow," whereas those for which PDI > 2 are said to be "broad," of course, such designations are highly subjective. As a very simple illustration, the two-molecule example given in the previous section consisting of a 10-mer and a 20-mer has a number-average degree of polymerization of 15 and a weight average of 16.7; thus its PDI $= 1.11$, which in polymer terms would be considered "narrow." This trivial example actually underscores an important point to bear in mind: polymer samples with "narrow" distributions will still contain molecules that are quite different in size (see Problem 1.8 for another instance).

In most fields of science, distributions are generally characterized by a *mean* and a *standard deviation*. We will now develop the relationships between these quantities and M_w and M_n, and in so doing justify the assertion that the PDI is a useful measure of the breadth of a distribution. The mean of any distribution of a variable i, $\langle i \rangle$, is defined as

$$\langle i \rangle = \frac{\sum_i i n_i}{\sum_i n_i} = \sum_i i x_i \approx \int_0^\infty i P(i)\, d i \tag{1.7.7}$$

where both discrete (x_i) and continuous $(P(i))$ versions are considered. From this definition, and Equation 1.7.2, we can see that $\langle i \rangle$ is nothing more than the number-average degree of polymerization, and thus M_n is just the mean molecular weight.

The standard deviation σ quantifies the width of the distribution. It is defined as

$$\sigma \equiv \left(\frac{\sum_i n_i (i - \langle i \rangle)^2}{\sum_i n_i} \right)^{1/2} = \left(\sum_i x_i (i - \langle i \rangle)^2 \right)^{1/2} \tag{1.7.8}$$

Note that σ^2 has the significance of being the mean value of the square of the deviations of individual values from the mean. Accordingly, σ is sometimes called the root mean square (rms) deviation.

From a computational point of view, the standard deviation may be written in a more convenient form by carrying out the following operations. First both sides of Equation 1.7.8 are squared, and then the difference $i-\langle i \rangle$ is squared to give

$$\sigma^2 = \frac{\sum_i n_i i^2}{\sum_i n_i} - 2\langle i \rangle \frac{\sum_i n_i i}{\sum_i n_i} + \langle i \rangle^2 \tag{1.7.9}$$

Recalling the definition of the mean, we recognize the first term on the right-hand side of Equation 1.7.9 to be the mean value of i^2 and write

$$\sigma^2 = \langle i^2 \rangle - 2\langle i \rangle^2 + \langle i \rangle^2 = \langle i^2 \rangle - \langle i \rangle^2 \tag{1.7.10}$$

It is, of course, important to realize that $\langle i^2 \rangle \neq \langle i \rangle^2$. An alternative to Equation 1.7.8 as a definition of standard deviation is, therefore,

$$\sigma = (\langle i^2 \rangle - \langle i \rangle^2)^{1/2} \tag{1.7.11}$$

Similarly, the standard deviation can be written

$$\sigma = \left[\sum_i x_i (M_i - M_n)^2 \right]^{1/2} \tag{1.7.12}$$

where in this case the standard deviation will have the units of molecular weight. If we expand Equation 1.7.12 we find

$$\sigma = \left[\sum_i x_i (M_i^2 - 2M_i M_n + M_n^2) \right]^{1/2} = \left[\left(\sum_i x_i M_i^2 \right) - M_n^2 \right]^{1/2} \tag{1.7.13}$$

We can factor out $M_n = \sum_i x_i M_i$ to obtain

$$\sigma = M_n \left[\frac{\sum_i x_i M_i^2}{(\sum_i x_i M_i)^2} - 1 \right]^{1/2} \tag{1.7.14}$$

and, finally, by recognizing from Equation 1.7.4 that

$$M_w = \frac{\sum_i n_i M_i^2}{\sum_i n_i M_i} = \frac{\sum_i x_i M_i^2}{\sum_i x_i M_i} \qquad (1.7.15)$$

we reach

$$\sigma = M_n \left[\frac{M_w}{M_n} - 1 \right]^{1/2} \qquad (1.7.16)$$

This result shows that the square root of the amount by which the ratio M_w/M_n exceeds unity equals the standard deviation of the distribution relative to the number-average molecular weight. Thus if a distribution is characterized by $M_n = 10,000$ and $\sigma = 3000$, then $M_w/M_n = 1.09$. Alternatively, if $M_w/M_n = 1.50$, then the standard deviation is 71% of the value of M_n. This shows that reporting the mean and standard deviation of a distribution or the values of M_n and M_w/M_n gives equivalent information.

We can define the quantities known as *moments* of a distribution, again either in the discrete or continuous forms. The k-th moment μ_k is given by

$$\mu_k = \sum_i x_i i^k \text{ or } \int_0^\infty i^k P(i) \, di \qquad (1.7.17)$$

In this definition both x_i and $P(i)$ are *normalized* distributions, which mean that $\sum_i x_i = 1$ and $\int_0^\infty P(i) \, di = 1$. From Equation 1.7.17 we can also see that the mean is equivalent to the first moment. From Equation 1.7.4 and Equation 1.7.5 it is apparent that M_w and M_z are proportional to the ratio of the second to the first moment and the third to the second, respectively. More generally, moments can be referred to a particular value, such as the k-th moment about the mean, ν_k:

$$\nu_k = \sum_i x_i (i - \langle i \rangle)^k \qquad (1.7.18)$$

From this expression we see that σ^2 is the second moment about the mean.

1.7.3 Examples of Distributions

First, consider the following numerical example in which we apply some of the equations of this section to hypothetical data.

Example 1.5

The first and second columns of Table 1.4 give the number of moles of polymer in six different molecular weight fractions. Calculate M_w/M_n for this polymer and evaluate σ using both Equation 1.7.12 and Equation 1.7.16.

Table 1.4 Some Molecular Weight Data for a Hypothetical Polymer Used in Example 1.5

n_i (mol)	M_i (g/mol)	m_i (g)	$m_i M_i \times 10^{-5}$ (g²/mol)	$(M_i - M_n)^2 \times 10^{-6}$ (g²/mol²)	$n_i(M_i - M_n)^2 \times 10^{-4}$ (g²/mol)
0.003	10,000	30	3.0	25	7.50
0.008	12,000	96	11.5	9	7.20
0.011	14,000	154	21.6	1	1.10
0.017	16,000	272	43.5	1	1.70
0.009	18,000	162	29.2	9	8.10
0.001	20,000	20	4.0	25	2.50
$\Sigma = 0.049$		$\Sigma = 734$	$\Sigma = 113$	$\Sigma = 70$	$\Sigma = 28.10$

Solution

Evaluate the product $n_i M_i = m_i$ for each class; this is required for the calculation of both M_n and M_w. Values of this quantity are listed in the third column of Table 1.4. From $\Sigma_i n_i M_i$ and $\Sigma_i n_i$, $M_n = 734/0.049 = 15{,}000$. (The matter of significant figures will not be strictly adhered to in this example. As a general rule, one has to work pretty hard to obtain more than two significant figures in an experimental determination of M.)

The products $m_i M_i$ are mass-weighted contributions and are listed in the fourth column of Table 1.4. From $\Sigma_i m_i$ and $\Sigma_i n_i M_i$, $M_w = 113 \times 10^5/734 = 15{,}400$.

The ratio M_w/M_n is found to be $15{,}400/15{,}000 = 1.026$ for these data. Using Equation 1.7.16, we have $\sigma/M_n = (1.026 - 1)^{1/2} = 0.162$ or $\sigma = 0.162(15{,}000) = 2430$.

To evaluate σ via Equation 1.7.12, differences between M_i and M_n must be considered. The fifth and sixth columns in Table 1.4 list $(M_i - M_n)^2$ and $n_i(M_i - M_n)^2$ for each class of data. From $\Sigma_i n_i$ and $\Sigma_i n_i(M_i - M_n)^2$, $\sigma^2 = 28.1 \times 10^4/0.049 = 5.73 \times 10^6$ and $\sigma = 2390$.

The discrepancy between the two values of σ is not meaningful in terms of significant figures; the standard deviation is 2400.

As polymers go, this is a very narrow molecular weight distribution.

When we consider particular polymerization schemes in Chapter 2 through Chapter 4, we will derive explicit expressions for the expected distributions x_i and w_i. For now, however, let us consider a particular mathematical function known as the *Schulz–Zimm distribution*. It has the virtue that by varying a single parameter, z, it is possible to obtain reasonable descriptions for typical narrow or moderately broad samples. We will use it to illustrate graphically how the distribution might appear for a given polydispersity. The Schulz–Zimm distribution can be expressed as

$$P(M_i) = \frac{z^{z+1}}{\Gamma(z+1)} \frac{M_i^{z-1}}{M_n^z} \exp\left(-\frac{zM_i}{M_n}\right) \tag{1.7.19}$$

where $\Gamma(z=1)$ is the so-called gamma function (which is tabulated in many mathematical references). For integer values of z, $\Gamma(z+1) = z!$, where $z!$ (z factorial) $= z \times (z-1) \times (z-2) \times \cdots \times 1$. From Equation 1.7.2 and Equation 1.7.3 we can see that $w_i = x_i M_i/M_n$, and thus

$$w_i = \frac{z^{z+1}}{\Gamma(z+1)} \frac{M_i^z}{M_n^{z+1}} \exp\left(-\frac{zM_i}{M_n}\right) \tag{1.7.20}$$

The utility of this distribution arises in part because of a very simple relationship between the parameter z and the polydispersity:

$$\frac{M_w}{M_n} = \frac{z+1}{z} \tag{1.7.21}$$

The proof of Equation 1.7.21 is left to Problem 1.9 at the end of the chapter. Now we can use Equation 1.7.19 and Equation 1.7.20 to generate distributions for specified values of M_n and M_w/M_n.

Figure 1.5a shows the mole fraction and weight fraction as a function of M, with the particular choice of $M_n = 10{,}000$, $M_0 = 100$, and $z = 1$. Thus, according to Equation 1.7.21, the PDI $= 2$ and $M_w = 20{,}000$. Both M_w and M_n are indicated by vertical lines on the plot. There are several remarkable features to point out. First, x_i is a continuously decreasing function of M (and therefore i). We shall see in Chapter 2 that this is to be expected in step-growth polymerizations. It means, for example, that there are more unreacted monomers ($i = 1$) than any other particular i-mer. The weight fraction, however, has a distinct but broad maximum. Notice also how many different values of M are present in significant amounts. For example, there is certainly a significant mass of the sample that is five times smaller than M_w, or five times larger than M_n.

Figure 1.5b shows the analogous curves, but now with PDI $=4$ and $z=0.333$. Although the mole fraction looks superficially similar to the previous case, the mass distribution is exceedingly broad, with a long tail on the high M side of M_n. (Note that the coincidence between M_n and the peak in w_i is a feature of this distribution, but not a general result in polymers; see Problem 1.10.)

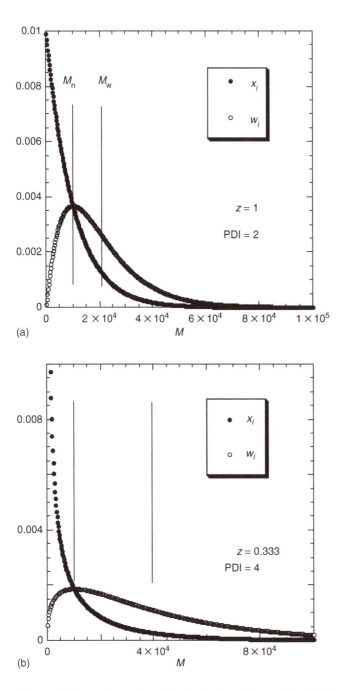

Figure 1.5 Number and weight distributions for the Schulz–Zimm distribution with the indicated polydispersities, and $M_n = 10,000$.

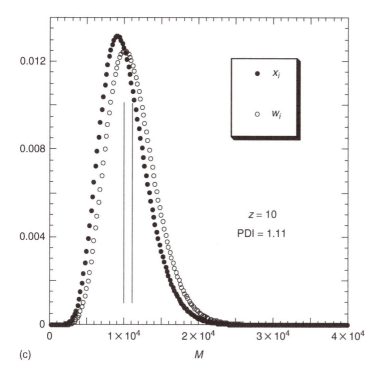

(c)

Figure 1.5 (continued)

Figure 1.5c shows the opposite extreme, or a narrow distribution with $z = 10$ and PDI $= 1.11$. Here both x_i and w_i are much narrower, and have distinct maxima that are also close to M_n and M_w, respectively. (Note also that the ordinate scale has been truncated.) Even though this is a narrow distribution by polymer standards, there are still significant numbers of molecules that are 50% larger and 50% smaller than the mean.

1.8   Measurement of Molecular Weight

1.8.1   General Considerations

The measurement of molecular weight is clearly the most important step in characterizing a polymer sample, although the previous sections have introduced many other aspects of polymer structure necessary for a full analysis. A rich variety of experimental techniques have been developed and employed for this purpose, and the major ones are listed in Table 1.5. As indicated in the rightmost column, we will describe end group analysis and matrix-assisted laser desorption/ ionization (MALDI) mass spectrometry in this section, while deferring treatment of size exclusion chromatography (SEC), osmotic pressure, light scattering, and intrinsic viscometry to subsequent chapters. Other techniques, such as those involving sedimentation, we will omit entirely. Before considering any technique in detail, some general comments about the entries in Table 1.5 are in order.

The first two techniques listed, SEC and MALDI, can provide information on the entire distribution of molecular weights. To the extent that either one can do this reliably, accurately, and conveniently, there is a diminished need for any other approach. Over the past 30 years, SEC has unquestionably emerged as the dominant method. Automated analysis of a few milligrams of sample dissolved in a good solvent can be achieved in half an hour. It is hard to imagine any serious polymer laboratory that does not have SEC capability. Nevertheless, as will be discussed in

Table 1.5 Summary of the Molecular Weight Averages Most Widely Encountered in Polymer Chemistry

Information	Definition	Methods	Sections
Full distribution	x_i, w_i	Size exclusion chromatography	9.8
		MALDI mass spectrometry	1.8
M_n	$\dfrac{\sum_i n_i M_i}{\sum_i n_i}$	Osmotic pressure	7.4
		Other colligative properties	—
		End group analysis	1.8
M_w	$\dfrac{\sum_i n_i M_i^2}{\sum_i n_i M_i}$	Light scattering	8
		Sedimentation velocity	—
M_z	$\dfrac{\sum_i n_i M_i^3}{\sum_i n_i M_i^2}$	Sedimentation equilibrium	—
M_v	$\left(\dfrac{\sum_i n_i M_i^{1+a}}{\sum_i n_i M_i}\right)^{1/a}$	Intrinsic viscosity	9.3

Chapter 9, SEC has some serious limitations; one of these is a lack of *resolution*. Resolution in this context refers to the difference in M values that can be determined; SEC would struggle to help you decide whether your sample was narrowly distributed with $M_n = 50,000$, or was bimodal with peak molecular weights at 40,000 and 60,000. In contrast, resolution is the real strength of MALDI, as we will see below. MALDI is a relative newcomer among the techniques listed in Table 1.5, but it will undoubtedly grow in importance as its scope expands.

The next group of techniques provides measurements of M_n. They do so by being sensitive to the *number* of solute molecules in solution, as is inherent in the so-called *colligative properties* (osmotic pressure, freezing point depression, boiling point elevation, etc.). Of these, osmotic pressure is the most commonly employed. It has the virtue (shared with light scattering) of being a technique based on equilibrium thermodynamics that can provide an absolute measurement without resorting to calibration against other polymer samples. End group analysis, to be discussed below, includes any of a number of analytical tools that can be used to quantify the presence of the unique structure of the polymer chain ends. For a linear chain, with two and only two ends, counting the number of ends is equivalent to counting the number of molecules.

Light scattering is sufficiently important that it merits a full chapter; determination of M_w is only one facet of the information that can be obtained. Sedimentation experiments will not be discussed further in this text, although they play a crucial role for biopolymer analysis in general, and proteins in particular. Similarly, gel electrophoresis (not listed in Table 1.5 or discussed further here) is an analytical method of central importance in biological sciences, and especially for the separation and sequencing of DNA in the Human Genome Project.

The intrinsic viscosity approach holds a place of particular historical importance; in the days before routine use of SEC (up to about 1970) it was by far the easiest way to obtain molecular weight information. The viscosity average molecular weight defined in Table 1.5 is not a simple moment of the distribution, but involves the *Mark–Houwink exponent a*, which needs to be known based on other information. As $0.5 \leq a \leq 0.8$ for most flexible polymers in solution, $M_n \leq M_v \leq M_w$. The relation between the viscosity of a dilute polymer solution and the molecular weight of the polymer actually rests on some rather subtle hydrodynamics, as we will explore in Chapter 9.

1.8.2 End Group Analysis

As indicated previously, the end groups of polymers are inherently different in chemical structure from the repeat units of the chain, and thus provide a possible means of counting the number of molecules in a sample. Any analytical technique that can reliably quantify the concentration of end groups can potentially be used in this manner, and over the years many have been so employed. It should be apparent from the discussion in the previous section that this approach will yield a measure of the number-average molecular weight, M_n. The experiment will involve preparing a known mass of sample, probably in solution, which given M_0 corresponds to a certain number of repeat units. The number of end groups is directly proportional to the number of polymers and the ratio of the number of repeat units to the number of polymers is the number-average degree of polymerization.

Several general principles apply to end group analysis:

1. The chemical structure of the end group must be sufficiently different from that of the repeat unit for the chosen analytical technique to resolve the two clearly.
2. There must be a well-defined number of end groups per polymer, at least on average. For a linear polymer, there will be two and only two end groups per molecule, which may or may not be distinct from each other. For branched polymers, the relation of the number of end groups to the number of polymers is ambiguous, unless the total number of branching points is also known.
3. The technique is limited to relatively low molecular weights, as the end groups become more and more dilute as N increases. This is an obvious corollary of the fact that we can ignore end groups in considering the structure of high molecular weight chains. How low is low in this context? The answer will depend on the particular system and analytical technique, but as a rule of thumb end groups present at the 1% level (corresponding to degrees of polymerization of 100 for a single end group, 200 for both end groups) can be reliably determined; those at the 0.1% level cannot.

As an example, consider condensation polymers such as polyesters and polyamides. They are especially well suited to this molecular weight determination, because they tend to have lower molecular weights than addition polymers, and because they naturally have unreacted functional groups at each end. Using polyamides as an example, we can readily account for the following possibilities:

1. A linear molecule has a carboxyl group at one end and an amino group at the other, such as poly(ε-caprolactam):

    In this case, there is one functional group of each kind per molecule and could be detected for example by titration with a strong base (for –COOH) or strong acid (for –NH$_2$).
2. If a polyamide is prepared in the presence of a large excess of diamine, the average chain will be capped by an amino group at each end:

    In this case, only the amine can be titrated, and two ends are counted per molecule.
3. Similarly, if a polyamide is prepared in the presence of a large excess of dicarboxylic acid, then the average chain will have a carboxyl group at each end:

In this case, the acid group can be titrated, and again two ends should be counted per molecule.

The preceding discussion illustrated how classical acid–base titration could be used for molecular weight determination. In current practice, nuclear magnetic resonance (NMR) spectroscopy is probably the most commonly used analytical method for end group analysis, especially proton (^{1}H) NMR. An additional advantage of this approach is the possibility of obtaining further information about the polymer structure from the same measurement, as illustrated in the following example.

Example 1.6

The ^{1}H NMR spectrum in Figure 1.6 corresponds to a sample of polyisoprene containing a *sec*-butyl initiating group and a hydroxyl terminating end group. The relative peak integrations are (a) 26.9, (b) 5.22, (c) 2.00, and (d) 5.95. What is M_n for this polymer? What is the relative percentage of 1,4 and 3,4 addition?

Solution

Peak (c) corresponds to the methylene protons adjacent to the hydroxyl end group; there are two such protons per polymer. Peak (a) reflects the single olefinic proton per 1,4 repeat unit, whereas peak (b) shows the two vinyl protons per 3,4 repeat unit. If we represent the integration of peak i as I_i, then the degree of polymerization is proportional to $(I_a + I_b/2) = 26.9 + 2.61 = 29.5$. The number of polymers is proportional to $I_c/2 = 1$. Thus the number-average degree of polymerization is 29.5, which gives an $M_n = 29.5 \times 68 = 2{,}000$. Peak (d) indicates the six methyl protons on the initiator fragment. The peak integration $I_d = 5.95$ should be $3I_c$, which is within experimental error. For this particular molecular structure, therefore, either end group could be used. An additional conclusion is that essentially 100% of the polymers were terminated with a hydroxyl group.

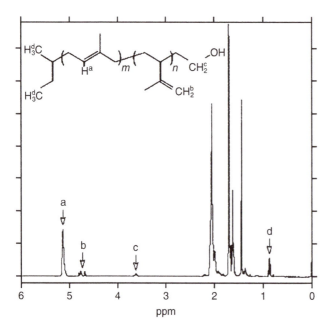

Figure 1.6 ^{1}H NMR spectrum of a polyisoprene sample, discussed in Example 1.6. (Data courtesy of N. Lynd and M.A. Hillmyer.)

The percent 1,4 addition can be computed as follows:

$$\%1,4 = \frac{I_a}{I_a + (I_b/2)} \times 100 = \frac{26.9}{29.5} \times 100 = 91.2\%.$$

This is a typical composition for polyisoprene prepared by anionic polymerization in a nonpolar solvent (see Chapter 4).

1.8.3 MALDI Mass Spectrometry

Mass spectrometry offers unprecedented resolution in the analysis of gas phase ions, and is a workhorse of chemical analysis. Its application to synthetic polymers has been limited until rather recently, primarily due to the difficulty of transferring high molecular weight species into the gas phase without degradation. However, progress in recent years has been quite rapid, with two general approaches being particularly productive. In one, electrospray ionization, a polymer solution is ejected through a small orifice into a vacuum environment; an electrode at the exit deposits a charge onto each drop of solution. The solvent then evaporates, leaving behind a charged macromolecule in the gas phase. In the other, the polymer sample is dispersed in a particular matrix on a solid substrate. An intense laser pulse is absorbed by the matrix, and the resulting energy transfer vaporizes both polymer and matrix. For uncharged synthetic polymers, the necessary charge is usually complexed with the polymer in the gas phase, after a suitable salt has been codissolved in the matrix. This technique, *matrix-assisted laser desorption/ionization mass spectrometry*, or MALDI for short, is already a standard approach in the biopolymer arena, and is making substantial inroads for synthetic polymers as well.

It is worth recalling the basic ingredients of a mass spectrometric experiment. A sample molecule of mass m is first introduced into the gas phase in a high vacuum, and at some point in the process it must acquire a net charge z. The resulting ion is accelerated along a particular direction by suitably placed electrodes, and ultimately collected and counted by a suitable analyzer. The ion acquires kinetic energy in the applied field, which depends on the net charge z and the applied voltage. This energy will result in a mass-dependent velocity v, as the kinetic energy $= mv^2/2$. This allows for the discrimination of different masses, by a variety of possible schemes. For example, if the ions experience an orthogonal magnetic field B, their trajectories will be curved to different extents, and it is possible to tune the magnitude of B to allow a particular mass to pass through an aperture before the detector. For polymers it turns out to be more effective to use time-of-flight (TOF) analysis. For a given applied field and flight path to the detector, larger masses will take longer to reach the detector. As long as all the molecules are introduced into the gas phase at the same instant in time, the time of arrival can be converted directly into a value of m/z.

The preceding discussion may give the misleading impression that MALDI is rather a straightforward technique. This is not, in fact, the case, especially for synthetic polymers. A great deal remains to be learned about both the desorption and ionization processes, and standard practice is to follow particular recipes (matrix and salt) that have been found to be successful for a given polymer. For example, polystyrene samples are most often dispersed in dithranol (1,8,9-anthracenetriol) with a silver salt such as silver trifluoroacetate. This mixture is co-dissolved in a volatile common good solvent such as tetrahydrofuran, to ensure homogeneity; after depositing a drop on a sample plate, the solvent is then allowed to evaporate. An intense pulse from a nitrogen laser ($\lambda = 337$ nm) desorbs some portion of the sample, and some fraction of the resulting gas-phase polystyrene molecules are complexed with a single silver cation.

Two examples of MALDI spectra on narrow distribution of polystyrene samples are shown in Figure 1.7a and Figure 1.7b. In the former, the average molecular weight is in the neighborhood of 5000, and different i-mers are clearly resolved. Each peak is separated by 104 g/mol, which is the repeat unit molecular weight. The absolute molecular weight of each peak should correspond to the following sum: $(104i + 108 + 1 + 57) = 104i + 166$, where 108, 1, and 57 are the contributions of the silver ion, terminal proton, and *sec*-butyl initiator fragment, respectively. Using the formulas given in Equation 1.7.2 and Equation 1.7.4, M_n and M_w can be calculated as 4620 and 4971, respectively, and the polydispersity is 1.076. This sample, which was prepared by living anionic polymerization as described in Chapter 4, is thus quite narrow. Nevertheless, the plot shows distinctly how many different *i*-mers are present, and in what relative proportion. The assumption is made that the height of each peak is proportional to its number concentration in the sample, and thus the *y*-axis corresponds to an unnormalized form of mole fraction x_i. This image, perhaps more than any other, underscores the point we have already made several times: even the best of polymer samples is quite heterogeneous. Recalling Example 1.4 and the tank car of polybutadiene, it is worth pointing out that each peak in Figure 1.7a corresponds to many structurally different molecules, in terms of the stereochemical sequence along the backbone.

The MALDI spectrum in Figure 1.7b corresponds to a sample about 10 times higher in molecular weight. At this point it is not possible to see any structure between different *i*-mers, although the expanded version shows that there is still a hint of resolution of distinct molecular weights. This serves to point out one limitation with MALDI, namely that its main attribute, high resolution, is diminished as M increases. Not apparent from this plot, but even more troublesome, is the fact that the absolute amplitude of the signal is greatly reduced compared to Figure 1.7a. It is

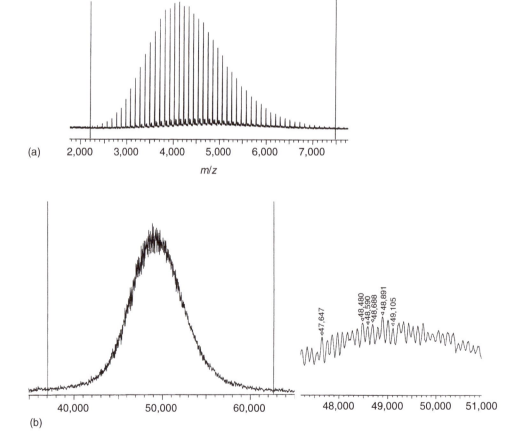

Figure 1.7   MALDI spectra for anionically prepared polystyrenes with (a) $M_w = 4{,}971$, $M_n = 4{,}620$, PDI = 1.076, and (b) $M_w = 49{,}306$, $M_n = 48{,}916$, PDI = 1.008. Inset shows that even in (b) there is some resolution of the different *i*-mers. (Data courtesy of K. Fagerquist, T. Chang, and T.P. Lodge.)

simply much harder to get higher molecular weight molecules into the gas phase. Nevertheless, the data in Figure 1.7b give $M_n = 48,916$, $M_w = 49,306$, and a polydispersity of 1.008. This turns out to be nearly as narrow as the theoretical limit for this class of polymerizations (see Chapter 4), yet it is still obviously quite heterogeneous.

We conclude this section with some further general observations about MALDI:

1. Generally, the more polar a polymer, the easier it is to analyze by MALDI. Thus poly(ethylene oxide) is relatively easy; poly(methyl methacrylate) is easier than polystyrene; polyethylene is almost impossible.
2. An important, unresolved issue is relating the amplitude of the signal of a particular peak to the relative abundance of that molecule in the sample. For example, are all molecular weights desorbed to the same extent within a given laser pulse (unlikely), and are all molecular weights equally likely to be ionized once in the gas phase (no)? Consequently, it can be dangerous to extract M_w and M_n as we did for the samples in Figure 1.7a and Figure 1.7b, because the signals have an unknown sensitivity to molecular weight. (In this instance, this problem is mitigated because the distributions are quite narrow.) In general, lower molecular weights have a much higher yield.
3. Multiply charged ions can present a problem, because one cannot distinguish between a molecule with a charge of 1 and a molecule of twice the molecular weight but with a charge of 2. In fact, if the technique were called "mass-to-charge ratio spectrometry" it would be a mouthful, but it would serve as a constant reminder of this important complication.
4. It is difficult to compare the amplitudes of peaks from one laser pulse to another, and from one sample drop to another. This presumably reflects the microscopic details of the spot on the sample that is actually at the focus of the laser beam. As a consequence any quantitative interpretation should be restricted to a given spectrum.

1.9   Preview of Things to Come

The contents of this book may be considered to comprise three sections, each containing four separate chapters. The first section, including Chapter 2 through Chapter 5, addresses the synthesis of polymers, the various reaction mechanisms and kinetics, the resulting molecular weight distributions, and some aspects of molecular characterization. In particular, Chapter 2 concerns step-growth (condensation) polymerization and Chapter 3 chain-growth (free radical) polymerization. Chapter 4 describes a family of particular polymerization schemes that permit a much higher degree of control over molecular weight, molecular weight distribution, and molecular architecture than those in the preceding two chapters. Chapter 5 addresses some of the factors that control the structural details within polymers, especially copolymers and stereoregular polymers, and aspects of their characterization.

The second section takes up the behavior of polymers dissolved in solution. The conformations of polymers, and especially random coils, are treated in Chapter 6. Solution thermodynamics are the subject of Chapter 7, including the concepts of solvent quality, osmotic pressure, and phase behavior. The technique of light scattering, which provides direct information about molecular weight, solvent quality, and chain conformations, is covered in detail in Chapter 8. Chapter 9 explores the various hydrodynamic properties of polymers in solution, and especially as they impact viscosity, diffusivity, and SEC.

The concluding section addresses the properties of polymers in the bulk, with a particular emphasis on the various solid states: rubber, glass, and crystal. Thus Chapter 10 considers polymer networks and their characteristic and remarkable elasticity. Chapter 11 treats the unusual visco-elastic behavior of polymer liquids, in a way that combines central concepts from both Chapter 9 and Chapter 10. Chapter 12 introduces the phenomenon of the glass transition, which is central to all polymer materials yet relatively unimportant in most atomic or small molecule-based materials. Finally, the rich crystallization properties of polymers are taken up in Chapter 13. The text concludes with an Appendix that reviews some of the mathematical manipulations encountered throughout the book.

1.10   Chapter Summary

In this chapter, we have introduced the central concept of chain molecules, and identified various ways in which polymers may be classified. The importance of molecular weight and its distribution was emphasized, and associated averages defined. Examples were given of the many possible structural variations that commonly occur in synthetic polymers:

1.   The most important feature of a polymer is its degree of polymerization or molecular weight. For example, even though polyethylene has the same chemical formula as the *n*-alkanes, it has remarkably different physical properties from its small molecule analogs.
2.   The statistical nature of polymerization schemes inevitably leads to a distribution of molecular weights. These can be characterized via specific averages, such as the number-average and weight-average molecular weights, or by the full distribution, which can be determined by SEC or MALDI mass spectrometry.
3.   Polymers can exhibit many different architectures, such as linear, randomly branched, or regularly branched chains, and networks. Homopolymers contain only one type of repeat unit, whereas copolymers contain two or more.
4.   There are many possible variations in local structure along a polymer chain, which we have classified as positional, stereochemical, or geometrical isomers. Given these possibilities, and those identified in the previous point, it is unlikely that any two polymer molecules within a particular sample have exactly the same chemical structure, even without considering differences in molecular weight.
5.   Natural polymers such as polysaccharides, proteins, and nucleic acids share many of the attributes of their synthetic analogs, and as such are an important part of the subject of this book. On the other hand, the specific biological functions of these macromolecules, especially proteins and nucleic acids, fall outside our scope.

Problems

To a significant extent the problems in this book are based on data from the original literature. In many instances the values given have been estimated from graphs, transformed from other functional representations, or changed in units. Therefore, these quantities do not necessarily reflect the accuracy of the original work, nor is the given number of significant figures always justified. Finally, the data may be used for purposes other than were intended in the original study.

1.   R.E. Cohen and A.R. Ramos[†] describe phase equilibrium studies of block copolymers of butadiene (B) and isoprene (I). One such polymer is described as having a 2:1 molar ratio of B to I with the following microstructure:
     B—45% *cis*-1,4; 45% *trans*-1,4; 10% vinyl.
     I—over 92% *cis*-1,4.
     Draw the structure of a portion of this polymer consisting of about 15 repeat units, and having approximately the composition of this polymer.
2.   Hydrogenation of polybutadiene converts both *cis* and *trans* isomers to the same linear structure, and vinyl groups to ethyl branches. A polybutadiene sample of molecular weight 168,000 was found by infrared spectroscopy to contain double bonds consisting of 47.2% *cis*,

---

[†] R.E. Cohen and A.R. Ramos *Macromolecules*, 12, 131 (1979).

44.9% *trans*, and 7.9% vinyl.[†] After hydrogenation, what is the average number of backbone carbon atoms between ethyl side chains?

3. Landel used a commercial material called Vulcollan 18/40 to study the rubber-to-glass transition of a polyurethane.[‡] This material is described as being "prepared from a low molecular weight polyester which is extended and cross-linked by reacting it with naphtha-lene-1,4,-diisocyanate and 1,4-butanediol. The polyester is prepared from adipic acid and a mixture of ethylene and propylene glycols." Draw the structural formula of a portion of the cross-linked polymer which includes the various possible linkages that this description includes. Remember that isocyanates react with active hydrogens; use this fact to account for the cross-linking.

4. Some polymers are listed below using either IUPAC (I) names or acceptable trivial (T) names. Draw structural formulas for the repeat units in these polymers, and propose an alternative name in the system other than the one given:
   Polymethylene (I)
   Polyformaldehyde (T)
   Poly(phenylene oxide) (T)
   Poly[(2-propyl-1,3-dioxane-4,6-diyl)methylene] (I)
   Poly(1-acetoxyethylene) (I)
   Poly(methyl acrylate) (T)

5. Star polymers are branched molecules with a controlled number of linear arms anchored to one central molecular unit acting as a branch point. Schaefgen and Flory[§] prepared poly (ε-caprolactam) four- and eight-arm stars using cyclohexanone tetrapropionic acid and dicyclohexanone octapropionic acid as branch points. The authors present the following stoichiometric definitions/relations to relate the molecular weight of the polymer to the concentration of unreacted acid groups in the product. Provide the information required for each of the following steps:

   (a) The product has the formula R-$\{-CO[-NH(CH_2)_5CO-]_y-OH\}_b$. What is the significance of R, y, and b?

   (b) If Q is the number of equivalents of multifunctional reactant which react per mole of monomer and L represents the number of equivalents of unreacted (end) groups per mole of monomer, then $<y> = (1-L)/(Q+L)$. Justify this relationship, assuming all functional groups are equal in reactivity.

   (c) If M_0 is the molecular weight of the repeat unit and M_b is the molecular weight of the original branch molecule divided by b, then the number-average molecular weight of the star polymer is

   $$M_n = b\left\{M_0\frac{1-L}{Q+L} + M_b\right\}$$

   Justify this result and evaluate M_0 and M_b for the $b=4$ and $b=8$ stars.

   (d) Evaluate M_n for the following molecules:

b	Q	L
4	0.2169	0.0018
8	0.134	0.00093

[†] W.E. Rochefort, G.G. Smith, H. Rachapudy, V.R. Raju, and W.W. Graessley, *J. Polym. Sci., Polym. Phys.*, 17, 1197 (1979).
[‡] R.F. Landel, *J. Colloid Sci.*, 12, 308 (1957).
[§] J.R. Schaefgen and P.J. Flory, *J. Am. Chem. Soc.*, 70, 2709 (1948).

6. Batzer reported the following data for a fractionated polyester made from sebacic acid and 1,6-hexanediol;[†] evaluate M_n, M_w, and M_z.

Fraction	1	2	3	4	5	6	7	8	9
Mass (g)	1.15	0.73	0.415	0.35	0.51	0.34	1.78	0.10	0.94
$M \times 10^{-4}$	1.25	2.05	2.40	3.20	3.90	4.50	6.35	4.10	9.40

7. The Mark–Houwink exponent a for poly(methyl methacrylate) at $25°C$ has the value 0.69 in acetone and 0.80 in chloroform. Calculate (retaining more significant figures than strictly warranted) the value of M_v that would be obtained for a sample with the following molecular weight distribution if the sample were studied by viscometry in each of these solvents.[‡] Compare the values of M_v with M_n and M_w.

$n_i \times 10^3$ (mol)	1.2	2.7	4.9	3.1	0.9
$M_i \times 10^{-5}$ (g/mol)	2.0	4.0	6.0	8.0	10.0

8. Consider a set consisting of 4–8 family members, friends, neighbors, etc. Try to select a variety of ages, genders, and other attributes. Take the mass of each individual (a rough estimate is probably wiser than asking directly) and calculate the number- and weight-average masses for this set. Does the resulting PDI indicate a rather "narrow" distribution? If you picture this group in your mind, do you imagine them all to be roughly the same size, as the PDI probably suggests?

9. Prove that the polydispersity of the Schulz–Zimm distribution is given by Equation 1.7.21. You may want to look up the general solution for integrals of the type $\int x^a e^{-bx} dx$.

10. In Figure 1.5a through Figure 1.5c it appears that the maximum in w_i corresponds closely to M_n. Differentiate Equation 1.7.20 with respect to M_i to show why this is the case.

11. The MALDI spectrum in Figure 1.7b resembles a Gaussian or normal distribution. One property of a Gaussian distribution is that the half-width at half-height of the peak is approximately equal to 1.2σ. Use this relation to estimate σ from the trace, and compare it to the value you would get from Equation 1.7.16.

12. Give the overall chemical reactions involved in the polymerization of these monomers, the resulting repeat unit structure, and an acceptable name for the polymer.

(a)

(b)

(c)

(d)

[†] H. Batzer, *Makromol. Chem.*, 5, 5 (1950).
[‡] S.N. Chinai, J.D. Matlock, A.L. Resnick, and R.J. Samuels, *J. Polym. Sci.*, 17, 391 (1955).

(e)

O=C=N

+ HO $\diagup$ OH
            4

N=C=O

(f)

Me

H$_2$N $\diagdown$ OH

O

13. A MALDI-TOF analysis of a polystyrene sample exhibited a peak (one of many) at 1206. The sample was prepared in dithranol, with silver nitrate as the salt. Assuming no head-to-head defects, how many distinct chemical structures could this peak represent? Propose structures for the end groups of the polymer as well.

14. Proton NMR is used to attempt to quantify the molecular weight of a poly(ethylene oxide) molecule with methyoxy end groups at each terminus. If the integration of the methyl protons relative to the methylene protons gave a ratio of 1:20, what can you say about the molecular weight?

15. What would be M_w and M_n for a sample obtained by mixing 10 g of polystyrene ($M_w = 100,000$, $M_n = 70,000$) with 20 g of another polystyrene ($M_w = 60,000$, $M_n = 20,000$)?

16. What would M_w and M_n be for an equimolar mixture of tetradecane and decane? (Ignore isotope effects.)

References

1. *Chemical & Engineering News*, 58, 29 (1980).
2. *The Graduate*, directed by Mike Nichols, Embassy Pictures Corporation, 1967.
3. No enterprise as rich as polymer science has only one "father." Herman Mark is one of those to whom the title could readily be applied. An interesting interview with Professor Mark appears in the *Journal of Chemical Education*, 56, 38 (1979).
4. H. Morawetz, *Polymers: The Origins and Growth of a Science*, Wiley, New York, 1985.
5. R.B. Seymour, *History of Polymer Science and Technology*, Marcel Dekker, New York, 1982.
6. C.S. Marvel, another pioneer in polymer chemistry, reminisced about the early days of polymer chemistry in the United States in the *Journal of Chemical Education*, 58, 535 (1981).
7. See IUPAC Macromolecular Nomenclature Commission, *Macromolecules*, 6, 149 (1973).

Further Readings

H.R. Allcock, F.W. Lampe, and J.E. Mark, *Contemporary Polymer Chemistry*, 3rd ed., Prentice Hall, Englewood Cliffs, NJ, 2003.

P.J. Flory, *Principles of Polymer Chemistry*, Cornell University Press, Ithaca, NY, 1953.

A.Y. Grosberg, and A.R. Khokhlov, *Giant Molecules, Here, There and Everywhere*, Academic Press, San Diego, CA, 1997.

A.L. Lehninger, *Biochemistry*, 2nd ed., Worth Publishers, New York, 1975.

P. Munk and T.M. Aminabhavi, *Introduction to Macromolecular Science*, 2nd ed., Wiley Interscience, New York, 2002.

G. Odian, *Principles of Polymerization*, 4th ed., Wiley, New York, 2004.

P.C. Painter and M.M. Coleman, *Fundamentals of Polymer Science*, Technomic, Lancaster, PA, 1994.

R.B. Seymour and C.E. Carraher, Jr., *Polymer Chemistry: An Introduction*, Marcel Dekker, New York, 1981.

L.H. Sperling, *Introduction to Physical Polymer Science*, Wiley Interscience, New York, 1986.

R.J. Young and P.A. Lovell, *Introduction to Polymers*, 2nd ed., Chapman and Hall, London, 1991.

2

Step-Growth Polymerization

2.1 Introduction

In Section 1.4, we discussed the classification of polymers into the categories of addition or condensation. At that time we noted that these classifications could be based on the following:

1. Stoichiometry of the polymerization reaction (small molecule eliminated?)
2. Composition of the backbone of the polymer (atoms other than carbon present?)
3. Mechanism of the polymerization (stepwise or chain reaction?)

It is the third of these criteria that offers the most powerful insight into the nature of the polymerization process for this important class of materials. We shall sometimes use the terms *step-growth* and *condensation polymers* as synonyms, although step-growth polymerization encompasses a wider range of reactions and products than either criteria (1) or (2) above would indicate.

The chapter is organized as follows. First, we examine how the degree of polymerization and its distribution vary with the progress of the polymerization reaction, with the latter defined both in terms of stoichiometry and time (Section 2.2 through Section 2.4). Initially we consider these topics for simple reaction mixtures, that is, those in which the proportions of reactants agree exactly with the stoichiometry of the reactions. After this, we consider two important classes of condensation or step-growth polymers: polyesters and polyamides (Section 2.5 and Section 2.6). Finally we consider nonstoichiometric proportions of reactants (Section 2.7). The important case of multifunctional monomers, which can introduce branching and cross-linking into the products, is deferred until Chapter 10.

2.2 Condensation Polymers: One Step at a Time

As the name implies, step-growth polymers are formed through a series of steps, and high molecular weight materials result from a large number of steps. Although our interest lies in high molecular weight, long-chain molecules, a crucial premise of this chapter is that these molecules can be effectively discussed in terms of the individual steps that lead to the formation of the polymer. Thus, polyesters and polyamides are substances that result from the occurrence of many steps in which ester or amide linkages are formed between the reactants. Central to our discussion is the idea that these steps may be treated in essentially the same way, whether they occur between small molecules or polymeric species. We shall return to a discussion of the implications and justification of this *assumption of equal reactivity* throughout this chapter.

2.2.1 Classes of Step-Growth Polymers

Here are examples of important classes of step-growth polymers:

1. Polyesters—successive reactions between diols and dicarboxylic acids:

$$n \; HO \overset{R}{\frown} OH \; + \; n \; HO \overset{O \quad O}{\underset{R'}{\frown}} OH \; \longrightarrow \; \left(O \underset{R}{\frown} O \overset{O}{\underset{O}{\frown}} \overset{R'}{\underset{O}{\frown}} \right)_n + 2n \; H_2O \tag{2.A}$$

2. Polyamides—successive reactions between diamines and dicarboxylic acids:

$$n \; H_2N{\sim}R{\sim}NH_2 \;+\; n \; HO-C(=O)-R'-C(=O)-OH \longrightarrow \left(N(H)-R-N(H)-C(=O)-R'-C(=O) \right)_n + 2n \; H_2O \tag{2.B}$$

3. General—successive reactions between difunctional monomer A–A and difunctional monomer B–B:

$$n \; A{-}A \;+\; n \; B{-}B \longrightarrow \left(a{\sim}a{\sim}b{\sim}b \right)_n + \cdots \tag{2.C}$$

Since the two reacting functional groups can be located in the same reactant molecule, we add the following:

4. Poly(amino acid)

$$n \; H_2N{\sim}C(R)(H){-}C(=O){-}OH \longrightarrow \left(N(H)-C(R)(H)-C(=O) \right)_n + n \; H_2O \tag{2.D}$$

5. General

$$n \; A{-}B \longrightarrow \left(a{\sim}b \right)_n + \cdots \tag{2.E}$$

Of course, in Reaction (2.A) and Reaction (2.B) the hydrocarbon sequences R and R' can be the same or different, contain any number of carbon atoms, be linear or cyclic, and so on. Likewise, the general reactions, Reaction (2.C) and Reaction (2.E), certainly involve hydrocarbon sequences between the reactive groups A and B. The notation involved in these latter reactions is particularly convenient, however, and we shall use it extensively in this chapter. It will become clear as we proceed that the stoichiometric proportions of reactive groups—A and B in the above notation—play an important role in determining the characteristics of the polymeric product. Accordingly, we shall confine our discussion for the present to reactions of the type given by Reaction (2.E), since equimolar proportions of A and B are assured by the structure of the monomer.

2.2.2 First Look at the Distribution of Products

Table 2.1 presents a hypothetical picture of how Reaction (2.E) might appear if we examined the distribution of product molecules in detail. Row 1 of Table 2.1 shows the initial pool of monomers,

Table 2.1 Hypothetical Step-Growth Polymerization of 10 AB Molecules[a]

Row	Molecular species present
1.	AB AB AB AB AB AB AB AB AB AB
2.	AbaB AbaB AbaB AbaB AB AB
3.	AbababaB AbaB AbabaB AB
4.	AbababaB AbabababaB AB
5.	AbababaB AbababababaB
6.	AbabababababababababaB

[a]A and B represent two different functional groups and ab is the product of their reaction with each other. Consult the text for a discussion of the line-by-line development of the reaction.

10 molecules in this example. Row 2 shows a possible composition after a certain amount of reaction has occurred. We shall see in Section 2.4 that the particular condensations that account for the differences between the first and second rows are not highly probable. Our objective here is not to assess the probability of certain reactions, but rather to consider some possibilities. Stoichiometrically, we can still account for the initial set of 10 A groups and 10 B groups; we indicate those that have reacted with each other as ab groups. The same conservation of atom groupings would be obtained if row 2 showed one trimer, two dimers, and three monomers instead of the four dimers and two monomers indicated in Table 2.1. Other combinations could also be assembled. These possibilities indicate one of the questions that we shall answer in this chapter: How do the molecules distribute themselves among the different possible species as the reaction proceeds?

Row 3 of Table 2.1 shows the mixture after two more reaction steps have occurred. Again, the components we have elected to show are an arbitrary possibility. For the monomer system we have chosen, the concentration of A and B groups in the initial monomer sample are equal to each other and equal to the concentration of monomer. In this case, an assay of either A groups or B groups in the mixture could be used to monitor the progress of the reaction. Choosing the number of A groups for this purpose, we see that this quantity drops from 10 to 6 to 4, respectively, as we proceed through rows 1, 2, and 3 of Table 2.1. What we wish to point out here is the fact that the 10 initial monomers are now present in four molecules, so the number average degree of polymerization is only 2.5, even though only 40% of the initial reactive groups remain. Another question is thus raised: In general, how does the average molecular weight vary with the extent of the reaction?

The reaction mixture in the fourth row of Table 2.1 is characterized by a number average degree of polymerization $N_n = 10/3 = 3.3$, with only 30% of the functional groups remaining. This means that 70% of the possible reactions have already occurred, even though we are still dealing with a very low average degree of polymerization. Note that the average degree of polymerization would be the same if the 70% reaction of functional groups led to the mixture AbababababababaB and two AB's. This is because the initial 10 monomers are present in three molecules in both instances, and we are using number averages to talk about these possibilities. The weight average would be different in the two cases. This poses still another question: How does the molecular weight distribution vary with the extent of reaction?

By the fifth row, the reaction has reached 80% completion and the number average value of the degree of polymerization N_n is 5. Although we have considered this slowly evolving polymer in terms of the extent of reaction, another question starts to be worrisome: How long is this going to take?

The sixth row represents the end of the reaction as far as linear polymer is concerned. Of the 10 initial A groups, 1 is still unreacted, but this situation raises the possibility that the decamer shown in row 6—or for that matter, some other i-mer, including monomer—might form a ring or cyclic compound, thereby eliminating functional groups without advancing the polymerization. Throughout this chapter we will assume that the extent of ring formation is negligible.

It is an easy matter to generalize the procedure we have been following and express the number average degree of polymerization in terms of the extent of reaction, regardless of the initial sample size. We have been dividing the initial number of monomers present by the total number of molecules present after any extent of reaction. Each molecular species—whether monomer or polymer of any length—contains just one A group. The total number of *monomers* is therefore equal to the initial (superscript o) number of A groups, ν_A^o; the total number of *molecules* at any extent of reaction (no superscript) is equal to the number of A groups, ν_A, present at that point. The number average degree of polymerization is therefore given by

$$N_n = \frac{\nu_A^o}{\nu_A} \tag{2.2.1}$$

It is convenient to define the fraction of reacted functional groups in a reaction mixture by a parameter p, called the *extent of reaction*. Thus, p is the fraction of A groups that have reacted at any stage of the process, and $1-p$ is the unreacted fraction:

$$1 - p = \frac{\nu_A}{\nu_A^o} \tag{2.2.2}$$

or

$$p = 1 - \frac{\nu_A}{\nu_A^o} \tag{2.2.3}$$

Comparison of Equation 2.2.1 and Equation 2.2.2 enables us to write very simply:

$$N_n = \frac{1}{1-p} \tag{2.2.4}$$

This expression is consistent with the analysis of each of the rows in Table 2.1 as presented above and provides a general answer to one of the questions posed there. It is often a relatively easy matter to monitor the concentration of functional groups in a reaction mixture; Equation 2.2.4 represents a quantitative summary of an end group method for determining N_n. We reiterate that Equation 2.2.4 assumes equal numbers of A and B groups, with none of either lost in nonpolymer reactions.

From row 6 in Table 2.1, we see that $N_n = 10$ when $p = 0.9$. The fact that this is also the maximum value for N is an artifact of the example. In a larger sample of monomers higher average degrees of polymerization are attainable. Equation 2.2.4 enables us to calculate that N_n becomes 20, 100, and 200 for extents of reaction 0.950, 0.990, and 0.995, respectively. These results reveal why condensation polymers are often of relatively modest molecular weight: it may be very difficult to achieve the extents of reaction required for very high molecular weights. As p increases the concentration of H_2O (or other small molecule product) will increase, and the law of mass action will oppose further polymerization. Consequently, steps must be taken to remove the small molecule as it is formed, if high molecular weights are desired.

2.2.3 A First Look at Reactivity and Reaction Rates

Most of the questions raised in the past few paragraphs will be answered during the course of this chapter, some for systems considerably more involved than the one considered here. Before proceeding further, we should reemphasize one premise that underlies the entire discussion of Table 2.1: How do the chemical reactivities of A and B groups depend on the degree of polymerization of the reaction mixture? In Table 2.1, successive entries were generated by simply linking together at random those species present in the preceding row. We have thus assumed that, as far as reactivity is concerned, an A reacts as an A and a B reacts as a B, regardless of the size of the molecule to which the group is attached. If this assumption of equal reactivity is valid, it results in a tremendous simplification; otherwise we shall have to characterize reactivity as a function of degree of polymerization, extent of reaction, and so on.

One of the most sensitive tests of the dependence of chemical reactivity on the size of the reacting molecules is the comparison of the rates of reaction for compounds that are members of a homologous series with different chain lengths. Studies by Flory and others on the rates of esterification and saponification of esters were the first investigations conducted to clarify the dependence of reactivity on molecular size [1]. The rate constants for these reactions are observed to converge quite rapidly to a constant value that is independent of molecular size, after an initial dependence on molecular size for small molecules. In the esterification of carboxylic acids, for example, the rate constants are different for acetic, propionic, and butyric acids, but constant for carboxylic acids with 4–18 carbon atoms. This observation on nonpolymeric compounds has been

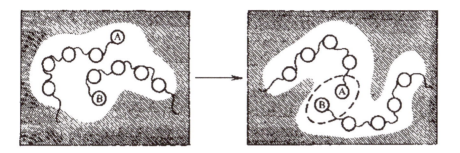

Figure 2.1   The reaction of A and B groups at the ends of two different chains. Note that rotations around only a few bonds will bring A and B into the same cage of neighboring groups, indicated by the dashed-line enclosure.

generalized to polymerization reactions as well. The latter are subject to several complications that are not involved in the study of simple model compounds, but when these complications are properly considered, the independence of reactivity on molecular size has been repeatedly verified.

The foregoing conclusion does not mean that a constant *rate* of reaction persists throughout Table 2.1. The rate of reaction depends on the concentrations of reactive groups, as well as on their reactivities. Accordingly, the rate of the reaction decreases as the extent of reaction progresses. When the rate law for the reaction is extracted from proper kinetic experiments, specific reactions are found to be characterized by fixed rate constants over a range of N_n values.

Among the further complications that can interfere with this conclusion is the possibility that the polymer becomes insoluble beyond a critical molecular weight or that the low molecular weight by-product molecules accumulate and thereby shift the equilibrium to favor reactants. It is also possible that the transport of reactants will be affected by the increasing viscosity of the polymerization medium, which is a very complicated issue.

Figure 2.1 suggests that reactive end-groups may be brought into contact by rotation around only a few bonds, an effect which is therefore independent of chain length. Once in close proximity, the A and B groups may be thought of as being in the same "cage" defined by near neighbors. It may take some time for the two reactive groups to diffuse together, but it will also take some time for them to diffuse apart; this provides the opportunity to react. The rate at which an A and a B group react to form an ab linkage therefore depends on the relative rates of three processes: the rate to diffuse together, the rate at which they diffuse apart, and the rate at which "trapped" A and B groups react. These considerations can be expressed more quantitatively by writing the process in terms of the following mechanism:

$$-A + -B \; \underset{k_o}{\overset{k_i}{\rightleftharpoons}} \; (-A + B-) \overset{k_r}{\rightarrow} -ab - \tag{2.F}$$

where the parentheses represent the caged pair, as in Figure 2.1, and the ks are the rate constants for the individual steps: k_i and k_o for diffusion into and out of the cage, respectively, and k_r for the reaction itself.

Since this is the first occasion we have had to examine the rates at which chemical reactions occur, a few remarks about mechanistic steps and rate laws seem appropriate. The reader who feels the need for additional information on this topic should consult any introductory physical chemistry text.

As a brief review we recall the following:

1.   The rate of a process is expressed by the derivative of a concentration (square brackets) with respect to time, $d[A]/dt$. If the concentration of reaction product is used, this quantity is

positive; if a reactant is used, it is negative and a minus sign must be included. Also, each derivative $d[A]/dt$ should be divided by the coefficient of that component in the chemical equation that describes the reaction, so that a single rate is described whichever component in the reaction is used to monitor it.

2. A *rate law* describes the rate of reaction as the product of a constant k, called the *rate constant*, and various concentrations, each raised to specific powers. The power of an individual concentration term in a rate law is called the *order* with respect to that component, and the sum of the exponents of all concentration terms gives the *overall order* of the reaction. Thus in the rate law Rate $= k[X]^1[Y]^2$, the reaction is first order in X, second order in Y, and third order overall.

3. A rate law is determined experimentally and the rate constant evaluated empirically. *There is no necessary connection between the stoichiometry of a reaction and the form of the rate law.*

4. A mechanism is a series of simple reaction steps that, when added together, account for the overall reaction. The rate law for the individual steps of the mechanism may be written by inspection of the mechanistic steps. The coefficients of the reactants in the chemical equation describing the step become the exponents of these concentrations in the rate law for that step.

5. Frequently it is possible to write more than one mechanism that is compatible with an observed rate law. Thus, the ability to account for an experimental rate law is a necessary but not a sufficient criterion for the correctness of the mechanism.

These ideas are readily applied to the mechanism described by Reaction (2.F). To begin with, the rate at which ab links are formed is first order with respect to the concentration of entrapped pairs. In this sense, the latter behaves as a reaction intermediate or transition state according to this mechanism. Therefore

$$\text{Rate of ab formation} = k_r[(-A + B-)] \tag{2.2.5}$$

These entrapped pairs, in turn, form at a rate given by the rate at which the two groups diffuse together minus the rate at which they either diffuse apart or are lost by reaction:

$$\frac{d[(-A + B-)]}{dt} = k_i[A][B] - k_o[(-A + B-)] - k_r[(-A + B-)] \tag{2.2.6}$$

The concentration of entrapped pairs is assumed to exist at some stationary-state (subscript s) level in which the rates of formation and loss are equal. In this stationary state $d[(-A + B-)]/dt = 0$ and Equation 2.2.6 becomes

$$[(-A + B-)]_s = \frac{k_i}{k_o + k_r}[A][B] \tag{2.2.7}$$

where the subscript s reminds us that this is the stationary-state value. Substituting Equation 2.2.7 into Equation 2.2.5 gives

$$\text{Rate of ab formation} = \frac{k_i k_r}{k_o + k_r}[A][B] \tag{2.2.8}$$

We shall return to this type of kinetic analysis in Chapter 3 where we discuss chain-growth polymerization.

According to the mechanism provided by Reaction (2.F) and the analysis given by Equation 2.2.8, the rate of polymerization is dependent upon the following:

1. The concentrations of both A and B, hence the reaction slows down as the conversion to polymer progresses.

2. The three constants associated with the rates of the individual steps in Reaction (2.F).

3. If the rate of chemical reaction is very slow compared to the rate of group diffusion ($k_r \ll k_i, k_o$), then Equation 2.2.8 reduces to

$$\text{Rate of ab formation} = \frac{k_i}{k_o} k_r [A][B] \tag{2.2.9}$$

4. The two constants k_i and k_o describe exactly the same kind of diffusional process, and differ only in direction. Hence they should have the same dependence on molecular size, whatever that might be, and that dependence therefore cancels out.
5. The mechanism in Reaction (2.F) is entirely comparable to the same reaction in low molecular weight systems. Such reactions involve considerably larger activation energies than physical processes like diffusion and, hence, do proceed slowly.
6. If $k_r \gg k_i, k_o$, then Equation 2.2.8 reduces to

$$\text{Rate of ab formation} = k_i [A][B] \tag{2.2.10}$$

Note that the rate law in this case depends only on k_i and any size dependence for this constant would not cancel out.

Both Equation 2.2.9 and Equation 2.2.10 predict rate laws that are first order with respect to the concentration of each of the reactive groups; the proportionality constant has a different significance in the two cases, however. The observed rate laws, which suggest a reactivity that is independent of molecular size and the a priori expectation cited in item 5 above regarding the magnitudes of different kinds of k values, lend credibility to the version presented in Equation 2.2.9.

Our objective in the preceding argument has been to justify the attitude that each ab linkage forms according to the same rate law, regardless of the extent of the reaction. While our attention is focused on the rate laws, we might as well consider the question, raised above, about the actual rates of these reactions. This is the topic of the next section.

2.3 Kinetics of Step-Growth Polymerization

In this section we consider the experimental side of condensation kinetics. The kinds of ab links that have been most extensively studied are ester and amide groups, although numerous additional systems could also be cited. In many of these systems the carbonyl group is present and believed to play an important role in stabilizing the actual chemical transition state involved in the reactions. The situation can be represented by the following schematic reaction:

$$\tag{2.G}$$

in which the intermediate is stabilized by coordination with protons, metal ions, or other Lewis acids. The importance of this is to emphasize that the kinds of reactions we are considering are often conducted in the presence of an acid catalyst, frequently something like a sulfonic acid or a metal oxide. The purpose of a catalyst is to modify the rate of a reaction, so we must be attentive to the situation with respect to catalysts. At present, we assume a constant concentration of catalyst and attach a subscript c to the rate constant to remind us of the assumption. Accordingly, we write

$$-\frac{d[A]}{dt} = k_c [A][B] \tag{2.3.1}$$

which is consistent with both Equation 2.2.9 and Equation 2.2.10. We expect the constant k_c to be dependent on the concentration of the catalyst in some way which means that Equation 2.3.1 may be called a pseudo-second-order rate law. We shall presently consider these reactions in the absence of external catalysts. For now it is easier to proceed with the catalyzed case.

2.3.1 Catalyzed Step-Growth Reactions

Equation 2.3.1 is the differential form of the rate law that describes the rate at which A groups are used up. To test a proposed rate law and evaluate the rate constant it is preferable to work with the integrated form of the rate law. The integration of Equation 2.3.1 yields different results, depending on whether the concentrations of A and B are the same or different:

1. We define [A] and [B] as the instantaneous concentrations of these groups at any time t during the reaction, and $[A]_0$ and $[B]_0$ as the concentrations of these groups at $t = 0$.
2. If $[A]_0 = [B]_0$, the integration of Equation 2.3.1 yields

$$\frac{1}{[A]} - \frac{1}{[A]_0} = k_c t \tag{2.3.2}$$

3. If $[A]_0 \neq [B]_0$, the integration yields

$$\frac{1}{[A]_0 - [B]_0} \ln\left(\frac{[A][B]_0}{[A]_0[B]}\right) = k_c t \tag{2.3.3}$$

Both of these results are readily obtained; we examine the less obvious relationship in Equation 2.3.3 in the following example. The consequences of different A and B concentrations on the molecular weight of the polymer will be discussed in Section 2.7.

Example 2.1

By differentiation, verify that Equation 2.3.3 is a solution to Equation 2.3.1 for the conditions given.

Solution

Neither $[A]_0$ nor $[B]_0$ are functions of t, although both [A] and [B] are. We write the latter two as $[A] = [A]_0 - x$ and $[B] = [B]_0 - x$. Substitute these results into Equation 2.3.3 and rearrange:

$$\ln\frac{[A]_0 - x}{[B]_0 - x} + \ln\frac{[B]_0}{[A]_0} = ([A]_0 - [B]_0)\, k_c t$$

now differentiate with respect to t, noting that only x is a function of t:

$$\left(\frac{[B]_0 - x}{[A]_0 - x}\right)\left(\frac{-([B]_0 - x) + ([A]_0 - x)}{([B]_0 - x)^2}\right)\frac{dx}{dt} = ([A]_0 - [B]_0)\, k_c$$

that after cancellation and rearrangement gives

$$\frac{dx}{dt} = k_c([A]_0 - x)([B]_0 - x) = k_c[A][B]$$

Since $d[A]/dt = -dx/dt$ by the definition of x, this proves Equation 2.3.3 to be a solution to Equation 2.3.1. Equation 2.3.3 is undefined in the event $[A]_0 = [B]_0$, but in this case the expression is anyhow inapplicable. Since A and B react in a 1:1 proportion, their concentrations are identical at all stages of reaction if they are equal initially. In this case, Equation 2.3.1 would reduce to a simpler second-order rate law, which integrates to Equation 2.3.2.

We shall now proceed on the assumption that $[A]_0$ and $[B]_0$ are equal. As noted above, having both reactive groups on the same molecule is one way of enforcing this condition. Accordingly, we rearrange Equation 2.3.2 to give the instantaneous concentrations of unreacted A groups as a function of time:

$$[A] = \frac{[A]_0}{1 + k_c[A]_0\, t} \tag{2.3.4}$$

At this point, it is convenient to recall the extent of reaction parameter, p, defined by Equation 2.2.3. If we combine Equation 2.2.2 and Equation 2.3.4, we obtain

$$1 - p = \frac{1}{1 + k_c[A]_0\, t} \tag{2.3.5}$$

or

$$\frac{1}{1-p} = N_n = 1 + k_c[A]_0\, t \tag{2.3.6}$$

where we incorporated Equation 2.2.4 into the present discussion. These last expressions provide two very useful views of the progress of a condensation polymerization reaction with time. Equation 2.3.4 describes how the concentration of A groups asymptotically approaches zero at long times; Equation 2.3.6 describes how the number average degree of polymerization increases linearly with time.

Equation 2.3.6 predicts a straight line when $1/(1-p)$ is plotted against t. Figure 2.2 shows such a plot for adipic acid reacted with 1,10-decamethylene glycol, catalyzed by p-toluene sulfonic acid. The reaction had already been run to consume 82% of the reactive groups before this experiment was conducted. Interpreting the slope of the line in terms of Equation 2.3.6 and in the light of actual initial concentrations gives a value of $k_c = 0.097$ kg eq^{-1} min^{-1}. Note that

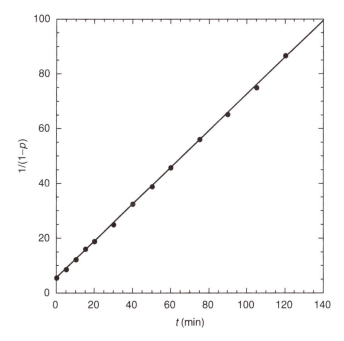

Figure 2.2 Plot of $1/(1-p)$ versus time for the late stages of esterification of adipic acid with 1,10-decamethylene glycol at 161°C, catalyzed by p-toluene sulfonic acid. The reaction time $t = 0$ corresponds to a previous extent of reaction in which 82% of the COOH groups had been consumed. (Data from Hamann, S.D., Solomon, D.H., and Swift, J.D., *J. Macromol. Sci. Chem.*, A2, 153, 1968.)

these units imply group concentrations expressed as equivalents per kilogram. Mass rather than volume units are often used for concentration, as substantial volume changes may occur during polymerization.

2.3.2 How Should Experimental Data Be Compared with Theoretical Rate Laws?

Although the results presented in Figure 2.2 appear to verify the predictions of Equation 2.3.6, this verification is not free from controversy. This controversy arises because various workers in this field employ different criteria in evaluating the success of the relationships we have presented in fitting experimental polymerization data. One school of thought maintains that an adequate kinetic description of a process must apply to the data over a large part of the *time* of the experiment. A second point of view maintains that a rate law correctly describes a process when it applies over a wide portion of the *concentration change* that occurs during a reaction. Each of these criteria seeks to maximize the region of fit, but the former emphasizes maximizing the range of t while the latter maximizes the range of p. Both standards tolerate deviations from their respective ideals at the beginning or the end of the experiment. Deviations at the beginning of a process are rationalized in terms of experimental uncertainties at the point of mixing or modelistic difficulties on attainment of stationary-state conditions.

The existence of these two different standards for success would be only of academic interest if the analysis we have discussed applied to experimental results over most of the time range and over most extents of reactions as well. Unfortunately, this is not the case in all of the systems that have been investigated. Ref. [2], for example, shows one particular set of data—adipic acid and diethylene glycol at $166°C$, similar reactants as the system in Figure 2.2—analyzed according to two different rate laws. This system obeys one rate law between $p = 0.50$ and 0.85 that represents 15% of the duration of the experiment, and another rate law between $p = 0.80$ and 0.93, which spans 45% of the reaction time. These would be interpreted differently by the two standards above. This sort of dilemma is not unique to the present problem, but arises in many situations where one variable undergoes a large percentage of its total change while the other variable undergoes only a small fraction of its change. In the present context one way out of the dilemma is to take the view that only the latter stages of the reaction are significant, as it is only beyond, say, $p = 0.80$, that it makes sense to consider the process as one of polymerization. Thus, it is only at large extents of reaction that polymeric products are formed and, hence, the kinetics of polymerization should be based on a description of this part of the process. This viewpoint intentionally focuses attention on a relatively modest but definite range of p values. Since the reaction is necessarily slow as the number of unreacted functional groups decreases, this position tends to maximize the time over which the rate law fits the data. Calculation from the ordinate of Figure 2.2 shows that the data presented there represent only about the last 20% of the range of p values. The zero of the timescale has thus been shifted to pick up the analysis of the reaction at this point.

We commented above that the deviations at the beginning or the end of kinetic experiments can be rationalized, although the different schools of thought would disagree as to what constitutes "beginning" and "end." Now that we have settled upon the polymer range, let us consider specifically why deviations occur from a simple second-order kinetic analysis in the case of catalyzed polymerizations. At the beginning of the experiment, say, up to $p \approx 0.5$, the concentrations of A and B groups change dramatically, even though the number average degree of polymerization has only changed from monomer to dimer. By ordinary polymeric standards, we are still dealing with a low molecular weight system that might be regarded as the solvent medium for the formation of polymer. During this transformation, however, 50% of the very-polar A groups and 50% of the very-polar B groups have been converted to the less-polar ab groups. Thus, a significant change in the polarity of the polymerization medium occurs during the first half of the change in p, even though an insignificant amount of true polymer has formed. In view of the role of ionic intermediates as suggested by Reaction (2.G), the polarity of the reaction medium might very well influence the rate law during this stage of the reaction.

At the other end of the reaction, deviations from idealized rate laws are attributed to secondary reactions such as degradation of acids, alcohols, and amines through decarboxylation, dehydration, and deamination, respectively. The step-growth polymers that have been most widely studied are simple condensation products such as polyesters and polyamides. Although we shall take up these classes of polymers—polyesters and polyamides—specifically in Section 2.5 and Section 2.6, respectively, it is appropriate to mention here that these are typically equilibrium reactions.

$$\text{(2.H)}$$

and

$$\text{(2.I)}$$

In order to achieve large p's and high molecular weights, it is essential that these equilibria be shifted to the right by removing the by-product molecule, water in these reactions. This may be accomplished by heating, imposing a partial vacuum, or purging with an inert gas, or some combination of the three. These treatments also open up the possibility of reactant loss due to volatility, which may accumulate to a significant source of error for reactions that are carried out to large values of p.

2.3.3 Uncatalyzed Step-Growth Reactions

Until now we have been discussing the kinetics of catalyzed reactions. Losses due to volatility and side reactions also raise questions as to the validity of assuming a constant concentration of catalyst. Of course, one way of avoiding this issue is to omit an outside catalyst; reactions involving carboxylic acids can be catalyzed by these compounds themselves. Experiments conducted under these conditions are informative in their own right and not merely as a means of eliminating errors in the catalyzed case. As noted in connection with the discussion of Reaction (2.G), the intermediate is stabilized by coordination with a proton from the catalyst. In the case of autoprotolysis by the carboxylic intermediate,

as this intermediate involves an additional equivalent of acid functional groups, the rate law for the disappearance of A groups becomes

$$-\frac{d[A]}{dt} = k_u[A]^2[B] \qquad (2.3.7)$$

on the assumption that A represents carboxyl groups. In this case, k_u is the rate constant for the uncatalyzed reaction. This differential rate law is the equivalent of Equation 2.3.1 for the catalyzed reaction. Equation 2.3.7 is readily integrated when $[A]_0 = [B]_0$, in which case it becomes

$$-\frac{1}{[A]^3}d[A] = k_u dt \qquad (2.3.8)$$

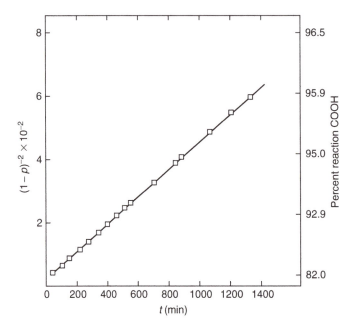

Figure 2.3 Plot of $1/(1-p)^2$ (left ordinate) and p (right ordinate) versus time for an uncatalyzed esterification. (Data from Hamann, S.D., Solomon, D.H., and Swift, J.D., *J. Macromol. Sci. Chem.*, A2, 153, 1968.)

This integrates to

$$\frac{1}{[A]^2} - \frac{1}{[A]_0^2} = 2k_u t \tag{2.3.9}$$

Thus for the uncatalyzed reaction, we have the following:

1. The rate law is third order.
2. Since $[A]/[A]_0 = 1-p$, Equation 2.3.9 may be rewritten as

$$\frac{1}{(1-p)^2} = 1 + 2k_u[A]_0^2 t \tag{2.3.10}$$

and this shows that a plot of $(1-p)^{-2}$ increases linearly with t.
3. Since $[A]/[A]_0 = 1/N_n$, Equation 2.3.10 becomes

$$N_n^2 = 1 + 2k_u[A]_0^2 t \tag{2.3.11}$$

which shows that N_n increases more gradually with t than in the catalyzed case, all other things being equal.

Figure 2.3 shows data for the uncatalyzed polymerization of adipic acid and 1,10-decamethylene glycol at 161°C plotted according to Equation 2.3.10. The various provisos of the catalyzed case apply here also, so it continues to be appropriate to consider only the final stages of the conversion to polymer. From these results, k_u is about 4.3×10^{-3} kg^2 eq^{-2} min^{-1} at 161°C.

We conclude this section with a numerical example that serves to review and compare some of the important relationships we have considered.

Example 2.2

Assuming that $k_c = 10^{-1}$ kg eq^{-1} min^{-1}, $k_u = 10^{-3}$ kg^2 eq^{-1} min^{-1}, and $[A]_0 = 10$ eq kg^{-1}, calculate the time required for p to reach values 0.2, 0.4, 0.6, and so on, for both catalyzed and uncatalyzed polymerizations, assuming that Equation 2.3.2 and Equation 2.3.9, respectively, apply to the entire reaction. Compare the results obtained in terms of both the degree of polymerization and the fraction of unreacted A groups as a function of time.

Solution

Since we are asked to evaluate t, N_n, and $[A]/[A]_0$ for specific values of p, it is convenient to summarize the following relationships:

1. Equation 2.2.4: $N_n = 1/(1-p)$.
2. Equation 2.2.2: $[A]/[A]_0 = 1-p$.
3. Equation 2.3.6: $t = (N_n-1)/k_c[A]_0 = N_n-1$ if catalyzed, since $10^{-1}(10) = 1$.
4. Equation 2.3.11: $t = (N_n^2-1)/2\,k_u[A]_0^2 = (N_n^2-1)(5)$ if uncatalyzed, since $2(10^{-3})\,(10)^2 = 0.2$.

Using these relationships, the accompanying table is developed.

p	$[A]/[A]_0$	N_n	Time (min) catalyzed	Time (min) uncatalyzed
0.2	0.8	1.25	0.25	2.8
0.4	0.6	1.67	0.67	8.9
0.6	0.4	2.50	1.5	26
0.8	0.2	5.00	4.0	120
0.9	0.1	10.0	9.0	500
0.95	0.05	20.0	19	2.0×10^3
0.99	0.01	100	99	5.0×10^4
0.992	0.008	120	119	7.2×10^4
0.998	0.002	500	499	1.3×10^6

A graphical comparison of the trends appearing here is presented in Figure 2.4. The importance of the catalyst is readily apparent in this hypothetical but not atypical system: To reach $N_n = 5$ requires 4 min in the catalyzed case and 120 min without any catalyst, assuming that the appropriate rate law describes the entire reaction in each case.

The question posed in Section 2.2—how long will it take to reach a certain extent of reaction or degree of polymerization?—is now answered. As is often the case, the answer begins, "It all depends...."

2.4   Distribution of Molecular Sizes

In this section we turn our attention to two other questions raised in Section 2.2, namely, how do the molecules distribute themselves among the different possible species, and how does this distribution vary with the extent of reaction? Since a range of species is present at each stage of polymerization, it is apparent that a statistical answer is required for these questions. This time, our answer begins, "On the average...."

We shall continue basing our discussion on the step-growth polymerization of the hypothetical monomer AB. In Section 2.7, we shall take a second look at this problem for the case of unequal concentrations of A and B groups. For now, however, we assure this equality by considering a monomer that contains one group of each type. In a previous discussion of the polymer formed from this monomer, we noted that remnants of the original functional groups are still recognizable, although modified, along the backbone of the polymer chain. This state of affairs is emphasized by the notation Ababa...abaB in which the a's and b's of the ab linkages are groups of atoms carried over the initial A and B reactive groups. In this type of polymer molecule, then, there are $i-1$ a's

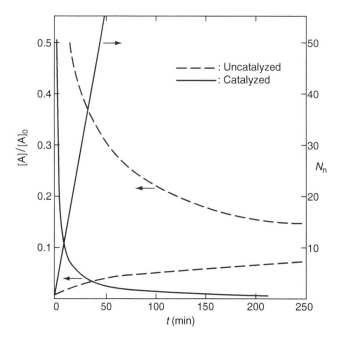

Figure 2.4 Comparison of catalyzed (solid lines) and uncatalyzed (dashed lines) polymerizations using results calculated in Example 2.2. Here $1-p$ (left ordinate) and N_n (right ordinate) are plotted versus time.

and 1 A if the degree of polymerization of the polymer is i. The a's differ from the A's precisely in that the former have undergone reaction whereas the latter have not. At any point during the polymerization reaction the fraction of the initial number of A groups that have reacted to become a's is given by p, and the fraction that remains as A's is given by $1-p$. In these expressions p is the same extent of reaction defined by Equation 2.2.3.

2.4.1 Mole Fractions of Species

We now turn to the question of evaluating the fraction of i-mers in a mixture as a function of p. The fraction of molecules of a particular type in a population is just another way of describing the probability of such a molecule. Hence our restated objective is to find the probability of an i-mer in terms of p; we symbolize this quantity as the mole fraction $x_i(p)$. Since the i-mer consists of $i-1$ a's and 1 A, its probability is the same as the probability of finding $i-1$ a's and 1 A in the same molecule. Recalling from Chapter 1 how such probabilities are compounded, we write

$$x_i(p) = p_a^{i-1} p_A = p^{i-1}(1 - p) \tag{2.4.1}$$

where p_a and p_A are the probabilities of individual a and A groups, respectively, and $p_a = p$ and $p_A = 1-p$. Equation 2.4.1 is known as the *most probable distribution*, and it arises in several circumstances in polymer science, in particular free radical polymerization (see Chapter 3). The probability of an i-mer can be converted to the number of i-mer molecules in the reaction mixture, n_i, by multiplying by the total number of molecules m in the mixture after the reaction has occurred to the extent p:

$$n_i = m\, p^{i-1}(1 - p) \tag{2.4.2}$$

Note that n_i/m gives the mole fraction of i-mers in a mixture at an extent of reaction p. As we have seen before, $m = (1-p)$ $[A]_0$, since each molecule in the mixture contains one unreacted A group. Incorporating this result into Equation 2.4.2 yields

$$n_i = p^{i-1}(1-p)^2 m_0 \qquad (2.4.3)$$

where m_0 is the total number of monomers present initially; $m_0 = [A]_0$ for AB monomers. This result may be used to evaluate the number of molecules of whatever degree of polymerization we elect to consider, in terms of p and m_0. As such, it provides the answer to one of the questions posed earlier.

Figure 2.5 is a plot of the ratio n_i/m versus i for several values of p. Several features are apparent from Figure 2.5 concerning the number distribution of molecules among the various species present:

1. On a number basis, the fraction of molecules always decreases with increasing i, regardless of the value of p. The distributions in Table 2.1 are unrealistic in this regard.
2. As p increases, the proportion of molecules with smaller i values decreases and the proportion with larger i values increases.
3. The combination of effects described in item (2) tends to flatten the curves as p increases, but not to the extent that the effect of item (1) disappears.

The number average degree of polymerization for these mixtures is easily obtained by recalling the definition of this average from Section 1.7. It is given by the sum of all possible i values, with each multiplied by its appropriate weighting factor provided by Equation 2.4.1:

$$N_n = \sum_{i=1}^{m_0} i\, x_i(p) = \sum_{i-1}^{\infty} i\, p^{i-1}(1-p) \qquad (2.4.4)$$

Note that the upper limit of the second summation has been shifted from m_0 to ∞ for mathematical reasons, namely that the answer is simple and known (see Appendix). The change is of little practical significance, since Equation 2.4.1 drops off for very large values of i. In particular, the result derived in the Appendix is

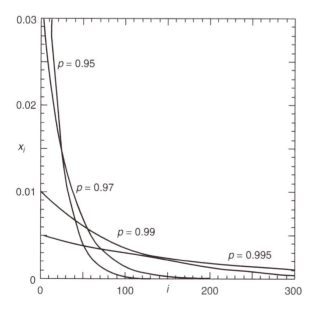

Figure 2.5   Mole fraction of i-mer as a function of i for several values of p.

$$\sum_{i=1}^{\infty} i\, p^{i-1} = \frac{1}{(1-p)^2}$$

Simplification of the summation in Equation 2.4.4 thus yields

$$N_n = \frac{1}{1-p} \tag{2.4.5}$$

Of course, this is the same result that was obtained more simply in Equation 2.2.4. The earlier result, however, was based on purely stoichiometric considerations and not on the detailed distribution as is the present result.

2.4.2 Weight Fractions of Species

Next we turn our attention to the distribution of the molecules *by weight* among the various species. This will lead directly to the determination of the weight average molecular weight and the ratio M_w/M_n.

We begin by recognizing that the weight fraction w_i of i-mers in the polymer mixture at any value of p equals the ratio of the mass of i-mer in the mixture divided by the mass of the total mixture. The former is given by the product $i\, n_i M_0$, where M_0 is the molecular weight of the repeat unit; the latter is given by $m_0 M_0$. Therefore we write

$$w_i = \frac{i\, n_i}{m_0} \tag{2.4.6}$$

into which Equation 2.4.3 may be substituted to give

$$w_i = i\, p^{i-1}(1-p)^2 \tag{2.4.7}$$

The weight fraction of i-mers is plotted as a function of i in Figure 2.6 for several large values of p. Inspection of Figure 2.6 and comparison with Figure 2.5 reveals the following:

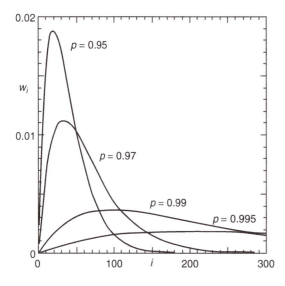

Figure 2.6   Weight fraction of i-mer as a function of i for several values of p.

1. At any p, very small and very large values of i contribute a lower weight fraction to the mixture than do intermediate values of i. This arises because of the product $i n_i$ in Equation 2.4.6: n_i is large for monomers, in which case i is low, and then n_i decreases as i increases. At intermediate values of i, w_i goes through a maximum.
2. As p increases, the maximum in the curves shifts to larger i values and the tail of the curve extends to higher values of i.
3. The effect in item (2) is not merely a matter of shifting curves toward higher i values as p increases, but reflects a distinct broadening of the distribution of i values as p increases.

The weight average degree of polymerization is obtained by averaging the contributions of various i values using weight fractions as weighting factors in the averaging procedure:

$$N_w = \frac{\sum_{i=1}^{m_0} i w_i}{\sum_{i=1}^{m_0} w_i} = \frac{\sum_{i=1}^{\infty} i^2 p^{i-1} (1-p)^2}{\sum_{i=1}^{\infty} i p^{i-1} (1-p)^2} \tag{2.4.8}$$

where the upper limit on i has been extended to infinity as before. The new summation that we need is also evaluated in the Appendix:

$$\sum_{i=1}^{\infty} i^2 p^{i-1} = \frac{1+p}{(1-p)^3}$$

Using this in Equation 2.4.8 gives

$$N_w = \frac{1+p}{1-p} \tag{2.4.9}$$

which is the desired result.

We saw in Chapter 1 that the ratio M_w/M_n, or polydispersity index, is widely used in polymer chemistry as a measure of the width of a molecular weight distribution. If the effect of chain ends is disregarded, this ratio is the same as the corresponding ratio of i values:

$$\frac{M_w}{M_n} = \frac{N_w}{N_n} = 1 + p \tag{2.4.10}$$

where the ratio of Equation 2.4.9 to Equation 2.2.4 has been used. Table 2.2 lists values of N_w, N_n, and N_w/N_n for a range of high p values. Note that $N_w/N_n \to 2$ as $p \to 1$; this is a characteristic result

Table 2.2 Values of N_n, N_w, and N_w/N_n for Various Large Values of p

p	N_n	N_w	N_w/N_n
0.90	10.0	19.0	1.90
0.92	12.5	24.0	1.92
0.94	16.7	32.3	1.94
0.96	25.0	49.0	1.96
0.98	50.0	99.0	1.98
0.990	100	199	1.990
0.992	125	249	1.992
0.994	167	332	1.994
0.996	250	499	1.996
0.998	500	999	1.998

of the most probable distribution. In light of Equation 1.7.16, the standard deviation of the molecular distribution is equal to M_n for the polymer sample produced by this polymerization. In a manner of speaking, the molecular weight distribution is as wide as the average is high. The broadening of the distribution with increasing p is dramatically shown by comparing the values in Table 2.2 with the situation at a low p value, say $p = 0.5$. At $p = 0.5$, $N_n = 2$, $N_w = 3$, and $N_w/N_n = 1.5$.

Since Equation 2.3.5 and Equation 2.3.11 give p as a function of time for the catalyzed and uncatalyzed polymerizations, respectively, the distributions discussed in the last few paragraphs can also be expressed with time as the independent variable instead of p.

The results we have obtained on the basis of the hypothetical monomer AB are also applicable to polymerizations between monomers of the AA and BB type, as long as the condition $[A] = [B]$ is maintained. In Section 2.7, we shall extend the arguments of this section to conditions in which $[A] \neq [B]$. In the meanwhile, we interrupt this line of reasoning by considering a few particular condensation polymers as examples of step-growth systems. The actual systems we discuss will serve both to verify and reveal the limitations of the concepts we have been discussing. In addition, they point out some of the topics that still need clarification. We anticipate some of the latter points by noting the following:

1. When $[A] \neq [B]$, both ends of the growing chain tend to be terminated by the group that is present in excess. Subsequent reaction of such a molecule involves reaction with the limiting group. The effect is a decrease in the maximum attainable degree of polymerization.
2. When a monofunctional reactant is present—one containing a single A or B group—the effect is also clearly a decrease in the average degree of polymerization. It is precisely because this type of reactant can only react once that it is sometimes introduced into polymer formulations, thereby eliminating the possibility of long-term combination of chain ends, and/or restricting the average molecular weight.

Polyesters and polyamides are two of the most-studied step-growth polymers, as well as substances of great commercial importance. We shall consider polyesters in Section 2.5 and polyamides in Section 2.6.

2.5 Polyesters

The preceding discussions of the kinetics and molecular weight distributions in the step-growth polymerizations of AB monomers are exemplified by esterification reactions between such monomers as glycolic acid and ω-hydroxydecanoic acid. Therefore one method of polyester synthesis is the following:

1. Esterification of a hydroxycarboxylic acid

    Several other chemical reactions are also widely used for the synthesis of these polymers. This list enumerates some of the possibilities, and Table 2.3 illustrates these reactions by schematic chemical equations.

2. Esterification of a diacid and a diol
3. Ester interchange with alcohol
4. Ester interchange with ester
5. Esterification of acid chlorides
6. Lactone polymerization

We have not attempted to indicate the conditions of temperature, catalyst, solvent, and so on, for these various reactions. For this type of information, references that deal specifically with synthetic polymer chemistry should be consulted. In the next few paragraphs we shall comment on the various routes to polyester formation in the order summarized above and followed in Table 2.3. The studies summarized in Figure 2.2 and Figure 2.3 are examples of Reaction 2 in Table 2.3.

Table 2.3 Some Schematic Reactions for the Formation of Polyesters

1. Esterification of a hydroxycarboxylic acid:

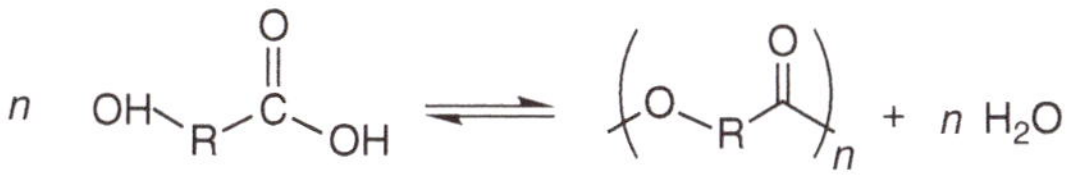

2. Esterification of diacid and diol:

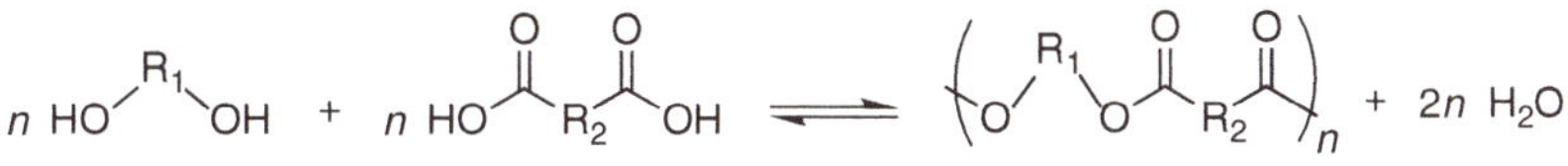

3. Ester interchange with alcohol:

4. Ester interchange with ester ("transesterification"):

5. Esterification of acid chlorides (Schotten–Baumann reaction):

6. Lactone polymerization:

While up to now we have emphasized bifunctional reactants, both monofunctional compounds and monomers with functionality greater than 2 are present in some polymerization processes, either intentionally or adventitiously. The effect of the monofunctional reactant is clearly to limit chain growth. As noted above, a functionality greater than 2 results in branching. A type of polyester that includes mono-, di-, and trifunctional monomers is the so-called alkyd resin. A typical example is based on the polymerization of phthalic acid (or anhydride), glycerol, and an unsaturated mono-carboxylic acid. The following suggests the structure of a portion of such a polyester:

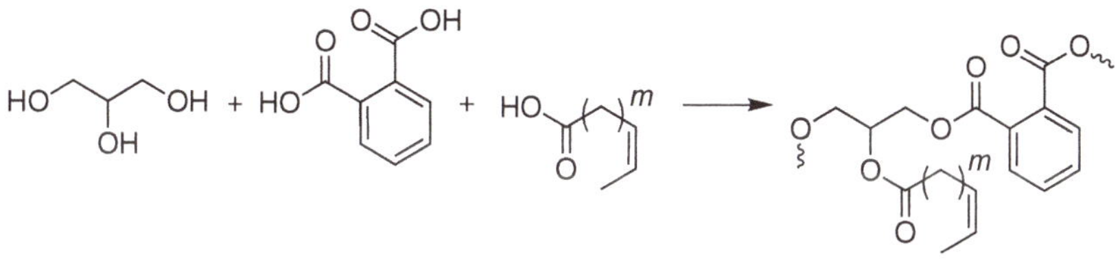

(2.J)

The presence of the unsaturated substituent along this polyester backbone gives this polymer cross-linking possibilities through a secondary reaction of the double bond. These polymers are used in paints, varnishes, and lacquers, where the ultimate cross-linked product results from the oxidation of the double bond as the coating cures. A cross-linked polyester could also result from Reaction (2.J) without the unsaturated carboxylic acid, but the latter would produce a gel in which the entire reaction mass solidified, and is therefore not as well suited to coating applications as a polymer that cross-links upon "drying."

Many of the reactions listed at the beginning of this section are acid catalyzed, although a number of basic catalysts are also employed. Esterifications are equilibrium reactions, and often carried out at elevated temperatures for favorable rate and equilibrium constants and to shift the equilibrium in favor of the polymer by volatilization of the by-product molecules. An undesired feature of higher polymerization temperatures is the increased possibility of side reactions, such as the dehydration of the diol or the pyrolysis of the ester. Basic catalysts tend to produce fewer undesirable side reactions.

Ester exchange reactions are valuable, since, say, methyl esters of dicarboxylic acids are often more soluble and easier to purify than the diacid itself. The methanol by-product is easily removed by evaporation. Poly(ethylene terephthalate) is an example of a polymer prepared by double application of Reaction 4 in Table 2.3. The first stage of the reaction is conducted at temperatures below 200°C and involves the interchange of dimethyl terephthalate with ethylene glycol.

$$(2.K)$$

The rate of this reaction is increased by using excess ethylene glycol, and removal of the methanol is assured by the elevated temperature. Polymer is produced in the second stage after the temperature is raised above the melting point of the polymer, about 260°C.

The ethylene glycol liberated by Reaction (2.L) is removed by lowering the pressure or purging with an inert gas. Because the ethylene glycol produced by Reaction (2.L) is removed, proper stoichiometry is assured by proceeding via the intermediate bis(2-hydroxyethyl) terephthalate; otherwise the excess glycol used initially would have a deleterious effect on the degree of polymerization. Poly(ethylene terephthalate) is more familiar by some of its trade names: Mylar as a film and Dacron, Kodel, or Terylene as fibers; it is also known by the acronym PET.

Ester interchange reactions like that shown in Reaction 4 in Table 2.3 (transesterification) can be carried out on polyesters themselves to produce a scrambling between the two polymers. Studies of this sort between high and low molecular weight prepolymers result in a single polymer with the same molecular weight distribution as would have been obtained from a similarly constituted diol–diacid mixture by direct polymerization. This is true when the time-catalyst conditions allow the randomization to reach equilibrium. If the two prepolymers are polyesters formed from different monomers, the product of the ester interchange reaction will be a copolymer of some sort. If the reaction conditions favor esterification, the two chains will merely link together and a block copolymer results. If the conditions favor the ester interchange reaction, then a scrambled copolymer molecule results. These possibilities underscore the idea that the derivations of the preceding sections are based on complete equilibrium among all molecular species present during the condensation reaction.

Example 2.3

It has been hypothesized that cross-linked polymers would have better mechanical properties if interchain bridges were located at the ends rather than the center of chains. To test this, low molecular weight polyesters were synthesized from a diol and two different diacids: one saturated and the other unsaturated. The synthetic procedure was such that the unsaturated acid units were located at either the center ("centrene") or the ends ("endene") of the chains. Some pertinent aspects of the overall experiment are listed below:

		Endene	Centrene
Step 1:	8 h at about 150°C–200°C		
	Maleic anhydride (mol)	0	2.0
	Succinic anhydride (mol)	2.0	0
	Diethylene glycol (mol)	3.0	3.0
Step 2:	About 1/2 h at about 120°C–130°C		
	Maleic anhydride (mol)	2.0	0
	Succinic anhydride (mol)	0	2.0
	Catalyst	0	0
Step 3:	30% Styrene + catalyst		
	16 h at 55°C + 1 h at 110°C		
	Elastic modulus (Pa)	21,550	16,500

On the basis of these facts, do the following:

1. Comment on the likelihood that the comonomers are segregated as the names of these polymers suggest.
2. Sketch the structure of the average endene and centrene molecules.
3. Comment on the results in terms of the initial hypothesis.

Solution

1. Since the reaction conditions are mild in step 2 (only 6% as much time allowed as in step 1 at a lower temperature) and no catalyst is present, it is unlikely that any significant amount of ester scrambling occurs. Isomerization of maleate to fumarate is also known to be insignificant under these conditions.
2. The idealized structures of these molecules are

Centrene

Endene

3. A cross-linked product with unsaturation at the chain ends does, indeed, have a higher modulus. This could be of commercial importance and indicates that industrial products might be formed by a nonequilibrium process precisely for this sort of reason. A fuller discussion of the factors that contribute to the modulus will be given in Chapter 10 and Chapter 12.

Acid chlorides are generally more reactive than the parent acids, so polyester formation via Reaction 5 in Table 2.3 can be carried out in solution and at lower temperatures, in contrast with the bulk reactions of the melt as described above. Again, the by-product molecules must be eliminated either by distillation or precipitation. The method of interfacial condensation, described in Section 2.6, can be applied to this type of reaction.

The formation of polyesters from the polymerization of lactones (Reaction 6 in Table 2.3) is a ring-opening reaction that may follow either a step-growth or chain mechanism, depending on conditions. For now our only concern is to note that the equilibrium representing this reaction in Table 2.3 describes polymerization by the forward reaction and ring formation by the back reaction. Rings clearly compete with polymers for monomer in all polymerizations. Throughout the chapter we have assumed that all competing side reactions, including ring formation, could be neglected.

2.6 Polyamides

The discussion of polyamides parallels that of polyesters in many ways. To begin with, polyamides may be formed from an AB monomer, in this case amino acids:

1. Amidation of amino acids

Additional synthetic routes that closely resemble polyesters are also available. Several more of these are listed below and are illustrated by schematic reactions in Table 2.4:

Table 2.4 Some Schematic Reactions for the Formation of Polyamides

1. Amidation of amino acids:

2. Amidation of diamine and diacid:

3. Interchange reactions:

4. Amidation of acid chlorides:

5. Lactam polymerization:

2. Amidation of a diacid and a diamine
3. Interchange reactions
4. Amidation of acid chlorides
5. Lactam polymerization

We only need to recall the trade name of synthetic polyamides, nylon, to recognize the importance of these polymers and the reactions employed to prepare them. Recall the system for naming these compounds (see Section 1.5): the first number after the name gives the number of carbon atoms in the diamine, and the second, the number of carbons in the diacid.

The diacid–diamine amidation described in Reaction 2 in Table 2.4 has been widely studied in the melt, solution, and the solid state. When equal amounts of two functional groups are present, both the rate laws and the molecular weight distributions are given by the treatment of the preceding sections. The stoichiometric balance between reactive groups is readily obtained by precipitating the 1:1 ammonium salt from ethanol:

(2.M)

This compound is sometimes called a nylon salt. The salt ⇌ polymer equilibrium is more favorable to the production of polymer than in the case of polyesters, so this reaction is often carried out in a sealed tube or autoclave at about 200°C until a fairly high extent of reaction is reached; then the temperature is raised and the water driven off to attain high molecular weight polymer. Also in contrast to polyesters, Reaction 1 and Reaction 2 in Table 2.4 can be conducted rapidly without an acid catalyst.

The process represented by Reaction 2 in Table 2.4 actually entails a number of additional equilibrium reactions. Some of the equilibria that have been considered include the following:

(2.N)

(2.O)

(2.P)

$2 \, H_2O \rightleftharpoons H_3O^+ + {}^-OH$
(2.Q)

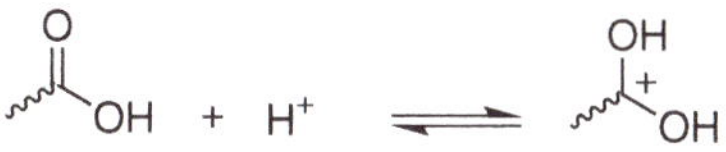

(2.R)

(2.S)

Reaction (2.N) describes the nylon salt ⇌ nylon equilibrium. Reaction (2.O) and Reaction (2.P) show proton transfer with water between carboxyl and amine groups. Since proton transfer equilibria are involved, the self-ionization of water, Reaction (2.Q), must also be included. Especially in the presence of acidic catalysts, Reaction (2.R) and Reaction (2.S) are the equilibria of the acid-catalyzed intermediate described in general in Reaction (2.G). The main point in including all of these equilibria is to indicate that the precise concentration of A and B groups in a diacid–diamine reaction mixture is a complicated function of the moisture content and the pH, as well as the initial amounts of reactants introduced. Because of the high affinity for water of the various functional groups present, the complete removal of water is impossible: the equilibrium moisture content of molten nylon-6,6 at 290°C under steam at 1 atm is 0.15%. Likewise, the various ionic possibilities mean that at both high and low pH values the concentration of unionized carboxyl or amine groups may be considerably different from the total concentration—without regard to state of ionization—of these groups. As usual, upsetting the stoichiometric balance of the reactive groups lowers the degree of polymerization attainable. The abundance of high-quality nylon products is evidence that these complications have been overcome in practice.

Amide interchange reactions of the type represented by Reaction 3 in Table 2.4 are known to occur more slowly than direct amidation; nevertheless, reactions between high and low molecular weight polyamides result in a polymer of intermediate molecular weight. The polymer is initially a block copolymer of the two starting materials, but randomization is eventually attained.

As with polyesters, the amidation reaction of acid chlorides may be carried out in solution because of the enhanced reactivity of acid chlorides compared with carboxylic acids. A technique known as interfacial polymerization has been employed for the formation of polyamides and other step-growth polymers, including polyesters, polyurethanes, and polycarbonates. In this method, the polymerization is carried out at the interface between two immiscible solutions, one of which contains one of the dissolved reactants, while the second monomer is dissolved in the other. Figure 2.7 shows a polyamide film forming at the interface between layers of an aqueous diamine solution and a solution of diacid chloride in an organic solvent. In this form, interfacial polymerization is part of the standard repertoire of chemical demonstrations. It is sometimes called the "nylon rope trick" because of the filament of nylon that can be produced by withdrawing the collapsed film.

The amidation of the reactive groups in interfacial polymerization is governed by the rates at which these groups can diffuse to the interface where the growing polymer is deposited. Accordingly, new reactants add to existing chains rather than interacting to form new chains. This is

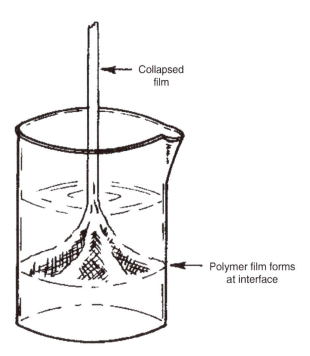

Collapsed
film

Polymer film forms
at interface

Figure 2.7 Sketch of an interfacial polymerization with the collapsed polymer film being withdrawn from the surface between the immiscible phases. (From Morgan, P.W. and Kwolek, S.L. *J. Chem. Educ.*, 36, 182, 1959. With permission.)

different than the bulk mechanism that we have discussed elsewhere in this chapter, and it is evident that a higher molecular weight polymer should result from this difference. The HCl by-product of the amidation reaction is neutralized by also dissolving an inorganic base in the aqueous layer in interfacial polymerization. The choice of the organic solvent plays a role in determining the properties of the polymer produced, probably because of the differences in solvent quality for the resulting polymer. Since this reaction is carried out at low temperatures, the complications associated with side reactions can be kept to a minimum. Polymer yield may be increased by increasing the area of the interface between the two solutions by stirring.

Lactam polymerization represented by Reaction 5 in Table 2.4 is another example of a ring-opening reaction, the reverse of which is a possible competitor with polymer for reactants. The various mechanical properties of polyamides may be traced in many instances to the possibility of intermolecular hydrogen bonding between the polymer molecules, and to the relatively stiff chains these substances possess. The latter, in turn, may be understood by considering still another equilibrium, this one among resonance structures along the chain backbone:

$$\text{(2.T)}$$

The combination of strong intermolecular forces and high chain-stiffness accounts for the high melting points of polyamides (see Chapter 13). The remarks of this section and Section 2.5 represent only a small fraction of what could be said about these important materials. We have commented on aspects of the polymerization processes and of the polymers themselves that have a direct bearing on the concepts discussed throughout this volume. This material provides an excellent example of the symbiosis between theoretical and application-oriented viewpoints. Each stimulates and reinforces the other with new challenges, although it must be conceded that many industrial processes reach a fairly high degree of empirical refinement before the conceptual basis is quantitatively developed.

2.7   Stoichiometric Imbalance

We now turn to one of the problems we have sidestepped until now—the polymerization of reactants in which a stoichiometric imbalance exists in the numbers of reactive groups A and B. In earlier sections dealing with the quantitative aspects of step-growth polymerization, we focused attention on monomers of the AB type to assure equality of reactive groups. The results obtained above also apply to AA and BB polymerizations, provided that the numbers of reactive groups are equal. There are obvious practical difficulties associated with the requirement of stoichiometric balance. Rigorous purification of monomers is difficult and adds to the cost of the final product. The effective loss of functional groups to side reactions imposes restrictions on the range of experimental conditions at best and is unavoidable at worst. These latter considerations apply even in the case of the AB monomer. We have already stated that the effect of the imbalance of A and B groups is to lower the eventual degree of polymerization of the product. A quantitative assessment of this limitation is what we now seek.

We define the problem by assuming that the polymerization involves AA and BB monomers and that the B groups are present in excess. We define ν_A and ν_B to be the numbers of A and B functional groups, respectively. The number of either of these quantities in the initial reaction mixture is indicated by a superscript o; the numbers at various stages of reaction have no superscript. The stoichiometric imbalance is defined by the ratio r, where

$$r \equiv \frac{\nu_A^o}{\nu_B^o} \tag{2.7.1}$$

By definition of the problem, this ratio cannot exceed unity.

As with other problems with stoichiometry, it is the less-abundant reactant that limits the product. Accordingly, we define the extent of reaction p to be the fraction of A groups that have reacted at any point. Since A and B groups react in a 1:1 proportion, the number of B groups that have reacted when the extent of reaction has reached p equals $p\nu_A^o$, which in turn equals $pr\nu_B^o$. The product pr gives the fraction of B groups that have reacted at any point. With these definitions in mind, the following relationships are readily obtained:

1. The number of unreacted functional groups after the reaction reaches extent p is

$$\nu_A = (1-p)\nu_A^o \tag{2.7.2}$$

and

$$\nu_B = (1-pr)\nu_B^o = (1-pr)\frac{\nu_A^o}{r} \tag{2.7.3}$$

2. The total number of chain ends is given by the sum of Equation 2.7.2 and Equation 2.7.3:

$$\nu_{ends} = \left(1-p+\frac{1-pr}{r}\right)\nu_A^o \tag{2.7.4}$$

3. The total number of chains is half the number of chain ends:

$$\nu_{chains} = \frac{1}{2}\left(1+\frac{1}{r}-2p\right)\nu_A^o \tag{2.7.5}$$

4. The total number of repeat units distributed among these chains is the number of monomer molecules present initially:

$$\nu_{repeat\ units} = \frac{1}{2}\nu_A^o + \frac{1}{2}\nu_B^o = \frac{1}{2}\left(1+\frac{1}{r}\right)\nu_A^o \tag{2.7.6}$$

The number average degree of polymerization is given by dividing the number of repeat units by the number of chains, or

$$N_n = \frac{1+1/r}{1+1/r-2p} = \frac{1+r}{1+r-2pr} \tag{2.7.7}$$

As a check that we have done this correctly, note that Equation 2.7.7 reduces to the previously established Equation 2.4.5 when $r=1$.

One distinction that should be pointed out involves the comparison of Equation 2.2.1 and Equation 2.7.7. In the former we considered explicitly the AB monomer, whereas the latter is based on the polymerization of AA and BB monomers. In both instances N_n is obtained by dividing the total number of monomer molecules initially present by the total number of chains after the reaction has occurred to extent p. Following the same procedure for different reaction mixtures results in a different definition of the repeat unit. In the case of the AB monomer, the repeat unit is the ab entity, which differs from AB by the elimination of the by-product molecule. In the case of the AA and BB monomers, the repeat unit in the polymer is the aabb unit, which differs from AA + BB by two by-product molecules. Equation 2.2.1 counts the number of ab units in the polymer directly. Equation 2.7.7 counts the number of aa + bb units. *The number of aa + bb units is twice the number of aabb units.* Rather than attempting to formalize this distinction by introducing more complex notation, we simply point out that application of the formulas of this chapter to specific systems must be accompanied by a reflection on the precise meaning of the calculated quantity for the system under consideration.

The distinction pointed out in the last paragraph carries over to the evaluation of M_n from N_n. We assume that the chain length of the polymer is great enough to render unnecessary any correction for the uniqueness of chain ends. In such a case, the molecular weight of the polymer is obtained from the degree of polymerization by multiplying the latter by the molecular weight of the repeat unit. The following examples illustrate the distinction under consideration:

1. Polymerization of an AB monomer is illustrated by the polyester formed from glycolic acid. The repeat unit in this polymer has the structure

   and $M_0 = 58$. Neglecting end groups, we have $M_n = 58\,N_n$ with N_n given by Equation 2.2.1.

2. Polymerization of AA and BB monomers is illustrated by butane-1,4-diol and adipic acid. The aabb repeat unit in the polymer has an M_0 value of 200. If Equation 2.2.4 is used to evaluate N_n, it gives the number of aa + bb units; therefore $M_n = (200\,N_n)/2$.

3. An equivalent way of looking at the conclusion of item (2) is to recall that Equation 2.7.7 gives the (number average) number of monomers of both kinds in the polymer; we should multiply this quantity by the average molecular weight of the two kinds of units in the structure: $(88 + 112)/2 = 100$.

Equation 2.7.7 also applies to the case when some of the excess B groups present are in the form of monofunctional reagents. In this latter situation the definition of r is modified somewhat (and labeled with a prime) to allow for the fact that some of the B groups are in BB-type monomers (unprimed) and some are in monofunctional (primed) molecules:

$$r' = \frac{\nu_A^o}{\nu_B^o + 2\nu_{B'}^o},$$
(2.7.8)

The parameter r' continues to measure the ratio of the number of A and B groups; the factor 2 enters since the monofunctional reagent has the same effect on the degree of polymerization as a difunctional molecule with two B groups, hence, is doubly effective compared to the latter. With this modification taken into account, Equation 2.7.7 enables us to evaluate quantitatively the effect of stoichiometric imbalance or monofunctional reagents, whether these are intentionally intro-duced to regulate N_n or whether they arise from impurities or side reactions.

The parameter r varies between 0 and 1; as such it has the same range as p. Although the quantitative effect of r and p on N_n is different, the qualitative effect is similar for each: the closer each of these fractions is to unity, higher degrees of polymerization are obtained. Table 2.5 shows some values of N_n calculated from Equation 2.7.7 for several combinations of (larger values of) r and p. Inspection of Table 2.5 reveals the following:

Table 2.5 Some Values of N_n Calculated by Equation 2.7.7 for Values of r and p Close to Unity

r	$p=0.95$	$p=0.97$	$p=0.99$	$p=1.00$
0.95	13.5	18.2	28.3	39.0
0.97	15.5	22.3	39.9	65.7
0.99	18.3	28.7	66.8	199
1.00	20.0	33.3	100	∞

1. For any value of r, N_n is greater for larger values of p; this conclusion is the same whether the proportions of A and B are balanced or not.
2. The final 0.05 increase in p has a bigger effect on N_n at r values that are closer to unity than for less-balanced mixtures.
3. For any value of p, N_n is greater for larger values of r; stoichiometric imbalance lowers the average chain-length for the preparation.
4. An 0.05 increase in r produces a much bigger increase in N_n at $p = 1$ than in mixtures that have reacted to a lesser extent.

An interesting special case occurs when $p = 1$; Equation 2.7.7 then becomes

$$N_n = \frac{1 + r}{1 - r} \qquad (2.7.9)$$

The following example illustrates some of the concepts developed in this section.

Example 2.4

It is desired to prepare a polyester with $M_n = 5000$ by reacting 1 mol of butane-1,4-diol with 1 mol of adipic acid.

1. Calculate the value of p at which the reaction should be stopped to obtain this polymer, assuming perfect stoichiometric balance and neglecting end group effects on M_n.
2. Assuming that 0.5 mol% of the diol is lost to polymerization by dehydration to olefin, what would be the value of M_n if the reaction were carried out to the same extent as in (1)?
3. How could the loss in (2) be offset so that the desired polymer is still obtained?
4. Suppose the total number of carboxyl groups in the original mixture is 2 mol, of which 1.0% is present as acetic acid to render the resulting polymer inert to subsequent esterification. What value of p would be required to produce the desired polymer in this case, assuming no other stoichiometric imbalance?

Solution

The various expressions we have developed in this section relating p to the size of the polymer are all based on N_n. Accordingly, we note that the average reactant molecule in this mixture has a molecular weight of 100 as calculated above. Therefore the desired polymer has a value of $N_n = 50$.

1. We use Equation 2.4.5 for the case of equal numbers of A and B groups and find that $p = 1 - 1/N_n = 0.980$. Even though Equation 2.4.5 was derived for an AB monomer, it applies to this case with the "average monomer" as the repeat unit.
2. Component AA is the diol in this case and $\nu_A^o = 0.995$ mol; therefore $r = 0.995/1.00 = 0.995$. We use Equation 2.7.7 and solve for N_n with $p = 0.980$ and $r = 0.995$:

$$N_n = \frac{1.995}{1.995 - 2\,(0.995)\,(0.980)} = 44.5$$

and therefore $M_n = 44.5 \times 100 = 4450$ g mol^{-1}.
3. The effect of the lost hydroxyl groups can be offset by carrying out the polymerization to a higher extent of reaction. We use Equation 2.7.7 and solve for p with $N_n = 50$ and $r = 0.995$:

$$p = \left(1 - \frac{1}{N_n}\right)\left(\frac{1 + r}{2r}\right) = \left(1 - \frac{1}{50}\right)\frac{1.995}{1.990} = 0.9825$$

4. The monofunctional reagent B$'$ is the acetic acid in this case and the number of monofunctional carboxyl groups is $2(0.010) = 0.020 = \nu_{B'}^o$. The number of B groups in BB monomers is

$1.980 = \nu_B$. We use Equation 2.7.8 to define r' for this situation, assuming the number of hydroxyl groups equals 2.00 mol:

$$r' = \frac{2.00}{1.980 + 2\,(0.020)} = 0.990$$

Equation 2.7.7 is now solved for p using $N_n = 50$ and $r' = 0.990$:

$$p = \left(1 - \frac{1}{N_n}\right)\left(\frac{1+r'}{2r'}\right) = \left(1 - \frac{1}{50}\right)\frac{1.990}{1.980} = 0.9849$$

Remember from Section 2.3 that a progressively longer period of time is required to shift the reaction to larger values of p. In practice, therefore, the effects of side reactions and monofunctional reactants are often not compensated by longer polymerization times, but are accepted in the form of lower molecular weight polymers.

2.8 Chapter Summary

In this chapter we have considered step-growth or condensation polymerization, one of the two main routes to synthetic polymers. Our emphasis has been on the description of the distribution of polymer sizes as a function of the extent of reaction and the concentration of reactants, and on the associated kinetics. In addition, we have given an introduction to the two major classes of commercial condensation polymers, polyesters and polyamides, and the different ways they may be produced. The principal results are as follows:

1. In the simplest case of stoichiometric balance, that is, equal numbers of A and B reactive groups, the number average degree of polymerization N_n is given by $1/(1-p)$, where p is the extent of reaction, equal to the fraction of A (or B) groups reacted. In general, therefore, the reaction must be driven far toward products ($p \rightarrow 1$) before appreciable molecular weights can be attained.
2. The resulting distribution of molecular sizes is called the most probable distribution and the associated polydispersity index approaches 2 as $p \rightarrow 1$. Two important features of this distribution are that there are always more i-mers present than $(i+1)$-mers, for any value of i, but there is an intermediate value of i for which the weight fraction w_i is maximum.
3. If the reaction is run in the presence of a catalyst (the usual situation), then N_n should grow linearly in time, whereas for the uncatalyzed case, N_n will grow with the square root of time.
4. In reality N will almost always be lower than the theoretical value for a given p, due to a combination of side reactions, including ring formation, contamination by monofunctional reagents, and stoichiometric imbalance.
5. The analysis of these reactions builds on the principle of equal reactivity, the assumption that the reactivity of a given functional group is independent of the molecular weight of the polymer to which it is attached. This assumption is quite reliable in most cases of interest.

Problems

1. Howard describes a model system used to test the molecular weight distribution of a condensation polymer.[†] "The polymer sample was an acetic acid–stabilized equilibrium nylon-6,6. Analysis showed it to have the following end group composition (in equivalents per 10^6 g): acetyl = 28.9, amine = 35.3, and carboxyl = 96.5. The number average degree of

---

polymerization is, therefore, 110 and the conversion degree ($=$ extent of reaction) $= 0.9909$."
Verify the self-consistency of those numbers.

2. Haward et al. have reported some research in which a copolymer of styrene and hydro-xyethylmethacrylate was cross-linked by hexamethylene di-isocyanate.[†] Draw the structural formula for a portion of this cross-linked polymer and indicate what part of the molecule is the result of a condensation reaction and what part results from addition polymerization. These authors indicate that the cross-linking reaction is carried out in sufficiently dilute solutions of copolymer that the cross-linking is primarily intramolecular rather than intermo-lecular. Explain the distinction between these two terms and why concentration affects the relative amounts of each.

3. The polymerization of β-carboxymethyl caprolactam has been observed to consist of initial isomerization via a second-order kinetic process followed by condensation of the isomer to polymer:

The rates of polymerization are thus of first order in ν_{NH_2} and in $\nu_{(CO)_2O}$ or second order overall. Since $\nu_{NH_2} = \nu_{(CO)_2O}$, the rate $= kc^2$, if catalyzed; third order is expected under uncatalyzed conditions. The indirect evaluation of c was accomplished by measuring the amount of monomer reacted, and the average degree of polymerization of the mixture was determined by viscosity at different times. The following data were obtained at 270°C; the early part of the experiment gives nonlinear results.[‡] Graphically test whether these data indicate catalyzed or uncatalyzed conditions, and evaluate the rate of constant for polymerization at 270°C. Propose a name for the polymer.

t (min)	c (Mole fraction)	t (min)	c (Mole fraction)
20	0.042	90	0.015
30	0.039	110	0.013
40	0.028	120	0.012
50	0.024	150	0.0096
60	0.021	180	0.0082
80	0.018		

4. Examination of Figure 2.5 shows that N_i/N is greater for $i = 40$ at $p = 0.97$ than at either $p = 0.95$ or $p = 0.99$. This is generally true: various i-mers go through a maximum in numerical abundance as p increases. Show that the extent of reaction at which this maximum occurs varies with i as follows: $p_{max} = (i - 1)/i$. For a catalyzed AB reaction, extend this expression to give a function for the time required for an i-mer to reach its maximum numerical abundance. If $k_c = 2.47 \times 10^{-4}$ L mol^{-1} s^{-1} at 160.5°C for the polymerization of 12-hydroxystearic acid,[§] calculate the time at which 15-mers show their maximum abundance if the initial concentration of monomer is 3.0 M.

[†] R.N. Haward, B.M. Parker, and E.F.T. White, *Adv. Chem.*, 91, 498 (1969).
[‡] H.K. Reimschuessel, *Adv. Chem.*, 91, 717 (1969).
[§] C.E.H. Bawn and M.B. Huglin, *Polymer*, 3, 257 (1962).

5. In the presence of pyridine-cuprous chloride catalyst, the following polymerization occurs:

In an investigation to examine the mechanism of this reaction, the dimer $(i=2)$

was used as a starting material. The composition of the mixture was studied as the reaction progressed and the accompanying results were obtained:[†]

Percent of theoretical O_2 absorbed	Weight percent composition in reaction mixture			
	Monomer	Dimer	Trimer	Tetramer
9	1	69	15	9
12	1.5	68	24	9
20	3	38.5	23	9
35	6	26	21	11
60	11	4	4	1
80	1	0	0	0

Plot a family of curves, each of different i, with composition as the y-axis and O_2 absorbed as the x-axis. Evaluate w_i by Equation 2.4.7 for $i=1$, 2, 3, and 4 and $0.1 \leq p \leq 0.9$ in increments of 0.1. Plot these results (w_i on the y-axis) on a separate graph drawn to the same scale as the experimental results. Compare your calculated curves with the experimental curves with respect to each of the following points: (1) coordinates used, (2) general shape of curves, and (3) labeling of curves.

6. The polymer described in the last problem is commercially called poly(phenylene oxide), which is not a "proper" name for a molecule with this structure. Propose a more correct name. Use the results of the last problem to criticize or defend the following proposition: The experimental data for dimer polymerization can be understood if it is assumed that one molecule of water and one molecule of monomer may split out in the condensation step. Steps involving incorporation of the monomer itself (with only water split out) also occur.

7. Taylor carefully fractionated a sample of nylon-6,6 and determined the weight fraction of different i-mers in the resulting mixture.[‡] The results obtained are given below. Evaluate N_w from these data, then use Equation 2.4.9 to calculate the corresponding value of p. Calculate the theoretical weight fraction of i-mers using this value of p and a suitable array of i values. Plot your theoretical curve and the above data points on the same graph. Criticize or defend the

[†] G.D. Cooper and A. Katchman, *Adv. Chem.*, 91, 660 (1969).
[‡] G.B. Taylor, *J. Am. Chem. Soc.*, 69, 638 (1947).

following proposition: although the fit of the data points is acceptable with this value of p, it appears that a slightly smaller value of p would give an even better fit.

i	$w_i \times 10^{-4}$	i	$w_i \times 10^{-4}$
12	6.5	311	15.2
35	19.6	334	14.1
58	29.4	357	13.0
81	33.0	380	11.5
104	35.4	403	11.0
127	36.5	426	9.1
150	33.0	449	7.2
173	27.6	472	6.5
196	25.2	495	4.9
219	22.9	518	4.3
242	19.4	541	3.9
265	18.5	564	3.3
288	16.8		

8. Paper chromatograms were developed for 50:50 blends of nylon-6,6 and nylon-6,10 after the mixture had been heated to 290°C for various periods of time.[†] The following observations describe the chromatograms after the indicated times of heating:

0 h—two spots with R_f values of individual polymers.
1/4 h—two distinct spots, but closer together than those of 0 h.
1/2 h—spots are linked together.
3/4 h—one long, diffuse spot.
11/2 h—one compact spot, intermediate R_f value.

On the basis of these observations, criticize or defend the following proposition: the fact that the separate spots fuse into a single spot of intermediate R_f value proves that block copolymers form between the two species within the blend upon heating.

9. Reimschuessel and Dege polymerized caprolactam in sealed tubes containing about 0.0205 mol H_2O per mole caprolactam.[‡] In addition, acetic acid (V), sebacic acid (S), hexamethylene diamine (H), and trimesic acid (T) were introduced as additives into separate runs. The following table lists (all data per mole caprolactam) the amounts of additive present and the analysis for end groups in various runs. Neglecting end group effects, calculate M_n for each of these polymers from the end group data. Are the trends in molecular weight qualitatively what would be expected in terms of the role of the additive in the reaction mixture? Explain briefly.

Additives	Moles additive	—COOH (mEq)	—NH$_2$ (mEq)
None	—	5.40	4.99
V	0.0205	19.8	2.3
S	0.0102	21.1	2.3
H	0.0102	1.4	19.7
T	0.0067	22.0	2.5

10. In the study described in the last problem, caprolactam was polymerized for 24 h at 225°C in sealed tubes containing various amounts of water. M_n and M_w were measured for the

[†] C.W. Ayers, *J. Appl. Chem.*, 4, 444 (1954).
[‡] H.L. Reimschuessel and G.J. Dege, *J. Polym. Sci.*, A-1, 2343 (1971).

resulting mixture by osmometry (see Chapter 7) and light scattering (see Chapter 8), respectively, and the following results were obtained:

Moles H_2O ($\times 10^3$)/mole Caprolactam	$M_n \times 10^{-3}$	$M_n \times 10^{-3}$
49.3	13.4	20.0
34.0	16.4	25.6
25.6	17.9	29.8
20.5	19.4	36.6

Use the molecular weight ratio to calculate the apparent extent of reaction of the caprolactam in these systems. Is the variation in p qualitatively consistent with your expectations of the effect of increased water content in the system? Plot p versus moisture content and estimate by extrapolation the equilibrium moisture content of nylon-6 at 255°C. Does the apparent equilibrium moisture content of this polymer seem consistent with the value given in Section 2.6 for nylon-6,6 at 290°C?

11. At 270°C adipic acid decomposes to the extent of 0.31 mol% after 1.5 h.[†] Suppose an initially equimolar mixture of adipic acid and diol achieves a value of $p = 0.990$ after 1.5 h, compare the expected and observed values of N_n in this experiment. Criticize or defend the following proposition: the difference between the observed and expected values would be even greater than calculated above if, instead of the extent of reaction being measured analytically, the value of p expected (neglecting decomposition) after 1.5 h was calculated by an appropriate kinetic equation.

12. Show the reaction sequence and the structure of the resulting polymer from the polycondensation of these two monomers; note that the reaction (a) has two distinct steps, and that (b) it is base-catalyzed.

13. A polyester is prepared under conditions of stoichiometric balance, but no attempt is made to remove water. Eventually, the reaction comes to equilibrium with equilibrium constant K. If $[COOH]_0$ is the initial concentration of carboxylic acid groups, show that the equilibrium water concentration is

$$[H_2O] = K \frac{[COOH]_0}{N_n(N_n - 1)}$$

14. For the most probable distribution, it is clear that there is always more i-mer present than $(i+1)$-mer, at any $0 < p < 1$. However, the absolute amount of an i-mer should go through a maximum with time, as the reaction progresses; there is zero to start, but at late enough stages i-mer will have mostly reacted to contribute to all the larger species. Use the chain rule and any

[†] V.V. Korshak and S.V. Vinogradova, *Polyesters*, Pergamon, Oxford, 1965.

suitable simplifications ($k[A]_0 t \gg 1$?) to find the degrees of conversion at which the mole fraction and the absolute concentration of i-mer have their maximum in time. Compare this to the number average degree of polymerization at the same conversion; does the answer make sense?

15. For the polymerization of succinic acid and 1,4-butanediol under stoichiometric balance in xylene:

    (a) Draw the chemical structures of the reactants, products, and important intermediates for both the strong acid–catalyzed and self-catalyzed case.

    (b) Generate a quantitative plot of N_n versus time for the self-catalyzed case up to 28,000 s, given $k = 6 \times 10^{-3} \, mol^{-2} \, L^2 \, s^{-2}$ and 3 mol L^{-1} starting concentration of each monomer. How many hours would it take to make a polymer with $N_n = 300$?

    (c) Do the same for the catalyzed case, with $k = 6 \times 10^{-2} \, mol^{-1} \, L \, s^{-1}$ and the same starting concentration. How many hours would it take to make a polymer with $N_n = 300$?

    (d) Qualitatively explain the origin of the different shapes of the curves in the two plots.

16. Hydrolysis of an aromatic polyamide with $M_n = 24,116$ gives 39.31% by weight m-amino-aniline, 59.81% terephthalic acid, and 0.88% benzoic acid. Draw the repeat unit structure of the polymer. Calculate the degree of polymerization and the extent of reaction. Calculate what the degree of polymerization would have been if the amount of benzoic acid were doubled.

17. Calculate the feed ratio of adipic acid and hexamethylene diamine necessary to achieve a molecular weight of approximately 10,000 at 99.5% conversion. What would the identity of the end groups be in the resulting polymer?

References

1. Flory, P.J., *Principles of Polymer Chemistry*, Cornell University Press, Ithaca, NY, 1953.
2. Solomon, D.H. (Ed.), *Step Growth Polymerization*, Marcel Dekker, New York, 1972.

Further Readings

Allcock, H.R. and Lampe, F.W., *Contemporary Polymer Chemistry*, 2nd ed., Prentice Hall, Englewood Cliffs, NJ, 1990.
Odian, G. *Principles of Polymerization*, 4th ed., Wiley, New York, 2004.
Rempp, P. and Merrill, E.W., *Polymer Synthesis*, 2nd ed., Hüthig & Wepf, Basel, 1991.

3

Chain-Growth Polymerization

3.1 Introduction

In Chapter 1 we indicated that the category of addition polymers is best characterized by the mechanism of the polymerization reaction rather than by the addition reaction itself. This is known to be a chain mechanism, so in the case of addition polymers we have chain reactions producing chain molecules. One thing to bear in mind is the two uses of the word *chain* in this discussion. The word chain continues to offer the best description of large polymer molecules. A chain reaction, on the other hand, describes a whole series of successive events triggered by some initial occurrence. We sometimes encounter this description of highway accidents in which one traffic mishap on a fogbound highway results in a pileup of colliding vehicles that can extend for miles. In nuclear reactors a cascade of fission reactions occurs, which is initiated by the capture of the first neutron. In both of these examples some initiating event is required. This is also true in chain-growth polymerization.

In the above examples the size of the chain can be measured by considering the number of automobile collisions that result from the first accident, or the number of fission reactions that follow from the first neutron capture. When we think about the number of monomers that react as a result of a single initiation step, we are led directly to the degree of polymerization of the resulting molecule. In this way the chain mechanism and the properties of the polymer chains are directly related.

Chain reactions do not go on forever. The fog may clear and the improved visibility ends the succession of accidents. Neutron-scavenging control rods may be inserted to shut down a nuclear reactor. The chemical reactions that terminate polymer chain growth are also an important part of the polymerization mechanism. Killing off the reactive intermediate that keeps the chain going is the essence of a termination reaction. Some interesting polymers can be formed when this termination process is suppressed; these are called *living* polymers, and will be discussed extensively in Chapter 4.

The kind of reaction that produces a "dead" polymer from a growing chain depends on the nature of the reactive intermediate. These intermediates may be free radicals, anions, or cations. We shall devote the rest of this chapter to a discussion of the free-radical mechanism, as it readily lends itself to a very general treatment. Furthermore, it is by far the most important chain-growth mechanism from a commercial point of view; examples include polyethylene (specifically, low-density polyethylene, LDPE), polystyrene, poly(vinyl chloride), and poly(acrylates) and poly(methacrylates). Anionic polymerization plays a central role in Chapter 4, where we discuss the so-called living polymerizations. In this chapter we deal exclusively with homopolymers. The important case of copolymers formed by chain-growth mechanisms is taken up in Chapter 4 and Chapter 5; block copolymers in the former, statistical or random copolymers in the latter.

3.2 Chain-Growth and Step-Growth Polymerizations: Some Comparisons

Our primary focus in this section is to point out some of the similarities and differences between step-growth and chain-growth polymerizations. In so doing we shall also have the opportunity to indicate some of the different types of chain-growth polymerization systems.

In Chapter 2 we saw that step-growth polymerizations occur, one step at a time, through a series of relatively simple organic reactions. By treating the reactivity of the functional groups as independent of the size of the molecule carrying the group, the entire course of the polymerization is described by the conversion of these groups to their condensation products. Two consequences of this are that both high yield and high molecular weight require the reaction to approach completion. In contrast, chain-growth polymerization occurs by introducing an active growth center into a reservoir of monomer, followed by the addition of many monomers to that center by a chain-type kinetic mechanism. The active center is ultimately killed off by a termination step. The (average) degree of polymerization that characterizes the system depends on the frequency of addition steps relative to termination steps. Thus high-molecular-weight polymer can be produced almost immediately. The only thing that is accomplished by allowing the reaction to proceed further is an increased yield of polymer. The molecular weight of the product is relatively unaffected. (This simple argument tends to break down at high extents of conversion. For this reason we shall focus attention in this chapter on low to moderate conversions to polymer, except where noted.)

Step-growth polymerizations can be schematically represented by one of the individual reaction steps A + B → ab, with the realization that the species so connected can be any molecules containing A and B groups. Chain-growth polymerization, by contrast, requires at least three distinctly different kinds of reactions to describe the mechanism. These three types of reactions will be discussed in the following sections in considerable detail; for now our purpose is just to introduce some vocabulary. The principal steps in the chain-growth mechanism are the following:

1. *Initiation.* An active species I* is formed by the decomposition of an initiator molecule I:

$$I \rightarrow I^* \tag{3.A}$$

2. *Propagation.* The initiator fragment reacts with a monomer M to begin the conversion to polymer; the center of activity is retained in the adduct. Monomers continue to add in the same way until polymers P_i are formed with the degree of polymerization i:

$$I^* + M \rightarrow IM^* \xrightarrow{M} IMM^* \rightarrow \rightarrow \rightarrow P_i^* \tag{3.B}$$

   If i is large enough, the initiator fragment—an endgroup—need not be written explicitly.

3. *Termination.* By some reaction, generally involving two polymers containing active centers, the growth center is deactivated, resulting in dead polymer:

$$P_i^* + P_j^* \rightarrow P_{i+j} \text{ (dead polymer)} \tag{3.C}$$

Elsewhere in this chapter we shall see that other reactions—notably, chain transfer and chain inhibition—also need to be considered to give a more fully developed picture of chain-growth polymerization, but we shall omit these for the time being. Most of this chapter examines the kinetics of these three mechanistic steps. We shall describe the rates of the three general kinds of reactions by the notation R_i, R_p, and R_t for initiation, propagation, and termination, respectively.

In the last chapter we presented arguments supporting the idea that reactivity is independent of molecular size. Although the chemical reactions are certainly different between this chapter and the last, we shall also adopt this *assumption of equal reactivity* for addition polymerization. For step-growth polymerization this assumption simplified the discussion tremendously and at the same time needed careful qualification. We recall that the equal reactivity premise is valid only after an initial size dependence for smaller molecules. The same variability applies to the propagation step of addition polymerizations for short-chain oligomers, although things soon level off and the assumption of equal reactivity holds. We are thus able to treat all propagation steps by the single rate constant k_p. Since the total polymer may be the product of hundreds or even thousands of such steps, no serious error is made in neglecting the variation that occurs in the first few steps.

In Section 2.3 we rationalized that, say, the first 50% of a step-growth reaction might be different from the second 50% because the reaction causes dramatic changes in the polarity of the reaction mixture. We shall see that, under certain circumstances, the rate of addition polymerization accelerates as the extent of conversion to polymer increases due to a composition-dependent effect on termination. In spite of these deviations from the assumption of equal reactivity at all extents of reaction, we continue to make this assumption because of the simplification it allows. We will then seek to explain the deviations from this ideal or to find experimental conditions—low conversions to polymer—under which the assumptions apply. This approach is common in chemistry; for example, most discussions of gases begin with the ideal gas law and describe real gases as deviating from the ideal at high pressures and approaching the ideal as pressure approaches zero.

In the last chapter we saw that two reactive groups per molecule are the norm for the formation of linear step-growth polymers. A pair of monofunctional reactants might undergo essentially the same reaction, but no polymer is produced because no additional functional groups remain to react. On the other hand, if a molecule contains more than two reactive groups, then branched or cross-linked products can result from step-growth polymerization. By comparison, a wide variety of unsaturated monomers undergo chain-growth polymerization. A single kind of monomer suffices—more than one yields a copolymer—and more than one double bond per monomer may result in branching or cross-linking. For example, the 1,2-addition reaction of butadiene results in a chain that has a substituent vinyl group capable of branch formation. Divinyl benzene is an example of a bifunctional monomer, which is used as a cross-linking agent in chain-growth polymerizations. We shall be primarily concerned with various alkenes or olefins as the monomers of interest; however, the carbon–oxygen double bond in aldehydes and ketones can also serve as the unsaturation required for addition polymerization. The polymerization of alkenes yields a carbon atom backbone, whereas the carbonyl group introduces carbon and oxygen atoms into the backbone, thereby illustrating the inadequacy of backbone composition as a basis for distinguishing between addition and condensation polymers.

It might be noted that most (but not all) alkenes are polymerizable by the chain mechanism involving free-radical intermediates, whereas the carbonyl group is generally not polymerized by the free-radical mechanism. Carbonyl groups and some carbon–carbon double bonds are polymerized by ionic mechanisms. Monomers display far more specificity where the ionic mechanism is involved than with the free-radical mechanism. For example, acrylamide will polymerize through an anionic intermediate but not a cationic one, N-vinyl pyrrolidones by cationic but not anionic intermediates, and halogenated olefins by neither ionic species. In all of these cases free-radical polymerization is possible.

The initiators used in addition polymerizations are sometimes called "catalysts," although strictly speaking this is a misnomer. A true catalyst is recoverable at the end of the reaction, chemically unchanged. This is not true of the initiator molecules in most addition polymerizations. Monomer and polymer are the initial and final states of the polymerization process, and these govern the thermodynamics of the reaction; the nature and concentration of the intermediates in the process, on the other hand, determine the rate. This makes initiator and catalyst synonyms for the same material. The former term stresses the effect of the reagent on the intermediate, and the latter its effect on the rate. The term catalyst is particularly common in the language of ionic polymerizations, but this terminology should not obscure the importance of the initiation step in the overall polymerization mechanism.

In the next three sections (Section 3.3 through Section 3.5) we consider initiation, termination, and propagation steps in the free-radical mechanism for addition polymerization. As noted above two additional steps, inhibition and chain transfer, are being ignored at this point. We shall take up these latter topics in Section 3.8.

3.3 Initiation

In this section we shall discuss the initiation step of free-radical polymerization. This discussion is centered around initiators and their decomposition behavior. The first requirement for an initiator is that it be a source of free radicals. In addition, the radicals must be produced at an acceptable rate

Table 3.1 Examples of Free-Radical Initiation Reactions

1. Organic peroxides or hydroperoxides

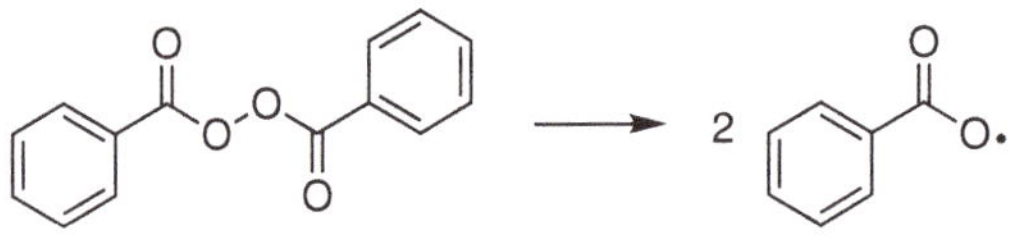

Benzoyl peroxide

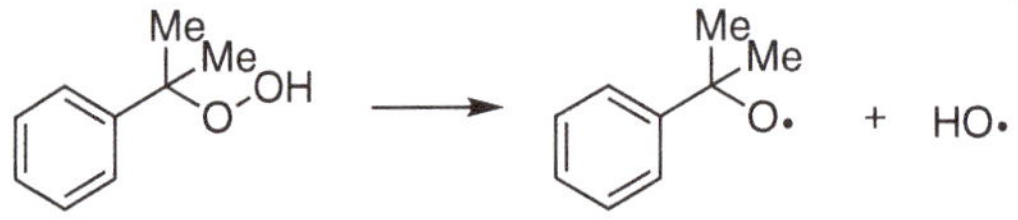

Cumyl hydroperoxide

2. Azo compounds

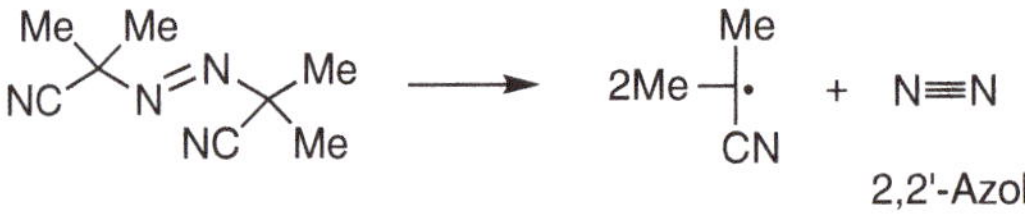

2,2'-Azobisisobutyronitrile (AIBN)

3. Redox systems

$$H_2O_2 \ + \ Fe^{2+} \longrightarrow \ {}^-OH \ + \ Fe^{3+} \ + \ {}^{\bullet}OH$$

$$S_2O_8^{2-} \ + \ Fe^{2+} \longrightarrow \ SO_4^{2-} \ + \ Fe^{3+} \ + \ SO_4^{-\bullet}$$

4. Electromagnetic radiation

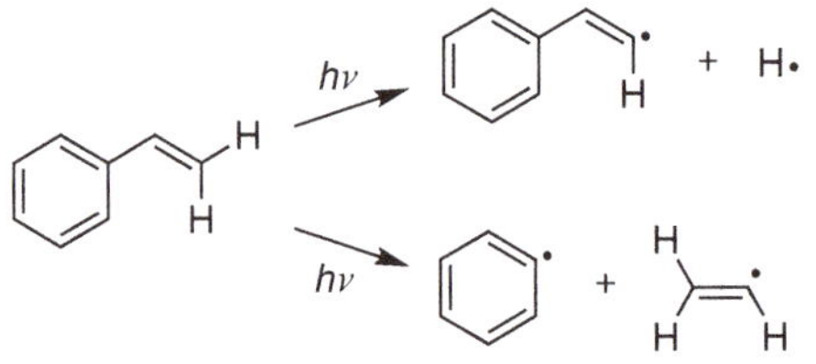

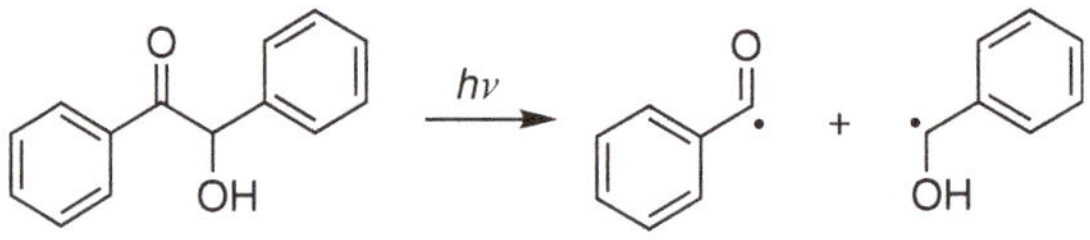

Benzoin

at convenient temperatures; have the required solubility behavior; transfer their activity to monomers efficiently; be amenable to analysis, preparation, purification, and so on.

3.3.1 Initiation Reactions

Some of the most widely used initiator systems are listed below, and Table 3.1 illustrates their behavior by typical reactions:

1. Organic peroxides or hydroperoxides
2. Azo compounds

3. Redox systems
4. Thermal or light energy

Peroxides and hydroperoxides are useful as initiators because of the low dissociation energy of the O−O bond. This very property makes the range of possible compounds somewhat limited because of the instability of these reagents. In the case of azo compounds the homolysis is driven by the liberation of the very stable N_2 molecule, despite the relatively high dissociation energy of the C−N bond. The redox systems listed in Table 3.1 have the advantage of water solubility, although redox systems that operate in organic solvents are also available. One advantage of redox reactions as a source of free radicals is the fact that these reactions often proceed more rapidly and at lower temperatures than the thermal homolysis of the peroxide and azo compounds.

The initiation reactions shown under the heading of electromagnetic radiation in Table 3.1 indicate two possibilities out of a large number of examples that might be cited. One mode of photochemical initiation involves the direct excitation of the monomer with subsequent bond rupture. The second example cited is the photolytic fragmentation of initiators such as alkyl halides and ketones. Because of the specificity of light absorption, photochemical initiators include a wider variety of compounds than those which decompose thermally. Photosensitizers can also be used to absorb and transfer radiation energy to either monomer or initiator molecules. Finally, we note that high-energy radiation such as x-rays and γ-rays and particulate radiation such as α or β particles can also produce free radicals. These latter sources of radiant energy are nonselective and produce a wider array of initiating species. Even though such high-energy radiations produce both ionic and free-radical species, the polymerizations that are so initiated follow the free-radical mechanism almost exclusively, except at very low temperatures, where ionic intermediates become more stable. We shall not deal further with these higher energy sources of initiating radicals, but we shall return to light as a photochemical initiator because of its utility in the evaluation of kinetic rate constants.

3.3.2 Fate of Free Radicals

All of the reactions listed in Table 3.1 produce free radicals, so we are presented with a number of alternatives for initiating a polymerization reaction. Our next concern is the fate of these radicals or, stated in terms of our interest in polymers, the efficiency with which these radicals initiate polymerization. Since these free radicals are relatively reactive species, there are a variety of processes they can undergo as alternatives to adding to monomers to form polymer.

In discussing mechanisms in the last chapter (Reaction 2.F) we noted that the entrapment of two reactive species in the same solvent cage may be considered a transition state in the reaction of these species. Reactions such as the thermal homolysis of peroxides and azo compounds result in the formation of two radicals already trapped together in a cage that promotes direct recombination, as with the 2-cyanopropyl radicals from 2,2′-azobisisobutyronitrile (AIBN),

(3.D)

or the recombination of degradation products of the initial radicals, as with acetoxy radicals from acetyl peroxide.

$$2 \; Me \overset{O}{\overset{\|}{C}} O\cdot \; \begin{cases} \longrightarrow \; Me \overset{O}{\overset{\|}{C}} O \diagdown Me \; + \; CO_2 \\ \\ \searrow \; Me-Me \; + \; 2CO_2 \end{cases}$$

(3.E)

In both of these examples, initiator is consumed, but no polymerization is started.

Once the radicals diffuse out of the solvent cage, reaction with monomer is the most probable reaction in bulk polymerizations, since monomers are the species most likely to be encountered. Reaction with polymer radicals or initiator molecules cannot be ruled out, but these are less important because of the lower concentration of the latter species. In the presence of solvent, reactions between the initiator radical and the solvent may effectively compete with polymer initiation. This depends very much on the specific chemicals involved. For example, carbon tetrachloride is quite reactive toward radicals because of the resonance stabilization of the solvent radical produced:

[I]

While this reaction with solvent continues to provide free radicals, these may be less reactive species than the original initiator fragments. We shall have more to say about the transfer of free-radical functionality to solvent in Section 3.8.

The significant thing about these, and numerous other side reactions that could be described, is the fact that they lower the efficiency of the initiator in promoting polymerization. To quantify this concept we define the *initiator efficiency f* to be the following fraction:

$$f = \frac{\text{Radicals incorporated into polymer}}{\text{Radicals formed by initiator}}$$

(3.3.1)

The initiator efficiency is not an exclusive property of the initiator, but depends on the conditions of the polymerization experiment, including the solvent. In many experimental situations, f lies in the range of $0.3-0.8$. The efficiency should be regarded as an empirical parameter whose value is determined experimentally. Several methods are used for the evaluation of initiator efficiency, the best being the direct analysis for initiator fragments as endgroups compared to the amount of initiator consumed, with proper allowances for stoichiometry. As an endgroup method, this procedure is difficult in addition polymers, where molecular weights are higher than in condensation polymers. Research with isotopically labeled initiators is particularly useful in this application. Since the quantity is so dependent on the conditions of the experiment, it should be monitored for each system studied.

Scavengers such as diphenylpicrylhydrazyl radicals [II] react with other radicals and thus provide an indirect method for analysis of the number of free radicals in a system:

(3.F)

[II]

The diphenylpicrylhydrazyl radical itself is readily followed spectrophotometrically, as it loses an intense purple color on reacting. Unfortunately, this reaction is not always quantitative.

3.3.3 Kinetics of Initiation

We recall some of the ideas of kinetics from the summary given in Section 2.2 and recognize that the rates of initiator decomposition can be developed in terms of the reactions listed in Table 3.1.

Using the change in initiator radical concentration $d[I\cdot]/dt$ to monitor the rates, we write the following:

1. For peroxides and azo compounds

$$\frac{d[I\cdot]}{dt} = 2k_d[I] \tag{3.3.2}$$

   where k_d is the rate constant for the homolytic decomposition of the initiator and $[I]$ is the concentration of the initiator. The factor 2 appears because of the stoichiometry in these particular reactions.

2. For redox systems

$$\frac{d[I\cdot]}{dt} = k[Ox][Red] \tag{3.3.3}$$

   where the bracketed terms describe the concentrations of oxidizing and reducing agents and k is the rate constant for the particular reactants.

3. For photochemical initiation

$$\frac{d[I\cdot]}{dt} = 2\phi' I_{abs} \tag{3.3.4}$$

where I_{abs} is the intensity of the light absorbed and the constant ϕ' is called the quantum yield. The factor 2 is again included for reasons of stoichiometry.

Since $(1/2)\,d[I\cdot]/dt = -d[I]/dt$ in the case of the azo initiators, Equation 3.3.2 can also be written as $-d[I]/dt = k_d[I]$ or, by integration, $\ln([I]/[I]_0) = -k_d t$, where $[I]_0$ is the initiator concentration at $t = 0$. Figure 3.1 shows a test of this relationship for AIBN in xylene at 77°C. Except for a short induction period, the data points fall on a straight line. The evaluation of k_d from these data is presented in the following example.

Example 3.1

The decomposition of AIBN in xylene at 77°C was studied by measuring the volume of N_2 evolved as a function of time. The volumes obtained at time t and $t = \infty$ are V_t and V_∞, respectively. Show that the manner of plotting used in Figure 3.1 is consistent with the integrated first-order rate law and evaluate k_d.

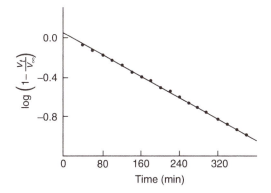

Figure 3.1 Volume of nitrogen evolved from the decomposition of AIBN at 77°C plotted according to the first-order rate law as discussed in Example 3.1. (Reprinted from Arnett, L.M., *J. Am. Chem. Soc.*, 74, 2027, 1952. With permission.)

Solution

The ratio $[I]/[I]_0$ gives the fraction of initiator remaining at time t. The volume of N_2 evolved is:

1. $V_0 = 0$ at $t = 0$, when no decomposition has occurred.
2. V_∞ at $t = \infty$, when complete decomposition has occurred.
3. V_t at time t, when some fraction of initiator has decomposed.

The fraction decomposed at t is given by $(V_t - V_0)/(V_\infty - V_0)$ and the fraction remaining at t is $1 - (V_t - V_0)/(V_\infty - V_0) = (V_\infty - V_t)/(V_\infty - V_0)$. Since $V_0 = 0$, this becomes $(V_\infty - V_t)/V_\infty$ or $[I]/[I]_0 = 1 - V_t/V_\infty$. Therefore a plot of $\ln(1 - V_t/V_\infty)$ versus t is predicted to be linear with slope $-k_d$. (If logarithms to base 10 were used, the slope would equal $-k_d/2.303$.)

From Figure 3.1,

$$\text{Slope} = \frac{-0.4 - (-0.8)}{160 - 320} = -2.5 \times 10^{-3} \text{ min}^{-1} = \frac{-k_d}{2.303}$$

$$k_d = 5.8 \times 10^{-3} \text{ min}^{-1}$$

Next we assume that only a fraction f of these initiator fragments actually reacts with monomer to transfer the radical functionality to monomer:

$$I\cdot + M \xrightarrow{f} IM\cdot \qquad (3.G)$$

As indicated in the last section, we regard the reactivity of the species $IP_i\cdot$ to be independent of the value of i. Accordingly, all subsequent additions to $IM\cdot$ in Reaction (3.G) are propagation steps and Reaction (3.G) represents the initiation of polymerization. Although it is premature at this point, we disregard endgroups and represent the polymeric radicals of whatever size by the symbol $P\cdot$. Accordingly, we write the following for the initiation of polymer radicals:

1. By peroxide and azo compounds

$$\frac{d[P\cdot]}{dt} = 2fk_d[I] \qquad (3.3.5)$$

2. By redox systems

$$\frac{d[P\cdot]}{dt} = fk[\text{Ox}][\text{Red}] \qquad (3.3.6)$$

3. By photochemical initiation

$$\frac{d[P\cdot]}{dt} = 2f\phi' I_{\text{abs}} = 2\phi I_{\text{abs}} \qquad (3.3.7)$$

where we have combined the factors of f and ϕ' into a composite quantum yield ϕ, since both of the separate factors are measures of efficiency.

Any one of these expressions gives the rate of initiation R_i for the particular catalytic system employed. We shall focus attention on the homolytic decomposition of a single initiator as the mode of initiation throughout most of this chapter, since this reaction typifies the most widely used free-radical initiators. Appropriate expressions for initiation that follow Equation 3.3.6 are readily derived.

3.3.4   Photochemical Initiation

An important application of photochemical initiation is in the determination of the rate constants that appear in the overall analysis of the chain-growth mechanism. Although we outline this method in Section 3.6, it is worthwhile to develop Equation 3.3.7 somewhat further at this point. It is not feasible to give a detailed treatment of light absorption here. Instead, we summarize some pertinent relationships and refer the reader who desires more information to standard textbooks of analytical or physical chemistry.

1. Intensity of light transmitted (subscript t) through a sample I_t depends on the intensity of the incident (subscript 0) light I_0, the thickness of the sample b, and the concentration $[c]$ of the absorbing species

$$I_t = I_0 e^{-\varepsilon[c]b} \tag{3.3.8}$$

where the proportionality constant ε is called the absorption coefficient (or molar absorptivity if $[c]$ is in moles/liter) and is a property of the absorber. The reader may recognize this equation as a form of the famous *Beer's law*.

2. Absorbance A as measured by spectrophotometers is defined as

$$A = \log_{10}\left(\frac{I_0}{I_t}\right) \tag{3.3.9}$$

The variation in absorbance with wavelength reflects the wavelength dependence of ε.

3. Since I_{abs} equals the difference $I_0 - I_t$

$$I_{abs} = I_0(1 - e^{-\varepsilon[c]b}) \tag{3.3.10}$$

If the exponent in Equation 3.3.10 is small—which in practice means dilute solutions, since most absorption experiments are done where ε is large—then the exponential can be expanded (see Appendix), $e^x \cong 1 + x + \cdots$, with only the leading terms retained to give

$$I_{abs} = I_0(\varepsilon[c]b) \tag{3.3.11}$$

4. Substituting this result into Equation 3.3.7 gives

$$\frac{d[P\bullet]}{dt} = 2\phi I_0 \varepsilon[c]b \tag{3.3.12}$$

where $[c]$ is the concentration of monomer or initiator for the two reactions shown in Table 3.1.

3.3.5 Temperature Dependence of Initiation Rates

Note that although Equation 3.3.5 and Equation 3.3.12 are both first-order rate laws, the physical significance of the proportionality factors is quite different in the two cases. The rate constants shown in Equation 3.3.5 and Equation 3.3.6 show a temperature dependence described by the Arrhenius equation:

$$k = Ae^{-E^*/RT} \tag{3.3.13}$$

where E^* is the activation energy, which is interpreted as the height of the energy barrier to a reaction, and A is the prefactor. Activation energies are evaluated from experiments in which rate constants are measured at different temperatures. Taking logarithms on both sides of Equation 3.3.13 gives $\ln k = \ln A - E^*/RT$. Therefore E^* is obtained from the slope of a plot of $\ln k$ against $1/T$. As usual, T is in kelvin and R and E^* are in (the same) energy units.

Since E^* is positive according to this picture, the form of the Arrhenius equation assures that k gets larger as T increases. This means that a larger proportion of molecules have sufficient energy to surmount the energy barrier at higher temperatures. This assumes, of course, that thermal energy is the source of E^*, something that is not the case in photoinitiated reactions. The effective first-order rate constants k and $I_0\varepsilon b$—for thermal initiation and photoinitiation, respectively—do

Table 3.2 Rate Constants (at the Indicated Temperature) and Activation Energies for Some Initiator Decomposition Reactions

Initiator	Solvent	T (°C)	k_d (s^{-1})	E_d^* (kJ mol^{-1})
2,2'-Azobisisobutyronitrile	Benzene	70	3.17×10^{-5}	123.4
	CCl$_4$	40	2.15×10^{-7}	128.4
	Toluene	100	1.60×10^{-3}	121.3
t-Butyl peroxide	Benzene	100	8.8×10^{-7}	146.9
Benzoyl peroxide	Benzene	70	1.48×10^{-5}	123.8
	Cumene	60	1.45×10^{-6}	120.5
t-Butyl hydroperoxide	Benzene	169	2.0×10^{-5}	170.7

Source: Data from Masson, J.C. in *Polymer Handbook*, 3rd ed., Brandrup, J. and Immergut, E.H. (Eds.), Wiley, New York, 1989.

not show the same temperature dependence. The former follows the Arrhenius equation, whereas the latter cluster of terms in Equation 3.3.12 is essentially independent of T.

The activation energies for the decomposition (subscript d) reaction of several different initiators in various solvents are shown in Table 3.2. Also listed are values of k_d for these systems at the temperature shown. The Arrhenius equation can be used in the form $\ln (k_{d,1}/k_{d,2}) = -(E^*/R)$ $(1/T_1 - 1/T_2)$ to evaluate k_d values for these systems at temperatures different from those given in Table 3.2.

3.4 Termination

The formation of initiator radicals is not the only process that determines the concentration of free radicals in a polymerization system. Polymer propagation itself does not change the radical concentration; it merely converts one radical to another. Termination steps also occur, however, and these remove radicals from the system. We shall discuss *combination* and *disproportionation* reactions as the two principal modes of termination.

3.4.1 Combination and Disproportionation

Termination by combination results in the simultaneous destruction of two radicals by direct coupling:

$$P_i\bullet + \bullet P_j \rightarrow P_{i+j} \tag{3.H}$$

The degrees of polymerization i and j in the two combining radicals can have any values, and the molecular weight of the product molecule will be considerably higher on the average than the radicals so terminated. The polymeric product molecule contains two initiator fragments per molecule by this mode of termination. Note also that for a vinyl monomer, such as styrene or methyl methacrylate, the combination reaction produces a single head-to-head linkage, with the side groups attached to adjacent backbone carbons instead of every other carbon.

Termination by disproportionation comes about when an atom, usually hydrogen, is transferred from one polymer radical to another:

$$\tag{3.I}$$

This mode of termination produces a negligible effect on the molecular weight of the reacting species, but it does produce a terminal unsaturation in one of the dead polymer molecules. Each polymer molecule contains one initiator fragment when termination occurs by disproportionation.

Kinetic analysis of the two modes of termination is quite straightforward, since each mode of termination involves a bimolecular reaction between two radicals. Accordingly, we write the following:

1. For general termination,

$$R_t = -\frac{d[P\bullet]}{dt} = 2k_t[P\bullet]^2 \tag{3.4.1}$$

   where R_t and k_t are the rate and rate constant for termination (subscript t) and the factor 2 enters (by convention) because two radicals are lost for each termination step.
2. The polymer radical concentration in Equation 3.4.1 represents the total concentration of all such species, regardless of their degree of polymerization; that is,

$$[P\bullet] = \sum_{\text{all i}} [P_i\bullet] \tag{3.4.2}$$

3. For combination,

$$R_t = -\frac{d[P\bullet]}{dt} = 2k_{t,c}[P\bullet]^2 \tag{3.4.3}$$

   where the subscript c specifically indicates termination by combination.
4. For disproportionation,

$$R_t = -\frac{d[P\bullet]}{dt} = 2k_{t,d}[P\bullet]^2 \tag{3.4.4}$$

   where the subscript d specifically indicates termination by disproportionation.
5. In the event that the two modes of termination are not distinguished, Equation 3.4.1 represents the sum of Equation 3.4.3 and Equation 3.4.4, or

$$k_t = k_{t,c} + k_{t,d} \tag{3.4.5}$$

Combination and disproportionation are competitive processes and do not occur to the same extent for all polymers, although in general combination is more prevalent. For poly(methyl methacrylate), both reactions contribute to termination, with disproportionation favored. Both rate constants for termination individually follow the Arrhenius equation, so the relative amounts of termination by the two modes are given by

$$\frac{\text{Termination by combination}}{\text{Termination by disproportionation}} = \frac{k_{t,c}}{k_{t,d}} = \frac{A_{t,c}e^{-E_{t,c}^*/RT}}{A_{t,d}e^{-E_{t,d}^*/RT}} = \frac{A_{t,c}}{A_{t,d}}\exp\left(\frac{-(E_{t,c}^* - E_{t,d}^*)}{RT}\right) \tag{3.4.6}$$

Since the disproportionation reaction requires bond breaking, which is not required for combination, $E_{t,d}^*$ is expected to be greater than $E_{t,c}^*$. This causes the exponential to be large at low temperatures, making combination the preferred mode of termination under these circumstances. Note that at higher temperatures this bias in favor of one mode of termination over another decreases as the difference in activation energies becomes smaller relative to the thermal energy RT. Experimental results on modes of termination show that this qualitative argument must be applied cautiously. The actual determination of the partitioning between the two modes of termination is best accomplished by analysis of endgroups, using the difference in endgroup distribution noted above.

Table 3.3 lists the activation energies for termination (these are overall values, not identified as to mode) of several different radicals. The rate constants for termination at 60°C are also given. We shall see in Section 3.6 how these constants are determined.

Table 3.3 Rate Constants at 60°C and Activation Energies for Some
Termination Reactions

Monomer	E_t^* (kJ mol^{-1})	$k_t \times 10^{-7}$ (L mol^{-1} s^{-1})
Acrylonitrile	1.5	78.2
Methyl acrylate	22.2	0.95
Methyl methacrylate	11.9	2.55
Styrene	8.0	6.0
Vinyl acetate	21.9	2.9
2-Vinyl pyridine	21.0	3.3

Source: Data from Korus, R. and O'Driscoll, K.F. in *Polymer Handbook*, 3rd ed.,
Brandrup, J. and Immergut, E.H. (Eds.), Wiley, New York, 1989.

3.4.2 Effect of Termination on Conversion to Polymer

The assumption that k values are constant over the entire duration of the reaction breaks down
for termination reactions in bulk polymerizations. Here, as in Section 2.2, we can consider the
termination process—whether by combination or disproportionation—to depend on the rates at
which polymer molecules can diffuse into (characterized by k_i) or out of (characterized by k_o) the
same solvent cage and the rate at which chemical reaction between them (characterized by k_r)
occurs in that cage. In Chapter 2, we saw that two limiting cases of Equation 2.2.8 could be readily
identified:

1. Rate of diffusion > rate of reaction (Equation 2.2.9):

$$k_t = \frac{k_i}{k_o} k_r \tag{3.4.7}$$

2. This situation seems highly probable for step-growth polymerization because of the high
    activation energy of many condensation reactions. The constants for the diffusion-dependent
    steps, which might be functions of molecular size or the extent of the reaction, cancel out.
3. Rate of reaction > rate of diffusion (Equation 2.2.10):

$$k_t = k_i \tag{3.4.8}$$

4. This situation is expected to apply to radical termination, especially by combination, because
    of the high reactivity of the trapped radicals. Only one constant appears that depends on the
    diffusion of the polymer radicals, so it cannot cancel out and may contribute to a dependence
    of k_t on the extent of reaction or the degree of polymerization.

Figure 3.2 shows how the percent conversion of methyl methacrylate to polymer varies with
time. These experiments were carried out in benzene at 50°C. The different curves correspond to
different concentrations of monomer. Up to about 40% monomer, the conversion varies smoothly
with time, gradually slowing down at higher conversions owing to the depletion of monomer. At
high concentrations, however, the polymerization starts to show an acceleration between 20% and
40% conversion. This behavior, known as the *Trommsdorff effect* [2], is attributed to a decrease in
the rate of termination with increasing conversion. This, in turn, is due to the increase in viscosity
that has an adverse effect on k_t through Equation 3.4.8. Considerations of this sort are important in
bulk polymerizations where high conversion is the objective, but this complication is something
we will avoid. Hence we shall be mainly concerned with solution polymerization and/or low
degrees of conversion where k_t may be justifiably treated as a true constant. We shall see in Section
3.8 that the introduction of solvent is accompanied by some complications of its own, but we shall
ignore this for now.

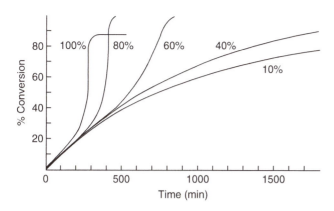

Figure 3.2 Acceleration of the polymerization rate for methyl methacrylate at the concentrations shown in benzene at 50°C. (Reprinted from Schulz, G.V. and Harborth, G., *Makromol. Chem.*, 1, 106, 1948. With permission.)

3.4.3 Stationary-State Radical Concentration

Polymer propagation steps do not change the total radical concentration, so we recognize that the two opposing processes, initiation and termination, will eventually reach a point of balance. This condition is called the *stationary state* and is characterized by a constant total concentration of free radicals. Under stationary-state conditions (subscript s) the net rate of initiation must equal the net rate of termination. Using Equation 3.3.5 for the rate of initiation (i.e., two radicals per initiator molecule) and Equation 3.4.1 for termination, we write

$$2 f k_d [I] = 2 k_t [P\bullet]_s^2 \qquad (3.4.9)$$

or

$$[P\bullet]_s = \left(\frac{f k_d}{k_t}\right)^{1/2} [I]^{1/2} \qquad (3.4.10)$$

This important equation shows that the stationary-state free-radical concentration increases with $[I]^{1/2}$ and varies directly with $k_d^{1/2}$ and inversely with $k_t^{1/2}$. The concentration of free radicals determines the rate at which polymer forms and the eventual molecular weight of the polymer, since each radical is a growth site. We shall examine these aspects of Equation 3.4.10 in the next section. We conclude this section with a numerical example illustrating the stationary-state radical concentration for a typical system.

Example 3.2

For an initiator concentration that is constant at $[I]_0$, the *nonstationary-state* radical concentration varies with time according to the following expression:

$$\frac{[P\bullet]}{[P\bullet]_s} = \frac{\exp\left[(4 f k_d k_t [I]_0)^{1/2} t\right] - 1}{\exp\left[(4 f k_d k_t [I]_0)^{1/2} t\right] + 1}$$

Calculate $[P\bullet]_s$ and the time required for the free-radical concentration to reach 99% of this value using the following as typical values for constants and concentrations: $k_d = 1.0 \times 10^{-4}$ s^{-1}, $k_t = 3 \times 10^7$ L mol^{-1} s^{-1}, $f = 1/2$, and $[I]_0 = 10^{-3}$ mol L^{-1}. Comment on the assumption $[I] = [I]_0$ that was made in deriving this nonstationary-state equation.

Solution

Use Equation 3.4.10 to evaluate $[P\cdot]_s$ for the system under consideration:

$$[P\cdot]_s = \left(\frac{fk_d}{k_t}[I]_0\right)^{1/2} = \left(\frac{(1/2)(1.0 \times 10^{-4})(10^{-3})}{3 \times 10^7}\right)^{1/2} = (1.67 \times 10^{-15})^{1/2}$$
$$= 4.08 \times 10^{-8} \text{ mol L}^{-1}$$

This low concentration is typical of free-radical polymerizations. Next we inquire how long it will take the free-radical concentration to reach 0.99 $[P\cdot]_s$, or 4.04×10^{-8} mol L^{-1} in this case. Let $a = (4fk_dk_t[I]_0)^{1/2}$ and rearrange the expression given to solve for t when $[P\cdot]/[P\cdot]_s = 0.99$: 0.99 $(e^{at} + 1) = e^{at} - 1$, or $1 + 0.99 = e^{at}(1 - 0.99)$. Therefore the product $at = \ln(1.99/0.01) = \ln 199 = 5.29$, and $a = [4(1/2)(1.0 \times 10^{-4})(3 \times 10^7)(10^{-3})]^{1/2} = 2.45$ s^{-1}. Hence $t = 5.29/2.45 = 2.16$ s.

This short period is also typical of the time required to reach the stationary state. The assumption that $[I] = [I]_0$ may be assessed by examining the integrated form of Equation 3.3.2 for this system and calculating the ratio $[I]/[I]_0$ after 2.16 s:

$$\ln\left(\frac{[I]}{[I]_0}\right) = -k_dt = -(1.0 \times 10^{-4})(2.16) = -2.16 \times 10^{-4}$$
$$\frac{[I]}{[I]_0} = 0.9998$$

Over the time required to reach the stationary state, the initiator concentration is essentially unchanged. As a matter of fact, it would take about 100 s for $[I]$ to reach 0.99 $[I]_0$ and about 8.5 min to reach 0.95 $[I]_0$, so the assumption that $[I] = [I]_0$ is entirely justified over the short times involved.

3.5 Propagation

The propagation of polymer chains is easy to consider under stationary-state conditions. As the preceding example illustrates, the stationary state is reached very rapidly, so we lose only a brief period at the start of the reaction by restricting ourselves to the stationary state. Of course, the stationary-state approximation breaks down at the end of the reaction also, when the radical concentration drops toward zero. We shall restrict our attention to relatively low conversion to polymer, however, to avoid the complications of the Trommsdorff effect. Therefore deviations from the stationary state at long times need not concern us.

It is worth taking a moment to examine the propagation step more explicitly in terms of the reaction mechanism itself. As an example, consider the case of styrene as a representative vinyl monomer. The polystyryl radical is stabilized on the terminal-substituted carbon by resonance delocalization:

Consequently, the addition of the next monomer is virtually exclusively in a "head-to-tail" arrangement, leading to an all-carbon backbone with substituents (X) on alternating backbone atoms:

$$(3.J)$$

This should be contrasted with the single head-to-head linkage that results from termination by recombination (recall Reaction (3.H)).

3.5.1 Rate Laws for Propagation

Consideration of Reaction (3.B) leads to

$$-\frac{d[M]}{dt} = k_p[M][P\cdot]$$

$$(3.5.1)$$

as the expression for the rate at which monomer is converted to polymer. In writing this expression, we assume the following:

1. The radical concentration has the stationary-state value given by Equation 3.4.10.
2. k_p is a constant independent of the size of the growing chain and the extent of conversion to polymer.
3. The rate at which monomer is consumed is equal to the rate of polymer formation R_p:

$$-\frac{d[M]}{dt} = \frac{d[\text{polymer}]}{dt} = R_p$$

$$(3.5.2)$$

Combining Equation 3.4.10 and Equation 3.5.1 yields

$$R_p = k_p[M]\left(\frac{fk_d}{k_t}\right)^{1/2}[I]^{1/2} = k_{app}[M][I]^{1/2}$$

$$(3.5.3)$$

in which the second form reminds us that an experimental study of the rate of polymerization yields a single *apparent* rate constant (subscript app), which the mechanism reveals to be a composite of three different rate constants. Equation 3.5.3 shows that the rate of polymerization is first order in monomer and half order in initiator and depends on the rate constants for each of the three types of steps—initiation, propagation, and termination—that make up the chain mechanism. Since the concentrations change with time, it is important to realize that Equation 3.5.3 gives an instantaneous rate of polymerization at the concentrations considered. The equation can be applied to the initial concentrations of monomer and initiator in a reaction mixture only to describe the initial rate of polymerization. Unless stated otherwise, we shall assume the initial conditions apply when we use this result.

The initial rate of polymerization is a measurable quantity. The amount of polymer formed after various times in the early stages of the reaction can be determined directly by precipitating the polymer and weighing. Alternatively, some property such as the volume of the system (or the density, the refractive index, or the viscosity) can be measured. Using an analysis similar to that followed in Example 3.1, we can relate the values of the property measured at t, $t = 0$ and $t = \infty$ to the fraction of monomer converted to polymer. If the rate of polymerization is measured under known and essentially constant concentrations of monomer and initiator, then the cluster of constants $(fk_p^2k_d/k_t)^{1/2}$ can be evaluated from the experiment. As noted earlier, f is best investigated by endgroup analysis. Even with the factor f excluded, experiments on the rate of polymerization still leave us with three unknowns. Two other measurable relationships among these unknowns must be found if the individual constants are to be resolved. In anticipation of this development, we list values of k_p and the corresponding activation energies for several common monomers in Table 3.4.

Table 3.4 Rate Constants at 60°C and Activation Energies for Some
Propagation Reactions

Monomer	E_p^* (kJ mol^{-1})	$k_p \times 10^{-3}$ (L mol^{-1} s^{-1})
Acrylonitrile	16.2	1.96
Methyl acrylate	29.7	2.09
Methyl methacrylate	26.4	0.515
Styrene	26.0	0.165
Vinyl acetate	18.0	2.30
2-Vinyl pyridine	33.0	0.186

Source: Data from Korus, R. and O'Driscoll, K.F. in *Polymer Handbook*, 3rd ed.,
Brandrup, J. and Immergut, E.H. (Eds.), Wiley, New York, 1989.

Equation 3.5.3 is an important result, which can be expressed in several alternate forms:

1. The variation in monomer concentration may be taken into account by writing the equation in the integrated form and treating the initiator concentration as constant at $[I]_0$ over the interval considered:

$$\ln\left(\frac{[M]}{[M]_0}\right) = -\left(\frac{fk_p^2 k_d}{k_t}[I]_0\right)^{1/2} t \tag{3.5.4}$$

   where $[M] = [M]_0$ at $t = 0$.

2. Instead of using $2fk_d[I]$ for the rate of initiation, we can simply write this latter quantity as R_i, in which case the stationary-state radical concentration is

$$[P\cdot]_s = \left(\frac{R_i}{2k_t}\right)^{1/2} \tag{3.5.5}$$

   and the rate of polymerization becomes

$$R_p = \left(\frac{k_p^2}{2k_t}\right)^{1/2} R_i^{1/2}[M] \tag{3.5.6}$$

   If the rate of initiation is investigated independently, the rate of polymerization measures a combination of k_p and k_t.

3. Alternatively, Equation 3.3.6 and Equation 3.3.7 can be used as expressions for R_i in Equation 3.5.6 to describe redox or photoinitiated polymerization.

Figure 3.3 shows some data that constitute a test of Equation 3.5.3. In Figure 3.3a, R_p and [M] are plotted on a log–log scale for a constant level of redox initiator. The slope of this line, which indicates the order of the polymerization with respect to monomer, is unity, showing that the polymerization of methyl methacrylate is first order in monomer. Figure 3.3b is a similar plot of the initial rate of polymerization—which essentially maintains the monomer at constant concentration—versus initiator concentration for two different monomer–initiator combinations. Each of the lines has a slope of 1/2, indicating a half-order dependence on [I] as predicted by Equation 3.5.3.

3.5.2 Temperature Dependence of Propagation Rates

The apparent rate constant in Equation 3.5.3 follows the Arrhenius equation and yields an apparent activation energy:

$$\ln k_{app} = \ln A_{app} - \frac{E_{app}^*}{RT} \tag{3.5.7}$$

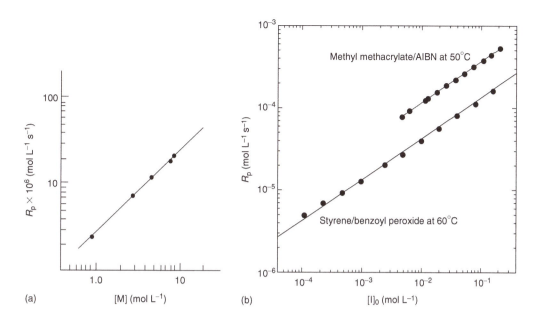

Figure 3.3 Log–log plots of R_p versus concentration that confirm the kinetic order with respect to the constituent varied. (a) Monomer (methyl methacrylate) concentration varied at constant initiator concentration. (Data from Sugimura, T. and Minoura, Y., *J. Polym. Sci.*, A-1, 2735, 1966.) (b) Initiator concentration varied: AIBN in methyl methacrylate (data from Arnett, L.M., *J. Am. Chem. Soc.*, 74, 2027, 1952) and benzoyl peroxide in styrene (data from Mayo, F.R., Gregg, R.A., and Matheson, M.S., *J. Am. Chem. Soc.*, 73, 1691, 1951).

The mechanistic analysis of the rate of polymerization and the fact that the separate constants individually follow the Arrhenius equations means that

$$\ln k_{app} = \ln k_p \left(\frac{fk_d}{k_t}\right)^{1/2}$$

$$= \ln A_p \left(\frac{fA_d}{A_t}\right)^{1/2} - \frac{E_p^* + E_d^*/2 - E_t^*/2}{RT}$$

(3.5.8)

This enables us to identify the apparent activation energy in Equation 3.5.7 with the difference in E^* values for the various steps:

$$E_{app}^* = E_p^* + \frac{E_d^*}{2} - \frac{E_t^*}{2}$$

(3.5.9)

Equation 3.5.9 allows us to conveniently assess the effect of temperature variation on the rate of polymerization. This effect is considered in the following example.

Example 3.3

Using typical activation energies from Table 3.2 through Table 3.4, estimate the percent change in the rate of polymerization with a 1°C change in temperature at 50°C, for both thermally initiated and photoinitiated polymerization.

Solution

Write Equation 3.5.3 in the form

$$\ln R_p = \ln k_{app} + \ln [M] + \frac{1}{2} \ln [I]$$

Taking the derivative, treating [M] and [I] as constants with respect to T while k is a function of T:

$$d \ln R_p = \frac{dR_p}{R_p} = d \ln k_{app}$$

Expand $d \ln k_{app}$ by means of the Arrhenius equation via Equation 3.5.8:

$$\frac{dR_p}{R_p} = d \ln A_{app} - d\left(\frac{E^*_{app}}{RT}\right) = \frac{E^*_{app}}{RT^2} dT$$

Substitute Equation 3.5.9 for E^*_{app}:

$$\frac{dR_p}{R_p} = \frac{E^*_p + E^*_d/2 - E^*_t/2}{RT^2} dT$$

Finally, we recognize that a $1°C$ temperature variation can be approximated as dT and that $(dR_p/R_p) \times 100$ gives the approximate percent change in the rate of polymerization. Taking average values of E^* from the appropriate tables, we obtain $E^*_d = 145$, $E^*_t = 16.8$, and $E^*_p = 24.9$ kJ mol^{-1}. For thermally initiated polymerization

$$\frac{dR_p}{R_p} = \frac{(24.9 + 145/2 - 16.8/2)(10^3)(1)}{(8.314)(323)^2} = 0.103$$

or 10.3% per $°C$.

For photoinitiation there is no activation energy for the initiator decomposition; hence

$$\frac{dR_p}{R_p} = \frac{(24.9 - 16.8/2)(10^3)(1)}{(8.314)(323)^2} = 1.90 \times 10^{-2}$$

or 1.90% per $°C$. Note that the initiator decomposition makes the largest contribution to E^*; therefore photoinitiated processes display a considerably lower temperature dependence for the rate of polymerization.

3.5.3 Kinetic Chain Length

Suppose we consider the ratio

$$R_p/R_i = \frac{-d[M]/dt}{-d[I]/dt}$$

under conditions where an initiator yields one radical, where $f = 1$, and where the final polymer contains one initiator fragment per molecule. For this set of conditions the ratio gives the number of monomers polymerized per chain initiated, which is the average degree of polymerization. A more general development of this idea is based on a quantity called the *kinetic chain length* $\bar{\nu}$. The kinetic chain length is defined as the ratio of the number of propagation steps to the number of initiation steps, regardless of the mode of termination:

$$\bar{\nu} = \frac{R_p}{R_i} = \frac{R_p}{R_t} \qquad (3.5.10)$$

where the second form of this expression uses the stationary-state condition $R_i = R_t$. The significance of the kinetic chain length is seen in the following statements:

1. For termination by disproportionation

$$\bar{\nu} = N_n \tag{3.5.11}$$

   where N_n is the number average degree of polymerization.
2. For termination by combination

$$\bar{\nu} = \frac{N_n}{2} \tag{3.5.12}$$

3. $\bar{\nu}$ is an average quantity—indicated by the overbar—since not all kinetic chains are identical any more than all molecular chains are.

   Using Equation 3.5.3 and Equation 3.4.4 for R_p and R_t, respectively, we write

$$\bar{\nu} = \frac{k_p[P\cdot][M]}{2k_t[P\cdot]^2} = \frac{k_p[M]}{2k_t[P\cdot]} \tag{3.5.13}$$

   This may be combined with Equation 3.4.10 to give the stationary-state value for $\bar{\nu}$:

$$\bar{\nu} = \frac{k_p[M]}{2k_t(fk_d[I]/k_t)^{1/2}} = \frac{k_p[M]}{2(fk_tk_d[I])^{1/2}} \tag{3.5.14}$$

As with the rate of polymerization, we see from Equation 3.5.14 that the kinetic chain length depends on the monomer and initiator concentrations and on the constants for the three different kinds of kinetic processes that constitute the mechanism. When the initial monomer and initiator concentrations are used, Equation 3.5.14 describes the initial polymer formed. The initial degree of polymerization is a measurable quantity, so Equation 3.5.14 provides a second functional relationship, distinct from Equation 3.5.3, among experimentally available quantities—N_n, [M], [I]—and theoretically important parameters—k_p, k_t, and k_d. Note that the mode of termination, which establishes the connection between $\bar{\nu}$ and N_n, and the value of f are both accessible through endgroup characterization. Thus we have a second equation with three unknowns; one more independent equation and the evaluation of the individual kinetic constants from experimental results will be feasible.

There are several additional points about Equation 3.5.14 that are worthy of comment. First it must be recalled that we have intentionally ignored any kinetic factors other than initiation, propagation, and termination. We shall see in Section 3.8 that another process, chain transfer, has significant effects on the molecular weight of a polymer. The result we have obtained, therefore, is properly designated as the kinetic chain length without transfer. A second observation is that $\bar{\nu}$ depends not only on the nature and concentration of the monomer, but also on the nature and concentration of the initiator. The latter determines the number of different sites competing for the addition of monomer, so it is not surprising that $\bar{\nu}$ is decreased by increases in either k_d or [I]. Finally, we observe that both k_p and k_t are properties of a particular monomer. The relative molecular weight that a specific monomer tends toward—all other things being equal—is characterized by the ratio $k_p/k_t^{1/2}$ for a monomer. Using the values in Table 3.3 and Table 3.4, we see that $k_p/k_t^{1/2}$ equals 0.678 for methyl acrylate and 0.0213 for styrene at 60°C. The kinetic chain length for poly(methyl acrylate) is thus expected to be about 32 times greater than for polystyrene if the two are prepared with the same initiator (k_d) and the same concentrations [M] and [I]. Extension of this type of comparison to the degree of polymerization requires that the two polymers compared show the same proportion of the modes of termination. Thus for vinyl acetate (subscript V) relative to acrylonitrile (subscript A) at 60°C, with the same provisos as above, $\bar{\nu}_V/\bar{\nu}_A = 6$ while $N_{n,V}/N_{n,A} = 3$ because of the differences in the mode of termination for the two.

The proviso "all other things being equal" in discussing the last point clearly applies to temperature as well, since the kinetic constants can be highly sensitive to temperature. To evaluate the effect of temperature variation on the molecular weight of an addition polymer, we follow the same sort of logic as was used in Example 3.3:

1. Take logarithms of Equation 3.5.14:

$$\ln \bar{\nu} = \ln k_p (k_t k_d)^{-1/2} + \ln \left(\frac{[M]}{2(f[I])^{1/2}} \right) \tag{3.5.15}$$

2. Differentiate with respect to T, assuming the temperature dependence of the concentrations is negligible compared to that of the rate constants:

$$\frac{d\bar{\nu}}{\bar{\nu}} = d \ln k_p - 1/2 d \ln (k_t k_d) \tag{3.5.16}$$

3. By the Arrhenius equation $d \ln k = -d(E^*/RT) = (E^*/RT^2) \, dT$; therefore

$$\frac{d\bar{\nu}}{\bar{\nu}} = \frac{E_p^* - E_t^*/2 - E_d^*/2}{RT^2} dT \tag{3.5.17}$$

It is interesting to compare the application of this result to thermally initiated and photoinitiated polymerizations as we did in Example 3.3. Again using the average values of the constants from Table 3.2 through Table 3.4 and taking $T = 50°C$, we calculate that $\bar{\nu}$ decreases by about 6.5% per °C for thermal initiation and increases by about 2% per °C for photoinitiation. It is clearly the large activation energy for initiator dissociation that makes the difference. This term is omitted in the case of photoinitiation, where the temperature increase produces a bigger effect on propagation than on termination. On the other hand, for thermal initiation an increase in temperature produces a large increase in the number of growth centers, with the attendant reduction of the average kinetic chain length.

Photoinitiation is not as important as thermal initiation in the overall picture of free-radical chain-growth polymerization. The foregoing discussion reveals, however, that the contrast between the two modes of initiation does provide insight into, and confirmation of, various aspects of addition polymerization. The most important application of photoinitiated polymerization is in providing a third experimental relationship among the kinetic parameters of the chain mechanism. We shall consider this in the next section.

3.6 Radical Lifetime

In the preceding section we observed that both the rate of polymerization and the degree of polymerization under stationary-state conditions can be interpreted to yield some cluster of the constants k_p, k_t, and k_d. The situation is summarized diagrammatically in Figure 3.4. The circles at the two bottom corners of the triangle indicate the particular grouping of constants obtainable from the measurement of R_p or N_n, as shown. By combining these two sources of data in the manner suggested in the boxes situated along the lines connecting these circles k_d can be evaluated, as well as the ratio k_p^2/k_t. Using this stationary-state data, however, it is not possible to further resolve the propagation and termination constants. Another relationship is needed to do this. A quantity called the *radical lifetime* $\bar{\tau}$ supplies the additional relationship and enables us to move off the base of Figure 3.4.

To arrive at an expression for the radical lifetime, we return to Equation 3.5.1, which may be interpreted as follows:

1. $-d[M]/dt$ gives the rate at which monomers enter polymer molecules. This, in turn, is given by the product of number of growth sites, $[P\cdot]$, and the rate at which monomers add to each

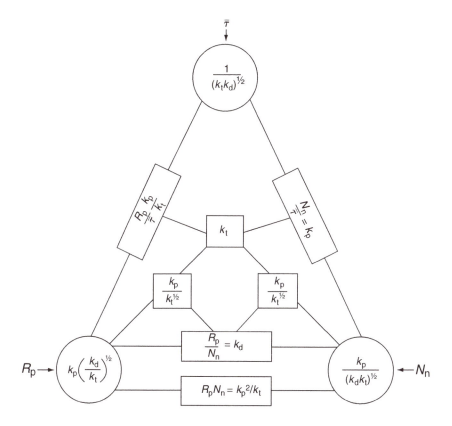

Figure 3.4 Schematic relationship among the various experimental quantities (R_p, N_n, and $\bar{\tau}$) and the rate constants k_d, k_p, and k_t derived therefrom.

growth site. On the basis of Equation 3.5.1, the rate at which monomers add to a radical is given by $k_p[M]$.

2. If $k_p[M]$ gives the number of monomers added per unit time, then $1/k_p[M]$ equals the time elapsed per monomer addition.

3. If we multiply the time elapsed per monomer added to a radical by the number of monomers in the average chain, then we obtain the time during which the radical exists. This is the definition of the radical lifetime. The number of monomers in a polymer chain is, of course, the degree of polymerization. Therefore we write

$$\bar{\tau} = \frac{N_n}{k_p[M]} \tag{3.6.1}$$

4. The degree of polymerization in Equation 3.6.1 can be replaced with the kinetic chain length, and the resulting expression simplified. To proceed, however, we must choose between the possibilities described in Equation 3.5.11 and Equation 3.5.12. Assuming termination by disproportionation, we replace N_n by $\bar{\nu}$, using Equation 3.5.14:

$$\bar{\tau} = \frac{k_p[M]}{2(fk_tk_d[I])^{1/2}} \frac{1}{k_p[M]} = \frac{1}{2(fk_tk_d[I])^{1/2}} \tag{3.6.2}$$

5. The radical lifetime is an average quantity, as indicated by the overbar.

We shall see presently that the lifetime of a radical can be measured. When such an experiment is conducted with a known concentration of initiator, then the cluster of constants $(k_t k_d)^{-1/2}$ can be evaluated. This is indicated at the apex of the triangle in Figure 3.4.

There are several things about Figure 3.4 that should be pointed out:

1. In going from the experimental quantities R_p, N_n, and $\bar{\tau}$ to the associated clusters of kinetic constants, it has been assumed that the monomer and initiator concentrations are known and essentially constant. In addition, the efficiency factor f has been left out, the assumption being that still another type of experiment has established its value.
2. By following the lines connecting two sources of circled information, the boxed result in the perimeter of the triangle may be established. Thus k_p is evaluated from $\bar{\tau}$ and N_n.
3. Here k_p can be combined with one of the various k_p/k_t ratios to permit the evaluation of k_t.

We can use the constants tabulated elsewhere in the chapter to get an idea of a typical radical lifetime. Choosing 10^{-3} mol L^{-1} AIBN as the initiator ($k_d = 0.85 \times 10^{-5}$ s^{-1} at 60°C) and vinyl acetate as the monomer (terminates entirely by disproportionation, $k_t = 2.9 \times 10^7$ L mol^{-1} s^{-1} at 60°C), and taking $f = 1$ for the purpose of calculation, we find $\bar{\tau} = 0.5[(1.0)(2.9 \times 10^7)(0.85 \times 10^{-5})(10^{-3})]^{-1/2} = 1.01$ s. This figure contrasts sharply with the time required to obtain high-molecular-weight molecules in step-growth polymerizations.

Since the radical lifetime provides the final piece of information needed to independently evaluate the three primary kinetic constants—remember, we are still neglecting chain transfer—the next order of business is a consideration of the measurement of $\bar{\tau}$. A widely used technique for measuring radical lifetime is based on photoinitiated polymerization using a light source, which blinks on and off at regular intervals. In practice, a rotating opaque disk with a wedge sliced out of it is interposed between the light and the reaction vessel. Thus the system is in darkness when the solid part of the disk is in the light path and is illuminated when the notch passes. With this device, called a rotating sector or chopper, the relative lengths of light and dark periods can be controlled by the area of the notch, and the frequency of the flickering by the velocity of rotation of the disk. We will not describe the rotating sector experiments in detail. It is sufficient to note that, with this method, the rate of photoinitiated polymerization is studied as a function of the time of illumination with the rapidly blinking light. The results show the rate of polymerization dropping from one plateau value at slow blink rates ("long" bursts of illumination) to a lower plateau at fast blink rates ("short" periods of illumination). A plot of the rate of polymerization versus the duration of an illuminated interval resembles an acid–base titration curve with a step between the two plateau regions. Just as the "step" marks the end point of a titration, the "step" in rotating sector data identifies the transition between relatively long and short periods of illumination. Here is the payoff: "long" and "short" times are defined relative to the average radical lifetime. Thus $\bar{\tau}$ may be read from the time axis at the midpoint of the transition between the two plateaus.

This qualitative description enables us to see that the radical lifetime described by Equation 3.6.2 is an experimentally accessible quantity. More precise values of $\bar{\tau}$ may be obtained by curve fitting since the nonstationary-state kinetics of the transition between plateaus have been analyzed in detail. To gain some additional familiarity with the concept of radical lifetime and to see how this quantity can be used to determine the absolute value of a kinetic constant, consider the following example:

Example 3.4

The polymerization of ethylene at 130°C and 1500 atm was studied using different concentrations of the initiator, 1-t-butylazo-1-phenoxycyclohexane. The rate of initiation was measured directly and radical lifetimes were determined using the rotating sector method. The following results were obtained.[†]

Run	$\bar{\tau}$ (s)	$R_i \times 10^9$ (mol L^{-1} s^{-1})
5	0.73	2.35
6	0.93	1.59
8	0.32	12.75
12	0.50	5.00
13	0.29	14.95

Demonstrate that the variations in the rate of initiation and $\bar{\tau}$ are consistent with free-radical kinetics, and evaluate k_t.

Solution

Since the rate of initiation is measured, we can substitute R_i for the terms $(2fk_d[I])^{1/2}$ in Equation 3.6.2 to give

$$\bar{\tau} = \frac{1}{(2k_tR_i)^{1/2}} \quad \text{or} \quad k_t = \frac{1}{2\bar{\tau}^2R_i}$$

If the data follow the kinetic scheme presented here, the values of k_t calculated for the different runs should be constant:

Run	$k_t \times 10^{-8}$ (L mol^{-1} s^{-1})
5	3.99
6	3.64
8	3.83
12	4.00
13	3.98
Average	3.89

Even though the rates of initiation span almost a 10-fold range, the values of k_t show a standard deviation of only 4%, which is excellent in view of the inevitable experimental errors. Note that the rotating sector method can be used in high-pressure experiments and other unusual situations, a highly desirable characteristic it shares with many optical methods in chemistry.

3.7 Distribution of Molecular Weights

Until this point in the chapter we have intentionally avoided making any differentiation among radicals on the basis of the degree of polymerization of the radical. Now we seek a description of the molecular weight distribution of addition polymer molecules. Toward this end it becomes necessary to consider radicals with different i values.

3.7.1 Distribution of *i*-mers: Termination by Disproportionation

We begin by writing a kinetic expression for the concentration of radicals of the degree of polymerization i, which we designate $[P_i\bullet]$. This rate law will be the sum of three contributions:

[†] Data from T. Takahashi and P. Ehrlich, *Polym. Prepr. Am. Chem. Soc. Polym. Chem. Div.*, 22, 203 (1981).

1. An increase that occurs by addition of monomer to the radical $P_{i-1}\cdot$
2. A decrease that occurs by addition of a monomer to the radical $P_i\cdot$
3. A decrease that occurs by the termination of $P_i\cdot$ with any other radical $P\cdot$

The change in $[P_i\cdot]$ under stationary-state conditions equals zero for all values of i; hence we can write

$$\frac{d[P_i\cdot]}{dt} = k_p[M][P_{i-1}\cdot] - k_p[M][P_i\cdot] - 2k_t[P_i\cdot][P\cdot] = 0 \tag{3.7.1}$$

which can be rearranged to

$$\frac{[P_i\cdot]}{[P_{i-1}\cdot]} = \frac{k_p[M]}{k_p[M] + 2k_t[P\cdot]} \tag{3.7.2}$$

Dividing the numerator and denominator of Equation 3.7.2 by $2k_t[P\cdot]$ and recalling the definition of $\bar{\nu}$ provided by Equation 3.5.13 enables us to express this result more succinctly as

$$\frac{[P_i\cdot]}{[P_{i-1}\cdot]} = \frac{\bar{\nu}}{1+\bar{\nu}} \tag{3.7.3}$$

Next let us consider the following sequence of multiplications:

$$\frac{[P_i\cdot]}{[P_{i-1}\cdot]}\frac{[P_{i-1}\cdot]}{[P_{i-2}\cdot]}\frac{[P_{i-2}\cdot]}{[P_{i-3}\cdot]}\cdots\frac{[P_{i-(i-2)}\cdot]}{[P_{i-(i-1)}\cdot]} = \frac{[P_i\cdot]}{[P_1\cdot]} \tag{3.7.4}$$

This shows that the number of i-mer radicals relative to the number of the smallest radicals is given by multiplying the ratio $[P_i\cdot]/[P_{i-1}\cdot]$ by $i-2$ analogous ratios. Since each of the individual ratios is given by $\bar{\nu}/(1+\bar{\nu})$, we can rewrite Equation 3.7.4 as

$$\frac{[P_i\cdot]}{[P_1\cdot]} = \frac{[P_i\cdot]}{[P_{i-1}\cdot]}\left(\frac{\bar{\nu}}{1+\bar{\nu}}\right)^{i-2} \tag{3.7.5}$$

or

$$[P_{i-1}\cdot] = [P_1\cdot]\left(\frac{\bar{\nu}}{1+\bar{\nu}}\right)^{(i-1)-1} \tag{3.7.6}$$

Since it is more convenient to focus attention on i-mers than $(i-1)$-mers, the corresponding expression for the i-mer is written by analogy:

$$[P_i\cdot] = [P_1\cdot]\left(\frac{\bar{\nu}}{1+\bar{\nu}}\right)^{i-1} \tag{3.7.7}$$

Dividing both sides of Equation 3.7.7 by $[P\cdot]$, the total radical concentration, gives the number (or mole) fraction of i-mer radicals in the total radical population. This ratio is the same as the number of i-mers n_i in the sample containing a total of n (no subscript) polymer molecules:

$$x_i = \frac{n_i}{n} = \frac{[P_i\cdot]}{[P\cdot]} = \frac{[P_1\cdot]}{[P\cdot]}\left(\frac{\bar{\nu}}{1+\bar{\nu}}\right)^{i-1} \tag{3.7.8}$$

The ratio $[P_1\cdot]/[P\cdot]$ in Equation 3.7.8 can be eliminated by applying Equation 3.7.1 explicitly to the P_1 radical:

1. Write Equation 3.7.1 for $P_1\cdot$, remembering in this case that the leading term describes initiation:

$$\frac{d[P_1\cdot]}{dt} = R_i - k_p[M][P_1\cdot] - 2k_t[P_1\cdot][P\cdot] = 0 \tag{3.7.9}$$

2. Rearrange under stationary-state conditions:

$$[P_1\bullet] = \frac{R_i}{k_p[M] + 2k_t[P\bullet]} \qquad (3.7.10)$$

The total radical concentration under stationary-state conditions can be similarly obtained.

3. Write Equation 3.4.9 using the same notation for initiation as in Equation 3.7.9:

$$\frac{d[P\bullet]}{dt} = R_i - 2k_t[P\bullet]^2 = 0 \qquad (3.7.11)$$

4. Rearrange under stationary-state conditions:

$$[P\bullet] = \frac{R_i}{2k_t[P\bullet]} \qquad (3.7.12)$$

5. Take the ratio of Equation 3.7.10 to Equation 3.7.12:

$$\frac{[P_1\bullet]}{[P\bullet]} = \frac{2k_t[P\bullet]}{k_p[M] + 2k_t[P\bullet]} = \frac{1}{1 + \bar{\nu}} \qquad (3.7.13)$$

Combining Equation 3.7.13 with Equation 3.7.8 gives

$$x_i = \frac{n_i}{n} = \frac{1}{1 + \bar{\nu}}\left(\frac{\bar{\nu}}{1 + \bar{\nu}}\right)^{i-1} = \frac{1}{\bar{\nu}}\left(\frac{\bar{\nu}}{1 + \bar{\nu}}\right)^{i} \qquad (3.7.14)$$

This expression gives the number fraction or mole fraction, x_i, of i-mers in the polymer and is thus equivalent to Equation 2.4.2 for step-growth polymerization.

The kinetic chain length $\bar{\nu}$ may also be viewed as merely a cluster of kinetic constants and concentrations, which was introduced into Equation 3.7.13 to simplify the notation. As an alternative, suppose we define for the purposes of this chapter a fraction p such that

$$p \equiv \frac{\bar{\nu}}{1 + \bar{\nu}} = \frac{k_p[M]}{k_p[M] + 2k_t[P\bullet]} \qquad (3.7.15)$$

It follows from this definition that $1/(1 + \bar{\nu}) = 1 - p$, so Equation 3.7.14 can be rewritten as

$$x_i = \frac{n_i}{n} = (1 - p)p^{i-1} \qquad (3.7.16)$$

This change of notation now expresses Equation 3.7.14 in exactly the same form as its equivalent in Section 2.4. In other words, the distribution of chain lengths is the most probable distribution, just as was the case for step-growth polymerization! Several similarities and differences should be noted in order to take full advantage of the parallel between this result and the corresponding material for condensation polymers in Chapter 2:

1. In Chapter 2, p was defined as the fraction (or probability) of functional groups that had reacted at a certain point in the polymerization. According to the current definition provided by Equation 3.7.15, p is the fraction (or probability) of propagation steps among the combined total of propagation and termination steps. The quantity $1 - p$ is therefore the fraction (or probability) of termination steps. An addition polymer with the degree of polymerization i has undergone $i - 1$ propagation steps and one termination step. Therefore it makes sense to describe its probability in the form of Equation 3.7.16.
2. It is apparent from Equation 3.7.15 that $p \to 1$ as $\bar{\nu} \to \infty$; hence those same conditions that favor the formation of a high-molecular-weight polymer also indicate p values close to unity.

3. In Chapter 2 all molecules—whether monomer or i-mers of any i—carry functional groups; hence the fraction described by Equation 2.4.1 applies to the entire reaction mixture. Equation 3.7.16, by contrast, applies *only to the radical population*. Since the radicals eventually end up as polymers, the equation also describes the polymer produced. Unreacted monomers are specifically excluded, however.
4. Only one additional stipulation needs to be made before adapting the results that follow from Equation 2.4.1 to addition polymers. The mode of termination must be specified to occur by disproportionation to use the results of Section 2.4 in this chapter, since termination by combination obviously changes the molecular weight distribution. We shall return to the case of termination by combination presently.
5. For termination by disproportionation (subscript d), we note that $p = k_p[M]/(k_p[M] + 2k_{t,d}[P\cdot])$, and therefore by analogy with Equation 2.4.5, Equation 2.4.9, and Equation 2.4.10,

$$(N_n)_d = \frac{1}{1-p} \tag{3.7.17}$$

$$(N_w)_d = \frac{1+p}{1-p} \tag{3.7.18}$$

$$\left(\frac{N_w}{N_n}\right)_d = 1 + p \rightarrow 2 \quad \text{as} \quad p \rightarrow 1 \tag{3.7.19}$$

By virtue of Equation 3.7.15, $(N_n)_d$ can also be written as $1 + \bar{\nu} \cong \bar{\nu}$ for large $\bar{\nu}$, which is the result already obtained in Equation 3.5.11. Figure 2.5 and Figure 2.6 also describe the distribution by number and weight of addition polymers, if the provisos enumerated above are applied.

3.7.2 Distribution of *i*-mers: Termination by Combination

To deal with the case of termination by combination, it is convenient to write some reactions by which an i-mer might be formed. Table 3.5 lists several specific chemical reactions and the corresponding rate expressions as well as the general form for the combination of an $(i-j)$-mer and a j-mer. On the assumption that all $k_{t,c}$ values are the same, we can write the total rate of change of $[P_i]$:

$$\left(\frac{d[P_i]}{dt}\right)_{tot} = k_{t,c} \sum_{j=1}^{i-1} [P_{i-j}\cdot][P_j\cdot] \tag{3.7.20}$$

The fraction of i-mers formed by combination may be evaluated by dividing $d[P_i]/dt$ by $\sum_i d[P_i]/dt$. Assuming that termination occurs exclusively by combination, then

$$\sum_i \frac{d[P_i]}{dt} = k_{t,c}[P\cdot]^2 \tag{3.7.21}$$

and the number or mole fraction of i-mers formed by combination (subscript c) is

$$\left(\frac{n_i}{n}\right)_c = \frac{d[P_i]/dt}{\sum_i d[P_i]/dt} = \frac{k_{t,c}\sum_{j=1}^{i-1}[P_{i-j}\cdot][P_j\cdot]}{k_{t,c}[P\cdot]^2} \tag{3.7.22}$$

Equation 3.7.16 can be used to relate $[P_{i-j}\cdot]$ and $[P_j\cdot]$ to the total radical concentration:

$$[P_{i-j}\cdot] = (1-p)p^{(i-j)-1}[P\cdot] \tag{3.7.23}$$

Table 3.5 Some Free-Radical Combination Reactions Which Yield i-mers and Their Rate Laws

Reaction	Rate law
$P_{i-1}\cdot + P_1\cdot \rightarrow P_i$	$\dfrac{d[P_i]}{dt} = k_{t,c}[P_{i-1}\cdot][P_1\cdot]$
$P_{i-2}\cdot + P_2\cdot \rightarrow P_i$	$\dfrac{d[P_i]}{dt} = k_{t,c}[P_{i-2}\cdot][P_2\cdot]$
$P_{i-3}\cdot + P_3\cdot \rightarrow P_i$	$\dfrac{d[P_i]}{dt} = k_{t,c}[P_{i-3}\cdot][P_3\cdot]$
$\vdots$	$\vdots$
$P_{i-j}\cdot + P_j\cdot \rightarrow P_i$	$\dfrac{d[P_i]}{dt} = k_{t,c}[P_{i-j}\cdot][P_j\cdot]$

and

$$[P_j\cdot] = (1-p)p^{(j-1)}[P\cdot] \tag{3.7.24}$$

Therefore

$$\left(\frac{n_i}{n}\right)_c = \frac{k_{t,c}\sum_{j=1}^{i-1}(1-p)p^{i-j-1}[P\cdot](1-p)p^{j-1}[P\cdot]}{k_{t,c}[P\cdot]^2} \tag{3.7.25}$$

$$= \sum_{j=1}^{i-1}(1-p)^2p^{i-2}$$

The index j drops out of the last summation; we compensate for this by multiplying the final result by $i-1$ in recognition of the fact that the summation adds up $i-1$ identical terms. Accordingly, the desired result is obtained:

$$\left(\frac{n_i}{n}\right)_c = x_i = (i-1)(1-p)^2p^{i-2} \tag{3.7.26}$$

This expression is plotted in Figure 3.5 for several values of p near unity. Although it shows the number distribution of polymers terminated by combination, the distribution looks quite different from Figure 2.5, which describes the number distribution for termination by disproportionation. In the latter x_i decreases monotonically with increasing i. With combination, however, the curves go through a maximum, which reflects the fact that the combination of two very small or two very large radicals is a less probable event than a more random combination.

Expressions for the various averages are readily derived from Equation 3.7.26 by procedures identical to those used in Section 2.4 (see Problem 6). We only quote the final results for the case where termination occurs exclusively by combination:

$$(N_n)_c = \frac{2}{1-p} \tag{3.7.27}$$

$$(N_w)_c = \frac{2+p}{1-p} \tag{3.7.28}$$

$$\left(\frac{N_w}{N_n}\right)_c = \frac{2+p}{2} \tag{3.7.29}$$

These various expressions differ from their analogs in the case of termination by disproportionation by the appearance of occasional 2's. These terms arise precisely because two chains are combined in this mode of termination. Again using Equation 3.7.15, we note that $(N_n)_c = 2(1+\bar{\nu}) \cong 2\bar{\nu}$ for large $\bar{\nu}$, a result that was already given as Equation 3.5.12.

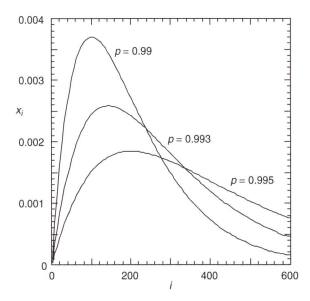

Figure 3.5 Mole fraction of i-mers as a function of i for termination by combination, according to Equation 3.7.26, for various values of p.

One rather different result that arises from the case of termination by combination is seen by examining the limit of Equation 3.7.29 for large values of p:

$$\frac{N_w}{N_n} \rightarrow \frac{2+1}{2} = 1.5 \quad \text{as} \quad p \rightarrow 1 \tag{3.7.30}$$

This contrasts with a limiting ratio of 2 for the case of termination by disproportionation. Since M_n and M_w can be measured, the difference is potentially a method for determining the mode of termination in a polymer system. In most instances, however, termination occurs by some proportion of both modes. Furthermore, other factors in the polymerization such as transfer, autoacceleration, etc., will also contribute to the experimental molecular weight distribution, so in general it is risky to draw too many conclusions about mechanisms from the measured distributions. Also, we have used p and $\bar{\nu}$ to describe the distribution of molecular weights, but it must be remembered that these quantities are defined in terms of various concentrations and therefore change as the reactions proceed. Accordingly, the results presented here are most simply applied at the start of the polymerization reaction when the initial concentrations of monomer and initiator can be used to evaluate p or $\bar{\nu}$.

3.8　Chain Transfer

The three-step mechanism for free-radical polymerization represented by Reaction (3.A) through Reaction (3.C) does not tell the whole story. Another type of free-radical reaction, called chain transfer, may also occur. This is unfortunate in the sense that it further complicates the picture presented so far. On the other hand, this additional reaction can be turned into an asset in actual polymer practice. One consequence of chain transfer reactions is a lowering of the kinetic chain length and hence the molecular weight of the polymer, without necessarily affecting the rate of polymerization. A certain minimum average molecular weight is often needed to achieve a desired physical property, but further increases in chain length simply make processing more difficult.

3.8.1 Chain Transfer Reactions

Chain transfer arises when hydrogen or some other atom X is transferred from a molecule in the system to the polymer radical. This terminates the growth of the original radical but replaces it with a new one: the fragment of the species from which X was extracted. These latter molecules will be designated by attaching the letter X to their symbol in this discussion. Thus if chain transfer involves an initiator molecule, we represent the latter as IX. Chain transfer can occur with any molecule in the system. The following reactions specifically describe transfer to initiator, monomer, solvent, and polymer molecules, respectively:

1. Transfer to initiator, IX:

$$P_i\cdot + IX \rightarrow P_iX + I\cdot \tag{3.K}$$

2. Transfer to monomer, MX:

$$P_i\cdot + MX \rightarrow P_iX + M\cdot \tag{3.L}$$

3. Transfer to solvent, SX:

$$P_i\cdot + SX \rightarrow P_iX + S\cdot \tag{3.M}$$

4. Transfer to polymer, P_jX:

$$P_i\cdot + P_jX \rightarrow P_iX + P_j\cdot \tag{3.N}$$

5. General transfer to RX:

$$P_i\cdot + RX \rightarrow P_iX + R\cdot \tag{3.O}$$

It is apparent from these reactions how chain transfer lowers the molecular weight of a chain-growth polymer. The effect of chain transfer on the rate of polymerization depends on the rate at which the new radicals reinitiate polymerization:

$$R\cdot + M \xrightarrow{k_R} RM\cdot \xrightarrow{k_p} \longrightarrow RP_i\cdot \tag{3.P}$$

If the rate constant k_R is comparable to k_p, the substitution of a polymer radical with a new radical has little or no effect on the rate of polymerization. If $k_R \ll k_p$, the rate of polymerization will be decreased, or even effectively suppressed by chain transfer.

The kinetic chain length acquires a slightly different definition in the presence of chain transfer. Instead of being simply the ratio R_p/R_t, it is redefined to be the rate of propagation relative to the rates of all other steps that compete with propagation; specifically, termination and transfer (subscript tr):

$$\bar{\nu}_{tr} = \frac{R_p}{R_t + R_{tr}} \tag{3.8.1}$$

The transfer reactions follow second-order kinetics, the general rate law being

$$R_{tr} = k_{tr}[P\cdot][RX] \tag{3.8.2}$$

where k_{tr} is the rate constant for chain transfer to a specific compound RX. Since chain transfer can occur with several different molecules in the reaction mixture, Equation 3.8.1 becomes

$$\bar{\nu}_{tr} = k_p[P\cdot][M]/\{2k_t[P\cdot]^2 + k_{tr,I}[P\cdot][IX] + k_{tr,M}[P\cdot][MX] + k_{tr,S}[P\cdot][SX] + k_{tr,P_i}[P\cdot][P_iX]\}$$

$$= \frac{k_p[M]}{2k_t[P\cdot] + \sum k_{tr,R}[RX]}$$

$$\tag{3.8.3}$$

where the summation is over all pertinent RX species. It is instructive to examine the reciprocal of this quantity:

$$\frac{1}{\bar{\nu}_{tr}} = \frac{2k_t[P\bullet]}{k_p[M]} + \frac{\sum k_{tr,R}[RX]}{k_p[M]} \tag{3.8.4}$$

Since the first term on the right-hand side is the reciprocal of the kinetic chain length in the absence of transfer, this becomes

$$\frac{1}{\bar{\nu}_{tr}} = \frac{1}{\bar{\nu}} + \frac{\sum k_{tr,R}[RX]}{k_p[M]} \tag{3.8.5}$$

This notation is simplified still further by defining the ratio of constants

$$\frac{k_{tr,R}}{k_p} = C_{RX} \tag{3.8.6}$$

which is called the *chain transfer constant* for the monomer in question to molecule RX:

$$\frac{1}{\bar{\nu}_{tr}} = \frac{1}{\bar{\nu}} + \sum_{\text{all RX}} C_{RX}\frac{[RX]}{[M]} \tag{3.8.7}$$

It is apparent from this expression that the larger the sum of chain transfer terms becomes, the smaller will be $\bar{\nu}_{tr}$.

The magnitude of the individual terms in the summation depends on both the specific chain transfer constants and the concentrations of the reactants under consideration. The former are characteristics of the system and hence quantities over which we have little control; the latter can often be adjusted to study a particular effect. For example, chain transfer constants are generally obtained under conditions of low conversion to polymer where the concentration of polymer is low enough to ignore the transfer to polymer. We shall return below to the case of high conversions where this is not true.

3.8.2 Evaluation of Chain Transfer Constants

If an experimental system is investigated in which only one molecule is significantly involved in transfer, then the chain transfer constant to that material is particularly straightforward to obtain. If we assume that species SX is the only molecule to which transfer occurs, Equation 3.8.7 becomes

$$\frac{1}{\bar{\nu}_{tr}} = \frac{1}{\bar{\nu}} + C_{SX}\frac{[SX]}{[M]} \tag{3.8.8}$$

This suggests that polymerizations should be conducted at different ratios of [SX]/[M] and the resulting molecular weight measured for each. Equation 3.8.8 indicates that a plot of $1/\bar{\nu}_{tr}$ versus [SX]/[M] should be a straight line with slope C_{SX}. Figure 3.6 shows this type of plot for the polymerization of styrene at 100°C in the presence of four different solvents. The fact that all show a common intercept as required by Equation 3.8.8 shows that the rate of initiation is unaffected by the nature of the solvent. The following example examines chain transfer constants evaluated in this situation.

Example 3.5

Estimate the chain transfer constants for styrene to isopropylbenzene, ethylbenzene, toluene, and benzene from the data presented in Figure 3.6. Comment on the relative magnitude of these constants in terms of the structure of the solvent molecules.

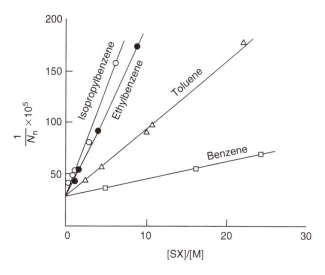

Figure 3.6 Effect of chain transfer to solvent according to Equation 3.8.8 for polystyrene at 100°C. Solvents used were ethylbenzene (●), isopropylbenzene (○), toluene (△), and benzene (□). (Data from Gregg, R.A. and Mayo, F.R., *Discuss. Faraday Soc.*, 2, 328, 1947. With permission.).

Solution

The chain transfer constants are given by Equation 3.8.8 as the slopes of the lines in Figure 3.6. These are estimated to be as follows (note that X = H in this case):

SX	i-$C_3H_7(C_6H_5)$	$C_2H_5(C_6H_5)$	$CH_3(C_6H_5)$	$H(C_6H_5)$
$C_{SX} \times 10^4$	2.08	1.38	0.55	0.16

The relative magnitudes of these constants are consistent with the general rule that benzylic hydrogens are more readily abstracted than those attached directly to the ring. The reactivity of the benzylic hydrogens themselves follows the order tertiary > secondary > primary, which is a well-established order in organic chemistry. The benzylic radical resulting from hydrogen abstraction is resonance stabilized. For toluene, as an example,

In certain commercial processes it is essential to regulate the molecular weight of the polymer either for ease of processing or because low molecular weight products are desirable for particular applications such as lubricants or plasticizers. In such cases the solvent or chain transfer agent is chosen and its concentration selected to produce the desired value of $\bar{\nu}_{tr}$. Certain mercaptans have particularly large chain transfer constants for many common monomers and are especially useful for molecular weight regulation. For example, styrene has a chain transfer constant for n-butyl mercaptan equal to 21 at 60°C. This is about 10^7 times larger than the chain transfer constant to benzene at the same temperature.

Chain transfer to initiator or monomer cannot always be ignored. It may be possible, however, to evaluate these transfer constants by conducting a similar analysis on polymerizations without added solvent or in the presence of a solvent for which C_{SX} is known to be negligibly small. Fairly extensive tables of chain transfer constants have been assembled on the basis of investigations of this sort. For example, the values of C_{MX} for acrylamide at 60°C is 6×10^{-5}, and that for vinyl chloride at 30°C is 6.3×10^{-4}. Likewise, for methyl methacrylate at 60°C, C_{IX} is 0.02 to benzoyl peroxide and 1.27 to t-butyl hydroperoxide.

3.8.3   Chain Transfer to Polymer

As noted above, chain transfer to polymer does not interfere with the determination of other transfer constants, since the latter are evaluated at low conversions. In polymer synthesis, however, high conversions are desirable and extensive chain transfer can have a dramatic effect on the properties of the product. This comes about since chain transfer to polymer introduces branching into the product:

$$(3.Q)$$

A moment's reflection reveals that the effect on $\bar{\nu}$ of transfer to polymer is different from the effects discussed above inasmuch as the overall degree of polymerization is not decreased by such transfers. Investigation of chain transfer to polymer is best handled by examining the extent of branching in the product. We shall not pursue the matter of evaluating the transfer constants, but shall consider describing two important specific examples of transfer to polymer.

Remember from Section 1.3 that graft copolymers have polymeric side chains that differ in the nature of the repeat unit from the backbone. These can be prepared by introducing a prepolymerized sample of the backbone polymer into a reactive mixture—i.e., one containing a source of free radicals—of the side-chain monomer. As an example, consider introducing 1,4-polybutadiene into a reactive mixture of styrene:

$$(3.R)$$

This procedure is used commercially to produce rubber-modified or high impact polystyrene (HIPS). The polybutadiene begins to segregate from the styrene as it polymerizes (see Chapter 7 to learn why), but is prevented from undergoing macroscopic phase separation due to the covalent linkages to polystyrene chains. Consequently, small (micron-sized) domains of polybutadiene rubber are distributed throughout the glassy polystyrene matrix. These "rubber balls" are able to dissipate energy effectively (see Chapter 10 and Chapter 12), and counteract the brittleness of polystyrene.

A second example of chain transfer to polymer is provided by the case of polyethylene. In this case the polymer product contains mainly ethyl and butyl side chains. At high conversions such side chains may occur as often as once every 15 backbone repeat units on the average. These short side chains are thought to arise from transfer reactions with methylene hydrogens along the same polymer chain. This process is called *backbiting* and reminds us of the stability of rings of certain sizes and the freedom of rotation around unsubstituted bonds:

$$(3.S)$$

However, transfer to polymer can also produce long-chain branches. The commercial product known as low density polyethylene is formed by a free-radical mechanism in a process conducted at high pressure. The presence of long-chain branches inhibits crystallization (see Chapter 13), and therefore results in a lower density product. These branches also have a profound effect on the flow properties of the material (see Chapter 11).

3.8.4 Suppressing Polymerization

We conclude this section by noting an extreme case of chain transfer, a reaction that produces radicals of such low reactivity that polymerization is effectively suppressed. Reagents that accomplish this are added to commercial monomers to prevent their premature polymerization during storage. These substances are called either *retarders* or *inhibitors*, depending on the degree of protection they afford. Such chemicals must be removed from monomers before use, and failure to achieve complete purification can considerably affect the polymerization reaction. Inhibitors and retarders differ in the extent to which they interfere with polymerization, but not in their essential activity. An inhibitor is defined as a substance that blocks polymerization completely until it is either removed or consumed. Thus failure to totally eliminate an inhibitor from purified monomer will result in an induction period in which an inhibitor is first converted to an inert form before polymerization can begin. A retarder is less efficient and merely slows down the polymerization process by competing for radicals.

Benzoquinone [III] is widely used as an inhibitor:

$$\text{(3.T)}$$

[III]

The resulting radical is stabilized by electron delocalization and eventually reacts with either another inhibitor radical by combination (dimerization) or disproportionation or with an inhibitor or other radical. Another commonly used inhibitor is 2,6-di-*tert*-butyl-4-methylphenol (butylated hydroxytoluene, or BHT):

which is also known as an *antioxidant*. Such free-radical scavengers often act as antioxidants, in that the first stage of oxidative attack generates a free radical.

Molecular oxygen contains two unpaired electrons and has the distinction of being capable of both initiating and inhibiting polymerization. Molecular oxygen functions in the latter capacity by forming the relatively unreactive peroxy radical:

$$O_2 + M\cdot \rightarrow M - O - O\cdot \qquad \text{(3.U)}$$

Inhibitors are characterized by inhibition constants, which are defined as the ratio of the rate constant for transfer to inhibitor to the propagation constants for the monomer, by analogy with Equation 3.8.6 for chain transfer constants. For styrene at $50°C$ the inhibition constant of *p*-benzoquinone is 518, and that for O_2 is 1.5×10^4. The *Polymer Handbook* [1] is an excellent source for these and most other rate constants discussed in this chapter.

3.9 Chapter Summary

In this chapter we have explored chain growth or addition polymerization, as exemplified by the free-radical mechanism. This particular polymerization route is the most prevalent from a commercial perspective, and is broadly applicable to a wide range of monomers, especially those containing carbon–carbon double bonds. The main points of the discussion may be summarized as follows:

1. In comparison with step-growth polymerization, free-radical polymerization can lead to much higher molecular weights and in much shorter times, although the resulting distributions of molecular weight are comparably broad.
2. There are three essential reaction steps in a chain-growth polymerization: initiation, propagation, and termination. A wide variety of free-radical initiators are available; the most common act by thermally induced cleavage of a peroxide or azo linkage. Propagation occurs by head-to-tail addition of a monomer to a growing polymer radical, and is typically very rapid. Termination occurs by reaction between two radicals, either by direct combination or by disproportionation.
3. A fourth class of reactions, termed transfer reactions, is almost always important in practice. The primary effect of transfer of a radical from a growing chain to another molecule is to reduce the average degree of polymerization of the resulting polymer chains, but in some cases it can also lead to interesting architectural consequences in the final polymer.
4. Kinetic analysis of the distribution of chain lengths is made tractable by three key assumptions. The steady-state approximation requires that the net rates of initiation and termination be equal; thus the total concentration of radicals is constant. The same approximation extends to the concentration of each radical species individually. The principle of equal reactivity asserts that a single rate constant describes each propagation step and each termination step, independent of the degree of polymerization of the radicals involved. Thirdly, transfer reactions are assumed to be absent.
5. The aforementioned assumptions are most successful in describing the early stages of polymerization, before a host of competing factors become significant, such as depletion of reactants, loss of mobility of chain radicals, etc. Under these assumptions explicit expressions for the number and weight distribution of polymer chains can be developed. In the case that termination occurs exclusively by disproportionation, the result is a most probable distribution of molecular weights, just as with step-growth polymerization. Termination by recombination, on the other hand, leads to a somewhat narrower distribution, with $M_w/M_n \approx 1.5$ rather than 2.

Problems

1. The efficiency of AIBN in initiating polymerization at $60°C$ was determined[†] by the following strategy. They measured R_p and $\bar{\nu}$ and calculated $R_i = R_p/\bar{\nu}$. The constant k_d was measured directly in the system, and from this quantity and the measured ratio $R_p/\bar{\nu}$ the fraction f could be determined. The following results were obtained for different concentrations of initiator:

[I] (g L^{-1})	$R_p/\bar{\nu} \times 10^8$ (mol L^{-1} s^{-1})
0.0556	0.377
0.250	1.57
0.250	1.72
1.00	6.77
1.50	10.9
2.50	17.1

Using $k_d = 0.0388$ h^{-1}, evaluate f from these data.

[†] J.C. Bevington, J.H. Bradbury, and G.M. Burnett, *J. Polym. Sci.*, 12, 469 (1954).

2. AIBN was synthesized using ^{14}C-labeled reagents and the tagged compound was used to initiate polymerization to methyl methacrylate and styrene. Samples of initiator and polymers containing initiator fragments were burned to CO_2. The radioactivity of uniform (in sample size and treatment) CO_2 samples was measured in counts per minute (cpm) by a suitable Geiger counter. A general formula for poly(methyl methacrylate) with its initiator fragments is $(C_5H_8O_2)_n(C_4H_6N)_m$, where n is the degree of polymerization for the polymer and m is either 1 or 2, depending on the mode of termination. The specific activity measured in the CO_2, resulting from combustion of the polymer relative to that produced by the initiator is

$$\frac{\text{Activity of C in polymer}}{\text{Activity of C in initiator}} = \frac{4m}{5n + 4m} \cong \frac{4m}{5n}$$

From the ratio of activities and measured values of n, the average number of initiator fragments per polymer can be determined.

Carry out a similar argument for the ratio of activities for polystyrene and evaluate the average number of initiator fragments per molecule for each polymer from the following data.[†] For both sets of data, the radioactivity from the labeled initiator gives 96,500 cpm when converted to CO_2.

Methyl methacrylate		Styrene	
$\overline{M}_n$	Counts per minute	$\overline{M}_n$	Counts per minute
444,000	20.6	383,000	25.5
312,000	30.1	117,000	86.5
298,000	29.0	114,000	89.5
147,000	60.5	104,000	96.4
124,000	76.5	101,000	113.5
91,300	103.4		
89,400	104.6		

3. In the research described in Example 3.4, the authors measured the following rates of polymerization:

Run number	$R_p \times 10^4$ (mol L^{-1} s^{-1})
5	3.40
6	2.24
8	6.50
12	5.48
13	7.59

They also reported a k_p value of 1.2×10^4 L mol^{-1} s^{-1}, but the concentrations of monomer in each run were not given. Use these values of R_p and k_p and the values of $\bar{\tau}$ and k_t given in Example 3.4 to evaluate [M] for each run. As a double check, evaluate [M] from these values of R_p (and k_p) and the values of R_i and k_t given in the example.

[†] J.C. Bevington, H.W. Melville, and R.P. Taylor, *J. Polym. Sci.*, 12, 449 (1954).

4. Arnett[†] initiated the polymerization of methyl methacrylate in benzene at 77°C with AIBN and measured the initial rates of polymerization for the concentrations listed:

[M] (mol L^{-1})	[I]$_0 \times 10^4$ (mol L^{-1})	$R_p \times 10^3$ (mol L^{-1} min^{-1})
9.04	2.35	11.61
8.63	2.06	10.20
7.19	2.55	9.92
6.13	2.28	7.75
4.96	3.13	7.31
4.75	1.92	5.62
4.22	2.30	5.20
4.17	5.81	7.81
3.26	2.45	4.29
2.07	2.11	2.49

Use these data to evaluate the cluster of constants $(fk_d/k_t)^{1/2}k_p$ at this temperature. Evaluate $k_p/k_t^{1/2}$ using Arnett's finding that $f = 1.0$ and assuming the k_d value determined in Example 3.1 for AIBN at 77°C in xylene also applies in benzene.

5. The lifetime of polystyrene radicals at 50°C was measured[‡] as a function of the extent of conversion to polymer. The following results were obtained:

Percent conversion	$\bar{\tau}$ (s)
0	2.29
32.7	1.80
36.3	9.1
39.5	13.1
43.8	18.8

Propose an explanation for the variation observed.

6. Derive Equation 3.7.27 and Equation 3.7.28.

7. The equations derived in Section 3.7 are based on the assumption that termination occurs exclusively by either disproportionation or combination. This is usually not the case; some proportion of each is more common. If α equals the fraction of chains for which termination occurs by disproportionation, it can be shown that

$$N_n = \frac{\alpha}{1-p} + \frac{(1-\alpha)2}{1-p} = \frac{2-\alpha}{1-p}$$

and

$$\frac{N_w}{N_n} = \frac{4 - 3\alpha - \alpha p + 2p}{(2-\alpha)^2}$$

From measurements of N_n and N_w/N_n it is possible in principle to evaluate α and p. May and Smith[*] have done this for a number of polystyrene samples. A selection of their data for which this approach seems feasible is presented below. Since p is very close to unity, it is

---

[†] M. Arnett, *J. Am. Chem. Soc.*, 74, 2027 (1952).

[‡] M.S. Matheson, E.E. Auer, E.B. Bevilacqua, and J.E. Hart, *J. Am. Chem. Soc.*, 73, 1700 (1951).

[*] J.A. May Jr. and W.B. Smith, *J. Phys. Chem.*, 72, 216 (1968).

adequate to assume this value and evaluate α from N_w/N_n and then use the value of α so obtained to evaluate a better value of p from N_n.

N_n	N_w/N_n
1129	1.60
924	1.67
674	1.73
609	1.74

8. Derive the two equations given in the previous problem. It may be helpful to recognize that for any distribution taken as a whole, $w_i = ix_i/N_n$.
9. In the research described in Problem 7, the authors determined the following distribution of molecular weights by a chromatographic procedure (w_i is the weight fraction of i-mer):

i	$w_i \times 10^4$	i	$w_i \times 10^4$
100	3.25	800	6.88
200	5.50	900	6.10
300	6.80	1200	4.20
400	7.45	1500	2.90
500	7.91	2000	1.20
600	7.82	2500	0.50
700	7.18	3000	0.20

They asserted that the points are described by the expression

$$w_i = \alpha i(1 - p)^2 p^{i-1} + 0.5(1 - \alpha)i(i - 1)(1 - p)^3 p^{i-2}$$

with $\alpha = 0.65$ and $p = 0.99754$. Calculate some representative points for this function and plot the theoretical and experimental points on the same graph. From the expression given extract the weight fraction i-mer resulting from termination by combination.

10. In fact, the expression in the previous problem is slightly incorrect. Derive the correct expression, and see if the implied values of α and p are significantly different. The solution to Problem 8 provides part of the answer.

11. Palit and Das[†] measured $\bar{\nu}_{tr}$ at 60°C for different values of the ratio [SX]/[M] and evaluated C_{SX} and $\bar{\nu}$ for vinyl acetate undergoing chain transfer with various solvents. Some of their measured and derived results are tabulated below (the same concentrations of AIBN and monomer were used in each run). Assuming that no other transfer reactions occur, calculate the values missing from the table. Criticize or defend the following proposition: The $\bar{\nu}$ values obtained from the limit [SX]/[M] $\to$ 0 show that the AIBN initiates polymerization identically in all solvents.

Solvent	$\bar{\nu}$	$\bar{\nu}_{tr}$	[SX]/[M]	$C_{SX} \times 10^4$
t-Butyl alcohol	6580	3709	—	0.46
Methyl isobutyl ketone	6670	510	0.492	—
Diethyl ketone	6670	—	0.583	114.4
Chloroform	—	93	0.772	125.2

[†] S.R. Palit and S.K. Das, *Proc. Roy. Soc. London*, 226A, 82 (1954).

12. Gregg and Mayo[†] studied the chain transfer between styrene and carbon tetrachloride at 60°C and 100°C. A sample of their data is given below for each of the temperatures.

At 60°C		At 100°C	
[CCl₄]/[Styrene]	$\bar{\nu}_{\text{tr}}^{-1} \times 10^5$	[CCl₄]/[Styrene]	$\bar{\nu}_{\text{tr}}^{-1} \times 10^5$
0.00614	16.1	0.00582	36.3
0.0267	35.9	0.0222	68.4
0.0393	49.8	0.0416	109
0.0704	74.8	0.0496	124
0.1000	106	0.0892	217
0.1643	156		
0.2595	242		
0.3045	289		

Evaluate the chain transfer constant (assuming that no other transfer reactions occur) at each temperature. By means of an Arrhenius analysis, estimate $E_{\text{tr}}^* - E_{\text{p}}^*$ for this reaction. Are the values of $\bar{\nu}$ in the limit of no transfer in the order expected for thermal polymerization? Explain.

13. Many olefins can be readily polymerized by a free-radical route. On the other hand, isobutylene is usually polymerized by a cationic mechanism. Explain.

14. Draw the mechanisms for the following processes in the radical polymerization of styrene in toluene: (a) initiation by cumyl peroxide; (b) propagation; (c) termination by disproportionation; and (d) transfer to solvent.

15. Show the mechanisms of addition of a butadiene monomer to a poly(butadienyl) radical, to give each of the three possible geometric isomers.

16. Consider the polymerization of styrene in toluene at 60°C initiated by di-t-butylperoxide for a solution containing 0.04 mol of initiator and 2 mol of monomer per liter. The initial rates of initiation, R_i, and propagation, R_p, are found to be 1.6×10^{-10} M s^{-1} and 6.4×10^{-7} M s^{-1}, respectively, at 60°C.

   (a) Calculate fk_d and $k_p/k_t^{1/2}$.
   (b) Assuming no chain transfer, calculate the initial kinetic chain length.
   (c) Assuming only disproportionation and under the conditions stated, the transfer constant of styrene, C_M, is 0.85×10^{-4}. How much does this transfer affect the molecular weight of the polymer?
   (d) The molecular weight of this polymer is too high. The desired molecular weight of this polymer is 40,000 g mol^{-1}. How much CCl₄ (in g L^{-1}) should be added to the reaction medium to attain the desired molecular weight? C_T of CCl₄ is 9×10^{-3}.
   (e) Under the conditions stated, the polymerization is too slow. What is the initial rate of polymerization if the temperature is raised to 100°C?
   (f) Calculate the conversion attained after the reaction has gone for 5 h at 100°C. Assume volume expansion does not change concentration significantly and that the initiator concentration is constant throughout the entire reaction.

References

1. Brandrup, J. and Immergut, E.H., Eds., *Polymer Handbook*, 3rd ed., Wiley, New York, 1989.
2. Trommsdorff, E., Kohle, H., and Lagally, P., *Makromol. Chem.*, 1, 169 (1948).

[†] R.A. Gregg and F.R. Mayo, *J. Am. Chem. Soc.*, 70, 2373 (1948).

Further Readings

Allcock, H.R. and Lampe, F.W., *Contemporary Polymer Chemistry*, 2nd ed., Prentice-Hall, Englewood Cliffs, NJ, 1990.

North, A.M., *The Kinetics of Free Radical Polymerization*, Pergamon, New York, 1966.

Odian, G., *Principles of Polymerization*, 4th ed., Wiley, New York, 2004.

Rempp, P. and Merrill, E.W., *Polymer Synthesis*, 2nd ed., Hüthig & Wepf, Basel, 1991.

4

Controlled Polymerization

4.1 Introduction

In the preceding chapters we have examined the two main classes of polymerization, namely step-growth and chain-growth polymerizations, with the latter exemplified by the free-radical mechanism. These are the workhorses of the polymer industry, permitting rapid and facile production of large quantities of useful materials. One common feature that emerged from the discussion of these mechanisms is the statistical nature of the polymerization process, which led directly to rather broad distributions of molecular weight. In particular, even in the simplest case (assuming the principle of equal reactivity, no transfer steps or side reactions, etc.), the product polymers of either a polycondensation or of a free-radical polymerization with termination by disproportionation would follow the most probable distribution, which has a polydispersity index (M_w/M_n) approaching 2. In commercial practice, the inevitable violation of most of the simplifying assumptions leads to even broader distributions, with polydispersity indices often falling between 2 and 10. In many cases the polymers have further degrees of heterogeneity, such as distributions of composition (e.g., copolymers), branching, tacticity, or microstructure (e.g., *cis* 1,4-, *trans* 1,4-, and 1,2-configurations in polybutadiene).

This state of affairs is rather unsatisfying, especially from the chemist's point of view. Chemists are used to the idea that every molecule of, say, ascorbic acid (vitamin C) is the same as every other one. Now we are confronted with the fact that a tank car full of the material called polybutadiene is unlikely to contain any two molecules with exactly the same chemical structure (recall Example 1.4). As polymers have found such widespread applications, we have obviously learned to live with this situation. However, if we could exert more control over the distribution of products, perhaps many more applications would be realized. In this chapter we describe several approaches designed to exert more control over the products of a polymerization. The major one is termed *living polymerization*, and can lead to much narrower molecular weight distributions. Furthermore, in addition to molecular weight control, living polymerization also enables the large-scale production of block copolymers, branched polymers of controlled architecture, and end-functionalized polymers.

A comparison between synthetic and biological macromolecules may be helpful at this stage. If condensation and free-radical polymerization represent the nadir of structural control, proteins and DNA represent the zenith. Proteins are "copolymers" that draw on 20 different amino acid monomers, yet each particular protein is synthesized within a cell with the identical degree of polymerization, composition, sequence, and stereochemistry. Similarly, DNAs with degrees of polymerization far in excess of those realized in commercial polymers can be faithfully replicated, with precise sequences of the four monomer units. One long-standing goal of polymer chemistry is to imitate nature's ability to exert complete control over polymerization. There are two ways to approach this. One is to begin with nature, and try to adapt its machinery to our purpose. This is exemplified by "training" cells into growing polymers that we want, for example, via recombinant DNA technology. The other approach, and the one described in this chapter, is to start with the polymerizations we already have, and try to improve them. Both approaches have merit, and we select the latter because it is currently much more established, and plays a central role in much of polymer research. It is worth noting that nature also makes use of many other macromolecular materials that are not so well-controlled as proteins and DNA; examples include polysaccharides such as cellulose, chitin, and starch. So in nature, as with commercial polymers, useful properties can still result from materials that are very heterogeneous at the molecular level.

The lack of control over molecular weight in polymerization arises directly from the random character of each step in the reaction. In a polycondensation, any molecule can react with any other at any time; the number of molecules is steadily decreasing, but the mole fraction of monomer is always larger than the mole fraction of any other species. In a free-radical polymerization, chains may be initiated at any time. Growing chains may also add monomer, or undergo a transfer or termination reaction at any time. The first requirement in controlling molecular weight is to fix the total number of polymers. This cannot be done in an unconstrained step-growth process, but it can be done in a chain-growth mechanism, through the concentration of initiators. The number of initiators will be equal to the number of polymers, assuming 100% initiation efficiency and assuming no transfer reactions that lead to new polymers. The second requirement is to distribute the total number of monomers as uniformly as possible among the fixed number of growing chains. If the polymerization then proceeds to completion, we could predict N_n precisely: it would simply be the ratio of the number of monomers to the number of initiators. To allow the reaction to proceed to completion, we would need to prevent termination steps, or at least defer them until we were ready. Now, suppose further that the reaction proceeds statistically, meaning that any monomer is equally likely to add to any growing chain at any time. If N_n was reasonably large, we could expect a rather narrow distribution of the number of monomers in each chain, just by probability. (This argument also assumes no transfer reactions, so that growing polymers are not terminated prematurely.) As an illustration, imagine placing an array of empty cups out in a steady rain; an empty cup is an "initiator" and a raindrop is a "monomer." As time goes on, the raindrops are distributed statistically among the cups, but after a lot of drops have fallen, the water level will be pretty much equal among the various cups. If a cup fell over, or a leaf fell and covered its top, that "polymer" would be "terminated," and its volume of water would not keep up with the others. Similarly, if you placed a cup outside a few minutes after the others, the delayed initiation would mean that it would never catch up with its neighbors. What we have just described is, in fact, the essence of a controlled polymerization: start with a fixed number of initiators, choose chemistry and conditions to eliminate transfer and termination reactions, and let the reaction start at a certain time and then go to completion. In order to control the local structural details, such as microstructure and stereochemistry, we have to influence the relative rates of various propagation steps. This can be achieved to some extent by manipulating the conditions at the active site at the growing end of the chain.

The remainder of this chapter is organized as follows. First we demonstrate how the kinetics of an ideal living polymerization leads to a narrow, Poisson distribution of chain lengths. Then, we consider chain-growth polymerization via an anionic propagating center; this has historically been the most commonly used controlled polymerization mechanism, and it can be conducted in such a way as to approach the ideal case very closely. In Section 4.4 we explore how the anionic mechanism can be extended to the preparation of block copolymers, end-functional polymers, and regular branched polymers of various architectures. We then turn our attention to other mechanisms that are capable of controlled polymerization, including cationic (Section 4.5), ring-opening (Section 4.8), and, especially, controlled radical polymerizations (Section 4.6). The concluding sections also address the concept of equilibrium polymerization, and a special class of controlled polymers called dendrimers.

4.2 Poisson Distribution for an Ideal Living Polymerization

In this section we lay out the kinetic scheme that describes a living polymerization, and thereby derive the resulting distribution of chain lengths. This scenario is most closely approached in the anionic case, but because it is not limited to anionic polymerizations, we will designate an active polymer of degree of polymerization i by P_i^*, and its concentration by $[P_i^*]$, where * represents the reactive end. *A living polymerization is defined as a chain-growth process for which there are no irreversible termination or transfer reactions.* There has been some controversy in the literature about the precise criteria for "livingness" [1], and whether they can ever be met in practice, but we will not dwell on this.

4.2.1 Kinetic Scheme

The concentration of unreacted monomer at time t will be denoted $[M]$. The initial concentrations of monomer and initiator are $[M]_0$ and $[I]_0$, respectively. The reaction steps can be represented as follows:

$$\text{Initiation:} \quad I + M \xrightarrow{k_i} P_1^* \tag{4.2.1}$$

$$\text{Propagation:} \quad P_1^* + M \xrightarrow{k_p} P_2^*$$

$$P_i^* + M \xrightarrow{k_p} P_{i+1}^* \tag{4.2.2}$$

Note that in using a single propagation rate constant, k_p, we are once again invoking the principle of equal reactivity.

We will now assume that initiation is effectively instantaneous relative to propagation ($k_i \gg k_p$), so that at time $t = 0$, $[P_1^*] = [I]_0$, and we will not worry about Equation 4.2.1 any further. Note that this criterion is not necessary to have a living polymerization, but it is necessary to achieve a narrow distribution of molecular weights. The concentration of unreacted monomer, $[M]$, will decrease in time as propagation takes over. The overall rate of polymerization, R_p, is the sum of the rates of consumption of monomer by all growing chains P_i^*. However, we know that, in the absence of termination or transfer reactions, the total concentration of P_i^* is always $[I]_0$: we have fixed the number of polymers. Therefore we can write

$$R_p = -\frac{d[M]}{dt} = k_p[M] \sum_i [P_i^*] = k_p[M][I]_0 \tag{4.2.3}$$

This is a linear, first-order differential equation for $[M]$, which has the solution

$$[M] = [M]_0 e^{-k_p[I]_0 t} \tag{4.2.4}$$

Therefore the concentration of monomer decreases exponentially to zero as time progresses. (Note that we are also assuming that propagation is irreversible, that is, there is no "back arrow" in Equation 4.2.2. The possibility of depolymerization reaction steps will be taken up in Section 4.7.)

At this stage it is very helpful to introduce a *kinetic chain length*, $\bar{\nu}$, analogous to the one we defined in Equation 3.5.10, as the ratio of the number of monomers incorporated into polymers to the number of polymers. The former is given by $[M]_0 - [M]$, and the latter by $[I]_0$, so we write

$$\bar{\nu} = \frac{[M]_0 - [M]}{[I]_0} \tag{4.2.5}$$

When the reaction has gone to completion, $[M]$ will be 0, and the kinetic chain length will be the number average degree of polymerization of the resulting polymer. It will also be helpful in the following development to differentiate Equation 4.2.5 with respect to time, and then incorporate Equation 4.2.3:

$$\frac{d\bar{\nu}}{dt} = -\frac{1}{[I]_0} \frac{d[M]}{dt} = k_p[M] \tag{4.2.6}$$

In order to obtain the distribution of chain lengths, we need to do a bit more work. We begin by writing an explicit equation for the rate of consumption of $[P_1^*]$:

$$-\frac{d[P_1^*]}{dt} = k_p[P_1^*][M] \tag{4.2.7}$$

We could insert Equation 4.2.4 into Equation 4.2.7 to replace [M], and thereby obtain an equation that can be solved. However, a simpler approach turns out to be to invoke the chain rule, as follows:

$$\frac{d[P_1^*]}{dt} = \frac{d[P_1^*]}{d\bar{\nu}}\frac{d\bar{\nu}}{dt} = \frac{d[P_1^*]}{d\bar{\nu}}k_p[M] \tag{4.2.8}$$

If we now compare Equation 4.2.7 and Equation 4.2.8 we can see that

$$-\frac{d[P_1^*]}{d\bar{\nu}} = [P_1^*] \tag{4.2.9}$$

and this equation is readily solved:

$$[P_1^*] = [P_1^*]_0 e^{-\bar{\nu}} = [I]_0 e^{-\bar{\nu}} \tag{4.2.10}$$

Now we repeat this process for $[P_2^*]$, beginning with the rate law. This is slightly more complicated, because $[P_2^*]$ grows by the reaction of P_1^* with monomer, but decreases by the reaction of P_2^* with monomer:

$$\begin{aligned}\frac{d[P_2^*]}{dt} &= k_p[P_1^*][M] - k_p[P_2^*][M] \\ &= k_p[M]\big([P_1^*] - [P_2^*]\big) = \frac{d\bar{\nu}}{dt}\big([P_1^*] - [P_2^*]\big)\end{aligned} \tag{4.2.11}$$

By invoking the chain rule once more

$$\frac{d[P_2^*]}{dt} = \frac{d[P_2^*]}{d\bar{\nu}}\frac{d\bar{\nu}}{dt} \tag{4.2.12}$$

and comparing with Equation 4.2.11 we obtain

$$\frac{d[P_2^*]}{d\bar{\nu}} + [P_2^*] = [P_1^*] = [I]_0 e^{-\bar{\nu}} \tag{4.2.13}$$

This equation has the solution

$$[P_2^*] = \bar{\nu}[I]_0 e^{-\bar{\nu}} \tag{4.2.14}$$

We can go through this sequence of steps once more, considering the concentration of trimer $[P_3^*]$:

$$\begin{aligned}\frac{d[P_3^*]}{dt} &= k_p[P_2^*][M] - k_p[P_3^*][M] \\ &= k_p[M]\big([P_2^*] - [P_3^*]\big) = \frac{d\bar{\nu}}{dt}\big([P_2^*] - [P_3^*]\big)\end{aligned} \tag{4.2.15}$$

leading to

$$\frac{d[P_3^*]}{d\bar{\nu}} + [P_3^*] = [P_2^*] = \bar{\nu}[I]_0 e^{-\bar{\nu}} \tag{4.2.16}$$

which has the solution (check it yourself)

$$[P_3^*] = \frac{1}{2}\bar{\nu}^2[I]_0 e^{-\bar{\nu}} \tag{4.2.17}$$

This pattern continues, and the result for the population of i-mer is

$$[P_i^*] = \frac{1}{(i-1)!}\bar{\nu}^{i-1}[I]_0 e^{-\bar{\nu}} \tag{4.2.18}$$

From this result we can obtain the desired distribution, namely the mole fraction of i-mer among all polymers, x_i, by dividing Equation 4.2.18 by the total number of polymers, $[I]_0$:

$$x_i = \frac{\bar{\nu}^{i-1}e^{-\bar{\nu}}}{(i-1)!} \tag{4.2.19}$$

This particular function, Equation 4.2.19, is called the *Poisson distribution*. Although we have obtained it from considering a specific kinetic scheme, in fact it will describe the situation whenever a larger number of objects (raindrops, or monomers) are distributed randomly among a small number of boxes (cups, or polymers). Once the polymerization reaction has gone to completion, and the polymers terminated by introduction of some appropriate reagent, the resulting molecular weight distribution should obey Equation 4.2.19, with $\bar{\nu}$ equal to $[M]_0/[I]_0$.

The following example illustrates some aspects of the kinetics of a living polymerization.

Example 4.1

The following data were reported for the living anionic polymerization of styrene.[†] The initial monomer concentration was 0.29 mol L^{-1}, and the initiator concentration was 0.00048 mol L^{-1}. The reactor was sampled at the indicated times, and the resulting polymer was terminated and analyzed for molecular weight and polydispersity. Use these data and Equation 4.2.4 and Equation 4.2.5 to answer the following questions: Does conversion of monomer to polymer follow the expected time dependence? What is the propagation rate constant under these conditions?

t (s)	M_n (g/mol)	N_n	PDI	$1-p$
238	3,770	36.3	1.06	0.940
888	20,600	198	1.02	0.672
1,626	33,700	324	1.02	0.463
2,296	43,000	413	1.01	0.316
3,098	49,800	479	1.008	0.207
4,220	54,900	528	1.006	0.127
14,345	61,700	593	1.005	0.018

Solution

We can equate the conversion of monomer to polymer with the familiar extent of reaction, p, as in Chapter 2 and Chapter 3:

$$p = \frac{[M]_0 - [M]}{[M]_0} = 1 - \frac{[M]}{[M]_0}$$

Using Equation 4.2.4 we see how p should evolve in time:

$$p = 1 - e^{-k_p[I]_0 t}$$

Therefore a plot of $\ln(1-p)$ versus t should give a straight line with slope equal to $-k_p[I]_0$. The data provided do not include [M] explicitly, but we can infer [M] and p from M_n. From Equation 4.2.5,

---

[†] W. Lee, H. Lee, J. Cha, T. Chang, K.J. Hanley, and T.P. Lodge, *Macromolecules*, 33, 5111 (2000).

the kinetic chain length is equal to $p[\text{I}]_0/[\text{M}]_0$, and it is also equal to N_n $(= M_n/M_0)$; thus $(1-p)$ in the table was obtained as

$$1 - p = 1 - \frac{[\text{I}]_0}{[\text{M}]_0}N_n = 1 - \frac{(0.00048 \text{ mol L}^{-1})}{(0.29 \text{ mol L}^{-1})}N_n$$

The suggested plot is shown below and the resulting slope from linear regression implies that $k_p \approx 1$ mol L^{-1} s^{-1}. (This is actually a rather low value, and in fact only an apparent value, due to a phenomenon to be described in Section 4.3 [also see Problem 3]. Also note that the last data point has been omitted from the fit, as it corresponds to essentially complete conversion, and thus is independent of t once the reaction is finished.)

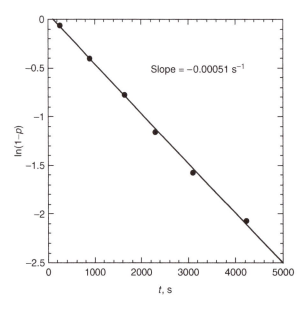

4.2.2 Breadth of the Poisson Distribution

Figure 4.1 illustrates the Poisson distribution for values of $\bar{\nu}$ equal to 100, 500, and 1000. For polystyrene with $M_0 = 104$, these would correspond to polymers with number average molecular weights of about 10^4, 5×10^4, and 10^5, respectively, which are moderate. The width of the distributions, although narrow, increases with $\bar{\nu}$, but as we shall see in a moment, the relative width (i.e., the width divided by $\bar{\nu}$) decreases steadily. It should be clear that these distributions are very narrow compared to the step-growth or free-radical polymerizations shown in Figure 2.5 and Figure 3.5, respectively. To underscore this, Figure 4.2 compares the theoretical distributions for free-radical polymerization with termination by combination (Equation 3.7.26) and for living polymerization, both with $\bar{\nu} = 100$. The difference is dramatic, and is made even more so when we recall that termination by combination leads to a relatively narrow distribution with M_w/M_n approaching 1.5 rather than 2.

For the Poisson distribution, the polydispersity index, M_w/M_n, in fact approaches unity as $\bar{\nu}$ increases indefinitely. The explicit relation for the Poisson distribution is

$$\frac{M_w}{M_n} = \frac{N_w}{N_n} = 1 + \frac{\bar{\nu}}{(1+\bar{\nu})^2} \approx 1 + \frac{1}{\bar{\nu}} \tag{4.2.20}$$

where the approximation applies for large $\bar{\nu}$. For $\bar{\nu} = 1000$ Equation 4.2.20 indicates that the polydispersity index will be 1.001, which is a far cry from 2.

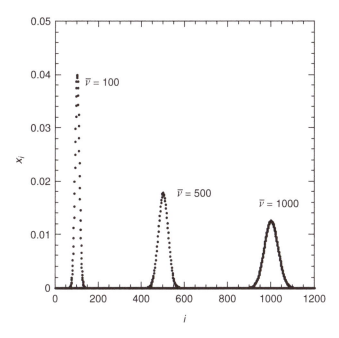

Figure 4.1   Mole fraction of i-mer for the Poisson distribution with the indicated kinetic chain lengths.

The derivation of Equation 4.2.20 is not too complicated, but it involves a couple of useful tricks, as we will show now. From Equation 1.7.2, we recall the definition of N_n, and insert Equation 4.2.19 to obtain

$$N_n = \sum_{i=1}^{\infty} i x_i = \sum_{i=1}^{\infty} \frac{i \, \bar{\nu}^{i-1} e^{-\bar{\nu}}}{(i-1)!} \qquad (4.2.21)$$

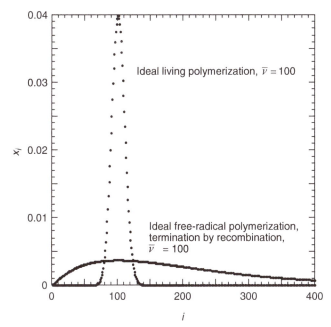

Figure 4.2   Comparison of Poisson distribution and distribution for free-radical polymerization with termination by combination.

To progress further with this, it is helpful to recall the infinite series expansion of e^x (see the Appendix if this is unfamiliar):

$$e^x = \sum_{i=0}^{\infty} \frac{x^i}{i!} = \sum_{i=1}^{\infty} \frac{x^{i-1}}{(i-1)!} \tag{4.2.22}$$

We will use this expansion to get rid of the factorials. Returning to Equation 4.2.21, we perform a series of manipulations, recognizing that $e^{-\bar{v}}$ does not depend on i and can be factored out of the sum, and that $i\bar{v}^{i-1}$ can be written as $d(\bar{v}^i)/d\bar{v}$:

$$N_n = \sum_{i=1}^{\infty} \frac{i\,\bar{v}^{i-1}e^{-\bar{v}}}{(i-1)!} = e^{-\bar{v}} \sum_{i=1}^{\infty} \frac{i\,\bar{v}^{i-1}}{(i-1)!}$$

$$= e^{-\bar{v}} \sum_{i=1}^{\infty} \frac{d}{d\bar{v}} \frac{\bar{v}^i}{(i-1)!} = e^{-\bar{v}} \frac{d}{d\bar{v}} \sum_{i=1}^{\infty} \frac{\bar{v}^i}{(i-1)!}$$

$$= e^{-\bar{v}} \frac{d}{d\bar{v}} \sum_{i=1}^{\infty} \bar{v} \frac{\bar{v}^{i-1}}{(i-1)!} = e^{-\bar{v}} \frac{d}{d\bar{v}} \left\{ \bar{v} \sum_{i=1}^{\infty} \frac{\bar{v}^{i-1}}{(i-1)!} \right\} = e^{-\bar{v}} \frac{d}{d\bar{v}} \{\bar{v} e^{\bar{v}}\} \tag{4.2.23}$$

This differentiation is straightforward, recalling the rule for differentiating the product of two functions, and that $d(e^x)/dx = e^x$:

$$e^{-\bar{v}} \frac{d}{d\bar{v}} \{\bar{v} e^{\bar{v}}\} = e^{-\bar{v}} \{e^{\bar{v}} + \bar{v} e^{\bar{v}}\} = 1 + \bar{v} \tag{4.2.24}$$

This relationship establishes that $N_n = 1 + \bar{v}$. (You may be wondering where the "1" came from. A glance at Equation 4.2.5 reveals the answer: before the reaction begins, when $[M] = [M]_0$, then $\bar{v} = 0$ when the degree of polymerization is actually 1. Of course, for any reasonable value of N_n, the difference between N_n and $N_n + 1$ is inconsequential.)

The development to obtain an expression for N_w follows a similar approach, beginning with the definition from Equation 1.7.4:

$$N_w = \sum_{i=1}^{\infty} iw_i = \frac{\sum_{i=1}^{\infty} i^2 x_i}{\sum_{i=1}^{\infty} i x_i} \tag{4.2.25}$$

We already know that the denominator on the right-hand side of Equation 4.2.25 is equal to $1 + \bar{v}$, so we just need to sort out the numerator

$$\sum_{i=1}^{\infty} i^2 x_i = \sum_{i=1}^{\infty} i^2 \frac{\bar{v}^{i-1}e^{-\bar{v}}}{(i-1)!} = e^{-\bar{v}} \sum_{i=1}^{\infty} i^2 \frac{\bar{v}^{i-1}}{(i-1)!}$$

$$= e^{-\bar{v}} \frac{d}{d\bar{v}} \left\{ \sum_{i=1}^{\infty} i \frac{\bar{v}^i}{(i-1)!} \right\} = e^{-\bar{v}} \frac{d}{d\bar{v}} \bar{v} \left\{ \sum_{i=1}^{\infty} i \frac{\bar{v}^{i-1}}{(i-1)!} \right\}$$

$$= e^{-\bar{v}} \frac{d}{d\bar{v}} \bar{v} \frac{d}{d\bar{v}} \left\{ \sum_{i=1}^{\infty} \frac{\bar{v}^i}{(i-1)!} \right\} = e^{-\bar{v}} \frac{d}{d\bar{v}} \bar{v} \frac{d}{d\bar{v}} \{\bar{v} e^{\bar{v}}\} \tag{4.2.26}$$

which leaves us with some more derivatives to take:

$$e^{-\bar{v}} \frac{d}{d\bar{v}} \left(\bar{v} \frac{d}{d\bar{v}} \{\bar{v} e^{\bar{v}}\} \right) = e^{-\bar{v}} \frac{d}{d\bar{v}} (\bar{v} \{e^{\bar{v}} + \bar{v} e^{\bar{v}}\})$$

$$= e^{-\bar{v}} \{\bar{v} e^{\bar{v}} + e^{\bar{v}} + 2\bar{v} e^{\bar{v}} + \bar{v}^2 e^{\bar{v}}\}$$

$$= 1 + 3\bar{v} + \bar{v}^2 \tag{4.2.27}$$

Finally, we can insert Equation 4.2.27 into Equation 4.2.25 to obtain N_w:

$$N_w = \frac{1 + 3\bar{\nu} + \bar{\nu}^2}{1 + \bar{\nu}} \tag{4.2.28}$$

It is now straightforward to obtain the result for the polydispersity index given in Equation 4.2.20, using Equation 4.2.24 and Equation 4.2.28:

$$\frac{N_w}{N_n} = \frac{1 + 3\bar{\nu} + \bar{\nu}^2}{(1 + \bar{\nu})^2} = \frac{(1 + \bar{\nu})^2 + \bar{\nu}}{(1 + \bar{\nu})^2} = 1 + \frac{\bar{\nu}}{(1 + \bar{\nu})^2} \tag{4.2.29}$$

The polydispersity data provided in Example 4.1 are compared with the Poisson distribution result (Equation 4.2.29) in Figure 4.3a. The experimental results are consistently larger than the prediction, but actually not by much. And, as the molecular weight increases, the experimental results seem to be approaching the Poisson result; the implications of this observation are considered in Problem 2. It is an interesting fact that this experimental test of Equation 4.2.29 was made possible only recently by advances in analytical techniques. To measure a polydispersity index below 1.01 would require an accuracy much better than 1% in the determination of M_w and M_n, and this is not yet possible using the standard techniques discussed in Chapter 1, Chapter 7, Chapter 8, and Chapter 9. In Figure 4.3b, the distribution for one particular sample obtained by matrix-assisted laser desorption/ionization (MALDI) mass spectrometry (and shown in Figure 1.7b) is compared with the Poisson distribution with the same mean; the agreement is excellent, with the experimental distribution being only slightly broader than the theoretical one.

We conclude this section with a summary of the requirements to achieve a narrow molecular weight distribution, and thereby draw an important distinction between "livingness" and the Poisson distribution. To recall the basic definition, a living polymerization is one that proceeds in the absence of transfer and termination reactions. Satisfying these two criteria is not sufficient to

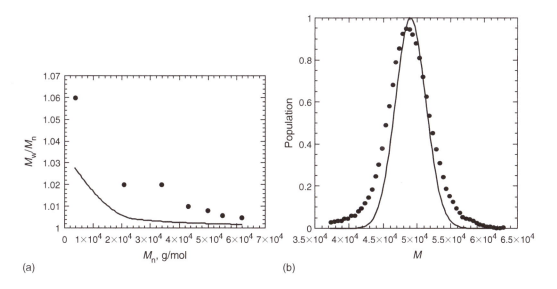

(a)   (b)

Figure 4.3   (a) Experimental polydispersities versus molecular weight for anionically polymerized polystyrenes, from the data in Example 4.1. (b) The distribution obtained by MALDI mass spectrometry for one particular sample. The smooth curves represent the results for the Poisson distribution, Equation 4.2.29 in (a) and Equation 4.2.19 in (b).

guarantee a narrow distribution, however. The additional requirements for approaching the Poisson distribution are:

1. All active chain ends must be equally likely to react with a monomer throughout the polymerization. This requires both the principle of equal reactivity, and good mixing of reagents at all times.
2. All active chain ends must be introduced at the same time. In practice, this means that the rate of initiation needs to be much more rapid than the rate of propagation, if all the monomer is added to the reaction mixture at the outset.
3. Propagation must be essentially irreversible, that is, the reverse "depolymerization" reaction does not occur to a significant extent. There are, in fact, cases where the propagation step is reversible, leading to the concept of an *equilibrium polymerization*, which we will take up in Section 4.8.

4.3   Anionic Polymerization

Anionic polymerization has been the most important mechanism for living polymerization, since its first realization in the 1950s [2]. Both modes of ionic polymerization (i.e., anionic and cationic) are described by the same vocabulary as the corresponding steps in the free-radical mechanism for chain-growth polymerization. However, initiation, propagation, transfer, and termination are quite different in ionic polymerization than in the free-radical case and, in fact, different in many ways between anionic and cationic mechanisms. In particular, termination by recombination is clearly not an option in ionic polymerization, a simple fact that underpins the development of living polymerization. In this section we will discuss some of the factors that contribute to a successful living anionic polymerization, and in the following section we will illustrate the extension of these techniques to block copolymers and controlled architecture branched polymers.

Monomers that are amenable to anionic polymerization include those with double bonds (vinyl, diene, and carbonyl functionality), and heterocyclic rings (see also Table 4.3). In the case of vinyl monomers $CH_2 = CHX$, the X group needs to have some electron withdrawing character, in order to stabilize the resulting carbanion. Examples include styrenes and substituted styrenes, vinyl aromatics, vinyl pyridines, alkyl methacrylates and acrylates, and conjugated dienes. The relative stabilities of these carbanions can be assessed by considering the pK_a of the corresponding conjugate acid. For example, the polystyryl carbanion is roughly equivalent to the conjugate base of toluene. The smaller the pK_a of the corresponding acid, the more stable the resulting carbanion. The more stable the carbanion, the more reactive the monomer in anionic polymerization. In the case of anionic ring-opening polymerization (ROP), the ring must be amenable to nucleophilic attack, as well as present a stable anion. Examples include epoxides, cyclic siloxanes, lactones, and carbonates. At the same time, there are many functionalities that will interfere with an anionic mechanism, especially those with an acidic proton (e.g., –OH, –NH$_3$, –COOH) or an electrophilic functional group (e.g., O_2, –C(O)–, CO_2). Anionic polymerization of monomers that include such functionalities can generally only be achieved if the functional group can be protected. As a corollary, the polymerization medium must be rigorously free of protic impurities such as water, as well as oxygen and carbon dioxide.

A wide variety of initiating systems have been developed for anionic polymerization. The first consideration is to choose an initiator that has a comparable or slightly higher reactivity than the intended carbanion. If the initiator is less reactive, the reaction will not proceed. If, on the other hand, it is too reactive, unwanted side reactions may result. As the pK_as of the conjugate acids for the many possible monomers span a wide range, so too must the pK_as of the conjugate acids of the initiators. Second, the initiator must be soluble in the same solvent as the monomer and resulting polymer. Common classes of initiators include radical anions, alkali metals, and especially alkyllithium compounds. We will illustrate two particular initiator systems: sodium naphthalenide, as an example of a radical anion, for the polymerization of styrene, and *sec*-butyllithium, as an alkyllithium, in the polymerization of isoprene.

The first living polymer studied in detail was polystyrene initiated with sodium naphthalenide in tetrahydrofuran (THF) at low temperatures:

1. The precursor to the initiator is prepared by the reaction of sodium metal with naphthalene and results in the formation of a radical ion:

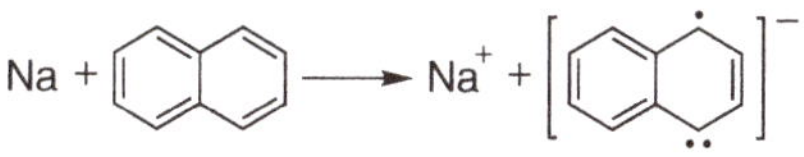

(4.A)

Of course the structure of the radical anion shown is just one of the several possible resonance forms.

2. These green radical ions react with styrene to produce the red styryl radical ion:

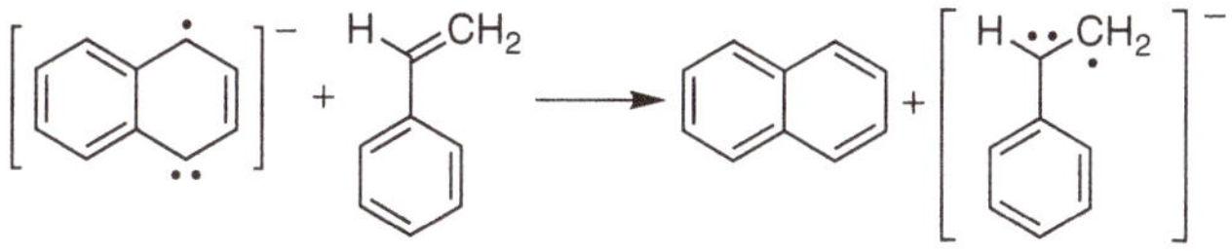

(4.B)

3. The latter undergoes radical combination to form the dianion, which subsequently initiates the polymerization:

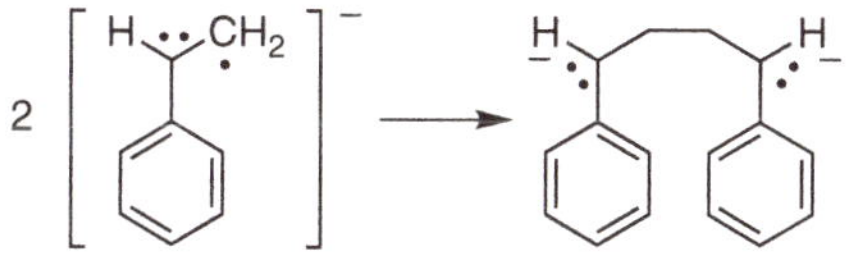

(4.C)

In this case, the degree of polymerization is $2\bar{\nu}$ because the initiator is difunctional; furthermore, there will be a single tail-to-tail linkage somewhere near the middle of each chain.

4. The propagation step at either end of the chain can be written as follows:

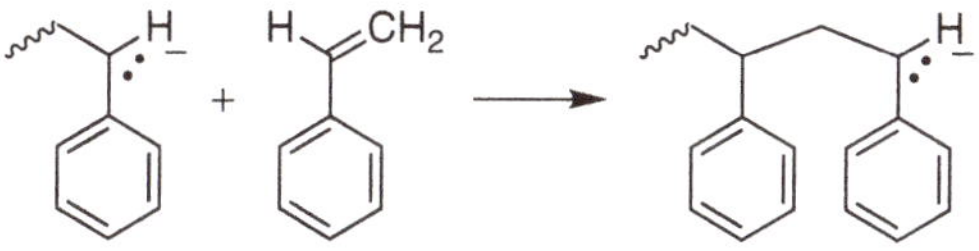

(4.D)

The carbanion attacks the more electropositive and less sterically hindered carbon to regenerate the more stable benzylic carbanion. Thus, the addition is essentially all head-to-tail in this case. Note also that the sodium counterions have not been written explicitly in Reactions (4.B)–(4.D), although of course they are present. As we will see below, the counterion can actually play a crucial role in the polymerization itself.

Now we consider the polymerization of isoprene by *sec*-butyllithium, in benzene at room temperature. In the first step, one monomer is added, but immediately there are many possibilities, as indicated:

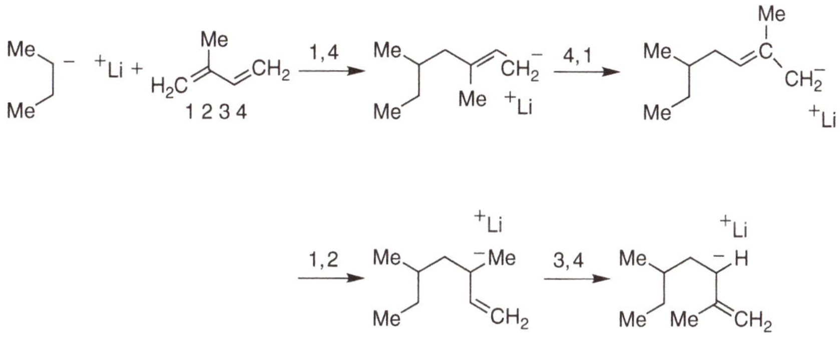

(4.E)

Which happens, and why? What happens when the next monomer adds? Is it the same configuration, or not? What does it all depend on? There is no simple answer to these questions, but we can gain a little insight into how to control the microstructure of a polydiene by looking at some data.

Table 4.1 gives the results of chemical analysis of the microstructure of polyisoprene after polymerization under the stated conditions. In the first two cases, there is a strong preference for 3,4 addition, with significant amounts of 1,2; relatively little 1,4 addition is found. The key feature here turns out to be the solvent polarity, as will be discussed below. When switching to heptane, a nonpolar solvent, the situation is reversed; now 1,4 *cis* is heavily favored. Interestingly, decreasing the initiator concentration by a factor of a thousand exerts a significant influence on the 1,4 *cis/ trans* ratio. At first glance this seems strange; the details of an addition step should not depend on the number of initiators. However, the answer lies in kinetics, as the propagation step is not as simple as one might naively expect. Finally, the last three entries show isoprene polymerized in bulk, which also corresponds to a nonpolar medium. In this case, we see that changing the counterion has a huge effect. Simply replacing lithium with sodium switches the product from almost all *cis* 1,4 to a mixture of *trans* 1,4 and 3,4.

The key factor that comes into play in nonpolar solvents is *ion pairing* or clustering of the living ends. Ionic species tend to be sparingly soluble in hydrocarbons, as the dielectric constant of the medium is too low. Consequently, the counterion is rather tightly associated with the carbanion, forming a dipole; these dipoles have a strong tendency to associate into a small cluster, with perhaps $n = 2$, 4, or 6 chains effectively connected as a star molecule. This equilibrium is illustrated in the cartoon below for the case $n = 4$:

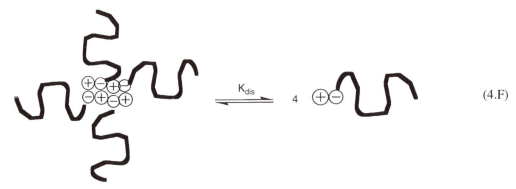

$$(4.F)$$

Addition steps occur primarily when the living chain end is not associated. This leads to an interesting dependence of the rate of polymerization, R_p, on the living chain concentration, as can readily be understood as follows (recall Equation 4.2.3):

Table 4.1 Polymerization of Polyisoprene under Various Conditions, and the Resulting Microstructure in %

Solvent	Counterion	T, °C	1,4 *cis*	1,4 *trans*	1,2	3,4
THF	Li	30	12 combined		29	59
Dioxane	Li	15	3	11	18	68
Heptane[a]	Li	−10	74	18	—	8
Heptane[b]	Li	−10	97	—	—	—
None	Li	25	94	—	—	6
None	Na	25	—	45	7	48
None	Cs	25	4	51	8	37

[a]Initiator concentration 6×10^{-3} M.
[b]Initiator concentration 8×10^{-6} M.

Source: From Hsieh, H.L. and Quirk, R.P., *Anionic Polymerization, Principles and Practical Applications*, Marcel Dekker, Inc., New York, NY, 1996.

$$R_p = -\frac{d[M]}{dt} = k_p[M][P^*]_{free} \tag{4.3.1}$$

where $[P^*]_{free}$ is the concentration of unassociated living chains. This concentration is set by the equilibrium between associated and free chains (see Reaction 4.F):

$$K_{dis} = \frac{([P^*]_{free})^4}{[(P^*)_4]} \tag{4.3.2}$$

Inserting Equation 4.3.2 into Equation 4.3.1 gives

$$R_p = k_p(K_{dis})^{1/4}[M][(P^*)_4]^{1/4} = k_{app}[M][P^*]^{1/4} \tag{4.3.3}$$

where we recognize that $[(P^*)_4] \approx (1/4)[P^*]$, as most of the chains are in aggregates, and that the apparent rate constant $k_{app} = k_p(K_{dis}/4)^{1/4}$. The rate of polymerization is therefore first order in monomer concentration, as one should expect, but has a $(1/n)$ fractional dependence on initiator concentration, where n is the average aggregate size. Accurate experimental determination of n is tricky, but a large body of data exists. It should also be noted that there is in all likelihood a distribution of states of association or ion clustering, so that the actual situation is considerably more complicated than implied by Reaction (4.F).

Increasing the size of the counterion increases the separation between charges at the end of the growing chain, thereby facilitating the insertion of the next monomer. The concentration of initiator can also influence n, presumably by the law of mass action. The dependence of the *cis* isomer concentration in heptane indicated in Table 4.1 is actually thought to be the result of a more subtle effect than this, however. It is generally accepted that the *cis* configuration is preferred immediately after addition of a monomer, but that isomerization to *trans* is possible, within an aggregate, given time. The rate of isomerization is proportional to the concentration of chains in aggregates and therefore proportional to $[P^*]$, whereas the rate of addition is proportional to a fractional power of $[P^*]$. Increasing the initiator concentration increases both rates, but favors isomerization relative to propagation.

Termination of an anionic polymerization is a relatively straightforward process; introduction of a suitable acidic proton source, such as methanol, will cap the growing chain and produce the corresponding salt, for example, $Li^+OCH_3^-$. Care must be taken that the termination is conducted under the same conditions of purity as the reaction itself, however. For example, introduction of oxygen along with the terminating agent can induce coupling of two living chains. However, in many cases it is desirable to introduce a particular chemical functionality at the end of the growing chain. One prime example is to switch to a second monomer, which is capable of continued polymerization to form a block copolymer. The second example is to use particular multifunctional terminating agents to prepare star-branched polymers. These cases, and other uses of end-functional chains, are the next subject we take up.

4.4 Block Copolymers, End-Functional Polymers, and Branched Polymers by Anionic Polymerization

The central importance of living anionic polymerization to current understanding of polymer behavior cannot be overstated. For example, throughout Chapter 6 through Chapter 13 we will derive a host of relationships between observable physical properties of polymers and their molecular weight. These relationships have been largely confirmed or established experimentally by measurements on narrow molecular weight distribution polymers, which were prepared by living anionic methods. However, it can be argued that even more important and interesting applications of living polymerization arise in the production of elaborate, controlled architectures; this section touches on some of these possibilities.

4.4.1 Block Copolymers

Before addressing the preparation of block copolymers by anionic polymerization, it is appropriate to consider some of the reasons why block copolymers are such an interesting class of macromolecules.

The importance of block copolymers begins with the fact that a single molecule contains two (or more) different polymers, and therefore may in some sense exhibit the characteristics of both components. This offers the possibility of tuning properties, or combinations of properties, between the extremes of the pure components. However, a random or statistical copolymer could also do that, without the effort required to prepare the block architecture. The important difference is that, for reasons that will be explored in Chapter 7, two different polymers will not usually mix; they tend to phase separate into almost pure components. The architecture of a block copolymer defeats this *macroscopic phase separation*, because of the covalent linkages between the different blocks. The consequence is that block copolymers undergo what is often called *microphase separation*; the blocks of one type segregate into domains that have dimensions on the lengthscale of the blocks themselves, i.e., 5–50 nm. In the current jargon, these polymers undergo *self-assembly* to produce particular nanostructures.

There are at least four broad arenas in which the self-assembly of block copolymers is useful, as illustrated in Figure 4.4:

1. *Micelles.* In a solvent that dissolves one block but not the other, copolymers will aggregate into micelles. A typical micelle is roughly spherical, about 20 nm in size, and contains 50–200 molecules. However, under appropriate conditions the micelles can be long, worm-like structures, or even flat bilayers that can curve around to form closed "bags" called *vesicles*. This behavior is analogous to that of small molecule surfactants or biological lipids. Micelles can be used to sequester, extract, or transport insoluble molecules through a solvent.

2. *Macromolecular surfactants.* Extending the analogy to small molecule surfactants, where the amphiphilic character of the molecule can stabilize dispersions of oil droplets in water (emulsions) or water in oil, an AB block copolymer could stabilize a dispersion of polymer A in a matrix of polymer B. This strategy is used to control the tendency of different polymers to phase separate on a macroscopic scale, and allows preparation of *compatibilized* polymer blends, with dispersed droplets on the micron scale.

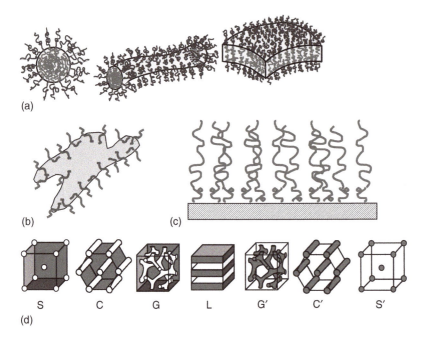

(a)

(b)   (c)

(d)   S   C   G   L   G′   C′   S′

Figure 4.4   Examples of block copolymer self-assembly: (a) as spherical, cylindrical, and bilayer micelles in a selective solvent for one block; (b) as surfactants in a dispersion of one polymer in an immiscible matrix polymer; (c) on surfaces, following adsorption of one block; (d) as bulk, nanostructured materials. Body-centered spherical micelles (S), hexagonally packed cylindrical micelles (C), bicontinuous double gyroid (G), lamellae (L).

3. *Tailored surfaces and thin films.* In a selective solvent, block copolymers can adsorb on a surface with the insoluble block forming a dense film, and the soluble block extending out into the solvent, forming a *brush*. Such brushes can impart colloidal stability to dispersed particles, or prevent protein adsorption in biomedical devices. Or, a thin film of copolymer can be allowed to self-assemble on a surface, forming nanoscale patterns such as stripes and spheres that are under consideration for lithographic applications.

4. *Nanostructured materials.* In the bulk state, or in concentrated solution, the self-assembly process can lead to structures with well-defined long-range order or symmetry. As illustrated in Figure 4.4, an AB diblock tends to adopt one of seven particular ordered phases, depending primarily on the relative lengths of the two blocks. For example, when the A fraction of the chain is small, perhaps 10%–20%, the A blocks collect in spherical domains, just like micelles in solution, and the micelles pack onto a body-centered cubic lattice. As the fraction of A is increased, the chains form cylindrical micelles on a hexagonal lattice, and then when the amounts of A and B are roughly equal, flat sheets or lamellae are formed. As A becomes the majority component, the same structures are seen, but now with the B blocks inside the cylinders and spheres. This sequence of interfacial curvature mirrors exactly that seen in solution micelles. However, one new feature is the presence of a bicontinuous cubic structure, the double gyroid, which intervenes between cylinders and lamellae.

In current commercial practice the most important block copolymer is the ABA triblock, where the A block is usually polystyrene and the B block is an elastomer such as isoprene, butadiene, or their saturated (i.e., hydrogenated) equivalents. Such polymers are known as *thermoplastic elastomers*, because at ambient temperatures they self-assemble in such a way that the small styrene domains, which are glassy, act as cross-links to form an extended, elastomeric network of the bridging B blocks. (We will discuss network elasticity in detail in Chapter 10 and the nature of the glass transition in Chapter 12.) At elevated temperatures (i.e., above 100°C), the polystyrene blocks can flow, and the network can be reformed into a new shape. These anionically prepared materials find use in such diverse applications as pressure-sensitive adhesives, hot melt adhesives, asphalt modifiers, sports footwear, and drug-releasing stents.

Block copolymers are usually prepared by sequential living anionic polymerization. This means that one block is polymerized to completion, but not terminated; the second monomer is then added to the reaction mixture. The living chains act as *macroinitiators* for the polymerization of the second block. After the second block is complete, a terminating agent can be introduced, or the monomer for a third block, and so on. The key requirements for this strategy to be successful include the following:

1. The most important criterion is that the carbanion of the first block be capable of initiating polymerization of the second block. Returning to the discussion in the previous section, this implies that the stability of the second block carbanion is greater than or equal to that of the first block, or equivalently that the pK_a of the conjugate acid is smaller. As an example, if it is desired to prepare polystyrene-*block*-poly(methyl methacylate), the polystyrene block must be prepared first. On the other hand, polystyryl, polybutadienyl, and polyisoprenyl anions can initiate one another, so in principle arbitrary sequences of these blocks are accessible.

2. The solvent system chosen must be suitable for all blocks, or it must be modified for the polymerization of the second block. For example, it is possible to prepare block copolymers of 1,4-polyisoprene and 1,2-polybutadiene, by adding a "polar modifier" in midstream. The first block microstructure calls for a nonpolar solvent, whereas the second requires a polar environment. Rather than switching solvents entirely, a polar modifier associates with the carbanion active site and directs the regiochemistry of addition in a similar fashion to a polar solvent. Examples of modifiers include Lewis bases such as triethylamine, N,N,N',N'-tetramethylethylenediamine (TMEDA), and 2,2'-bis(4,4,6-trimethyl-1,3-dioxane).

3. The counterion must also be suitable for polymerization of both the blocks.

These requirements, and especially the first, might appear to be rather limiting. For example, how could either poly(methyl methacrylate)-*block*-polystyrene-*block*-poly(methyl methacrylate) or polystyrene-*block*-poly(methyl methacrylate)-*block*-polystyrene triblocks be prepared? The answer in both cases is actually rather straightforward. In the first case, a difunctional initiating system such as sodium naphthalenide could be used; then the triblock would be grown from the middle out. In the second case, a *coupling agent* can be used, which would link two equivalent living polystyrene-*block*-poly(methyl methacrylate) diblocks together. The coupling agent is usually a difunctional molecule, in which each functional group is equally capable of terminating an anionic polymerization. This is illustrated by α,α'-dibromo-*p*-xylene in the following reaction:

$$(styrene)_n\text{—}(methyl\ methacrylate)_{2m}\text{—}(styrene)_n$$
$$+\ 2\ LiBr$$

(4.G)

Note that the resulting methyl methacrylate midblock will have one phenyl linkage in the middle. This coupling strategy has several potential advantages over sequential monomer addition. In addition to achieving otherwise inaccessible block sequences, the total polymerization time is roughly cut in half. Furthermore, the second "crossover" step is avoided, which is desirable in that each addition of monomer brings with it the possibility of contamination or less than complete initiation of the subsequent blocks. The primary limitation of coupling is the inevitability of incomplete conversion of diblock to triblock. If Reaction (4.G) is run with either excess living chain or coupling agent, there will be some remaining diblock. If run under stoichiometric conditions, incomplete coupling is still probable. Any excess diblock can be removed by fractionation, if necessary.

There are still many block copolymers, even diblocks, that simply cannot be prepared by sequential monomer addition: the conditions required for the polymerization of one block are not compatible with the other. In this case, one general strategy is to prepare batches of the two homopolymers, each functionalized at one end with a reactive group that can couple to the other. This potentially enables preparation of any conceivable diblock, and each block could be prepared by any suitable living polymerization scheme, not just the anionic one. However, this approach is usually the last resort, because polymer–polymer coupling reactions are notoriously inefficient, even assuming a common solvent can be found. Coupling reactions are practical in the anionic triblock case because the two reacting chains are already present in the reactor, and the carbanions are highly reactive; this might not be the case with, say, a hydroxyl-terminated polymer A and a carboxylic acid–terminated polymer B. A more efficient strategy is to terminate the polymerization of the first block in such a way as to leave a functional group that can subsequently be used to initiate living polymerization of the second monomer; this is the macroinitiator approach, but where the reaction conditions are completely changed in midstream. As an example, polystyrene and poly(ethylene oxide) are both amenable to living anionic polymerization, but often not under the same conditions. If ethylene oxide monomer is introduced to the polystyryl anion with a lithium counterion, it turns out that one monomer adds but no propagation occurs. Termination with a proton therefore generates a polystyrene molecule with a terminal hydroxyl group. This can then act as a macroinitiator; titration of the endgroup with the strong base potassium naphthanelide produces the terminal alkoxide with a potassium counterion, which can initiate ethylene oxide polymerization.

(4.H)

4.4.2 End-Functional Polymers

The previous illustration of the macroinitiator approach is an excellent example of the utility of an end-functional polymer, by which we mean a polymer with a well-defined, reactive chemical functionality at one end, or at both ends. Such polymers are also referred to as *telechelic*. It should be apparent that most condensation polymers have reactive groups at each end, and thus fall in this class. However, we are concerned here with polymers that have narrow molecular weight distributions as a result of a living polymerization. In essence, an end-functional polymer is a macromolecular reagent. It can be carefully characterized and then stored on the shelf until needed for a particular application. The following is a list of a few of the many examples of possible uses for end-functional polymers:

1. *Macroinitiators.* As illustrated in the previous section, a macroinitiator is an end-functional polymer in which the functional group can be used to initiate polymerization of a second monomer. In this way, block copolymers can be prepared that are not readily accessible by sequential monomer addition. Indeed, the second block could be polymerized by an entirely different mechanism than the first; other living polymerization schemes will be discussed in subsequent sections.
2. *Labeled polymers.* It is sometimes desired to attach a "label" to a particular polymer, such as a fluorescent dye or radioactive group, which will permit subsequent tracking of the location of the polymer in some process. By attaching the label to the end of the chain, the number of labels is well-defined, and labeled chains can be dispersed in otherwise equivalent unlabeled chains in any desired proportion.
3. *Chain coupling.* Both block copolymers and regular branched architectures can be accessed by coupling reactions between complementary functionalities on different chains.
4. *Macromonomers.* If the terminal functional group is actually polymerizable, such as a carbon–carbon double bond, polymerization through the double bond can produce densely branched comb or "bottlebrush" copolymers.
5. *Grafting to surfaces.* As mentioned in the context of copolymer adsorption to a surface, a densely packed layer of polymer chains emanating from a surface forms a brush. Such brushes can also be prepared by the grafting of end-functional chains, where the functionality is tailored to react with the surface. High grafting densities are hard to achieve by this strategy, however, due to steric crowding; the first chains anchored to the surface make it progressively harder for further chain ends to react.
6. *Controlled-branched and cyclic architectures.* Examples of branched structures will be given in the following section. Cyclic polymers can be prepared by intramolecular reaction of an "α,ω-heterotelechelic" linear precursor, where the two distinct end groups can react. Such ring-closing reactions have to be run at extreme dilution, to suppress interchain end linking.
7. *Network precursors.* Telechelic polymers can serve as precursors to network formation, when combined with suitable multifunctional linkers or catalysts. For example, some silicone adhesives contain poly(dimethylsiloxane) chains with vinyl groups at each end. In the unreacted form, these polymers form a low-viscosity fluid that can easily be mixed with catalyst and spread on the surfaces to be joined; the subsequent reaction produces an adhesive, three-dimensional network in situ.
8. *Reactive compatibilization.* As noted previously, block copolymers can act as macromolecular surfactants to stabilize dispersions of immiscible homopolymers. However, direct mixing of block copolymers during polymer processing is not always successful, as the copolymers have a tendency to aggregate into micelles and never reach the interface between the two polymers. One effective way to overcome this is to form the block copolymer at the targeted interface, by in situ reaction of suitable functional chains. Note that in this case it is not absolutely necessary that the reactive groups be at the chain ends.

There are two general routes to end-functional chains: use a functional initiator or use a functional terminating group. The use of a functional terminating agent proves to be the more flexible strategy for a rather straightforward reason. Any functional group present in the initiator must be inert to the polymerization, which can be problematic in the case of anionic polymerization. Thus, the functional group in the initiator must be protected in some way. In contrast, for the terminating agent all that is required are two functionalities: the desired one and another electrophilic one to terminate the polymerization. However, the functionality that is designed to terminate polymerization must be substantially more reactive to carbanions than the other functionality, or more than one chain end structure will result. Consequently, in most cases a protection strategy is also employed for the terminating agent. Nevertheless, in the termination case the demands on the protecting group are much reduced relative to initiation; in the former, the protecting group only needs to be significantly less reactive than the electrophile, whereas in the latter the protecting group must be substantially less reactive than the monomer.

For the living anionic polymerization of styrene, butadiene, and isoprene, an effective terminating strategy is to use alkanes that have bromo functionality at one end and the protecting group at the other. The halide is very reactive to the carbanion, readily eliminating the LiBr salt as the chain is terminated. Of course, the protecting group must then be removed in a separate step. Examples of protecting groups and the desired functionalities are given in Table 4.2. Some of the same protecting groups illustrated in Table 4.2 can also be used in functional initiators. For example, the *tert*-butyl dimethylsilyl moiety used to protect the thiol group can also be used to protect a hydroxyl group in the initiator, as in (3-(*tert*-butyl dimethylsilyloxy-1-propyllithium)).

Another powerful strategy for preparing end-functional polymers by anionic polymerization was implicitly suggested in the previous section, where addition of a nominally polymerizable monomer (ethylene oxide in that instance) to a growing polystyryl anion resulted in the addition of only one new monomer. It turns out that 1,1-diphenylethylene and derivatives thereof will only react with organolithium salts to form the associated relatively stable carbanion; no further propagation occurs:

Table 4.2 Examples of Protection Strategies for Preparing End-Functional Polymers by Living Anionic Polymerization of Styrenes and Dienes

Functional Group	Protected functionality
−OH	$\diagup O \diagdown Si(CH_3)_3$
−NH$_2$	$Si(CH_3)_3$ $\diagup N \diagdown Si(CH_3)_3$
−SH	$\diagup S \diagdown Si((CH_3)_2 t\text{-Bu})$
−COOH	$\diagup C(OCH_3)_3$
−C≡CH	$-C≡C-Si(CH_3)_3$

Termination by short alkanes with a halide at one end and the protected functionality at the other.
Source: From Hirao, A. and Hoyashi, M., *Acta Polymerica*, 50, 219, 1999.

$$(4.I)$$

In this structure R' and R'' could be any of a variety of protected or even unprotected functionalities. Even more interesting is the fact that this carbanion can be used to initiate anionic polymerization of a new monomer (such as methyl methacrylate, dienes, etc.) or even to reinitiate the polymerization of styrene. In this way, diphenylethylene derivatives can be used to place particular functional groups at desired locations along a homopolymer or copolymer, not just at the terminus.

4.4.3 Regular Branched Architectures

The kinds of synthetic methodology suggested in the previous section have been adapted to the preparation of a wide range of polymer structures with controlled branching [4]. The first architecture to consider is that of the regular star, in which a predetermined number of equal length arms are connected to a central core. There are two general strategies to prepare such a polymer by living anionic polymerization: use a multifunctional initiator, and grow the arms outwards simultaneously, or use a multifunctional terminating agent to link together premade arms. The first route is an example of an approach known as *grafting from*, whereas the second is termed *grafting to*. Or, in anticipation of the discussion of dendrimers in Section 4.9, grafting from and grafting to are analogous to divergent and convergent synthetic strategies. Although both have been used extensively, grafting to is more generally applicable to anionic polymerization due to the difficulty in preparing and dissolving small molecules with multiple alkyllithium functionalities. Furthermore, in order to achieve uniform arm lengths, it is essential that each initiation site be equally reactive and equally accessible to monomers in the reaction medium. If it is desired to terminate each star arm with a functional group, however, then grafting from may be preferred. Should the anionic polymerization be initiated by a potassium alkoxide group, as for example with the polymerization of ethylene oxide suggested in the context of Reaction (4.H), then preparation of initiators with multiple hydroxyl groups is quite feasible (see Reaction 4.EE for a specific example). Similarly, if other living polymerization routes are employed, such as controlled radical polymerization to be discussed in Section 4.6, then grafting from is more convenient than in the anionic case.

The preparation of an eight-arm polystyrene star by grafting to is illustrated in the following scheme. The most popular terminating functionality in this context is a chlorosilane, which reacts rapidly and cleanly with many polymeric carbanions, and which can be prepared with functionalities up to at least 32 without extraordinary effort. An octafunctional chlorosilane can be prepared starting with tetravinylsilane and dichloromethylsilane, using platinum as a catalyst:

$$(4.J)$$

This multifunctional terminating agent is then introduced directly into the reaction vessel containing the living polystyryl chains. The chains should be in stoichiometric excess to minimize the formation of a mixture of stars with different numbers of arms. This will necessitate separation of the unattached arms from the reaction mixture, but this is feasible. Moreover, an additional advantage of the grafting to approach is thus exposed: the unattached arms can be characterized (for molecular weight, polydispersity, etc.) independently of the stars themselves, a desirable step that is not possible when grafting from.

The scheme just outlined is not quite as straightforward as it might appear. The key issue is to make all eight terminating sites accessible to the polystyryl chains. As the number of attached arms grows, it becomes harder and harder for new chain ends to find their way into the reactive core. In order to reduce these steric effects, more methylene groups can be inserted into the terminating agent to spread out the chlorosilanes. In some cases, polystyryl chains have been capped with a few butadienyl units to reduce the steric bulk of the chain end. Clearly, all of these issues grow in importance as the number of arms increases. Note, however, that it is not necessary that all the chlorosilanes be equally reactive in order to preserve a narrow molecular weight distribution; it is only necessary that the attachment of the narrowly distributed arms be driven to completion (which may take some time).

As the desired number of arms increases, it is practical to surrender some control over the exact number of arms in favor of a simpler method for termination. A scheme that has been refined to a considerable extent is to introduce a difunctional monomer, such as divinylbenzene, as a polymerizable linking agent. The idea is illustrated in the following reaction:

$$(4.K)$$

One divinylbenzene molecule can thus couple two polystyryl chains and leave two anions for further reaction. Each anion might add one more divinylbenzene, each of which could then add one more polystyryl chain. At that point, the growing star molecule would have four arms, emanating from a core containing three divinylbenzene moieties and four anions. This process can continue until the divinylbenzene is consumed and the anions terminated. Clearly, there is potential for a great deal of variation in the resulting structures, both in the size of the core and in the number of arms. However, by carefully controlling the reaction conditions, and especially the ratio of divinylbenzene to living chains, reasonably narrow distributions of functionality can be obtained, with average numbers of arms even exceeding 100.

The preceding strategy can actually be classified as *grafting through*, a third approach that is particularly useful for the preparation of comb polymers. A comb polymer consists of a backbone to which a number of polymeric arms are attached; combs can be prepared by grafting from, grafting to, and grafting through. In the first case, the backbone must contain reactive sites that can used to initiate polymerization. The backbone can be characterized independently of the arms, but the arms themselves cannot. In grafting to, the backbone must contain reactive sites such as chlorosilanes that can act to terminate the polymerization of the arms. Clearly in this case, as with stars, the arms and the backbone can be characterized independently. The grafting through strategy takes advantage of what we previously termed macromonomers: the arms are polymers terminated with a polymerizable group. These groups can be copolymerized with the analogous monomers to generate the backbone. By varying the ratio of macromonomer to comonomer, the spacing of the "teeth" of the comb can be tuned. Note that this process is not necessarily straightforward. In Chapter 5 we will consider copolymerization in great detail, but a key concept is that of *reactivity ratio*. This refers to the relative probability of adding one monomer to a growing chain, depending on the identity of the previous monomer that attached. It is generally the case that there are significant preferences (i.e., the reactivity ratios of the two monomers are not unity), which means that the two monomers will not add completely randomly. These factors need to be understood before regular comb molecules with variable branching density can be prepared by grafting through.

The grafting through approach can be illustrated through the following sequence [5]. Polystyryl chains can be capped with one ethylene oxide unit (Reaction 4.H) followed by termination with methacryloyl chloride:

$$(4.L)$$

This macromonomer can then be copolymerized with methyl methacrylate, to produce a comb or graft copolymer, with a poly(methyl methacrylate) backbone and polystyrene arms:

$$PMMA-g-PS$$
$$(4.M)$$

This last example reminds us that the variety of possible controlled branched architectures is greatly enhanced when different chemistries are used for different parts of the molecule. If we confine ourselves to the case of stars, a molecule in which any two arms differ in a deliberate and significant way has been termed a *miktoarm star*, from the Greek word for mixed [4]. A whole host of different structures have been prepared in this manner. For example, an A_2B miktoarm star contains two equal length arms of polymer A and one arm of polymer B. Among the structures that have been reported are A_2B, A_3B, A_2B_2, A_4B_4, and a variety of ABC miktoarm terpolymers. It is even possible to produce asymmetric stars, in which the arms consist of the same polymer but differ in length.

4.5 Cationic Polymerization

Just as anionic polymerization is a chain-growth mechanism that shares important parallels with the free-radical route, so too cationic polymerizations can be discussed within the same framework: initiation, propagation, termination, and transfer. However, there are important differences between anionic and cationic polymerizations that have direct impact on the suitability of the latter for living polymerization. The principal differences between the two ionic routes are the following:

1. A single initiator species is often not sufficient in cationic polymerizations; frequently a second ingredient (or *cocatalyst*) is required.
2. Total dissociation of the cationic initiator is rather rare, which has implications for the ability to start all the chains growing at the same time.
3. Although both ionic mechanisms clearly eliminate termination by direct recombination of growing chains, cationic species are much more prone to transfer reactions than their anionic counterparts. Consequently, living cationic polymerization is much less prevalent than living anionic polymerization.
4. Most monomers that can be readily polymerized by anionic mechanisms are also amenable to free-radical polymerization. Thus, in commercial practice the rather more demanding anionic route is only employed when a higher degree of control is required, for example, in the preparation of styrene–diene block copolymers.
5. In contrast, although most monomers that can be polymerized by cationic mechanisms are also amenable to free-radical polymerization, there are important exceptions. The most significant from a total production point of view is polyisobutylene (butyl rubber), which is produced commercially by (both living and nonliving) cationic polymerization.

Table 4.3 General Summary of Polymerizability of Various Monomer Types by the Indicated Chain-Growth Modes

Monomer	Radical	Anionic	Cationic
Ethylene	✓	(✓)	×
α-Olefins	×	✓	(✓)
1,1-Dialkyl alkenes	×	×	✓
Halogenated alkenes	✓	×	×
1,3-Dienes	✓	✓	✓
Styrenes	✓	✓	✓
Acrylates, methacrylates	✓	✓	×
Acrylonitrile	✓	✓	×
Acrylamide, methacrylamide	✓	✓	×
Vinyl esters	✓	×	✓
Vinyl ethers	×	×	✓
Aldehydes, ketones	×	✓	✓

Note: Parentheses indicate not readily polymerized by this route.

Source: Adapted from Odian, G., *Principles of Polymerization*, 4th ed, Wiley-Interscience, Hoboken, NJ, 2004.

A brief summary of the applicability of the three chain-growth mechanisms—radical, anionic, cationic—to various monomer classes is presented in Table 4.3. In the remainder of this section we describe general aspects of cationic polymerization and introduce some of the transfer reactions that inhibit living polymerization. Then, we conclude by discussing the strategies that have been used to approach a living cationic polymerization.

4.5.1 Aspects of Cationic Polymerization

In cationic polymerization, the active species is the ion formed by the addition of a proton from the initiator system to a monomer (partly for this reason the initiator species is often called a catalyst, because it is not incorporated into the chain). For vinyl monomers the substituents which promote this type of polymerization are electron donating, to stabilize the propagating carbocation; examples include alkyl, 1,1-dialkyl, aryl, and alkoxy. Isobutylene, α-methylstyrene, and vinyl alkyl ethers are examples of monomers commonly polymerized via cationic intermediates.

The initiator systems are generally Lewis acids, such as BF_3, $AlCl_3$, and $TiCl_4$, or protonic acids, such as H_2SO_4, $HClO_4$, and HI. In the case of the Lewis acids, a proton-donating coinitiator (often called a cocatalyst) such as water or methanol is typically used:

$$H_2O + BF_3 \rightleftharpoons H^+ + F_3BOH^-$$

$$H_2O + AlCl_3 \rightleftharpoons H^+ + Cl_3AlOH^-$$

$$CH_3OH + TiCl_4 \rightleftharpoons H^+ + Cl_4TiOCH_3^-$$

(4.N)

With insufficient cocatalyst these equilibria lie too far to the left, while excess cocatalyst can terminate the chain or destroy the catalyst. Thus, the optimum proportion of catalyst and cocatalyst varies with the specific monomer and polymerization solvent. In the case of protonic acids, the concentration of protons depends on the position of the standard acid–base equilibria, but in the chosen organic solvent:

$$H_2SO_4 \rightleftharpoons H^+ + HSO_4^-$$

$$HClO_4 \rightleftharpoons H^+ + ClO_4^-$$

$$HI \rightleftharpoons H^+ + I^-$$

(4.O)

If we write the general formula for the initiator system as H^+B^-, then the initiation and propagation steps for a vinyl monomer $CH_2 = CHR$ can be written as follows. The proton adds to the more electronegative carbon atom in the olefin to initiate chain growth:

$$H_2C = \underset{H}{\overset{R}{<}} + H^+ B^- \longrightarrow Me - \underset{H}{\overset{R}{<}} {}^+ \; B^- \qquad (4.P)$$

The electron-donating character of the R group helps to stabilize this cation. As with anionic polymerization, the separation of the ions and the possibility of ion pairing play important roles in the ease of subsequent monomer insertion. The propagation proceeds in a head-to-tail manner:

$$Me - \underset{H}{\overset{R}{<}} {}^+ B^- + H_2C = \underset{H}{\overset{R}{<}} \longrightarrow \underset{Me \quad H \quad H}{R \qquad R} {}^+ \; B^- \qquad (4.Q)$$

Aldehydes can also be polymerized in this fashion, with the corresponding reactions for formaldehyde being

$$H^+ \; B^- + O = \underset{H}{\overset{H}{<}} \longrightarrow HO - \underset{H}{\overset{H}{<}} {}^+ \; B^- \longrightarrow \qquad (4.R)$$

One of the side reactions that can complicate cationic polymerization is the possibility of the ionic repeat unit undergoing rearrangement during the polymerization. The following example illustrates this situation.

Example 4.2

It has been observed that poly(1,1-dimethyl propane) is the product when 3-methylbutene-1 $(CH_2 = CH - CH(CH_3)_2)$ is polymerized with $AlCl_3$ in ethyl chloride at $-130°C$.[†] Draw structural formulas for the expected and observed repeat units, and propose an explanation.

Solution

The structures expected and found are sketched here:

Expected

Found

The conversion of the cationic intermediate of the monomer to the cation of the product occurs by a *hydride shift* between adjacent carbons:

[†] J.P. Kennedy and R.M. Thomas, *Makromol. Chem.*, 53, 28 (1962).

This is a well-known reaction that is favored by the greater stability of the tertiary compared to the secondary carbocation.

The preceding example illustrates one of the potential complications encountered in cationic polymerization, but it is not in itself an impediment to living polymerization. There are several other potential transfer reactions, however, that collectively do impede a living cationic polymerization. Four of these are the following:

1. *β-Proton transfer.* This is exemplified by the case of polyisobutylene. Protons on carbons adjacent (β) to the carbocation are electropositive, due to a phenomenon known as *hyperconjugation*; we can view this as partial electron delocalization through σ bonds, in contrast to resonance, which is delocalization through π bonds. Consequently there is a tendency for β-protons to react with any base present, such as a vinyl monomer:

$$(4.S)$$

The activated monomer can now participate in propagation reactions, whereas the previous chain is terminated. Note that in isobutylene there are two distinct β-protons, and thus two possible structures for the terminal unsaturation of the chain. There is also a possibility that these double bonds can react subsequently.

2. *Hydride transfer from monomer.* In this case, the transfer proceeds in the opposite direction, but has the same detrimental net effect from the point of view of achieving a living polymerization:

$$(4.T)$$

In the particular case of isobutylene the resulting primary carbocation is less stable than the tertiary one on the chain, so Reaction (4.T) is less of an issue than Reaction (4.S).

3. *Intermolecular hydride transfer.* This is an example of transfer to polymer, and can be written generally as

$$(4.U)$$

4. *Spontaneous termination.* This process, also known as chain transfer to counterion, is essentially a reversal of the initiation step, as a β-proton is transferred back to the anion (e.g., as in Reaction 4.P, but with a growing chain rather than the first monomer).

4.5.2 Living Cationic Polymerization

The preceding discussion provides some insight into the obstacles to achieve a living cationic polymerization. Nevertheless, living cationic polymerization is by now a relatively common tool, and many of the controlled architectures (block copolymers, end-functional chains, regular branched molecules) that we discussed in the context of anionic polymerization have been accessed [7]. In this section, we briefly describe the general strategy behind living cationic polymerization; recall that the essential elements are the absence of termination or transfer reactions:

1. Clearly, the reaction must be conducted in the absence of nucleophilic species that are capable of irreversible termination of the growing chain.
2. Similarly, the reaction should be conducted in the absence of bases that can participate in β-proton transfer. As discussed above, the monomer itself is such as base, and therefore

cationic polymerization always has a "built-in" transfer reaction. The key step, therefore, is to choose reaction conditions to maximize the rate of propagation relative to transfer, given that transfer probably cannot be completely eliminated.

3. Generally, both propagation and transfer are very rapid reactions, with transfer having the higher activation energy. Lower temperatures therefore favor propagation relative to transfer, as well as have the advantage of bringing both reactions under better control.

4. Cationic polymerization, like most chain-growth polymerization, is often highly exothermic. With the additional feature of very rapid reaction, it becomes important to reduce the rate of polymerization in order to remove the excess heat. Low temperature is the first option in this respect, followed by lower concentrations of growing chains.

5. Another way to view control in this context is to aim to extend the lifetime of the growing chain. As a point of reference, a living polystyryl carbanion can persist for years in a sealed reaction vessel; a polyisobutyl carbocation will probably not last for an hour under equivalently pristine conditions. While low temperature certainly aids in increasing the lifetime, another useful strategy is to make the growing center inactive or dormant for a significant fraction of the elapsed reaction time. This is done via the process of reversible termination, as illustrated by the sequence in Reaction (4.V):

$$
\begin{aligned}
HCl + TiCl_4 &\rightleftarrows TiCl_5^- + H^+ \\
H^+ + M &\longrightarrow P_1^+ \\
P_i^+ + M &\longrightarrow P_{i+1}^+ \\
P_i^+ + TiCl_5^- &\rightleftarrows P_iCl + TiCl_4
\end{aligned}
\tag{4.V}
$$

In this sequence, the first reaction generates the initiating proton, and the second and third reactions correspond to standard irreversible initiation and propagation steps involving monomer M. The fourth reaction is the key. The growing cationic i-mer P_i^+ is converted to a dormant, covalent species P_iCl by a reversible reaction. While the growing chain is in this form, it does not undergo transfer or propagation reactions, thereby extending its lifetime. The reversible activation/deactivation reaction must be sufficiently rapid to allow each chain to have many opportunities to add monomer during the polymerization, and the relative length of time spent in the active and dormant states can be controlled by the position of the associated equilibrium. This, in turn, offers many opportunities to tune a particular chemical system. For example, decreasing the polarity of the solvent or adding an inert salt that contains a common ion (chloride in this case) both push the equilibrium toward the dormant state.

We will revisit this idea of a dormant reactive species in the next section on controlled radical polymerization, where it plays the central role. We conclude this section with a specific example of a successful living cationic polymerization scheme. Isobutyl vinyl ether (and other vinyl ethers) can be polymerized by a combination of HI and ZnI_2 [8]. The hydrogen iodide "initiator" adds across the double bond, but forms an essentially unreactive species:

$$\tag{4.W}$$

The carbon–iodide bond is then activated by the relatively weak Lewis acid ZnI_2 to allow insertion of the next monomer. The transition state for propagation may be represented schematically as

$$\tag{4.X}$$

Experimentally, the system exhibits many of the characteristics associated with a living polymerization: polydispersities consistently below 1.1; M_n increasing linearly with conversion; the ability

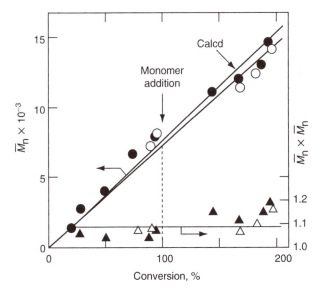

Figure 4.5 Living cationic polymerization of isobutyl vinyl ether in methylene chloride at –40°C. The calculated curve indicates the expected molecular weight assuming 100% initiator (HI) efficiency. After 100% conversion a new charge of monomer was added, demonstrating the ability of the chain ends to resume propagation. (Reproduced from Sawamoto, M., Okamoto, C., and Higashimura, T., *Macromolecules*, 20, 2693, 1987. With permission.)

to resume polymerization after addition of a new charge of monomer. These aspects are illustrated in Figure 4.5. The mechanism implied by Reaction (4.W) and Reaction (4.X) is consistent with the experimental observation that M_n is inversely proportional to the concentration of HI, but independent of the concentration of ZnI_2. On the other hand, the polymerization rate increases with added ZnI_2. ZnI_2 is apparently a sufficiently mild activator that the polymerization is still living at room temperature when conducted in toluene, whereas in the more polar solvent methylene chloride lower temperatures are required.

4.6   Controlled Radical Polymerization

In this section we take up the topic of *controlled radical polymerization*, which represents one of the most active fields in polymer synthesis in recent years. The combination of the general advantages of radical polymerization (a wide range of suitable monomers, tolerance to many functional groups, characteristically rapid reactions, relatively relaxed polymerization conditions) with the unique features of a living polymerization (narrow molecular weight distributions, controlled molecular weights, end functionality, block copolymers, and other complex architectures) has tremendous appeal in many different areas of polymer science. In this section we outline first in general terms how this combination is achieved, and then give some specific examples of the mechanistic details. We choose the term "controlled" rather than "living" in this section, because irreversible termination reactions cannot be rigorously excluded.

4.6.1   General Principles of Controlled Radical Polymerization

The first task is to resolve the apparent paradox: given that radicals can always combine to undergo termination reactions, how do we approximate a living polymerization? To develop the answer, it is helpful to start by summarizing once again the essence of a chain-growth polymerization in terms of initiation, propagation, and termination rates:

$$R_i = k_i[I][M] \tag{4.6.1}$$

$$R_p = k_p[P\bullet][M] \tag{4.6.2}$$

$$R_t = k_t[P\bullet][P\bullet] \tag{4.6.3}$$

where as before [I] is the concentration of initiating species, [M] is the concentration of unreacted monomer, and [P•] is the total concentration of radicals of any length. The key to a living polymerization is that $R_t \rightarrow 0$, or equivalently in practice, that $R_p \gg R_t$. From Table 3.3 and Table 3.4, we can see that typical values of k_t are about four orders of magnitude larger than k_p. Therefore, if we want R_p to be, say, 10^4 times larger than R_t we will need [M] to be 10^8 times larger than [P•]. Given that [M] could be on the order of 1–10 mol L^{-1} (i.e., in bulk or concentrated solution), which means the concentration of radicals, and therefore growing chains, will have to be 10^{-7}–10^{-8} mol L^{-1}. This is quite small, but from the calculations in Section 3.4.3 we know that it is quite feasible.

We can do even better than this, however, by a nifty trick already suggested in the context of living cationic polymerization. Suppose that the absolute concentration of radical forming species is not that small, but that each molecule spends the vast majority of its time in an unreactive, dormant form. This is illustrated schematically below:

$$PX \rightleftharpoons P\bullet + X \tag{4.Y}$$

where PX is the dormant species and X is a group (atom or molecular fragment) that can leave and reattach to the radical rapidly. (Note that radical species are apparently not conserved in the way Reaction (4.Y) is expressed, but as we will see subsequently this is not actually the case. Usually X is also a radical species, but one that is not capable of propagating.) If the equilibrium constant for the activation process, K_{act}, is small, then the instantaneous [P•] will be small even if [PX] is reasonably large:

$$[P\bullet] = K_{act}\frac{[PX]}{[X]} \tag{4.6.4}$$

The process of controlled radical polymerization can now be seen to take place as follows. The dormant species PX spontaneously dissociates into the active radical and the inert partner X. The exposed radical may then undergo propagation steps, or simply recombine with X so that no net reaction takes place. If each radical spends most of its time in the dormant state, the instantaneous concentration of radicals is small, and termination is very unlikely (but never impossible). During an average active period a given radical may add many new monomers, about one new monomer, or essentially no new monomers. It is actually the last situation that is most desirable, because it means that over time all radicals are equally likely to propagate, one monomer at a time. We can understand this concept in the following way. After the polymerization has proceeded for a reasonable time, so that each chain on average has experienced many active periods, the number of active periods per chain will follow the Poisson distribution (Equation 4.2.19). That is because we are randomly distributing a large number of items (active periods) into a smaller number of boxes (growing chains). In the limit where the likelihood of adding a monomer per active period is small, the average number of monomers added per chain will be directly proportional to the number of active periods, and thus follow the Poisson distribution as well. Of course, we are neglecting any termination and transfer reactions.

In contrast, if radicals tend to add monomers in a burst during each active period, the molecular weight distribution will not be as narrow unless the total degree of polymerization involves many such bursts. In fact, the length distribution of the "bursts" will be the most probable distribution, which (recall Equation 2.4.10 and Equation 3.7.19) has a polydispersity approaching 2. We can actually rationalize an approximate expression for the polydispersity of the resulting polymers, based on what we already know. Suppose the mean number of monomers added per active period is $q \geq 1$. The distribution of active periods from chain to chain still follows the Poisson distribution, so it is almost as though we were adding one q-length block per active period. Thus, the polydispersity index becomes that for the Poisson distribution (recall Equation 4.2.20) with a new "effective monomer" of molecular weight qM_0:

$$\frac{M_\text{w}}{M_\text{n}} \approx 1 + \frac{qM_0}{M_\text{n}} \qquad\qquad (4.6.5)$$

Equation 4.6.5 suggests that even if q is 10, a polydispersity of 1.1 is achievable if the total degree of polymerization exceeds 100. A more detailed analysis yields equations similar to Equation 4.6.5, when the average degree of polymerization is sufficiently large [9]. It is worth noting that there are several complications to this analysis, such as the fact that the value of q will actually change during the polymerization, as [M] decreases.

It should be evident from the preceding discussion that termination processes are not rigorously excluded in controlled radical polymerization, only significantly suppressed. Nevertheless, polydispersities $M_\text{w}/M_\text{n} < 1.1$–$1.2$ are routinely obtained by this methodology. It should also be evident that the higher the average chain length, the more likely termination steps become. This can be seen directly from Equation 4.6.2 and Equation 4.6.3; as time progresses, R_t remains essentially constant, whereas R_p decreases because [M] decreases with time. Consequently, the relative likelihood of a termination event increases steadily as the reaction progresses. In fact, there is really a three-way competition in designing a controlled radical polymerization scheme, among average molecular weight, polydispersity, and "efficiency," where we use efficiency to denote a combination of practical issues. For example, the higher the desired molecular weight, the broader the distribution will become, due to termination reactions. This could be mitigated to some extent by running at even higher dilution, but this costs time and generates large volumes of solvent waste. Or, the reaction vessel could be replenished with monomer, to keep [M] high even as the reaction progresses, but this wastes monomer, or at least necessitates a recovery process.

4.6.2 Particular Realizations of Controlled Radical Polymerization

A rich variety of systems that fall under the umbrella of Reaction (4.Y) have been reported. Three general schemes have so far emerged as the most prevalent, although there is no a priori reason why others may not become more popular in the years ahead. Each has particular advantages and disadvantages relative to the others, but for the purposes of this discussion we are really only interested in their evident success. All three have been the subject of extensive review articles; see for example Refs. [10,11,12].

4.6.2.1 Atom Transfer Radical Polymerization (ATRP)

In this approach the leaving group X in Reaction (4.Y) is a halide, such as a chloride or bromide, and it is extracted by a suitable metal, such as copper, nickel, iron, or ruthenium. The metal is chelated by ligands such as bipyridines, amines, and trialklyphosphines that can stabilize the metal in different oxidation states. A particular example of the activation/deactivation equilibrium using copper bromide/2,2′-bipyridine (bipy) can thus be written:

$$P_i\text{Br} + \text{CuBr(bipy)}_2 \rightleftharpoons P_i\bullet + \text{CuBr}_2\text{(bipy)}_2 \qquad\qquad (4.\text{Z})$$

where the copper atom is oxidized from Cu(I)Br to Cu(II)Br$_2$. Reaction (4.Z) suggests that the polymerization could be initiated by the appropriate halide of the monomer in question, such as 1-phenylethyl bromide when styrene is the monomer. Alternatively, a standard free-radical initiator such as AIBN could be employed (recall Section 3.3). A particularly appealing aspect of ATRP is the wide variety of monomers that are amenable to this approach: styrene and substituted styrenes, acrylates and methacrylates, and other vinyl monomers. Dienes and amine or carboxylic acid–containing monomers are more challenging.

The following example illustrates some of the quantitative aspects of ATRP of styrene.

Example 4.3

From a linear plot of ln $([M]_0/[M])$ versus time, it has been reported that apparent propagation rate constant for the ATRP of styrene in bulk is on the order of 10^{-4} s^{-1}, where the apparent rate constant k_p^{app} is defined by $R_p = k_p^{app}[M]$.[†] What is the order of magnitude of the concentration of active radicals at any time?

Solution

From Equation 4.6.2, we can see that k_p^{app} thus defined is actually equal to k_p [P•]. From Table 3.4 in Chapter 3 we know that a typical value for k_p for free-radical polymerization is 10^2–10^3 L mol^{-1} s^{-1}, and on this basis direct substitution tells us that [P•] is about 10^{-4} s^{-1}/10^{2-3} L mol^{-1} s^{-1} = 10^{-6}–10^{-7} mol L^{-1}. This is in line with the estimate given in the previous section, of the target concentration of active radicals needed to make the rate of termination small with respect to the rate of propagation.

4.6.2.2   Stable Free-Radical Polymerization (SFRP)

In this variant, the leaving group X in Reaction (4.Y) is a free-radical, but a sufficiently stable one that it does not initiate polymerization. The prime example of this class is the nitroxide radical, usually embedded in the (2,2,6,6-tetramethylpiperidinyloxy) "TEMPO" group [13]. When attached to a monomer analog or a growing polymer chain terminus through the alkoxyamine C–ON bond, homolytic cleavage of the C–O bond produces the stable TEMPO radical and an active radical species. This reaction is illustrated below for the case of styrene:

(4.AA)

The adduct of styrene and TEMPO on the left-hand side of Reaction (4.AA) can be prepared rather readily, purified, and stored indefinitely. In contrast to other controlled radical polymerization schemes, this approach is based on a single initiating species; no cocatalyst or transfer agent is needed. Even in the presence of a large excess of styrene monomer, it is not until the system is brought to an elevated temperature such as 125°C that polymerization proceeds directly. The reaction can be run under nitrogen, and the rigorous purification necessary for living ionic polymerizations is not required. Molecular weights well in excess of 10^5, with polydispersities in the range of 1.1–1.2, have been achieved. The range of accessible monomers is so far more restricted than with ATRP or reversible addition-fragmentation transfer (RAFT), with styrene, acrylate, and methacrylate derivatives being the monomers of choice. However, the polymerization is relatively tolerant of functional groups, and many functionalized initiators with TEMPO adducts have been designed. This makes SFRP an appealing alternative to living ionic polymerization for the production of end-functional polymers (recall Section 4.4), and by extension block copolymers and branched architectures, once the initiator is available.

[†] K. Matyjaszewski, T.E. Patten, and J. Xia, *J. Am. Chem. Soc.*, 119, 674 (1997).

Example 4.4

An interesting question arises upon examination of Reaction (4.AA): does each TEMPO radical remain associated with the same chain during the polymerization, or does it migrate freely through the reaction medium? In the case of anionic polymerization in a nonpolar solvent, the counterion is certainly closely associated with the chain end, due to the requirement of electrical neutrality. In the case, of conventional free-radical polymerization, we considered the "caging effect" that can severely limit the efficiency of an initiator (see Section 3.3). In this case, the relatively high temperature should enhance both the mobility of the individual species and the ability to escape from whatever attractive interaction would hold the two radical species in proximity. How could one test this intuition experimentally?

Solution

The unimolecular nature of the TEMPO-based initiator, plus its susceptibility to functionalization, offers a convenient solution, as has been demonstrated.[†] These authors prepared the styrene–TEMPO adduct shown in Reaction (4.AA), plus a dihydroxy-functionalized variant:

A 1:1 mixture of the two initiators was added to styrene monomer and heated to the polymerization temperature. At various times, the reaction mixture could be cooled, and analyzed. If the exchange of TEMPO groups was rapid, then one would expect four distinct chain populations, with roughly equal proportions: one with no hydroxyls, one with a hydroxyl at each end, one with a hydroxyl at the terminus, and one with a hydroxyl at the initial monomer. On the other hand, if there was little exchange, there should be just two populations: one with no hydroxyls and one with two. Liquid chromatographic analysis gave results that were fully consistent with the former scenario.

4.6.2.3 Reversible Addition-Fragmentation Transfer (RAFT) Polymerization

The principal distinction between RAFT polymerization on the one hand and ATRP or SFRP on the other is that RAFT polymerization involves a *reversible chain transfer*, whereas the other two involve *reversible chain termination*. The key player in the RAFT process is the chain transfer agent itself; the radicals are generally provided by conventional free-radical initiators such as AIBN. Dithioesters (RCSSR′) such as cumyl dithiobenzoate are often used; in this instance R is a phenyl ring and R′ is a cumyl group. The growing radical chain $P_i\cdot$ reacts with the transfer agent, and the cumyl group departs with the radical:

$$\text{(4.BB)}$$

A different growing radical $P_j\cdot$ can also react in the analogous manner:

$$\text{(4.CC)}$$

[†] C.J. Hawker, G.G. Barclay, and J. Dao, *J. Am. Chem. Soc.*, 118, 11467 (1996).

In this way, the dithioester end group is transferred from chain to chain. An important feature of this reaction scheme, which is important for achieving narrow molecular weight distributions, is that the ease of transfer of the dithioester is essentially independent of the length of the associated chain, or between chains and the R′ group.

There is an important feature of this scheme that is different from the other two controlled radical approaches. Namely, the number of chains is not determined by the number of initiators, but by the combination of conventional initiators (e.g., AIBN) and those from the RAFT agent, e.g., cumyl radicals (R′• in Reaction 4.BB). In fact, given that the decomposition of AIBN cannot be controlled, it is advantageous to use an excess of the RAFT agent, thereby dictating the number of chains initiated via R′•, which in turn is proportional to the concentration of RCSSR′. This process is facilitated by the fact that such dithioesters have very large chain transfer constants (recall Section 3.8), and thus a chain initiated by AIBN or by R′• is rapidly transformed into a dormant form, before achieving a significant degree of polymerization. The RAFT approach has been successful with a very wide variety of different monomers.

4.7 Polymerization Equilibrium

Up to this point we have tended to write chain-growth propagation steps as one-way reactions, with a single arrow pointing to the product:

$$P_i^* + M \xrightarrow{k_p} P_{i+1}^*$$

In fact, as a chemical reaction, there must be a reverse depropagation or *depolymerization* step, and the possibility of chemical equilibrium:

$$P_i^* + M \xleftrightarrow{K_{poly}} P_{i+1}^*$$

This equilibrium constant for polymerization, K_{poly}, can be written as the ratio of the forward and reverse rate constants, and as the appropriate ratio of species concentrations *at equilibrium*:

$$K_{poly} = \frac{k_p}{k_{dep}} = \frac{[P_{i+1}^*]}{[P_i^*][M]} \approx \frac{1}{[M]_{eq}} \tag{4.7.1}$$

The last term indicates that the equilibrium constant is the inverse of the equilibrium monomer concentration, because the concentrations of i-mer and $(i+1)$-mer must be nearly equal (recall Equation 3.7.3). The reason we have not emphasized the possibility of equilibrium so far is that almost all polymerization reactions are run under conditions where the equilibrium lies far to the right, in favor of products; the residual monomer concentration is very small. This is not always the case, however, as we shall now discuss.

The state of equilibrium is directly related to the Gibbs free energy of polymerization:

$$\Delta G_{poly} = \Delta G_{poly}^0 + RT \ln Q_{poly} \tag{4.7.2}$$

where the *reaction quotient*, Q_{poly}, is the same ratio of product and reactant concentrations as K, but not necessarily at equilibrium. The free energy of polymerization is the difference between the free energies of the products and the reactants, in kJ/mol, where for polymeric species we consider moles of repeat units. The superscript 0 indicates the standard quantity, where all species are at some specified standard state (e.g., pure monomer and repeat unit, or perhaps at 1 mol L^{-1} in solution). For the polymerization reaction to proceed spontaneously, $\Delta G_{poly} < 0$. When the reaction is allowed to come to equilibrium, $Q_{poly} = K_{poly}$ and $\Delta G_{poly} = 0$. Thus, we have the well-known relation

$$\Delta G^0_{poly} = -RT \ln K_{poly} \tag{4.7.3}$$

The free energy change per repeat unit upon polymerization may be further resolved into enthalpic (H) and entropic (S) contributions:

$$\Delta G^0_{poly} = \Delta H^0_{poly} - T\Delta S^0_{poly} \tag{4.7.4}$$

From Equation 4.7.3, we can see that the statement that K_{poly} is large, favoring products, is equivalent to saying that ΔG^0_{poly} is large and negative. From Equation 4.7.4, we can see that facile polymerization requires that either ΔH^0_{poly} is large and negative, that is, the reaction is exothermic, or that ΔS^0_{poly} is large and positive. In fact, ΔS^0_{poly} is usually negative; the monomers lose translational entropy when bonded together in a polymer. However, ΔH^0_{poly} is exothermic, because the extra energy of a carbon–carbon double bond relative to a single bond is released. In fact, we should have anticipated this conclusion from the outset: polymers could not be made inexpensively in large quantities if we had to put in energy for each propagation step.

Table 4.4 provides examples of the standard enthalpy and entropy of polymerization for a few common vinyl monomers. In all cases both the enthalpy and the entropy changes are negative, as expected; furthermore, ΔG^0_{poly} is negative at room temperature (300 K). Starting with ethylene as the reference, the relative enthalpies of polymerization can be understood in terms of two general effects. The first is the possibility of resonance stabilization of the double bond in the monomer that is lost upon polymerization. This results in lower exothermicity for butadiene, isoprene, styrene, and α-methylstyrene, for example. The second is steric hindrance in the resulting polymer. For example, disubstituted carbons in the polymer can lead to significant interactions between substituents on every other carbon, which therefore destabilize the polymer, as in the case of isobutylene, α-methylstyrene, and methyl methacrylate. Tetrafluoroethylene, with its unusually large exothermicity, is included in this short table in part to remind us that there are examples where we may not have a simple explanation. Further discussion of these issues is provided in Section 5.4.

Equation 4.7.4 indicates that as the polymerization temperature increases, the relative importance of entropy increases as well. As ΔS favors depolymerization, it is possible to reach a

Table 4.4 Values of the Standard Enthalpy and Entropy of Polymerization, as Reported in Odian [6]

Monomer	ΔH^0_{poly} (kJ mol^{-1})	ΔS^0_{poly} (J K^{-1} mol^{-1})	ΔG^0_{poly} at 300 K (kJ mol^{-1})
Ethylene	−93	−155	−47
Propylene	−84	−116	−49
Isobutylene	−48	−121	−12
1,3-Butadiene	−73	−89	−46
Isoprene	−75	−101	−45
Styrene	−73	−104	−42
α-Methylstyrene	−35	−110	−2
Tetrafluoroethylene	−163	−112	−130
Vinyl acetate	−88	−110	−55
Methyl methacrylate	−56	−117	−21

Note: The enthalpy corresponds to the conversion of liquid monomer (gas in the case of ethylene) to amorphous (or slightly crystalline) polymer. The entropy corresponds to conversion of a 1 mol L^{-1} solution of monomer to polymer.

temperature above which polymerization will not be spontaneous under standard conditions. This special temperature is referred to as the *ceiling temperature*, T_c. From Equation 4.7.4, we have

$$\Delta G^0_{poly} = 0 = \Delta H^0_{poly} - T_c \Delta S^0_{poly} \qquad (4.7.5)$$

and combining this relation with Equation 4.7.1 through Equation 4.7.3 we find

$$T_c = \frac{\Delta H^0_{poly}}{\Delta S^0_{poly} + R \ln [M]_{eq}} \qquad (4.7.6)$$

Using this relation, the data in Table 4.4, and assuming $[M] = 1$ mol L^{-1}, the ceiling temperature is 45°C for poly(α-methylstyrene) and 206°C for poly(methyl methacrylate). Note the important fact that according to Equation 4.7.6, T_c will depend on the monomer concentration and will therefore be different for a polymerization in dilute solution compared to one in bulk monomer.

Figure 4.6 illustrates the application of Equation 4.7.6 to poly(α-methylstyrene). The equilibrium monomer concentration is plotted against temperature according to Equation 4.7.6 and the indicated values of ΔH^0 and ΔS^0. The smooth curve corresponds to the ceiling temperature and the horizontal line indicates a 1 mol L^{-1} solution. Any solution of monomer that falls to the right of, or below, the curve (i.e., any combination of [M] and T) will simply not polymerize. Any solution to the left of, or above, the curve can polymerize, but only until the equilibrium monomer concentration is reached. For example, a 1 mol L^{-1} solution at 0°C could polymerize, but as the equilibrium monomer concentration is about 0.1 mol L^{-1}, the maximum conversion would only be 90%.

Interestingly, there are a few instances in which polymerization is driven by an *increase* in entropy, and where the enthalpy gain is almost negligible. Examples include the polymerization of cyclic oligomers of dimethylsiloxane, such as the cyclic trimer and tetramer, which we will discuss in the next section. In this case, the bonds that are broken and reformed are essentially the same,

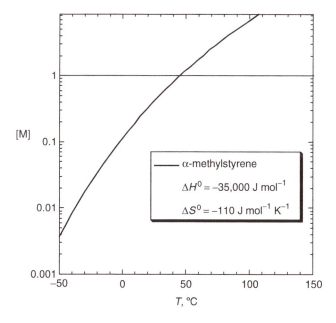

Figure 4.6 Illustration of the relation between equilibrium monomer concentration and ceiling temperature for poly(α-methylstyrene).

hence $\Delta H_{poly}^0 \approx 0$. On the other hand, possibly because of the greater conformational freedom in the linear polymer versus the small cycles, ΔS_{poly}^0 is positive. In such a case when ΔH_{poly}^0 is slightly positive, it is possible in principle to have a *floor temperature*, below which polymerization cannot occur at equilibrium. Furthermore, for the living polymerization of "D_3," the cyclic trimer of dimethylsiloxane, for example using an anionic initiator, ΔG_{poly} is never particularly favorable compared to the typical vinyl monomer case. Consequently, one has to be aware of the law of mass action, just as in a polycondensation. In other words, the reaction cannot be allowed to go to completion, because rather than achieving essentially 100% conversion to polymer, the reaction mixture stabilizes at the equilibrium monomer concentration. Then, random propagation and depropagation steps will degrade the narrow molecular weight distribution that was initially sought. This problem can be circumvented by adding more monomer than is necessary to achieve the target molecular weight, and using trial and error to determine the time (and fractional conversion) at which the desired average molecular weight has been achieved.

4.8   Ring-Opening Polymerization (ROP)

4.8.1   General Aspects

Cyclic molecules, in which the ring contains a modest number of atoms, say 3–8, can often be polymerized by a ring-opening reaction, in which a particular bond in the cycle is ruptured, and then reformed between two different monomers in a linear sequence. This process is illustrated in the following schematic reaction:

$$\sim\!\!B^* + \underset{A-B}{\boxed{}} \longrightarrow \sim\!\!B - A \qquad B^* \tag{4.DD}$$

In this instance the ring contains four atoms, and the A–B bond is the one that is preferentially cleaved. The propagating chain is shown with B containing the active center; it is often the case that ROP proceeds by an ionic mechanism. Comparison of monomer and repeat unit structures in Reaction (4.DD) reveals that the bonding sequence is the same in both cases, in marked contrast to either a chain-growth polymerization through, for example, a carbon–carbon double bond or a step-growth polymerization through, for example, condensation of acid and alcohol groups. In light of the previous section, where we considered the thermodynamics of polymerization, a basic question immediately arises: if the bonding is the same in monomer and polymer, what is the primary driving force for polymerization? The answer is *ring strain*. The linkage of the atoms into a ring generally enforces distortion of the preferred bond angles and even bond lengths, effects that are grouped together under the title ring strain. The amount of ring strain is a strong function of the number of atoms in the ring, r. For example, ethylene oxide, with $r = 3$, is quite a explosive gas at room temperature. On the other hand, cyclohexane, with $r = 6$, is almost inert. In fact, $r = 6$ represents a special case, at least for all carbon rings, as the natural sp^3 bond angles and lengths can be almost perfectly matched. The following example illustrates the effect of r on the thermodynamics of polymerization, for cyclic alkanes.

Example 4.5

The following values of ΔH_{lc}^0 and ΔS_{lc}^0 per methylene unit have been estimated at room temperature (as reported in Ref. [10]) for the process

$$-(CH_2)_r - \text{(liquid ring)} \quad \Leftrightarrow \quad -(CH_2)_n - \text{(crystalline linear polymer)}$$

The subscript "lc" denotes the liquid-to-crystal aspect of the process; as we will see in Chapter 13, high molecular weight linear polyethylene, the product of the hypothetical polymerization reaction, is crystalline at equilibrium at room temperature. Evaluate ΔG_{lc}^0 and K for this process and interpret the results:

r	ΔH_{lc}^0 (kJ mol^{-1})	ΔS_{lc}^0 (J mol^{-1} K^{-1})
3	−113	−69.1
4	−105	−55.3
5	−21.2	−42.7
6	+2.9	−10.5
7	−21.8	−15.9
8	−34.8	−3.3

Solution

We use the relationships $\Delta G^0 = \Delta H^0 - T\Delta S^0$ (Equation 4.7.4) and $\Delta G^0 = -RT \ln K$ (Equation 4.7.3), with $T = 298$ K and $R = 8.314$ J mol^{-1} K^{-1}, to obtain the following table:

r	ΔG_{lc}^0 (kJ mol^{-1})	K
3	−92	2×10^{16}
4	−89	3×10^{15}
5	−8.5	30
6	+6.0	0.09
7	−17	10^3
8	−34	8×10^5

The results indicate that for all but $r = 6$, polymerization is favored, consistent with the known stability of six-membered carbon rings. From the point of view of polymerization, the driving force should be ranked according to $r = 3$, $4 > r = 8 > r = 5$, 7. These are general trends, and different substituents or heteroatoms within the ring can change the numerical values significantly. Finally, while ΔH^0 shows a distinct maximum at $r = 6$, ΔS^0 decreases monotonically with r, while remaining consistently negative. However, we need to recognize that these values incorporate the relief of the ring strain, the incorporation of monomer into polymer, and the changes associated with crystallization of the liquid polymer. The opening of the ring affords more degrees of freedom to the molecule, increasing the entropy, but both subsequent polymerization and crystallization reduce it. Consequently, it is dangerous to overinterpret particular values of ΔS^0.

The preceding example nicely illustrates the importance of ring strain, but the fact is that the primary utility of ROP is not to produce polyethylene from cyclic alkanes. Rather, it is to produce interesting polymer structures from readily accessible cyclic monomers, structures that cannot be prepared more conveniently by classical step-growth or chain-growth polymerization. Examples of seven different classes of cyclic monomers and the resulting polymer structures are given in Table 4.5. In all cases the ring contains one or more heteroatoms, such as O, N, and Si. These participate in the bond-breaking process that is essential to ROP; in contrast, it is actually rather difficult in practice to polymerize cyclic alkanes, even when free energy considerations favor it. In Chapter 1 we suggested that the presence of a heteroatom in the backbone was often characteristic of a step-growth polymerization. One of the beauties of ROP is that it is a chain-growth mechanism, enabling the ready preparation of high molecular weight materials. For example, Entry 5 (a polyamide, poly(ε-caprolactam)) and Entry 6 (a polyester, polylactide) are polymers that could be prepared by condensation of the appropriate AB monomer. However, by using the cyclic monomer, the condensation step has already taken place, and the small molecule byproduct is

Table 4.5 Examples of Monomers Amenable
to Ring-Opening Polymerization

Monomer class	Example	Repeat unit
1. Epoxides	(epoxide structure)	(repeat unit with O)
2. Cyclic ethers	(tetrahydrofuran structure with O)	$\{CH_2)_4 O\}$
3. Cyclic acetals	(cyclic acetal, $()_5$)	$\{O \sim O (CH_2)\}_5$
4. Imines (cyclic amines)	(aziridine, N–H)	$\{\sim N(H)\}$
5. Lactams (cyclic amides)	(caprolactam, NH, =O)	$\{(CH_2)_5 N(H) C(=O)\}$
6. Lactones (cyclic esters)	(lactide structure, O, =O)	$\{O \overset{Me}{-} C(=O)\}$
7. Siloxanes	(cyclic siloxane, Si–O–Si)	$\{Si(Me)(Me) O\}$

removed. Thus, the law of mass action that typically limits the molecular weight of a step-growth polymerization is overcome. (This is not to say that polymerization equilibrium will not be an issue. In fact, Entry 6 and Entry 7 both indicate six-membered rings, and the correspondingly lower ring-strain does bring equilibration into play.)

From the perspective of this chapter the main point of ROP is not just its chain-growth character, but the fact that in many cases living ROP systems have been designed. As indicated above, most ROPs proceed via an ionic mechanism, which certainly invites attempts to achieve a living polymerization. We will briefly present specific examples of such systems, for three disparate but rather interesting and important polymers in Table 4.5: poly(ethylene oxide) (Entry 1), polylactide (Entry 6), and poly(dimethylsiloxane) (Entry 7). We will also consider a class of ROP that can produce all carbon backbones, via olefin metathesis.

4.8.2 Specific Examples of Living Ring-Opening Polymerizations

4.8.2.1 Poly(ethylene oxide)

Poly(ethylene oxide) represents one of the most versatile polymer structures for both fundamental studies and in commercial applications. It is readily prepared by living anionic polymerization, with molecular weights ranging all the way up to several millions. It is water soluble, a highly desirable yet relatively unusual characteristic of nonionic, controlled molecular weight polymers. Furthermore, it appears to be more or less benign in humans, thereby allowing its use in many consumer products, biomedical formulations, etc. In fact, a grafted layer of short chain poly(ethylene oxide)s (see Section 4.4) can confer long-term stability against protein adsorption or deposition of other biomacromolecules. In the biochemical arena short chain poly(ethylene oxide)s (with hydroxyl groups at both ends) are more commonly referred to as poly(ethylene glycol)s or PEGs. The grafting of PEG molecules onto a biomacromolecule or other substrate has become such a useful procedure that it has earned a special name: *PEGylation*. Poly(ethylene oxide) crystallizes rather readily, with a typical melting temperature near 65°C, which has led to its use in many fundamental studies of polymer crystallization (see Chapter 13). The first block copolymers to become commercially available were the so-called *polyoxamers*, diblocks and triblocks of poly(ethylene oxide) and poly(propylene oxide). Historically, it was the polymerization of ethylene oxide that Flory used as an example in proposing the Poisson distribution for chain-growth polymers prepared from a fixed number of initiators with rapid propagation [14].

In practice, the living anionic polymerization of ethylene oxide has been achieved by a variety of initiator systems, following the general principles laid out in Section 4.3. Examples include metal hydroxides, alkoxides, alkyls, and aryls. In contrast to styrenes and dienes, however, lithium is not an effective counter ion in this case. (As noted in the context of Reaction 4.H, this feature is actually convenient when it is desired to use ethylene oxide to end-functionalize such polymers.) The following scheme represents an example of a three-arm star prepared by grafting from; the initiator is trimethylol propane, which has three equivalent primary alcohols that can be activated by diphenylmethyl potassium. The addition of ethylene oxide monomer is straightforward, and the resulting polymer is terminated with acidic methanol to yield a terminal hydroxyl-functionality on each arm.

$$(4.EE)$$

An interesting feature of this reaction is that it can be carried out in THF, which like ethylene oxide is a cyclic ether. This again illustrates the importance of ring-strain in facilitating, or, in this case suppressing, polymerization. In fact, cyclic ethers including THF are usually polymerized only by a cationic ring-opening mechanism; the high ring-strain of ethylene oxide makes it the exception to this rule.

4.8.2.2 Polylactide

Polylactide is biodegradable and can be derived from biorenewable feedstocks such as corn. The former feature enables a long-standing application as resorbable sutures; after a period of days to weeks, the suture degrades and is metabolized by the body. The latter property underscores recent interest in the large-scale commercial production of polylactide for a wide variety of thermoplastic applications.

The structure of lactic acid is

Clearly, as it contains both a hydroxyl and a carboxylic acid, it could be polymerized directly to the corresponding polyester via condensation. However, if the starting material is lactide, the cyclic dimer of the corresponding ester, then ROP produces the same polymer structure but by a chain-growth process:

$$\text{(structure)} \qquad \text{(4.FF)}$$

Note that the central carbon in lactic acid is chiral, and therefore the corresponding two carbons in both lactide and the polylactide repeat unit are stereocenters. Consequently, there are three possibilities for the lactide monomer, according to the absolute configuration of these carbons: R,R; S,S; and R,S. Polymerization of either of the first two leads to the corresponding stereochemically pure PDLA and PLLA, which are crystallizable; the meso dyad leads to an atactic polymer. In this case, therefore, the responsibility of producing a particular tacticity is largely transferred from the catalyst (see Chapter 5) to the purification of the starting material. Of course, the catalyst has to guide the polymerization in such a way that the stereochemistry is not scrambled or epimerized.

From the point of view of designing a living polymerization of polylactide, there are two general issues to confront. First, as polylactide is a polyester, it is susceptible to transesterification reactions. This constraint favors lower reaction temperatures, conditions that are neither too basic nor acidic, and acts against the normal desire to make the catalyst more "active." Ironically, the advantageous degradability of polylactide through the hydrolysis of the ester linkage is thus a disadvantage from the point of view of molecular weight control. The second problem is equilibration. Recall from Example 4.5 that for cyclic alkanes, the six-membered ring has no ring-strain to speak of, and is therefore not polymerizable. Although lactide is a six-membered ring, it does possess sufficient ring-strain, but not a lot. Consequently, narrow molecular weight distributions are usually obtained only by terminating well before the reaction has approached completion.

A typical catalyst is based on a metal alkoxide, such as LMOR', where L is a "spectator" ligand and R' is a small alkyl group. The initiation step can be written as

$$\text{(structure)} + \text{LMOR'} \rightleftharpoons \text{(structure)} \qquad \text{(4.GG)}$$

where the lactide ring is cleaved at the bond between the oxygen and the carbonyl carbon. The subsequent propagation steps involve the same bond cleavage, with addition of the new monomer into the oxygen–metal bond at the growing chain end. In fact, this kind of polymerization has been classified as "anionic coordination," in distinction to anionic polymerization, as the crucial step is coordination of the metal with the carbonyl oxygen, followed by insertion of the alkoxide into the polarized C–O bond. The most commonly employed catalyst for polylactide is tin ethylhexanoate, but more success in terms of achieving living conditions has been realized with aluminum alkoxides. Interestingly, these aluminum species have a tendency to aggregate in solution, with the result that the reaction kinetics can become rather complicated; different aggregation states can exhibit very different propagation rates. This situation is reminiscent of the aggregation of carbanions in anionic polymerization in nonpolar solvents discussed in Section 4.3.

4.8.2.3 Poly(dimethylsiloxane)

Poly(dimethylsiloxane) has one of the lowest glass transition temperatures (T_g, see Chapter 12) of all common polymers. This is due in part to the great flexibility of the backbone structure (see Chapter 6), which reflects the longer Si–O bond compared to the C–C bond, the larger bond angle, and the absence of substituents on every other backbone atom. It is also chemically quite robust. It is used in any number of lubrication and adhesive applications ("silicones"), as well as a variety of rubber materials. Living anionic polymerization of the cyclic trimer hexamethylcyclotrisiloxane, D_3, has been achieved by a number of routes. Given that this monomer is a six-membered ring, we can anticipate that the polymerization is not strongly favored by thermodynamics. Consequently, narrow molecular weight distributions are achieved by terminating the reaction well before the consumption of all the monomer.

An example of a successful protocol is the following. A modest amount of cyclic trimer is initiated with potassium alkoxide, in cyclohexane solution. Under these conditions initiation is rather slow, but propagation is almost nonexistent, thereby allowing for complete initiation. Presumably, the lack of propagation is due to ion clustering as discussed in Section 4.3. The addition of THF, as a polar modifier, plus more monomer allows propagation to proceed for an empirically determined time interval. Termination is achieved with trimethylchlorosilane (TMSCl):

$$\text{(4.HH)}$$

4.8.2.4   Ring-Opening Metathesis Polymerization (ROMP)

Reactions in this class of ROP are distinct from those previously considered, both in the fact that the mechanism does not involve ionic intermediates and in the creation of all carbon backbones (albeit ones that contain double bonds). An *olefin metathesis* reaction is one in which two carbon–carbon double bonds are removed and two new ones are created. Generically, this can be represented by the following scheme, whereby $R_1HC = CHR_2$ reacts with $R_3HC = CHR_4$ to produce, for example, $R_1HC = CHR_3$ and $R_2HC = CHR_4$.

$$\text{(4.II)}$$

Although Reaction (4.II) illustrates the net outcome, it says nothing about the mechanism. Metathesis reactions are catalyzed by transition metal centers, with the associated ligand package providing tunability of reaction characteristics such as rate, selectivity, and stereochemistry of addition. In the case of ROMP, the metal forms a double bond with a carbon at one end of the chain; thus in Reaction (4.II) the R_1CH group would be replaced by the metal and its ligands (ML_n). For a propagation step, R_2 would denote a previously polymerized chain, P_i. In the monomer to be inserted into the chain, between the metal and the end of the previously polymerized chain, R_3 and R_4 are covalently linked to form a ring. Thus, the ring plays two key roles: the ring-strain provides the driving force for polymerization, and the ring structure provides the permanent connectivity between the two carbon atoms whose double bond is broken. The ROMP analog to Reaction (4.II) can thus be described schematically as

$$\text{(4.JJ)}$$

In this case, the monomer is cyclooctene. After the monomer insertion or propagation step, an active metal carbene remains at the chain terminus, and one carbon–carbon double bond remains in the backbone for each repeat unit.

There is a large literature on metathesis catalysts and associated mechanisms, many of which incorporate multiple components beyond the active metal. However, for the purposes of controlled ROMP, there are currently two families of single species catalysts that are highly successful. One, based on tungsten or molybdenum, is known as a Schrock catalyst [I], and the other, based on ruthenium, is a Grubbs catalyst [II]. These investigators were corecipients of the 2005 Nobel Prize in chemistry (with Y. Chauvin) for their work on the metathesis reaction. The structures of representative examples are given below, where the symbol Cy denotes a cyclohexyl ring:

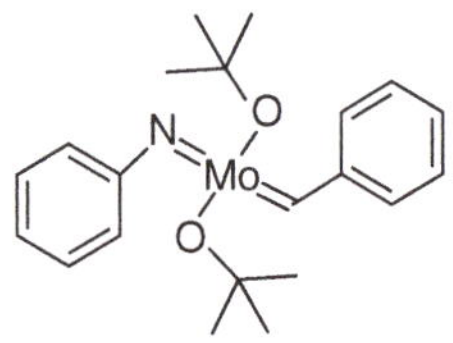

[I]

Schrock catalyst

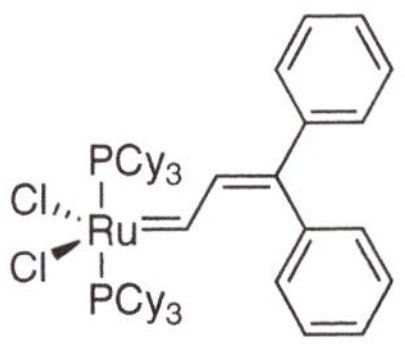

[II]

Grubbs catalyst

Note that the substituent on the metal–carbon double bond will become attached to the nonpropagating terminus of the chain. Collectively, catalysts in these two families have proven capable of achieving controlled polymerization of a wide variety of cyclic olefins, including those containing functional groups. In particular, while the Schrock catalysts tend to be more active, the Grubbs catalysts are more tolerant of functional groups, oxygen, and protic solvents. Reaction conditions are often mild, that is, near room temperature, and in some cases the polymerization can be conducted in water. Overall control is often quite good, and many block copolymers have been prepared by ROMP. Some ROMP systems have even been commercialized, including the polymerization of norbornene:

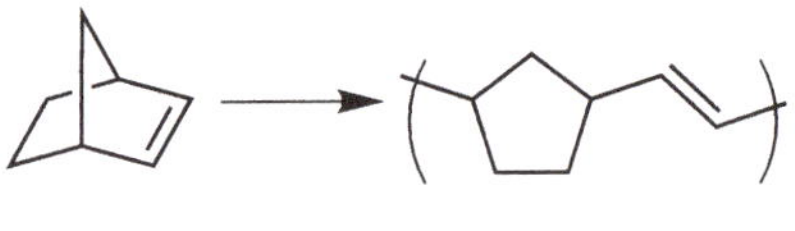

(4.KK)

4.9 Dendrimers

Dendrimers are an interesting, unique class of polymers with controlled structures. For example, they can have precisely defined molecular weights, even though the elementary addition steps are usually of the condensation variety. From an application point of view it is the structure of the dendrimer, rather than its molecular weight per se that is the source of its appeal. A cartoon example of a dendrimer is provided in Figure 1.2. The term comes from the Greek word *dendron*, or tree, and indeed a dendrimer is a highly branched polymer molecule. In particular, a dendrimer is usually an approximately spherical molecule with a radius of a few nanometers. Thus, a dendrimer is both a covalently assembled molecule and also a well-defined nanoparticle. The outer surface of the dendrimer is covered with a high density of functional groups that govern the interactions between the dendrimer and its environment. These exterior groups have the advantages of being numerous, and readily accessible for chemical transformation. The interior of the dendrimer can incorporate a distinct functionality that can endow the molecule with desirable properties. For example, the dendrimer might incorporate a highly absorbing group, for "light harvesting," or a fluorophore, for efficient emission. Other possibilities include catalytic centers or electrochemically active groups. By being housed within the dendrimer, this functional unit can be protected from unwanted interactions with the environment. The functional unit may be covalently bound within the dendrimer, or it may simply be encapsulated. The possibility of controlling uptake and release of specific agents by the dendrimer core also makes them appealing as possible delivery vehicles for pharmaceuticals or other therapeutic agents. As nanoparticles, dendrimers share certain attributes with other objects of similar size, such as globular proteins, surfactant and block copolymer micelles, hyperbranched polymers, and colloidal nanoparticles. Although beyond the scope of this chapter, it is interesting to speculate on the possible advantages and disadvantages of these various structures (see Problem 16).

There are two distinct, primary synthetic routes to prepare a dendrimer, termed *divergent* and *convergent*. In a divergent approach, the dendrimer is built-up by successive additions of monomers to a central, branched core unit, whereas in the convergent approach branched structures called *dendrons* are built-up separately, and then ultimately linked together to form the dendrimer in a final step. The divergent approach was conceived first, and is the more easily visualized. The process is illustrated schematically in Figure 4.7. The core molecule in this case has three functional groups denoted by the open circles. These are reacted with three equivalents of another three-functional monomer, but in this case two of the functional groups are protected (filled circles). After this reaction is complete, the growing molecule has six functional groups that are

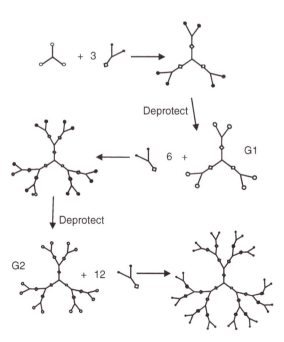

Figure 4.7 Schematic illustration of the divergent synthesis of a third-generation dendrimer from a trifunctional core. The open circles denote reactive groups and the filled circles protected groups.

then deprotected. At this stage the molecule is termed a *first generation* (G1) dendrimer. Another addition reaction is then performed, but now six equivalents of the protected monomer are required to complete the next generation. After deprotection the resulting G2 dendrimer has 12 functional groups. It is straightforward to see that the number of functional groups on the surface grows geometrically with the number of generations, g:

$$\text{Number of functional groups} = 3 \times 2^g \qquad (4.9.1)$$

Thus, a perfect G5, G6, and G7 dendrimer would have 96, 192, and 384 functional groups, respectively. Note that Equation 4.9.1 would need to be modified in the case of, for example, a tetrafunctional core. We have introduced the term "perfect" here to emphasize that it is certainly possible for a dendrimer molecule to have defects, or missing functional groups, which will propagate through all subsequent generations. For the first few generations it is usually not too difficult to approach perfection, but for G5 and above, the functional groups become rather congested, which makes complete addition of the next generation difficult. It also becomes harder to separate out defective structures. It is typically not practical to go beyond G8.

Further consideration of the divergent approach in Figure 4.7 reveals that, compared with most polymerization reactions, it is rather labor intensive. For example, the addition of each generation requires both an addition step and a deprotection step. The addition will typically be conducted in the presence of a substantial excess of protected monomer, to drive the completion of the new layer. The resulting products will need to be separated, in order to isolate the perfect dendrimer structure from all other reaction products and reagents. Similarly, the deprotection step needs to be driven to completion, and the pure product isolated. Thus, in the end there are two reaction steps and two purification steps required for each generation. This requires a significant amount of time, and it is challenging to prepare commercial-scale quantities of perfect, high generation products.

As a specific example of a divergent synthesis, we will consider the formation of the polyamidoamine (PAMAM) system. In this case, there are two monomers to be added sequentially in each generation, rather than one addition and one deprotection step. The core molecule and one of the monomers is typically ethylene diamine and the other monomer is methyl acrylate. The first step is addition of four methyl acrylate molecules to ethylene diamine in a solvent such as methanol. The Michael addition-type mechanism involves nucleophilic attack of the electron pair on the nitrogen to the double bond of the acrylate, which is activated by the electron withdrawing character of the ester group:

(4.LL)

The next second step involves amidation of each ester group by nucleophilic attack of the nitrogen on the electropositive carbonyl carbon, with release of methanol:

(4.MM)

The structure of the resulting G1 PAMAM dendrimer is therefore the following:

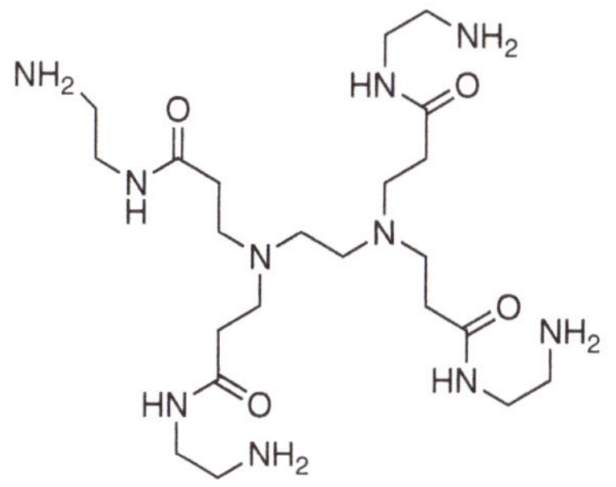

An alternative general strategy for preparing dendrimers and dendritic fragments, or dendrons, is the so-called convergent approach. This is illustrated schematically in Figure 4.8. As the name implies, these molecules are made "from the outside in," that is, the eventual surface group, denoted by "x" in Figure 4.8, is present in the initial reactants. The first reaction produces a molecule with two surface groups and one protected reactive group. The second step, after deprotection, doubles the number of surface groups, and so on. At any stage, a suitable multifunctional core molecule can be used, to stitch the appropriate number of dendrons (usually 3 or 4) together. Each growing wedge-shaped dendron possesses only one reactive group, which presents a significant advantage in terms of purification. At each growth step, a dendron either reacts or it does not, but the product and reactant are significantly different in molecular weight. By contrast, in the divergent approach, the surface of the dendrimer has many reactive groups, and it may not be easy to separate a G3 dendrimer with 24 newly added monomers from one with only 23 monomers. Furthermore, because there are so many more reactive groups on the dendrimer than on the added monomers, the monomers must be present in huge molar excess to drive each reaction to completion. In the convergent approach shown, there are only twice as many dendrons as new coupling molecules at stoichiometric equivalence, so a large excess of dendrons is not necessary.

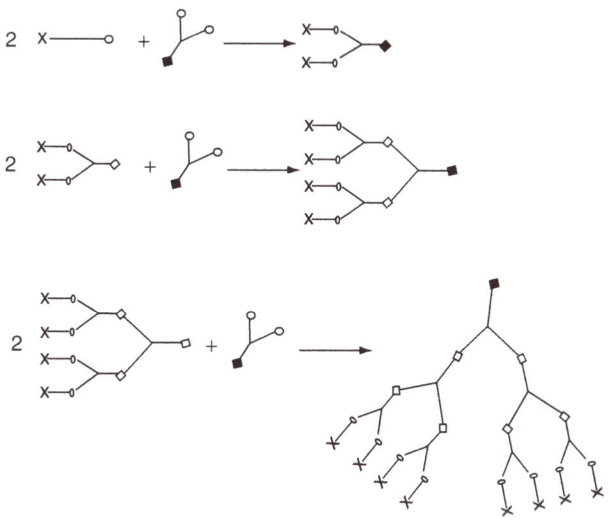

Figure 4.8 Illustration of the convergent approach to dendrimer synthesis. Each dendron is built-up by successive 2:1 reactions, before the final coupling step.

The initial demonstration of this approach was based on the following scheme [15]. The building blocks were 3,5-dihydroxybenzyl alcohol and a benzylic bromide. The first reaction, conducted in acetone, coupled two of the bromides with one alcohol. The surviving benzylic alcohol was then transformed back to a bromide functionality with carbon tetrabromide in the presence of triphenyl phosphine. Introduction of more 3,5-dihydroxybenzyl alcohol began the formation of the next generation dendron, and the process continued.

(4.NN)

4.10 Chapter Summary

In this chapter, we have considered a wide variety of synthetic strategies to exert greater control over the products of a polymerization, compared to the standard step-growth and chain-growth approaches. Although access to much narrower molecular weight distributions has been the primary focus, production of block copolymers, end-functional polymers, and controlled-branched architectures has also been explored. The central concept of the chapter is that of a living polymerization, defined as a chain-growth process that proceeds in the absence of irreversible chain termination or chain transfer:

1. When a living polymerization is conducted such that the rate of initiation is effectively instantaneous compared to propagation, it is possible to approach a Poisson distribution of molecular weights, where the polydispersity is $1 + (1/N_n)$.
2. Anionic polymerization is the most established method for approaching the ideal living polymerization, and effective protocols for a variety of monomers have been established.
3. Cationic polymerization can also be living, although it is generally harder to do so than for the anionic case, in large part due to the prevalence of transfer reactions, including transfer to monomer.
4. Using the concept of a reversibly dormant or inactive species, free-radical polymerizations have also been brought under much greater control. Three general flavors of controlled radical polymerization, known as ATRP, SFRP, and RAFT, are currently undergoing rapid development.
5. Living polymerization in general, and anionic polymerization in particular, can be used to produce block copolymers, end-functional polymers, and well-defined star and graft polymers, for a variety of possible uses.
6. Through basic thermodynamic considerations the concepts of equilibrium polymerization, ceiling temperature, and floor temperature have been explored.
7. The utility of ROPs has been established, where the thermodynamic driving force for chain growth relies on ring-strain. Specific systems of nearly living ring-opening polymerizations have been introduced, including important metal-catalyzed routes such as ROMP.

8. A particular class of highly branched, precisely controlled polymers called dendrimers can be prepared by either of two step-growth routes, referred to as convergent and divergent, respectively.

Problems

1. Experimental data cited in Example 4.1 for the anionic polymerization of styrene do not really test the relationship between conversion, p and time; why not? What additional experimental information should have been obtained if that were the object?
2. Although the polydispersities described in Example 4.1 are very low, they consistently exceed the theoretical Poisson limit. List four assumptions that are necessary for the Poisson distribution to apply, and then identify which one is most likely not satisfied. Justify your answer, based on the data provided.
3. For the living anionic polymerization of styrene discussed in Example 4.1, the solvent used was cyclohexane, and the kinetics are known to be 0.5 order with respect to initiator. What is the predominant species in terms of ion pairing, and what is the approximate dissociation constant for this cluster if k_p is actually 1000 mol L^{-1} s^{-1}?
4. One often-cited criterion for judging whether a polymerization is living is that M_n should increase linearly with conversion. Why is this not, in fact, a robust criterion?
5. A living polymerization of 2-vinyl pyridine was conducted using benzyl picolyl magnesium as the initiator.[†] Values of M_n were determined for polymers prepared with different initiator concentrations and different initial concentrations of monomer, as shown below. Calculate the expected M_n assuming complete conversion and 100% initiator efficiency; how well do the theoretical and experimental values agree?

[I] (mmol L^{-1})	[M]$_0$ (mmol L^{-1})	M_n (kg mol^{-1})
0.48	82	20
0.37	85	25
0.17	71	46
0.48	71	17
0.58	73	14
0.15	150	115

6. The following table shows values of ΔH^0 at 298 K for the gas phase reactions $X(g) + H^+(g) \rightarrow HX^+(g)$, where X is an olefin.[‡] Use these data to comment quantitatively on each of the following points:
    1. The cation is stabilized by electron-donating alkyl substituents.
    2. The carbonium ion rearrangement of n-propyl ions to i-propyl ions is energetically favored.
    3. With the supplementary information that ΔH_f^0 of 1-butene and cis-2-butene are $+1.6$ and -5.8 kJ mol^{-1}, respectively, evaluate the ΔH for the rearrangement n-butyl to sec-butyl ions and compare with the corresponding isomerization for the propyl cation.
    4. Of the monomers shown, only isobutene undergoes cationic polymerization to any significant extent. Criticize or defend the following proposition: the data explain this fact by showing that this is the only monomer listed that combines a sufficiently negative ΔH for protonation, with the freedom from interfering isomerization reactions.

[†] A. Soum and M. Fontanille, in *Anionic Polymerization*, J.E. McGrath, Ed., ACS Symposium Series, Vol 166, 1981.
[‡] P.H. Plesch, Ed., *Cationic Polymerization*, Macmillan, New York, 1963.

X	HX^+	ΔH^0 at 298 K (kJ mol^{-1})
$CH_2=CH_2$	$CH_3CH_2^+$	-640
$CH_3CH=CH_2$	$CH_3CH_2CH_2^+$	-690
$CH_3CH=CH_2$	$CH_3C^+HCH_3$	-757
$CH_3CH_2CH=CH_2$	$CH_3CH_2CH_2CH_2^+$	-682
$CH_3CH=CHCH_3$	$CH_3CH_2C^+HCH_3$	-782
$(CH_3)_2C=CH_2$	$(CH_3)_2CHCH_2^+$	-695

7. In the study discussed in Example 4.3, a solution ATRP of styrene gave an apparent propagation rate constant of 3.9×10^{-5} s^{-1}. Given that the initial monomer concentration was 4.3 mol L^{-1}, and that the initial concentrations of initiator and CuBr were 0.045 mol L^{-1}, estimate the equilibrium constant K for activation of the chain end radical.

8. For the solution polymerization of lactide with [M] = 1 mol L^{-1}, Duda and Penczek determined $\Delta H^0 = -22.9$ kJ mol^{-1} and $\Delta S^0 = -41.1$ J mol^{-1} K^{-1}[†]. What is the associated ceiling temperature for an equilibrium monomer concentration of 1 mol L^{-1}? Does the value you obtain suggest that equilibration is an issue in controlled polymerization of lactide? Compare these thermodynamic quantities with those for the cyclic alkanes in Table 4.4; how do you account for the differences between the six-membered alkane and lactide?

9. For the polymerization system in Problem 8, calculate the equilibrium monomer concentration that would actually be obtained, and the conversion to polymer, at 80°C and 120°C.

10. A typical propagation rate constant, k_p, for the anionic ROP of hexamethylcyclotrisiloxane (D$_3$) is 0.1 L mol^{-1} s^{-1}. Design a polymerization system (initial monomer and initiator concentrations) to obtain a narrow distribution poly(dimethylsiloxane) with $M_n = 50{,}000$, assuming that the reaction will be terminated at 50% conversion. At what time should the polymerization be terminated?

11. Given that ROP is often conducted under conditions in which reverse reactions are possible, do we need to worry about cyclization of the entire growing polymer? Why or why not?

12. Suggest a scheme to test the hypothesis that in lactide polymerization it is the acyl carbon–oxygen bond that is cleaved, rather than the alkyl carbon–oxygen bond.

13. Both ethyleneimine and ethylene sulfide are amenable to ROP. The former proceeds in the presence of acid, whereas the latter can follow either anionic or cationic routes. Propose structures for the three propagating chain ends and the resulting polymers.

14. Draw repeat unit structures for polymers made by ROMP of the following three monomers:

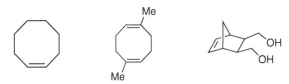

[†] A. Duda and S. Penczek *Macromolecules*, 23, 1636 (1990).

15. Suggest monomer structures that will lead to the following repeat unit structures following ROMP:

16. Compare and contrast dendrimers with block copolymer micelles, globular proteins, inorganic nanoparticles in terms of attributes and likely utility in the following applications: (a) drug delivery; (b) homogenous catalysis; and (c) solubilization.

17. In an ideal living polymerization, how should M_n and M_w/M_n vary with conversion of monomer to polymer? How should M_n of the formed polymer vary with time? Compare these to a radical polymerization with termination by disproportionation, and no transfer.

18. The following criteria have all been suggested and/or utilized as diagnostics for whether a polymerization is living or not. For each one, explain why it might be useful, and then decide whether or not it is a robust criterion, that is, can you think of a situation in which the criterion is satisfied but the polymerization is not living? (See also Problem 4.)

    1. Polymerization proceeds until all monomer is consumed. Polymerization continues if more monomer is then added.
    2. The number of polymer molecules is constant, and independent of conversion.
    3. Narrow molecular weight distributions are produced.
    4. The concentration of monomer decreases to zero, exponentially with time.

References

1. T.R. Darling, T.P. Davis, M. Fryd, A.A. Gridnev, D.M. Haddleton, S.D. Ittel, R.R. Matheson, Jr., G. Moad, and E. Rizzardo, *J. Polym. Sci., Polym. Chem. Ed.*, 38, 1706 (2000).
2. M. Szwarc, *Nature*, 178, 1168 (1956).
3. H.L. Hsieh and R.P. Quirk, *Anionic Polymerization, Principles and Practical Applications*, Marcel Dekker, Inc., New York, NY, 1996.
4. N. Hadjichristidis, M. Pitsikalis, S. Pispas, and H. Iatrou, *Chem. Rev.*, 101, 3747 (2001).
5. Y. Tsukahara, K. Tsutsumi, Y. Yamashita, and S. Shimada, *Macromolecules*, 23, 5201 (1990).
6. G. Odian, *Principles of Polymerization*, 4th ed., Wiley-Interscience, Hoboken, NJ, 2004.
7. J.E. Puskas, P. Antony, Y. Kwon, C. Paulo, M. Kovar, P.R. Norton, G. Kaszas, and V. Altstädt, *Macromol. Mater. Eng.*, 286, 565 (2001).
8. M. Sawamoto, C. Okamoto, and T. Higashimura, *Macromolecules*, 20, 2693 (1987).
9. J.E. Puskas, G. Kaszas, and M. Litt, *Macromolecules*, 24, 5278 (1991).
10. K. Matyjaszewski, (Ed.), *Controlled/Living Polymerization: Progress in ATRP, NMP, and RAFT*, ACS Symposium Series 768, Oxford University Press (2000).
11. K. Matyjaszewski, *Chem. Rev.*, 101, 2921 (2001); C.J. Hawker, A.W. Bosman, and E. Harth, *Chem. Rev.*, 101, 3661 (2001).
12. G. Moad, E. Rizzardo, S.H. Thang, *Aust. J. Chem.*, 58, 379 (2005).
13. C.J. Hawker, *Acc. Chem. Res.*, 30, 373 (1997).
14. P.J. Flory, *Nature (London)*, 62, 1561 (1940).
15. C.J. Hawker and J.M.J. Fréchet, *J. Am. Chem. Soc.*, 112, 7639 (1990).

Further Readings

N. Hadjichristidis, S. Pispas, and G. Floudas, *Block Copolymers: Synthetic Strategies, Physical Properties, and Applications*, John Wiley & Sons, Inc., New York, 2003.

H.L. Hsieh and R.P. Quirk, *Anionic Polymerization, Principles and Practical Applications*, Marcel Dekker, Inc., New York, NY, 1996.

G. Odian, *Principles of Polymerization*, 4th ed., Wiley-Interscience, Hoboken, NJ, 2004.

P. Rempp and E.W. Merrill, *Polymer Synthesis*, 2nd ed., Hüthig & Wepf, Basel, 1991.

M. Szwarc and M. van Beylen, *Ionic Polymerization and Living Polymers*, Chapman and Hall, New York, 1993.

5

Copolymers, Microstructure, and Stereoregularity

5.1 Introduction

All polymer molecules have unique features of one sort or another at the level of the individual repeat units. Occasional head-to-head or tail-to-tail orientations, random branching, and the distinctiveness of chain ends are all examples of such details. In this chapter, we shall focus attention on two other situations that introduce structural variation at the level of the repeat unit: the presence of two different monomers, and the regulation of configuration of successive repeat units. In the former case copolymers are produced, and in the latter case polymers with differences in tacticity. In the discussion of these combined topics, we use statistics extensively because the description of microstructure requires this kind of approach. This is the basis for merging a discussion of copolymers and stereoregular polymers into a single chapter. In other respects these two classes of materials and the processes that produce them are very different, and their description leads us into some rather diverse areas. Copolymerization offers a facile means to tune material properties, as the average composition of the resulting polymers can often be varied across the complete composition range. Similarly, control of stereoregularity plays an essential role in dictating the crystallinity of the resulting material, which in turn can exert a profound influence on the resulting physical properties.

The formation of copolymers involves the reaction of (at least) two kinds of monomers. This means that each must be capable of undergoing the same propagation reaction, but it is apparent that quite a range of reactivities are compatible with this broad requirement. We shall examine such things as the polarity of monomers, the degree of resonance stabilization they possess, and the steric hindrance they experience in an attempt to understand these differences in reactivity. There are few types of reactions for which chemists are successful in explaining all examples with general concepts such as these, and polymerization reactions are no exception. Even for the specific case of free-radical copolymerization, we shall see that reactivity involves the interplay of all these considerations.

To achieve any sort of pattern in configuration among successive repeat units in a polymer chain, the tendency toward random addition must be overcome. Although temperature effects are pertinent here—remember that high temperature is the great randomizer—real success in regulating the pattern of successive addition involves the use of catalysts that "pin down" both the monomer and the growing chain so that their reaction is biased in favor of one mode of addition or another. We shall discuss the Ziegler–Natta catalysts that accomplish this, and shall discover these to be complicated systems for which no single mechanism is entirely satisfactory. We shall also compare these to the more recently developed "single-site" catalysts, which offer great potential for controlling multiple aspects of polymer structure.

For both copolymers and stereoregular polymers, experimental methods for characterizing the products often involve spectroscopy. We shall see that nuclear magnetic resonance (NMR) spectroscopy is particularly well suited for the study of tacticity. This method is also used for the analysis of copolymers.

In spite of the assortment of things discussed in this chapter, there are also related topics that could be included but which are not owing to space limitations. We do not discuss copolymers formed by the step-growth mechanism, for example, or the use of Ziegler–Natta catalysts to regulate geometrical isomerism in, say, butadiene polymerization. Some other important omissions are noted in passing in the body of the chapter.

5.2   Copolymer Composition

We begin our discussion of copolymers by considering the free-radical polymerization of a mixture of two monomers, M_1 and M_2. This is already a narrow view of the entire field of copolymers, since more than two repeat units can be present in copolymers and, in addition, mechanisms other than free-radical chain growth can be responsible for copolymer formation. The essential features of the problem are introduced by this simpler special case, and so we shall restrict our attention to this system.

5.2.1   Rate Laws

The polymerization mechanism continues to include initiation, termination, and propagation steps, and we ignore transfer reactions for simplicity. This time, however, there are four distinctly different propagation reactions:

$$-M_1\cdot + M_1 \xrightarrow{\ k_{11}\ } -M_1M_1\cdot \tag{5.A}$$

$$-M_1\cdot + M_2 \xrightarrow{\ k_{12}\ } -M_1M_2\cdot \tag{5.B}$$

$$-M_2\cdot + M_1 \xrightarrow{\ k_{21}\ } -M_2M_1\cdot \tag{5.C}$$

$$-M_2\cdot + M_2 \xrightarrow{\ k_{22}\ } -M_2M_2\cdot \tag{5.D}$$

Each of these reactions is characterized by a propagation constant, which is labeled by a two-digit subscript: the first number identifies the terminal repeat unit in the growing radical and the second identifies the added monomer. The rate laws governing these four reactions are:

$$R_{p,11} = k_{11}[M_1\cdot][M_1] \tag{5.2.1}$$

$$R_{p,12} = k_{12}[M_1\cdot][M_2] \tag{5.2.2}$$

$$R_{p,21} = k_{21}[M_2\cdot][M_1] \tag{5.2.3}$$

$$R_{p,22} = k_{22}[M_2\cdot][M_2] \tag{5.2.4}$$

In writing Equation 5.2.1 through Equation 5.2.4 we make the customary assumption that the kinetic constants are independent of the size of the radical, and we indicate the concentration of all radicals ending with the M_1 repeat unit, whatever their chain length, by the notation $[M_1\cdot]$. This formalism therefore assumes that only the nature of the radical chain end influences the rate constant for propagation. We refer to this as the *terminal control mechanism*. If we wished to consider the effect of the next-to-last repeat unit in the radical, each of these reactions and the associated rate laws would be replaced by two alternatives. Thus Reaction (5.A) becomes

$$-M_1M_1\cdot + M_1 \xrightarrow{\ k_{111}\ } -M_1M_1M_1\cdot \tag{5.E}$$

$$-M_2M_1\cdot + M_1 \xrightarrow{\ k_{211}\ } -M_2M_1M_1\cdot \tag{5.F}$$

and Equation 5.2.1 becomes

$$R_{p,111} = k_{111}[M_1M_1\cdot][M_1] \tag{5.2.5}$$

$$R_{p,211} = k_{211}[M_2M_1\cdot][M_1] \tag{5.2.6}$$

where the effect of the next-to-last, or *penultimate*, unit is considered. For now we shall restrict ourselves to the simpler case where only the terminal unit determines behavior, although systems in which the penultimate effect is important are well known.

The magnitudes of the various k values in Equation 5.2.1 through Equation 5.2.4 describe the intrinsic differences between the various modes of addition, and the ks plus the concentrations of the different species determine the rates at which the four kinds of additions occur. It is the proportion of different steps that determines the composition of the copolymer produced.

Monomer M_1 is converted to polymer by Reaction (5.A) and Reaction (5.C); therefore the rate at which this occurs is the sum of $R_{p,11}$ and $R_{p,21}$:

$$-\frac{d[M_1]}{dt} = k_{11}[M_1\bullet][M_1] + k_{21}[M_2\bullet][M_1] \tag{5.2.7}$$

Likewise, Reaction (5.B) and Reaction (5.D) convert M_2 to polymer, and the rate at which this occurs is the sum of $R_{p,12}$ and $R_{p,22}$:

$$-\frac{d[M_2]}{dt} = k_{12}[M_1\bullet][M_2] + k_{22}[M_2\bullet][M_2] \tag{5.2.8}$$

The ratio of Equation 5.2.7 and Equation 5.2.8 gives the relative rates of the two monomer additions and, hence, the ratio of the two kinds of repeat units in the copolymer:

$$\frac{d[M_1]/dt}{d[M_2]/dt} = \frac{k_{11}[M_1\bullet][M_1] + k_{21}[M_2\bullet][M_1]}{k_{12}[M_1\bullet][M_2] + k_{22}[M_2\bullet][M_2]} \tag{5.2.9}$$

We saw in Chapter 3 that the stationary-state approximation is applicable to free-radical homopolymerizations, and the same is true of copolymerizations. Of course, it takes a brief time for the stationary-state radical concentration to be reached, but this period is insignificant compared to the total duration of a polymerization reaction. If the total concentration of radicals is constant, this means that the rate of crossover between the different types of terminal units is also equal, or that $R_{p,12} = R_{p,21}$:

$$k_{12}[M_1\bullet][M_2] = k_{21}[M_2\bullet][M_1] \tag{5.2.10}$$

or

$$\frac{[M_1\bullet]}{[M_2\bullet]} = \frac{k_{21}[M_1]}{k_{12}[M_2]} \tag{5.2.11}$$

Combining Equation 5.2.9 and Equation 5.2.11 yields one form of the important *copolymer composition equation* or *copolymerization equation*:

$$\frac{d[M_1]/dt}{d[M_2]/dt} = \frac{[M_1]}{[M_2]} \frac{(k_{11}/k_{12})[M_1] + [M_2]}{(k_{22}/k_{21})[M_2] + [M_1]} \tag{5.2.12}$$

Although there are a total of four different rate constants for propagation, Equation 5.2.12 shows that the relationship between the relative amounts of the two monomers incorporated into the polymer and the composition of the monomer feedstock involves only two ratios of different pairs of these constants. Accordingly, we simplify the notation by defining *reactivity ratios*:

$$r_1 = \frac{k_{11}}{k_{12}} \tag{5.2.13}$$

and

$$r_2 = \frac{k_{22}}{k_{21}} \tag{5.2.14}$$

With these substitutions, Equation 5.2.12 becomes

$$\frac{d[M_1]/dt}{d[M_2]/dt} = \frac{[M_1]}{[M_2]} \frac{r_1[M_1] + [M_2]}{r_2[M_2] + [M_1]} = \frac{1 + r_1[M_1]/[M_2]}{1 + r_2[M_2]/[M_1]} \tag{5.2.15}$$

The ratio $(d[M_1]/dt)/(d[M_2]/dt)$ is the same as the ratio of the numbers of each kind of repeat unit in the polymer formed from the solution containing M_1 and M_2 at concentrations $[M_1]$ and $[M_2]$, respectively. Since the composition of the monomer solution changes as the reaction progresses, Equation 5.2.15 applies to the feedstock as prepared only during the initial stages of the polymerization. Subsequently, the instantaneous concentrations in the prevailing mixture apply unless monomer is added continuously to replace that which has reacted and maintain the original composition of the feedstock. We shall assume that it is the initial product formed that we describe when we use Equation 5.2.15 so as to remove uncertainty as to the monomer concentrations.

5.2.2   Composition versus Feedstock

As an alternative to Equation 5.2.15, it is convenient to describe the composition of both the polymer and the feedstock in terms of the mole fraction of each monomer. Defining F_i as the mole fraction of the ith component in the polymer and f_i as the mole fraction of component i in the monomer solution, we observe that

$$F_1 = 1 - F_2 = \frac{d[M_1]/dt}{d[M_1]/dt + d[M_2]/dt} \tag{5.2.16}$$

and

$$f_1 = 1 - f_2 = \frac{[M_1]}{[M_1] + [M_2]} \tag{5.2.17}$$

Combining Equation 5.2.15 and Equation 5.2.16 into Equation 5.2.17 yields another form of the copolymer composition equation

$$F_1 = \frac{r_1 f_1^2 + f_1 f_2}{r_1 f_1^2 + 2 f_1 f_2 + r_2 f_2^2} \tag{5.2.18}$$

This equation relates the composition of the copolymer formed to the instantaneous composition of the feedstock and to the reactivity ratios r_1 and r_2 that characterize the specific system.

Figure 5.1 shows a plot of F_1 versus f_1—the mole fractions of monomer 1 in the copolymer and in the mixture, respectively—for several values of the reactivity ratios. Inspection of Figure 5.1 brings out the following points:

1. If $r_1 = r_2 = 1$, the copolymer and the feed mixture have the same composition at all times. In this case Equation 5.2.18 becomes

$$F_1 = \frac{f_1(f_1 + f_2)}{(f_1 + f_2)^2} = f_1 \tag{5.2.19}$$

2. If $r_1 = r_2$, the copolymer and the feed mixture have the same composition at $f_1 = 0.5$. In this case Equation 5.2.18 becomes $F_1 = (r + 1)/2(r + 1) = 0.5$.
3. If $r_1 = r_2$, with both values less than unity, the copolymer is richer in component 1 than the feed mixture for $f_1 < 0.5$, and richer in component 2 than the feed mixture for $f_1 > 0.5$.
4. If $r_1 = r_2$, with both values greater than unity, an S-shaped curve passing through the point $(0.5, 0.5)$ would also result, but in this case reflected across the $45°$ line compared to item (3).
5. If $r_1 \neq r_2$, with both values less than unity, the copolymer starts out richer in monomer 1 than the feed mixture and then crosses the $45°$ line, and is richer in component 2 beyond this

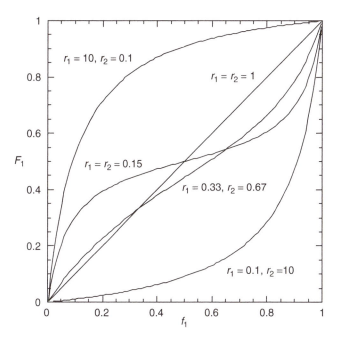

Figure 5.1 Mole fraction of component 1 in the copolymer as a function of feedstock composition for various reactivity ratios.

crossover point. At the crossover point the copolymer and feed mixture have the same composition. The monomer ratio at this point is conveniently solved by Equation 5.2.15:

$$\left(\frac{[M_1]}{[M_2]}\right)_{cross} = \frac{1 - r_2}{1 - r_1} \tag{5.2.20}$$

For the case of $r_1 = 0.33$ and $r_2 = 0.67$ shown in Figure 5.1, $[M_1]/[M_2]$ equals 0.5 and $f_1 = 0.33$. This mathematical analysis shows that a comparable result is possible with both r_1 and r_2 greater than unity, but is not possible for $r_1 > 1$ and $r_2 < 1$.

6. When $r_1 = 1/r_2$, the copolymer composition curve will be either convex or concave when viewed from the F_1 axis, depending on whether r_1 is greater or less than unity. The further removed from unity r_1 is, the farther the composition curve will be displaced from the 45° line. This situation where $r_1 r_2 = 1$ is called an *ideal copolymerization*. The example below explores the origin of this terminology.

There is a parallel between the composition of a copolymer produced from a certain feed and the composition of a vapor in equilibrium with a two-component liquid mixture. The following example illustrates this parallel when the liquid mixture is an ideal solution and the vapor is an ideal gas.

Example 5.1

An ideal gas obeys Dalton's law; that is, the total pressure is the sum of the partial pressures of the components. An ideal solution obeys Raoult's law; that is, the partial pressure of the ith component above a solution is equal to the mole fraction of that component in the solution times the vapor pressure of pure component i. Use these relationships to relate the mole fraction of component 1 in the equilibrium vapor to its mole fraction in a two-component solution and relate the result to the ideal case of the copolymer composition equation.

Solution

We define F_1 to be the mole fraction of component 1 in the vapor phase and f_1 to be its mole fraction in the liquid solution. Here p_1 and p_2 are the vapor pressures of components 1 and 2 in equilibrium with an ideal solution, and p_1^0 and p_2^0 are the vapor pressures of the two pure liquids. By Dalton's law, $p_{tot} = p_1 + p_2$ and $F_1 = p_1/p_{tot}$, since these are ideal gases and p is proportional to the number of moles. By Raoult's law, $p_1 = f_1 p_1^0$, $p_2 = f_2 p_2^0$, and $p_{tot} = f_1 p_1^0 + f_2 p_2^0$. Combining the two gives

$$F_1 = \frac{f_1 p_1^0}{f_1 p_1^0 + f_2 p_2^0} = \frac{f_1 (p_1^0/p_2^0)}{f_1 (p_1^0/p_2^0) + f_2}$$

Now examine Equation 5.2.18 for the case of $r_1 = 1/r_2$:

$$F_1 = \frac{r_1 f_1^2 + f_1 f_2}{r_1 f_1^2 + 2 f_1 f_2 + (1/r_1) f_2^2} = \frac{r_1 f_1 (r_1 f_1 + f_2)}{(r_1 f_1 + f_2^2)} = \frac{r_1 f_1}{r_1 f_1 + f_2}$$

This is identical to the ideal liquid–vapor equilibrium if r_1 is identified with p_1^0/p_2^0.

The vapor pressure ratio measures the intrinsic tendency of component 1 to enter the vapor phase relative to component 2. Likewise r_1 measures the tendency of M_1 to add to $M_1\cdot$ relative to M_1 adding to $M_2\cdot$. In this sense there is a certain parallel, but it is based on $M_1\cdot$ as a reference radical and hence appears to be less general than the vapor pressure ratio. Note, however, that $r_1 = 1/r_2$ means $k_{11}/k_{12} = k_{21}/k_{22}$. In this case the ratio of rate constants for monomer 1 relative to monomer 2 is the same regardless of the reference radical examined. This shows the parallelism to be exact.

Because of the analogy with liquid–vapor equilibrium, copolymers for which $r_1 = 1/r_2$ are said to be ideal. For those nonideal cases in which the copolymer and feedstock happen to have the same composition, the reaction is called an *azeotropic polymerization*. Just as in the case of azeotropic distillation, the composition of the reaction mixture does not change as copolymer is formed if the composition corresponds to the azeotrope. The proportion of the two monomers at this point is given by Equation 5.2.20.

In this section we have seen that the copolymer composition depends to a large extent on the four propagation constants, although it is sufficient to consider these in terms of the two reactivity ratios r_1 and r_2. In the next section we shall examine these ratios in somewhat greater detail.

5.3   Reactivity Ratios

The parameters r_1 and r_2 are the vehicles by which the nature of the reactants enter the copolymer composition equation. We shall call these radical reactivity ratios simply reactivity ratios, although similarly defined ratios also describe copolymerizations that involve ionic intermediates. There are several important things to note about reactivity ratios:

1. The single subscript used to label r is the index of the radical.
2. r_1 is the ratio of two propagation constants involving radical 1: The ratio always compares the propagation constant for the same monomer adding to the radical relative to the propagation constant for the addition of the other monomer. Thus, if $r_1 > 1$, $M_1\cdot$ adds M_1 in preference to M_2; if $r_1 < 1$, $M_1\cdot$ adds M_2 in preference to M_1.
3. Although r_1 is descriptive of radical $M_1\cdot$, it also depends on the identity of monomer 2; the pair of parameters r_1 and r_2 are both required to characterize a particular system, and the product $r_1 r_2$ is used to quantify this by a single parameter.
4. The reciprocal of a radical reactivity ratio can be used to quantify the reactivity of monomer M_2 by comparing its rate of addition to radical $M_1\cdot$ relative to the rate of M_1 adding $M_1\cdot$.

5. As the ratio of two rate constants, a radical reactivity ratio follows the Arrhenius equation with an apparent activation energy equal to the difference in the activation energies for the individual constants. Thus for r_1, $E^*_{app} = E^*_{p,11} - E^*_{p,12}$. Since the activation energies for propagation are not large to begin with, their difference is even smaller. Accordingly, the temperature dependence of r is relatively small.

5.3.1 Effects of *r* Values

The reactivity ratios of a copolymerization system are the fundamental parameters in terms of which the system is described. Since the copolymer composition equation relates the compositions of the product and the feedstock, it is clear that values of r can be evaluated from experimental data in which the corresponding compositions are measured. We shall consider this evaluation procedure in Section 5.6, where it will be found that this approach is not as free of ambiguity as might be desired. For now we shall simply assume that we know the desired r values for a system; in fact, extensive tabulations of such values exist. An especially convenient source of this information is the *Polymer Handbook* [1]. Table 5.1 lists some typical r values at 60°C.

Although Table 5.1 is rather arbitrarily assembled, note that it contains no system for which r_1 and r_2 are *both* greater than unity. Indeed, such systems are very rare. We can understand this by recognizing that, at least in the extreme case of very large r's, these monomers would tend to simultaneously homopolymerize. Because of this preference toward homopolymerization, any copolymer that does form in systems with r_1 and r_2 both greater than unity will be a block-type polymer with very long sequences of a single repeat unit. Since such systems are only infrequently encountered, we shall not consider them further.

Table 5.1 also lists the product r_1r_2 for the systems included. These products typically lie in the range between zero and unity, and it is instructive to consider the character of the copolymer produced toward each of these extremes.

In the extreme case where $r_1r_2 = 0$ because both r_1 and r_2 equal zero, the copolymer adds monomers with perfect alternation. This is apparent from the definition of r, which compares the addition of the same monomer to the other monomer for a particular radical. If both r's are zero, there is no tendency for a radical to add a monomer of the same kind as the growing end,

Table 5.1 Values of Reactivity Ratios r_1 and r_2 and the Product r_1r_2 for a Few Copolymers at 60°C

M_1	M_2	r_1	r_2	r_1r_2
Acrylonitrile	Methyl vinyl ketone	0.61	1.78	1.09
	Methyl methacrylate	0.13	1.16	0.15
	α-Methyl styrene	0.04	0.20	0.008
	Vinyl acetate	4.05	0.061	0.25
Methyl methacrylate	Styrene	0.46	0.52	0.24
	Methacrylic acid	1.18	0.63	0.74
	Vinyl acetate	20	0.015	0.30
	Vinylidene chloride	2.53	0.24	0.61
Styrene	Vinyl acetate	55	0.01	0.55
	Vinyl chloride	17	0.02	0.34
	Vinylidene chloride	1.85	0.085	0.16
	2-Vinyl pyridine	0.55	1.14	0.63
Vinyl acetate	1-Butene	2.0	0.34	0.68
	Isobutylene	2.15	0.31	0.67
	Vinyl chloride	0.23	1.68	0.39
	Vinylidene chloride	0.05	6.7	0.34

Source: Data from Young, L.J. in *Polymer Handbook*, 3rd ed., Brandrup, J. and Immergut, E.H. (Eds.), Wiley, New York, 1989.

whichever species is the terminal unit. When only one of the r's is zero, say r_1, then alternation occurs whenever the radical ends with an $M_1\cdot$ unit. There is thus a tendency toward alternation in this case, although it is less pronounced than in the case where both r's are zero. Accordingly, we find increasing tendency toward alternation as $r_1 \rightarrow 0$ and $r_2 \rightarrow 0$, or, more succinctly, as the product $r_1 r_2 \rightarrow 0$.

On the other end of the commonly encountered range we find the product $r_1 r_2 \rightarrow 1$. As noted above, this limit corresponds to ideal copolymerization and means the two monomers have the same relative tendency to add to both radicals. Thus if $r_1 \rightarrow 10$, monomer 1 is 10 times more likely to add to $M_1\cdot$ than monomer 2. At the same time $r_2 = 0.1$, which also means that monomer 1 is 10 times more likely to add to $M_2\cdot$. In this case the radicals exert the same influence, so the monomers add at random in the proportion governed by the specific values of the r's.

Recognition of these differences in behavior points out an important limitation on the copolymer composition equation. The equation describes the overall composition of the copolymer, but gives no information whatsoever about the *distribution* of the different kinds of repeat units within the polymer. While the overall composition is an important property of the copolymer, the detailed microstructural arrangement is also a significant feature of the molecule. It is possible for copolymers with the same overall composition to have very different properties because of the differences in microstructure. Reviewing the three categories presented in Chapter 1, we see the following:

1. Alternating structures are promoted by $r_1 \rightarrow 0$ and $r_2 \rightarrow 0$:

    $M_1 M_2 M_1 M_2 M_1 M_2 M_1 M_2 M_1 M_2 M_1 M_2 M_1 M_2 M_1 M_2 M_1 M_2 M_1 M_2$

2. Random structures are promoted by $r_1 r_2 \rightarrow 1$:

    $M_1 M_2 M_2 M_2 M_1 M_1 M_2 M_1 M_1 M_2 M_1 M_2 M_1 M_2 M_1 M_2 M_2 M_1 M_1 M_2$

3. "Blocky" structures are promoted by $r_1 r_2 > 1$:

    $M_1 M_1 M_1 M_1 M_1 M_1 M_1 M_1 M_1 M_1 M_2 M_2 M_2 M_2 M_2 M_2 M_2 M_2 M_2 M_2$

    Each of these polymers has a 50:50 proportion of the two components, but the products probably differ in properties. As examples of such differences, we note the following:

4. Alternating copolymers, while relatively rare, are characterized by combining the properties of the two monomers along with structural regularity. We will see in Chapter 13 that a very high degree of regularity—extending all the way to stereoregularity in the configuration of the repeat units—is required for crystallinity to develop in polymers.

5. Random copolymers tend to average the properties of the constituent monomers in proportion to the relative abundance of the two comonomers.

6. Block copolymers are closer to blends of homopolymers in properties, but without the latter's tendency to undergo phase separation. Diblock copolymers can be used as surfactants to bind immiscible homopolymer blends together and thus improve their mechanical properties. Block copolymers are generally prepared by sequential addition of monomers to living polymers, rather than by depending on the improbable $r_1 r_2 > 1$ criterion in monomers, as was discussed in Chapter 4.

Returning to the data of Table 5.1, it is apparent that there is a good deal of variability among the r values displayed by various systems. We have already seen the effect this produces on the overall copolymer composition; we shall return to this matter of microstructures in Section 5.5. First, however, let us consider the obvious question. What factors in the molecular structure of the two monomers govern the kinetics of the different addition steps? This question is considered in the following sections; for now we look for a way to systematize the data as the first step toward an answer.

5.3.2 Relation of Reactivity Ratios to Chemical Structure

We noted above that the product r_1r_2 can be used to locate a copolymer along an axis between alternating and random structures. It is by means of this product that some values from Table 5.1, supplemented by other results for additional systems, have been organized in Figure 5.2. Figure 5.2 has been constructed according to the following general principles:

1. Various monomers are listed along the base of the triangle.
2. The triangle is subdivided into an array of diamonds by lines drawn parallel to the two sides of the triangle.
3. The spacing of the lines is such that each monomer along the base serves as a label for a row of diamonds.
4. Each diamond marks the intersection of two such rows and therefore corresponds to two comonomers.
5. The r_1r_2 product for the various systems is the number entered in each diamond.
6. Individual monomers have been arranged in such a way as to achieve to the greatest extent possible the values of r_1r_2 that approach zero toward the apex of the triangle and values of r_1r_2 that approach unity toward the base of the triangle.

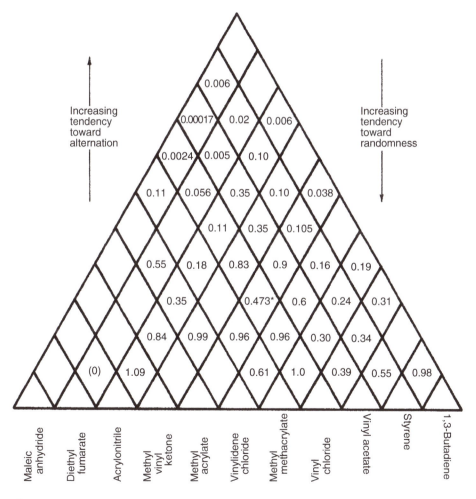

Figure 5.2 The product r_1r_2 for copolymers whose components define the intersection where the numbers appear. The value marked* is determined in Example 5.4; other values are from Ref. [1].

Before proceeding with a discussion of this display, it is important to acknowledge that the criteria for monomer placement can be met only in part. For one thing, there are combinations for which data are not readily available. Incidentally, not all of the $r_1 r_2$ values in Figure 5.2 were measured at the same temperature, but, as noted above, temperature effects are expected to be relatively unimportant. Also, there are outright exceptions to the pattern sought: Generalizations about chemical reactions always seem to be plagued by these. In spite of some reversals of ranking, the predominant trend moving upward from the base along any row of diamonds is a decrease in $r_1 r_2$ values.

From the geometry of this triangular display, it follows immediately—if one overlooks the exceptions—that the more widely separated a pair of comonomers are in Figure 5.2, the greater is their tendency toward randomness. We recognize a parallel here to the notion that widely separated elements in the periodic table will produce more polar bonds than those which are closer together, and vice versa. This is a purely empirical and qualitative trend. The next order of business is to seek an explanation for its origin in terms of molecular structure. If we focus attention on the electron-withdrawing or electron-donating attributes of the substituents on the double bond, we find that the substituents of monomers that are located toward the right-hand corner of the triangle in Figure 5.2 are recognized as electron donors. Likewise, the substituents in monomers located toward the left-hand corner of the triangle are electron acceptors. The demarcation between the two regions of behavior is indicated in Figure 5.2 by reversing the direction of the lettering at this point. Pushing this point of view somewhat further, we conclude that the sequence acetoxy < phenyl < vinyl is the order of increase in electron-donating tendency. Chloro < carbonyl < nitrile is the order of increase in electron-withdrawing tendency. The positions of diethyl fumarate and vinylidene chloride relative to their monosubstituted analogs indicates that "more is better" with respect to these substituent effects. The location of methyl methacrylate relative to methyl acrylate also indicates additivity, this time with partial compensation of opposing effects.

The reactivity ratios are kinetic in origin, and therefore reflect the mechanism or, more specifically, the transition state of a reaction. The transition state for the addition of a vinyl monomer to a growing radical involves the formation of a partial bond between the two species, with a corresponding reduction of the double-bond character of the vinyl group in the monomer:

$$\text{(5.G)}$$

If substituent X is an electron donor and Y an electron acceptor, then the partial bond in the transition state is stabilized by a resonance form (5.I), which attributes a certain polarity to the emerging bond:

[5.I]   [5.II]

The contribution of this polar structure to the bonding lowers the energy of the transition state. This may be viewed as a lower activation energy for the addition step and thus a factor that promotes this particular reaction. The effect is clearly larger the greater the difference in the donor–acceptor properties of X and Y. The transition state for the successive addition of the same monomer (whether X or Y substituted) is Structure (5.II). This involves a more uniform distribution of charge because of the identical substituents and thus lacks the stabilizing effect of the polar resonance form. The activation energy for this mode of addition is greater than that for alternation, at least when X and Y are sufficiently different.

Although we use the term *resonance* in describing the effect of polarity in stabilizing the transition state in alternating copolymers, the emphasis of the foregoing is definitely on polarity rather than resonance per se. It turns out, however, that resonance plays an important role in free-radical polymerization, even when polarity effects are ignored. In Section 5.4 we examine some evidence for this and consider the origin of this behavior.

5.4 Resonance and Reactivity

The tendency toward alternation is not the only pattern in terms of which copolymerization can be discussed. The reactivities of radicals and monomers may also be examined as a source of insight into copolymer formation. The reactivity of radical 1 copolymerizing with monomer 2 is measured by the rate constant k_{12}. The absolute value of this constant can be determined from copolymerization data (r_1) and studies yielding absolute homopolymerization constants (k_{11}):

$$k_{12} = \frac{k_{11}}{r_1} \tag{5.4.1}$$

Table 5.2 lists a few cross-propagation constants calculated by Equation 5.4.1. Far more extensive tabulations than this have been prepared by correlating copolymerization and homopolymerization data for additional systems.

Examination of Table 5.2 shows that the general order of increasing *radical* activity is styrene < acrylonitrile < methyl acrylate < vinyl acetate. An additional observation is that any one of these species shows the reverse order of reactivity for the corresponding monomers. As *monomers*, the order of reactivity in Table 5.2 is styrene > acrylonitrile > methyl acrylate > vinyl acetate. These and similar rankings based on more extensive comparisons are summarized in terms of substituents in Table 5.3.

An important pattern to recognize among the substituents listed in Table 5.3 is this: Those that have a double bond conjugated with the double bond in the olefin are the species that are more stable as radicals and more reactive as monomers. The inverse relationship between the stability of *monomers* and radicals arises precisely because monomers gain (or lose) stability by converting to the radical: The greater the gain (or loss), the greater (or less) the incentive for the monomer to react. It is important to realize that the ability to form conjugated structures is associated with a substituent, whether it is in a monomer or a radical. Conjugation allows greater electron delocalization, which, in turn, lowers the energy of the system that possesses this feature.

Comparison of the range of k_{12} along rows and columns in Table 5.2 suggests that resonance stabilization produces a bigger effect in the radical than in the monomer. After all, the right- and left-hand columns in Table 5.2 (various radicals) differ by factors of 100–1000, whereas the top

Table 5.2 Values of the Cross-Propagation Constants k_{12} (L mol^{-1} s^{-1}) for Four Monomer–Radical Combinations

Monomer	Radical			
	Styrene	Acrylonitrile	Methyl acrylate	Vinyl acetate
Styrene	145	49,000	14,000	230,000
Acrylonitrile	435	1,960	2,510	46,000
Methyl acrylate	203	1,310	2,090	23,000
Vinyl acetate	2.9	230	230	2,300

Source: From Brandrup, J. and Immergut, E.H. (Eds.), *Polymer Handbook*, 3rd ed., Wiley, New York, 1989.

Table 5.3 List of Substituents Ranked in Terms of Their Effects on
Monomer and Radical Reactivity

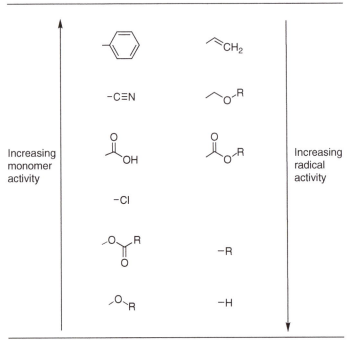

and bottom rows (various monomers) differ only by the factors of 50–100. In order to examine this effect in more detail, consider the addition reaction of monomer M to a reactant radical $R\bullet$ to form a product radical $P\bullet$. What distinguishes these species is the presence or absence of resonance stabilization (subscript rs). If the latter is operative, we must also consider which species benefit from its presence. There are four possibilities:

1. Unstabilized monomer converts stabilized radical to unstabilized radical:

$$R_{rs}\bullet + M \rightarrow P\bullet \tag{5.H}$$

There is an overall loss of resonance stabilization in this reaction. Since it is a radical which suffers the loss, the effect is larger than in the reaction in which....

2. Stabilized monomer converts stabilized radical to another stabilized radical:

$$R_{rs}\bullet + M_{rs} \rightarrow P_{rs}\bullet \tag{5.I}$$

Here too there is an overall loss of resonance stabilization, but it is monomer stabilization which is lost, and this is energetically less costly than Reaction (5.H).

3. Unstabilized monomer converts unstabilized radical to another unstabilized radical:

$$R\bullet + M \rightarrow P\bullet \tag{5.J}$$

This reaction suffers none of the reduction in resonance stabilization that is present in Reaction (5.H) and Reaction (5.I). It is energetically more favored than both of these, but not as much as the reaction in which....

4. Stabilized monomer converts unstabilized radical to stabilized radical:

$$R\bullet + M_{rs} \rightarrow P_{rs}\bullet \tag{5.K}$$

This reaction converts the less effective resonance stabilization of a monomer to a more effective form of radical stabilization. This is the most favorable of the four reaction possibilities.

In summary, we can rank these reactions in terms of their propagation constants as follows:

$$R_{rs} \cdot + M < R_{rs} \cdot + M_{rs} < R \cdot + M < R \cdot + M_{rs}$$

Systems from Table 5.2 which correspond to these situations are the following:

Radical	Styrene		Styrene		Vinyl acetate		Vinyl acetate
	+	<	+	<	+	<	+
Monomer	Vinyl acetate		Styrene		Vinyl acetate		Styrene

Note that this inquiry into copolymer propagation rates also increases our understanding of the differences in free-radical homopolymerization rates. Recall that in Chapter 3 a discussion of this aspect of homopolymerization was deferred until copolymerization was introduced. The trends under consideration enable us to make some sense out of the rate constants for propagation in free-radical homopolymerization as well. For example, in Table 3.4 we see that k_p values at 60°C for vinyl acetate and styrene are 2300 and 165 L mol^{-1} s^{-1}, respectively. The relative magnitude of these constants can be understood in terms of the sequence above.

Resonance stabilization energies are generally assessed from thermodynamic data. If we define ε_1 to be the resonance stabilization energy of species i, then the heat of formation of that species will be less by an amount ε_1 than for an otherwise equivalent molecule without resonance. Likewise, the change in enthalpy ΔH for a reaction that is influenced by resonance effects is less by an amount $\Delta \varepsilon$ (Δ is the usual difference: products minus reactants) that the ΔH for a reaction which is otherwise identical except for resonance effects:

$$\Delta H_{rs} = \Delta H_{no\ rs} - \Delta \varepsilon \tag{5.4.2}$$

Thus if we consider the homopolymerization of ethylene (no resonance possibilities),

$$-CH_2-CH_2 \cdot + CH_2{=}CH_2 \rightarrow -CH_2CH_2CH_2CH_2 \cdot \tag{5.L}$$

$$\Delta H_{no\ rs} = -88.7 \text{ kJ mol}^{-1}$$

as a reference reaction, and compare it with the homopolymerization of styrene (resonance effects present),

$$\tag{5.M}$$

$$\Delta H_{rs} = -69.9 \text{ kJ mol}^{-1}$$

we find a value of $\Delta \varepsilon = -19$ kJ mol^{-1}, according to Equation 5.4.2. Reaction (5.M) is a specific example of the general Reaction (5.I), and the negative value of $\Delta \varepsilon$ in this example indicates the overall loss of resonance stabilization, which is characteristic of Reaction (5.I).

Although it is not universally true that the activation energies of reactions parallel their heats of reaction, this is approximately true for the kind of addition reaction we are discussing. Accordingly, we can estimate $E^* = \alpha \Delta H$, with α an appropriate proportionality constant. If we consider the difference between two activation energies by combining this idea with Equation 5.4.2, the contribution of the nonstabilized reference reaction drops out of Equation 5.4.2 and we obtain

$$E_{11}^* - E_{12}^* = \alpha[-\Delta \varepsilon_{11} - (-\Delta \varepsilon_{12})]$$
$$= -(\varepsilon_{P_1 \cdot} - \varepsilon_{R_1 \cdot} - \varepsilon_{M_1}) + (\varepsilon_{P_2 \cdot} - \varepsilon_{R_1 \cdot} - \varepsilon_{M_2}) \tag{5.4.3}$$

In writing the second version of this relation, the proportionality constant has been set equal to unity as a simplification. Note that the resonance stabilization energy of the reference radical $R_1\cdot$ also cancels out of this expression.

The temperature dependence of the reactivity ratio r_1 also involves the $E_{11}^* - E_{12}^*$ difference through the Arrhenius equation; hence

$$r_1 \propto \exp\left(\frac{\varepsilon_{P_1^\bullet} - \varepsilon_{M_1}}{RT}\right) \exp\left(\frac{-(\varepsilon_{P_2^\bullet} - \varepsilon_{M_2})}{RT}\right) \tag{5.4.4}$$

An analogous expression can be written for r_2:

$$r_2 \propto \exp\left(\frac{\varepsilon_{P_2^\bullet} - \varepsilon_{M_2}}{RT}\right) \exp\left(\frac{-(\varepsilon_{P_1^\bullet} - \varepsilon_{M_1})}{RT}\right) \tag{5.4.5}$$

According to this formalism, the following applies:

1. The reactivity ratios are proportional to the product of two exponentials.
2. Each exponential involves the difference between the resonance stabilization energy of the radical and monomer of a particular species.
3. A positive exponent is associated with the same species as identifies the r (i.e., for r_1, $M_1 \rightarrow P_1\cdot$), whereas the negative exponent is associated with the other species (for r_1, $M_2 \rightarrow P_2\cdot$).

We might be hard-pressed to estimate the individual resonance stabilization energies in Equation 5.4.4 and Equation 5.4.5, but the quantitative application of these ideas is not difficult. Consider once again the styrene–vinyl acetate system:

1. Define styrene to be monomer 1 and vinyl acetate to be monomer 2.
2. The difference in resonance stabilization energy $\varepsilon_{P_1^\bullet} - \varepsilon_{M_1} > 1$, since styrene is resonance stabilized and the effect is larger for the radical than the monomer.
3. The difference $\varepsilon_{P_2^\bullet} - \varepsilon_{M_2} \cong 0$, since neither the radical nor the monomer of vinyl acetate shows appreciable stabilization.
4. Therefore, according to Equation 5.4.4 and Equation 5.4.5, $r_1 > 1$ while $r_2 < 1$.
5. Experimental values for this system are $r_1 = 55$ and $r_2 = 0.01$.

Although this approach does correctly rank the parameters r_1 and r_2 for the styrene–vinyl acetate system, this conclusion was already reached qualitatively above using the same concepts and without any mathematical manipulations. One point that the quantitative derivation makes clear is that explanations of copolymer behavior based exclusively on resonance concepts fail to describe the full picture. All that we need to do is examine the product $r_1 r_2$ as given by Equation 5.4.4 and Equation 5.4.5, and the shortcoming becomes apparent. According to these relationships, the product $r_1 r_2$ always equals unity, yet we saw in the last section that experimental $r_1 r_2$ values generally lie between zero and unity. We also saw that polarity effects could be invoked to rationalize the $r_1 r_2$ product.

The situation may be summarized as follows:

1. If resonance effects *alone* are considered, it is possible to make some sense of the ranking of various propagation constants.
2. In this case only random microstructure is predicted.
3. If polarity effects *alone* are considered, it is possible to make some sense out of the tendency toward alternation.
4. In this case homopolymerization is unexplained.

The way out of this apparent dilemma is easily stated, although not easily acted upon. It is not adequate to consider any one of these approaches for the explanation of something as complicated as these reactions. Polarity effects and resonance are both operative, and, if these still fall short of explaining all observations, there is another old standby to fall back on: steric effects. Resonance, polarity, and steric considerations are all believed to play an important role in copolymerization chemistry just as in the other areas of organic chemistry. Things are obviously simplified if only one of these is considered, but it must be remembered that doing this necessarily reveals only one facet of the problem. Nevertheless, there are times, particularly before launching an experimental investigation of a new system, when some guidelines are very useful. The following example illustrates this point.

Example 5.2

It is proposed to polymerize the vinyl group of the hemin molecule with other vinyl comonomers to prepare model compounds to be used in hemoglobin research. Considering hemin and styrene to be species 1 and 2, respectively, use the resonance concept to rank the reactivity ratios of r_1 and r_2.

Solution

Hemin is the complex between protoporphyrin and iron in the $+3$ oxidation state. Iron is in the $+2$ state in the heme of hemoglobin. The molecule has the following structure:

It is apparent from the size of the conjugated system here that numerous resonance possibilities exist in this species in both the radical and the molecular form. Styrene also has resonance structures in both forms. On the principle that these effects are larger for radicals than monomers, we conclude that the difference $\varepsilon_{P\bullet} - \varepsilon_M > 0$ for both hemin and styrene. On the principle that greater resonance effects result from greater delocalization, we expect the difference to be larger for hemin than for styrene. According to Equation 5.4.4, $r_1 \propto e^{\text{larger}} e^{-\text{smaller}} > 1$. According to Equation 5.4.5, $r_2 \propto e^{\text{smaller}} e^{-\text{larger}} < 1$. Experimentally, the values for these parameters turn out to be $r_1 = 65$ and $r_2 = 0.18$.

5.5 A Closer Look at Microstructure

In Section 5.3 we noted that variations in the product $r_1 r_2$ led to differences in the polymer microstructure, even when the overall compositions of two systems are the same. In this section we shall take a closer look at this variation, using the approach best suited for this kind of detail, statistics.

5.5.1　Sequence Distributions

Suppose we define as p_{ij} the probability that a unit of type i is followed in the polymer by a unit of type j, where both i and j can be either 1 or 2. Since an i unit must be followed by either an i or a j, the fraction of ij sequences (pairs) out of all possible sequences (pairs) defines p_{ij}:

$$p_{ij} = \frac{\text{Number of } ij \text{ sequences}}{\text{Number of } ij \text{ sequences} + \text{Number of } ii \text{ sequences}} \tag{5.5.1}$$

This equation can also be written in terms of the propagation rates of the different types of addition steps which generate the sequences:

$$p_{ij} = \frac{R_{ij}}{R_{ij} + R_{ii}} = \frac{k_{ij}[\text{M}_i\bullet][\text{M}_j]}{k_{ij}[\text{M}_i\bullet][\text{M}_j] + k_{ii}[\text{M}_i\bullet][\text{M}_i]} \tag{5.5.2}$$

For the various possible combinations in a copolymer, Equation 5.5.2 becomes

$$p_{11} = \frac{k_{11}[\text{M}_1\bullet][\text{M}_1]}{k_{11}[\text{M}_1\bullet][\text{M}_1] + k_{12}[\text{M}_1\bullet][\text{M}_2]} = \frac{r_1[\text{M}_1]}{r_1[\text{M}_1] + [\text{M}_2]} \tag{5.5.3}$$

$$p_{12} = \frac{[\text{M}_2]}{r_1[\text{M}_1] + [\text{M}_2]} \tag{5.5.4}$$

$$p_{22} = \frac{k_{22}[\text{M}_2\bullet][\text{M}_2]}{k_{22}[\text{M}_2\bullet][\text{M}_2] + k_{21}[\text{M}_2\bullet][\text{M}_1]} = \frac{r_2[\text{M}_2]}{r_2[\text{M}_2] + [\text{M}_1]} \tag{5.5.5}$$

$$p_{21} = \frac{[\text{M}_1]}{r_2[\text{M}_2] + [\text{M}_1]} \tag{5.5.6}$$

Note that $p_{11} + p_{12} = p_{22} + p_{21} = 1$. In writing these expressions we make the assumption that only the terminal unit of the radical influences the addition of the next monomer. This same assumption was made in deriving the copolymer composition equation. We shall have more to say below about this particular assumption.

Next let us consider the probability of finding a sequence of repeat units in a copolymer, which is exactly ν units of M_1 in length. This may be represented as $\text{M}_2(\text{M}_1)_\nu\text{M}_2$. Working from left to right in this sequence, we note the following:

1.　If the addition of monomer M_1 to a radical ending with M_2 occurs L times in a sample, then there will be a total of L sequences, of unspecified length, of M_1 units in the sample.
2.　If $\nu - 1$ consecutive M_1 monomers add to radicals capped by M_1 units, the total number of such sequences is expressed in terms of p_{11} to be $Lp_{11}^{\nu-1}$.
3.　If the sequence contains exactly ν units of type M_1, then the next step must be the addition of an M_2 unit. The probability of such an addition is given by p_{12}, and the number of sequences is $Lp_{11}^{\nu-1}p_{12}$.

Since L equals the total number of M_1 sequences of any length, the fraction of sequences of length ν, ϕ_ν, is given by

$$\phi_\nu = p_{11}^{\nu-1}p_{12} \tag{5.5.7}$$

The similarity of this derivation to those in Section 2.4 and Section 3.7 should be apparent. Substitution of the probabilities given by Equation 5.5.3 and Equation 5.5.4 leads to

$$\phi_\nu = \left(\frac{r_1[\text{M}_1]}{r_1[\text{M}_1] + [\text{M}_2]}\right)^{\nu-1}\left(\frac{[\text{M}_2]}{r_1[\text{M}_1] + [\text{M}_2]}\right) \tag{5.5.8}$$

A similar result can be written for ϕ_μ, where μ denotes the length of a sequence of M_2 units. These expressions give the fraction of sequences of specified length in terms of the reactivity ratios of the copolymer system and the composition of the feedstock. Figure 5.3 illustrates by means of a bar graph how ϕ_ν varies with ν for two polymer systems prepared from equimolar solutions of monomers. The shaded bars in Figure 5.3 describe the system for which $r_1 r_2 = 0.03$ and the unshaded bars describe $r_1 r_2 = 0.30$.

Table 5.4 shows the effect of variations in the composition of the feedstock for the system $r_1 r_2 = 1$. The following observations can be made concerning Figure 5.3 and Table 5.4:

1. In all situations, the fraction ϕ_ν decreases with increasing ν.
2. Figure 5.3 shows that for $r_1 r_2 = 0.03$, about 85% of the M_1 units are sandwiched between two M_2's. We have already concluded that low values of the $r_1 r_2$ product indicate a tendency toward alternation.
3. Figure 5.3 also shows that the proportion of alternating M_1 units decreases, and the fraction of longer sequences increases, as $r_1 r_2$ increases. The 50 mol% entry in Table 5.4 shows that the distribution of sequence lengths gets flatter and broader for $r_1 r_2 = 1$, the ideal case.
4. Table 5.4 also shows that increasing the percentage of M_1 in the monomer solution flattens and broadens the distribution of sequence lengths. Similar results are observed for lower values of $r_1 r_2$, but the broadening is less pronounced when the tendency toward alternation is high.

Next we consider the average value of a sequence length of M_1, $\bar{\nu}$. Combining Equation 1.7.7 and Equation 5.5.7 gives

$$\bar{\nu} = \frac{\sum_{\nu=1}^{\infty} \nu \phi_\nu}{\sum_{\nu=1}^{\infty} \phi_\nu} = \frac{\sum_{\nu=1}^{\infty} \nu p_{11}^{\nu-1} p_{12}}{\sum_{\nu=1}^{\infty} p_{11}^{\nu-1} p_{12}} \qquad (5.5.9)$$

Simplifying this result involves the same infinite series that we examined in connection with Equation 2.4.5; therefore we can write immediately

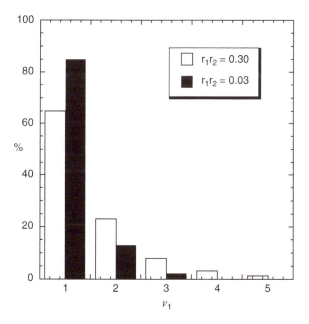

Figure 5.3 Fraction of sequences of the indicated length for copolymers prepared from equimolar feedstocks with $r_1 r_2 = 0.03$ (shaded) and $r_1 r_2 = 0.30$ (unshaded). (Data from Tosi, C., *Adv. Polym. Sci.*, 5, 451, 1968.)

Table 5.4 Percentage of Sequences of Length ν for Copolymers Prepared from Different Feedstocks f_1 with $r_1 r_2 = 1$

$\nu \backslash f_1$	0.1	0.2	0.3	0.4	0.5	0.6	0.7	0.8	0.9
1	90	80	70	60	50	40	30	20	10
2	9	16	21	24	25	24	21	16	9
3	0.9	3.2	6.3	9.6	12.5	14.4	14.7	12.8	8.1
4	0.09	0.64	1.89	3.84	6.25	8.64	10.3	10.2	7.29
5		0.13	0.57	1.54	3.13	5.18	7.20	8.19	6.56
6			0.17	0.62	1.56	3.11	5.04	6.55	5.90
7			0.05	0.25	0.78	1.87	3.53	5.24	5.31
8				0.10	0.39	1.12	2.47	4.19	4.78
9				0.04	0.20	0.67	1.73	3.36	4.30
10					0.10	0.40	1.21	2.68	3.87
11					0.05	0.24	0.85	2.15	3.59
12						0.14	0.59	1.72	3.23

$$\bar{\nu} = \frac{1}{1 - p_{11}} = \frac{1}{p_{12}} \tag{5.5.10}$$

By combining Equation 5.5.4 and Equation 5.5.10, we obtain

$$\bar{\nu} = 1 + r_1 \frac{[M_1]}{[M_2]} \tag{5.5.11}$$

A value of $\bar{\mu}$ is obtained by similar operations:

$$\bar{\mu} = 1 + r_2 \frac{[M_2]}{[M_1]} \tag{5.5.12}$$

The following example demonstrates the use of some of these relationships pertaining to microstructure.

Example 5.3

The hemoglobin molecule contains four heme units. It is proposed to synthesize a hemin (molecule 1)–styrene (molecule 2) copolymer such that $\bar{\nu} = 4$ in an attempt to test some theory concerning hemoglobin. As noted in Example 5.2, $r_1 = 65$ and $r_2 = 0.18$ for this system. What should be the proportion of monomers to obtain this average hemin sequence length? What is the average styrene sequence length at this composition? Does this system seem like a suitable model if the four hemin clusters are to be treated as isolated from one another in the theory being tested? Also evaluate ϕ_ν for several ν bracketing $\bar{\nu}$ to get an idea of the distribution of these values.

Solution

Use Equation 5.5.11 to evaluate $[M_1]/[M_2]$ for $r_1 = 65$ and $\bar{\nu} = 4$:

$$\frac{[M_1]}{[M_2]} = \frac{\bar{\nu} - 1}{r_1} = \frac{4 - 1}{65} = 0.046 \quad \text{and} \quad \frac{[M_2]}{[M_1]} = 21.7$$

Use this ratio of concentration in Equation 5.5.12 to evaluate $\bar{\mu}$:

$$\bar{\mu} = 1 + r_2 \frac{[M_2]}{[M_1]} = 1 + 0.18(21.7) = 4.9$$

The number of styrene units in an average sequence is a little larger than the length of the average hemin sequence. It is not unreasonable to describe the hemin clusters as isolated, on the average, in this molecule. The product $r_1 r_2 = 11.7$ in this system, which also indicates a tendency toward block formation. Use Equation 5.5.8 with $[M_1]/[M_2] = 0.046$ and the r_1 and r_2 values to evaluate ϕ_ν:

$$\phi_\nu = \left(\frac{65(0.046)}{65(0.046) + 1} \right)^{\nu-1} \left(\frac{1}{65(0.046) + 1} \right)$$

$$= \left(\frac{2.99}{3.99} \right)^{\nu-1} \left(\frac{1}{3.99} \right) = (0.0749)^{\nu-1}(0.251)$$

Solving for several values of ν, we conclude that the distribution of sequence length is quite broad:

ν	1	2	3	4	5	6
ϕ_ν	0.251	0.188	0.140	0.105	0.079	0.059

For the systems represented in Figure 5.3 and the equimolar case in Table 5.4, the average lengths are $\bar{\nu} = 1.173$ for $r_1 r_2 = 0.03$, $\bar{\nu} = 1.548$ for $r_1 r_2 = 0.30$, and $\bar{\nu} = 2.000$ for $r_1 r_2 = 1.0$.

Equation 5.5.11 and Equation 5.5.12 suggest a second method for the experimental determination of reactivity ratios, in addition to the copolymer composition equation. If the average sequence length can be determined for a feedstock of known composition, then r_1 and r_2 can be evaluated. We shall return to this possibility in the next section. In anticipation of applying this idea, let us review the assumptions and limitation to which Equation 5.5.11 and Equation 5.5.12 are subject:

1. The instantaneous monomer concentration must be used. Except at the azeotrope, this changes as the conversion of monomers to polymer progresses. As in Section 5.2, we assume that either the initial conditions apply (little change has taken place) or that monomers are continuously being added (replacement of reacted monomer).
2. The kinetic analysis described by Equation 5.5.3 and Equation 5.5.4 assumes that no repeat unit in the radical other than the terminal unit influences the addition. The penultimate unit in the radical as well as those still further from the growing end are assumed to have no effect.
3. Item (2) requires that each event in the addition process be independent of all others. We have consistently assumed this throughout this chapter, beginning with the copolymer composition equation. Until now we have said nothing about testing this assumption. Consideration of copolymer sequence length offers this possibility.

5.5.2 Terminal and Penultimate Models

We have suggested earlier that both the copolymer composition equation and the average sequence length offer possibilities for experimental evaluation of the reactivity ratios. Note that in so doing we are finding parameters which fit experimental results to the predictions of a model. Nothing about this tests the model itself. It could be argued that obtaining the same values of r_1 and r_2 from the fitting of composition and microstructure data would validate the model. It is not likely, however, that both types of data would be available and of sufficient quality to make this unambiguous. We shall examine the experimental side of this in the next section.

Statistical considerations make it possible to test the assumption of independent additions. Let us approach this topic by considering an easier problem: coin tossing. Under conditions where two events are purely random—as in tossing a fair coin—the probability of a specific sequence of outcomes is given by the product of the probabilities of the individual events. The probability of tossing a head followed by a head—indicated HH—is given by

$$p_{HH} = p_H p_H \tag{5.5.13}$$

If the events are not independent, provision must be made for this, so we define a quantity called the *conditional probability*. For the probability of a head *given the prior event of a head*, this is written $p_{H/H}$, where the first quantity in the subscript is the event under consideration and that following the slash mark is the prior condition. Thus $p_{T/H}$ is the probability of a tail following a head. If the events are independent, $p_{H/H} = p_H$; if not, then $p_{H/H}$ must be evaluated as a separate quantity. If the coin being tossed were biased, that is, if successive events are not independent, Equation 5.5.13 would become

$$p_{HH} = p_{H/H} p_H \tag{5.5.14}$$

We recall that the fraction of times a particular outcome occurs is used to estimate probabilities. Therefore we could evaluate $p_{H/H}$ by counting the number of times N_H the first toss yielded a head and the number of times N_{HH} two tosses yielded a head followed by a head and write

$$p_{H/H} = \frac{p_{HH}}{p_H} = \frac{N_{HH}}{N_H} \tag{5.5.15}$$

This procedure is readily extended to three tosses. For a fair coin the probability of three heads is the cube of the probability of tossing a single head:

$$p_{HHH} = p_H p_H p_H \tag{5.5.16}$$

If the coin is biased, conditional probabilities must be introduced:

$$p_{HHH} = p_{H/HH} p_{H/H} p_H \tag{5.5.17}$$

Using Equation 5.5.15 to eliminate $p_{H/H}$ from the last result gives

$$p_{HHH} = p_{H/HH} \left(\frac{p_{HH}}{p_H}\right) p_H \tag{5.5.18}$$

or

$$p_{H/HH} = \frac{p_{HHH}}{p_{HH}} = \frac{N_{HHH}}{N_{HH}} \tag{5.5.19}$$

If we were testing whether a coin were biased or not, we would use ideas like these as the basis for a test. We could count, for example, HHH and HH sequences and divide them according to Equation 5.5.19. If $p_{H/HH} \neq p_H$, we would be suspicious.

A similar logic can be applied to copolymers. The story is a bit more complicated to tell, so we only outline the method. If penultimate effects operate, then the probabilities p_{11}, p_{12}, etc., defined in Equation 5.5.3 through Equation 5.5.6 should be replaced by conditional probabilities. As a matter of fact, the kind of conditional probabilities needed must be based on the two preceding events. Thus Reaction (5.E) and Reaction (5.F) are two of the appropriate reactions, and the corresponding probabilities are $p_{1/11}$ and $p_{1/21}$. Rather than work out all of the probabilities in detail, we summarize the penultimate model as follows:

1. A total of eight different reactions are involved, since each reaction like Reaction (5.A) is replaced by a pair of reactions like Reaction (5.E) and Reaction (5.F).
2. There are eight different rate laws and rate constants associated with these reactions. Equation 5.2.1, for example, is replaced by Equation 5.2.5 and Equation 5.2.6.

3.   Eight rate constants are clustered in four ratios, which define new reactivity ratios. Thus r_1 as defined in Equation 5.2.13 is replaced by $r_1' = k_{111}/k_{112}$ and $r_1'' = k_{211}/k_{212}$ whereas r_2 is replaced by $r_2' = k_{222}/k_{221}$ and $r_2'' = k_{122}/k_{121}$.

4.   The probability p_{11} as given by Equation 5.5.3 is replaced by the conditional probability $p_{1/11}$, which is defined as

$$p_{1/11} = \frac{k_{111}[M_1M_1\cdot][M_1]}{k_{111}[M_1M_1\cdot][M_1] + k_{112}[M_1M_1\cdot][M_2]} = \frac{r_1'[M_1]/[M_2]}{1 + r_1'\{[M_1]/[M_2]}} \tag{5.5.20}$$

There are eight of these conditional probabilities, each associated with the reaction described in item (1).

5.   The probability p_{11} can be written as the ratio $N_{M_1M_1}/N_{M_1}$ using Equation 5.5.15. This is replaced by $p_{1/11}$, which is given by the ratio $N_{M_1M_1M_1}/N_{M_1M_1}$ according to Equation 5.5.19.

6.   Equation 5.5.4 shows that p_{11} is constant for a particular copolymer if the terminal model applies; therefore the ratio $N_{M_1M_1}/N_{M_1}$ also equals this constant. Equation 5.5.20 shows that $p_{1/11}$ is constant for a particular copolymer if the penultimate model applies; therefore the ratio $N_{M_1M_1M_1}/N_{M_1M_1}$ also equals this constant, but the ratio $N_{M_1M_1}/N_{M_1}$ does not have the same value.

These observations suggest how the terminal mechanism can be proved to apply to a copoly-merization reaction if experiments exist which permit the number of sequences of a particular length to be determined. If this is possible, we should count the number of M_1's (this is given by the copolymer composition) and the number of M_1M_1 and $M_1M_1M_1$ sequences. Specified sequences, of any definite composition, of two units are called dyads; those of three units, triads; those of four units, tetrads; those of five units, pentads; and so on. Next we examine the ratio $N_{M_1M_1}/N_{M_1}$ and $N_{M_1M_1M_1}/N_{M_1M_1}$. If these are the same, then the mechanism is *shown* to have terminal control; if not, it *may* be penultimate control. To prove the penultimate model it would also be necessary to count the number of M_1 tetrads. If the tetrad–triad ratio were the same as the triad–dyad ratio, the penultimate model is established.

This situation can be generalized. If the ratios do not become constant until the ratio of pentads to tetrads is considered, then the unit before the next to last—called the antepenultimate unit—plays a role in the addition. This situation has been observed, for example, for propylene oxide–maleic anhydride copolymers. The foregoing discussion has been conducted in terms of M_1 sequences. Additional relationships of the sort we have been considering also exist for dyads, triads, and so forth, of different types of specific composition. Thus an ability to investigate microstructure experimentally allows some rather subtle mechanistic effects to be studied. In the next section we shall see how such information is obtained.

5.6   Copolymer Composition and Microstructure: Experimental Aspects

As we have already seen, the reactivity ratios of a particular copolymer system determine both the composition and microstructure of the polymer. Thus it is important to have reliable values for these parameters. At the same time it suggests that experimental studies of composition and microstructure can be used to evaluate the various r's.

5.6.1   Evaluating Reactivity Ratios from Composition Data

Evaluation of reactivity ratios from the copolymer composition equation requires only composition data—that is, relatively straightforward analytical chemistry—and has been the method most widely used to evaluate r_1 and r_2. As noted in the last section, this method assumes terminal control and seeks the best fit of the data to that model. It offers no means for testing the model, and as we shall see, is subject to enough uncertainty to make even self-consistency difficult to achieve. Microstructure studies, by contrast, offer both a means to evaluate the reactivity ratios and also to test the model. The capability to investigate this level of structural detail was virtually nonexistent until the advent of modern instrumentation, and even now is limited to sequences of rather modest length.

In this section we shall use the evaluation of reactivity ratios as the unifying theme; the experimental methods constitute the new material introduced. The copolymer composition Equation 5.2.18 relates the r's to the mole fractions of the monomers in the feedstock and in the copolymer. To use the equation to evaluate r_1 and r_2, the composition of a copolymer resulting from a feedstock of known composition must be measured. The composition of the feedstock itself must also be known, but we assume this poses no problems. The copolymer specimen must be obtained by proper sampling procedures and purified of extraneous materials. Remember that monomers, initiators, and possibly solvents and soluble catalysts are involved in these reactions also, even though we have been focusing attention on the copolymer alone. The proportions of the two kinds of repeat unit in the copolymer are then determined by either chemical or physical methods. Elemental analysis is a widely used chemical method, but spectroscopic analysis (UV–visible, IR, NMR, and mass spectrometry) for functional groups is commonly employed.

Since the copolymer equation involves both r_1 and r_2 as unknowns, at least two polymers prepared from different feedstocks must be analyzed. It is preferable to use more than this minimum number of observations, and it is helpful to rearrange the copolymer composition equation into a linear form so that simple graphical methods can be employed to evaluate the r's. Several ways to linearize the equation exist:

1. Rearrange Equation 5.2.18 to give

$$\frac{f_1(1-2F_1)}{F_1(1-f_1)} = r_1\left(\frac{f_1^2(F_1-1)}{F_1(1-f_1)^2}\right) + r_2 \tag{5.6.1}$$

    This is the equation of a straight line, so r_1 and r_2 can be evaluated from the slope and intercept of an appropriate plot.

2. In terms of ratios rather than fractions, Equation 5.6.1 may be written as

$$\frac{[M_1]/[M_2]}{n_1/n_2}\left(\frac{n_1}{n_2}-1\right) = r_1\frac{([M_1]/[M_2])^2}{n_1/n_2} - r_2 \tag{5.6.2}$$

    where n_1 refers to the number of repeat units in the polymer. This expression is also of the form $y = mx + b$ if $x = ([M_1]/[M_2])^2/(n_1/n_2)$ and $y = ([M_1]/[M_2])/(n_1/n_2)/(n_1/n_2-1)$, so the slope and intercept yield r_1 and r_2, respectively. This type of analysis is known as a *Finemann–Ross plot*.

3. This last expression can be rearranged in several additional ways, which yield linear plots:

$$\frac{y}{x} = -r_2\frac{1}{x} + r_1 \tag{5.6.3}$$

$$x = \frac{1}{r_1}y + \frac{r_2}{r_1} \tag{5.6.4}$$

$$\frac{x}{y} = \frac{r_1}{r_2}\frac{1}{y} + \frac{1}{r_1} \tag{5.6.5}$$

Each of these forms weigh the errors in various data points differently, so some may be more suitable than others, depending on the precision of the data. Ideally all should yield the same values of the reactivity ratios. The following example illustrates the use of Equation 5.6.1 to evaluate r_1 and r_2.

Table 5.5 Values of F_1 as a Function for f_1 for the Methyl Acrylate (M_1)–Vinyl Chloride (M_2) System

f_1	F_1	f_1	F_1
0.075	0.441	0.421	0.864
0.154	0.699	0.521	0.900
0.237	0.753	0.744	0.968
0.326	0.828	0.867	0.983

Note: These data are also plotted in Figure 5.4.

Source: Data from Chapin, E.L., Ham, G., and Fordyce, R., *J. Am. Chem. Soc.*, 70, 538, 1948.

Example 5.4

The data in Table 5.5 list the mole fraction of methyl acrylate in the feedstock and in the copolymer for the methyl acrylate (M_1)–vinyl chloride (M_2) system. Use Equation 5.6.1 as the basis for the graphical determination of the reactivity ratios, which describe this system.

Solution

We calculate the variables to be used as ordinate and abscissa for the data in Table 5.5 using Equation 5.6.1:

$f_1(1-2F_1)/F_1(1-f_1)$	0.0217	−0.1036	−0.2087	−0.3832	−0.6127	−0.9668	−2.8102	−6.4061
$f_1^2(F_1-1)/F_1(1-f_1)^2$	−0.0083	−0.0143	−0.0316	−0.0486	−0.0832	−0.1315	−0.2792	−0.7349

Least-square analysis of these values gives a slope $r_1 = 8.929$ and an intercept $r_2 = 0.053$. Figure 5.4b shows these data plotted according to Equation 5.6.1. The line is drawn with the least-squares slope and intercept. The last point on the left in Figure 5.4b, which this line passes through, corresponds to $F_1 = 0.983$ and $f_1 = 0.867$. Because the functional form plotted involves the small differences $F_1 - 1$ and $1 - f_1$, this point is also subject to the largest error. This illustrates the value of having alternate methods for analyzing the data. The authors of this research carried out several different analyses of the same data; the values they obtained for r_1 and r_2 averaged over the various methods were $r_1 = 9.616 \pm 0.603$ and $r_2 = 0.0853 \pm 0.0239$. The standard deviations

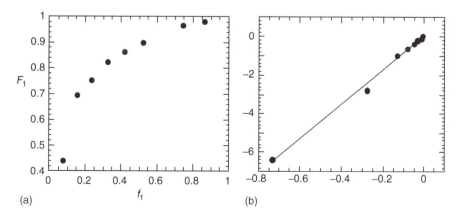

Figure 5.4 (a) Mole fraction of methyl acrylate in copolymers with vinyl chloride as a function of feedstock composition, and (b) Finemann–Ross plot to extract reactivity ratios, as described in Example 5.4.

of about 6% and 28% in r_1 and r_2 *analyzed from the same data* indicate the hazards of this method for determining r values.

5.6.2 Spectroscopic Techniques

In spite of the compounding of errors to which it is subject, the foregoing method was the best procedure for measuring reactivity ratios until the analysis of microstructure became feasible. Let us now consider this development. Most of the experimental information concerning copolymer microstructure has been obtained by modern instrumental methods. Techniques such as UV-visible, IR, NMR, and mass spectroscopy have all been used to good advantage in this type of research. Advances in instrumentation have made these physical methods particularly suitable to answer the question we pose: With what frequency do particular sequences of repeat units occur in a copolymer? The choice of the best method for answering this question is governed by the specific nature of the system under investigation. Few general principles exist beyond the importance of analyzing a representative sample of suitable purity. Our approach is to consider some specific examples. In view of the diversity of physical methods available and the number of copolymer combinations which exist, a few samples barely touch the subject. They will suffice to illustrate the concepts involved, however. The simpler question—What is the mole fraction of each repeat unit in the polymer sample?—can usually be answered via the same instrumental techniques.

Spectroscopic techniques based on the absorption of UV, visible, or IR radiation depend on the excitation from one quantum state to another. References in physical or analytical chemistry should be consulted for additional details, but a brief summary will be sufficient for our purposes:

1. The excitation energy ΔE reflects the separation between the final (subscript f) and initial (subscript i) quantum states:

    $$\Delta E = E_f - E_i \tag{5.6.6}$$

    The difference is positive for absorbed energy.
2. The energy absorbed is proportional to the frequency of the radiation via Planck's constant ($h = 6.63 \times 10^{-34}$ J s):

    $$\Delta E = h\nu = h\frac{c}{\lambda} \tag{5.6.7}$$

    In the second version of this equation c is the speed of light and λ the wavelength of the radiation.
3. The more widely separated two states are in energy, the shorter the wavelength of the radiation absorbed. Transitions between electronic states have higher energies, and correspond to UV–visible wavelengths, whereas vibrational quantum states are more closely spaced and are induced by IR radiation.
4. Different light-absorbing groups, called *chromophores*, absorb characteristic wavelengths, opening the possibility of qualitative analysis based on the location of an absorption peak.
5. If there is no band overlap in a spectrum, the absorbance at a characteristic wavelength is proportional to the concentration of chromophores present. This is the basis of quantitative analysis using spectra. With band overlap, things are more complicated but still possible.
6. The proportionality between the concentration of chromophores and the measured absorbance is given by Beer's law (recall the discussion in Section 3.3.4):

    $$A = \varepsilon b c \tag{5.6.8}$$

    where A is the (dimensionless) absorbance, b is the sample thickness, c is the chromophore concentration, and ε is the absorptivity. Usually quantitative measurements are facilitated by calibration with standards of known concentrations, so that ε, b, and various other instrumental parameters need not be determined individually.

7. For copolymers, or any other mixture of chromophores, the measured absorbance is given by the sum of individual Beer's law terms:

$$A = \varepsilon_1 bc_1 + \varepsilon_2 bc_2 + \varepsilon_3 bc_3 + \cdots \tag{5.6.9}$$

Recalling that ε depends on the chromophore and on the wavelength, measurements at different wavelengths can be used to extract the concentrations of each component. For a copolymer with two monomers, at least two wavelengths would be needed, and ideally they should be chosen to such that if ε_1 is large at λ_1, then ε_2 is large at λ_2.

$$A(\lambda_1) = \varepsilon_1(\lambda_1)bc_1 + \varepsilon_2(\lambda_1)bc_2$$
$$A(\lambda_2) = \varepsilon_1(\lambda_2)bc_1 + \varepsilon_2(\lambda_2)bc_2 \tag{5.6.10}$$

These relations amount to a system of two equations with two unknowns, c_1 and c_2, which can be solved in a straightforward manner.

NMR spectroscopy is especially useful for microstructure studies, because of the sensitivity to the chemical environment of a particular nucleus. We shall consider its application to copolymers now, and to questions of stereoregularity in Section 5.7. NMR has become such an important technique (actually a family of techniques) in organic chemistry that contemporary textbooks in the subject discuss its principles quite thoroughly, as do texts in physical and analytical chemistry, so here also we note only a few pertinent highlights:

1. Nuclei with an odd number of protons plus neutrons—especially ^{1}H and ^{13}C—possess magnetic moments and show two quantum states (spin up and spin down) in a strong magnetic field.
2. If energy of the proper frequency is supplied, a transition between these quantum states occurs with the absorption of an amount of energy equal to the separation of the states, just as in UV–visible and IR absorption. For NMR the frequency of the absorbed radiation lies in the radio frequency range and depends on the local magnetic field at the atom in question.
3. Electrons in a molecule also have magnetic moments and set up secondary magnetic fields, which partly screen each atom from the applied field. Thus atoms in different chemical environments display resonance at slightly different magnetic fields.
4. The displacement δ of individual resonances from that of a standard is small, and is measured in parts per million (ppm) relative to the applied field. These so-called *chemical shifts* are characteristic of a proton or carbon in a specific environment.
5. The interaction between nuclei splits resonances into multiple peaks, the number and relative intensity of which also assist in qualitative identification of the proton responsible for the absorption. Proton splitting is most commonly caused by the interaction of protons on adjacent carbons with the proton of interest. If there are m equivalent hydrogens on an adjacent carbon, the proton of interest produces $m + 1$ peaks by this coupling.
6. More distant coupling is revealed in high magnetic fields. Unresolved fine structures in a field of one strength may be resolved at higher field where more subtle long-range influences can be probed. The use of NMR spectroscopy to characterize copolymer microstructure takes advantage of this last ability to discern environmental effects that extend over the length of several repeat units. This capability is extremely valuable in analyzing the stereoregularity of a polymer, and we shall have more to say about it in that context in Section 5.7.
7. In NMR spectroscopy the "absorptivities" are, in essence, all the same, so that the integrated area under a peak is directly proportional to the number of nuclei of that type in the sample. Thus if different repeat units have identifiably different peaks, as is almost always the case, the relative abundance of each type can be extracted by peak integration without any additional calibration.

5.6.3 Sequence Distribution: Experimental Determination

As suggested in the foregoing, the analysis for overall composition in a copolymer sample is by now a relatively straightforward affair. The analysis for sequence distribution, however, is not. The primary difficulty is that the energy of a particular transition, be it electronic, vibrational, or nuclear, is determined primarily by the immediate chromophore of interest, and only weakly influenced by chemically bonded neighbors. NMR offers the most promise in this respect, especially with the advent of higher magnetic fields; this feature can provide sufficient resolution to detect the influence of repeat units up to about five monomers down the chain. Nevertheless, there are cases where UV–visible spectroscopy can help. An elegant example is the copolymer of styrene (molecule 1) and 1-chloro-1,3-butadiene (molecule 2). These molecules quantitatively degrade with the loss of HCl upon heating in base solution. This restores 1,3-unsaturation to the butadiene repeat unit:

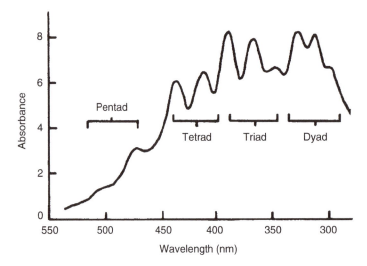

$$\text{(5.N)}$$

It is these conjugated double bonds that are the chromophores of interest in this system. What makes this particularly useful is the fact that the absorption maximum for this chromophore is displaced to longer wavelengths the more conjugated bonds there are in a sequence. Qualitatively, this can be understood in terms of a one-dimensional particle in a box model for which the energy level spacing is inversely proportional to the square of the length of the box. In this case the latter increases with the length of the conjugated polyene system. This in turn depends on the number of consecutive butadiene repeat units in the copolymer. For an isolated butadiene molecule dehalogenation produces one pair of conjugated double bonds; two adjacent butadienes, four conjugated double bonds; three adjacent butadienes, six conjugated double bonds; and so on. Sequences of these increasing lengths are expected to absorb at progressively longer wavelengths.

Figure 5.5 shows the appropriate portion of the spectrum for a copolymer prepared from a feedstock for which $f_1 = 0.153$. It turns out that each polyene produces a set of three bands: the

Figure 5.5 Ultraviolet–visible spectrum of dehydrohalogenated copolymers of styrene–1-chloro-1,3-butadiene. (Redrawn from Winston, A. and Wichacheewa, P., *Macromolecules*, 6, 200, 1973. With permission.)

dyad is identified with the peaks at $\lambda = 298$, 312, and 327 nm; the triad with the peaks at $\lambda = 347$, 367, and 388 nm; and the tetrad with the peaks at $\lambda = 412$ and 437 nm. Apparently one of the tetrad bands overlaps that of the triad and is not resolved. Likewise, only one band (at 473 nm) is observed for the pentad. The identification of these features can be confirmed with model compounds and the location and relative intensities of the peaks have been shown to be independent of copolymer composition. Once these features have been identified, the spectra can be interpreted in terms of the numbers of dyads, triads, tetrads, and maybe pentads of the butadiene units and compared with predicted sequences of various lengths. Further consideration of this system is left for Problem 3 through Problem 5 at the end of the chapter.

We now illustrate the application of NMR to gather copolymer sequence information. Suppose we consider the various triads of repeat units. There are six possibilities: $M_1M_1M_1$, $M_1M_1M_2$, $M_2M_1M_2$, $M_2M_2M_2$, $M_2M_2M_1$, and $M_1M_2M_1$. These can be divided into two groups of three, depending on the identity of the central unit. Thus the center of a triad can be bracketed by two monomers identical to itself, different from itself, or by one of each. In each of these cases the central repeat unit is in a different environment, and a characteristic proton in that repeat unit is in a different location, depending on the effect of that environment. As a specific example, consider the methoxy group in poly(methyl methacrylate). The hydrogens in the group are magnetically equivalent and hence produce a single resonance at $\delta = 3.74$ ppm. Now suppose we look for the same resonance feature in the copolymer of methyl methacrylate (M_1) and acrylonitrile (M_2). Figure 5.6 shows that 60 MHz spectrum of several of these copolymers in the neighborhood of the methoxy resonance. Three resonance peaks rather than one are observed. Figure 5.6 also lists the methyl methacrylate content of each of these polymers. As the methyl methacrylate content decreases, the peak on the right decreases and the left increases. We therefore identify the peak on the right-hand peak with the $M_1M_1M_1$ sequence, the left-hand peak with $M_2M_1M_2$, and the peak in the center with $M_1M_1M_2$. The $M_1M_1M_1$ peak occurs at the same location as in the methyl methacrylate homopolymer.

The areas under the three peaks give the relative proportions of three sequences. In the following example we consider some results on dyad sequences determined by comparable procedures in vinylidene chloride–isobutylene copolymers.

Example 5.5

The mole fractions of various dyads in the vinylidine chloride (M_1)–isobutylene (M_2) system were determined[†] by NMR spectroscopy. A selection of the values obtained are listed below, as well as the compositions of the feedstocks from which the copolymers were prepared; assuming terminal control, evaluate r_1 from each of the first three sets of data, and r_2 from each of the last three.

	Mole fraction of dyads		
f_1	11	12	22
0.584	0.68	0.29	—
0.505	0.61	0.36	—
0.471	0.59	0.38	—
0.130	—	0.67	0.08
0.121	—	0.66	0.10
0.083	—	0.64	0.17

[†] J.B. Kinsinger, T. Fischer, and C.W. Wilson, *Polym. Lett.*, 5, 285 (1967).

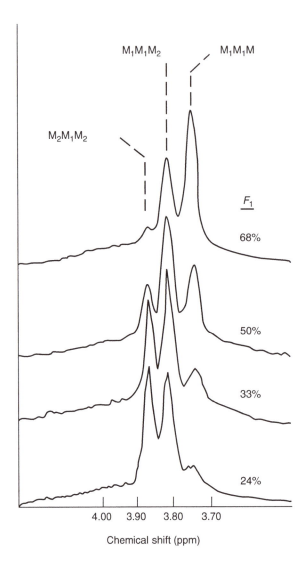

Figure 5.6 Chemical shift (from hexamethyldisiloxane) for acrylonitrile–methyl methacrylate copolymers of the indicated methyl methacrylate (M_1) content. Methoxyl resonances are labeled as to the triad source. (From Chujo, R., Ubara, H., and Nishioka, A., *Polym. J.*, 3, 670, 1972. With permission.)

Solution

Equation 5.5.3 and Equation 5.5.5 provide the method for evaluating the r's from the data given. We recognize that a 12 dyad can come about from 1 adding to 2 as well as from 2 adding to 1; therefore we use half the number of 12 dyads as a measure of the number of additions of monomer 2 to chain end 1. Accordingly, by Equation 5.5.1,

$$p_{11} = \frac{N_{11}}{N_{11} + (1/2)N_{12}} = \frac{2N_{11}}{2N_{11} + N_{12}} \quad \text{and} \quad p_{22} = \frac{2N_{22}}{2N_{22} + N_{12}}$$

Since $[M_1]/[M_2] = f_1/(1-f_1)$, Equation 5.5.2 can be written

$$p_{11} = \frac{r_1[f_1/(1-f_1)]}{1 + r_1[f_1/(1-f_1)]} \quad \text{and} \quad p_{22} = \frac{r_2[(1-f_1)/f_1]}{1 + r_2[(1-f_1)/f_1]}$$

From		From	
$\dfrac{2N_{11}}{2N_{11}+N_{12}}=\dfrac{r_1[f_1/(1-f_1)]}{1+r_1[f_1/(1-f_1)]}$		$\dfrac{2N_{22}}{2N_{22}+N_{12}}=\dfrac{r_2[(1-f_1)/f_1]}{1+r_2[(1-f_1)/f_1]}$	
f_1	r_1	f_1	r_2
0.584	3.33	0.130	0.036
0.505	3.32	0.121	0.042
0.471	3.48	0.083	0.048
Average	3.38	Average	0.042

Particularly when r values are close to zero, this method for evaluating small r's is superior to the graphical analysis of composition data (compare Example 5.4 and Figure 5.4).

By making measurements at higher magnetic fields, it is possible to resolve spectral features arising from still longer sequences. As a matter of fact, the authors of the research described in the last example were able to measure the fractions of tetrads of different composition in the same vinylidene chloride–isobutylene copolymer. Based on the longer sequences, they concluded that the penultimate model describes this system better than the terminal model, although the short-comings of the latter are not evident in the example. Problem 6 and Problem 7 at the end of the chapter also refer to this system.

5.7 Characterizing Stereoregularity

We introduced the concept of stereoregularity in Section 1.6. Figure 1.3 illustrates isotactic, syndiotactic, and atactic structures of a vinyl polymer in which successive repeat units along the fully extended chain lie, respectively, on the same side, alternating sides, or at random with respect to the backbone. It is important to appreciate the fact that these different structures—different configurations—have their origin in the bonding of the polymer, and no amount of rotation around bonds—changes in conformation—will convert one structure into another.

Our discussion of stereoregularity in this chapter is primarily concerned with polymers of monosubstituted ethylene repeat units. We shall represent these by

Monosubstituted ethylene

In this representation the X indicates the substituent; other bonds involve only hydrogens. This formalism also applies to 1,1-disubstituted ethylenes in which the substituents are different. With these symbols, the isotactic, syndiotactic, and atactic structures shown in Figure 1.3 are represented by Structure (5.III) through Structure (5.V), respectively:

(5.III)

(5.IV)

(5.V)

The carbon atoms carrying the substituents are not truly asymmetric, since the two chain sections—while generally of different length—are locally the same on either side of any carbon atom, except near the ends of the chain. As usual, we ignore any uniqueness associated with chain ends.

There are several topics pertaining to stereoregularity that we shall not cover to simplify the presentation:

1. Stereoregular copolymers. We shall restrict our discussion to stereoregular homopolymers.
2. Complications arising from other types of isomerism. Positional and geometrical isomerism, also described in Section 1.6, will be excluded for simplicity. In actual polymers these are not always so easily ignored.
3. Polymerization of 1,2-disubstituted ethylenes. Since these introduce two different "asymmetric" carbons into the polymer backbone (second substituent Y), they have the potential to display ditacticity. Our attention to these is limited to the illustration of some terminology, which is derived from carbohydrate nomenclature (Structure (5.VI) through Structure (5.IX)).

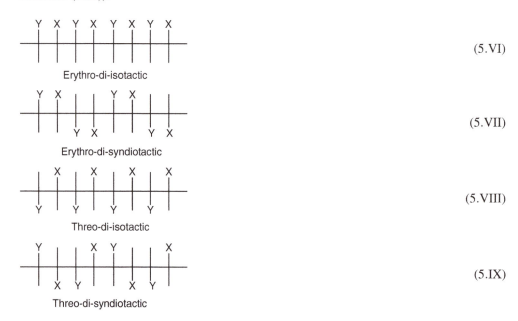

$$(5.VI)$$

Erythro-di-isotactic

$$(5.VII)$$

Erythro-di-syndiotactic

$$(5.VIII)$$

Threo-di-isotactic

$$(5.IX)$$

Threo-di-syndiotactic

The successive repeat units in Structure (5.III) through Structure (5.V) are of two different kinds. If they were labeled M_1 and M_2, we would find that, as far as microstructure is concerned, isotactic polymers are formally the same as homopolymers, syndiotactic polymers are formally the same as alternating copolymers, and atactic polymers are formally the same as random copolymers. The analog of block copolymers, stereoblock polymers, also exist. Instead of using M_1 and M_2 to differentiate between the two kinds of repeat units, we shall use the letters D and L as we did in Chapter 1.

The statistical nature of polymers and polymerization reactions has been illustrated at many points throughout this volume. It continues to be important in the discussion of stereoregularity. Thus it is generally more accurate to describe a polymer as, say, predominantly isotactic rather than perfectly isotactic. More quantitatively, we need to be able to describe a polymer in terms of the percentage of isotactic, syndiotactic, and atactic sequences.

Certain bulk properties of polymers also reflect differences in stereoregularity. We will see in Chapter 13 that crystallinity is virtually impossible unless a high degree of stereoregularity is present in a polymer. Since crystallinity plays such an important part in determining the mechanical properties of polymers, stereoregularity manifests itself in these other behaviors also. These gross, bulk properties provide qualitative evidence for differences in stereoregularity, but, as with copolymers, it is the microstructural detail that quantitatively characterizes the tacticity of a polymer. We shall examine the statistics of this situation in the next section, and the application of NMR in Section 5.9.

The analogy between stereoregular polymers and copolymers can be extended still further. We can write chemical equations for propagation reactions leading to products that differ in configuration along with the associated rate laws. We do this without specifying anything—at least for now—about the mechanism. There are several things that need to be defined to do this:

1. These are addition polymerizations in which chain growth is propagated through an active center. The latter could be a free radical or an ion; we shall see that a coordinated intermediate is the more usual case.
2. The active-center chain end is open to front or rear attack in general; hence the configuration of a repeat unit is not fixed until the next unit attaches to the growing chain.
3. The reactivity of a growing chain is, as usual, assumed to be independent of chain length. In representing this schematically, as either DM* or LM*, the M* indicates the terminal active center, and the D or L, the penultimate units of fixed configuration. From a kinetic point of view, we ignore what lies further back along the chain.
4. As in Chapter 3 and Chapter 4, the monomer is represented by M.

With these definitions in mind, we can write

$$—DM^* + M \nearrow\searrow \begin{matrix}DDM^*\\DLM^*\end{matrix} \quad \text{or} \quad —LM^* + M \nearrow\searrow \begin{matrix}LLM^*\\LDM^*\end{matrix} \tag{5.O}$$

What is significant about these reactions is that only two possibilities exist: addition with the same configuration (D → DD or L → LL) or addition with the opposite configuration (D → DL or L → LD). We shall designate these isotactic (subscript i) or syndiotactic (subscript s) additions, respectively, and shall define the rate constants for the two steps k_i and k_s. Therefore the rates of isotactic and syndiotactic propagation become

$$R_{p,i} = k_i[M^*][M] \tag{5.7.1}$$

and

$$R_{p,s} = k_s[M^*][M] \tag{5.7.2}$$

and, since the concentration dependences are identical, the relative rate of the two processes is given by the ratio of the rate constants. This same ratio also gives the relative number of dyads having the same or different configurations:

$$\frac{R_{p,i}}{R_{p,s}} = \frac{k_i}{k_s} = \frac{\text{Number dyads with same configuration}}{\text{Number dyads with different configurations}} \tag{5.7.3}$$

The Arrhenius equation enables us to expand on this still further:

$$\frac{\text{Iso dyads}}{\text{Syndio dyads}} = \frac{A_i}{A_s} e^{-(E_i^* - E_s^*)/RT} \tag{5.7.4}$$

The main conclusion we wish to draw from this line of development is that the difference between E_i^* and E_s^* could vary widely, depending on the nature of the active center. If the active center in a polymerization is a free radical unencumbered by interaction with any surrounding species, we would expect $E_i^* - E_s^*$ to be small. Experiment confirms this expectation; for vinyl chloride it is on the order of 1.3 kJ mol^{-1}. Thus at the temperatures usually encountered in free-radical polymerizations (ca. 60°C), the exponential in Equation 5.7.4 is small and the proportions of isotactic and syndiotactic dyads are roughly equal. This is the case for poly(vinyl chloride), for which $k_i/k_s = 0.63$ at 60°C. The preference for syndiotactic addition is greater than this (i.e., $E_i^* - E_s^*$ is larger) in some systems, apparently because there is less repulsion between substituents when they are staggered in the transition state. In all cases, whatever difference in activation energies exists manifests itself in product composition to a greater extent at low temperatures. At high temperatures small differences in E^* value are leveled out by the high average thermal energy available.

The foregoing remarks refer explicitly to free-radical polymerizations. If the active center is some kind of associated species—an ion pair or a coordination complex—then predictions based on unencumbered intermediates are irrelevant. It turns out that the Ziegler–Natta catalysts—which won their discoverers the Nobel Prize—apparently operate in this way. The active center of the chain coordinates with the catalyst in such a way as to block one mode of addition. High levels of stereoregularity are achieved in this case. Although these substances also initiate the polymerization, the term *catalyst* is especially appropriate in the present context, since the activation energy for one mode of addition is dramatically altered relative to the other by these materials. We shall discuss the chemical makeup of Ziegler–Natta catalysts and some ideas about how they work in Section 5.10. For now it is sufficient to recognize that these catalysts introduce a real bias into Equation 5.7.4 and thereby favor one pattern of addition.

In the next section we take up the statistical description of various possible sequences.

5.8   A Statistical Description of Stereoregularity

Since it is unlikely that a polymer will possess perfect stereoregularity, it is desirable to assess this property quantitatively, both to describe the polymer and to evaluate the effectiveness of various catalysts in this regard. In discussing tacticity in terms of microstructure, it has become conventional to designate a dyad as *meso* if the repeat units have the same configuration, and as *racemic* if the configuration is reversed. The terminology is derived from the stereochemistry of small molecules; its basis is seen by focusing attention on the methylene group in the backbone of the vinyl polymer. This methylene lies in a plane of symmetry in the isotactic molecule [5.X],

$$(5.X)$$

and thereby defines a meso (subscript m) structure as far as the dyad is concerned. Considering only the dyad, we see that these two methylene protons are in different environments. Therefore each will show a different chemical shift in an NMR spectrum. In addition, each proton splits the resonance of the other into a doublet, so a quartet of peaks appears in the spectrum. Still considering only the dyad, we see that the methylene in a syndiotactic grouping [5.XI] contains two protons in identical environments:

$$(5.XI)$$

These protons show a single chemical shift in the NMR spectrum. This is called a racemic (subscript r) structure, since it contains equal amounts of D and L character. In the next section we shall discuss the NMR spectra of stereoregular polymers in more detail.

If we define p_m and p_r as the probability of addition occurring in the meso and racemic modes, respectively, then $p_m + p_r = 1$, since there are only two possibilities. The probability p_m is the analog of p_{ij} for copolymers; hence, by analogy with Equation 5.5.1, this equals the fraction of isotactic dyads among all dyads. In terms of the kinetic approach of the last section, p_m is equal to the rate of an iso addition divided by the combined rates of iso and syndio additions:

$$p_m = \frac{k_i}{k_i + k_s} \qquad (5.8.1)$$

This expression is the equivalent of Equation 5.5.2 for copolymers.

The system of notation we have defined can readily be extended to sequences of greater length. Table 5.6 illustrates how either m or r dyads can be bracketed by two additional repeat units to form a tetrad. Each of the outer units is either m or r with respect to the unit it is attached to, so the meso dyad generates three tetrads. Note that the tetrads mmr and rmm are equivalent and are not distinguished. A similar set of tetrads is generated from the r dyad.

The same system of notation can be extended further by focusing attention on the backbone substituents rather than on the methylenes. Consider bracketing a center substituent with a pair of monomers in which the substituents have either the same or opposite configurations as the central substituent. Thus the probabilities of the resulting triads are obtained from the probabilities of the respective m or r additions. The following possibilities exist:

Table 5.6 The Splitting of Meso and Racemic Dyads into Six Tetrads

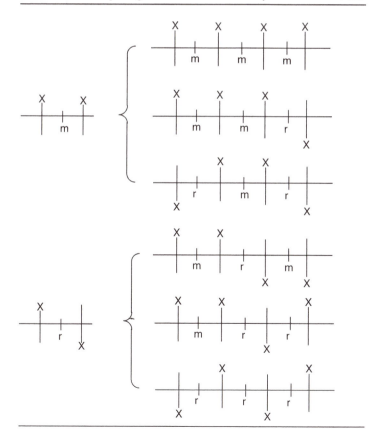

1. An isotactic triad (5.XII) is generated by two successive meso additions:

$$(5.XII)$$

The probability of the isotactic triad is

$$p_i = p_m^2 \tag{5.8.2}$$

2. A syndiotactic triad (5.XIII) is generated by two successive racemic additions:

$$(5.XIII)$$

The probability of the syndiotactic triad is given by p_r^2, which becomes

$$p_s = (1 - p_m)^2 \tag{5.8.3}$$

3. A heterotactic triad (5.XIV) is generated by mr and rm sequences of additions:

$$(5.XIV)$$

The probability of a heterotactic (subscript h) triad is

$$p_h = 2p_m(1 - p_m) \tag{5.8.4}$$

The factor 2 arises because this particular sequence can be generated in two different orders.

These triads can also be bracketed by two more units to generate 10 different pentads following the pattern established in Table 5.6. It is left for the reader to verify this number by generating the various structures.

The probabilities of the various dyad, triad, and other sequences that we have examined have all been described by a single probability parameter p_m. When we used the same kind of statistics for copolymers, we called the situation one of terminal control. We are considering similar statistics here, but the idea that the stereochemistry is controlled by the terminal unit is inappropriate. The active center of the chain end governs the *chemistry* of the addition, but not the *stereochemistry*. Equation 5.7.1 and Equation 5.7.2 merely state that an addition must be of one kind or another, but that the rates are not necessarily identical.

A mechanism in which the stereochemistry of the growing chain does exert an influence on the addition might exist, but at least two repeat units in the chain are required to define any such stereochemistry. Therefore this possibility is equivalent to the penultimate mechanism in copolymers. In this case the addition would be described in terms of conditional probabilities, just as Equation 5.5.20 does for copolymers. Thus the probability of an isotactic triad controlled by the stereochemistry of the growing chain would be represented by the reaction.

$$(5.P)$$

and described by the probability

$$p_{control} = p_m p_{m/m} \tag{5.8.5}$$

where the conditional probability $p_{m/m}$, is the probability of an m addition, given the fact of a prior m addition. As with copolymers, triads must be considered in order to test whether the simple statistics apply. Still longer sequences need to be examined to test whether stereochemical control is exerted by the chain. Although such situations are known, we shall limit our discussion to the simple case where the single probability p_m is sufficient to describe the various additions. The latter, incidentally, may be called *zero-order Markov* (or *Bernoulli*) *statistics* to avoid the vocabulary of terminal control. The case where the addition is influenced by whether the last linkage in the chain is m or r is said to follow a first-order Markov process.

The number of m or r linkages in an "*n*-ad" is $n-1$. Thus dyads are characterized by a single linkage (either m or r), triads by two linkages (either mm, mr, or rr), and so forth. The m and r notation thus reduces by 1 the order of the description from what is obtained when the repeat units themselves are described. For this reason the terminal control mechanism for copolymers is a first-order Markov process and the penultimate model is a second-order Markov process. Note that the compound probabilities which describe the probability of an *n*-ad in terms of p_m are also of order $n-1$. In the following example we calculate the probability of various triads on the basis of zero-order Markov statistics.

Example 5.6

Use zero-order Markov statistics to evaluate the probability of isotactic, syndiotactic, and heterotactic triads for the series of p_m values spaced at intervals of 0.1. Plot and comment on the results.

Solution

Evaluate Equation 5.8.2 through Equation 5.8.4 for p_m between zero and unity; these results are plotted in Figure 5.7.

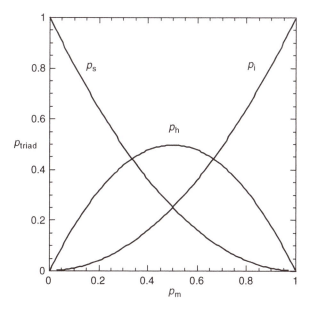

Figure 5.7 Fractions of iso, syndio, and hetero triads as a function of p_m, calculated assuming zero-order Markov statistics in Example 5.6.

p_m	p_m^2	$(1-p_m)^2$	$2p_m(1-p_m)$
0	0	0	0
0.1	0.01	0.81	0.18
0.2	0.04	0.64	0.32
0.3	0.09	0.49	0.42
0.4	0.16	0.36	0.48
0.5	0.25	0.25	0.50
0.6	0.36	0.16	0.48
0.7	0.49	0.09	0.42
0.8	0.64	0.04	0.32
0.9	0.81	0.01	0.18
1.0	1.0	0	0

The following observations can be made from these calculations:

1. The probabilities give the fractions of the three different types of triads in the polymer.
2. If the fractions of triads could be measured, they either would or would not lie on a single vertical line in Figure 5.7. If they did occur at a single value of p_m, this would not only give the value of p_m (which could be obtained from the fraction of one kind of triad), but would also prove the statistics assumed. If the fractions were not consistent with a single p_m value, higher-order Markov statistics are indicated.
3. The fraction of isotactic sequences increases as p_m increases, as required by the definition of these quantities.
4. The fraction of syndiotactic sequences increases as $p_m \rightarrow 0$, which corresponds to $p_r \rightarrow 1$.
5. The fraction of heterotactic triads is a maximum at $p_m = p_r = 0.5$ and drops to zero at either extreme.
6. For an atactic polymer the proportions of isotactic, syndiotactic, and heterotactic triads are 0.25:0.25:0.50.

To investigate the triads by NMR, the resonances associated with the chain substituent are examined, since Structure (5.XII) through Structure (5.XIV) show that it is these that experience different environments in the various triads. If dyad information is sufficient, the resonances of the methylenes in the chain backbone are measured. Structure (5.X) and Structure (5.XI) show that these serve as probes of the environment in dyads. In the next section we shall examine in more detail how this type of NMR data is interpreted.

5.9　Assessing Stereoregularity by Nuclear Magnetic Resonance

It is not the purpose of this book to discuss in detail the contributions of NMR spectroscopy to the determination of molecular structure. This is a specialized field in itself and a great deal has been written on the subject. In this section we shall consider only the application of NMR to the elucidation of stereoregularity in polymers. Numerous other applications of this powerful technique have also been made in polymer chemistry, including the study of positional and geometrical isomerism (Section 1.6) and copolymers (Section 5.7). We shall also make no attempt to compare the NMR spectra of various different polymers; instead, we shall examine primarily the NMR spectra of different poly(methyl methacrylate) preparations to illustrate the capabilities of the method using the first system that was investigated by this technique as the example.

Figure 5.8 shows the 60 MHz spectra of poly(methyl methacrylate) prepared with different catalysts so that predominantly isotactic, syndiotactic, and atactic products are formed. The three spectra in Figure 5.8 are identified in terms of this predominant character. It is apparent that the

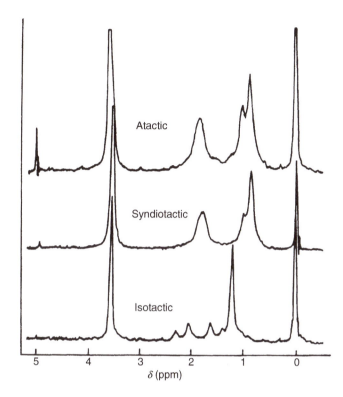

Figure 5.8 Nuclear magnetic resonance spectra of three poly(methyl methacrylate) samples. Curves are labeled according to the predominant tacticity of samples. (From McCall, D.W. and Slichter, W.P., in *Newer Methods of Polymer Characterization*, Ke, B. (Ed.), Interscience, New York, 1964. With permission.)

spectra are quite different, especially in the range of δ values between 1 and 2 ppm. Since the atactic polymer has the least regular structure, we concentrate on the other two to make the assignment of the spectral features to the various protons.

Several observations from the last section provide the basis of interpreting these spectra:

1.   Hydrogens of the methylene group in the backbone of the poly(methyl methacrylate) produce a single peak in a racemic dyad, as illustrated by Structure (5.XIII).
2.   The same group of hydrogens in a meso dyad (5.X) produces a quartet of peaks: two different chemical shifts, each split into two by the two hydrogens in the methylene.
3.   The peaks centered at $\delta = 1.84$ ppm—a singlet in the syndiotactic and a quartet in the isotactic polymers—are thus identified with these protons. This provides an unambiguous identification of the predominant stereoregularity of these samples.
4.   The features that occur near $\delta = 1.0$ ppm are associated with the protons of the α-methyl group. The location of this peak depends on the configurations of the nearest neighbors.
5.   Working from the methylene assignments, we see that the peak at $\delta = 1.22$ ppm in the isotactic polymer arises from the methyl in the center of an isotactic triad, the peak at $\delta = 0.87$ ppm from a syndiotactic triad, and the peak at $\delta = 1.02$ ppm from a homotactic triad.
6.   The peak at $\delta = 3.5$ ppm is due to the methoxy group.

Once these assignments are made, the areas under the various peaks can be measured to determine the various fractions:

1.   The area under the methylene peaks is proportional to the dyad concentration: The singlet gives the racemic dyads and the quartet gives the meso dyads.

2. The area under one of the methyl peaks is proportional to the concentration of the corresponding triad.
3. It is apparent that it is not particularly easy to determine the exact areas of these features when the various contributions occur together to any significant extent. This is clear from the atactic spectrum, in which slight shoulders on both the methylene and methyl peaks are the only evidence of meso methylenes and iso methyls.

The spectra shown in Figure 5.8 were early attempts at this kind of experiment, and the measurement of peak areas in this case was a rather subjective affair. We shall continue with an analysis of these spectra, even though improved instrumentation has resulted in greatly enhanced spectra. One development that has produced better resolution is the use of higher magnetic fields. As the magnetic field increases, the chemical shifts for the various features are displaced proportionately. The splitting caused by spin–spin coupling, on the other hand, is unaffected. This can produce a considerable sharpening of the NMR spectrum. Other procedures such as spin decoupling, isotopic substitution, computerized stripping of superimposed spectra, and ^{13}C-NMR also offer methods for identifying and quantifying NMR spectra.

Table 5.7 lists the estimated fractions of dyads of types m and r and the fractions of triads of types i, s, and h from Figure 5.8. These fractions represent the area under a specific peak (or four peaks in the case of the meso dyads) divided by the total area under all of the peaks in either the dyad or triad category. As expected for the sample labeled isotactic, 89% of the triads are of type i and 87% of the dyads are of type m. Likewise, in the sample labeled syndiotactic, 68% of the triads are s and 83% of the dyads are r.

The sample labeled atactic in Figure 5.8 was prepared by a free-radical mechanism and is expected to follow zero-order Markov statistics. As a test for this, we examine Figure 5.7 to see whether the values of p_i, p_s, and p_h, which are given by the fractions in Table 5.7, agree with a single set of p_m values. When this is done, it is apparent that these proportions are consistent with this type of statistics within experimental error and that $p_m \cong 0.25$ for poly(methyl methacrylate). Under the conditions of this polymerization, the free-radical mechanism is biased in favor of syndiotactic additions over isotactic additions by about 3:1, according to Equation 5.8.1. Presumably this is due to steric effects involving the two substituents on the α-carbon.

With this kind of information it is not difficult to evaluate the average lengths of isotactic and syndiotactic sequences in a polymer. As a step toward this objective, we define the following:

1. The number of isotactic sequences containing n_i iso repeat units is N_{n_i}.
2. The number of syndiotactic sequences containing n_s syndio repeat units is N_{n_s}.
3. Since isotactic and syndiotactic sequences must alternate, it follows that:

$$\sum N_{n_i} = \sum N_{n_s} \tag{5.9.1}$$

Table 5.7 The Fractions of Meso and Racemic Dyads and Iso, Syndio, and Hetero Triads for the Data in Figure 5.8

Sample	Dyads		Triads		
	Meso	Racemic	Iso	Syndio	Hetero
Atactic	0.22	0.78	0.07	0.55	0.38
Syndiotactic	0.17	0.83	0.04	0.68	0.28
Isotactic	0.87	0.13	0.89	0.04	0.07

Source: Data from McCall, D.W. and Slichter, W.P. in *Newer Methods of Polymer Characterization*, Ke, B. (Ed.), Interscience, New York, 1964.

4. The number of iso triads in a sequence of n_i iso repeat units is n_i-1, and the number of syndio triads in a sequence of n_s syndio repeat units is n_s-1. We can verify these relationships by examining a specific chain segment:

    $$-DDLDLDLDLD*DDDDDDDDL-$$

    In this example both the iso and syndio sequences consist of eight repeat units, with seven triads in each. The repeat unit marked * is counted as part of each type of triad, but is itself the center of a hetero triad.
5. The number of racemic dyads in a sequence is the same as the number of syndiotactic units n_s. The number of meso dyads in a sequence is the same as the number of iso units n_i. These can also be verified from structure above.

    With these definitions in mind, we can immediately write expressions for the ratio of the total number of iso triads, ν_i, to the total number of syndio triads, ν_s:

$$\frac{\nu_i}{\nu_s} = \frac{\sum N_{n_i}(n_i - 1)}{\sum N_{n_s}(n_s - 1)} = \frac{\sum N_{n_i}(n_i) - \sum N_{n_i}}{\sum N_{n_s}(n_s) - \sum N_{n_s}} \tag{5.9.2}$$

In this equation the summations are over all values of n of the specified type. Also remember that the ν's and n's in this discussion (with subscript i or s) are defined differently from the ν's and n's defined earlier in the chapter for copolymers. Using Equation 5.9.1 and remembering the definition of an average provided by Equation 1.7.7, we see that Equation 5.9.2 becomes

$$\frac{\nu_i}{\nu_s} = \frac{\bar{n}_i - 1}{\bar{n}_s - 1} \tag{5.9.3}$$

where the overbar indicates the average length of the indicated sequence.

A similar result can be written for the ratio of the total number (ν) of dyads of the two types (m and r), using item (5) above:

$$\frac{\nu_m}{\nu_r} = \frac{\sum N_{n_i}(n_i)}{\sum N_{n_s}(n_s)} = \frac{\bar{n}_i}{\bar{n}_s} \tag{5.9.4}$$

Equation 5.9.3 and Equation 5.9.4 can be solved simultaneously for $\bar{n}_i$ and $\bar{n}_s$ in terms of the total number of dyads and triads:

$$\bar{n}_i = \frac{1 - \nu_i/\nu_s}{1 - (\nu_i/\nu_s)(\nu_r/\nu_m)} \tag{5.9.5}$$

and

$$\bar{n}_s = \frac{1 - \nu_i/\nu_s}{(\nu_m/\nu_r) - (\nu_i/\nu_s)} \tag{5.9.6}$$

Use of these relationships is illustrated in the following example.

Example 5.7

Use the dyad and triad fractions in Table 5.7 to calculate the average lengths of isotactic and syndiotactic sequences for the polymers of Figure 5.8. Comment on the results.

Solution

Since the total numbers of dyads and triads always occur as ratios in Equation 5.9.3 and Equation 5.9.4, both the numerators and denominators of these ratios can be divided by the total number of dyads or triads to convert these total numbers into fractions, i.e.,

$$\nu_i/\nu_s = (\nu_i/\nu_{tot})/(\nu_s/\nu_{tot}) = p_i/p_s$$

Thus the fractions in Table 5.7 can be substituted for the ν's in Equation 5.9.3 and Equation 5.9.4. The values of $\bar{n}_i$ and $\bar{n}_s$ so calculated for the three polymers are:

	$\bar{n}_i$	$\bar{n}_s$
Atactic	1.59	5.64
Syndiotactic	1.32	6.45
Isotactic	9.14	1.37

This analysis adds nothing new to the picture already presented by the dyad and triad probabilities. It is somewhat easier to visualize an average sequence, however, although it must be remembered that the latter implies nothing about the distribution of sequence lengths.

We conclude this section via Figure 5.9, which introduces the use of ^{13}C-NMR obtained at 100 MHz for the analysis of stereoregularity in polypropylene. This spectrum shows the carbons on the pendant methyl groups for an atactic polymer. Individual peaks are resolved for all the possible pentad sequences. Polypropylene also serves as an excellent starting point for the next section, in which we examine some of the catalysts that are able to control stereoregularity in such polymers.

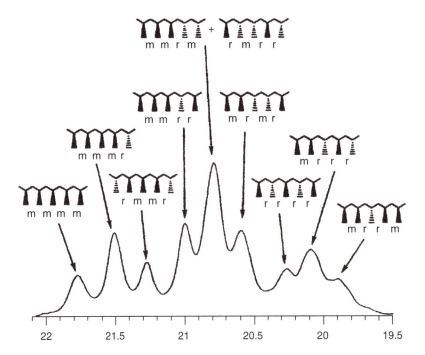

Figure 5.9 ^{13}C-NMR assignments for polypropylene. (From Bruce, M.D. and Waymouth, R.M., *Macromolecules*, 31, 2707, 1998. With permission.)

5.10   Ziegler–Natta Catalysts

In this discussion we consider Ziegler–Natta catalysts and their role in achieving stereoregularity. This is a somewhat restrictive view of the situation, since there are other catalysts—such as phenyl magnesium bromide, a Grignard reagent—which can produce stereoregularity; the Ziegler–Natta catalysts are also used to produce polymers—unbranched polyethylene to name one—which lack stereoregularity. However, Ziegler–Natta catalysts are historically the most widely used and best-understood stereoregulating systems, so the loss of generality in this approach is not of great consequence.

The fundamental Ziegler–Natta recipe consists of two components: a halide or other compound of a transition metal from among the group IVB to VIIIB elements, and an organometallic compound of a representative metal from groups IA to IIIA. Some of the transition metal compounds studied include $TiCl_4$, $TiCl_3$, VCl_4, VCl_3, $ZrCl_4$, $CrCl_3$, $MoCl_5$, and $CuCl$. Representative organometallics include $(C_2H_5)_3Al$, $(C_2H_5)_2Mg$, C_4H_9Li, and $(C_2H_5)_2Zn$. These are only a few of the possible compounds, so the number of combinations is very large.

The individual components of the Ziegler–Natta system can separately account for the initiation of some forms of polymerization reactions, but not for the fact of stereoregularity. For example, butyl lithium can initiate anionic polymerization (see Section 4.3) and $TiCl_4$ can initiate cationic polymerization (see Section 4.5). In combination, still another mechanism for polymerization, coordination polymerization, is indicated. When the two components of the Ziegler–Natta system are present together, complicated exchange reactions are possible. Often the catalyst must "age" to attain maximum effectiveness; presumably this allows these exchange reactions to occur. Some possible exchange equilibria are

$$2Al(C_2H_5)_3 \Leftrightarrow Al_2(C_2H_5)_6 \Leftrightarrow [Al(C_2H_5)_2]^+[Al(C_2H_5)_4]^-$$
$$TiCl_4 + [Al(C_2H_5)_2]^+ \Leftrightarrow C_2H_5TiCl_3 + [Al(C_2H_5)Cl]^+ \tag{5.Q}$$

The organotitanium halide can then be reduced to $TiCl_3$:

$$C_2H_5TiCl_3 \rightarrow TiCl_3 + C_2H_5 \cdot \tag{5.R}$$

Among other possibilities in these reactions, these free radicals can initiate ordinary free-radical polymerization. The Ziegler–Natta systems are thus seen to encompass several mechanisms for the initiation of polymerization. Neither ionic nor free-radical mechanisms account for stereoregularity, however, so we must look further for the mechanism whereby the Ziegler–Natta systems produce this interesting effect.

The stereoregulating capability of Ziegler–Natta catalysts is believed to depend on a coordination mechanism in which both the growing polymer chain and the monomer coordinate with the catalyst. The addition then occurs by insertion of the monomer between the growing chain and the catalysts by a concerted mechanism (5.XV):

$$\tag{5.XV}$$

Since the coordination almost certainly involves the transition metal atom, there is a resemblance here to anionic polymerization. The coordination is an important aspect of the present picture, since it is this feature that allows the catalyst to serve as a template for stereoregulation.

The assortment of combinations of components is not the only variable to consider in describing Ziegler–Natta catalysts. Some other variables include the following:

1. Catalyst solubility. Polymerization systems may consist of one or two phases. Titanium-based catalysts are the most common of the heterogeneous systems; vanadium-based catalysts are the most common homogeneous systems. Since the catalyst functions as a template for the formation of a stereoregular product, it follows that the more extreme orienting effect of solid surface (i.e., heterogeneous catalysts) is required for those monomers that interact only weakly with the catalyst. The latter are nonpolar monomers. Polar monomers interact more strongly with catalysts, and dissolved catalysts are able to exert sufficient control for stereoregularity.

2. Crystal structure of solids. The α-crystal form of $TiCl_3$ is an excellent catalyst and has been investigated extensively. In this particular crystal form of $TiCl_3$, the titanium ions are located in an octahedral environment of chloride ions. It is believed that the stereoactive titanium ions in this crystal are located at the edges of the crystal, where chloride ion vacancies in the coordination sphere allow coordination with the monomer molecules.

3. Tacticity of products. Most solid catalysts produce isotactic products. This is probably because of the highly orienting effect of the solid surface, as noted in item (1). The preferred isotactic configuration produced at these surfaces is largely governed by steric and electrostatic interactions between the monomer and the ligands of the transition metal. Syndiotacticity is mostly produced by soluble catalysts. Syndiotactic polymerizations are carried out at low temperatures, and even the catalyst must be prepared at low temperatures; otherwise specificity is lost. With polar monomers syndiotacticity is also promoted by polar reaction media. Apparently the polar solvent molecules compete with monomer for coordination sites, and thus indicate more loosely coordinated reactive species.

4. Rate of polymerization. The rate of polymerization for homogeneous systems closely resembles anionic polymerization. For heterogeneous systems the concentration of alkylated transition metal sites on the surface appears in the rate law. The latter depends on the particle size of the solid catalyst and may be complicated by sites of various degrees of activity. There is sometimes an inverse relationship between the degree of stereoregularity produced by a catalyst and the rate at which polymerization occurs.

The catalysts under consideration both initiate the polymerization and regulate the polymer formed. There is general agreement that the mechanism by which these materials exert their regulatory role involves coordination of monomer with the transition metal atom, but proposed details beyond this are almost as numerous and specific as the catalysts themselves. We shall return to a description of two specific mechanisms below. The general picture postulates an interaction between monomer and catalyst such that a complex is formed between the π electrons of the olefin and the d orbitals of the transition metal. Figure 5.10 shows that the overlap between the filled orbitals of the monomer can overlap with vacant $d_{x^2-y^2}$ orbitals of the metal. Alternatively, hybrid orbitals may be involved on the metal. There is a precedent for such bonding in simple model compounds. It is known, for example, that Pt^{2+} complexes with ethylene by forming a dsp^2 hybrid–π sigma bond and a dp hybrid–π* pi bond. A crucial consideration in the coordination is maximizing the overlap of the orbitals involved. Titanium(III) ions seem ideally suited for this

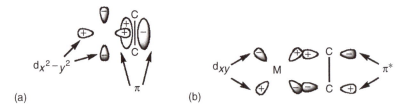

(a)   (b)

Figure 5.10 Possible orbital overlaps between a transition metal and an olefin.

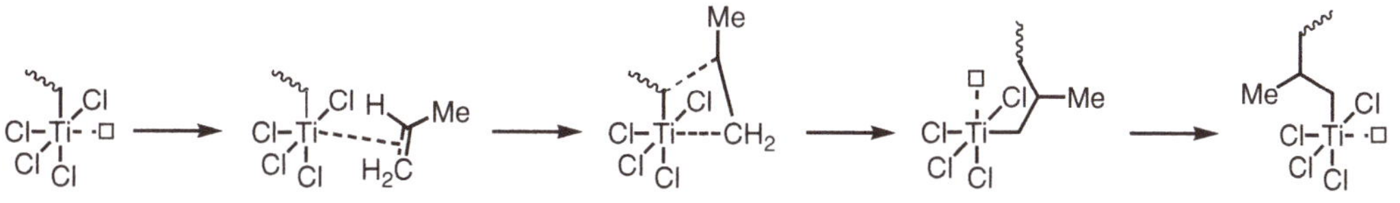

Figure 5.11 Monometallic mechanism. The square indicates a vacant ligand site.

function; higher effective nuclear charge on the metal results in less spatial extension of d orbitals and diminished overlap.

Many mechanisms have been proposed that elaborate on this picture. These are often so specific that they cannot be generalized beyond the systems for which they are proposed. Two schemes that do allow some generalization are presented here. Although they share certain common features, these mechanisms are distinguished by the fact that one—the monometallic model—does not include any participation by the representative metal in the mechanism. The second—the bimetallic model—does assume the involvement of both metals in the mechanism.

The monometallic mechanism is illustrated by Figure 5.11. It involves the monomer coordinating with an alkylated titanium atom. The insertion of the monomer into the titanium–carbon bond propagates the chain. As shown in Figure 5.11 this shifts the vacancy—represented by the square—in the coordination sphere of the titanium to a different site. Syndiotactic regulation occurs if the next addition takes place via this newly created vacancy. In this case the monomer and the growing chain occupy alternating coordination sites in successive steps. For the more common isotactic growth the polymer chain must migrate back to its original position.

The bimetallic mechanism is illustrated in Figure 5.12; the bimetallic active center is the distinguishing feature of this mechanism. The precise distribution of halides and alkyls is not spelled out because of the exchanges described by Reaction (5.Q). An alkyl bridge is assumed based on observations of other organometallic compounds. The π coordination of the olefin with the titanium is followed by insertion of the monomer into the bridge to propagate the reaction.

At present it is not possible to determine which of these mechanisms or their variations most accurately represents the behavior of Ziegler–Natta catalysts. In view of the number of variables in these catalyzed polymerizations, both mechanisms may be valid, each for different specific systems. In the following example the termination step of coordination polymerizations is considered.

Example 5.8

Polypropylene polymerized with triethyl aluminum and titanium trichloride has been found to contain various kinds of chain ends. Both terminal vinylidene unsaturation and aluminum-bound chain ends have been identified. Propose two termination reactions to account for these observations. Do the termination reactions allow any discrimination between the monometallic and bimetallic propagation mechanisms?

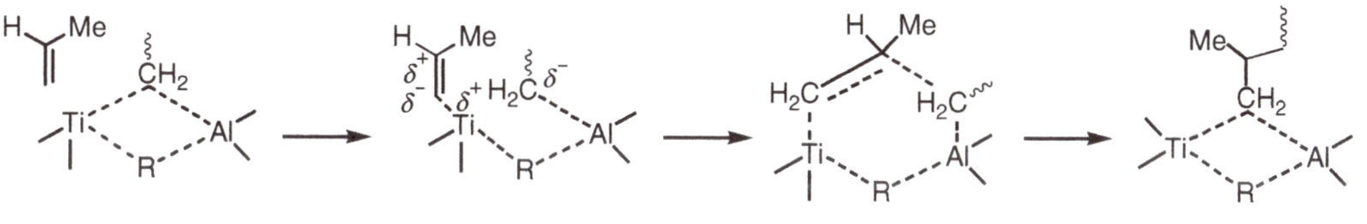

Figure 5.12 The bimetallic mechanism.

Solution

A reaction analogous to the alkylation step of Reaction (4.Q) can account for the association of an aluminum species with chain ends:

$$(5.8a)$$

The transfer of a tertiary hydrogen between the polymer chain and a monomer can account for the vinylidene group in the polymer:

$$(5.8b)$$

These reactions appear equally feasible for titanium in either the monometallic or bimetallic intermediate. Thus they account for the different types of end groups in the polymer, but do not differentiate between propagation intermediates. In the commercial process for the production of polypropylene by Ziegler–Natta catalysts, hydrogen is added to terminate the reaction, so neither of these reactions is pertinent in this case.

5.11   Single-Site Catalysts

The discussion in the preceding section indicates that Ziegler–Natta catalysts represent a rather complicated subject. This complexity is often reflected in the structure of the polymers produced. For example, the different ways that the two metal centers may or may not interact during an addition step suggest that there are, in fact, multiple catalytic sites active in a given polymerization. This can lead to sites with greatly different propagation rates, different stereoselectivity, and different propensities to incorporate any comonomers present. The net result is that polymer materials produced by Ziegler–Natta catalysts, especially under commercial conditions, tend to be highly heterogeneous at the molecular level. A broad strategy to overcome this limitation is based on the concept of a *single-site catalyst*, i.e., one that has a single, well-defined catalytic geometry that can control the desired aspect of propagation. In this section, we briefly consider some examples of such catalysts for stereochemical control in the polymerization of α-olefins. We begin with a little more consideration of catalysis in general.

The majority of catalysts in commercial use are *heterogeneous*. In this usage, the term heterogeneous means that the phase of the catalyst (e.g., solid) is distinct from that of the reagents and products (usually gases and liquids). When the catalyst is a relatively small molecule, it is retained in the solid phase by immobilization on some kind of inert, robust support. The reaction of interest therefore takes place at the solid–liquid or solid–gas interface. The fact of immobilization can itself contribute to the multiple site nature of heterogeneous catalysts, for example by exposing different faces of the catalytically active metal center, by restricting accessibility of reagents to catalyst particles deep within a porous support, and by presenting a distribution of different cluster sizes of catalytic particles. Given these disadvantages, one might ask why heterogeneous catalysis is the norm. The answer is simple: It is much easier to separate (and possibly regenerate) heterogeneous catalysts from products and unreacted reagents. Note that if the activity of a catalyst is sufficiently high (in terms of grams polymer produced per gram catalyst employed), then separation and recovery of the catalyst may not be necessary. In contrast, single-site polymerization catalysts are usually *homogeneous*: they are molecularly dispersed within the reaction medium. This situation leads to better defined products, and is much more amenable to detailed studies of mechanism. Furthermore, strategies for immobilizing such catalysts are available, making them also of potential commercial interest.

Most single-site catalysts have the general formula [L_nMP], where L_n represents a set of ligands, M is the active metal center, and P is the growing polymer. Furthermore, a common motif is for two of the ligands to contain cyclopentadienyl (Cp) rings, which may themselves be covalently linked or bridged. The example shown below (5.XVI) was one of the first such *metallocene* systems and produces highly isotactic polypropylene.

ZrX$_2$ (5.XVI)

However, this representation is not complete. Just as Ziegler–Natta catalysts always involve a mixture of at least two active ingredients, single-site catalysts involve another component. The most common is a partially hydrolyzed trimethyl aluminum species—methylaluminoxane (MAO). The active center is more properly denoted $[L_nMP]^+[X]^-$, where the metal site is cationic by virtue of being coordinatively unsaturated, and the counterion contains MAO and a displaced ligand, such as chloride.

The choice of metal, ligands, and design of the overall constraining geometry provide a rich palette from which catalysts may be designed. In general, the stereoregulation of monomer addition can be achieved through one of two modes. Under *chain-end control*, the addition of a monomer is influenced mostly by the configuration of the previous repeat unit, which is reminiscent of the terminal model of copolymerization. To appreciate how this can happen, it is important to realize that the growing polymer remains bound to the metal center during the addition step. Alternatively, under *site control* the ligand set may be chosen to provide a chiral confining environment, which exerts a dominant influence on the stereochemistry of addition. The symmetry of the catalyst is often strongly correlated with the mode and effect of stereocontrol. This is summarized in Figure 5.13. Catalysts with a plane of symmetry, or C_s, tend to produce syndiotactic polymers under site control, but either iso- or syndiotactic polymers under chain-end control. When the symmetry is C_2, i.e., identical after rotation by $180°$ about a single axis, the addition is

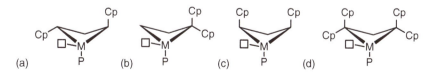

Figure 5.13 Role of catalyst symmetry in stereocontrol. The open square represents the unsaturated site for monomer addition, and the Cp rings are represented by the pendant lines. A catalyst of type (a) is isospecific and (b) is syndiospecific, when under site control; (c) and (d) can be either iso- or syndiospecific, under chain-end control. (From Coates, G.W., *Chem. Rev.*, 100, 1223, 2000.)

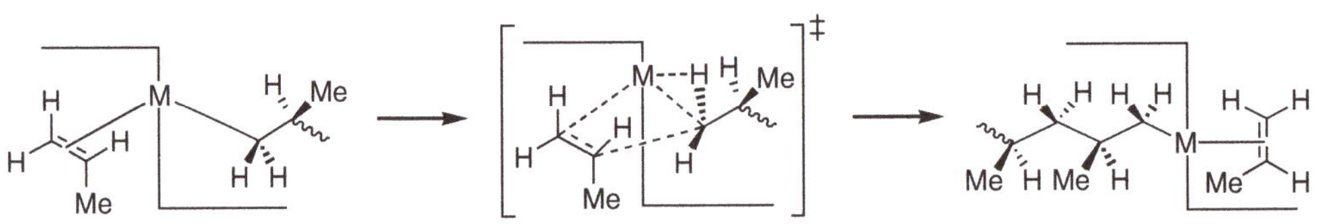

Figure 5.14 Proposed mechanism of isospecific polymerization of polypropylene. (From Coates, G.W., *Chem. Rev.*, 100, 1223, 2000.)

isospecific under site control. When a further mirror plane exists, in C_{2v} symmetry, chain-end control leads to either iso- or syndiospecific addition.

We now illustrate these phenomena with two particular catalysts and a cartoon sketch of the mechanism of monomer insertion. The monomer in question is polypropylene, the commercially most important stereogenic polyolefin and the most studied model system. However, it should be noted that the flexibility of design for single-site catalysts offers the possibility of more tolerance toward monomer polarity or functionality than in the Ziegler–Natta analogs, thereby enabling stereocontrol of many different monomers or comonomers. The catalyst (5.XVI) has C_2 symmetry and is isospecific under site control. The mechanism is illustrated in Figure 5.14, where for simplicity the Cp-containing ligands are represented by horizontal lines. The polymer chain is bound to the metal through the unsubstituted backbone carbon and the orientation of the incoming monomer is influenced by the location of the Cp ligand. In the transition state the unsubstituted carbon of the new monomer coordinates with the metal and will become the new terminal carbon of the growing chain. A key role is thought to be played by a so-called "α-agostic" interaction between the metal and the hydrogen on the terminal carbon of the polymer chain, which stabilizes the particular geometry of the transition state. After the incorporation of the monomer, the polymer chain (or a last few repeat units thereof) has "flipped" to the other side of the metal center, in a process which is often compared to the action of a windshield wiper.

In contrast, the following zirconocene (5.XVII) is syndiospecific, consistent with its C_s symmetry. The mechanism is analogous to that illustrated in Figure 5.14, except that the inversion of the position of the bulky ligand inverts the preferred orientation of the incoming monomer.

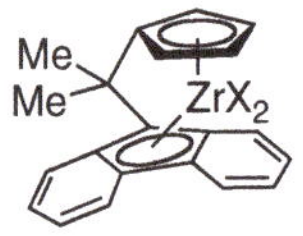

(5.XVII)

The range of possibilities afforded by this class of catalysts is vast. As one last example, consider the following zirconocene (5.XVIII), developed by Coates and Waymouth [2]:

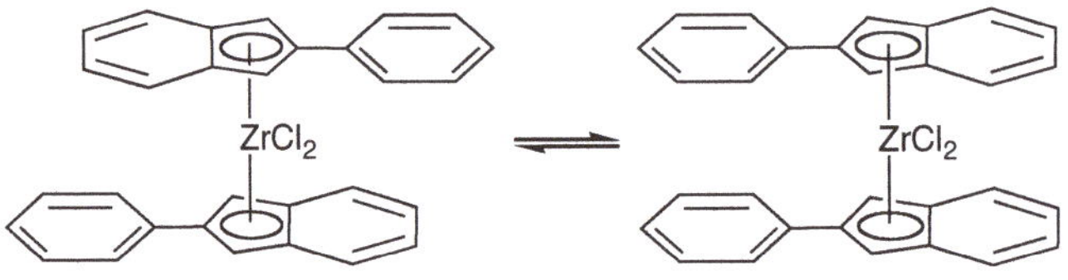

(5.XVIII)

As indicated by the double arrows, the catalyst actually oscillates between two isomeric structures. The structure on the left is chiral with C_2 symmetry, and gives isotactic polypropylene (note that the chloride ligands are not in the plane of the page). The structure on the right, however, is achiral, and actually leads to random stereochemistry, i.e., atactic polypropylene. Now, consider the interesting situation where the rate of monomer insertion is more rapid than the rate of exchange between the two structures, say by a factor of 20. In such a case the resulting polymer would be a "stereoblock copolymer," with alternating sequences of isotactic and atactic polypropylene, where the average sequence length would be about 20. Such a polymer has some very appealing properties. The isotactic blocks can crystallize, as will be discussed in detail in Chapter 13, whereas the atactic blocks cannot. The result is that for temperatures above the glass transition of the atactic block (about $-10°C$, see Chapter 12) but below the melting temperature of the stereoregular block (about $140°C$) the material acts as a crosslinked elastomer (see Chapter 10). The crystallites tie the different molecules together, imparting mechanical strength, but the atactic blocks can be stretched appreciably without breaking, like a rubbery material. The mechanical response is sensitive to the relative lengths of the two blocks, which can be tuned through monomer concentration and polymerization temperature. The result is an appealing situation in which an inexpensive monomer can be used to produce a variety of different products by straightforward modification to the reaction conditions.

5.12 Chapter Summary

This chapter has covered a broad range of issues relating to the structure of polymer chains at the level of a few repeat units. The two main topics have been copolymerization and stereoregularity. These topics share many features in common, including (i) the importance of the relative reactivity of a growing chain end to addition of a particular monomer, or a monomer in a particular configuration; (ii) the use of statistics in describing composition, average sequence lengths, and sequence length distribution; (iii) the central role of spectroscopic methods, and especially NMR, in characterizing structural details.

1. The key parameters in copolymerization are the reactivity ratios, which influence the relative rates at which a given radical will add the same monomer versus a comonomer. Thus a given reactivity ratio is specific to a particular pair of monomers, and copolymerization of two monomer system requires specification of two reactivity ratios.
2. The copolymerization equation relates the mole fraction of monomers in polymer to the composition of the feedstock via the reactivity ratios. Different classes of behavior may be assigned based on the product of the reactivity ratios, including an "ideal" copolymerization when the two reactivity ratios are reciprocals of one another.
3. The relative magnitudes of reactivity ratios can be understood, at least qualitatively, by considering the contributions of resonance stabilization, polarity differences, and possible steric effects.
4. Statistical considerations give predictions for the average sequence length and sequence length distributions in a copolymer on the basis of reactivity ratios and feedstock composition. However, the probability of adding a given monomer to a growing chain end may be determined by the last, the last plus next-to-last, or even the last, next-to-last and second-to-last monomers added. These mechanisms are referred to as terminal, penultimate, and antepenultimate control, respectively.
5. Stereoregularity may be viewed as a subset of copolymerization, in which addition of a monomer with an asymmetric center may follow the same stereochemistry as the previous repeat unit, thereby forming a meso dyad, or by the opposite stereochemistry, forming a racemic dyad. Isotactic, syndiotactic, and atactic polymers thus correspond to predominantly meso dyads, predominantly racemic dyads, or random mixtures of the two, respectively.
6. Copolymer sequence lengths (dyads, triads, tetrads, etc.) can be determined by NMR methods. These in turn may be used to discriminate among terminal, penultimate, and antepenultimate

control mechanisms. Similarly NMR gives access to stereochemical information, being sensitive to sequences of meso dyads, racemic dyads, and even longer sequences.

7. Stereoregularity is obtained by coordination polymerization in the presence of particular catalysts. The most commonly used systems for the polymerization of α-olefins are referred to as Ziegler–Natta catalysts, a class which actually spans a large variety of particular compounds. The mechanisms of action of these catalysts are typically rather complicated. More recently there have been rapid advances in the development of single-site catalysts, which are usually based on metallocenes: a metal center coordinated to one or more cyclopentadienyl ligands. The terminology refers to the presence of a well-defined catalytic site throughout the polymerization medium, leading to more homogeneous products. These systems are capable of being fine-tuned to regulate a variety of structural features, including stereochemistry and comonomer addition.

Problems

1. Write structural formulas for maleic anhydride (M_1) and stilbene (M_2). Neither of these monomers homopolymerize to any significant extent, presumably owing to steric effects. These monomers form a copolymer, however, with $r_1 = r_2 = 0.03$.[†] Criticize or defend the following proposition: The strong tendency toward alternation in this copolymer suggests that polarity effects offset the steric hindrance and permit copolymerization of these monomers.

2. Styrene and methyl methacrylate have been used as comonomers in many investigations of copolymerization. Use the following list of r_1 values for each of these copolymerizing with the monomers listed below to rank the latter with respect to reactivity. To the extent that the data allow, suggest where these substituents might be positioned in Table 5.3.

M_2	Styrene as M_1	Methyl methacrylate as M_1
Acrylonitrile	0.41	1.35
Allyl acetate	90	23
1,2-Dichloropropene-2	5	5.5
Methacrylonitrile	0.30	0.67
Vinyl chloride	17	12.5
Vinylidene chloride	1.85	2.53
2-Vinyl pyridine	0.55	0.395

3. As part of the research described in Figure 5.5, Winston and Wichacheewa measured the weight percentages of carbon and chlorine in copolymers of styrene (molecule 1) and 1-chloro-1,3-butadiene (molecule 2) prepared from various feedstocks. A portion of their data is given below. Use these data to calculate F_1, the mole fraction of styrene in these copolymers.

f_1	Percent C	Percent Cl
0.892	81.80	10.88
0.649	71.34	20.14
0.324	64.95	27.92
0.153	58.69	34.79

4. Additional data from the research of the last problem yield the following pairs of f_1, F_1 values (remember that styrene is component 1 in the styrene–1-chloro-1,3-butadiene system). Use the

[†] F.M. Lewis and F.R. Mayo, *J. Am. Chem. Soc.*, 70, 1533 (1948).

form suggested by Equation 5.6.1 to prepare a graph based on these data and evaluate r_1 and r_2.

f_1	F_1	f_1	F_1
0.947	0.829	0.448	0.362
0.861	0.688	0.247	0.207
0.698	0.515	0.221	0.200
0.602	0.452		

5. The reactivity ratios for the styrene (M_1)–1-chloro-1,3-butadiene (M_2) system were found to be $r_1 = 0.26$ and $r_2 = 1.02$ by the authors of the research described in the last two problems, using the results of all their measurements. Use these r values and the feed compositions listed below to calculate the fraction expected in the copolymer of 1-chlorobutadiene sequences of lengths $\nu = 2$, 3, or 4. From these calculated results, evaluate the ratios N_{222}/N_{22} and N_{2222}/N_{222}. Copolymers prepared from these feedstocks were dehydrohalogenated to yield the polyenes like that whose spectrum is shown in Figure 5.5. The absorbance at the indicated wavelengths was measured for 1% solutions of the products after HCl elimination.

	Absorbance		
f_1	$\lambda = 312$ nm	$\lambda = 367$ nm	$\lambda = 412$ nm
0.829	74	13	—
0.734	71	19	—
0.551	154	77	20
0.490	151	78	42

As noted in Section 5.6, these different wavelengths correspond to absorbance by sequences of different lengths. Compare the appropriate absorbance ratios with the theoretical sequence length ratios calculated above and comment briefly on the results.

6. Use the values determined in Example 5.5 for the vinylidene chloride (M_1)–isobutylene (M_2) system to calculate F_1, for various values of f_1, according to the terminal mechanism. Prepare a plot of the results. On the same graph, plot the following experimentally measured values of f_1 and F_1. Comment on the quality of the fit.

f_1	F_1	f_1	F_1
0.548	0.83	0.225	0.66
0.471	0.79	0.206	0.64
0.391	0.74	0.159	0.61
0.318	0.71	0.126	0.58
0.288	0.70	0.083	0.52

7. Some additional dyad fractions from the research cited in the last problem are reported at intermediate feedstock concentrations $(M_1 = \text{vinylidene chloride}; M_2 = \text{isobutylene})$.[†] Still assuming terminal control, evaluate r_1 and r_2 from these data. Criticize or defend the following proposition: The copolymer composition equation does not provide a very sensitive test for

---

[†] J.B. Kinsinger, T. Fischer, and C.W. Wilson, *Polym. Lett.*, 5, 285 (1967).

the terminal control mechanism. Dyad fractions are more sensitive, but must be examined over a wide range of compositions to provide a valid test.

	Mole fraction of dyads		
f_1	11	12	22
0.418	0.55	0.43	0.03
0.353	0.48	0.49	0.04
0.317	0.44	0.52	0.04
0.247	0.38	0.58	0.04
0.213	0.34	0.62	0.04
0. 198	0.32	0.64	0.05

8. Fox and Schnecko carried out the free-radical polymerization of methyl methacrylate between $-40°C$ and $250°C$. By analysis of the α-methyl peaks in the NMR spectra of the products, they determined the following values of β, the probability of an isotactic placement in the products prepared at different temperatures.

T (°C)	250	150	100	95	60	30	0	-20	-40
β	0.36	0.33	0.27	0.27	0.24	0.22	0.20	0.18	0.14

Evaluate $E_i^* - E_s^*$ by means of an Arrhenius plot of these data using $\beta/(1-\beta)$ as a measure of k_i/k_s. Briefly justify this last relationship.

9. A hetero triad occurs at each interface between iso and syndio triads. The total number of hetero triads, therefore, equals the total number of sequences of all other types:

$$\nu_h = \sum N_{n_i} + \sum N_{n_s}$$

Use this relationship and Equation 5.9.1 to derive the expression

$$p_h = \frac{\nu_h}{\nu_h + \nu_i + \nu_s} = \frac{2}{\bar{n}_i + \bar{n}_s}$$

Criticize or defend the following proposition: The sequence DL– is already two thirds of the way to becoming a hetero triad, whereas the sequence DD– is two thirds of the way toward an iso triad. This means that the fraction of heterotactic triads is larger when the average length of syndio sequences is greater than the average length of iso sequences.

10. Randall[†] used ^{13}C-NMR to study the methylene spectrum of polystyrene. In 1,2,4-trichlorobenzene at $120°C$, nine resonances were observed. These were assumed to arise from a combination of tetrads and hexads. Using m and r notation, extend Table 5.6 to include all 20 possible hexads. Criticize or defend the following proposition: Assuming that none of the resonances are obscured by overlap, there is only one way that nine methylene resonances can be produced, namely, by one of the tetrads being split into hexads whereas the remaining tetrads remain unsplit.

11. In the research described in the preceding problem, Randall was able to assign the five peaks associated with tetrads in the ^{13}C-NMR spectrum on the basis of their relative intensities, assuming zero-order Markov statistics with $p_m = 0.575$. The five tetrad intensities and their chemical shifts from TMS are as follows:

[†] J.C. Randall, *J. Polym. Sci., Polym. Phys. Ed.*, 13, 889 (1975).

^{13}C δ_{TMS} (ppm)	Relative area under peak
45.38	0.10
44.94	0.28
44.25	0.13
43.77	0.19
42.84	0.09

The remaining 21% of the peak area is distributed among the remaining hexad features. Use the value of p_m given to calculate the probabilities of the unsplit tetrads (see Problem 10) and on this basis assign the features listed above to the appropriate tetrads. Which of the tetrads appears to be split into hexads?

12. The fraction of sequences of the length indicated below have been measured for a copolymer system at different feed ratios.[†] From appropriate ratios of these sequence lengths, what conclusions can be drawn concerning terminal versus penultimate control of addition?

$[M_1]/[M_2]$	$P(M_1)$	$P(M_1M_1)$	$P(M_1M_1M_1)$
3	0.168	0.0643	0.0149
4	0.189	0.0563	0.0161
9	0.388	0.225	0.107
19	0.592	0.425	0.278

13. The following are experimental tacticity fractions of polymers prepared from different monomers and with various catalysts. On the basis of Figure 5.7, decide whether these preparations are adequately described by a single parameter p_m or whether some other type of statistical description is required (remember to make some allowance for experimental error). On the basis of these observations, criticize or defend the following proposition: Regardless of the monomer used, zero-order Markov statistics apply to all free-radical, anionic, and cationic polymerizations, but not to Ziegler–Natta catalyzed systems.

Catalyst	Solvent	T (°C)	Iso	Hetero	Syndio
			Fraction of polymer		
Methyl methacrylate[a]					
Thermal	Toluene	60	8	33	59
n-Butyl lithium	Toluene	−78	78	16	6
n-Butyl lithium	Methyl isobutyrate	−78	21	31	48
α-Methyl styrene[b]					
$TiCl_4$	Toluene	−78	—	19	81
$Et_3Al/TiCl_4$	Benzene	25	3	35	62
n-Butyl lithium	Cyclohexane	4	—	31	69

[a]Methyl methacrylate data from K. Hatada, K. Ota, and H. Yuki, Polym. Lett., 5, 225 (1967).
[b]α-Methyl styrene data from S. Brownstein, S. Bywater, and O.J. Worsfold, Makromol. Chem., 48, 127 (1961).

14. Replacing one of the alkyl groups in R_3Al with a halogen increases the stereospecificity of the Ziegler–Natta catalyst in the order I > Br > Cl > R. Replacement of a second alkyl by halogen decreases specificity. Criticize or defend the following proposition on the basis of these observations: The observed result of halogen substitution is consistent with the effect on the

[†] K. Ito and Y. Yamashita, J. Polym. Sci., 3A, 2165 (1965).

ease of alkylation produced by substituents of different electronegativity. This evidence thus adds credence to the monometallic mechanism, even though the observation involves the organometallic.

15. The weight percent propylene in ethylene–propylene copolymers for different Ziegler–Natta catalysts was measured for the initial polymer produced from identical feedstocks.[†] The following results were obtained. Interpret these results in terms of the relative influence of the two components of the catalyst on the product found.

Catalyst components	Weight percent propylene	Catalyst components	Weight percent propylene
VCl_4, plus		$Al(i\text{-}Bu)_3$, plus	
$Al(i\text{-}Bu)_3$	4.5	$HfCl_4$	0.7
CH_3TiCl_3	4.5	$ZrCl_4$	0.8
$Zn(C_2H_5)_2$	4.5	$VOCl_3$	2.4
$Zn(n\text{-}Bu)_2$	4.5		

16. Imagine a given single-site catalyst for polypropylene introduced a stereodefect on average once every 10 monomer additions. Furthermore, assume the catalyst was supposed to be highly isospecific. Explain how measurements of triad populations (e.g., mmm, mmr, etc.) could be used to distinguish between chain-end control and site control. (Hint: consider the sequences of D and L in the two cases.)

References

1. Brandrup, J. and Immergut, E.H., Eds., *Polymer Handbook*, 3rd ed., Wiley, New York, 1989.
2. Coates, G.W. and Waymouth, R.M., *Science*, 267, 217 (1995).

Further Readings

Allcock, H.R. and Lampe, F.W., *Contemporary Polymer Chemistry*, 2nd ed., Prentice-Hall, Englewood Cliffs, NJ, 1990.
Bovey, F.W., *High Resolution NMR of Macromolecules*, Academic Press, New York, 1972.
Coates, G.W., Precise control of polyolefin stereochemistry using single-site metal catalysts, *Chem. Rev.*, 100, 1223 (2000).
Koenig, J.L., *Chemical Microstructure of Polymer Chains*, Wiley, New York, 1980.
North, A.M., *The Kinetics of Free Radical Polymerization*, Pergamon Press, New York, 1966.
Odian, G., *Principles of Polymerization*, 4th ed., Wiley, New York, 2004.
Rempp, P. and Merrill, E.W., *Polymer Synthesis*, 2nd ed., Hüthig & Wepf, Basel, 1991.

[†] F.J. Karol and W.L. Carrick, *J. Am. Chem. Soc.*, 83, 585 (1960).

6

Polymer Conformations

6.1 Conformations, Bond Rotation, and Polymer Size

The remarkable properties of polymers derive from their size. As pointed out in Chapter 1, it is not the high molecular weight per se that gives polymers mechanical strength, flexibility, elasticity, etc., but rather their large spatial extent. In this chapter, we will learn how to describe the three-dimensional shape of polymers in an average sense, and how the average size of the object in space will depend on molecular weight. We will also explore the equilibrium distribution of sizes.

To gain an appreciation of the possibilities, consider a polyethylene molecule with $M = 280,000$ g/mol (this is a reasonable value for a randomly selected molecule in some commercial grades of polyethylene). As the monomer ($-CH_2-CH_2-$) molecular weight is 28 g/mol, the degree of polymerization, N, is 10,000, and there are 20,000 C—C backbone bonds. Assuming a perfectly linear structure (actually not likely for a commercial polyethylene), the contour length L of this molecule would be roughly $20,000 \times 1.5$ Å $= 30,000$ Å or about 3 μm, because 1.5 Å is approximately the average length of a C—C bond. This is simply huge. If stretched out to its full extent, L would be half the size of a red blood cell, and possibly visible under a high-power optical microscope. Some commercial polymers are 10 times bigger than this one, and some DNA molecules have molecular weights in excess of 10^9 g/mol. However, as we will see, it is very rare indeed for a chain to be so extended, and the contour length is not usually the most useful measure of size. Now consider the opposite extreme, where the same polyethylene molecule collapses into a dense ball or *globule*. The density of bulk polyethylene is about 0.9 g/mL. The volume occupied by this 280,000 g/mol molecule would be $(280,000/0.9)/(6 \times 10^{23})$ mL $= 520,000$ Å^3, and if we assume it is a sphere, the radius would be $((3/4\pi)\text{volume})^{1/3} \approx 50$ Å. The range from 50 Å at the smallest to 3 μm at the largest covers three orders of magnitude; it is a remarkable fact that such a mundane molecule could adopt conformations with sizes varying over that range. If the dense sphere were a tennis ball, the chain contour would be the length of a football field.

Polyethylene, and most carbon chain polymers, is not likely to adopt either of these extreme conformations. The reason is easy to see. Select a C—C bond anywhere along the chain; we can represent the structure as R'CH$_2$—CH$_2$R''. There is rotation about this bond, with three energetically preferred relative orientations of R' and R'' called *trans* (t), *gauche plus* (g$^+$), and *gauche minus* (g$^-$) (see Figure 6.1a). For the chain of 20,000 bonds, there are three possible conformations for each bond, and therefore $3^{20,000} \approx 10^{10,000}$ possible conformations. This number is effectively infinite. If our molecule were in a high temperature liquid state, and if we assume it takes 1 ps to change one bond conformation, then the molecule would not even approach sampling all possible conformations over the history of the universe. Similarly, it would be highly improbable for it to even visit any given conformation twice. We can now see why the chance of being fully extended, in the all-*trans* state, is unlikely to say the least; the probability is about 1 in $10^{10,000}$. The dense sphere state might be marginally more probable, as there are many sequences of t, g$^+$, and g$^-$ that might produce something close to that, but it is still essentially impossible without the action of some external force. What the polymer does instead is form what is called a *random coil* (Figure 6.2). The different sequences of t, g$^+$, and g$^-$ cause the chain to wander about haphazardly in space, with a typical size intermediate between the dense sphere and the extended chain. With so

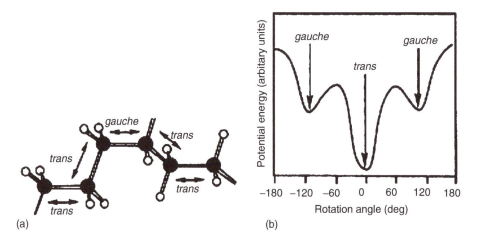

(a)   (b)

Figure 6.1   Backbone bond conformations for polyethylene. (a) Illustration of *trans* and *gauche* arrangements of the backbone bonds. (b) Schematic plot of the potential energy as a function of rotation angle about a single backbone bond.

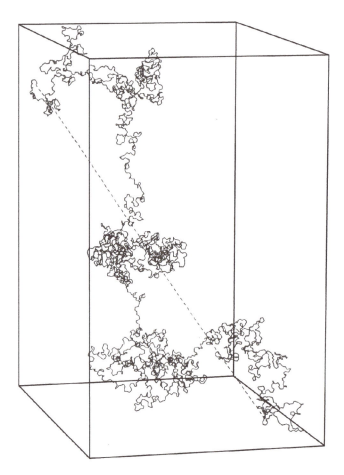

Figure 6.2   Illustration of a random walk in three dimensions. The walk has 4000 steps, and the walk touches each face of the box. (From Lodge, A.S., *An Introduction to Elastomer Molecular Network Theory*, Bannatek Press, Madison, 1999. With permission.)

many possibilities, we obviously cannot predict any instantaneous conformation or size, but we will be able to say a great deal about the average size.

The preceding argument, although fundamentally sound, neglects a very important aspect of chain conformations. The energies of the t, g^+, and g^- states are not equal; for polyethylene t is energetically favorable relative to g^+ or g^- by about $\Delta E = 3$ kJ/mol (or 0.7 kcal/mol). Therefore, at equilibrium, the population of t states will exceed that of g^+ or g^- by the appropriate Boltzmann factor, $\exp(-\Delta E/RT)$, which in this case is about 2 at $T = 500$ K ($RT \approx 4.2$ kJ/mol or 1 kcal/mol). (We have injected some statistical mechanics here. For a large collection or *ensemble* of molecules at equilibrium, the relative populations of any two possible states are given by this Boltzmann factor.) This bias is still not enough to put much of a dent into the vast number of possible conformations, but it does matter when computing the detailed conformational statistics for a given polymer chain. There is a further issue of importance, namely, how high are the energy barriers among the t, g^+, and g^- states? If these are too high, conformational rearrangements will not occur rapidly. Figure 6.1b shows a schematic plot of the potential energy as a function of rotation about the C—C bond in polyethylene. The barrier heights are on the order of 10 kJ/mol (2.5 kcal/mol), which corresponds to about 2.5 times RT, the available thermal energy per mole at 500 K. Thus rotation should be relatively facile for polyethylene. (To put this energy barrier on a chemist's scale, it is comparable to the energy of a weak hydrogen bond.) However, for polymers with larger side-groups or with more complicated backbone structures, these barriers can become substantial. For example, in poly(n-hexyl isocyanate) ($-N(C_6H_{13})-C(O)-$) the n-hexyl side chain forces the backbone to favor a helical conformation, and the molecule becomes relatively extended.

At this point it might look like a very daunting task to calculate the probable conformation of a given polymer, and it will require some detailed information about bond rotational potentials, etc., for each structure. However, it turns out that we can go a long way without any such knowledge. What we will calculate first is the average distance between the ends of a chain, as a function of the number of steps in a chain. We will show that this is given by a simple formula, and that all the details about chemical structures, bond potentials, etc., can be grouped into a single parameter. We will also consider the distribution of possible values of this end-to-end distance. The average could, in principle, be taken in two different ways. One would be to follow a single chain as it samples many different conformations—a time average. Another would be to look at a large collection of structurally identical chains at a given instant in time—an ensemble average. In this example, these two averages should be the same; when this occurs, we say the system is *ergodic*. In a real polymer sample, a measurement will also average over a distribution of chain lengths or molecular weights; that is a different average, which we will have to reckon with when we consider particular experimental techniques.

6.2   Average End-to-End Distance for Model Chains

In this section, we calculate the root-mean-square (rms) end-to-end distance $\langle h^2 \rangle^{1/2}$ for an imaginary chain, made up of n rigid links, each with length ℓ. The model is sketched in Figure 6.3. At this stage there is no need to worry about whether the link is meant to represent a real C—C bond or not; we will make the correspondence to real polymers later. If we arbitrarily select one end as the starting point, each link can be represented by a vector, $\vec{\ell}_i$, with $i = 1, 2, 3, \ldots n$. The instantaneous end-to-end vector, $\vec{h}$, is simply the sum of the link vectors:

$$\vec{h} = \sum_{i=1}^{n} \vec{\ell}_i \qquad\qquad\qquad (6.2.1)$$

If we have a chain that wiggles around over time, or if we look at an ensemble of similar chains, there is no reason for $\vec{h}$ to point in any one direction more than any other, and the average $\langle \vec{h} \rangle = 0$; we say the sample is *isotropic*. What we really care about is the average end-to-end *distance*, which

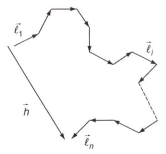

Figure 6.3 Model chain consisting of n links of length ℓ. Each link is represented by a vector, $\vec{\ell}_i$, and the end-to-end vector $\vec{h}$ extends from the start of the first link to the end of the last one.

we can calculate remembering that the length of a vector is obtained by taking the dot product of the vector with itself:

$$\langle h^2 \rangle^{1/2} = \langle \vec{h} \cdot \vec{h} \rangle^{1/2} = \left\langle \sum_{i=1}^{n} \vec{\ell}_i \cdot \sum_{j=1}^{n} \vec{\ell}_j \right\rangle^{1/2} \tag{6.2.2}$$

The double sum can be broken into two parts, remembering also that the summations can be taken outside the average as follows:

$$\langle h^2 \rangle = \left\langle \sum_{i=1}^{n} \vec{\ell}_i \cdot \sum_{j=1}^{n} \vec{\ell}_j \right\rangle = \sum_{i=1}^{n} \sum_{j=1}^{n} \left\langle \vec{\ell}_i \cdot \vec{\ell}_j \right\rangle$$

$$= \sum_{i=1}^{n} \left\langle \vec{\ell}_i \cdot \vec{\ell}_i \right\rangle + \sum_{i=1}^{n} \sum_{j \neq i}^{n} \left\langle \vec{\ell}_i \cdot \vec{\ell}_j \right\rangle$$

$$= n\ell^2 + \sum_{i=1}^{n} \sum_{j \neq i}^{n} \left\langle \vec{\ell}_i \cdot \vec{\ell}_j \right\rangle \tag{6.2.3}$$

where the first term, $n\ell^2$, accounts for the "self-terms," i.e., each of the n link vectors dotted into itself, and the second term accounts for the "cross terms," i.e., each link vector dotted into the $n-1$ other link vectors.

We now develop explicit results for $\langle h^2 \rangle^{1/2}$, with three different rules for how the orientation of a given link, $\vec{\ell}_i$, is constrained by the orientation of its predecessor, $\vec{\ell}_{i-1}$.

Case 6.2.1   The Freely Jointed Chain

In this simplest possible case, the orientation of link i is unaffected by link $i-1$, and is equally likely to point in any direction. It can even lie on top of link $i-1$ by pointing in the opposite direction. (Remember we are dealing with imaginary links, not real chemical bonds, so this is permissible.) Mathematically we represent this approximation using the relation

$$\vec{\ell}_i \cdot \vec{\ell}_{i-1} = \ell^2 \cos \theta \tag{6.2.4}$$

where θ is the angle between $\vec{\ell}_i$ and $\vec{\ell}_{i-1}$. For the freely jointed chain, θ ranges freely from $0°$ to $180°$. Thus on average

$$\left\langle \vec{\ell}_i \cdot \vec{\ell}_{i-1} \right\rangle = \ell^2 \langle \cos \theta \rangle = 0 \tag{6.2.5}$$

When the orientations of two links are *uncorrelated*, then $\langle \cos \theta \rangle = 0$, because $\cos \theta$ ranges from -1 to $+1$, with $+$ or $-$ values equally probable. If we consider the relative orientations of any two different links along the freely jointed chain, they must all be uncorrelated, so that

$$\left\langle \vec{\ell}_i \cdot \vec{\ell}_j \right\rangle = 0 \tag{6.2.6}$$

whenever $i \neq j$. In other words if the orientation of a given link is unaffected by its nearest neighbor, it must also be unaffected by more distant neighbors. From Equation 6.2.3 we now obtain the simple but tremendously important result

$$\langle h^2 \rangle = n\ell^2 \quad \text{or} \quad \langle h^2 \rangle^{1/2} = \sqrt{n}\ell \tag{6.2.7}$$

because all the cross terms vanish. This is the classic result for the so-called *random walk* (or *random flight*): the root-mean-square excursion is given by the step length times the square root of the number of steps. We will invoke this result repeatedly in subsequent chapters.

At this point you may be thinking "Fine, but even if the link is not a C—C bond, a real polymer chain cannot reverse its direction 180° at a joint, so how can this result be relevant?" Good question. Be patient for a bit.

Case 6.2.2 The Freely Rotating Chain

Now we make the model a bit more realistic. We will constrain the angle between adjacent links to be a fixed value, θ, but still allow free rotation of the link around the cone defined by θ (see Figure 6.4). What happens? Now, $\vec{\ell}_i \cdot \vec{\ell}_{i-1} = \ell^2 \cos \theta$ does not average to zero because θ is fixed. For simplicity, we will define $\alpha = \cos \theta$ just to avoid writing $\cos \theta$ over and over. Returning to Equation 6.2.3, what we need to calculate is the double sum over all possible $\langle \vec{\ell}_i \cdot \vec{\ell}_j \rangle$, i.e., the cross terms as well as the self-terms. We now know $\langle \vec{\ell}_i \cdot \vec{\ell}_j \rangle = \ell^2 \alpha$ when $|i - j| = 1$, but what about $|i - j| = 2, 3$, etc.? This is a little sneaky: $\vec{\ell}_i$ has a component parallel to $\vec{\ell}_{i-1}$, with length $\ell \alpha$, but it also has a component perpendicular to $\vec{\ell}_{i-1}$ (with length $\ell \sin \theta$). However, because of the free rotation, over time the perpendicular part will average to zero (see Figure 6.4). So from the point of view of bond $\vec{\ell}_{i-2}$, *on average* $\vec{\ell}_i$ looks just like a bond of length $\ell \alpha$ pointing in the same direction as $\vec{\ell}_{i-1}$, and thus $\langle \vec{\ell}_{i-2} \cdot \vec{\ell}_i \rangle = \ell^2 \alpha^2$. The same argument can be extended to any pair of bonds i, j:

$$\left\langle \vec{\ell}_i \cdot \vec{\ell}_j \right\rangle = \ell^2 \alpha^{|i-j|} \tag{6.2.8}$$

Now we define a new summation index $k = |i-j|$ and write

$$\sum_{i=1}^{n} \sum_{j \neq i}^{n} \left\langle \vec{\ell}_i \cdot \vec{\ell}_j \right\rangle = \sum_{i=1}^{n} \sum_{j \neq i}^{n} \ell^2 \alpha^{|i-j|}$$

$$2 \sum_{k=1}^{n-1} \ell^2 \alpha^k (n - k) = 2n\ell^2 \sum_{k=1}^{n-1} \alpha^k - 2\ell^2 \sum_{k=1}^{n-1} k\alpha^k \tag{6.2.9}$$

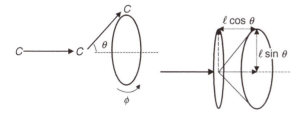

Figure 6.4 Definition of the angles θ and ϕ for the freely rotating and hindered rotation chains; the correspondence to a polyethylene molecule is suggested by the locations of the carbons.

The factor of 2 comes about because there are two ways to get each value of k (one with $i > j$ and one with $i < j$). The term $(n - k)$ arises because there are $n - 1$ nearest neighbors $(k = 1)$, $n - 2$ next nearest neighbors $(k = 2)$, etc. Note also that careful attention has to be paid to the limits of the sums.

The relevant summations have algebraic answers that are derived in the Appendix, namely:

$$\sum_{k=1}^{n-1} \alpha^k = \alpha \frac{1 - \alpha^{n-1}}{1 - \alpha} \approx \frac{\alpha}{1 - \alpha} \tag{6.2.10a}$$

for $n \to \infty$, assuming $|\alpha| < 1$ (i.e., $\theta \neq 90°$), and

$$\sum_{k=1}^{n-1} k\alpha^k = \alpha \frac{1 - \alpha^n}{(1 - \alpha)^2} \approx \frac{\alpha}{(1 - \alpha)^2} \tag{6.2.10b}$$

where again we let $n \to \infty$ to reach the last expression. Now we insert Equation 6.2.10a and Equation 6.2.10b into Equation 6.2.9, and we recall Equation 6.2.3 to obtain

$$\langle h^2 \rangle = n\ell^2 + 2n\ell^2 \left(\frac{\alpha}{1 - \alpha} \right) - 2\ell^2 \frac{\alpha}{(1 - \alpha)^2}$$

$$= n\ell^2 \left\{ \frac{1 + \alpha}{1 - \alpha} - \frac{2}{n} \cdot \frac{\alpha}{(1 - \alpha)^2} \right\} \approx n\ell^2 \left\{ \frac{1 + \alpha}{1 - \alpha} \right\}$$

$$= n\ell^2 \left\{ \frac{1 + \cos\theta}{1 - \cos\theta} \right\} \tag{6.2.11}$$

where again we assume n is large in the penultimate step. (We also reinserted $\cos\theta$ for α.) We can learn three important things from this result.

1. $\langle h^2 \rangle$ is *larger* than the freely jointed chain result if $\theta < 90°$ ($\cos\theta > 0$). This is very reasonable; if each link has some preference for heading in the same direction as the previous link, the chain will double back on itself less often. As an example, for C–C single bonds, θ is close to 70.5° (the complement to the tetrahedral angle) and for this value $\langle h^2 \rangle \approx 2n\ell^2$.

2. $\langle h^2 \rangle$ is still *proportional* to $n\ell^2$; the proportionality factor is just a number that depends on the details of the local constraints placed on link orientation. In particular, therefore, $\langle h^2 \rangle^{1/2}$ is still proportional to $\sqrt{n}$.

3. The previous statement applies strictly only in the large n limit, i.e., when the term proportional to $1/n$ in Equation 6.2.11 is negligible and when α^{n-1} vanishes. This is a commonly encountered caveat in polymer science: we can derive relatively simple expressions, but they will often be valid only in the large n limit. The answer to "How large is large enough?" will depend on the particular property, but when the correction is proportional to $1/n$, as it is in Equation 6.2.11, it will drop to the order of 1% when $n \approx 100$, which is not a particularly large number of backbone bonds.

Case 6.2.3   Hindered Rotation Chain

In a real polyethylene chain, the rotation about the cone is not free; there are three preferred conformations (t, g$^+$, g$^-$) as discussed in Section 6.1. Furthermore, all values of the rotation angle ϕ are possible to some extent (see Figure 6.1b). The derivation of $\langle h^2 \rangle$ is more complicated for this case, but it is similar in spirit to that for the freely rotating chain; it may be found in Flory's second book [1]. The large n result is

$$\langle h^2 \rangle = n\ell^2 \left\{ \frac{1 + \cos\theta}{1 - \cos\theta} \right\} \left\{ \frac{1 + \langle\cos\phi\rangle}{1 - \langle\cos\phi\rangle} \right\} \tag{6.2.12}$$

where $\langle \cos \phi \rangle$ is the average of $\cos \phi$ over the appropriate potential energy curve (Figure 6.1b). However, the important message is that $\langle h^2 \rangle$ is still proportional to $n\ell^2$; all that has changed is a numerical prefactor that depends on specific local constraints. This point, in fact, can be stated as a theorem:

If we take the limit $n \to \infty$, and if we consider "phantom" chains that can double back on themselves, then $\langle h^2 \rangle = Cn\ell^2$, where C is a numerical constant that depends only on local constraints and not on n.

The physical content of this theorem can be summarized as follows. If we have a chain of links with any degree of conformational freedom, no matter how limited that freedom may be, and if we track the conformation over enough links, the orientation of the last link will have lost all memory of the orientation of the first. At this point we could replace that entire subset of links with one new link, and it would be freely jointed with respect to the next set and the previous set. In other words, for any chain of n links whose relative orientations are constrained, we can always generate an equivalent chain with a new (bigger) link that is freely jointed, so that the original chain and the new chain have the same $\langle h^2 \rangle$. We will illustrate this concept in the next section.

The issue of how large n needs to be for this theorem to be useful was mentioned before; n of a hundred or so is usually more than adequate. The fact that real polymers occupy real volume means that polymer chains cannot double back on themselves, or have two or more links at the same point in space. This so-called "excluded volume" problem is actually very serious, and makes an exact solution for $\langle h^2 \rangle$ of a real polymer much more complicated. However, it turns out that there are two practical situations in which we can make this problem essentially go away; one is in a molten polymer, and the other is in a particular kind of solvent, called a theta solvent. Under these circumstances, a polymer is said to exhibit "unperturbed" dimensions. Thus this theorem is of tremendous practical importance. Further discussion of the excluded volume effect will be deferred to Section 6.8, and then it will be revisited in more detail in Chapter 7.7.

6.3   Characteristic Ratio and Statistical Segment Length

We can define a quantity C_n, called the *characteristic ratio*, which for any polymer structure describes the effect of local constraints on the chain dimensions:

$$C_n \equiv \frac{\langle h^2 \rangle_0}{n\ell^2} \tag{6.3.1}$$

In this equation, $\langle h^2 \rangle_0$ is the actual mean-square end-to-end distance of the polymer chain, and the subscript 0 reminds us that we are referring to unperturbed dimensions. In Equation 6.3.1, n denotes the number of chemical bonds along the polymer backbone, and ℓ is the actual length of a backbone bond, e.g., 1.5 Å for C—C. (For polymers containing different kinds of backbone bonds, such as polyisoprene or poly(ethylene oxide), it is appropriate simply to add $n_1\ell_1^2 + n_2\ell_2^2 + \cdots$ where n_i and ℓ_i are the number and length of bonds of type i, respectively.) C_n is a measure of chain flexibility: the larger the value of C_n, the more the local constraints have caused the chain to extend in one direction. As defined in Equation 6.3.1, C_n depends on n, but it approaches a constant value at large n; this is often denoted C_∞. For the freely rotating chain, C_∞ is $(1 + \cos \theta)/(1 - \cos \theta)$ from Equation 6.2.11. The dependence of C_n on n is shown in Figure 6.5 for several theoretical chains. The values of C_∞ for several common polymers are listed in Table 6.1. For polymers that have primarily C—C or C—O single bonds along the backbone, C_∞ ranges from about 4 to about 12. Using these values, or those provided in reference books, it is straightforward to estimate $\langle h^2 \rangle_0$ for any polymer of known structure and molecular weight.

Although calculating $\langle h^2 \rangle_0$ for a given polymer is thus a solved problem, this approach using C_∞ is not always the most convenient. For example, it requires remembering particular bond lengths

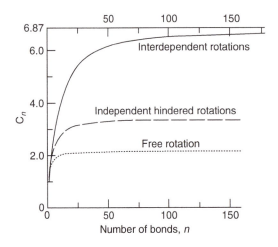

Figure 6.5 Characteristic ratio as a function of the number of bonds for three model chains. The dotted curve represents the freely rotating chain with $\theta = 68°$. The long dashed curve corresponds to a particular hindered rotation chain with the preferred values of ϕ 120° apart, but in which values of ϕ for neighboring bonds are independent. The smooth curve applies to an interdependence among values of ϕ on neighboring bonds. (Reproduced from Flory, P.J., *Statistical Mechanics of Chain Molecules*, Wiley-Interscience, New York, 1969. With permission.)

and the number of bonds per repeat unit. A more popular approach was suggested in the previous section. We could rewrite Equation 6.3.1 in the following way:

$$\langle h^2 \rangle_0 = C_\infty n\ell^2 = n\ell_{\text{eff}}^2 = Nb^2 \tag{6.3.2}$$

where $\ell_{\text{eff}} = \ell\sqrt{C_\infty}$ is a new effective bond length with the following meaning: the real chain with local constraints has an end-to-end distance, which is the same as that of a freely jointed chain with the same number of links n, but with a different (larger) step length ℓ_{eff}. Continuing in this vein, we can replace the number of bonds, n, with the number of monomers or repeat units, N, and subsume the proportionality factor between n and N into a new effective step length, b. The quantity b, defined by Equation 6.3.2, is called the *statistical segment length*. The calculation of $\langle h^2 \rangle_0$ through

Table 6.1 Representative Values of the Characteristic Ratio, Statistical Segment Length, and Persistence Length for Various Flexible Polymers, Calculated from the Experimental Quantities, $\langle h^2 \rangle_0/M$ (Å² mol/g), via Equation 6.3.2 and Equation 6.4.5b

Polymer	C_∞	b (Å)	ℓ_p (Å)	$\langle h^2 \rangle_0/M$ (Å² mol/g)	T (°C)
Poly(ethylene oxide)	5.6	6.0	4.1	0.80$_5$	140
1,4-Polybutadiene	5.3	6.9	4.0	0.87$_6$	140
1,4-Polyisoprene	4.8	6.5	3.5	0.62$_5$	140
Poly(dimethylsiloxane)	6.6	5.8	5.3	0.45$_7$	140
Polyethylene	7.4	5.9	5.7	1.2$_5$	140
Polypropylene	5.9	5.3	4.6	0.67	140
Polyisobutylene	6.7	5.6	5.2	0.57	140
Poly(methyl methacrylate)	9.0	6.5	6.9	0.42$_5$	140
Poly(vinyl acetate)	8.9	6.5	6.8	0.49	25
Polystyrene	9.5	6.7	7.3	0.43$_4$	140

The experimental chain dimensions were obtained by small-angle neutron scattering, as compiled in Fetters, L.J., Lohse, D.J., Witten, T.A., and Zirkel, A., *Macromolecules*, 27, 4639, 1994. The uncertainties in $\langle h^2 \rangle_0$ and M are typically a few percent. The temperature of the measurement is indicated, because the distribution of chain conformations depends on temperature.

N and b treats the real chain as though it were a freely jointed chain with N links of length b. This proves to be a useful computational scheme, but it is important to realize that b has no simple correspondence with the physical chain; it is not a measure of the real size of a real monomer. Furthermore, it contains no new information beyond that embodied in C_∞. Values of b are also given in Table 6.1. Note that although b varies monotonically with C_∞, it is not as simple a measure of flexibility. For example, b for polystyrene (6.7 Å) and polyisoprene (6.5 Å) differ by only a few percent, whereas polyisoprene is considered to be a relatively flexible polymer, and polystyrene a relatively stiff one. The resolution of this apparent paradox is left to Problem 6.2. Values of C_∞ and b may be determined experimentally in various ways, but the most direct route is by scattering measurements of the chain dimensions; this method will be described in detail in Chapter 8.

The approach taken above, and indeed for the remainder of the book, is that C_∞ or b are structure-specific parameters that we can look up as needed; the dependence of $\langle h^2 \rangle_0$ on molecular weight is universal and therefore more important to understand. However, it is of considerable interest to ask whether the techniques of computational statistical mechanics can be used to calculate C_∞ from first principles, i.e., from knowledge of bond angles, rotational potentials, etc. A highly successful scheme for doing so, called the rotational isomeric state approach, was developed by Flory [1]. It is beyond the scope of this book to describe it, but it is worth mentioning that even today it is not a trivial matter to execute such calculations, and that controversy exists about the correct values of C_∞ for some relatively simple chemical structures. These controversies are also not easily resolved by experiment; combined uncertainties in measured molecular weights and chain dimensions often exceed 10%.

Example 6.1

It is an interesting fact that bulk polyethylene has a positive coefficient of thermal expansion (about 2.5×10^{-4} per °C at room temperature), whereas the individual chain dimensions have a negative coefficient (d ln$\langle h^2 \rangle_0$/d$T = -1.2 \times 10^{-3} \text{deg}^{-1}$). In other words, when a piece of polyethylene is heated, the volume increases while the individual chain dimensions shrink. How does this come about?

Solution

Thermal expansion corresponds to a decrease in the density of the material, which reflects primarily an increase in the average distance *between* molecules; the radii of the individual atoms and the bond lengths also tend to increase, but to a much smaller extent. In contrast, the reduction in $\langle h^2 \rangle_0$ is primarily of intramolecular origin. From Equation 6.3.1, we can see that as n is independent of T and ℓ, if anything, increases with T, then there must be a decrease in the characteristic ratio, C_∞. The origin of this effect can be seen from Figure 6.1. As temperature increases, the Boltzmann factors that dictate the relative equilibrium populations of *trans* and *gauche* conformations change, and the *gauche* states become relatively more populated. As the *trans* conformations favor larger $\langle h^2 \rangle_0$, the net result is a reduction in C_∞. Note that this simple relation between C_∞ and the relative populations of *trans* and *gauche* states does not necessarily extend to more complicated backbone structures. For example, d ln$\langle h^2 \rangle_0$/dT is positive for 100% *cis*-1,4-polybutadiene, but negative for the all-*trans* versions (see Problem 6.5). This observation is not easy to anticipate based on the molecular structure.

6.4 Semiflexible Chains and the Persistence Length

For many macromolecules, the backbone does not consist of a string of single bonds with facile rotations, but rather some combination of bonds that tend to make the backbone continue in one direction. Such chains are called semiflexible, and examples (see Figure 6.6) include polymers with mostly aromatic rings along the backbone, such as poly(*p*-phenylene); polymers with large

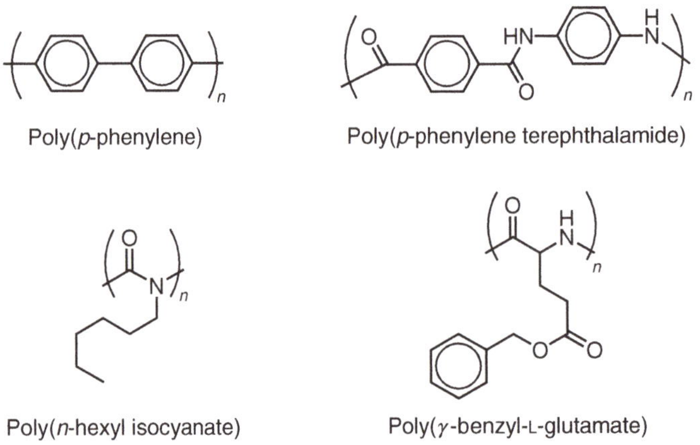

Poly(*p*-phenylene)   Poly(*p*-phenylene terephthalamide)

Poly(*n*-hexyl isocyanate)   Poly(*γ*-benzyl-L-glutamate)

Figure 6.6 Examples of polymer structures that are semiflexible or stiff chains.

side-groups that for steric reasons induce the backbone to adopt a helical conformation, such as poly(*n*-hexyl isocyanate) and poly(*γ*-benzyl-L-glutamate); biopolymers such as DNA and collagen that involve intertwined double or triple helices. Short versions of these molecules are essentially "rigid rods," but very long versions will wander about enough to be random coils. The description of chain dimensions in terms of either C_∞ or b turns out not to be as useful for this class of macromolecules. Therefore it is desirable to have a method to calculate the dimensions of such molecules, and particularly to understand the crossover from rod-like to coil-like behavior. Such a scheme is provided by the so-called *worm-like chain* of Kratky and Porod [2]; the fundamental concept is that of the *persistence length*, ℓ_p, which is a measure of how far along the backbone one has to go before the orientation changes appreciably. A garden hose provides a good everyday analogy to a semiflexible polymer, with a persistence length on the order of 1 ft. A 2 in. section of hose is relatively stiff or rigid, whereas the full 50 ft hose can be wrapped around and tangled with itself many times like a random coil. We will first define ℓ_p for flexible chains and see how it is simply related to C_∞. Then we will develop in terms of ℓ_p an expression for $\langle h^2 \rangle$ that can be used to describe flexible, semiflexible, and rigid chains.

The persistence length represents the tendency of the chain to continue to point in a particular direction as one moves along the backbone. It can be calculated by taking the projection of the end-to-end vector on the direction of the first bond ($\vec{\ell}_1/\ell$ is a unit vector in the direction of $\vec{\ell}_1$):

$$\ell_p \equiv \left\langle \frac{1}{\ell}\vec{\ell}_1 \cdot \vec{h} \right\rangle = \left\langle \frac{1}{\ell}\vec{\ell}_1 \cdot \sum_{i=1}^{n} \vec{\ell}_i \right\rangle$$

$$= \frac{1}{\ell}\left\{ \left\langle \vec{\ell}_1 \cdot \vec{\ell}_1 \right\rangle + \left\langle \vec{\ell}_1 \cdot \vec{\ell}_2 \right\rangle + \ldots + \left\langle \vec{\ell}_1 \cdot \vec{\ell}_n \right\rangle \right\} \qquad (6.4.1)$$

For the freely jointed chain, as discussed above, all the terms in the expansion in Equation 6.4.1 are zero except the first, and thus $\ell_p = \ell$. For chains with more and more conformational constraints that encourage the backbone to straighten out, more and more terms in the expansion will contribute positively, and ℓ_p increases. In the limit that every bond points in the same direction, the persistence length tends to infinity. When $\ell_p > L$, where $L = n\ell$ is the contour length of the chain, such a molecule is called a rigid rod.

6.4.1 Persistence Length of Flexible Chains

We now seek a relation between ℓ_p and C_∞ for long, flexible chains. We can rewrite Equation 6.4.1 as

$$\ell_p = \frac{1}{\ell} \sum_{j=1}^{n} \left\langle \vec{\ell}_1 \cdot \vec{\ell}_j \right\rangle = \frac{1}{\ell} \sum_{j=x}^{n} \left\langle \vec{\ell}_x \cdot \vec{\ell}_j \right\rangle \tag{6.4.2}$$

where x is any arbitrary bond in the chain. We can make this substitution because for a flexible chain, only a few terms j with small $|j - x|$ will contribute. Now we change the limits of the sum over j to extend over the complete chain; in other words we look in both directions from bond x. This amounts to a double counting, so we multiply ℓ_p by 2:

$$2\ell_p = \frac{1}{\ell} \sum_{j=1}^{n} \left\langle \vec{\ell}_x \cdot \vec{\ell}_j \right\rangle + \ell \tag{6.4.3}$$

Where did the extra ℓ on the right hand side of Equation 6.4.3 come from? Well, the double counting was not quite complete; there were two terms with $|x - j| = 1$, two with $|x - j| = 2$, etc., but only one with $x = j$. We need this contribution of ℓ in order to obtain $2\ell_p$, and therefore the missing "self" term is appended to Equation 6.4.3.

In order to remove the arbitrary choice of bond x, we sum over all possible choices of x, and assume that we get the same answer for each x; this approximation neglects the effects of chain ends, so it is valid only in the limit of large n.

$$\ell_p = \frac{1}{2n\ell} \sum_{j=1}^{n} \sum_{x=1}^{n} \left\langle \vec{\ell}_x \cdot \vec{\ell}_j \right\rangle + \frac{\ell}{2} \tag{6.4.4}$$

Here we divided by n in front of the double sum because we added n identical terms through the sum over x. All of these manipulations finally pay off: we recognize this double sum as exactly $\langle h^2 \rangle$ from Equation 6.2.2, and hence

$$\ell_p = \frac{1}{2n\ell} \langle h^2 \rangle + \frac{\ell}{2} = \frac{1}{2n\ell} C_\infty n\ell^2 + \frac{\ell}{2} = (C_\infty + 1)\frac{\ell}{2} \tag{6.4.5a}$$

In some derivations of this relation, the joint limit $n \to \infty$ and $\ell \to 0$ is taken (see following section). In this case, the extra ℓ on the right hand side of Equation 6.4.3 would vanish, and Equation 6.4.5a would become

$$\ell_p = \frac{1}{2n\ell} \langle h^2 \rangle = \frac{1}{2n\ell} C_\infty n\ell^2 = C_\infty \frac{\ell}{2} \tag{6.4.5b}$$

The difference between Equation 6.4.5a and Equation 6.4.5b is not particularly important, especially for stiffer chains where $C_\infty \gg 1$; we will use the latter form below, because it is simpler.

A related quantity in common use is the *Kuhn length* [3], ℓ_k, which is defined as twice the persistence length:

$$\ell_k \equiv 2\ell_p = C_\infty \ell \tag{6.4.6}$$

We thus have three different, but fully equivalent expressions for the mean-square unperturbed end-to-end distance of a flexible chain:

$$\langle h^2 \rangle_0 = C_\infty n\ell^2 = Nb^2 = L\ell_k \tag{6.4.7}$$

All three are useful and frequently employed, so they are worth remembering. Estimates of the persistence lengths of flexible polymers are also listed in Table 6.1.

Returning to Equation 6.4.1, we can actually extract a very appealing physical meaning for ℓ_p and ℓ_k. The terms in the expansion become progressively smaller as the average orientation of bond i becomes less correlated with that of the first bond. In fact, when bond i is on average perpendicular to the first bond (i.e., is uncorrelated), $\langle \vec{\ell_1} \cdot \vec{\ell_j} \rangle \approx 0$. All higher-order terms will also vanish. Thus the persistence length measures how far we have to travel along the chain before it will, on average, bend 90°. Similarly, the Kuhn length tells us how far we have to go along the chain contour before it will, on average, reverse direction completely. Equation 6.4.6 also provides a simple interpretation for C_∞: it is the number of backbone bonds needed for the chain to easily bend 180°.

6.4.2 Worm-Like Chains

So far, all that the persistence length has given us is a new way to express $\langle h^2 \rangle_0$ for flexible chains. To obtain a useful result for semiflexible and stiff chains, we will return to the freely rotating chain of Section 6.2 and transform it into a continuous worm-like chain. This we do by taking a special limit alluded to earlier; we will let the number of bonds, n, go to infinity, but the length of each bond, ℓ, will go to zero, while maintaining the contour length $L = n\ell$ constant. We begin by relating ℓ_p to $\alpha = \cos\theta$ of the freely rotating chain, starting from Equation 6.4.1:

$$\ell_p = \frac{1}{\ell}\left\{ \langle \vec{\ell_1} \cdot \vec{\ell_1} \rangle + \langle \vec{\ell_1} \cdot \vec{\ell_2} \rangle + \cdots + \langle \vec{\ell_1} \cdot \vec{\ell_n} \rangle \right\}$$
$$= \frac{1}{\ell}\left\{ \ell^2 + \ell^2\alpha + \ell^2\alpha^2 + \cdots + \ell^2\alpha^{n-1} \right\}$$
$$= \ell\left\{ 1 + \alpha + \alpha^2 + \cdots + \alpha^{n-1} \right\} = \ell\left(\frac{1-\alpha^n}{1-\alpha}\right) = \frac{\ell}{1-\alpha} \tag{6.4.8}$$

The last transformation utilized the summation results in Equation 6.2.10a, and the large n limit. Thus we can write

$$\alpha = 1 - \frac{\ell}{\ell_p} \approx \exp(-\ell/\ell_p) \quad \text{for} \quad \ell \to 0 \tag{6.4.9}$$

where we invoke the series expansion (see the Appendix):

$$e^x = 1 + x + \frac{x^2}{2!} + \cdots$$

Now we recall Equation 6.2.11 for the freely rotating chain, but retaining the term in α^n:

$$\langle h^2 \rangle = n\ell^2\frac{1+\alpha}{1-\alpha} - 2\ell^2\alpha\frac{(1-\alpha^n)}{(1-\alpha)^2}$$
$$= n\ell^2\left(\frac{2-\ell/\ell_p}{\ell/\ell_p}\right) - 2\ell^2(1-\ell/\ell_p)\left(\frac{1-\exp[-L/\ell_p]}{(\ell/\ell_p)^2}\right)$$
$$= L\ell_p(2-\ell/\ell_p) - 2\ell_p^2(1-\ell/\ell_p)(1-\exp[-L/\ell_p]) \tag{6.4.10}$$

and thus

$$\langle h^2 \rangle = 2\ell_p L - 2\ell_p^2(1-\exp[-L/\ell_p]) \quad \text{as} \quad \ell \to 0 \tag{6.4.11}$$

This expression is the result for the worm-like chain obtained by Kratky and Porod [2]. (Note that we took $(n-1)\ell \approx n\ell = L$ in Equation 6.4.10.) It is left as an exercise (Problem 6.6) to show that in the coil limit ($L \gg \ell_p$) this expression reverts to Equation 6.4.7 and that in the rod limit ($L \ll \ell_p$), $\langle h^2 \rangle = L^2$, as it should. Examples of experimental persistence lengths for semiflexible polymers are given in Table 6.2.

Table 6.2 Representative Values of Persistence Lengths for Semiflexible Polymers. Note That Values May Depend Either Weakly or Strongly on the Choice of Solvent[a]

Polymer	Solvent	ℓ_p (Å)
Hydroxypropyl cellulose	Dimethylacetamide	65
Poly(p-phenylene)[b]	Toluene	130
Poly(p-phenylene terephthalamide)	Methane sulfonic acid[c]	100
	96% Sulfuric acid	180
Poly(n-hexyl isocyanate)	Hexane[d]	420
	Dichloromethane	210
DNA (double helix)	0.2 M NaCl	600
Xanthan (double helix)	0.1 M NaCl	1200
Poly(γ-benzyl-L-glutamate)	Dimethylformamide	1500
Schizophyllan	Water	2000

[a]Data are summarized in Sato, T. and Teramoto, A., *Adv. Polym. Sci.*, 126, 85, 1996, except for [b]Vanhee, S., Rulkens, R., Lehmann, U., Rosenauer, C., Schulze, M., Koehler, W., and Wegner, G., *Macromolecules*, 29, 5136, 1996; [c]Chu, S.G., Venkatraman, S., Berry, G.C., and Einaga, Y., *Macromolecules*, 14, 939, 1981; and [d]Murakami, H., Norisuye, T., and Fujita, H., *Macromolecules*, 13, 345, 1980.

Figure 6.7 shows a plot of a dimensionless form of Equation 6.4.11, obtained by dividing through by ℓ_p^2 and plotting against L/ℓ_p. This independent variable is the number of persistence lengths in the chain, i.e., an effective degree of polymerization. The curve illustrates the smooth crossover from the rod-like behavior at small L/ℓ_p, with $\langle h^2 \rangle_0 \sim M^2$, to the coil-like behavior at large L/ℓ_p, with $\langle h^2 \rangle_0 \sim M$. Thus the worm-like chain model is able to describe both flexible and semiflexible chains with one expression. The double logarithmic format of Figure 6.7 is often employed in polymer science, when both the independent variable (such as M) and the dependent

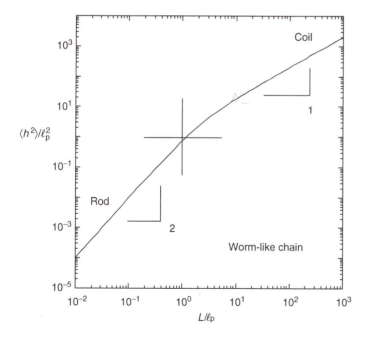

Figure 6.7 The mean-square end-to-end distance, normalized by the squared persistence length, as a function of the number of persistence lengths per chain (given by the ratio of contour and persistence lengths), according to the Kratky–Porod worm-like chain. The asymptotic slopes of 2 (rod limit) and 1 (coil limit) are also shown, as is the location of a chain with length equal to one persistence length.

variable can range over several orders of magnitude. If the functional relation is a power law, then in this format the plot will be a straight line and the slope gives the power law exponent.

6.5   Radius of Gyration

So far we have considered chain dimensions solely in terms of the average end-to-end distance. However, there are two severe limitations to this approach. First, the end-to-end distance is generally very difficult to measure experimentally. Second, for many interesting polymer structures (e.g., stars, rings, combs, dendrimers, etc.) it cannot even be defined unambiguously. The end-to-end distance assigns particular significance to the first and last monomers, but all monomers are of importance. A useful way to incorporate this fact is to calculate the average distance of all monomers from the center of mass. We denote the instantaneous vector from the center of mass to monomer i as $\vec{s}_i$, as shown in Figure 6.8. The center of mass at any instant in time for any polymer structure is the point in space such that

$$\sum_{i=1}^{N} m_i \vec{s}_i = 0 \tag{6.5.1}$$

where m_i is the mass of monomer i. Note that the center of mass does not need to be actually on the chain (in fact, it is unlikely to be). The root-mean-square, mass-weighted average distance of monomers from the center of mass is called the *radius of gyration*, R_g, or $\langle s^2 \rangle^{1/2}$, and is determined by

$$R_g = \langle s^2 \rangle^{1/2} \equiv \left\{ \frac{\sum_{i=1}^{N} m_i \langle s_i^2 \rangle}{\sum_{i=1}^{N} m_i} \right\}^{1/2} = \left\{ \frac{1}{N} \sum_{i=1}^{N} \langle s_i^2 \rangle \right\}^{1/2} \tag{6.5.2}$$

Here, just as with the end-to-end vector, it is useful to take $\langle s_i^2 \rangle = \langle \vec{s}_i \cdot \vec{s}_i \rangle$ in order to obtain an average distance rather than an average vector (which would zero by isotropy). In the second transformation we have assumed equal masses, i.e., a homopolymer, and $m_i = m$ cancels out. (Note that the summations run up to N, the number of monomers, and not n, the number of backbone bonds.) It is worth mentioning that the term radius of gyration is unfortunate, in that it invites confusion with the radius of gyration in mechanics; the latter refers to the mass-weighted, root-mean-square distance from an axis of rotation, not from a single point. However, the term radius of gyration in reference to Equation 6.5.2 appears to be firmly entrenched in polymer

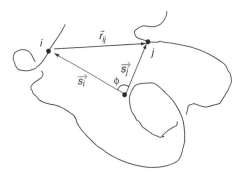

Figure 6.8   Illustration of the vectors from the center of mass to monomers i and j, $\vec{s}_i$ and $\vec{s}_j$, respectively, and the vector from monomer i to monomer j, $\vec{r}_{ij}$.

science. A more correct description will emerge in Section 6.7, namely that $\langle s^2 \rangle$ is the "second moment of the monomer distribution about the center of mass," but this terminology is rather unwieldy for daily use.

It is clear that R_g can be defined for any polymer structure, and thus avoids the second objection to $\langle h^2 \rangle$ listed above. It can be measured directly by light scattering techniques, as will be described in Chapter 8, and indirectly through various solution dynamics properties, as explained in Chapter 9, thereby avoiding the first objection. However, we went to some trouble to calculate $\langle h^2 \rangle_0$ for various chains, and to establish the utility of C_∞, b, and ℓ_p. Is that all out the window? No, it is not. We will now show that R_g is, in fact, very simply related to $\langle h^2 \rangle_0$ for an unperturbed linear chain, namely

$$R_g^2 = \frac{\langle h^2 \rangle_0}{6} = \frac{Nb^2}{6} = \frac{\ell_p L}{3} \tag{6.5.3}$$

Consider the dot product of the vectors from the center of mass to any two monomers i and j. By the law of cosines

$$\vec{s}_i \bullet \vec{s}_j = |\vec{s}_i||\vec{s}_j| \cos \phi = \frac{1}{2}\left[s_i^2 + s_j^2 - r_{ij}^2 \right] \tag{6.5.4}$$

where ϕ is the angle between $\vec{s}_i$ and $\vec{s}_j$, and r_{ij}^2 is the square distance between monomers i and j (see Figure 6.8). Now we take the average of each term in Equation 6.5.4, and then double sum over i and j:

$$\sum_{i=1}^{N}\sum_{j=1}^{N}\left\langle \vec{s}_i \bullet \vec{s}_j \right\rangle = \frac{1}{2}\sum_{i=1}^{N}\sum_{j=1}^{N}\left\langle s_i^2 \right\rangle + \frac{1}{2}\sum_{i=1}^{N}\sum_{j=1}^{N}\left\langle s_j^2 \right\rangle - \frac{1}{2}\sum_{i=1}^{N}\sum_{j=1}^{N}\left\langle r_{ij}^2 \right\rangle = 0 \tag{6.5.5}$$

This expression equals zero because

$$\sum_{i=1}^{N}\sum_{j=1}^{N}\left\langle \vec{s}_i \bullet \vec{s}_j \right\rangle = \left\langle \sum_{i=1}^{N}\vec{s}_i \bullet \sum_{j=1}^{N}\vec{s}_j \right\rangle = \langle 0 \bullet 0 \rangle = 0 \tag{6.5.6}$$

from Equation 6.5.1 (assuming all masses are equal). Returning to the second part of Equation 6.5.5, and utilizing Equation 6.5.2, we obtain

$$\frac{1}{2}\sum_{i=1}^{N}\sum_{j=1}^{N}\left\langle s_i^2 \right\rangle = \frac{1}{2}\sum_{i=1}^{N}\sum_{j=1}^{N}\left\langle s_j^2 \right\rangle = \frac{N^2}{2}R_g^2 \tag{6.5.7}$$

which can be rearranged to

$$R_g^2 = \frac{1}{2N^2}\sum_{i=1}^{N}\sum_{j=1}^{N}\left\langle r_{ij}^2 \right\rangle \tag{6.5.8}$$

This equation turns out to be a useful alternative definition of R_g. It expresses R_g in terms of the average distances between all pairs of monomers in the molecule; the location of the center of mass is not needed. Furthermore, Equation 6.5.8 is valid for any structure; it need not be a linear chain, and it need not have unperturbed dimensions.

Now we can derive the specific result Equation 6.5.3 for the freely jointed chain by realizing that

$$\left\langle r_{ij}^2 \right\rangle = |i - j|b^2 \tag{6.5.9}$$

or, in other words, $\langle r_{ij}^2 \rangle$ represents the end-to-end mean-square distance between any pair of monomers i and j separated by $k = |i - j|$ links. We can thus write

$$R_g^2 = \frac{b^2}{2N^2} \sum_{i=1}^{N} \sum_{j=1}^{N} |i - j| = \frac{b^2}{2N^2} \sum_{k=1}^{N-1} 2k(N - k) \qquad (6.5.10)$$

which comes from the fact that there are $2(N-1)$ terms where $|i - j| = 1$; $2(N - 2)$ terms where $|i - j| = 2$; $2(N - 3)$ terms where $|i - j| = 3$; etc. (If this seems mysterious, draw an $N \times N$ matrix, where the rows are numbers $i = 1 \ldots N$ and the columns are $j = 1 \ldots N$. For each matrix element, enter $|i - j|$. There will be N 0's along the main diagonal, $N - 1$ 1's immediately adjacent to [and on both sides of] the main diagonal, $N-2$ 2's in the next place over, etc.). Therefore

$$R_g^2 = \frac{b^2}{N^2} \left\{ N \sum_{k=1}^{N-1} k - \sum_{k=1}^{N-1} k^2 \right\} = \frac{b^2}{N^2} \left\{ \frac{N^2(N - 1)}{2} - \frac{N(N - 1)(2N - 1)}{6} \right\}$$

$$= \frac{b^2}{N^2} \left\{ \frac{3N^2(N - 1) - (N - 1)(2N^2 - N)}{6} \right\} \qquad (6.5.11)$$

$$= \frac{Nb^2}{6} - \frac{b^2}{6N} \approx \frac{Nb^2}{6}$$

where once again the last formula applies in the high N limit.

Example 6.2

A useful rule of thumb for polymers is that R_g is about 100 Å when $M = 10^5$ g/mol. This number can be used to estimate R_g for any other M by recalling the proportionality of R_g to $M^{1/2}$. Use the data in Table 6.1 to assess the reliability of this rule of thumb.

Solution

We can take the values for $\langle h^2 \rangle_0/M$ directly from the fifth column and multiply each one by 10^5. The largest will be 125,000 Å^2 for polyethylene, and the smallest will be 42,500 Å^2 for poly(methyl methacrylate). Then we need to divide by 6 and take the square root to obtain R_g:

For polyethylene, $R_g = (125,000/6)^{1/2} = 144$ Å

For poly(methyl methacrylate), $R_g = (42,500/6)^{1/2} = 84$ Å

All of the other polymers in Table 6.1 will give values between these two. We may conclude that the rule of thumb is reliable to at least one significant figure, and is better than that for many polymers.

Example 6.3

Use the experimental data for R_g for polystyrenes dissolved in cyclohexane in Figure 6.9 to estimate C_∞, ℓ_p, and b. Note that these data are for remarkably large molecular weights.

Solution

The straight line fit to the data gives $R_g = 0.25\ M^{0.51}$. To make things convenient, we can choose $N = 10^5$, for which $M = 104 \times 10^5 = 1.04 \times 10^7$ g/mol. From the fitting equation we obtain $R_g = 948$ Å. The number of backbone bonds $n = 2 \times 10^5$, and we use a more precise estimate of the bond length of 1.53 Å. From Equation 6.3.2 and Equation 6.5.3, then

$$C_\infty = \frac{\langle h^2 \rangle_0}{n\ell^2} = \frac{6 \times (948)^2}{2 \times 10^5 (1.53)^2} = 11.5$$

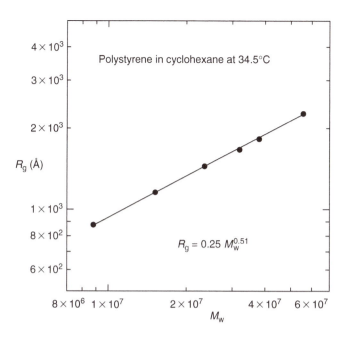

Polystyrene in cyclohexane at 34.5°C

R_g (Å)

$R_g = 0.25\, M_w^{0.51}$

M_w

Figure 6.9 Radius of gyration for very high molecular weight polystyrene in cyclohexane at the theta temperature. (Data from Miyake, Y., Einaga, Y., and Fujita, M., *Macromolecules*, 11, 1180, 1978.)

Also from Equation 6.3.2 and Equation 6.5.3,

$$b = \frac{\sqrt{6}}{\sqrt{N}} R_g = \frac{\sqrt{6}}{\sqrt{1 \times 10^5}} \times 948 = 7.3 \text{ Å}$$

Finally, from Equation 6.4.5b we have

$$\ell_p = C_\infty \frac{\ell}{2} = 11.5 \times \frac{1.53}{2} = 8.8 \text{ Å}$$

These values are systematically larger than those given in Table 6.1. Part of this difference may be attributed to experimental uncertainty, but most of the difference stems from the fact that the polystyrene data in Table 6.1 were obtained from molten polystyrenes, whereas the data in Figure 6.9 are for dilute solutions. Although the chain dimensions in a dilute theta solution and in the bulk both increase as $M^{1/2}$, the prefactor (e.g., C_∞) can be slightly different. In fact, the dimensions of a given polymer may differ by as much as 10% between two different theta solvents.

The worm-like chain of Section 6.4.2 also has an expression for R_g^2, which is

$$R_g^2 = \frac{1}{3}\ell_p L - \ell_p^2 + \frac{2\ell_p^4}{L^2}(\exp[-L/\ell_p] - 1) + \frac{2\ell_p^3}{L} \tag{6.5.12}$$

This result can be derived from the expression for $\langle h^2 \rangle$, Equation 6.4.11, by way of the relation 6.5.8 and a transformation of sums to integrals; this is left as Problem 6.8. An example of the application of the worm-like chain model is shown in Figure 6.10. The material is poly(n-hexyl isocyanate) (see Figure 6.6) dissolved in hexane and the coil dimensions were measured by light scattering (see Chapter 8). The smooth curve corresponds to Equation 6.5.12 with a persistence length of 42 nm, and the contour length determined as L (nm) $= M$ (g/mol)/715 (g/mol/nm). The factor of 715 therefore reflects the molar mass per nanometer of contour length. The correspondence between the data and the model is extremely good, except for the two very highest M samples.

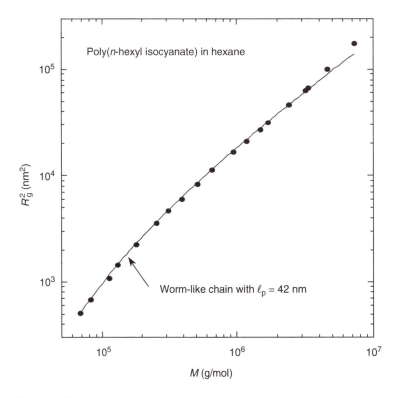

Figure 6.10   Radius of gyration versus molecular weight for poly(*n*-hexyl isocyanate) in hexane. The curve corresponds to the worm-like chain model with a persistence length of 42 nm. (Data from Murakami, H., Norisuye, T., and Fujita, H., *Macromolecules*, 13, 345, 1980.)

This deviation may be attributed to the onset of excluded volume effects, whereby the coil conformations are larger than anticipated by the freely jointed chain (Gaussian) limit.

6.6   Spheres, Rods, and Coils

We have derived results for $\langle h^2 \rangle_0$ for coils, rods, and everything in between. The conformation is assumed to be determined by the steric constraints induced by connecting the monomers chemically. But we could also think about the molecule more abstractly, as a series of freely jointed effective subunits. Now suppose we have the ability to "dial in" a through-space interaction between these subunits, an interaction that might be either attractive or repulsive. The former could arise naturally through dispersion forces, for example; all molecules attract one another in this way, and if we put our chain in a vacuum, those forces might dominate. If the attractive interaction were sufficiently strong, the chain could collapse into a dense, roughly spherical ball or globule. Repulsion could arise if each subunit bore a charge of the same sign, a so-called polyelectrolyte. This is commonly encountered in biological macromolecules, DNA for example. If the repulsive interaction were sufficiently strong, the chain could extend out to be a rod. It is instructive to think of the globule, coil, and rod as the three archetypical possible conformations of a macromolecule, and for many systems coil ↔ globule and coil ↔ rod transitions are experimentally accessible. For example, proteins in their native state are often globular, but upon denaturing the attractive interactions that cause them to fold are released, and the molecule becomes more coil-like. Similarly, a synthetic, neutral polymer dispersed in a bad solvent will collapse into a globule when it precipitates out of solution. A relatively short DNA double helix is reasonably rod-like ($\ell_p \approx 600$ Å), but if the double helix is denatured or "melted," the two separated strands can become coils. A poly(carboxylic acid) such as poly(methacrylic acid) in water will have a different density of charges along the chain as the pH is varied. At high pH, virtually every monomer will bear a negative charge and although the chain will not straighten out into a completely rigid, all-*trans*-backbone chain, it will show a size that scales almost linearly with M rather than as $\sqrt{M}$. In short, polymers can adopt conformations varying from dense spheres through flexible coils to rigid rods.

The scaling of the size with molecular weight is quite different in each case. We can encompass the various possibilities by writing a proportionality between the size and the degree of polymerization:

$$R_g \sim N^\nu \tag{6.6.1}$$

For the globule or dense object, $\nu = 1/3$. The volume occupied by the molecule is proportional to N and thus the radius goes as (volume)$^{1/3}$. For the unperturbed coil $\nu = 1/2$, as we have seen. However, we will find out in Chapter 7 that in a good solvent $\nu \approx 3/5$ due to the excluded volume effect. For a rod, clearly $\nu = 1$. Equation 6.6.1 is an example of a scaling relation; it expresses the most important aspect of chain dimensions, namely how the size varies with the degree of polymerization, but provides no numerical prefactors. The value of the exponent is universal, in the sense that any particular value of ν (1/3, 1/2, 3/5, or 1) will apply to all molecules in the same class.

As illustrated at the very beginning of this chapter, the size of these various structures (globule, coil, and rod) would be very different for a given polymer. Then we considered a polyethylene molecule, which in the liquid state will always be a coil. Now consider a representative DNA from bacteriophage T2. It has a contour length of 60 μm or almost 0.1 mm. As a coil, it should have $R_g = (L_k/6)^{1/2} \approx 1$ μm or 10,000 Å. With $M \approx 10^8$ and assuming a density of 1 g/mL, it would form a dense sphere with a radius of 340 Å. It is an amazing fact that the bacteriophage actually packages the DNA molecule to almost this extent. As shown in Figure 6.11, upon experiencing an osmotic shock the bacteriophage releases the DNA, which had been tightly wound up inside its head. The mechanism by which the DNA is packed so tightly remains incompletely understood. It is particularly remarkable, given that DNA carries negative charges all along its contour, which should create a strong repulsion between two portions of helix. This example also underscores another important point about chain dimensions: they can be very sensitive to the environment of the molecule, and not only to the intramolecular bonding constraints.

6.7   Distributions for End-to-End Distance and Segment Density

So far in this chapter we have only considered the *average* size and conformation of a polymer. Now we will figure out how to describe the *distribution* of sizes or conformations for a particular chain. We seek an expression for the probability $P(N, \vec{h})\, d\vec{h}$ that a random walk of N steps of length b will have an end-to-end vector, $\vec{h}$, lying between $\vec{h}$ and $\vec{h} + d\vec{h}$, as illustrated in Figure 6.12a. In other words, if the start of the chain defines the origin, we want the probability that the other end falls in an infinitesimal box with coordinates between x and $x + dx$, y and $y + dy$, and z and $z + dz$. From such a function, we will be able to obtain related functions for the probability $P(N, h)\, dh$ that the same walk has an end-to-end distance, $h = |\vec{h}|$, lying between h and $h + dh$, and the probability $\rho(N, r)\, dr$ that a monomer will be found between a distance r and $r + dr$ from the center of mass. It turns out that all of these distributions are approximately Gaussian functions, just like the familiar normal distribution for error analysis. In particular, the answer for $P(N, \vec{h})$ is

$$P(N, \vec{h}) = \left[\frac{3}{2\pi N b^2}\right]^{3/2} \exp\left[-\frac{3|h|^2}{2N b^2}\right] \tag{6.7.1}$$

a result that we will now derive.

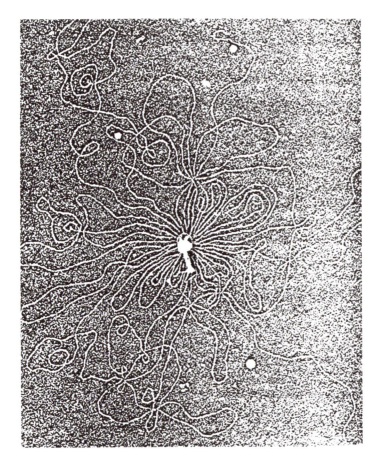

Figure 6.11   The DNA within a single T2 bacteriophage is released by "osmotic shock." (Reproduced from Kleinschmidt, A.K., et al., *Biochim. Biophys. Acta*, 61, 857, 1962. With permission.)

6.7.1   Distribution of the End-to-End Vector

We begin with a one-dimensional random walk with N steps of length b. In other words, at each step we go a distance b in either the $+x$ or the $-x$ direction, with the probability of $+$ or $-$ each

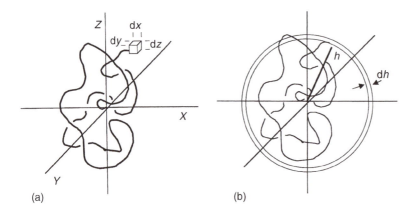

Figure 6.12   A flexible coil with one end at the origin and the other in (a) a volume element $dx\,dy\,dz$ and (b) a spherical shell of volume $4\pi h^2\,dh$.

being 1/2. At the end of the N steps, let p be the total number of $+$ steps and q be the total number of $-$ steps. Clearly $N = p + q$ and the net distance traveled will be $x = b(p - q)$. The probability of any given outcome for this kind of process is given by the binomial theorem:

$$P(N,x) = \left(\frac{1}{2}\right)^p \left(\frac{1}{2}\right)^q \frac{N!}{p!q!} \tag{6.7.2}$$

This expression arises as follows. The probability of a sequence of events that are independent is equal to the product of the probabilities of each event; this gives the factors of $(1/2)^p$ and $(1/2)^q$. However, this would underestimate what we want, namely a net displacement of x. This is because $(1/2)^p(1/2)^q$ assumes we have all $p+$ steps in succession, followed by $q-$ steps, whereas in fact the order of the individual steps does not matter, only the total p and q. There are N possible choices for the first step, then $N - 1$ for the next and so on, which increases the total probability by a factor of $N!$. This, however, now overestimates the answer, because all of the $p+$ steps are indistinguishable, as are all of the $q-$ steps. There are $p!$ possible permutations of the $+$ steps, all of which would give the same answer, and similarly $q!$ permutations of the $-$ steps; both of these factors are counted in $N!$, so we have to divide by them out. Thus we arrive at Equation 6.7.2.

Now we make the simple substitutions

$$p = \frac{1}{2}\left(N + \frac{x}{b}\right), \quad q = \frac{1}{2}\left(N - \frac{x}{b}\right) \tag{6.7.3}$$

to obtain

$$P(N,x) = \left(\frac{1}{2}\right)^N N! \frac{2!}{\left(N + \frac{x}{b}\right)!} \frac{2!}{\left(N - \frac{x}{b}\right)!} \tag{6.7.4}$$

The expression is simplified by means of Stirling's large N approximation, namely

$$\ln N! \approx N \ln N - N \tag{6.7.5}$$

which gets rid of the nasty factorials. (It is worth noting that Stirling's approximation is excellent when N is on the order of Avogadro's number, but it is not quantitatively accurate for $N \approx 100$; nevertheless these errors largely cancel in deriving the Gaussian distribution.) Utilizing this, Equation 6.7.4 can be expanded:

$$\ln P(N,x) = -\frac{1}{2}\left(N + \frac{x}{b}\right)\ln\left(1 + \frac{x}{bN}\right) - \frac{1}{2}\left(1 - \frac{x}{bN}\right)\ln\left(1 - \frac{x}{bN}\right) \tag{6.7.6}$$

after some algebra. Now we recall the expansion of $\ln(1 + x)$ when $x \ll 1$, namely

$$\ln(1 + x) = x - \frac{1}{2}x^2 + \frac{1}{3}x^3 \cdots \tag{6.7.7}$$

and realize that $(x/bN) \to 0$ as N gets very big. Thus $\ln(1 + x) \approx x$ applies

$$\ln P(N,x) = -\frac{1}{2}\left(N + \frac{x}{b}\right)\left(\frac{x}{Nb}\right) - \frac{1}{2}\left(N - \frac{x}{b}\right)\left(\frac{x}{Nb}\right)$$
$$= \frac{-x^2}{2Nb^2} \tag{6.7.8}$$

or

$$P(N, x) \sim \exp\left[\frac{-x^2}{2Nb^2}\right] \tag{6.7.9}$$

We insert a proportional sign here to allow for the appropriate normalization (see below). Now to convert to a three-dimensional N-step random walk, we take $N/3$ steps along x, $N/3$ along y, and $N/3$ along z, and recognize that the probabilities along the three directions are independent.

Therefore

$$P(N,\vec{h}) = P\left(\frac{N}{3},x\right)P\left(\frac{N}{3},y\right)P\left(\frac{N}{3},z\right) \sim \exp\left[\frac{-3|h|^2}{2Nb^2}\right] \tag{6.7.10}$$

where we have inserted $|h|^2 = x^2 + y^2 + z^2$. The final step is the normalization, which accounts for the fact that if we look over all space for the end of our walk, we must find it exactly once. This is expressed by

$$\int\limits_{-\infty}^{\infty}\int\limits_{-\infty}^{\infty}\int\limits_{-\infty}^{\infty} P(N,\vec{h})\,dx,dy,dz = 1 \tag{6.7.11}$$

You can find in a table of integrals that

$$\int\limits_{-\infty}^{\infty} \exp(-kx^2)\,dx = \sqrt{\frac{\pi}{k}}$$

and thus from the triple integral of Equation 6.7.11 we find that the result is $[3/2\pi Nb^2]^{-3/2}$. Therefore we need to multiply our exponential factor by $[3/2\pi Nb^2]^{3/2}$ to satisfy Equation 6.7.11 and arrive at the result given in Equation 6.7.1.

This Gaussian distribution function for $\vec{h}$ is plotted against $|\vec{h}|$ in Figure 6.13a. Although it is a distribution function for a vector quantity, it only depends on the length of h; this is a natural consequence of assuming that x, y, and z steps are equally probable. It is peaked at the origin, which means that the single most probable outcome is $\vec{h} = 0$, i.e., the walk returns to the origin. However, because of the prefactor, the probability of this particular outcome shrinks as $N^{-3/2}$, even though it is more likely than any other single outcome. Finally, this expression for $P(N,\vec{h})$ was obtained by assuming large N (no surprise here). How large does N have to be for the Gaussian function to be useful? It turns out that even for $N \approx 10$, the real distribution for a random walk looks reasonably Gaussian. Already for very small N it is symmetric and peaked at the origin, but it

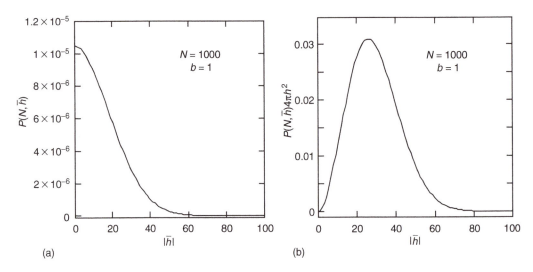

Figure 6.13 Gaussian probability distributions for a chain of N steps of length b, plotted as (a) the probability of an end-to-end vector $\vec{h}$ versus $|\vec{h}|$ and (b) the probability of an end-to-end distance h.

will be "bumpy" because b has a discrete value. As N gets larger, the fact that b is discrete becomes less and less important, although the fact that we are representing a discrete function by a continuous one never goes away for finite N. Indeed, we can assert that even though the Gaussian distribution may provide numerical answers that are very accurate, it can never be exactly correct for a real chain. That is because a real chain has a finite contour length that $|\vec{h}|$ can never exceed, whereas Equation 6.7.1 provides a finite probability for any value of $|\vec{h}|$ all the way out to infinity.

6.7.2 Distribution of the End-to-End Distance

We now turn to obtaining $P(N, h)$ for the distance h. The transformation is illustrated in Figure 6.12b. We consider a spherical shell at a distance h from the origin. It has a surface area $4\pi h^2$ and a thickness $\mathrm{d}h$, so its volume is $4\pi h^2\,\mathrm{d}h$. Any walk whose end-to-end vector $\vec{h}$ lies in this shell will have the same length h, so the answer we seek is just

$$P(N, h)\,\mathrm{d}h = 4\pi h^2 P(N, \vec{h})\,\mathrm{d}h$$

$$= 4\pi h^2 \left[\frac{3}{2\pi Nb^2}\right]^{3/2} \exp\left[\frac{-3\left|\vec{h}\right|^2}{2Nb^2}\right]\mathrm{d}h \qquad (6.7.12)$$

This function is already normalized correctly. It should satisfy the one-dimensional integral

$$\int_0^\infty P(N, h)\,\mathrm{d}h = 1 \qquad (6.7.13)$$

and you can show that it does, armed with the knowledge that

$$\int_0^\infty x^2 \exp(-kx^2)\,\mathrm{d}x = \frac{1}{4k}\sqrt{\frac{\pi}{k}}$$

Note that the normalization integral goes from 0 to ∞, as h cannot be negative.

This distribution function is plotted in Figure 6.13b and is rather different from $P(N, \vec{h})$. In particular, it vanishes at the origin and has a peak at a finite value of h before decaying to zero as $N \to \infty$. The fact that it vanishes at the origin is due to multiplying the exponential decay by h^2. There is thus a big difference between finding the most probable *vector* position (which is the origin) and finding the most probable *distance* (the position of the peak in $P(N, h)$, see Problem 6.10). You have probably encountered this contrast before, for example, in the radial distribution function for the s electrons of a hydrogen-like atom, or for the Maxwell distribution of molecular velocities in a gas. In the former, the most probable position of the electron is at the nucleus, but the most probable distance is a finite quantity, the Bohr radius. In the latter, the most probable velocity in the gas is zero, but the most probable speed is finite.

As a simple application of this distribution function, we can ask what is the mean square value of h?

$$\langle h^2 \rangle = \int_0^\infty h^2 P(N, h)\,\mathrm{d}h$$

$$= \int_0^\infty h^2 4\pi h^2 \left[\frac{3}{2\pi Nb^2}\right]^{3/2} \exp\left[\frac{-3\left|\vec{h}\right|^2}{2Nb^2}\right]\mathrm{d}h \qquad (6.7.14)$$

where now we need to know

$$\int_0^\infty x^4 \exp(-kx^2)\, dx = \frac{3}{8k^2}\sqrt{\frac{\pi}{k}}$$

Applying this in Equation 6.7.14 we obtain

$$\langle h^2 \rangle = 4\pi \left[\frac{3}{2\pi Nb^2}\right]^{3/2} \frac{3}{8}\left(\frac{3}{2Nb^2}\right)^{-2}\sqrt{\pi}\left(\frac{2Nb^2}{3}\right)^{1/2} = Nb^2 \qquad (6.7.15)$$

as expected. The quantity $\langle h^2 \rangle$ is also called the *second moment* of the distribution $P(N, h)$. (You may recall the discussion of moments in the context of molecular weights in Chapter 1.7.)

6.7.3 Distribution about the Center of Mass

The last thing we consider in this section is the related distribution $\rho(N, r)\, dr$, the probability that a monomer is between r and $r + dr$ from the center of mass. It turns out that there is no simple analytical expression for this distribution, even for a Gaussian chain [4], but the resulting distribution is well approximated by a Gaussian:

$$\rho(N, r) = N4\pi r^2 \left[\frac{3}{2\pi\langle s^2\rangle}\right]^{3/2}\exp\left[\frac{-3r^2}{2\langle s^2\rangle}\right] \qquad (6.7.16)$$

where we use the second moment $\langle s^2\rangle = Nb^2/6$ explicitly in the expression (compare Equation 6.7.1, where Nb^2 could have been replaced by $\langle h^2\rangle$). One important point is the factor of N in front. This is simply a new normalization so that

$$\int_0^\infty \rho(N, r)\, dr = N \qquad (6.7.17)$$

which reflects the fact that when we look for monomers over all space, we must find all N of them.

The segment density distribution can also be used for a solid object, where there is no need to identify N separate subunits. In such a case $\rho(r)$ can be used to find $R_g = \langle s^2 \rangle^{1/2}$, from

$$\langle s^2 \rangle = \frac{\int_0^\infty r^2 \rho(r)\, dr}{\int_0^\infty \rho(r)\, dr} = \frac{\int_0^\infty 4\pi r^4 \rho(\vec{r})\, dr}{\int_0^\infty 4\pi r^2 \rho(\vec{r})\, dr} \qquad (6.7.18)$$

The integral in the denominator provides the necessary normalization. As an example, consider a solid sphere with radius R_0 and uniform density ρ_0. The distribution function for $\rho(\vec{r})$ is just a constant, ρ_0, for $0 \leq r \leq R_0$, and 0 for $r > R_0$. (Note that this simple function is $\rho(\vec{r})$, not $\rho(r)$ because the latter must increase as r^2 for $r \leq R_0$; there are more monomers near the surface of the sphere than at the center.) Substituting $\rho(\vec{r})$ into Equation 6.7.18 we obtain

$$\langle s^2 \rangle = \frac{\int_0^{R_0} \rho_0 4\pi r^4\, dr}{\int_0^{R_0} \rho_0 4\pi r^2\, dr} = \frac{R_0^5/5}{R_0^3/3} = \frac{3}{5}R_0^2 \qquad (6.7.19)$$

Thus for a solid sphere

Table 6.3 Formulae for the Radii of Gyration for Various Shapes

Structure	R_g^2	Parameters
Gaussian coil	$\dfrac{Nb^2}{6}$	Degree of polymerization N
Gaussian star	$\dfrac{3f - 2}{f}\dfrac{N_{arm}b^2}{6}$	Statistical segment length b Arm degree of polymerization N_{arm} Number of arms f
Gaussian ring	$\dfrac{Nb^2}{12}$	Statistical segment length b Degree of polymerization N
Solid sphere	$\dfrac{3}{5}R^2$	Statistical segment length b Sphere radius R
Solid ellipsoid	$\dfrac{1}{5}(R_1^2 + R_2^2 + R_3^2)$	Ellipsoid principal radii R_1, R_2, R_3
Thin rod	$\dfrac{1}{12}L^2$	Rod length L
Cylinder	$\dfrac{1}{12}L^2 + \dfrac{1}{2}r^2$	Cylinder radius r, length L
Thin disk	$\dfrac{1}{2}r^2$	Disk radius r

$$R_g = \sqrt{\frac{3}{5}}R_0 \qquad\qquad (6.7.20)$$

This result serves to emphasize an important point that R_g reflects an *average* of the monomer distribution and not the *total spatial extent* of the object. There are always monomers further than R_g from the center of mass as well as monomers closer than R_g to the center of mass. Expressions for R_g for other shapes are listed in Table 6.3.

6.8 Self-Avoiding Chains: A First Look

We mentioned at the end of Section 6.2 that there is a further important issue in chain conformations, that of excluded volume. The simple fact is that a real polymer occupies real space and no two monomers in the chain can share the same location concurrently. This means that the statistics of the conformation are no longer those of a random walk, but rather a *self-avoiding walk*. This difference might appear subtle, which it is, but the consequences are profound.

Most importantly, we can no longer write down an analytical expression for any of the desired distribution functions such as $P(N, h)$. Nor can we calculate in any simple yet rigorous way the dependence of R_g on N. The mathematical reason for this difficulty is the way the problem becomes more complicated as N increases. In the case of the random walk, the orientation of any link or step is dictated by random chance and its position in space is determined only by where the previous link is. For the self-avoiding walk, in contrast, we would need to ask where every previous link is in order to establish whether a particular orientation would be allowed for the link in question. It would not be allowed if it intersected any previous link. Consequently, the calculation becomes more and more complicated as N increases. Some very sophisticated mathematics has been employed on this problem, but we will not discuss this at all. We can draw an important qualitative conclusion, however: the excluded volume effect will tend to make the average coil size *larger*, as the chain seeks conformations without self-intersections. From the most sophisticated analysis, $R_g \sim N^{0.589}$ instead of $N^{1/2}$; in other words the exponent ν from Equation 6.6.1 is 0.589 (although most people use 0.6 as a reasonable approximation).

As also mentioned in Section 6.2, there are two experimental situations in which this complication goes away, and $\nu = 0.5$ again. One case is a molten polymer. Here a chain still cannot have two monomers occupying the same place, but there is no benefit to expanding the coil. The reason is that space is full of monomers, and a monomer on one chain cannot tell if its immediate neighbor in space belongs to a different chain, or is attached by many bonds to the same chain. Consequently, it does not gain anything by expanding beyond the Gaussian distribution. The second case is for a chain dissolved in a particular kind of solvent called a theta solvent. A theta solvent is actually a not-very-good solvent, in the sense that for energetic reasons monomers would much prefer to be next to other monomers than next to solvent molecules. This has a tendency to shrink the chain, and a theta solvent refers to a particular solvent at a particular temperature where the expansion due to the self-avoiding nature of the chain is exactly canceled by shrinking due to unfavorable polymer–solvent interactions. We will explore this in more detail in the next chapter, where we consider the thermodynamics of polymer solutions.

6.9 Chapter Summary

In this chapter, we have examined the spatial extent of polymer chains as a function of molecular weight and chemical structure. The principal results are the following:

1. A single chain can adopt an almost infinite number of possible conformations; we must settle for describing the average size.
2. For any chain with some degree of conformational freedom the average size will grow as the square root of the degree of polymerization: this is the classic result for a random walk. Furthermore, the distribution of chain sizes is approximately Gaussian.
3. The prefactor that relates size to molecular weight is a measure of local flexibility; three interchangeable schemes for quantifying the prefactors are the characteristic ratio, the statistical segment length, and the persistence length.
4. Chemical structures for which the chain orientation can reverse direction in about 20 backbone bonds or less are called "flexible"; much stiffer polymers are termed "semiflexible." This latter class is best considered through the worm-like chain model using the persistence length as the key measure of local flexibility.
5. The radius of gyration is the most commonly employed measure of size; it can be defined for any chemical structure and it is directly measurable.
6. Because of excluded volume, real chains dissolved in a good solvent are not random walks, but self-avoiding walks. The corresponding size grows with a slightly larger power of molecular weight. In theta solvents or in the melt, the excluded volume effect is canceled out and the random walk result applies.

Problems

1. Experimental chain dimensions for poly(ethylene terephthalate) (PET) at 275°C are given by $\langle h^2 \rangle_0 / M \approx 0.90$ Å^2 mol/g. Calculate C_∞, the statistical segment length, and the persistence length for this polymer. Based on these numbers, is PET a flexible polymer, or not? What would you expect based on the molecular structure?
2. Resolve the paradox noted in Section 6.3: polystyrene is considered to be relatively stiff, and polyisoprene relatively flexible, yet their statistical segment lengths are almost identical.
3. For freely jointed copolymers with n_A steps of length ℓ_A and n_B steps of length ℓ_B find $\langle h^2 \rangle$ (large n limit) for strictly alternating, random, and diblock architectures. Are the answers the same or different? Why?

4. Assume a freely rotating, strictly alternating copolymer chain, with alternating bond angles θ_A and θ_B and step lengths ℓ_A and ℓ_B; find $\langle h^2 \rangle$.
5. The unperturbed dimensions of polymer chains depend weakly on temperature. Interestingly, the sign of the temperature coefficient can be either positive or negative. For example, for 100% *cis*-polybutadiene, $d(\ln \langle h^2 \rangle)/dT \approx 0.0004$ deg^{-1}, whereas for 100% *trans*-polybutadiene, the same quantity is -0.0006 deg^{-1}. Provide an explanation for the observation that this quantity can, in general, be either positive or negative, and speculate how your explanation might apply in this particular case. Note that it would require a very careful calculation to actually show why these two coefficients have different signs.
6. Show that the expression for the mean-square end-to-end distance of the worm-like chain given in Equation 6.4.11 reduces to the expected answers for a random coil and a rigid rod, when the limits $L \gg \ell_p$ and $\ell_p \gg L$ are taken, respectively.
7. From the following data of Kirste[†] estimate the persistence length of calf thymus DNA and the mass per persistence length M_p:

M_w	R_g (Å)	M_w	R_g (Å)
3.5×10^5	450	3.45×10^6	1700
4.6×10^5	480	4.6×10^6	2000
6.9×10^5	650	6.3×10^6	2300
1.15×10^6	900	6.9×10^6	2500
1.6×10^6	1100	9.2×10^6	3000
2.3×10^6	1390	1.35×10^7	3600
2.8×10^6	1550		

Equation 6.5.12 is a good place to start. One approach is graphical, i.e., to compare the data (plotted logarithmically) against a theoretical plot of the dimensional quantities R_g/ℓ_p versus M/M_p. Alternatively, the data can be fit to Equation 6.5.12 using a nonlinear regression routine. However, some care must be taken in weighting the data in the fit. Is there a reason why the highest molecular weight data should be accorded less significance?

8. Derive the expression for the mean square radius of gyration of the worm-like chain, Equation 6.5.12, from the corresponding expression for the mean-square end-to-end distance, Equation 6.4.11. First show that this relation is equivalent to Equation 6.5.8.

$$R_g^2 = \frac{1}{n^2} \sum_{j=2}^{n} \sum_{i=1}^{j-1} \langle h_{ij}^2 \rangle$$

Then make the correspondences $L = n\ell$, $x = i\ell$, and $y = j\ell$ when using Equation 6.4.11 for $\langle h_{ij}^2 \rangle$. The final step is to equate the double sum above with the following integrals and carry out the integration:

$$\sum_{j=2}^{n} \sum_{i=1}^{j-1} \Rightarrow \int_0^L \frac{1}{\ell} dy \int_0^y \frac{1}{\ell} dx$$

9. Miyaki, Einaga, and Fujita[‡] reported measurements of R_g for very high M polystyrenes in benzene at 25°C (a good solvent) and cyclohexane at 34.5°C (theta conditions). Find the relation between R_g and M_w in the case of benzene (the cyclohexane data were analyzed in Example 6.3). How does the exponent compare with expectations? If you extrapolate to lower

[†] R.G. Kirste, *Disc. Farad. Soc.* 49, 51 (1970).
[‡] Y. Miyaki, Y. Einaga, and H. Fujita, *Macromolecules* 11, 1180 (1978).

M, by what degree of polymerization would excluded volume increase R_g (i) by only 10%?
(ii) by a factor of 2?

$M_w \times 10^{-6}$	R_g (nm), Benzene	R_g (nm), Cyclohexane
56.2 ± 1	506 ± 10	228 ± 5
39.5 ± 1	392 ± 8	183 ± 4
32.0 ± 0.6	353 ± 7	167 ± 4
23.5 ± 0.5	297 ± 9	145 ± 3
15.1 ± 0.5	227 ± 7	116 ± 2
8.77 ± 0.3	164 ± 4	87.9 ± 2

10. For the Gaussian distribution function for the end-to-end distance, calculate the most probable distance, the mean distance, and the root-mean-square distance. Generate a good plot of this distribution function for some value of N and indicate where these three characteristic distances fall on the plot.

11. Find the mean square radius of gyration for an infinitely thin rod of length L, with mass density (per unit length) ρ in the center $L/2$ section and mass density (per unit length) 2ρ for the $L/4$ sections at each end.

12. Find the radius of gyration of a sphere with an outer shell of different density, where the inner spherical core has a radius of R_1 and a density of ρ, and the outer shell (the "corona") extends to a radius of $R_2 = 2R_1$ and has a density equal to 2ρ. (This geometry is a reasonable representation of a spherical micelle, which can be formed in a solution of block copolymers when one block is relatively insoluble; see Chapter 4.4.)

13. Consider the coil dimensions of an A–B diblock copolymer, with total degree of polymerization N and the fraction of A monomers given by f. Assuming that the mass of each A and B monomers is the same, show that

$$\langle R_g^2 \rangle = f \langle R_g^2 \rangle_A + (1 - f) \langle R_g^2 \rangle_B + f(1 - f) \langle Z^2 \rangle_{AB}$$

where $\langle R_g^2 \rangle_i$ is the mean square radius of gyration for block i and $\langle Z^2 \rangle_{AB}$ is the mean square separation of block centers-of-mass.

14. Consider a statistical A–B copolymer, with total degree of polymerization N and the fraction of A monomers given by f. Derive a simple expression for $\langle h^2 \rangle$ in terms of N, f, b_A, and b_B, assuming Gaussian statistics. This equation will probably not be exactly correct, in practice, even for large N. One reason is that thermodynamic interactions between A and B monomers (usually effectively repulsive) will tend to expand the chain. However, there is another reason, connected to the nature of the statistical length; what is it?

15. A distinguished polymer scientist is reputed to have remarked "an infinite steel girder would be a random coil." Explain the important point that this striking comment is intended to illustrate.

16. How does the average polymer concentration inside an individual coil vary with M in good and theta solvents? Estimate this concentration in g/cm^3 for polystyrene with $M = 10^6$ under both conditions, using Miyake et al.'s data given in Problem 9. Explain why this concentration is often referred to as the coil overlap concentration, c^*.

References

1. Flory, P.J., *Statistical Mechanics of Chain Molecules*, Wiley-Interscience, New York, 1969.
2. Kratky, O. and Porod, G., *Rec. Trav. Chim.*, 68, 1106 (1949).
3. Kuhn, W., *Kolloid Z.*, 68, 2 (1934).
4. Yamakawa, H., *Modern Theory of Polymer Solutions*, Harper & Row, New York, 1971.

Further Readings

Doi, M. and Edwards, S.F., *The Theory of Polymer Dynamics*, Clarendon Press, Oxford, 1986.

Flory, P.J., *Statistical Mechanics of Chain Molecules*, Wiley-Interscience, New York, 1969.

Graessley, W.W., *Polymeric Liquids and Networks: Structure and Properties*, Garland Science, New York, 2003.

Rubinstein, M. and Colby, R.H., *Polymer Physics*, Oxford University Press, New York, 2003.

Volkenstein, M.V., *Configurational Statistics of Polymeric Chains*, Wiley-Interscience, London, 1963.

Yamakawa, H., *Modern Theory of Polymer Solutions*, Harper & Row, New York, 1971.

7

Thermodynamics of Polymer Solutions

7.1 Review of Thermodynamic and Statistical Thermodynamic Concepts

In this chapter we shall consider some thermodynamic properties of solutions in which a polymer is the solute and some low-molecular-weight species is the solvent. Our special interest is in the application of solution thermodynamics to problems of phase equilibrium.

An important fact to remember about the field of thermodynamics is that it is blind to details concerning the structure of matter. Thermodynamics is concerned with observable quantities and the relationships among them, although there is a danger of losing sight of this fact in the somewhat abstract mathematical formalism of the subject. For example, we will take the position that entropy is often more intelligible from a statistical, atomistic point of view than from purely phenomenological perspective. It is the latter that is *pure* thermodynamics; the former is the approach of *statistical* thermodynamics. In this chapter we shall make extensive use of the statistical point of view to understand the molecular origin of certain phenomena. The treatment of heat capacity in physical chemistry provides an excellent example of the relationship between pure and statistical thermodynamics. Heat capacity is defined experimentally as the heat required to change the temperature of a sample in, say, a constant-pressure experiment. Various equations relate the heat capacity to other thermodynamic quantities such as enthalpy, H and entropy, S. The alternative approach to heat capacity would be to account for the storage of energy in molecules in terms of the various translational, rotational, and vibrational degrees of freedom. *Doing* thermodynamics does not even require knowledge that molecules exist, much less how they store energy, whereas *understanding* thermodynamics benefits considerably from the molecular point of view.

One drawback of the statistical approach is that it depends on models, and models are bound to oversimplify. Nevertheless, we can learn a great deal from the attempt to evaluate thermodynamic properties from molecular models, even if the effort falls short of quantitative success.

There is probably no area of science that is as rich in mathematical relationships as thermodynamics. This makes thermodynamics very powerful, but such an abundance of riches can also be intimidating. In this chapter we assume that the reader is familiar with basic chemical and statistical thermodynamics at the level that these topics are treated in undergraduate physical chemistry textbooks. This premise notwithstanding, a brief review of some pertinent relationships will be a useful way to get started.

Notation frequently poses problems in science and this chapter is an example of such a situation. Our problem at present is that we have too many things to count: they cannot all be designated n. In thermodynamics n is widely used to designate the number of moles and so we will adhere to this convention. Since we deal with (at least) two-component systems in this chapter, any count of the number of moles will always carry a subscript to indicate the component under consideration. We shall use the subscript 1 to designate the solvent and 2 to designate the solute. We have consistently used N to designate the degree of polymerization and shall continue with this notation, although the definition of "monomer" will be modified slightly.

To describe the state of a two-component system at equilibrium, we must specify the number of moles n_1 and n_2 of each component, as well as the pressure p and the absolute temperature T. It is the Gibbs free energy, G, that provides the most familiar access to a discussion of equilibrium. The increment of G associated with increments in the independent variables mentioned above is given by the equation

$$dG = V\,dp - S\,dT + \sum_{i=1,2} \mu_i\,dn_i \qquad (7.1.1)$$

where V is the volume, S is the entropy, and μ_i is the chemical potential of component i. An important aspect of thermodynamics is the fact that the state variables (in the present context, especially H, G, and the internal energy U) can be expanded as partial derivatives of the fundamental variables. Hence we can also write

$$dG = \left(\frac{\partial G}{\partial p}\right)_{T,n_1,n_2} dp + \left(\frac{\partial G}{\partial T}\right)_{p,n_1,n_2} dT + \left(\frac{\partial G}{\partial n_1}\right)_{p,T,n_2} dn_1 + \left(\frac{\partial G}{\partial n_2}\right)_{p,T,n_1} dn_2 \qquad (7.1.2)$$

Comparing Equation 7.1.1 and Equation 7.1.2 gives

$$V = \left(\frac{\partial G}{\partial p}\right)_{T,n_1,n_2} \qquad (7.1.3)$$

$$S = -\left(\frac{\partial G}{\partial T}\right)_{p,n_1,n_2} \qquad (7.1.4)$$

and

$$\mu_i = \left(\frac{\partial G}{\partial n_i}\right)_{p,T,n_{j\neq i}} \qquad (7.1.5)$$

The chemical potential is an example of a *partial molar* quantity: μ_i is the partial molar Gibbs free energy with respect to component i. Other partial molar quantities exist and share the following features:

1. We may define, say, partial molar volume, enthalpy, or entropy by analogy with Equation 7.1.5:

$$\overline{Y}_i = \left(\frac{\partial Y}{\partial n_i}\right)_{p,T,n_{j\neq i}} \qquad (7.1.6)$$

    where $Y = V, H$, or S, respectively. Except for the partial molar Gibbs free energy, we shall use the notation $\overline{Y}_i$ to signify a partial molar quantity, where Y stands for the symbol of the appropriate variable.
2. Partial molar quantities have per mole units, and for Y_i this is understood to mean per mole of component i. The value of this coefficient depends on the overall composition of the mixture. Thus $\overline{V}_{H_2O}$ is not the same for a water–alcohol mixture that is 10% water as for one that is 90% water.
3. For a pure component the partial molar quantity is identical to the molar (superscript $^\wedge$) value of the pure substance. Thus for pure component i

$$\mu_i = \widehat{G}_i \qquad (7.1.7)$$

4. A useful feature of the partial molar properties is that the property of a mixture (subscript m) can be written as the sum of the mole-weighted contributions of the partial molar properties of the components:

$$Y_m = n_1\overline{Y}_1 + n_2\overline{Y}_2 \qquad (7.1.8)$$

In this expression n_1 and n_2 are the numbers of moles of components 1 and 2 in the mixture under consideration.

5. To express the value of property Y_m on a per mole basis, it is necessary to divide Equation 7.1.8 by the total number of moles, $n_1 + n_2$. The mole fraction x_i of component i is written

$$x_i = \frac{n_i}{\sum_{i=1,2} n_i} \tag{7.1.9}$$

therefore

$$\frac{Y_m}{n_1 + n_2} = x_1 \overline{Y}_1 + x_2 \overline{Y}_2 \tag{7.1.10}$$

6. Relationships that exist among ordinary thermodynamic variables also apply to the corresponding partial molar quantities. Two such relationships are

$$\mu_i = \overline{H}_i - T\overline{S}_i \tag{7.1.11}$$

and

$$\overline{V}_i = \left(\frac{\partial \mu_i}{\partial p}\right)_{T, n_{j \neq i}} \tag{7.1.12}$$

As noted above, all of the partial molar quantities are, in general, concentration-dependent.

It is convenient to define a thermodynamic concentration called the *activity* a_i in terms of which the chemical potential is given by the relationship

$$\mu_i = \mu_i^o + RT \ln a_i \tag{7.1.13}$$

The quantity μ_i^o is called the *standard state* (superscript o) value of μ_i; it is the value of μ_i when $a_i = 1$. Neither μ_i nor G can be measured absolutely; we deal with differences in the quantities and the standard state value disappears when differences are taken. Although the standard state is defined differently in various situations, we shall generally take the pure component as the standard state and $\mu_i^o = \widehat{G}_i^o$.

7.2 Regular Solution Theory

Regular solution theory illustrates how a relatively simple statistical model can provide a useful expression for the free energy of mixing for a binary solution of two components. Furthermore, the Flory–Huggins theory, which we will develop in the next section and which is the starting point for discussions of both polymer solutions and polymer blends, is really nothing more than regular solution theory extended to polymers. We will derive these results using a lattice approach because the physical picture is particularly clear, but it should be realized that the expression for ΔG_m that results could be obtained without such a seemingly artificial assumption.

We will designate the number of *molecules* of the two species m_1 and m_2 and the number of *moles* n_1 and n_2. We assume that the two molecules have equal volumes and also that their partial molar volumes are equal (and concentration-independent): $\overline{V}_1 = \overline{V}_2$. The cell size of the lattice is chosen to equal the molecular volume, and the lattice has a coordination number z. The total number of molecules $m = m_1 + m_2$ is also the total number of lattice sites. A section of a two-dimensional square lattice is illustrated in Figure 7.1.

7.2.1 Regular Solution Theory: Entropy of Mixing

The entropy of mixing is obtained from the Boltzmann definition of entropy:

$$S = k \ln \Omega \tag{7.2.1}$$

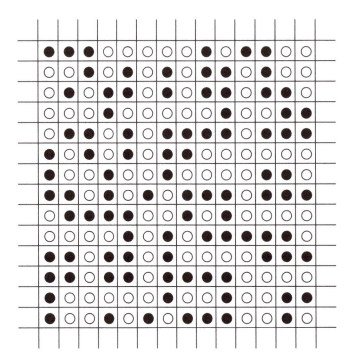

Figure 7.1   Section of a two-dimensional square lattice with each site occupied by either of two species.

where Ω is the total number of possible configurations that the system can adopt. For placing m_1 objects of type 1 and m_2 objects of type 2 on a lattice of m sites the total number of configurations is

$$\Omega = \frac{m!}{m_1! \, m_2!} \tag{7.2.2}$$

where the terms in the denominator take account of the fact that the m_1 molecules of type 1 are indistinguishable from one another, as are the m_2 molecules of type 2 (compare this with Equation 6.7.2). Stirling's approximation ($\ln N! \approx N \ln N - N$ for large N) is used to get rid of the factorials, just as in Section 6.7:

$$
\begin{aligned}
S_{\mathrm{m}} &= k\{\ln m! - \ln m_1! - \ln m_2!\} \\
&= k\{m \ln m - m - m_1 \ln m_1 + m_1 - m_2 \ln m_2 + m_2\} \\
&= k\{(m_1 + m_2) \ln(m_1 + m_2) - m_1 \ln m_1 - m_2 \ln m_2\} \\
&= -k\{m_1(\ln m_1 - \ln m) + m_2(\ln m_2 - \ln m)\} \\
&= -k\{m_1 \ln x_1 + m_2 \ln x_2\} \tag{7.2.3}
\end{aligned}
$$

Now the configurational entropy for each pure component (and we are assuming that configurational entropy is the only source of entropy) is zero:

$$
\begin{aligned}
S_1 &= k \ln \Omega_1 = k \ln \frac{m_1!}{m_1!} = 0 \\[6pt]
S_2 &= k \ln \Omega_2 = k \ln \frac{m_2!}{m_2!} = 0
\end{aligned}
\tag{7.2.4}
$$

so Equation 7.2.3 is actually the change in entropy with mixing

$$\Delta S_m = S_m - S_1 - S_2 = -k(m_1 \ln x_1 + m_2 \ln x_2) \tag{7.2.5a}$$

This expression applies to the entire mixture, and could also be written

$$\Delta S_m = -R\{n_1 \ln x_1 + n_2 \ln x_2\} \tag{7.2.5b}$$

where we have factored out N_{av}. It is often useful to have an intensive expression, i.e., the entropy of mixing per lattice site (and therefore per molecule)

$$\Delta S_m = -k(x_1 \ln x_1 + x_2 \ln x_2) \tag{7.2.5c}$$

or per mole of lattice sites (and therefore, in this case, per mole of molecules)

$$\Delta S_m = -R\{x_1 \ln x_1 + x_2 \ln x_2\} \tag{7.2.5d}$$

It is important to recognize that Equation 7.2.5a through Equation 7.2.5d are all the same in physical content, but different in units, and one must be careful to use the appropriate form of ΔS_m in calculations.

There are three important features to this expression for ΔS_m.

1. As x_1 and x_2 are always between 0 and 1 in a mixture, the natural logarithm terms are always negative and the overall $\Delta S_m > 0$. Therefore, configurational entropy always favors spontaneous mixing.
2. The expression is symmetric with respect to exchange of 1 and 2, which is a consequence of the assumption of equal molecular sizes. In real mixtures this condition will be hard to satisfy.
3. This calculation of the entropy assumed that all configurations on the lattice were equally probable, i.e., there was no energetic benefit or price for having 1 next to 1 and 2 next to 2, versus having 1 next to 2. If there were such an energy term, each configuration ought to be further weighted by the appropriate Boltzmann factor $\exp[-E/RT]$, where E is the total energy of that configuration.

7.2.2 Regular Solution Theory: Enthalpy of Mixing

We now compute the enthalpy of mixing for this model, ΔH_m, and will assume that the resulting energy term is sufficiently small that it does not matter in ΔS_m. First, we assume that $\Delta H_m = \Delta U_m$, in other words no $p - V$ terms contribute to H; this is consistent with the lattice approach and emphasizes that we assume there is no volume change on mixing. We introduce the interaction energies in the pure state, w_{11} and w_{22}, which act between two molecules of type 1 and between two molecules of type 2, respectively; this is shown schematically in Figure 7.2a. All molecules *attract* one another by dispersion forces, so w_{11} and w_{22} are *negative*. Under these assumptions the enthalpy in the pure state is given by

$$H_1 = \frac{1}{2}m_1 z w_{11}$$

$$H_2 = \frac{1}{2}m_2 z w_{22} \tag{7.2.6}$$

which means each molecule has z neighbors and therefore z interactions, but we divide by 2 because there is only w_{11} worth of interaction energy per *pair* of molecules. We are also assuming that molecules only interact with their nearest neighbors.

In the mixture we assume that the placement of 1 and 2 is completely random, as in the calculation of ΔS_m. Thus the probability that a neighboring site is occupied by a molecule of type

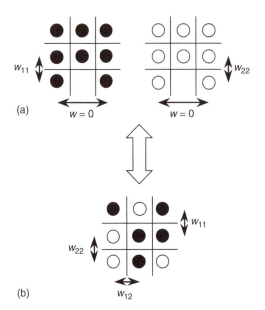

Figure 7.2 Illustration of pairwise nearest-neighbor interaction energies, w_{ij}, in (a) the pure components and (b) a mixture.

1 or 2 is given by x_1 or x_2, respectively. If we call the interaction energy between molecules 1 and 2 w_{12} (w_{12} is also negative), then (see Figure 7.2b)

$$H_m = \frac{1}{2}m_1(zx_1w_{11} + zx_2w_{12}) + \frac{1}{2}m_2(zx_1w_{12} + zx_2w_{22}) \tag{7.2.7}$$

and thus

$$\Delta H_m = H_m - H_1 - H_2$$

$$= \frac{1}{2}z\{w_{12}(m_1x_2 + m_2x_1) + w_{11}(m_1x_1 - m_1) + w_{22}(m_2x_2 - m_2)\}$$

$$= \frac{1}{2}z\left\{w_{12}\left(\frac{2m_1m_2}{m}\right) - w_{11}\left(\frac{m_1m_2}{m}\right) - w_{22}\left(\frac{m_1m_2}{m}\right)\right\}$$

$$= \frac{m_1m_2}{m}z\left(w_{12} - \frac{w_{11}}{2} - \frac{w_{22}}{2}\right) = \frac{m_1m_2}{m}z\Delta w \tag{7.2.8}$$

We define an *exchange energy* $\Delta w = (w_{12} - w_{11}/2 - w_{22}/2)$, which represents the difference between the attractive cross-interaction of 1 and 2 and the average self-interaction of 1 with 1 and 2 with 2. For dispersion forces, and for the regular solution theory, $\Delta w \geq 0$, as we will discuss in Section 7.6; this is a manifestation of "like prefers like," indicating that the self-interactions are more attractive than cross-interactions. For now, however, we can just take it as an energy parameter that is probably positive.

We make a further definition, that of the interaction parameter, χ:

$$\chi \equiv \frac{z\Delta w}{kT} \tag{7.2.9}$$

which is the exchange energy per molecule, normalized by the thermal energy kT. In other words, χ is the fraction of kT you must pay in order to lift one molecule of type 1 out of its beaker, one molecule of type 2 out of its beaker, and exchange them. Note that although χ is dimensionless, its value does depend on the chosen size of the lattice site, through Δw. Using this definition we can write

$$\Delta H_m = m_1 x_2 \chi kT = n_1 x_2 \chi RT \tag{7.2.10a}$$

for the whole system, or

$$\Delta H_m = x_1 x_2 \chi kT \tag{7.2.10b}$$

per lattice site or molecule, and

$$\Delta H_m = x_1 x_2 \chi RT \tag{7.2.10c}$$

per mole. When χ is positive ΔH_m is positive, and therefore opposes spontaneous mixing. We can now express the free energy of mixing, $\Delta G_m = \Delta H_m - T\Delta S_m$ as

$$\frac{\Delta G_m}{RT} = (n_1 \ln x_1 + n_2 \ln x_2 + n_1 x_2 \chi) \tag{7.2.11a}$$

for the system as a whole, or

$$\frac{\Delta G_m}{kT} = (x_1 \ln x_1 + x_2 \ln x_2 + x_1 x_2 \chi) \tag{7.2.11b}$$

per site. The first two terms represent the entropy, and, as noted above, they favor mixing, whereas the last enthalpic term is assumed to be positive and therefore opposes mixing. The second form of ΔG_m, Equation 7.2.11b, is probably the easier one to remember because of its obvious symmetry. However, when we subsequently compute chemical potentials, we will need to take partial derivatives with respect to n_1 and n_2, and the form in Equation 7.2.11a will be more appropriate. The implications of regular solution theory, particularly in terms of the predictions for the phase behavior, will be discussed in Section 7.5.

Example 7.1

Assuming that w_{11}, w_{12}, and w_{22} are approximately -1.17×10^{-20}, -1.08×10^{-20}, and -1.01×10^{-20} J for toluene and cyclohexane, respectively, estimate the free energy of mixing for 1 mol of toluene with 1 mol of cyclohexane at room temperature. (We will see in Section 7.6 how these interaction energies can be estimated from experimental measurements.)

Solution

We will first calculate a value for χ, and then substitute into Equation 7.2.11a. We need to assume a value of the coordination number, z; it is typically about 10 for small molecule liquids (recall that in a close-packed lattice of spheres there are 12 nearest neighbors).

$$\chi = \frac{z\Delta w}{kT} \approx \frac{10}{(1.4 \times 10^{-23})(298)} \left\{ -1.08 + \frac{1.17 + 1.01}{2} \right\} \times 10^{-20} = 0.24$$

We have $n_1 = n_2 = 1$, and $x_1 = x_2 = 0.5$, so

$$\frac{\Delta G_m}{RT} = \ln(0.5) + \ln(0.5) + (0.5 \times 0.24) = -1.3$$

For the mixture of 2 mol the total free energy of mixing is therefore $(-1.3 \times 2 \times 8.3 \times 298) \approx -6.5$ kJ. Although we should not take the exact numbers too seriously, the overall negative sign confirms our expectation that toluene and cyclohexane should be quite happy to mix, even though χ is positive.

Before proceeding to polymer solutions, one further comment about this derivation of ΔG_m is in order. In computing ΔH_m we assumed completely random mixing, as in Equation 7.2.7. But, if χ is nonzero, there will presumably be some finite preference for "clustering," e.g., 1 with 1 and 2 with 2 if $\chi > 0$. Thus the probability that a lattice site immediately adjacent to a type 1 molecule is occupied by another 1 may actually be larger than x_1 and the same for component 2. This possibility is simply not accommodated in the model and represents a fundamental limitation. A theory that assumes that the local interactions are determined solely by the bulk average composition, i.e., x_1 and x_2 in this case, is called a "mean-field" theory. We will encounter other examples of mean-field theories in this book. They are popular, as one might expect, because they are relatively tractable; it also turns out that in many polymer problems, especially in undiluted polymers, they are remarkably reliable.

7.3 Flory–Huggins Theory

Flory and Huggins independently considered ΔG_m for polymer solutions, and the essence of their model is developed here [1,2]. As noted in the previous section, the Flory–Huggins theory is a natural extension of regular solution theory to the case where at least one of the components is polymeric. To proceed, we will adopt the same lattice model as before, with one important difference. We choose the lattice site to have the volume of one solvent molecule (subscript 1) and each polymer (subscript 2) occupies N lattice sites. This is illustrated in Figure 7.3. Thus N is proportional to the degree of polymerization, but the monomer unit is now defined to have the same volume as the solvent. (Equivalently, N is the ratio of the molar volume of the polymer to that of the solvent.) We will refer to this subunit of the polymer as a "segment." We will also switch from mole fractions to volume fractions, ϕ_1 and ϕ_2, in describing composition. We do this because in order to use moles accurately with polymers, one needs to know the molecular weight precisely, and one also has to know the full molecular weight distribution, whereas the volume fraction is easily obtained from the measured mass and known densities. (However, we will need to be a little careful in some thermodynamic manipulations in subsequent sections, where for example the

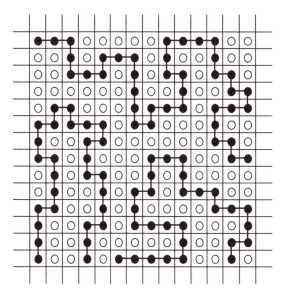

Figure 7.3 Section of a two-dimensional square lattice with each site occupied either by a solvent molecule or a polymer segment.

chemical potential should be calculated by taking partial derivatives with respect to the number of moles.) Accordingly, the volume fractions are defined as

$$\phi_1 = \frac{m_1}{m_1 + Nm_2}$$
$$\phi_2 = \frac{Nm_2}{m_1 + Nm_2}$$

(7.3.1)

where the total number of lattice sites m is now given by $m_1 + Nm_2$.

There are at least two ways to obtain the Flory–Huggins entropy of mixing; one is rather informal, but is very simple, whereas the other is more careful in the application of the model, but algebraically tedious. We shall go through both, but you may wish to skip the longer version on the first pass.

7.3.1 Flory–Huggins Theory: Entropy of Mixing by a Quick Route

The polymer molecule has many internal degrees of freedom, as described in Chapter 6. However, we can assume that the number of conformational degrees of freedom is the same in the solution as in the pure polymer; in other words, the polymer is the same random coil with the same segment distribution function in the bulk as in solution. Thus the only configurational entropy of mixing comes from the increased possibilities for placing the center of mass. The polymers occupy Nm_2 sites in the bulk polymer, and the number of possible locations for the center of mass is thus proportional to Nm_2. On the other hand, in the mixture the number of locations is proportional to $m_1 + Nm_2$. Thus for one polymer molecule (recall Equation 7.2.1)

$$\Delta S_{m,2} \sim k \ln(m_1 + Nm_2) - k \ln Nm_2 = -k \ln \phi_2$$

(7.3.2)

where the unspecified proportionality factor drops out in the ratio. A similar argument holds for each solvent molecule, and therefore

$$\Delta S_{m,1} = -k \ln \phi_1$$

(7.3.3)

To obtain ΔS_m for the entire system, we multiply the respective terms by the number of molecules of each type, and sum:

$$\Delta S_m = -k(m_1 \ln \phi_1 + m_2 \ln \phi_2) = -R(n_1 \ln \phi_1 + n_2 \ln \phi_2)$$

(7.3.4a)

If we compare this expression with Equation 7.2.5 for regular solution theory, the only change is the switch to volume fractions. Furthermore, if we let $N = 1$, then $\phi_1 = x_1$ and $\phi_2 = x_2$, and regular solution theory can be recovered as a special case of the Flory–Huggins model. To obtain the expression for ΔS_m per site, we divide by $m = m_1 + Nm_2$:

$$\Delta S_m = -k\left(\frac{m_1}{m} \ln \phi_1 + \frac{m_2}{m} \ln \phi_2\right) = -k\left(\phi_1 \ln \phi_1 + \frac{\phi_2}{N} \ln \phi_2\right)$$

(7.3.4b)

This expression has the same physical content as Equation 7.3.4a, but it brings out the importance of N. Because N is usually a large number, the contribution to the entropy of mixing from the polymer is very small, often almost negligible. Thus immediately we can predict that for systems with $\chi > 0$, where spontaneous mixing is driven only by ΔS_m, it will be harder to mix a solvent with a polymer than with another small molecule.

7.3.2 Flory–Huggins Theory: Entropy of Mixing by a Longer Route

Herewith a more detailed derivation is presented. Let us suppose we have added i polymers onto the lattice and we inquire about the number of possibilities for placing the $(i + 1)$th chain. At this moment, the probability that a lattice site does not contain a monomer is $f_i = 1 - (iN/m)$, where m is again the total number of sites. Now we add the next polymer segment by segment (and assume f_i is constant throughout the process). The first segment has $m - iN$ choices, the second has zf_i, and the third and all subsequent segments have $(z - 1)f_i$ choices. The number of configurations available to the $(i + 1)$th chain, Ω_i, is thus

$$\Omega_{i+1} = (m - iN)(z)(f_i)(z - 1)^{N-2}(f_i)^{N-2}$$
$$\approx (m - iN)(z - 1)^{N-1}\left(\frac{m - iN}{m}\right)^{N-1} \tag{7.3.5}$$

where we neglect the difference between z and $z-1$ in one term. The total number of configurations for adding m_2 polymers in succession is therefore

$$\Omega_{tot} = \frac{1}{m_2!}\prod_{i=0}^{m_2-1}\Omega_{i+1} = \frac{1}{m_2!}\prod_{i=0}^{m_2-1}(m - iN)(z - 1)^{N-1}\left(\frac{m - iN}{m}\right)^{N-1}$$
$$= \frac{1}{m_2!}\left(\frac{z-1}{m}\right)^{m_2(N-1)}\prod_{i=0}^{m_2-1}(m - iN)^N \tag{7.3.6}$$

where the factor of $1/m_2!$ takes care of the fact that the polymers are indistinguishable. Now we need to play a trick to deal with the term inside the product. Consider the following ratio of factorials:

$$\frac{(m - iN)!}{(m - (i + 1)N)!} = \frac{(m - iN)(m - iN - 1)\cdots(m - iN - N)!}{(m - iN - N)!}$$
$$= (m - iN)(m - iN - 1)\cdots(m - iN - N + 1) \approx (m - iN)^N \tag{7.3.7}$$

where the last transformation invokes a dilute solution approximation: $iN \ll m$ always, and $iN \gg N$ for all but a few values of i. This now allows us to replace the product of powers in the last version of Equation 7.3.6 with a ratio of factorials:

$$\frac{1}{m_2!}\left(\frac{z-1}{m}\right)^{m_2(N-1)}\prod_{i=0}^{m_2-1}\frac{(m - iN)!}{(m - iN - N)!} = \frac{1}{m_2!}\left(\frac{z-1}{m}\right)^{m_2(N-1)}$$
$$\times \left\{\frac{m!(m - N)!\cdots(m - (m_2 - 1)N)!}{(m - N)!(m - 2N)!\cdots(m - m_2N)!}\right\}$$
$$= \frac{1}{m_2!}\left(\frac{z-1}{m}\right)^{m_2(N-1)}\cdot\frac{m!}{(m - m_2N)!} = \frac{m!}{m_1!m_2!}\left(\frac{z-1}{m}\right)^{m_2(N-1)} \tag{7.3.8}$$

where the last step recognizes that $m_1 = m - m_2N$. Now we are ready for help from Mr. Stirling:

$$\ln\Omega_{tot} = \ln m! - \ln m_1! - \ln m_2! + m_2(N - 1)\ln\left(\frac{z-1}{m}\right)$$
$$= m\ln m - m - m_1\ln m_1 + m_1 - m_2\ln m_2 + m_2 + m_2(N - 1)\ln(z - 1) - m_2(N - 1)\ln m$$
$$= (m_1 + m_2)\ln(m_1 + Nm_2) - m_1\ln m_1 - m_2\ln m_2 - m_2(N - 1)(1 - \ln(z - 1)) \tag{7.3.9}$$

where the last line took a little algebra. Anyway, this now gives us $S_m = k\ln\Omega_{tot}$. To obtain S for the pure polymer and pure solvent, we can set either m_1 or $m_2 = 0$ in Equation 7.3.9, respectively:

$$\ln\Omega_1 = m_1\ln m_1 - m_1\ln m_1 = 0$$
$$\ln\Omega_2 = m_2\ln m_2N - m_2\ln m_2 - m_2(N - 1)(1 - \ln(z - 1)) \tag{7.3.10}$$

Finally, then

$$\Delta S_m = k\{(m_1 + m_2)\ln(m_1 + m_2 N) - m_1 \ln m_1 - m_2 \ln(m_2 N)\}$$

$$= k\left\{m_1 \ln\left(\frac{m_1 + m_2 N}{m_1}\right) + m_2 \ln\left(\frac{m_1 + m_2 N}{m_2 N}\right)\right\}$$

$$= -k\{m_1 \ln\phi_1 + m_2 \ln\phi_2\}$$

$$= -R\{n_1 \ln\phi_1 + n_2 \ln\phi_2\} \tag{7.3.11}$$

which is exactly the result obtained in the previous section. Next time, you will probably just settle for the simple argument!

7.3.3 Flory–Huggins Theory: Enthalpy of Mixing

The expression for the enthalpy of mixing in Flory–Huggins theory is exactly that for regular solution theory, once the substitutions of ϕ_1 and ϕ_2 for x_1 and x_2 have been made. This can be seen because the enthalpy was computed on a lattice site basis with only local interactions and the calculation would not be changed by linking the monomers together. (This is not strictly true, because now for each monomer there are only $z-2$ neighboring sites that could be occupied by either monomer or solvent; two sites are required to be other monomers by covalent attachment. However, as z and Δw do not appear independently in the final expression for ΔG_m, but are subsumed into the parameter χ, and we do not know Δw exactly anyway, we ignore this complication.) Thus we can write

$$\Delta H_m = m_1 \phi_2 \chi kT = n_1 \phi_2 \chi RT \tag{7.3.12a}$$

for the system, and

$$\Delta H_m = \phi_1 \phi_2 \chi kT \tag{7.3.12b}$$

per site. Combining the expressions for ΔS_m and ΔH_m we arrive at the final result:

$$\frac{\Delta G_m}{RT} = n_1 \ln\phi_1 + n_2 \ln\phi_2 + n_1 \phi_2 \chi \tag{7.3.13a}$$

for the system, and

$$\frac{\Delta G_m}{kT} = \phi_1 \ln\phi_1 + \frac{\phi_2}{N} \ln\phi_2 + \phi_1 \phi_2 \chi \tag{7.3.13b}$$

per site.

The main features of Equation 7.3.13a and Equation 7.3.13b are that the entropy terms always favor mixing, the enthalpy opposes mixing when $\chi > 0$, and the big difference from regular solution theory is the factor of N reducing the polymer contribution to the entropy of mixing. These expressions are very powerful, as they can be used to calculate many thermodynamic quantities of interest. For example, in the next two sections we will develop the explicit predictions of the model for two experimentally important quantities, the osmotic pressure and the phase diagram, respectively.

One further point to bring out now, however, is how problematic the mean-field assumption can be for dilute polymer solutions. This is illustrated schematically in Figure 7.4, where a dilute solution is pictured, along with a trajectory through the solution that happens to pass through two coils. Also shown is the "local" monomer concentration along that trajectory. It has two regions where the trajectory passes through the coils, and the local concentration of monomers is significantly higher than the solution average, ϕ_2. However, there are also substantial regions between coils where the actual monomer concentration is zero. In other words, chain connectivity

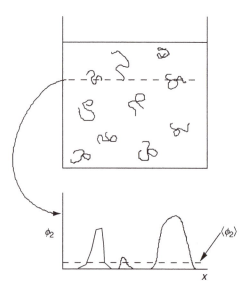

Figure 7.4 Illustration of the failure of the mean-field assumption in dilute solutions: The local concentration along some arbitrary trajectory fluctuates between higher than average, inside a polymer coil, and lower than average, outside a coil.

guarantees that monomers are clustered in space, whereas the model assumes that the local concentration is uniformly ϕ_2 throughout the sample. This turns out to be a major limitation to the quantitative application of Flory–Huggins theory to dilute solutions; however, it also suggests that the theory should get progressively better when the concentration is increased, and the coils begin to interpenetrate. This turns out to be the case.

7.3.4 Flory–Huggins Theory: Summary of Assumptions

At this point it is worthwhile to summarize the main assumptions employed in order to arrive at the expression for the free energy of mixing, Equation 7.3.13.

1. There is no volume change on mixing, and $\overline{V}_1 = \hat{V}_1, \overline{V}_2 = \hat{V}_2$ are independent of concentration.
2. ΔS_m is entirely the ideal combinatorial entropy of mixing.
3. ΔH_m is entirely the internal energy of mixing.
4. Both ΔS_m and ΔH_m are computed assuming entirely random mixing.
5. The interactions are short-ranged (nearest neighbors only), isotropic, and pairwise additive.
6. The local concentration is always given by the bulk average composition (the mean-field assumption).

7.4 Osmotic Pressure

In this section we consider the osmotic pressure, Π, of a dilute, uncharged polymer solution. First, we will develop the virial expansion for Π, which is based on general thermodynamic principles and therefore completely model-independent. We will learn how measurements of Π can be used to determine the number-average molecular weight of a polymer, and how the so-called second virial coefficient, B, is a diagnostic of the quality of a solvent for a given polymer. Then we will return to the Flory–Huggins theory, and see what it predicts for Π and B. This will lead us to our first working definition of a theta solvent, a very important concept in polymer solutions. Finally, the concept of osmotic pressure will turn out later to be central to understanding scattering experiments (see Chapter 8). In short, this is a very important section.

7.4.1 Osmotic Pressure: General Case

The experiment is illustrated schematically in Figure 7.5. A thermostated chamber at pressure p_0 and temperature T is divided into two compartments by a semipermeable membrane; the membrane passes solvent easily, but is impermeable to polymers (e.g., because of their size). There is a thin tube emerging from the top of each compartment and the height of the fluid in each tube reflects the pressure in that compartment. The compartment on the left is full of pure solvent, which therefore has its standard state chemical potential $\mu_1^o(T, p_0)$. The compartment on the right has a dilute polymer solution of known concentration, c (in g/mL), and therefore at the instant the solution is introduced into its compartment, the solvent component has a different chemical potential, μ_1. To reach equilibrium, so that the solvent chemical potential is equal on both sides of the membrane, there must be a net flow of solvent from the left compartment to the right. This can be easily seen, in that the simplest way to equalize the two chemical potentials would be to have equal polymer concentrations on each side. As the polymer cannot move from right to left, some solvent must move from left to right. However, because the solution in the right is contained, the influx of solvent increases the column of solvent in the tube, i.e., the pressure goes up. This increase in pressure defines the osmotic pressure, Π, and ultimately Π will oppose further solvent transfer. At equilibrium, then,

$$\mu_1^o(T, p_0) = \mu_1(T, p_0 + \Pi) = \mu_1(T, p_0) + \int_{p_0}^{p_0+\Pi} \left(\frac{\partial \mu_1}{\partial p} \right)_T dp \qquad (7.4.1)$$

From Equation 7.1.12 we recall that the integrand in Equation 7.4.1 is just the partial molar volume of the solvent, $\overline{V}_1$. We can assume this is constant over the relatively small pressure change, Π, and thus

$$\mu_1^o(T, p_0) = \mu_1(T, p_0) + \Pi \overline{V}_1 \qquad (7.4.2)$$

Rearranging, we obtain

$$\Pi \overline{V}_1 = \mu_1^o - \mu_1 = -RT \ln a_1 = -RT \ln \gamma_1 x_1 \approx -RT \ln x_1 \qquad (7.4.3)$$

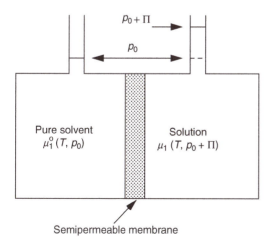

Figure 7.5 Schematic diagram of the osmotic pressure experiment, and the operational definition of the osmotic pressure, Π.

where we have recalled the definition of the solvent activity, a_1, from Equation 7.1.13, and recognized that γ_1, the activity coefficient, will approach 1 in sufficiently dilute solution. Converting to the polymer concentration, $x_2 = 1 - x_1$, gives

$$\frac{\Pi}{RT} = -\frac{1}{\overline{V}_1} \ln(1 - x_2) \approx \frac{x_2}{\overline{V}_1} \tag{7.4.4}$$

recalling again that $\ln(1 - x) \approx -x$ for small x. In order to allow for the effects of finite solute concentration, it is customary to expand the right-hand side of Equation 7.4.4 in powers of x_2, a *virial expansion*:

$$\frac{\Pi}{RT} = \frac{x_2}{\overline{V}_1} + \beta x_2^2 + \beta_3 x_2^3 + \cdots \tag{7.4.5}$$

To replace x_2 with "practical" polymer units, such as c, we recall that $c = n_2 M/V$ where M is the molecular weight and V is the solution volume. Thus

$$x_2 = \frac{n_2}{n_1 + n_2} = \frac{cV/M}{(V/\overline{V}_1) + (cV/M)} \cong \frac{\overline{V}_1 c}{M} \tag{7.4.6}$$

The last simplification in Equation 7.4.6 is equivalent to ignoring n_2 relative to n_1 in the denominator. For a solution with $c = 0.01$ g/mL, $M = 100,000$ g/mol, and V in mL, n_2/V will be 10^{-7}, whereas n_1/V will be about 10^{-2} (if the solvent density ≈ 1 g/mL and molecular weight ≈ 100 g/mol), so this is a very reasonable approximation. Substituting Equation 7.4.6 into Equation 7.4.5 leads to

$$\frac{\Pi}{RT} = \frac{c}{M} + \beta \overline{V}_1^2 \left(\frac{c}{M} \right)^2 + \cdots = \frac{c}{M} + Bc^2 + B_3 c^3 + \cdots \tag{7.4.7a}$$

where B is called the *second virial coefficient*: it has units of cm^3 mol/g^2. The quantity B_3, the third virial coefficient, reflects ternary or three-body interactions, and is important when c is large enough and when B is small enough; we will not consider it further here. Equation 7.4.7a is the central result of this section. The quantity on the left-hand side is measurable; of the quantities on the right, c is determined by solution preparation, and M and B are determined by examining the c dependence of Π. In Figure 7.6a Π/RT is plotted against c for three different values of B: $B > 0$,

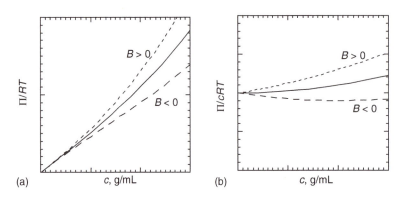

Figure 7.6 Generic plots of osmotic pressure versus concentration for dilute polymer solutions: (a) Π/RT versus c and (b) Π/cRT versus c, for the indicated values of B. Note the curvature at high c due to three-body interactions (the c^3 term in Equation 7.4.7a).

$B = 0$, and $B < 0$. In Figure 7.6b we choose a second format, obtained by dividing Equation 7.4.7a through by c:

$$\frac{\Pi}{cRT} = \frac{1}{M} + Bc + \cdots \tag{7.4.7b}$$

This format is often employed in practice and reveals clearly that the molecular weight is determined from the y-intercept of the plot and the sign of B corresponds to the sign of the initial slope.

So, the experimental approach is clear enough: Measure Π for a series of solutions of known c, with c sufficiently low that higher order terms in the virial expansion are not too important. But, what does Equation 7.4.7a mean physically?

1. The first term in Equation 7.4.7a, c/M, is the number of molecules per unit volume (in mol/cm^3). Thus, in the very dilute limit, the osmotic pressure is determined only by the *number* of solute molecules, whatever they may be. This is characteristic of all *colligative properties*, including Π, freezing point depression, and boiling point elevation. However, as c increases, there will be binary solute–solute interactions. The second term, proportional to c^2, accounts for these. Now there are three possibilities:
2. If the polymers are in a good solvent, monomers on different chains are happy enough to be surrounded by solvent; when two coils approach one another, there is steric repulsion (i.e., excluded volume) and the chains separate. Therefore, the coil–coil interaction is effectively repulsive. This corresponds to $B > 0$; it means there is an even greater drive for solvent to flow into the right compartment and dilute the solution.
3. Conversely, if the solvent is poor, such that the polymer is barely able to stay dissolved in solution, monomers find it energetically favorable to be close to other monomers. Thus when two coils approach one another, there is a tendency to cluster. In this case, $B < 0$ and the effectively attractive solute–solute interactions resist the uptake of further solvent.
4. There is a special intermediate case where $B = 0$. This corresponds to a "not-very-good" solvent where the excluded volume and relatively unfavorable solvent–solute interactions cancel one another in the net Π. This case is given a special name; following Flory, it is called a *theta solvent* [3]. We will explore the significance of a theta solvent in more detail subsequently.
5. It is worth pointing out that there is an analogy with the van der Waals equation of state for 1 mol of an imperfect gas:

$$\left(p + \frac{\alpha}{V^2}\right)(V - \beta) = RT \tag{7.4.8}$$

in which β accounts for the excluded volume and α the intermolecular interactions. When this equation is expanded in a virial series in the density $(1/V)$

$$\frac{p}{RT} = \frac{1}{V} + \frac{\beta - \alpha/RT}{V^2} + \frac{\beta^2}{V^3} + \cdots \tag{7.4.9}$$

the second virial coefficient is $\beta - (\alpha/RT)$. This virial coefficient vanishes at a special temperature, known as the Boyle point, when the excluded volume (β) and interaction (α/RT) terms exactly cancel one another. Similarly, the theta temperature for a particular polymer–solvent system is the temperature at which B vanishes due to a cancelation of effects from excluded volume and net polymer–polymer interactions. It is important to realize, however, that it is not the excluded volume itself that vanishes, but just its effect on the osmotic pressure.

7.4.1.1 Number-Average Molecular Weight

We noted previously that moles were a troublesome unit for polymers, in part because of the inevitable molecular weight distribution. Now we can address the important issue of what average

M will the osmotic pressure experiment measure for a polydisperse sample? The answer is appealingly simple and rigorous; as a colligative property, Π at infinite dilution depends only on the number of solute molecules per unit volume, and therefore should determine the number-average molecular weight, M_n. To see that this is indeed so, we can write for very dilute solutions:

$$\Pi = \frac{RTc}{M} = RT \sum \frac{c_i}{M_i} \tag{7.4.10}$$

for a distribution of molecular weights, M_i. Next we form the ratio

$$\lim_{c \to 0} \left(\frac{\Pi}{cRT} \right) = \frac{\sum c_i/M_i}{\sum c_i} = \frac{\sum n_i M_i (1/M_i)/V}{\sum n_i M_i/V}$$

$$= \frac{\sum n_i}{\sum n_i M_i} = \frac{1}{M_n} \tag{7.4.11}$$

where we recall the definition of M_n from Chapter 1. Thus a properly conducted osmotic pressure measurement can determine the absolute value of M_n.

Example 7.2

To illustrate the typical magnitude of the osmotic pressure and the extraction of values of M_n and B, consider the following data for Π (in atm) for a polystyrene sample in cyclohexane at three temperatures. The data are also plotted in Figure 7.7a.

c (g/mL)	20.0°C	34.5°C	50.0°C
0.005	0.0061	0.0063	0.0067
0.010	0.0114	0.0124	0.0136
0.015	0.0173	0.0187	0.0203
0.020	0.0215	0.0255	0.0276
0.025	0.0276	0.0316	0.0354
0.030	0.0332	0.0381	0.0421
0.040	0.0405	0.0502	0.0580

Solution

For each data set, divide Π by cRT, where $R = 82.1$ cm^3 atm/K mol and T is in kelvin. The results are plotted in Figure 7.7b. Then fit each data set to a straight line by linear regression (a hand calculator is sufficient). The results are

$$\text{20.0°C:} \quad \text{Slope} = B = -2.0 \times 10^{-4} \text{ cm}^3 \text{ mol/g}^2$$
$$\text{1/intercept} = M_n = 1.97 \times 10^4 \text{ g/mol}$$
$$\text{34.5°C:} \quad \text{Slope} = B = 1.7 \times 10^{-5} \text{ cm}^3 \text{ mol/g}^2$$
$$\text{1/intercept} = M_n = 2.02 \times 10^4 \text{ g/mol}$$
$$\text{50.0°C:} \quad \text{Slope} = B = 1.2 \times 10^{-4} \text{ cm}^3 \text{ mol/g}^2$$
$$\text{1/intercept} = M_n = 2.00 \times 10^4 \text{ g/mol}$$

The three values of M_n are very comparable, as they should be. The values of B are consistent with the notion that cyclohexane is a theta solvent for polystyrene at 34.5°C. The experimental value of B is essentially zero at this temperature, considering the experimental uncertainty. Note also how the plotting format of Figure 7.7b accentuates the scatter in the data.

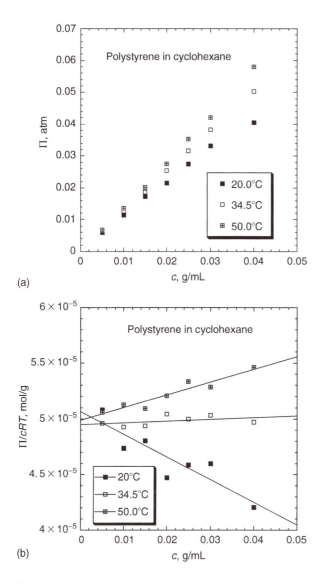

(a)

(b)

Figure 7.7   Osmotic pressure for polystyrene in cyclohexane plotted as (a) Π versus c and (b) Π/cRT versus c, at the indicated temperatures. The straight lines are linear regression fits. The data are provided in Example 7.2.

7.4.2   Osmotic Pressure: Flory–Huggins Theory

To conclude this section, we return to the Flory–Huggins expression for the free energy of mixing for a polymer solution, Equation 7.3.13, and see what it predicts for Π. From Equation 7.1.5 we can write

$$-\Pi\overline{V}_1 = \mu_1 - \mu_1^{\circ} = \left(\frac{\partial}{\partial n_1}\Delta G_m\right)_{p,T,n_2} \tag{7.4.12}$$

so we need to take the derivative with respect to n_1 of Equation 7.3.13, i.e.,

$$\frac{\partial}{\partial n_1}\left\{RT(n_1 \ln\phi_1 + n_2 \ln\phi_2 + n_1\phi_2\chi)\right\}$$

To find the answer, we recall $\phi_1 = n_1/(n_1 + Nn_2)$, and so

$$\frac{\partial \phi_1}{\partial n_1} = \frac{(n_1 + Nn_2) - n_1}{(n_1 + Nn_2)^2} = \frac{\phi_2}{(n_1 + Nn_2)}$$

$$= \frac{\phi_2^2}{Nn_2} = \frac{\phi_2 \phi_1}{n_1} \tag{7.4.13a}$$

similarly,

$$\frac{\partial \phi_2}{\partial n_1} = \frac{\partial(1 - \phi_1)}{\partial n_1} = -\frac{\partial \phi_1}{\partial n_1} = \frac{-\phi_2^2}{Nn_2} = \frac{-\phi_2 \phi_1}{n_1} \tag{7.4.13b}$$

Proceeding with the differentiation of ΔG_m, we have

$$\frac{\Pi \bar{V}_1}{RT} = -\left\{ \ln \phi_1 + \frac{n_1}{\phi_1} \frac{\phi_1 \phi_2}{n_1} + \frac{n_2}{\phi_2} \left(\frac{-\phi_2^2}{Nn_2} \right) + (\phi_2 - \phi_1 \phi_2) \chi \right\}$$

$$= -\left\{ \ln(1 - \phi_2) + \phi_2 \left(1 - \frac{1}{N} \right) + \chi \phi_2^2 \right\} \tag{7.4.14}$$

where we choose to write everything in terms of the polymer concentration. The expansion of $\ln(1 - \phi_2) = -\phi_2 - \phi_2^2/2 - \cdots$ is used to get rid of the logarithm (we keep the $\phi_2^2/2$ term here because of the virial expansion to second order), and we take $\bar{V}_1$ over to the other side:

$$\frac{\Pi}{RT} = \frac{1}{\bar{V}_1} \frac{\phi_2}{N} - \frac{1}{\bar{V}_1} \left(\chi - \frac{1}{2} \right) \phi_2^2 + \cdots \tag{7.4.15}$$

Finally, we convert from ϕ_2 to c:

$$\phi_2 = \frac{c\bar{V}_2}{M} = \frac{cN\bar{V}_1}{M} \tag{7.4.16}$$

and obtain

$$\frac{\Pi}{RT} = \frac{c}{M} + \left(\frac{1}{2} - \chi \right) \bar{V}_1 \frac{N^2}{M^2} c^2 + \cdots \tag{7.4.17}$$

and thus for the Flory–Huggins model,

$$B = \left(\frac{1}{2} - \chi \right) \bar{V}_1 \frac{N^2}{M^2} = \left(\frac{1}{2} - \chi \right) \frac{\bar{V}_2^2}{\bar{V}_1} \frac{1}{M^2} \tag{7.4.18}$$

This equation has two important features. First, when $\chi = 1/2$, then $B = 0$ and we have a theta solvent. Thus $\chi = 1/2$ represents a second operational definition of the theta point. For $\chi > 1/2$, $B < 0$, and the solvent is poor, whereas for $\chi < 1/2$, the solvent is good. Figure 7.8 shows data for $B(T)$ for several polystyrenes in cyclohexane and the theta temperature is determined to be 34.5°C. Second, B is predicted to be independent of molecular weight (note that $N \sim M$ in Equation 7.4.17 and thus the M dependence cancels out). It turns out experimentally that this is not quite true; for example B varies approximately as $M^{-0.2}$ in good solvents. This incorrect prediction is a direct consequence of the mean-field assumption that we discussed in the previous section.

7.5 Phase Behavior of Polymer Solutions

In this section we examine the phase behavior of a polymer solution, or, more precisely, we consider the temperature-composition plane at fixed pressure and locate the regions where a

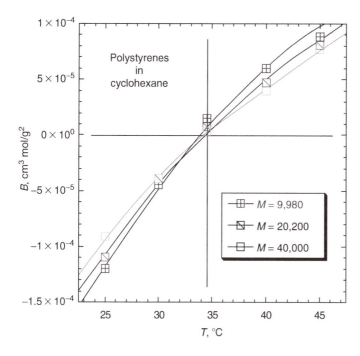

Figure 7.8 Second virial coefficient for three polystyrenes in cyclohexane as a function of temperature. The theta temperature based on these data alone lies between 34°C and 35°C. (From Yamakawa, H., Abe, F., and Einaga, Y., *Macromolecules*, 27, 5704, 1994.)

one-phase solution is stable, and where the mixture will undergo liquid–liquid phase separation into two phases. We will do this first for regular solution theory, as a means to illustrate the various concepts and steps in the procedure. Then we will return to Flory–Huggins theory and see the consequences of having one component substantially larger in molecular weight than the other.

7.5.1 Overview of the Phase Diagram

The phase diagram for a regular solution is shown schematically in Figure 7.9. It has the following important features, which we will see how to calculate:

1. A *critical point* (T_c, x_c) such that for $T > T_c$ a one-phase solution is formed for all compositions.
2. A *coexistence curve*, or *binodal*, which describes the compositions of the two phases x_1' and x_1'' that coexist at equilibrium, after liquid–liquid separation at some fixed $T < T_c$. Any solution prepared such that (T, x_1) lies under the binodal will be out of equilibrium until it has undergone phase separation.
3. A *stability limit*, or *spinodal*, which divides the two-phase region into a *metastable* window, between the binodal and the spinodal, and an *unstable* region, below the spinodal. The significance of the terms metastable and unstable will be explained subsequently. Note that the binodal and spinodal curves meet at the critical point.

Qualitatively, of course, we should expect one-phase behavior at high T because $\Delta S_m > 0$, and therefore $-T\Delta S_m$ contributes an increasingly negative term to ΔG_m. However, although $\Delta G_m < 0$ is the criterion for spontaneous mixing, it by no means guarantees a *single* mixed phase, as we shall

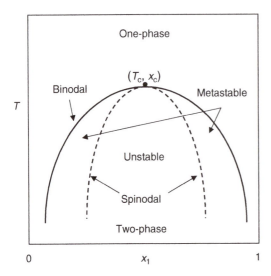

Figure 7.9 Phase diagram (temperature versus mole fraction of component 1) for regular solution theory. The binodal (coexistence curve) separates the one-phase region at high temperature from the two-phase region at low temperature. The spinodal curve (stability limit) separates the unstable and metastable windows within the two-phase region. The binodal and spinodal curves meet at a critical point.

see. To begin the analysis, we resolve the two contributions to $\Delta G_m/RT$ from regular solution theory (Equation 7.2.11):

$$-\frac{\Delta S_m}{R} = x_1 \ln x_1 + x_2 \ln x_2$$
$$\frac{\Delta H_m}{RT} = x_1 x_2 \chi \tag{7.5.1}$$

and recall from its definition (Equation 7.2.9) that $\chi \sim 1/T$. These two functions are plotted in Figure 7.10a and Figure 7.10b, respectively. Note that both are symmetric about $x_1 = 1/2$, and that in this format the entropy term is independent of T, whereas the enthalpy term is not (due to χ). Furthermore, we take $\chi > 0$, as expected by the theory. In Figure 7.10c we combine the two terms, at two generic temperatures, one "high" and one "low." At the higher T, χ is so small that ΔG_m looks much like the ΔS_m term; it is always concave up. However, at the lower T, the larger χ in the enthalpy term produces a "bump," or local maximum in the free energy. This will turn out to have profound consequences. Note that even at the lower T, $\Delta G_m < 0$ for all compositions considered in this example.

Phase separation will occur whenever the system can lower its total free energy by dividing into two phases. If we prepare a solution with overall composition $\langle x_1 \rangle$, and then ask will it prefer to separate into phases with compositions x_1' and x_1'', we can find the answer simply by drawing a line connecting the corresponding points on the ΔG_m curve (i.e., $\Delta G_m(x_1')$ to $\Delta G_m(x_1'')$), as shown in Figure 7.11a. Because ΔG_m is an extensive property, this line represents the hypothetical free energy of a combination of two phases, x_1' and x_1'', for any overall composition $\langle x_1 \rangle$ that lies in between. (Note that the relative proportions of the two phases with compositions x_1' and x_1'' are determined once $\langle x_1 \rangle$ is selected, by the so-called lever rule.)

What we now realize is that, so long as ΔG_m is concave up, this straight line will lie *above* ΔG_m at $\langle x_1 \rangle$ for any choice of x_1' and x_1'', and therefore phase separation would *increase* the free energy.

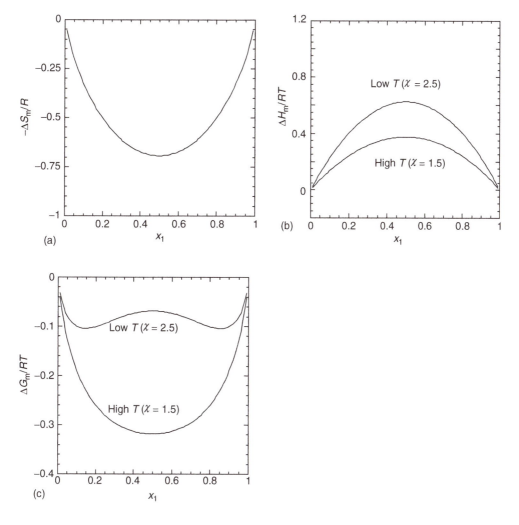

Figure 7.10 Predictions of regular solution theory for (a) entropy of mixing, plotted as $-\Delta S_m/R$; (b) enthalpy of mixing, plotted as $\Delta H_m/RT$, for two temperatures; (c) free energy of mixing obtained by combining panels (a) and (b), plotted as $\Delta G_m/RT$.

Thus "concave up" gives us the criterion for *stability* of the one-phase solution; the mathematical expression of concave up is

$$\left(\frac{\partial^2 \Delta G_m}{\partial x_i^2}\right)_{T,p} > 0 \tag{7.5.2}$$

where the second derivative can be taken with respect to the mole fraction of any component. The meaning of stability is this: In any mixture at a finite temperature, there will be spontaneous, small local fluctuations in concentration δx, such that there are small regions that have x_1 bigger than the average, and some regions where it is smaller. Now, by the argument given above, any such fluctuation will actually *increase* the free energy; the straight line connecting $\langle x_1 \rangle - \delta x_1$ and $\langle x_1 \rangle + \delta x_1$ will fall above $\Delta G_m(x_1)$. Consequently all these fluctuations will relax back to $\langle x_1 \rangle$. The importance of these spontaneous fluctuations will be taken up again in Chapter 8, where we will show how they are the origin of light scattering.

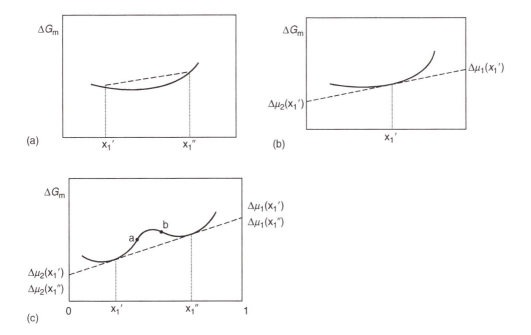

Figure 7.11 Generic free energy of mixing versus composition curves. (a) If a solution with overall composition between x_1' and x_1'' were to separate into two phases with compositions x_1' and x_1'', the resulting free energy (the dashed line) would lie above the one-phase case (smooth curve). (b) Tangent construction showing how the chemical potentials of the two components may be obtained for a given composition x_1'. (c) Tangent construction finds the compositions of the two phases x_1' and x_1'' that would coexist at equilibrium, for a system with overall composition between x_1' and x_1''. Points a and b denote the inflection points of ΔG_m, which separate the metastable ($x_1' < x_1 < a, b < x_1 < x_1''$) and unstable regions ($a < x_1 < b$).

7.5.2 Finding the Binodal

Now consider the lower T curve in Figure 7.10c, where ΔG_m shows the bump. Here we can see that if we prepared a solution with $\langle x_1 \rangle$ somewhere near the local maximum of ΔG_m, we could find an x_1' and x_1'' such that the straight line between them would fall below the ΔG_m curve for our $\langle x_1 \rangle$, and phase separation should occur. In fact, there are many such pairs x_1' and x_1'' that would lower ΔG_m, so which pair is chosen? We recall the criteria for phase equilibria: T and p must be identical in the two phases, and

$$\mu_1(x_1') = \mu_1(x_1''), \quad \mu_2(x_1') = \mu_2(x_1'') \tag{7.5.3}$$

The chemical potential of component 1 is the same in both phases and the chemical potential of component 2 is equal in both phases. (Be careful with this; both relations must be satisfied simultaneously, but it is *not* an equality between μ_1 and μ_2.) It turns out that there will be only one solution (x_1', x_1'') for both of these relations at a particular T, which we can identify by the *common tangent* construction. We can write the free energy as the mole-weighted sum of the chemical potentials (which are the partial molar free energies, Equation 7.1.5):

$$\Delta G_m = n_1 \Delta \mu_1 + n_2 \Delta \mu_2$$

or   (7.5.4)

$$\Delta G_m = x_1 \Delta \mu_1 + (1 - x_1) \Delta \mu_2 = \Delta \mu_2 + x_1 (\Delta \mu_1 - \Delta \mu_2)$$

where we have divided by the total number of moles to get to mole fractions, and where $\Delta \mu_i = \mu_i - \mu_i^\circ$. Now imagine we draw a straight line that is tangent to ΔG_m at some composition, x_1', as shown in Figure 7.11b. This line can be written generically as

$$y = kx_1 + b \tag{7.5.5}$$

where k is the slope and b is the $x_1 = 0$ intercept. But we chose $y = \Delta G_m$ for $x_1 = x_1'$, so inserting Equation 7.5.4 into Equation 7.5.5 we find

$$kx_1' + b = \Delta\mu_2(x_1') + x_1'\left[\Delta\mu_1(x_1') - \Delta\mu_2(x_1')\right] \tag{7.5.6}$$

But this relation holds whatever x_1' we choose, so we can match the intercepts and slopes to obtain

$$\begin{aligned}
b &= \Delta\mu_2(x_1') \\
k &= \Delta\mu_1(x_1') - \Delta\mu_2(x_1')
\end{aligned} \tag{7.5.7}$$

In other words, if we follow the tangent to the $x_1 = 0$ intercept, we obtain $b = \Delta\mu_2(x_1')$, and if we follow it to the $x_1 = 1$ intercept, $k + b = \Delta\mu_1(x_1')$.

The argument so far applies for any ΔG_m curve. Now if we have a ΔG_m curve with a bump as in Figure 7.11c, we can draw one straight line that is tangent to ΔG_m at *two* particular points, call them x_1' and x_1''. From the argument above, the $x_1 = 0$ intercept gives us *both* $\Delta\mu_2(x_1')$ and $\Delta\mu_2(x_1'')$, so these two chemical potentials must be equal. By the same reasoning the other intercept gives $\Delta\mu_1(x_1') = \Delta\mu_1(x_1'')$, and therefore we have shown that x_1' and x_1'' defined by the common tangent are indeed the compositions of the two coexisting phases. (Warning: for regular solution theory, where the ΔG_m curve is symmetric, x_1' and x_1'' coincide with the local minima in the ΔG_m curve, but this is not generally true.) So, in summary, one can locate the coexistence concentrations by geometrical construction on a plot of ΔG_m versus composition, or one could do it from the analytical expressions for the two chemical potentials. However, the latter is algebraically a little tricky, particularly because of the natural logarithm terms (see, for example, Equation 7.4.14).

7.5.3 Finding the Spinodal

The next issue to address is the location of the spinodal, or *stability limit*. We have already indicated the condition for stability, namely Equation 7.5.2. The stability limit, then, is found where the second derivative of ΔG_m changes sign, which defines an *inflection point*:

$$\left(\frac{\partial^2 \Delta G_m}{\partial x_i^2}\right)_{T,p} = 0 \quad \text{on the spinodal} \tag{7.5.8}$$

Returning to Figure 7.11c, we see that there are two inflection points, marked a and b, on each side of the bump. Between these two compositions, the free energy is concave down, and we say the solution for that (x_1, T) is *unstable*. What does this mean? For any small local fluctuation in concentration δx_1, the straight line connecting $x_1 - \delta x_1$ and $x_1 + \delta x_1$ will fall below $\Delta G_m(x_1)$. These fluctuations will therefore grow in amplitude and spatial extent; the mixture will spontaneously phase separate into two phases with compositions x_1' and x_1''. Thus in a region where the ΔG_m curve is concave down, the solution is unstable with respect to any fluctuation in concentration. The mechanism by which this phase separation occurs is called *spinodal decomposition*, and it is quite interesting in its own right. However, in this chapter we are concerned with thermodynamics, not kinetics, so we will not pursue this here.

You may have noticed that there are two regions on the curve in Figure 7.11c, between x_1' and a, and between b and x_1'', where the curve is locally concave up, indicating stability, yet we already know that the equilibrium state in these intervals should be liquid–liquid coexistence with concentrations x_1' and x_1''. What does this mean? These regions fall between the binodal and spinodal, and are termed *metastable*. They are stable against small, spontaneous fluctuations, but not globally stable against phase separation. Consequently, a system in the metastable region may remain there indefinitely; it requires *nucleation* of a region of the new phase before separation

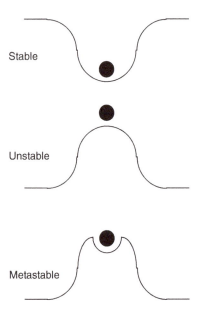

Stable

Unstable

Metastable

Figure 7.12 Schematic illustration of the difference between stable, unstable, and metastable states.

proceeds. Nucleation, and the ensuing process of domain growth, is another interesting kinetic process that we will not discuss here. However, metastability can be a wonderful thing; diamond is metastable with respect to graphite, the equilibrium phase of carbon at room T and p, but no one worries about diamonds transforming to graphite in their lifetime. A mechanical analogy is helpful in distinguishing among stable, metastable, and unstable systems, as shown in Figure 7.12. The ball in panel (a) may rattle around near the bottom of the bowl, but it will never come out; the system is stable. The ball in panel (b) is precariously perched on top of the inverted bowl, and the slightest breeze or vibration will knock it off; the system is unstable. The ball in panel (c) can rattle around in the small depression, and may appear to be stable for long periods of time, but with a sufficiently large impulse it will roll over the barrier and downhill to a lower energy state; the system is metastable. Only state (a) is an equilibrium state, but state (c) might not change in our lifetime.

7.5.4 Finding the Critical Point

The final feature to locate in the phase diagram is the critical point. We know it lies on the spinodal, so it must satisfy Equation 7.5.8. But, we need another condition to make it a single, special point. The easiest way to visualize this is to return to Figure 7.10c and the plots of ΔG_m at different temperatures. Phase separation occurs only when we have the bump in ΔG_m, so the critical point marks the temperature where the bump first appears. This also corresponds to the temperature where the two inflection points merge into one and this is determined by

$$\left(\frac{\partial^3 \Delta G_m}{\partial x_i^3}\right)_{T,p} = 0 \quad \text{at the critical point} \tag{7.5.9}$$

We can understand this by realizing that as T approaches T_c from below, one inflection point moves to the right, and one to the left. The rate of change of the inflection point, $\partial/\partial x_i(\partial^2 \Delta G_m/\partial x_i^2)$, vanishes when the two meet.

Algebraic expressions for the spinodal and the critical point of regular solution theory can be directly obtained as follows. The chemical potential for component 1 (and of course, by symmetry, we could equally well use component 2) comes from differentiating ΔG_m:

$$\frac{\mu_1}{RT} = \frac{\partial}{\partial n_1} \{n_1 \ln x_1 + n_2 \ln(1 - x_1) + n_1(1 - x_1)\chi\}$$

$$= \ln x_1 + \frac{n_1}{x_1}\frac{\partial x_1}{\partial n_1} - \frac{n_2}{1 - x_1}\frac{\partial x_1}{\partial n_1} + (1 - x_1)\chi - n_1\frac{\partial x_1}{\partial n_1}\chi \tag{7.5.10}$$

Now

$$\frac{\partial x_1}{\partial n_1} = \frac{n_1 + n_2 - n_1}{(n_1 + n_2)^2} = \frac{x_1 x_2}{n_1} = \frac{x_2^2}{n_2} \tag{7.5.11}$$

so

$$\frac{\mu_1}{RT} = \ln x_1 + (1 - x_1) - (1 - x_1) + \chi(1 - x_1)^2$$

$$= \ln x_1 + \chi(1 - x_1)^2 \tag{7.5.12}$$

The stability limit can now be obtained by taking the derivative with respect to x_1:

$$\frac{\partial}{\partial x_1}\left(\frac{\mu_1}{RT}\right) = \frac{1}{x_1} - 2\chi_s(1 - x_1) = 0 \tag{7.5.13}$$

where the subscript s denotes the value of χ on the spinodal. (You should convince yourself that if we followed the prescription for the stability limit given by Equation 7.5.8, and took the second derivative of $\Delta G_m/kT$ from Equation 7.2.11b with respect to x_1 instead of first obtaining μ_1, we would get the same relation.) This equation is a quadratic in x_1:

$$x_1^2 - x_1 + \frac{1}{2\chi_s} = 0 \tag{7.5.14a}$$

Note that this relation can be rewritten in the appealingly symmetric form

$$\frac{1}{x_1} + \frac{1}{x_2} - 2\chi_s = 0 \tag{7.5.14b}$$

The critical point requires that we differentiate Equation 7.5.13 once more:

$$\frac{\partial}{\partial x_1}\left(\frac{1}{x_1} - 2\chi(1 - x_1)\right) = -\frac{1}{x_1^2} + 2\chi_c = 0 \tag{7.5.15}$$

Equation 7.5.14a and Equation 7.5.15 constitute two simultaneous equations that can be solved to obtain the critical point (see Problem 9): The result is $x_{1,c} = 1/2$ (which we could have guessed from the outset, due to symmetry) and $\chi_c = 2$. This means that unless it costs at least $2kT$ to exchange one molecule of type 1 with one molecule of type 2, there will be no phase separation. To obtain the critical temperature for a particular system, T_c, we need to know the value of χ (i.e., $z\Delta w$):

$$T_c = \frac{z\Delta w}{k\chi_c} = \frac{z\Delta w}{2k} \tag{7.5.16}$$

Generically, however, we can see that the larger Δw, the larger T_c will be, and therefore the larger the two-phase window. If, perhaps due to some specific interactions, Δw happens to be negative, there will be no critical point according to regular solution theory; the system will be completely miscible at all temperatures and in all proportions.

7.5.5 Phase Diagram from Flory–Huggins Theory

Now we can repeat this entire procedure for the Flory–Huggins theory. The main difference will be that the value of N breaks the symmetry of the ΔS_m expression and will produce an asymmetric phase diagram. We already found the expression for $\Delta\mu_1/RT$, in Equation 7.4.14, so we can start from there. We should take derivatives with respect to x_1, but in fact we can get away with the much easier task of differentiating with respect to ϕ_2. This is because $\partial/\partial\phi_2 = (\partial x_2/\partial\phi_2)\partial/\partial x_2$ and as we will be setting the expressions equal to zero, $\partial x_2/\partial\phi_2$ will divide out. Also, $\partial/\partial\phi_2 = -\partial/\partial\phi_1$, so we can work with μ_1. Thus the spinodal curve can be found from

$$\frac{\partial}{\partial\phi_2}\left(\frac{\Delta\mu_1}{RT}\right) = \frac{\partial}{\partial\phi_2}\left\{\ln(1-\phi_2) + \phi_2\left(1-\frac{1}{N}\right) + \chi\phi_2^2\right\}$$

$$= \frac{-1}{1-\phi_2} + 1 - \frac{1}{N} + 2\chi_s\phi_2 = \frac{1}{\phi_1} + \frac{1}{\phi_2 N} - 2\chi_s = 0 \qquad (7.5.17)$$

The critical point comes from

$$\frac{\partial^2}{\partial\phi_2^2}\left(\frac{\Delta\mu_1}{RT}\right) = \frac{-1}{(1-\phi_{2,c})^2} + 2\chi_c = 0 \qquad (7.5.18)$$

or

$$\chi_c = \frac{1}{2}\frac{1}{(1-\phi_{2,c})^2} \qquad (7.5.19)$$

We can now substitute this relation into Equation 7.5.17

$$\frac{-1}{1-\phi_{2,c}} + \left(1 - \frac{1}{N}\right) + \frac{\phi_{2,c}}{(1-\phi_{2,c})^2} = 0 \qquad (7.5.20)$$

which is again a quadratic:

$$\phi_{2,c}^2\left(1-\frac{1}{N}\right) + \frac{2\phi_{2,c}}{N} - \frac{1}{N} = 0 \qquad (7.5.21)$$

$$\phi_{2,c} = \frac{-2/N + \sqrt{\frac{4}{N^2} + \frac{4}{N}\left(1-\frac{1}{N}\right)}}{2\left(1-\frac{1}{N}\right)} = \frac{\sqrt{N}-1}{N-1} = \frac{\sqrt{N}-1}{(\sqrt{N}-1)(\sqrt{N}+1)}$$

$$= \frac{1}{1+\sqrt{N}} \approx \frac{1}{\sqrt{N}} \qquad (7.5.22)$$

Thus the critical polymer concentration depends inversely on $\sqrt{N}$. For larger and larger N, ϕ_c will become lower and lower, approaching 0 as $N\to\infty$. Now to complete the analysis, we return to Equation 7.5.19 and find χ_c:

$$\chi_c = \frac{1}{2}\frac{1}{(1-\phi_{2,c})^2} = \frac{1}{2}\frac{(1+\sqrt{N})^2}{(\sqrt{N})^2} = \frac{1}{2}\left(\frac{1}{N} + \frac{2}{\sqrt{N}} + 1\right) \qquad (7.5.23)$$

So, as N increases, χ_c approaches 1/2. Recall from regular solution theory (which we can obtain from Flory–Huggins theory by setting $N=1$) that $\chi_c = 2$ and $\phi_c = 1/2$. Thus, as we increase N, the critical concentration moves more and more toward solutions dilute in polymer, and χ_c becomes smaller. For a given $z\Delta w$, therefore, T_c increases as N increases, meaning that polymers become less likely to form a homogeneous solution at a given T as N increases. This is illustrated in Figure 7.13 for various values of N.

Recall an important result from the previous section: At the theta temperature $B=0$ and $\chi=1/2$. Now we have a third definition of the theta temperature; it is the critical temperature for a given polymer–solvent system in the limit of infinite molecular weight. A polymer will be completely miscible with a solvent above the theta temperature, but anywhere below the theta temperature, there

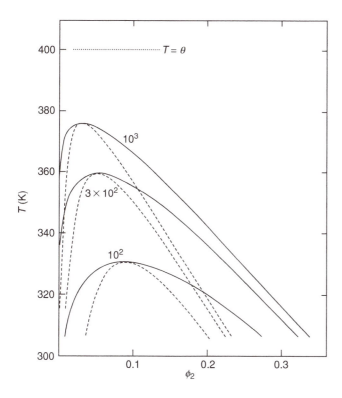

Figure 7.13 Phase diagrams for a polymer–solvent system with a theta temperature of 400 K, and N values of 10^2, 3×10^2 and 10^3. Smooth curves are binodals, dashed curves are spinodals. (Reproduced from Munk, P. and Aminabhavi, T.M. in *Introduction to Macromolecular Science*, 2nd ed., Wiley, New York, 2002. With permission.)

is a danger of phase separation. Furthermore, as a polydisperse solution is cooled below the theta point, the higher molecular weight chains will tend to phase separate first, a feature which can be used to advantage in fractionation.

Examples of the phase behavior of polymer solutions are presented in Figure 7.14. The classic results of Shultz and Flory for polystyrene in cyclohexane and polyisobutylene in diisobutyl ketone are reproduced in Figure 7.14a and Figure 7.14b, respectively. The data are experimental estimates of the coexistence concentrations (binodal), with smooth curves drawn to guide the eye. The dashed lines correspond to the predictions of the Flory–Huggins theory. It is clear from these figures that the theory indeed captures the main features of the data, namely a critical concentration that is small and decreases with increasing M, and a critical temperature that increases with increasing M. Neither the shape of the binodal nor the exact concentration dependence of the critical composition is correct, however. The critical temperatures for these two systems are plotted as a function of M in Figure 7.14c, in a format suggested by Equation 7.5.23. The plots are linear, and permit reliable determination of the theta temperatures, with values that are in good agreement with those determined by locating $B(T) = 0$ (see Figure 7.8).

A third system, polystyrene in acetone, is illustrated in Figure 7.15. Here we see phenomena that are not described by the theory at all, namely phase separation upon heating for certain values of M. For example, for $M = 10,300$ there appear to be two critical temperatures, one just below 0°C and the other just above 140°C. The solution would be two-phase for temperatures below the former and above the latter and one-phase at intermediate temperatures. For $M = 19,800$, there is no temperature at which a solution with $0.1 < \phi_2 < 0.15$ would be one-phase. What are we to make of this? First, there is nothing in thermodynamics that forbids this kind of behavior. The Flory–Huggins theory, however, cannot

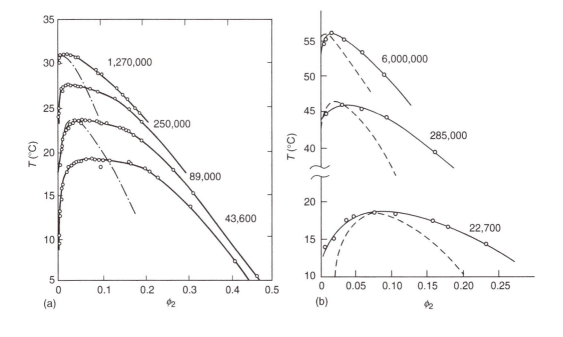

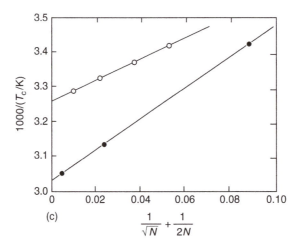

Figure 7.14 Experimental coexistence curves compared to the Flory–Huggins theory (dashed curves) for (a) polystyrene in cyclohexane and (b) polyisobutylene in diisobutylketone. (c) Resulting critical temperatures versus degree of polymerization, plotted as suggested by Equation 7.5.23. (From Shultz, A.R. and Flory, P.J., *J. Am. Chem. Soc.*, 74, 4760, 1952. With permission.)

describe it (at least with $\chi > 0$, see the next section) and thus this system illustrates some of the qualitative limitations of the theory. The critical point on a phase boundary that separates a two-phase region at low temperature from a one-phase region at high temperature is called an *upper critical solution temperature* (UCST), whereas a critical point on a phase boundary that separates a two-phase region at high temperature from a one-phase region at low temperature is called a *lower critical solution temperature* (LCST). Thus for polystyrene in acetone with $M = 10{,}300$, both a UCST and an

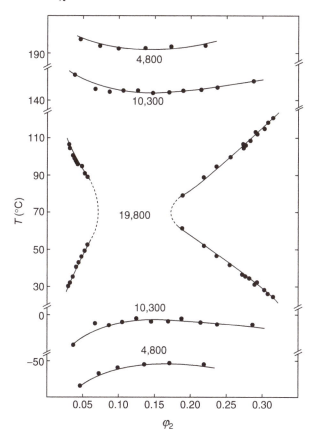

Figure 7.15 Experimental phase diagrams for polystyrene in acetone. (From Siow, K.S., Delmas, G., and Patterson, D., *Macromolecules*, 5, 29, 1972. With permission.)

LCST are observed; the Flory–Huggins theory is only capable of predicting UCST behavior. (Note a possible source of confusion: the lower critical temperature lies *above* the upper critical temperature.)

7.6 What's in χ?

As we have seen, the phase behavior of a polymer solution is determined largely by the interaction parameter χ and by N. In this section we inspect some of the various ways to look at χ. This will give us some insight into how intermolecular interactions determine phase behavior. Furthermore, it turns out that χ is used in the research literature in a range of different specific ways and it is important to see how that comes about.

7.6.1 χ from Regular Solution Theory

We begin by returning to regular solution theory, and the initial definition of χ in terms of $z\Delta w$ in Equation 7.2.9. If we know something about the intermolecular interactions, we ought to be able to say something more specific about w_{ij} rather than just leaving it as a parameter. For London (dispersion) interactions, which are caused by spontaneous dipolar fluctuations on one molecule inducing a dipole on another, the interaction energy is

$$w_{ij} \sim -\frac{\alpha_i \alpha_j}{r_{ij}^6} \tag{7.6.1}$$

where α_i is the polarizability of molecule i, and r_{ij} is the distance between molecules i and j. Because w_{ij} falls off with distance so rapidly, we are justified in considering only nearest-neighbor interactions (as in regular solution or Flory–Huggins theory). All atoms and molecules experience this attractive (and fundamentally quantum mechanical) interaction, and for many molecules without strong dipole moments or hydrogen bonds, it is the only interaction that matters. For example, simple alkanes and inert gases form condensed phases primarily because of London forces. Now if we have a lattice where we have imposed $r_{ii} = r_{jj} = r_{ij}$ by design, then

$$w_{12} = -\frac{\alpha_1 \alpha_2}{r_{12}^6} = -\frac{\alpha_1}{r_{11}^3}\frac{\alpha_2}{r_{22}^3} = -\sqrt{w_{11}w_{22}} \tag{7.6.2}$$

This particular "mixing rule" for w_{ij} is called the Berthelot rule, and is certainly a plausible starting point. Under this assumption, we have

$$
\begin{aligned}
z\Delta w &= z\left(w_{12} - \frac{1}{2}w_{11} - \frac{1}{2}w_{22}\right) \\
&= \frac{z}{2}\left(-2\sqrt{w_{11}w_{22}} + \left[\left(\sqrt{|w_{11}|}\right)^2 + \left(\sqrt{|w_{22}|}\right)^2\right]\right) \\
&= \frac{z}{2}\left[\sqrt{|w_{11}|} - \sqrt{|w_{22}|}\right]^2
\end{aligned}
\tag{7.6.3}
$$

Now we can see that because Δw can be written as a perfect square, it must be greater than or equal to zero, which is why $\chi \geq 0$ in regular solution theory and the Flory–Huggins theory.

We can go further with this approach. The molar heat of vaporization for a pure substance, $\widehat{U}_{vap}$, should be directly related to w, i.e.,

$$\widehat{U}_{i,\text{vap}} = -N_{av}z\frac{w_{ii}}{2} \tag{7.6.4}$$

meaning that $zw_{ii}/2$ is the interaction energy lost by removing one i molecule from the pure substance. The *cohesive energy density* (CED) is defined by dividing $\widehat{U}_{i,\text{vap}}$ by the molar volume, $\widehat{V}_i$, which in regular solution theory is just Avogadro's number times the lattice site volume:

$$\text{CED} = \frac{\widehat{U}_{i,\text{vap}}}{\widehat{V}_i} = -N_{av}z\frac{w_{ii}}{2\widehat{V}_i} \equiv \delta_i^2 \tag{7.6.5}$$

Here we have introduced the *solubility parameter*, δ, defined as $\sqrt{\text{CED}}$. The CED is usually given in units of cal/cm^3, and thus δ has the unusual units of $(\text{cal/cm}^3)^{1/2}$. Now we can rewrite Equation 7.6.3 as

$$\chi = \frac{\widehat{V}_1}{RT}(\delta_1 - \delta_2)^2 \tag{7.6.6}$$

This is a very simple expression for χ that is directly related to fundamental physical quantities ($\widehat{U}_{vap}$, CED, or δ). Note that in the Flory–Huggins theory we have assumed no volume change on mixing, so $\widehat{V}_1 = \overline{V}_1$ and the lattice size is defined by the solvent size. By examining a table of solubility parameters (see examples in Table 7.1), an estimate of χ can be quickly obtained. To obtain a good solvent for a polymer, one could begin by seeking solvents with similar solubility parameters. For small molecules the heat of vaporization can be measured precisely, but for polymers this is not the case; thus solubility parameter values are obtained indirectly, and can be quite uncertain.

7.6.2 χ from Experiment

It turns out that the solubility parameter approach to χ does not work well as a predictive method for polymer solutions. As a first example, consider polystyrene in cyclohexane, which is a theta solvent at 34.5°C (see Figure 7.7 and Figure 7.8). We can calculate χ by Equation 7.6.6, using

Table 7.1 Solubility Values for Common Polymers and Solvents (Values for Polymers Are Representative Only)

Polymer	δ $(cal/cm^3)^{1/2}$	Solvent	δ $(cal/cm^3)^{1/2}$
Poly(tetrafluoroethylene)	6.2	n-Hexane	7.3
Poly(dimethylsiloxane)	7.4	Cyclohexane	8.2
Polyisobutylene	7.9	Carbon tetrachloride	8.6
Polyethylene	7.9	Toluene	8.9
Polyisoprene	8.1	Ethyl acetate	9.1
1,4-Polybutadiene	8.3	Tetrahydrofuran	9.1
Polystyrene	9.1	Chloroform	9.3
Atactic polypropylene	9.2	Carbon disulfide	10.0
Poly(methyl methacrylate)	9.2	Dioxane	10.0
Poly(vinyl acetate)	9.4	Ethanol	12.7
Poly(vinyl chloride)	9.7	Methanol	14.5
Poly(ethylene oxide)	9.9	Water	23.4

Source: E.A. Grulke in *Polymer Handbook*, Brandrup, J. and Immergut, E.H. (Eds.), 3rd ed., Wiley, New York, 1989.

$\widehat{V}_1 \approx M/\rho = (84 \text{ g/mol})/(0.78 \text{ g/cm}^3) = 108 \text{ cm}^3/\text{mol}$ and $R = 1.987$ cal/K mol; the result is $\chi \approx 0.14$. But, we know that $\chi = 0.50$ at the theta temperature, so we are off by almost a factor of 4! To see that this is not an isolated case, consider the following example.

Example 7.3

Three other reported theta systems are polyisoprene in dioxane at 34°C, poly(methyl methacrylate) in carbon tetrachloride at 27°C, and poly(vinyl acetate) in ethanol at 19°C. Estimate χ for each of these cases, using Equation 7.6.6 and Table 7.1.

Solution

We will also need molecular weights and densities for the three solvents; they are approximately 88 g/mol and 1.03 g/cm^3 for dioxane; 154 g/mol and 1.59 g/cm^3 for carbon tetrachloride; 46 g/mol and 0.79 g/cm^3 for ethanol:

Polyisoprene/dioxane: $\chi \approx \dfrac{(88/1.03)}{(1.987 \times 307)}(10 - 8.1)^2 = 0.51$

Poly(methyl methacrylate)–carbon tetrachloride: $\chi \approx \dfrac{(154/1.59)}{(1.987 \times 300)}(8.6 - 9.2)^2 = 0.06$

Poly(vinyl acetate)–ethanol: $\chi \approx \dfrac{(46/0.79)}{(1.987 \times 292)}(12.7 - 9.4)^2 = 1.1$

According to the theory, all three χ values should be 0.5. In fact, one value is spot on, one is much too small, and one is much too big.

What should we conclude from these calculations? First, the solubility parameter approach is not quantitatively reliable for polymer solutions; the predicted value of χ can be greater than, less than, or very close to an experimental value. For at least one of the examples above, poly(vinyl acetate) in ethanol, the possibility of hydrogen bonding invalidates the basic assumptions of the theory, so we should not be too surprised by this result. For poly(methyl methacrylate) in carbon tetrachloride, the predicted χ is much too small. Based on comparisons of experiments with many polymers and solvents, the following empirical equation is found to be a much more reliable route to estimate χ when the predicted value is less than about 0.3:

$$\chi = 0.34 + \frac{\widehat{V}_1}{RT}(\delta_1 - \delta_2)^2 \tag{7.6.7}$$

In the poly(methyl methacrylate)–carbon tetrachloride and polystyrene–cyclohexane cases Equation 7.6.7 does a much better job, albeit still not perfect. From Equation 7.6.7 there appears to be a nearly constant (and substantial) temperature-independent contribution to χ in polymer–solvent systems, which is not anticipated by the regular solution theory approach. The fact that the 0.34 term does not have an explicit temperature dependence suggests that it reflects an additional entropy of mixing contribution, rather than the purely enthalpic χ anticipated by the model.

This is but one example of the limitations of this theory; another was provided by the polystyrene–acetone phase diagrams in Figure 7.15, where the observed LCST behavior cannot be explained by an interaction parameter that follows Equation 7.6.6. Still other problems are apparent from the solubility parameter values in Table 7.1. For example, poly(ethylene oxide) is actually water-soluble, even though the difference between the two solubility parameters is huge. This particular case involves hydrogen bonding and the rather unusual properties of water. In fact, the assumptions of the theory are not consistent with any kind of strong or directional interaction, such as those involving permanent dipoles or hydrogen bonds. The packing of dipoles can influence the entropy of mixing through the various orientational degrees of freedom, which are not included in the purely combinatorial entropy of the lattice model. Similarly, the energy of interaction between two dipoles is very sensitive to the relative orientation, and the orientation of one dipole will be sensitive to the positions and orientations of all neighbors. In short, there are many ways in which the basic theory fails to incorporate important features of real systems, especially when the interactions are strong and directional.

7.6.3 Further Approaches to χ

Two questions should immediately come to mind. First, have we wasted our time examining the Flory–Huggins theory in such detail, given that it fails to describe the thermodynamic properties of polymer solutions quantitatively, and, in many cases, even qualitatively? Second, can we do anything to rectify the situation? The answer to the first question is simple: no, we have not wasted our time. All of the developments that we went through (finding expressions for Π, μ_i, and mapping out the phase diagram, etc.) were model-independent thermodynamics. The model really only entered through the explicit expression for ΔG_m that we used (i.e., Equation 7.3.13), and all differences between experiment and model predictions are directly attributable to inadequacies of Equation 7.3.13. The answer to the second question is not so clear-cut. There are three general strategies employed in the polymer community. One is to try and improve the model, for example by identifying further interaction terms, additional sources of entropy, etc. This approach has been pursued for many years. It has the virtue that it is possible to keep track of both the intended meaning and quantitative effect of any added term in the expression for ΔG_m. It has the drawback that it is not yet generally successful if we try to restrict ourselves to only a small number of new terms. A second strategy is known as the "equation of state" approach. In this case, the ΔG_m expression is recast in a general form, with a finite number of parameters. In many cases the behavior of mixtures can be well predicted based on knowledge of the parameters of the pure components, so it can be a very practical strategy. One disadvantage is that it is difficult to gain much physical insight into the underlying molecular processes from the parameter values. The third approach is the one most favored by experimentalists in polymer science, and so we will examine this one a little more carefully. In essence it amounts to using χ as a fitting *function*, not a number, so that χ takes on whatever attributes are necessary to describe the data.

In general the free energy of mixing can be divided into two parts, an *ideal* part (superscript id) and an *excess* part (superscript ex):

$$\Delta G_m = \Delta G_m^{id} + \Delta G_m^{ex} = \Delta H_m^{id} - T\Delta S_m^{id} + \Delta H_m^{ex} - T\Delta S_m^{ex} \tag{7.6.8}$$

Furthermore, an ideal solution is defined as one for which

$$\Delta S_m^{id} = -k(x_1 \ln x_1 + x_2 \ln x_2), \quad \Delta H_m^{id} = 0 \tag{7.6.9}$$

In other words, the entropy of mixing of an ideal solution is the purely combinatorial entropy of mixing, and there is no enthalpy of mixing (i.e., it is an *athermal* solution). Now we can view regular solution theory, or the Flory–Huggins theory, as making specific predictions for the excess quantities, namely

$$\Delta S_m^{ex} = 0, \quad \Delta H_m^{ex} = x_1 x_2 \chi kT \tag{7.6.10}$$

The approach that is often adopted is to define an *effective interaction parameter* χ_{eff} in terms of the experimentally accessible excess free energy of mixing (i.e., the experimentally measured ΔG_m less the ideal part from Equation 7.6.9):

$$\chi_{eff} \equiv \frac{\Delta G_m^{ex}}{\phi_1 \phi_2 kT} \tag{7.6.11}$$

This χ_{eff} will therefore include all the ingredients that pertain to the system under study, but it is an *experimental result* not a *model prediction*. The scientific literature is often very confusing on this point. For example, relatively few authors will actually mention Equation 7.6.11, even though that is what they are doing when they fit the data. Furthermore, quite a few will refer to this χ_{eff} as the "Flory–Huggins interaction parameter," which it is not; the Flory–Huggins χ is correctly given by Equation 7.2.9.

Empirically, the χ_{eff} function obtained by fitting data usually follows the form

$$\chi_{eff} = \frac{\alpha}{T} + \beta = \chi_h + \chi_s \tag{7.6.12}$$

where χ_{eff} is sometimes resolved into two components, the "enthalpic part" χ_h and the "entropic part" χ_s. Interestingly, the parameters α and β may each be positive or negative. A negative α implies some kind of specific attractive interaction between the components, such as might occur if one component was a hydrogen bond donor and the other an acceptor. The sign of β is harder to interpret, but presumably reflects details of molecular packing. (Recall that in the lattice model we assume that each molecule has the same shape, and fits neatly into the lattice site, so there is no entropy associated with how the molecule is oriented within a site. For real molecules anisotropy of shape is almost inevitable, but it is not necessarily obvious whether there will be more or less packing possibilities per unit volume in the mixture compared to in the pure components, and hence the sign of the excess entropy can be hard to predict.) These various possibilities for the signs of α and β give some insight into the various kinds of phase diagram that arise; for example, an LCST system would be the natural consequence of having $\alpha < 0$ and $\beta > 0$.

One final point about this experimental χ_{eff} parameter: it very often exhibits a concentration dependence, which is equivalent to saying that the excess free energy of mixing is not entirely quadratic and symmetric with respect to components 1 and 2 (i.e., not $\Delta G_m^{ex} \propto \phi_1 \phi_2$). This should not be too surprising, given the great disparity in size and shape between a polymer and a solvent molecule. This feature does have an important practical consequence, however. We can write the free energy of mixing as

$$\frac{\Delta G_m}{kT} = \phi_1 \ln \phi_1 + \frac{\phi_2}{N} \ln \phi_2 + \chi_{eff} \phi_1 \phi_2 \tag{7.6.13a}$$

and if we actually measure ΔG_m in an experiment, we can extract χ_{eff} directly. However, it is not easy to measure ΔG_m; more often we make a measurement that is really sensing a chemical potential, such as the osmotic pressure. In this case, if we take Equation 7.4.14 and substitute χ_{eff} for χ, we would have

$$\frac{\Pi \bar{V}_1}{RT} = -\left\{ \ln(1 - \phi_2) + \phi_2 \left(1 - \frac{1}{N} \right) + \chi_{eff} \phi_2^2 \right\} \tag{7.6.13b}$$

The resulting χ_{eff} would be *different* from that obtained from Equation 7.6.13a. The reason is that in obtaining Equation 7.6.13b from Equation 7.6.13a we took a derivative with respect to a concentration variable, but under the assumption that χ was not a function of concentration. When χ is a function of concentration, the derivative will yield an additional term involving $\partial \chi_{\text{eff}}/\partial \phi$. As we will see in Chapter 8, a scattering experiment actually measures the second derivative of ΔG_{m}, and thus the answer includes both first and second derivatives of χ_{eff}. Consequently, one must be very careful in comparing the values of χ_{eff} obtained by different experiments. This issue is explored further in Problem 7.10.

7.7   Excluded Volume and Chains in a Good Solvent

We now return to this rather tricky problem, armed with enough information to say something useful. In Chapter 6 we examined in detail the mean-square radius of gyration, and the segment distribution function, for flexible chains. As pointed out then, we left out one very important physical feature, namely that a real polymer chain cannot intersect itself. We modeled the chain as a random walk, whereas in reality it is a *self-avoiding walk*. The current best estimate of the exact result for a self-avoiding walk is $R_{\text{g}} \sim N^{0.589}$, but we follow common practice and approximate this relation as $R_{\text{g}} \sim N^{0.6}$. The problem of a polymer in solution has another aspect we did not consider in Chapter 6, namely that there will generally be some interaction energy between a polymer segment and a solvent molecule. In an athermal solvent ($\Delta H_{\text{m}} = 0$), $\chi = 0$, and there is no energetic price to pay for having a solvent molecule next to a polymer segment. In such a case the full self-avoiding walk statistics apply, and the chain is larger than its unperturbed dimensions described by Equation 6.5.3. Often this coil is said to be "swollen," and the degree of swelling or "coil expansion" is quantified by the expansion factor, α:

$$\alpha \equiv \frac{R_{\text{g}}}{R_{\text{g,0}}} \tag{7.7.1}$$

where the subscript "0" denotes the unperturbed dimensions, as in Chapter 6. However, as χ increases, there is an increasing penalty for having solvent next to polymer and this will begin to favor monomer–monomer contacts. Consequently, the coil starts to become more compact. If χ continues to increase, the coils eventually give up, collapse, and precipitate out of solution. It turns out that there is a special value of χ for which the unfavorable segment–solvent interactions counteract the self-avoiding nature of the chain, and the random-walk conformation is recovered. This value, as you may have guessed, is $\chi = 1/2$: a theta solvent. Thus theta solvents are of particular value because they provide an environment in which we have a very full description of the conformational statistics. In Section 7.4 we saw how the second virial coefficient, B, vanished in a theta solvent, and this was attributed to the canceling of excluded volume effects *between coils* by the unfavorable polymer–solvent interactions. In the discussion here we have been talking about excluded volume interactions *within a single coil*, but unless one examines the theoretical issues very deeply, this is a subtle distinction that can be ignored.

The preceding discussion gives an overview of the problem, but the fact that the exponent ν (defined by $R_{\text{g}} \sim N^{\nu}$ in Equation 6.6.1) takes on the approximate value 3/5 was just stated, as was the fact that intramolecular excluded volume effects are canceled out at the theta temperature. In the remainder of this section we will give a derivation of these results which, though far from rigorous or detailed, at least provides some rationale.

Imagine a single polymer coil in a very good solvent, confined within a hypothetical spherical semipermeable membrane which allows solvent in or out, but not polymer segments. Thus we are imagining a "single solute molecule" osmotic pressure experiment, as shown in Figure 7.16. The question is, as we increase N, how much does the radius of this spherical membrane, R, increase? Solvent can flow in to swell (i.e., dilute) the segments, but the chain will be distorted away from

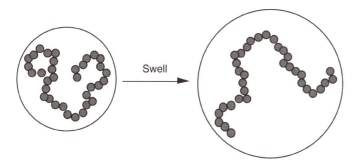

Figure 7.16 Schematic illustration of the swelling of a coil in a good solvent as a "single chain osmotic pressure" experiment.

random-walk statistics, and the larger the chain dimensions, the fewer the possible conformations. Thus entropy resists unlimited coil expansion. The balance between the osmotic drive to swell and the entropic drive to stay coiled up sets the ultimate dimensions. We estimate the "free energy" per unit volume of this chain by a scaling argument, as follows [4]. (A scaling argument means that we leave out unimportant numerical factors, and emphasize the dependence of R on the main variable, in this case N. Also, we assume R will have the same dependence on N as does R_g.) The osmotic part is driven by the segment–segment interactions, which we can write as $v(T)c^2$ per unit volume:

$$\frac{F_{os}}{kT} \sim v(T)c^2 \tag{7.7.2}$$

where c is the number of segments per unit volume, and $v(T)$ is the strength of the excluded volume interaction between any two segments. (In particular, $v \sim B$, and the c^2 dependence is just a reflection of the probability of two segments coming into contact. Basically Equation 7.7.2 is the second term of the virial expansion, Equation 7.4.7a; the first term does not matter because we can let N become very large.) Now $c \sim N/R^3$ and when we integrate F_{os} over the volume of the coil (we assume F_{os} is the same everywhere, i.e., a mean-field approximation within the coil), we gain another factor of R^3:

$$\frac{F_{os}}{kT} \sim \int_{coil} v(T)\frac{N^2}{R^6} d(\text{volume}) = v(T)\frac{N^2}{R^6}R^3 = v(T)\frac{N^2}{R^3} \tag{7.7.3}$$

Now we turn to the resistance to swelling, often referred to as an elastic or stretching penalty. We will consider the elastic force that resists deforming a flexible chain in detail in Chapter 10; it is the reason why rubber bands work. However, we can extract the main relation that we need now. The elastic energy, F_{el}, is proportional to $-TS$, where S is the entropy lost on stretching. If we assume that we begin with a Gaussian chain, we can write

$$F_{el} \sim -TS = -kT \ln P(N,R)$$
$$\sim -kT \ln\left(\exp\left[\frac{-3R^2}{2Nb^2}\right]\right) = \frac{3}{2}kT\frac{R^2}{Nb^2} \tag{7.7.4}$$

Thus the scaling argument says

$$\frac{F_{el}}{kT} \sim \frac{R^2}{N} \tag{7.7.5}$$

and combining Equation 7.7.3 and Equation 7.7.5 we have

$$\frac{F_{tot}}{kT} \sim v(T)\frac{N^2}{R^3} + \frac{R^2}{N} \tag{7.7.6}$$

To find the desired relation between R and N, we minimize F_{tot} with respect to R:

$$\frac{d}{dR}\left(\frac{F_{tot}}{kT}\right) \sim -3v(T)\frac{N^2}{R^4} + \frac{2R}{N} = 0 \tag{7.7.7}$$

and so

$$R \sim v(T)^{1/5}N^{3/5} \sim B(T)^{1/5}N^{3/5} \tag{7.7.8}$$

We can also recast this equation in terms of the expansion factor, α, by dividing each side by $R_{g,0} \sim N^{1/2}$, and then raising each side to the fifth power:

$$\alpha^5 \sim v(T)N^{1/2} \tag{7.7.9}$$

The preceding argument is not quite right, but it actually succeeds by a partial cancelation of errors. The osmotic term reflects a mean-field argument—the probability of segment–segment contact is taken to be uniformly c^2 across the coil—which overestimates its importance. On the other hand, the elastic part presupposes a Gaussian chain, which is also not quite correct. Nevertheless, the argument is reasonably simple and it captures the essence of the problem: a balance between excluded volume and coil distortion.

Flory and Krigbaum worked out a much more detailed version of this calculation [5], with the result

$$\alpha^5 - \alpha^3 = 2C_M\left(\frac{1}{2} - \chi\right)\sqrt{M} \tag{7.7.10}$$

where the prefactor C_M is given by

$$C_M = \left(\frac{27}{\sqrt{32\pi^3}}\right)\left(\frac{\widehat{V}_2^2}{M^2 N_{av}\,\widehat{V}_1}\right)\left(\frac{\langle h^2\rangle_0}{M}\right)^{-3/2} \tag{7.7.11}$$

The quantities $\widehat{V}_2$ and $\widehat{V}_1$ are the molar volumes of the polymer and solvent, respectively (which are equal to the partial molar quantities in the Flory–Huggins theory). The differences between Equation 7.7.9 and Equation 7.7.10 are the term in α^3, which arises from an additional entropy change for the chain due to the increased volume accessible to it, and the explicit expression for the prefactor. The factor of $(1/2 - \chi)$ in Equation 7.7.10 is proportional to B (recall Equation 7.4.18), just as was $v(T)$ in Equation 7.7.9.

Now we can briefly examine the implications of Equation 7.7.10.

1. In a theta solvent, $\chi = 1/2$, and so $\alpha = 1$; there is no swelling. Thus in a theta solvent the chain behaves as a random walk for all N large enough to be random walks.
2. In a good solvent, $\chi < 1/2$, the chain swells ($\alpha > 1$). In the limit of a very good solvent, $\alpha^5 \gg \alpha^3$, and thus $\alpha^5 \sim N^{1/2}$ (Equation 7.7.9). Therefore,

$$\alpha^5 = \left(\frac{R_g}{R_{g,o}}\right)^5 \propto \left(\frac{N^\nu}{N^{1/2}}\right)^5 \propto N^{1/2}$$

$$\frac{N^\nu}{N^{1/2}} \propto N^{1/10}$$

$$N^\nu \propto N^{5/10}N^{1/10} = N^{3/5} \tag{7.7.12}$$

This result, that $\nu = 3/5$, is the classic result for the excluded volume (self-avoiding walk) exponent that we cited before. The high N behavior in good and theta solvents is illustrated in Figure 7.17, for polystyrene in benzene (a good solvent), cyclohexane (theta), and *trans*-decalin (theta).

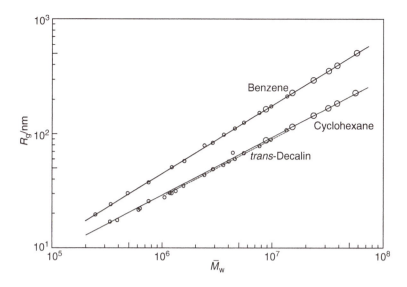

Figure 7.17 Experimental radii of gyration for polystyrenes in benzene, a good solvent, and cyclohexane and *trans*-decalin, both theta solvents. Data obtained by light scattering by several authors. (From Miyake, A., Einaga, Y., and Fujita, H., *Macromolecules*, 11, 1180, 1978. With permission.)

3. The crossover from theta-like behavior ($\alpha \approx 1$) to fully developed excluded volume, $\alpha \gg 1$, is very broad. It depends on both N and χ (and therefore T). For example, for a given N the chain will swell progressively as T is increased above $T = \Theta$, and for a given $T > \Theta$, a larger N chain will swell more than a shorter one. Because of this broad crossover, it is very common in experiments to find apparent values of the exponent ν falling between 1/2 and 3/5. This will become important particularly in the context of the intrinsic viscosity, as we shall see in Chapter 9.

4. Although designed for polymer solutions, Equation 7.7.10 actually hints at a very important result for molten polymers. Suppose the solvent were a chain of the same monomer, but with a different degree of polymerization, P. Presumably $\chi \approx 0$, so the chain should swell. However, $\widehat{V}_1 \sim P$, so if P exceeds $\sqrt{N}$ in length there should be little or no swelling. In other words, chains in their own melt should be Gaussian. Flory made this very important prediction in the 1950s, but it was not until the advent of small-angle neutron scattering in the early 1970s that this fundamental result could be confirmed.

7.8 Chapter Summary

In this chapter we have covered a great deal of material concerning the thermodynamic properties of polymer solutions. We have interwoven results that are strictly thermodynamic with those that depend on a particular model, the Flory–Huggins theory. The main points are as follows:

1. Using thermodynamics alone we were able to show how osmotic pressure measurements on dilute solutions can be used to determine M_n and the second virial coefficient, B. The latter gives direct information about the solvent quality: $B > 0$ corresponds to a good solvent, and $B < 0$ to a poor solvent.

2. Thermodynamic arguments were also sufficient to show how the complete phase diagram (T, ϕ) for a binary system could be constructed from an expression for the free energy of mixing. The key features of the phase diagram are the critical point, the coexistence curve (binodal), and the stability limit (spinodal).

3. We developed an expression for the free energy of mixing based on the Flory–Huggins theory. This is a mean-field theory and reduces to the standard regular solution theory when the degree of polymerization of the polymer component is set equal to 1. The numerous assumptions of the model were identified.

4. The Flory–Huggins model was explored in detail, in terms of its predictions for B and for the phase diagram. Comparison with experiments reveals that in some systems the Flory–Huggins theory captures the phenomenology in a qualitative manner, but in others it does not. It does not provide a quantitative description for any dilute polymer solution.

5. The concept of a theta solvent emerges as a central feature of polymer solutions. It has four equivalent operational definitions: (a) the temperature where $B = 0$, (b) the temperature where the interaction parameter $\chi = 1/2$, (c) the limit of the critical temperature, T_c, as $M \to \infty$, and (d) a solvent in which $R_g \sim M^{1/2}$. Physically, a theta solvent is one in which the polymer–solvent interactions are rather unfavorable, so that the chain shrinks to its random-walk dimensions. This contraction cancels the effect of the excluded volume interactions, which otherwise swell the chain to self-avoiding conformations: $R_g \sim M^{3/5}$.

6. The Flory–Huggins parameter χ can be related to thermodynamic quantities such as the heat of vaporization and the cohesive energy density. Only for systems with very weak intermolecular interactions can one hope to estimate χ reliably based on tabulated thermodynamic quantities, and even then there needs to be a substantial, empirical correction term for polymer solutions. This motivates the use of an effective interaction function that can be used to fit experimental data; the resulting χ_{eff} has become a commonly used scheme to describe the excess free energy of mixing.

Problems

1. In the derivation of Equation 7.3.12 it is assumed that each polymer segment is surrounded by z sites which are occupied at random by either solvent molecules or polymer segments. Actually, this is true of only $(z - 2)$ of the sites in the coordination sphere—$(z - 1)$ for chain ends—since two of the sites are occupied by polymer segments which are covalently bound to other polymer segments. Criticize or defend the following proposition concerning this effect: The kinds of physical interactions that we identify as London or spontaneous dipole–dipole attractions can also operate between segments which are covalently bonded together, so the w_{22} contribution continues to be valid. A slight error in counting is made—to allow for simplification of the resulting function—but this is a tolerable approximation in concentrated solutions. In dilute solutions the approximation introduces more error, but the model is in trouble in such solutions anyhow, so another approximation makes little difference.

2. Show that

$$T_c = (2/R)\,\widehat{V}_1 (\delta_1 - \delta_2)^2$$

where T_c is the critical temperature for phase separation. For polystyrene with $M \cong 3 \times 10^6$, Shultz and Flory observed T_c values of 68°C and 84°C, respectively, for cyclohexanone and cyclohexanol. Values of $\widehat{V}_1$ for these solvents are about 108 and 106 cm^3/mol, and δ_1 values are 9.9 and 11.4 (cal/cm^3)$^{1/2}$, respectively. Use each of these T_c values to form separate estimates of δ_2 for polystyrene and compare the calculated values with each other and with the value of δ_2 from Table 7.1. Comment on the agreement or lack thereof for the calculated and accepted δ's in terms of the assumptions inherent in this method. If Equation 7.6.7 was used instead of Equation 7.6.6, does the agreement among the δ_2 values improve?

3. The term $(1/2 - \chi)$ that appears, for example, in the Flory–Huggins expression for B (Equation 7.4.18) is sometimes replaced by

$$\frac{1}{2} - \chi = \psi \left(1 - \frac{\Theta}{T} \right)$$

where Θ is the theta temperature (K) and ψ is a new parameter. This substitution has the effect of replacing one parameter (χ) with two (ψ and Θ). Why might one do this? How do ψ and Θ fit into the discussion of effective χ parameters in Section 7.6.3?

4. The osmotic pressure of polystyrene fractions in toluene and methyl ethyl ketone was measured by Bawn et al.[†] at 25°C, and the following results were obtained. Make plots of Π/c versus c, and evaluate M_n for the three fractions in an appropriate way. Do the results make sense? (Hint: If they do not, perhaps a more sophisticated analysis would help.) What can you say about the quality of these solvents?

Fraction	Toluene c	Toluene Π	MEK c	MEK Π
I	4.27	0.22	2.67	0.04
	6.97	0.58	6.12	0.14
	9.00	1.00	8.91	0.31
	10.96	1.53		
II	1.55	0.16	3.93	0.40
	2.56	0.28	8.08	0.95
	2.93	0.32	10.13	1.30
	3.80	0.47		
	5.38	0.77		
	7.80	1.36		
	8.68	1.60		
III	1.75	0.31	1.41	0.23
	2.85	0.53	2.90	0.48
	4.35	0.88	6.24	1.11
	6.50	1.49	8.57	1.63
	8.85	2.36		

The units of c are mg/mL, and Π is in g/cm^2. (This unit of g/cm^2 is a bit old fashioned; to convert Π to dyn/cm^2, multiply by the acceleration due to gravity, $g = 980$ cm/s^2.)

5. Write down by inspection the Flory–Huggins theory prediction for the free energy of mixing of a ternary solution (polymer A, polymer B, and solvent).

6. The osmotic pressure of solutions of polystyrene in cyclohexane was measured at several different temperatures, and the following results were obtained:[‡]

	$T = 24°C$	
Fraction	c (g/cm^3)	$\Pi/RTc \times 10^6$ (mol/g)
II	0.0976	8.0
	0.182	6.0
	0.259	8.7

	$T = 34°C$	
Fraction	c (g/cm^3)	$\Pi/RTc \times 10^6$ (mol/g)
II	0.0081	13.3
	0.0201	14.2
	0.0964	14.2
	0.180	18.7
	0.257	26.2

[†] C. Bawn, R. Freeman, and A. Kamaliddin, *Trans. Faraday Soc.*, 46, 862 (1950).
[‡] W.R. Krigbaum and D.O. Geymer, *J. Am. Chem. Soc.*, 81, 1859 (1959).

III	0.0156	2.46
	0.0482	2.24
	0.0911	3.42
	0.126	4.96
	0.139	6.05

		$T = 44°C$	
Fraction	c (g/cm^3)	$\Pi/RTc \times 10^6$ (mol/g)
II	0.0959	18.6
	0.178	28.1
	0.255	40.0
III	0.0478	5.50
	0.125	11.0
	0.138	13.2

Plot all of these data on a single graph as Π/RTc versus c, connecting the points so as to present as coherent a display of the results as possible. Evaluate the molecular weights of the two polystyrene fractions. Criticize or defend the following proposition: These data show that the Θ temperature for this system is about 34°C. As expected, the range of concentrations which are adequately described by the first two terms of the virial equation is less for sample III than for sample II. Above this range other contributions to nonideality contribute positive deviations from the two-term osmotic pressure equation. It would be interesting to see how this last effect appears for sample III at 24°C, but this measurement was probably impossible to carry out owing to phase separation.

7. By combining Equation 7.5.23 and Problem 3, the following relationship can be obtained between the critical temperature T_c for phase separation and the degree of polymerization:

$$\frac{1}{T_c} = \frac{1}{\Theta} + \frac{1}{\Theta\psi}\left(\frac{1}{\sqrt{N}} + \frac{1}{2N}\right)$$

Derive this relationship and explain the graphical method it suggests for evaluating Θ and ψ. The critical temperatures for precipitation for the data shown in Figure 7.14b are the following:

M (g/mol)	22,700	285,000	6,000,000
T_c (°C)	18.2	45.9	56.2

Use the graphical method outlined above to evaluate Θ, ψ, and χ for polyisobutylene in diisobutylketone.

8. Assume χ for a polymer–solvent system followed Equation 7.6.12. There are four possible cases according to whether the parameters α and β were positive or negative. What possible phase diagrams could be observed in each case, i.e., UCST only, LCST only, both, neither, etc.

9. Fill in the steps omitted in the text to derive the relations for the critical mole fraction and the critical χ for regular solution theory (see Equation 7.5.15).

10. Find the Flory–Huggins expression for $\Delta\mu_2(\phi_1)$, and then find the range of χ for which there can be two physically meaningful values of ϕ_1 for which $\Delta\mu_2$ is the same. (Hint: Take the large N limit, but at the correct moment.) What is the significance of this range of χ?

11. A polymer–solvent mixture has a critical temperature of 300 K. The polymer is monodisperse and has a molar volume equal to three times that of the solvent. If the solvent were replaced

by its dimer, what does Flory–Huggins theory predict for the new critical temperature? (Hints: It is a good idea to figure out which way T_c should change before you do the details. Also, remember that although χ is dimensionless, it does depend on the chosen site volume.)

12. Suppose you have just developed a synthesis method for a new, flexible polymer, capable of producing high-molecular-weight samples. Provide a step-by-step outline of how you would go about finding a theta solvent for this polymer, and then locating the theta temperature precisely. Note that although there are many ways to proceed, some will be much more labor-intensive than others. Try to minimize the amount of effort required.

13. The Flory–Huggins theory may be extended to a binary mixture of different polymers, i.e., an A/B blend. The resulting free energy of mixing can be written

$$\Delta G_m = kT \left\{ \frac{\phi_A}{N_A} \ln \phi_A + \frac{\phi_B}{N_B} \ln \phi_B + \chi \phi_A \phi_B \right\}$$

where N_A and N_B are the respective degrees of polymerization. Find the critical point (χ_c and ϕ_c) in terms of N_A and N_B. Show that your results reduce to the solution case when $N_B \rightarrow 1$. Compare the solution and symmetric blend (i.e., $N_A = N_B$) results in the limit of infinite molecular weight; what is the crucial difference?

14. The results from Problem 13 imply that the critical temperature for a symmetric polymer blend should increase linearly with N. Confirm this result. What would the N dependence be for the critical temperature if in fact χ followed Equation 7.6.12?

15. Prove that for a polymer solution in the athermal limit ($\Delta H_m = 0$), no phase separation can occur in the Flory–Huggins model. (Hint: The sign of ΔG_m is not enough.)

16. Consider the Flory–Huggins theory as applied to an AB statistical copolymer dissolved in solution. It is possible to apply the same expression for the free energy of mixing as for a homopolymer, if a new effective χ parameter is employed. Develop a simple relation for χ_{eff} in terms of f_A, χ_A, χ_B, and χ_{AB}, where f_A is the volume fraction of A units in the copolymer, and the three χ's refer to polymer A–solvent, polymer B–solvent, and polymer A–polymer B interactions, respectively. Use mean-field approximations in the same spirit as in the original theory. The important feature is that there are unfavorable A–B interactions in the pure statistical copolymer that are diminished in dilute solution. If your result is correct, you can use it to show that, in principle, it should be possible to find a solvent that dissolves the statistical copolymer but neither of the constituent homopolymers, at constant M.

17. Extend the derivation for the excluded volume exponent ($\nu = 0.6$) given in Section 7.7 to a chain in a space of dimensionality d. What is ν in two dimensions; why is it different from 0.6? What does the result for four dimensions tell you?

18. Repeat Problem 8 for a polymer–polymer blend at high molecular weight.

19. For a polydisperse sample, what average degree of polymerization would apply in the corresponding Flory–Huggins expression for the spinodal? Prove your answer; one way to do this is to find the spinodal for a binary mixture of two degrees of polymerization.

20. Recast Equation 7.7.10 in terms of the second virial coefficient, $B(T)$.

References

1. Flory, P.J., *J. Chem. Phys.*, 10, 51 (1942).
2. Huggins, M.L., *J. Am. Chem. Soc.*, 64, 1712 (1942).
3. Flory, P.J., *Principles of Polymer Chemistry*, Cornell University Press, Ithaca, NY, 1953.
4. de Gennes, P.G., *Scaling Concepts in Polymer Physics*, Cornell University Press, Ithaca, NY, 1979.
5. Krigbaum, W.R. and Flory, P.J., *J. Am. Chem. Soc.*, 75, 1775 (1953).

Further Readings

Flory, P.J., *Principles of Polymer Chemistry*, Cornell University Press, Ithaca, NY, 1953.
Graessley, W.W., *Polymeric Liquids and Networks: Structure and Properties*, Garland Science, New York, 2003.
Koningsfeld, R., Stockmayer, W.H., and Nies, E., *Polymer Phase Diagrams*, Oxford University Press, New York, 2001.
Kurata, M., *Thermodynamics of Polymer Solutions*, Harwood, New York, 1982.
Rubinstein, M. and Colby, R.H., *Polymer Physics*, Oxford University Press, New York, 2003.
Tanford, C., *Physical Chemistry of Macromolecules*, Wiley, New York, 1961.
Yamakawa, H., *Modern Theory of Polymer Solutions*, Harper & Row, New York, 1971.

8

Light Scattering by Polymer Solutions

8.1 Introduction: Light Waves

In this chapter we will explore the phenomenon of light scattering from dilute polymer solutions. Light scattering is an important experimental technique for polymers for several reasons. First, it provides a direct, absolute measurement of the weight average molecular weight, M_w. Second, it gives information about polymer–polymer and polymer–solvent interactions, through the second virial coefficient, B (introduced in Chapter 7). Third, under many circumstances light scattering can be used to determine the radius of gyration, R_g (described in Chapter 6) without any prior knowledge about the shape of the molecule (e.g., coil, rod, globule). Fourth, the description of the scattering process that we will develop in this chapter may be readily adapted to x-ray and neutron scattering, two other techniques in common use in polymer science. In short, the wealth of information that may be obtained from light scattering more than justifies the effort we will need to expend in this chapter to understand how it works. As a final introductory comment, the fact that light scattering can determine M_w and B tells us that it is a thermodynamic measurement. The fact that it can measure R_g tells us that it is a structural tool as well.

A light beam may be described as a traveling electromagnetic wave. For our purposes (i.e., polymers and solvents containing mostly C, H, O, and N atoms), the magnetic component of the wave is of no consequence, and so we can represent the wave as

$$\vec{E} = \vec{E}_o \cos\left(\omega t - \vec{k} \cdot \vec{r}\right) \tag{8.1.1}$$

where $\vec{E}_o$ is the *amplitude* of the electric field, ω is the *frequency* in rad/s, t is time, $\vec{r}$ is the position, and $\vec{k}$ is the *wavevector*. This is illustrated in Figure 8.1. Several comments about Equation 8.1.1 are appropriate.

1. The amplitude $\vec{E}_o$ is itself a vector. If we take the wave to be travelling along the x direction, $\vec{E}_o$ lies in the y–z plane. If $\vec{E}_o$ lies exclusively along one axis, say z, then the beam is said to be z-polarized or *vertically polarized*; on the other hand, if $\vec{E}_o$ is uniformly distributed in the y–z plane, then the wave is *unpolarized*.
2. The frequency can also be written

$$\omega = 2\pi\nu = \frac{2\pi c}{\lambda_0} \tag{8.1.2}$$

where ν is the frequency in cycles per second (Hz), c is the speed of light in vacuum, and λ_0 is the wavelength in vacuum.
3. In the photon picture of light, the energy carried by each photon $E = h\nu$, where h is Planck's constant (6.63×10^{-34} Js).

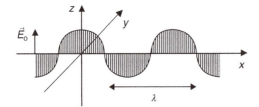

Figure 8.1 Schematic of the electric field component of a z-polarized light wave propagating in the x direction.

4. The phase factor, $\vec{k} \cdot \vec{r}$, is the projection of the wavevector $\vec{k}$ onto the particular position vector of interest. The wavevector points in the direction of propagation, x in this instance, and has amplitude $2\pi/\lambda$. Thus, our wave could be rewritten as

$$\vec{E}(x,t) = \vec{E}_o(y,z)\cos(\omega t - kx) = \vec{E}_o \cos\left(2\pi\left[\nu t - \frac{x}{\lambda}\right]\right) \tag{8.1.3}$$

The essence of the traveling wave is that if we consider a particular instant in time (ωt fixed), then the wave oscillates in space with wavelength λ; and if we consider a particular point in space (kx fixed) the wave oscillates in time with frequency ω. It will turn out that it is the spatial dependence, $\vec{k} \cdot \vec{r}$ or kx, that will play the crucial role in scattering experiments.

For visible light $\lambda_0 \approx 3500$–7000 Å, and consequently $\nu = c/\lambda_0 \approx 10^{15}$ Hz (recall $c = 3.0 \times 10^8$ m/s). This means that the oscillations in the electric field amplitude are much too rapid for detectors to follow: photomultiplier tubes, photodiodes, and other photodetectors typically have time constants on the order of nanoseconds. Instead, the detector actually integrates in time the incident intensity, I (energy/area/time), which is proportional to $|\vec{E} \cdot \vec{E}| = |\vec{E}|^2$; such detectors are called *square-law detectors*. We will not worry about the proportionality factors between intensity and electric field squared, because in scattering experiments we will always use the ratio of the scattered intensity to the incident intensity, thereby canceling out these prefactors. (Even if we did not do that, the detectors actually generate an electrical current proportional to the incident intensity, and it is rare that anyone goes through the trouble of converting this signal into the actual incident intensity. Rather, we can calibrate the electrical signal with a reference intensity, as will be described in Section 8.7.)

We could continue to use the cosine representation of the light wave as in Equation 8.1.3, but it turns out the arithmetic is much more convenient if we use complex notation. If you do not recall how complex numbers work, they are reviewed in the Appendix. The advantages stem primarily from the fact that complex numbers allow us to factor out the temporal and spatial dependences of the wave amplitude, and that squaring the amplitude becomes much easier. Accordingly, the wave can be written as

$$\vec{E} = \text{Re}\left\{\vec{E}_o \exp\left[i\left(\omega t - \vec{k} \cdot \vec{r}\right)\right]\right\} = \text{Re}\left\{\vec{E}_o \exp[i\omega t] \exp\left[-i\vec{k} \cdot \vec{r}\right]\right\} \tag{8.1.4}$$

where Re{...} means the *real part* of the complex argument. The intensity is proportional to

$$I \sim \left|\vec{E}^* \cdot \vec{E}\right| = \left|\vec{E}_o \cdot \vec{E}_o\left\{e^{i\omega t}e^{-i\omega t}e^{-i\vec{k} \cdot \vec{r}}e^{i\vec{k} \cdot \vec{r}}\right\}\right| = \left|\vec{E}_o \cdot \vec{E}_o\right| \tag{8.1.5}$$

where $\vec{E}^*$ is the *complex conjugate* of $\vec{E}$, obtained by replacing i ($= \sqrt{-1}$) with $-i$. The use of the complex conjugate guarantees that I is a real number, because $\exp(ix)\exp(-ix) = 1$. Generally, we forget about saying the wave is the real part of the complex form, and just remember that at the end of a calculation we convert to intensities by Equation 8.1.5.

Table 8.1   Refractive Indices of Common Polymers and Solvents

Polymer	n	Solvent	n
Poly(tetrafluoroethylene)	1.41	Methanol	1.326
Poly(dimethylsiloxane)	1.43	Water	1.333
Poly(ethylene oxide)	1.46	Ethanol	1.359
Atactic polypropylene	1.47	Ethyl acetate	1.370
Amorphous polyethylene	1.49	n-Hexane	1.372
Poly(vinyl acetate)	1.49	Tetrahydrofuran	1.404
Poly(methyl methacrylate)	1.49	Cyclohexane	1.424
1,4-Polyisoprene	1.50	Chloroform	1.444
Polyisobutylene	1.51	Toluene	1.494
1,4-Polybutadiene	1.52	Bromobenzene	1.557
Polystyrene	1.59	Carbon disulfide	1.628

Refractive indices for $\lambda_o = 5893$ Å, at 20°C or 25°C.

Source: From Brandrup, J. and Immergut, E.H. (Eds.), *Polymer Handbook*, 3rd ed., Wiley, New York, 1989.

Equation 8.1.1 represents a solution to Maxwell's equations in a homogeneous medium. The frequency of the wave (and also the energy of the equivalent photon, $h\nu$) is independent of the medium, but the wavelength is not. The wavelength in the material, λ, relative to the wavelength in vacuum, λ_0, is determined by a material property called the *refractive index n*:

$$n = \frac{\lambda_0}{\lambda} = \frac{c}{v} \tag{8.1.6}$$

where v is the speed of light in the material. Thus the amplitude of the wavevector, $k = |\vec{k}|$, is often written as $2\pi n/\lambda_0$. The refractive index is determined, in turn, by the polarizability of the constituent molecules α and their spatial arrangement (density and orientation). Qualitatively, the polarizability reflects the ability of the incident electric field to distort the electronic distribution within a molecule, and this distortion, in turn, reduces c to v. Consequently, more polarizable chemical moieties, such as aromatic rings, generally lead to higher refractive indices than less polarizable groups, such as $-CH_3$ or $-CH_2-$. In the liquid state, all orientations of the individual molecules are equally probable, and we say the liquid is *isotropic*. (In this case we need not worry about the fact that most molecules are actually *anisotropic*: the polarizability is different along different molecular axes). The magnitude of n ranges from about 1.3 to 1.6 for common polymers and aqueous or organic solvents as can be seen from the representative values in Table 8.1. The relevant equation that relates the material property, n, to the molecular property, α, is a version of the Lorentz–Lorenz equation for an ideal gas (also known as the Clausius–Mosotti equation):

$$\alpha \approx \frac{n^2 - n_s^2}{4\pi} \Psi \tag{8.1.7}$$

where n is the refractive index of the solution, n_s is the refractive index of the pure solvent, and $1/\Psi$ is the number of solute particles per unit volume.

8.2   Basic Concepts of Scattering

Scattering is the reradiation of a traveling wave due to a change in the character of the medium in which the wave is propagating. The mathematical description of scattering could be largely developed without specifying whether we are talking about light waves, sound waves, electrical signals, or waves on a lake, as there is a great deal of commonality to all these phenomena. For light, scattering will be caused by local changes in refractive index or polarizability, due to, for example, a dust particle in the air or to a polymer in the solvent. We will only consider nonabsorbing media here, so the incident light intensity will either be transmitted or scattered, but not absorbed. In general, the scattered wave propagates in all directions, i.e., it is a spherical wave in three dimensions or a circular wave in two. If the incident and scattered frequencies are the same, i.e., no energy is exchanged between the medium and the wave, the scattering process is classified as *elastic*. We will only consider elastic scattering in this chapter. (There is also *quasielastic* scattering, in which very small differences in energy are detected; this is the basis of the powerful technique of dynamic light scattering, which will be described briefly in Chapter 9. Raman and Brillouin scattering are examples of *inelastic* processes, where the light exchanges vibrational or rotational quanta with the molecules in the former, and exchanges energy with traveling density waves or *phonons* in the latter.) Lastly, scattering is termed *incoherent* if the intensity is independent of the scattering angle, and *coherent* if it depends on the scattering angle. The origin of this terminology will become apparent in the following discussion, but generically coherence implies that there exists some particular phase relationship among different waves.

8.2.1 Scattering from Randomly Placed Objects

Imagine that we have scattering from randomly placed, noninteracting objects; we should get a total scattered intensity, I_s, proportional to the number of scattering objects:

$$I_s \sim (\text{number of scatterers}) \times (\text{scattering power of each object}) \qquad (8.2.1)$$

for the case of molecules in solution. We anticipate that bigger molecules will scatter proportionally more than smaller ones. On the other hand, suppose we scatter light from objects that are connected to one another, e.g., monomers within one polymer. Now the phases of the waves scattered from different monomers should be related, because the distance r_{jk} between any two monomers j and k has some preferred value or range of values. This will lead to some interference between these waves and a net loss of scattered intensity. Recall that interference between two waves is *constructive* only if the difference in distance that the two waves travel to the detector is some integral number of wavelengths, m; in other words, when $kr = m\lambda$. The interference will be completely *destructive* if the path difference is an integral number plus one half of λ: $kr = (m + 1/2)\lambda$. We can expect, therefore, that in order to see significant destructive interference we need the distance between scatterers to be a significant fraction of λ, say a few percent. This will turn out to be the case. For visible light with $\lambda_0 \approx 5000$ Å, only for polymers that are larger than at least 100 Å do we have to worry about this kind of interference. You might recall from Example 6.2 that this size typically corresponds to molecular weights in excess of 10^5 g/mol.

8.2.2 Scattering from a Perfect Crystal

Imagine we have an absolutely perfect, regular array of scatterers (atoms, if you like) placed every 1 Å apart, and we shine $\lambda = 5000$ Å light on them (Figure 8.2a). What will the scattering look like? Each atom will radiate a scattered wave with the same amplitude and wavelength, but the net scattered intensity will be zero. Why? If we select any particular scatterer and particular detection angle, we can always find another scatterer that is exactly $\lambda/2$ further away from the detector. The two waves from these scatterers will cancel each other. We can do this because the array is perfectly regular, and because the array is essentially a continuum ($1 \ll 5000$ Å). The only direction where this argument does not apply is forward, i.e., the scattering angle $\theta = 0°$. Here the phase shift of the scattered waves between two atoms is exactly canceled by the phase shift between the incident waves arriving at the two atoms. Thus the light beam propagates happily straight through the material. The important lesson is this: *there is no scattering from a perfectly homogeneous material*.

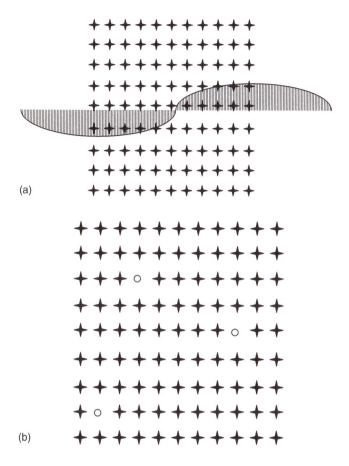

Figure 8.2   (a) A light wave propagating through a perfectly regular array of scatterers will not be scattered if the distance between neighbors is much less than λ. (b) Random fluctuations in the perfect array lead to incoherent scattering.

8.2.3   Origins of Incoherent and Coherent Scattering

There are two ways by which we could generate scattering from our hypothetical array. One would be to remove a few scatterers at random (Figure 8.2b). Then the "pairing-off" argument fails because each atom we remove used to cancel a scattered wave from some other atom, but now it cannot. Thus we conclude that *random fluctuations* in an otherwise homogeneous medium give rise to scattering. This scattering is incoherent: because the fluctuations are random by construction, on average there can be no phase relation between them.

The second way of obtaining scattering from our hypothetical array would be to make λ close to the distance between the scatterers. For our example, if we use x-rays, where $\lambda = 1.54$ Å is a typical value, then the pairing-off argument fails again. There will be particular angles in which planes of atoms are separated by integral multiples of λ and the scattering from different atoms will be in phase, i.e., coherent. This process is called *Bragg diffraction*, as illustrated in Figure 8.3 and described below. The point we want to emphasize now is that there will be coherent scattering whenever there is a spatial correlation between scattering objects on a distance scale comparable to λ. Bragg diffraction from atomic crystals gives sharp scattering peaks because the spatial coherence is very high; the crystal structure is regular over large distances (see Chapter 13). In scattering from polymers, we will find that the spatial coherence is lower, but not vanishing; for example,

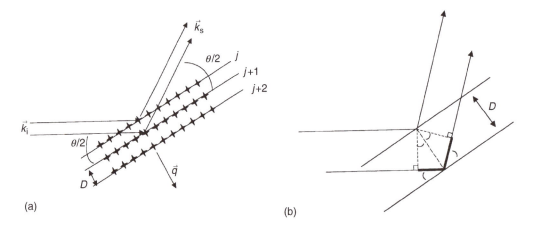

Figure 8.3 Illustration of Bragg's law. (a) The incident light wave and the diffracted light wave each makes an angle $\theta/2$ with each plane in the crystal. The planes are separated by a distance D, and the direction of the scattering vector $\vec{q}$ is indicated. (b) The bold line segments correspond to the extra distance the wave travels in being scattered from the second plane. All four indicated angles are equal to $\theta/2$, and thus each of the bold line segments has length $D \sin(\theta/2)$.

within a coil, the likelihood of two monomers being separated by a certain distance might be given by the Gaussian distribution (Equation 6.7.12). This lower degree of spatial coherence will lead to scattering that is a smooth function of the scattering angle, rather than sharp peaks, but the underlying phenomenon is the same.

The main relationship we will derive in this chapter, known as the *Zimm equation* (Equation 8.5.18), reflects these two sources of scattering. A polymer solution would be a homogeneous medium, except for two things. There are random fluctuations in concentration that give rise to incoherent scattering, with intensity proportional to the product of concentration and molecular weight (Equation 8.4.24b). Then, large polymers have monomers with partially correlated spatial separations that are a significant fraction of λ. This correlation leads to interference and coherent scattering, which can be used to determine R_g (Equation 8.5.18). The development will proceed in three stages. First we will consider scattering from a single isolated atom or molecule, a result generally attributed to Lord Rayleigh. In the second stage we will apply this result to a dilute solution of small polymers, to obtain an expression for the incoherent scattering. Finally, we will consider larger molecules and the resulting coherent contribution. To conclude this overview section, we develop Bragg's equation and define the scattering vector $\vec{q}$.

8.2.4 Bragg's Law and the Scattering Vector

Consider a wave incident on a series of equally spaced, parallel planes of scatterers (which could be atoms, but need not be). The planes are separated by a distance D, and the angle between the incident wave direction and the scattering planes is $\theta/2$, as shown in Figure 8.3a. What is the relation among D, λ, and θ for there to be scattering (or diffraction, as it is called in this context) in the direction an angle θ away from the incident direction? The distance the light travels from the source to the detector is the same for all atoms in a given plane j, so there is no problem there. What is essential is that the distance traveled by waves scattered from the next further plane $j+1$ be $m\lambda$ larger, where m is an integer. Then the waves scattered from one plane will be in phase with the waves scattered from the next, and there will be constructive interference at the detector. By extension, waves from plane $j+2$ will travel $2m\lambda$ further than from plane j, and therefore will still be in phase, and so will waves from all the planes in the array. This condition is the basis of Bragg's law. To state this as an equation, consider the enlarged diagram in Figure 8.3b. By constructing the indicated perpendiculars, it is possible to see that there are two extra segments that the wave must travel when it is scattered off plane $j + 1$. The length of each of these segments is $D \sin(\theta/2)$. Thus Bragg's law can be written as

$$m\lambda = 2D \sin\left(\frac{\theta}{2}\right) \tag{8.2.2}$$

A central quantity in any scattering process is the *scattering vector*, defined by the difference between the incident and scattered wave vectors:

$$\vec{q} \equiv \vec{k}_i - \vec{k}_s \tag{8.2.3}$$

This is illustrated in Figure 8.4, where again θ is the angle between the scattered and incident waves. For elastic scattering, both incident and scattered wave vectors have magnitude $2\pi/\lambda = 2\pi n/\lambda_0$, and thus the magnitude of $\vec{q}$ can be obtained (see Figure 8.4):

$$|\vec{q}| \equiv q = 2\left(\frac{2\pi}{\lambda}\right) \sin\left(\frac{\theta}{2}\right) = \frac{4\pi}{\lambda} \sin\left(\frac{\theta}{2}\right) \tag{8.2.4}$$

What is the significance of $\vec{q}$? There are three aspects worth bringing out now:

1. It is also known as the *momentum transfer vector*. Although no energy is exchanged in elastic scattering, photons carry momentum and there must be a change in momentum when the propagation direction changes. The law of conservation tells us that momentum is transferred into the medium along the direction of $\vec{q}$. In fact, at sufficiently high intensity the resulting "radiation pressure" can cause particles to move.

2. The direction of $\vec{q}$, shown in Figure 8.3a, corresponds to the normal to the planes of scatterers. A vector pointing in this direction with amplitude $2\pi/D$ is called a *reciprocal lattice vector* (reciprocal because D is in the denominator), and when the magnitudes of these two vectors coincide

$$\frac{2\pi}{D} = \frac{4\pi}{\lambda} \sin\left(\frac{\theta}{2}\right) \tag{8.2.5}$$

we can see that Bragg's law (Equation 8.2.2) is automatically satisfied. In other words, the criterion for Bragg diffraction is that the scattering vector (determined by the apparatus) coincides with a reciprocal lattice vector (determined by the material).

3. The magnitude q has dimensions of inverse length. As will become clearer in Section 8.5 and Section 8.6, the scattering will be sensitive to structure in the solution on the length scale $1/q$, so the choice of q (which is fixed by choice of λ and θ) ultimately determines what kind of structural information a given scattering experiment can provide.

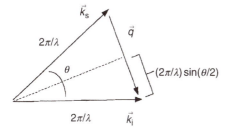

Figure 8.4   The length of the scattering vector is calculated in terms of the scattering angle θ and the magnitude of the incident and scattered wavevectors, $2\pi/\lambda$.

In the next two sections we will be concerned with incoherent scattering, and will not need to be concerned with the role of q, but then in Section 8.5 and Section 8.6 it will be of central importance.

8.3 Scattering by an Isolated Small Molecule

We begin with the fact that an incident light wave $\vec{E}_i$ will induce a dipole moment $\vec{\mu}$ in an atom or molecule, where

$$\vec{\mu} = \alpha \vec{E}_i = \alpha \vec{E}_o \exp[\mathrm{i}(\omega t - \vec{k} \cdot \vec{r})] \tag{8.3.1}$$

The induced dipole moment oscillates at the same frequency ω as the field. An oscillating dipole involves an accelerating charge, and will therefore radiate an oscillating electric field that we will call the scattered wave $\vec{E}_s$. The magnitude of this wave depends on the direction of observation (through an angle ϕ defined below), the speed of light, the distance from the dipole (r), the electron charge (e), and the acceleration of the charge in the dipole ($\vec{a}$):

$$\left|\vec{E}_s\right| = \frac{e|\vec{a}|}{rc^2} \sin \phi \tag{8.3.2}$$

The derivation of this equation comes from basic electromagnetism, but it involves some rather hairy vector calculus, so we omit it; it can be found in many introductory physics texts. However, we can understand the important inverse-dependence on r. The total energy of the scattered wave should be conserved. From Equation 8.3.2, we see that the intensity will be proportional to $|\vec{E}_s^* \cdot \vec{E}_s| \sim r^{-2}$. The total energy will be proportional to the intensity integrated over the surface of a sphere of radius r. As we go further away from the dipole and integrate $|\vec{E}_s^* \cdot \vec{E}_s|$ over the surface of the sphere, the area will increase as r^2, and thus the total energy will be independent of r.

Note that $(\vec{\mu}/e)$ has the dimensions of length, since the induced dipole is essentially the product of a charge and the distance of charge separation. Therefore we can write the magnitude of the acceleration as

$$|\vec{a}| = \left|\frac{\mathrm{d}^2}{\mathrm{d}t^2}\left(\frac{\vec{\mu}}{e}\right)\right| = \frac{\alpha}{e}\left|\frac{\mathrm{d}^2}{\mathrm{d}t^2}\vec{E}_i\right| = \frac{\alpha \omega^2}{e}\left|\vec{E}_i\right| \tag{8.3.3}$$

The ratio of the scattered intensity, I_s, to the incident intensity, I_o, is given by

$$\begin{aligned}
\frac{I_s}{I_o} &= \frac{\left|\vec{E}_s^* \cdot \vec{E}_s\right|}{\left|\vec{E}_i^* \cdot \vec{E}_i\right|} = \frac{e^2|\vec{a}|^2}{r^2 c^4} \sin^2 \phi \frac{1}{\left|\vec{E}_i^* \cdot \vec{E}_i\right|} \\
&= e^2 \left(\frac{\alpha^2 \omega^4}{e^2}\right) \frac{\left|\vec{E}_i^* \cdot \vec{E}_i\right| \sin^2 \phi}{r^2 c^4 \left|\vec{E}_i^* \cdot \vec{E}_i\right|} \\
&= \frac{16\pi^4 \alpha^2}{r^2 \lambda_0^4} \sin^2 \phi
\end{aligned} \tag{8.3.4}$$

where we have used $\omega = 2\pi\nu = 2\pi c/\lambda_0$, and as before the dot product is required in taking the square of the electric field vector.

If the incident wave travels along x, and is polarized along z, then ϕ is the angle of detection relative to the z-axis (see Figure 8.5a). There is no angular dependence in the $x-y$ plane since $\sin \phi = 1$, and there is no scattered wave in the vertical direction z ($\sin \phi = 0$). If we use vertically polarized light and detect scattering in the horizontal plane (currently the most commonly employed geometry), then $\sin \phi = 1$ and life is easy. The scattered intensity, normalized to the incident vertically polarized intensity, is given by

$$\frac{I_s}{I_{o,v}} = \frac{16\pi^4 \alpha^2}{r^2 \lambda_0^4} \tag{8.3.5}$$

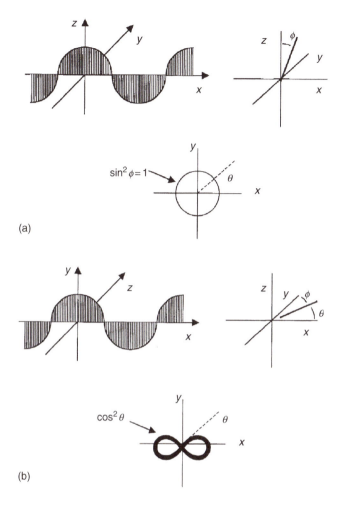

Figure 8.5 Illustration of the effect of incident beam polarization for a light wave propagating along x, and with scattering being detected in the $x-y$ plane. (a) Vertically (z) polarized beam; ϕ is the angle of scattering with respect to z, and $\sin\phi = 1$ for all θ in the $x-y$ plane. (b) Horizontally (y) polarized beam; ϕ is the angle of scattering with respect to y, and therefore $\sin\phi = \sin(\pi/2 - \theta) = \cos\theta$ in the $x-y$ plane. The dependence of the magnitude of $\sin\phi$ on θ is also illustrated for both cases.

Suppose, however, that we are dealing with unpolarized incident light, which is sometimes the case. This we can view as equal parts of vertically and horizontally polarized light. Now we define the scattering angle, θ, between the direction of the transmitted (unscattered) wave (x in this case) and the scattered wave to the detector in the $x-y$ plane. For the vertically polarized component of the incident beam, the scattered wave is the same for all θ. For the horizontally polarized part, which is y-polarized, there will be no scattering along y. Figure 8.5b shows that the $\sin\phi$ factor becomes $\sin(\pi/2 - \theta) = \cos\theta$ in our nomenclature. Thus the scattered field from the horizontal part varies as $\cos\theta$, and the scattered intensity as $\cos^2\theta$, whereas for the vertical part it is constant. The incident intensity was 50% of each, so the net intensity varies as $(1 + \cos^2\theta)/2$, and we insert this in Equation 8.3.4 to obtain

$$\frac{I_s}{I_{0,u}} = \frac{8\pi^4\alpha^2(1 + \cos^2\theta)}{r^2\lambda_0^4} \tag{8.3.6}$$

where the subscript u denotes unpolarized incident light. This equation is associated with the name of Lord Rayleigh, and such scattering from independent polarizable objects is called *Rayleigh scattering* [2]. One interesting feature of Equation 8.3.5 and Equation 8.3.6 is the dependence on λ_0^{-4}. This strong dependence means that shorter wavelength, higher frequency, or higher energy waves are scattered considerably more. In the atmosphere, molecules and particles scatter blue light down to the earth preferentially over red light, and the sky overhead appears blue. On the other hand, when the sun is low in the sky, the blue light is scattered away from the observer and the remaining, transmitted light has a reddish hue.

We will use Equation 8.3.5 throughout the rest of this chapter, but it is important to remember that we are assuming vertically polarized incident light, and that the scattering is detected in the horizontal plane.

Example 8.1

Can we measure the light scattering from a single water molecule in empty space?

Solution

We begin by applying Rayleigh's result, Equation 8.3.5, assuming polarized light for simplicity. We need to know the distance to the detector, r, the wavelength of light, λ_0, and the polarizability of water, α. Let us assume we place our detector 1 m from the molecule, and that we are using a 1 W argon laser with $\lambda_0 = 488$ nm. The average polarizability of water is given as 1.65×10^{-24} cm^3 [3]. However, it is also instructive to estimate it via Equation 8.1.7, the Lorentz–Lorenz equation. The refractive index of water is about 1.333 (see Table 8.1), its molecular weight is 18 g/mol, and its density is about 1.0 g/cm^3, so Equation 8.1.7 gives

$$\alpha \cong \frac{(1.333)^2 - 1^2}{4 \times 3.14} \times \frac{18}{1.0 \times 6 \times 10^{23}} = 1.86 \times 10^{-24} \text{ cm}^3$$

which is pretty close to the tabulated value. Using the tabulated value in Equation 8.3.5 we have

$$\frac{I_s}{I_0} \cong \frac{16 \times (3.14)^4 \times (1.65 \times 10^{-24})^2}{(100)^2 \times (488 \times 10^{-7})^4} = 7.5 \times 10^{-32}$$

after taking care to put all the lengths in centimeters. This is a very small number, which does not look promising. But, even though it is a very small fraction, how much light is going in? Here it is easiest to adopt the photon picture. A 1 W laser emits 1 J/s; how many photons is that? Here we recall the discussion following Equation 8.1.2, and find that 1 photon at this wavelength gives

$$E = \frac{hc}{\lambda_0} = \frac{6.63 \times 10^{-34} \times 3 \times 10^8}{488 \times 10^{-9}} = 4.1 \times 10^{-19} \text{ J/photon}$$

Taking 1 J/s $\div$ (4.1×10^{-19}) J/photon $= 2.5 \times 10^{18}$ photons/s. This is a lot of photons, when we bear in mind that sensitive photodetectors can count just a few photons per second; but it clearly is not enough: $2.5 \times 10^{18} \times 7.5 \times 10^{-32} \approx 2 \times 10^{-13}$ is still a very small number. Furthermore, the calculation gives the total scattered intensity over the surface of a sphere of radius 1 m, whereas our detector would only collect some small fraction of that. On the other hand, it is worth remembering that if our laser beam encountered a mole of water molecules, the story would be quite different. And, in particular, sunlight traverses great distances through the atmosphere and encounters many moles of gas phase species, so atmospheric scattering is far from negligible.

8.4   Scattering from a Dilute Polymer Solution

In this section we adapt the Rayleigh scattering equation for an isolated object of polarizability α, Equation 8.3.5, to the case of a dilute, nonabsorbing polymer solution. Our vertically polarized incident light beam will illuminate some region of the sample solution, and the detection optics will be arranged to collect light from some portion of the illuminated region; we call this portion the *scattering volume*. Now we divide the scattering volume into a large number of imaginary "cells" of volume Ψ, with the following criteria:

1. $\Psi^{1/3} \ll \lambda$, so that each cell is effectively a point scatterer.
2. Each cell contains many monomers, with a concentration c subject to statistical fluctuations.
3. The cells are statistically independent, i.e., the fluctuations in any one cell are uncorrelated with those in any other.

These assumptions are not too restrictive, except that they collectively imply that R_g (and therefore the molecular weight) of the polymer is not too big. We will deal with the large molecule case in the subsequent sections. Now each cell will have an instantaneous polarizability, which can be expressed as the sum of an average value, $\langle \alpha \rangle$, plus a fluctuation, $\delta \alpha$:

$$\alpha = \langle \alpha \rangle + \delta \alpha \tag{8.4.1}$$

The average could be either the time average for one cell, or the ensemble average for the scattering volume, as these two should be equivalent (recall from Chapter 6 that such a system is said to be *ergodic*). In the actual measurement, we will record the signal for a finite amount of time, therefore performing a time average, and we will record scattering from all the cells in the scattering volume, therefore performing an ensemble average as well. From Equation 8.3.5, we know that the average scattering from any cell will depend on the average of the squared polarizability:

$$\langle \alpha^2 \rangle = \left\langle \left((\langle \alpha \rangle + \delta \alpha)^2 \right) \right\rangle$$
$$= \left\langle \langle \alpha \rangle^2 \right\rangle + 2\langle \alpha \rangle \langle \delta \alpha \rangle + \left\langle (\delta \alpha)^2 \right\rangle \tag{8.4.2}$$

There are three terms on the right hand side of Equation 8.4.2. The first will not contribute to the net scattering from the solution, because it is the same for every cell. By the argument given in Section 8.2, a completely uniform material does not scatter. The second term is identically zero, because by definition $\langle \delta \alpha \rangle = 0$; the fluctuations are equally likely to be positive or negative. Consequently, we reach the very important conclusion that the scattering is determined entirely by the *mean-square fluctuations in polarizability*, $\left\langle (\delta \alpha)^2 \right\rangle$.

Now we need to relate $\delta \alpha$ to fluctuations in the thermodynamic variables p, T, and c:

$$\delta \alpha = \left(\frac{\partial \alpha}{\partial p} \right)_{T,c} \delta p + \left(\frac{\partial \alpha}{\partial T} \right)_{p,c} \delta T + \left(\frac{\partial \alpha}{\partial c} \right)_{T,p} \delta c \tag{8.4.3}$$

and we simplify this through a very important assumption, that the scattering from fluctuations in pressure (δp) and temperature (δT) is the same in the neat solvent as in the dilute solution. Consequently, when we consider the *excess* scattered intensity, $I_{ex} = I_s^{solution} - I_s^{solvent}$, only the concentration fluctuations matter. With these developments Equation 8.3.5 can be transformed into

$$\frac{I_{ex}}{I_0} = \frac{16\pi^4}{r^2 \lambda_0^4} \left\langle (\delta \alpha)^2 \right\rangle \frac{1}{\Psi} = \frac{16\pi^4}{r^2 \lambda_0^4} \left(\frac{\partial \alpha}{\partial c} \right)_{T,p}^2 \left\langle (\delta c)^2 \right\rangle \frac{1}{\Psi} \tag{8.4.4}$$

where $1/\Psi$ is the number of cells per unit volume. At this stage we need to work on two terms, $(\partial \alpha / \partial c)^2$ and $\langle (\delta c)^2 \rangle$. The former is transformed using the Lorentz–Lorenz equation (Equation 8.1.7):

$$\alpha = \Psi \frac{n^2 - n_s^2}{4\pi} \tag{8.4.5}$$

Therefore the concentration derivative is

$$\left(\frac{\partial \alpha}{\partial c} \right)_{T,p} = \frac{n\Psi}{2\pi} \left(\frac{\partial n}{\partial c} \right)_{T,p} \tag{8.4.6}$$

The key part of this relation is the so-called *refractive index increment*, $\partial n/\partial c$ (where we drop the reminder about constant T and p from now on), which can either be measured precisely using a differential refractometer, or looked up in tables, as will be discussed in Section 8.7.

Now we return to the concentration fluctuation term, $\langle(\delta c)^2\rangle$. We can say that

$$\langle(\delta c)^2\rangle = \frac{\int\limits_{-\infty}^{\infty}(\delta c)^2 P(\delta c)\,\mathrm{d}\delta c}{\int\limits_{-\infty}^{\infty}P(\delta c)\,\mathrm{d}\delta c} \tag{8.4.7}$$

where $P(\delta c)$ is the probability of a given fluctuation δc. As positive and negative fluctuations are equally probable, P must be symmetric about zero (just like the Gaussian distribution for the end-to-end vector in Equation 6.7.1). The size of a fluctuation is related to the associated fluctuation in free energy, δG, through the Boltzmann factor:

$$P(\delta c) = A\exp\left[\frac{-\delta G}{kT}\right] \tag{8.4.8}$$

and we can expand δG as a Taylor series (see Appendix):

$$\delta G = \left(\frac{\partial G}{\partial c}\right)_{T,p}\delta c + \frac{1}{2!}\left(\frac{\partial^2 G}{\partial c^2}\right)_{T,p}(\delta c)^2 + \cdots \tag{8.4.9}$$

Note that the first term in Equation 8.4.9 is not symmetric about $\delta c = 0$, and therefore does not contribute. If we insert the remaining $\partial^2 G/\partial c^2$ term into the exponential and perform the integrals in Equation 8.4.7 using the formula cited in Section 6.7, we arrive at the very simple relation (see also Problem 3)

$$\langle(\delta c)^2\rangle = \frac{kT}{(\partial^2 G/\partial c^2)_{T,p}} \tag{8.4.10}$$

which relates the concentration fluctuations to the associated free-energy penalty. The larger the cost in free energy, the smaller the average fluctuations will be. (Recall from Section 7.5 that for a system at equilibrium, i.e., one that is stable, $\partial^2 G/\partial c^2$ must be positive.) However, if the solution approaches a stability limit or spinodal, the denominator tends to zero and the fluctuations, and therefore the scattering, can get very big indeed. The numerator of Equation 8.4.10 just acknowledges the fact that the more thermal energy is available, the larger the fluctuations will tend to be.

We can now recast Equation 8.4.4 as

$$\frac{I_{ex}}{I_o} = \frac{4\pi^2 n^2 \Psi}{r^2 \lambda_0^4}\left(\frac{\partial n}{\partial c}\right)^2 \frac{kT}{(\partial^2 G/\partial c^2)_{T,p}} \tag{8.4.11}$$

by incorporating Equation 8.4.6 and Equation 8.4.10. The last transformation we need is to relate $\partial^2 G/\partial c^2$ to a virial expansion appropriate for dilute solutions (see Section 7.4). A slight complication arises because we have been working with c as the concentration variable, and we need to return to numbers of moles, n_i, in order to handle the chemical potential. The solution volume, V, can be written in terms of the partial molar volumes (see Equation 7.1.8):

$$V = n_1\bar{V}_1 + n_2\bar{V}_2 \tag{8.4.12a}$$

and because we need not worry about fluctuations in volume, $\mathrm{d}V = 0$, and therefore

$$\mathrm{d}n_1 = -\frac{\bar{V}_2}{\bar{V}_1}\mathrm{d}n_2 \tag{8.4.12b}$$

Since

$$c = \frac{n_2 M}{V} \qquad (8.4.13a)$$

then

$$\frac{dn_2}{dc} = \frac{V}{M} \qquad (8.4.13b)$$

At constant T and p (recall Equation 7.1.1)

$$dG = \mu_1 dn_1 + \mu_2 dn_2 = \left(\mu_2 - \frac{\bar{V}_2}{\bar{V}_1}\mu_1\right) dn_2 \qquad (8.4.14)$$

Inserting 8.4.13b into 8.4.14 we find

$$\frac{\partial G}{\partial c} = \left(\mu_2 - \frac{\bar{V}_2}{\bar{V}_1}\mu_1\right)\frac{V}{M} \qquad (8.4.15a)$$

and

$$\frac{\partial^2 G}{\partial c^2} = \left(\frac{\partial \mu_2}{\partial c} - \frac{\bar{V}_2}{\bar{V}_1}\frac{\partial \mu_1}{\partial c}\right)\frac{V}{M} \qquad (8.4.15b)$$

The Gibbs–Duhem relation $n_1 d\mu_1 + n_2 d\mu_2 = 0$ at fixed T and p gives

$$\frac{\partial \mu_2}{\partial c} = -\frac{n_1}{n_2}\frac{\partial \mu_1}{\partial c} \qquad (8.4.16)$$

so

$$\frac{\partial^2 G}{\partial c^2} = -\frac{V}{M}\left(\frac{n_1}{n_2} + \frac{\bar{V}_2}{\bar{V}_1}\right)\frac{\partial \mu_1}{\partial c} = -\frac{V}{M}\left(\frac{n_1 \bar{V}_1 + n_2 \bar{V}_2}{n_2 \bar{V}_1}\right)\frac{\partial \mu_1}{\partial c}$$

$$= \frac{-V}{c\bar{V}_1}\left(\frac{\partial \mu_1}{\partial c}\right) \qquad (8.4.17)$$

From Equation 7.4.2 for the osmotic pressure,

$$\Pi = \left(\frac{\mu_1 - \mu_1^{\circ}}{\bar{V}_1}\right)$$

we have

$$\frac{\partial \mu_1}{\partial c} = -\bar{V}_1\frac{\partial \Pi}{\partial c} = -\bar{V}_1 RT\left(\frac{1}{M} + 2Bc + \cdots\right) \qquad (8.4.18)$$

where we have used Equation 7.4.7, the virial expansion for Π/RT. To conclude,

$$\frac{kT}{(\partial^2 G/\partial c^2)} = \frac{-kT}{(V/\bar{V}_1 c)\left(\dfrac{\partial \mu_1}{\partial c}\right)} = \frac{ckT}{VRT(1/M + 2Bc + \cdots)}$$

$$= \frac{c}{VN_{av}}\frac{1}{(1/M + 2Bc + \cdots)} \qquad (8.4.19)$$

and as we have been working with the fluctuations in a cell of volume Ψ, we set $V = \Psi$ in Equation 8.4.11. The final result is therefore

$$\frac{I_{ex}}{I_o} = \frac{4\pi^2 n^2 (\partial n/\partial c)^2}{r^2 \lambda_0^4 N_{av}} \frac{c}{\left(\frac{1}{M} + 2Bc + \cdots\right)} \tag{8.4.20}$$

where the volume of the fictional cell, Ψ, has happily disappeared.

We now regroup some terms:

$$\frac{I_{ex} r^2}{I_o} \equiv R_\theta \tag{8.4.21}$$

where R_θ is the so-called *Rayleigh ratio*. It is the normalized excess scattered intensity per unit volume, with the purely geometrical quantity r factored out; therefore R_θ should depend only on the solution and λ_0, and not the instrument used to measure it. Note that R_θ has units of cm^{-1}, because I_{ex} is the scattered intensity per unit volume, whereas I_o is just the incident intensity. Similarly, the purely optical factors can be grouped

$$K \equiv \frac{4\pi^2 n^2 (\partial n/\partial c)^2}{\lambda_0^4 N_{av}} \tag{8.4.22}$$

to reach the result

$$R_\theta = \frac{Kc}{\dfrac{1}{M} + 2Bc + \cdots} \tag{8.4.23}$$

This expression is usually rearranged in either of two ways, as follows:

$$\frac{Kc}{R_\theta} = \frac{1}{M} + 2Bc + \cdots \tag{8.4.24a}$$

or

$$R_\theta = KcM\{1 - 2BcM \cdots\} \tag{8.4.24b}$$

where the second version is obtained by recalling that $1/(1 + x) = 1 - x + x^2 + \cdots$. The former version is more often used when plotting data, because the right hand side should be linear in c, whereas the latter is more transparent in terms of the physical content. These equations tell us several important things.

1. In the limit of low concentration Equation 8.4.24b reduces to $I_{ex} \sim$ (number of scatterers)$\times$(size of scatterer)$\sim cM$, just as advertised for incoherent scattering in Section 8.2. Note that Equation 8.4.24a, somewhat perversely, has the experimental signal—the scattered intensity—in the denominator, so this result is not so obvious.

2. In the limit of infinite dilution, light scattering measures M, as does the osmotic pressure. These two experiments are intimately related, because the light scattering is determined by concentration fluctuations, and their amplitude is related to the associated osmotic cost. In other words, one could imagine a semipermeable membrane around each of our fictional cells. Any extra polymer in a particular cell will drive up Π, whereas a lower-than-average concentration would necessarily cause Π to increase in some other cells. Thermal energy drives random fluctuations, but the osmotic compressibility resists them.

3. The virial coefficient is obtained from the concentration dependence of the scattering, just as in the osmotic pressure experiment. Equation 8.4.24b shows directly that in a good solvent, when $B > 0$, the intensity will first increase linearly with c, but the rate of increase will drop

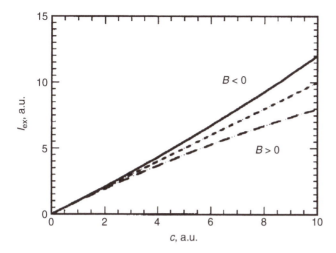

Figure 8.6 The excess scattered intensity as a function of concentration for different values of the second virial coefficient, B.

when the B term contributes appreciably. On the other hand, in a poor solvent, with $B \approx 0$ or even negative, the intensity will increase more rapidly with further increase in c. This is illustrated schematically in Figure 8.6.

4. We reiterate that Equation 8.4.24a and Equation 8.4.24b are valid for "small" polymers only; we will explore this restriction more quantitatively in the following section.
5. Although we have stressed the intimate connection between light scattering and osmotic pressure, there is one important difference. For a dilute but polydisperse sample, we can see that

$$\lim_{c \to 0} R_\theta = KcM = K \sum c_i M_i \tag{8.4.25}$$

and so

$$\lim_{c \to 0} \frac{Kc}{R_\theta} = \frac{K \sum c_i}{K \sum c_i M_i} = \frac{\sum \frac{n_i M_i}{V}}{\sum \frac{n_i M_i^2}{V}} = \frac{1}{M_w} \tag{8.4.26}$$

Thus light scattering measures the absolute weight average molecular weight, whereas the osmotic pressure experiment gives the number average (recall Equation 7.4.11). The reason for this difference is that the intensity scattered from an individual polymer is proportional to M, so that although the fluctuations are determined by the number of molecules per unit volume, the scattering signal is weighted by an additional factor of M.

Example 8.2

The following light scattering data were obtained on solutions of polystyrene in toluene at 25°C. A polarized Helium–Neon laser was used ($\lambda_0 = 633$ nm); $\partial n/\partial c$ was found to be 0.108 mL/g, and $n = 1.494$. Calculate the weight average molecular weight and the second virial coefficient.

c (g/mL)	0.0011	0.0026	0.0039	0.0052	0.0067	0.0089
$R_\theta \times 10^5$ (cm^{-1})	1.87	3.75	5.09	6.06	7.02	8.04

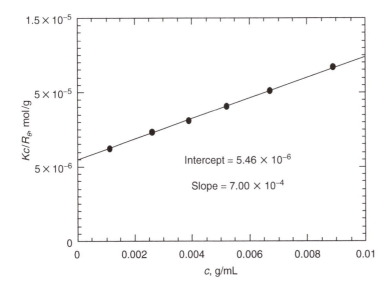

Figure 8.7 Plot of data for Example 8.2 according to Equation 8.42a.

Solution

First we need to calculate K by Equation 8.4.22:

$$K = \frac{4 \times \pi^2 \times (1.494)^2 \times (0.108)^2}{(6.33 \times 10^{-5})^4 \times 6.02 \times 10^{23}} = 1.06 \times 10^{-7} \text{ cm}^2 \text{ mol}/\text{g}^2$$

Then we plot Kc/R_θ versus c, as shown in Figure 8.7. Linear regression gives an intercept of 5.46×10^{-6} mol/g and a slope of 7.00×10^{-4} mL mol/g^2. Thus we obtain

$$M_w = 1/(5.46 \times 10^{-6}) = 183,000 \text{ g}/\text{mol}$$
$$B = 7.00 \times 10^{-4}/2 = 3.5 \times 10^{-4} \text{ mL mol}/\text{g}^2$$

8.5   The Form Factor and the Zimm Equation

We now take into account the finite size of a polymer. Unfortunately, the mathematics is a little tedious, but there is no shortcut. On the other hand, the main result will turn out to be relatively simple. We will restrict ourselves to the limit of a very dilute solution, i.e., we only consider one polymer at a time. The basic idea was outlined at the beginning of the chapter, namely that if the distance, r_{jk}, between two monomers j and k on a chain is a significant fraction of the radiation wavelength, λ, then the waves scattered from each monomer will have a phase difference at the detector. This leads to some destructive interference and a net reduction in the scattered intensity. This phase difference will depend on the scattering angle, θ, and therefore the intensity will also depend on θ; thus we are dealing with coherent scattering. Operationally, we can define a *form factor* for a single polymer, $P(\theta)$, as

$$P(\theta) = \frac{\text{Actual } I_{\text{ex}}(\theta)}{\text{Rayleigh } I_{\text{ex}}(\theta)} \tag{8.5.1}$$

where the Rayleigh scattering is given by Equation 8.4.24. From this relation we can see that $0 \leq P(\theta) \leq 1$. In general, we should worry about interference between waves scattered from different polymers as well as different monomers on one polymer, and then Equation 8.5.1 would become the operational definition of what is called the solution *structure factor*, $S(\theta)$. However, because we are restricting ourselves to very dilute solutions, there should be no correlation between the positions of different polymers and the intermolecular interference terms do not contribute.

8.5.1 Mathematical Expression for the Form Factor

To develop a mathematically precise definition of $P(\theta)$ we consider each monomer to be a Rayleigh scatterer. The electric field scattered by the molecule, $\vec{E}_{s,\text{tot}}$, is given by the superposition of the fields scattered by each monomer:

$$\vec{E}_{s,\text{tot}} = \sum_{j=1}^{N} \vec{E}_{s,j} = \sum_{j=1}^{N} \vec{E}_{o,s} \exp\left[i(\omega t + \delta_j)\right] = \vec{E}_{o,s} e^{i\omega t} \sum_{j=1}^{N} e^{i\delta_j} \tag{8.5.2}$$

where $\vec{E}_{o,s}$ is the amplitude of the field scattered from each monomer and δ_j is the phase of the wave from monomer j. The summation runs over the N monomers of the chain. (We choose to use sums over j and k to avoid confusion with $i = \sqrt{-1}$.) In Equation 8.5.2 we have assumed that each monomer is identical in scattering power (i.e., a homopolymer). The scattered intensity requires the squaring of the total field:

$$\begin{aligned} I_s &\sim \left|\vec{E}_{s,\text{tot}}^* \cdot \vec{E}_{s,\text{tot}}\right| \sim \left(\vec{E}_{s,\text{tot}}^* e^{-i\omega t} \sum_{j=1}^{N} e^{-i\delta_j}\right) \cdot \left(\vec{E}_{o,s} e^{i\omega t} \sum_{k=1}^{N} e^{i\delta_k}\right) \\ &= \left|\vec{E}_{o,s}\right|^2 \sum_{j=1}^{N}\sum_{k=1}^{N} e^{i(\delta_k - \delta_j)} \end{aligned} \tag{8.5.3}$$

For Rayleigh scattering we assumed that each polymer was very small compared to λ, and therefore the phase is the same for each monomer along the chain: $\delta_j = \delta_k$. With this simplification we can write

$$P(\theta) = \frac{\left|\vec{E}_{o,s}\right|^2 \sum_{j=1}^{N}\sum_{k=1}^{N} e^{i(\delta_k - \delta_j)}}{\left|\vec{E}_{o,s}\right|^2 \sum_{j=1}^{N}\sum_{k=1}^{N} e^{i0}} = \frac{1}{N^2}\sum_{j=1}^{N}\sum_{k=1}^{N}\left\langle e^{i(\delta_k - \delta_j)}\right\rangle \tag{8.5.4}$$

We have also taken the average $\left\langle e^{i(\delta_k - \delta_j)}\right\rangle$ because that is the experimentally important quantity. Equation 8.5.4 is an alternative definition of $P(\theta)$, and again we emphasize that it is the single chain form factor; the double summation is over the N monomers of one chain. For the general case, we would just extend the double summation over all pairs of monomers in the scattering volume, and then Equation 8.5.4 would define the structure factor.

The next step is to develop an expression for $\delta_k - \delta_j$, the phase difference between the waves scattered from any pair of monomers k and j. Here the scattering vector comes to the rescue, because this phase difference is very simply expressed as the dot product of $\vec{q}$ with the vector separating monomers j and k:

$$\delta_k - \delta_j = \vec{q} \cdot \vec{r}_{jk} \tag{8.5.5}$$

If we recall the derivation of Bragg's law in Section 8.2, two particles in a given lattice plane automatically scatter waves in phase with one another. In this case $\vec{q} \cdot \vec{r}_{jk} = 0$, because $\vec{q}$ is perpendicular to the lattice planes. In other words, the phase difference is determined solely by the component of $\vec{r}_{jk}$ that is parallel to the scattering vector.

Now we insert this result into Equation 8.5.4:

$$P(\theta) = P(q) = \frac{1}{N^2} \sum_{j=1}^{N} \sum_{k=1}^{N} \langle \exp[i\vec{q} \cdot \vec{r}_{jk}] \rangle \tag{8.5.6}$$

This equation represents the standard definition of the form factor. It may appear rather complicated, because it involves a complex number and two vectors. However, remember that the complex numbers are just a convenience and they will disappear when we calculate anything observable. Furthermore, the dot product of two vectors is a scalar, so we should be able to get rid of the vectors as well. The average in Equation 8.5.6 is

$$\langle \exp[i\vec{q} \cdot \vec{r}_{jk}] \rangle = \int_{-\infty}^{\infty} P(\vec{r}_{jk}) \exp[i\vec{q} \cdot \vec{r}_{jk}] \, d\vec{r}_{jk} \tag{8.5.7}$$

where $P(\vec{r}_{jk})$ is the probability of monomers j and k being separated by a vector $\vec{r}_{jk}$. For the case of a chain in a theta solvent (recall Chapter 6 and Chapter 7), we already know this probability function: it is a Gaussian function (Equation 6.7.1).

8.5.2 Form Factor for Isotropic Solutions

Now, if we restrict our attention to isotropic samples, such that in spherical coordinates (r,θ,ϕ) $P(\vec{r}_{jk})$ has no θ or ϕ dependence, then we can get rid of the vectors altogether. First, we recognize that

$$d\vec{r}_{jk} = dx_{jk}\, dy_{jk}\, dz_{jk} = r_{jk}^2 \sin\theta \, d\theta \, d\phi \, dr_{jk} \tag{8.5.8}$$

using the transformation to spherical coordinates (if this is unfamiliar, see the Appendix). Then we eliminate the dot product:

$$\exp[i\vec{q} \cdot \vec{r}_{jk}] = \exp[iqr_{jk}\cos\theta] \tag{8.5.9}$$

in which $q = (4\pi/\lambda)\sin(\theta/2)$ from Equation 8.2.4. If we expand the exponential as a power series this becomes

$$\exp[i\vec{q} \cdot \vec{r}_{jk}] = 1 + iqr_{jk}\cos\theta + \frac{i^2 q^2 r_{jk}^2 \cos^2\theta}{2!} + \cdots$$

Then we can write

$$\langle \exp[i\vec{q} \cdot \vec{r}_{jk}] \rangle = \int_0^{\infty} P(\vec{r}_{jk}) r_{jk}^2 \, dr_{jk} \int_0^{2\pi} d\phi \int_0^{\pi} d\theta \sin\theta \left(1 + iqr_{jk}\cos\theta + \cdots \right)$$

$$= \int_0^{\infty} P(\vec{r}_{jk}) r_{jk}^2 \, dr_{jk} 2\pi \int_0^{\pi} \left(\sin\theta + iqr_{jk}\cos\theta\sin\theta + \frac{i^2 q^2 r_{jk}^2 \cos^2\theta\sin\theta + \cdots}{2} \right) d\theta$$

$$= \int_0^{\infty} P(\vec{r}_{jk}) 2\pi r_{jk}^2 \, dr_{jk} \left[2 + 0 - \frac{q^2 r_{jk}^2}{2!} \frac{2}{3} + 0 + \frac{q^4 r_{jk}^4}{4!} \frac{2}{5} + \cdots \right] \tag{8.5.10}$$

where we have looked up the integrals for $\sin\theta\cos^k\theta$. The resulting series in square brackets is, in fact, nothing more than

$$2\left[1 - \frac{(qr_{jk})^2}{3!} + \frac{(qr_{jk})^4}{5!} + \cdots \right] = 2\frac{\sin qr_{jk}}{qr_{jk}} \tag{8.5.11}$$

so in the end,

$$\langle \exp(i\vec{q} \cdot \vec{r}_{jk}) \rangle = \int_0^\infty P(r_{jk}) \left(\frac{\sin qr_{jk}}{qr_{jk}} \right) dr_{jk} = \left\langle \frac{\sin qr_{jk}}{qr_{jk}} \right\rangle \tag{8.5.12}$$

where $P(r_{jk}) = 4\pi r_{jk}^2 \, P(\vec{r}_{jk})$.

This assumption of isotropy is not at all restrictive for a dilute polymer solution at rest. Note that it is *not* an assumption that the *shape* of the molecule is spherical, only that *on average* the intramolecular bond vectors point equally in all directions. Thus, even a solution of rod-like particles can be isotropic in this sense. We can use Equation 8.5.12 whenever we have information about $P(r_{jk})$. However, it turns out we can also extract something very useful even if we do not know anything particular about this distribution.

8.5.3 Form Factor as $qR_g \to 0$

The power series in qr_{jk} given in Equation 8.5.11, combined with Equation 8.5.6, gives for $P(q)$:

$$P(q) = \frac{1}{N^2} \sum_{j=1}^N \sum_{k=1}^N \left\{ 1 - \frac{q^2}{6} \left\langle r_{jk}^2 \right\rangle + \frac{q^4}{120} \left\langle r_{jk}^4 \right\rangle \cdots \right\} \tag{8.5.13}$$

But, the average $\sum_{j=1}^N \sum_{k=1}^N \left\langle r_{jk}^2 \right\rangle$ is directly related to $\langle s^2 \rangle \left(= R_g^2 \right)$ for any shape, as shown in Equation 6.5.8:

$$R_g^2 = \frac{1}{2N^2} \sum_{j=1}^N \sum_{k=1}^N \left\langle r_{jk}^2 \right\rangle \tag{8.5.14}$$

and therefore we have the important result that

$$P(q) = 1 - \frac{q^2}{3} R_g^2 + \cdots \tag{8.5.15}$$

independent of the shape of the particle. Thus if the experiment is designed such that $q^2 R_g^2 < 1$, the higher order terms in the expansion Equation 8.5.15 can be neglected, and R_g can be determined without any prior knowledge of the average conformation. It turns out that for flexible and semi-flexible polymers this condition is quite often satisfied (see Example 8.4 and Problem 6).

8.5.4 Zimm Equation

We now return to the end of the previous section, and insert $P(q) = P(\theta)$ into the scattering equation (Equation 8.4.24a):

$$\frac{Kc}{R_\theta} = \frac{1}{M_w} \frac{1}{P(\theta)} + 2Bc + \cdots \tag{8.5.16}$$

in the limit of $c \to 0$. Now it is traditional to manipulate $1/P(\theta)$ as follows:

$$\frac{1}{P(\theta)} = \frac{1}{1 - \frac{q^2}{3} R_g^2 + \cdots} = 1 + \frac{1}{3} q^2 R_g^2 + \cdots \tag{8.5.17}$$

because $1/(1-x) = 1 + x + x^2 + \cdots$. We now use this result to obtain the *Zimm equation*:

$$\frac{Kc}{R_\theta} = \frac{1}{M_w} \left(1 + \frac{q^2}{3} R_g^2 + \cdots \right) + 2Bc + \cdots \tag{8.5.18}$$

This is the fundamental result for light scattering from dilute polymer solutions [4].

If we recast this relation from Equation 8.4.24b, we obtain

$$R_\theta = KcM_w P(\theta)\{1 - 2BcM_w P(\theta)\ldots\}$$

$$= KcM_w\left(1 - \frac{q^2}{3}R_g^2\cdots\right)\left\{1 - 2BcM_w\left(1 - \frac{q^2}{3}R_g^2\cdots\right)\cdots\right\} \tag{8.5.19}$$

You should confirm that if you start from Equation 8.5.18, and transform it appropriately you will recover Equation 8.5.19. It is important to point out a subtlety here. You might well ask, if c is high enough that polymer–polymer terms contribute to the concentration fluctuations (i.e., through B), why don't we have to worry about interference effects between monomers on nearby chains? The answer is that actually we do, and that is the source of the second $P(\theta)$ term multiplying B in Equation 8.5.19. In the simplest approach (which is not simple), the *single contact approximation* introduced by Zimm [4] (Equation 8.5.19) is the result. We should also point out that several texts write the right hand side of Equation 8.5.18 incorrectly, as $(1/M_w + 2Bc + \cdots)(1 + q^2/3R_g^2 + \cdots)$.

8.5.5 Zimm Plot

Equation 8.5.18 provides the basis for a particular method to analyze light-scattering data, the so-called *Zimm plot*. We need to perform two extrapolations, to zero scattering angle ($\theta = 0°$ or $q = 0$) and to zero concentration ($2Bc = 0$), and the result will be $1/M_w$. Furthermore, the slopes of the angle extrapolation and the concentration extrapolation, respectively, will provide values of R_g and B, assuming that the range of the independent variable is such that only the first term of the relevant expansion is important. Accordingly, we would plot Kc/R_θ versus $\sin^2(\theta/2) + \gamma c$, where γ is an arbitrary constant; the $\sin^2(\theta/2)$ term is proportional to q^2. The value of γ is chosen to spread the data out. An example is shown in Figure 8.8 for a solution of methylcellulose in water. One

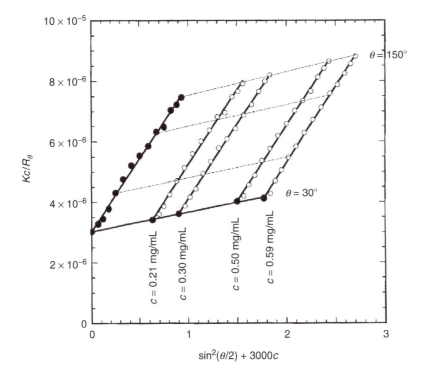

Figure 8.8 Zimm plot for a sample of methylcellulose in water. (Reproduced from Kobayashi, K., Huang, C.-I., and Lodge, T.P., *Macromolecules*, 32, 7070, 1999. With permission.)

choice of γ is approximately $1/\Delta c$, where Δc is the difference between successive concentrations. In this case because $\sin^2(\theta/2) \leq 1$, all of the angle data for a given concentration lie to the left of all of the angle data for the next higher concentration. In any event, the choice of γ is essentially cosmetic. The next step is to perform linear regression on each set of data, as follows. For each shifted concentration γc, Kc/R_θ should be fit to a straight line against $\sin^2(\theta/2)$, and the $\theta = 0°$ intercept recorded (and plotted, see the filled circles in Figure 8.8). Similarly, for each scattering angle θ, Kc/R_θ should be fit to a straight line against c, and the $c = 0$ intercept recorded (and plotted). Then, the $\theta = 0°$ intercepts should be fit to a straight line against c, and the $c = 0$ intercepts to a straight line against $\sin^2(\theta/2)$. Both of these lines should meet on the Kc/R_θ axis, and the resulting value is $1/M_w$. The slopes of these lines are proportional to R_g^2 and B, respectively, with the proportionality depending on exactly how the data were treated. This process may be understood in detail by working through Problem 10, or the following example.

Example 8.3

The scattering data for the methylcellulose sample presented in Figure 8.8 are given in the following table. The data were taken at $20°C$, with an argon ion laser operating at 488 nm. Under these conditions, n for water is 1.33, and $\partial n/\partial c$ was determined to be 0.137 mL/g. Calculate M_w, R_g, and B.

Solution

The horizontal axis of the Zimm plot requires a choice of the shift factor γ. As the different concentrations are separated by $\Delta c \approx 0.1$ or 0.2 mg/mL, a reasonable choice for γ would be in the range $1/\Delta c \approx 5000$–$10,000$; in fact, a value of 3000 was selected for Figure 8.8. The next step is to perform the first extrapolation to zero angle for each concentration. This may be done directly from the data in the table, using $\sin^2(\theta/2)$ as the x axis, by linear regression ("least squares"). The resulting $\theta = 0°$ data are listed below in the second table, and these extrapolations are shown in Figure 8.9a, with the $\theta = 0°$ data highlighted as solid circles. Note that to plot the data in Figure 8.9a, each value of $\sin^2(\theta/2)$ has to be added to the appropriate value of γc.

θ (°)	$\sin^2(\theta/2)$	$Kc/R_\theta \times 10^6$			
		0.21 mg/mL	0.30 mg/mL	0.50 mg/mL	0.59 mg/mL
30	0.0670	3.61	3.87	4.21	4.29
40	0.117	3.86	4.16	4.57	4.71
50	0.179	4.25	4.43	4.89	5.07
60	0.250	4.71	4.95	5.38	5.48
70	0.329	5.14	5.34	5.76	5.84
80	0.413	5.60	5.78	6.20	6.29
90	0.500	6.01	6.08	6.59	6.72
100	0.587	6.37	6.43	7.02	7.14
110	0.671	6.83	6.89	7.35	7.59
120	0.750	6.96	7.23	7.66	7.88
130	0.821	7.49	7.66	8.14	8.25
140	0.883	7.66	8.00	8.42	8.57
150	0.933	7.92	8.20	8.65	8.82

c (mg/mL)	0.21	0.30	0.50	0.59
$Kc/R_\theta \times 10^6$, $\theta = 0°$	3.42	3.62	4.03	4.13

The second extrapolation is to $c = 0$ for each angle. The resulting linear regression curves and intercepts are shown in Figure 8.9b, and the $c = 0$ values are listed in the following table.

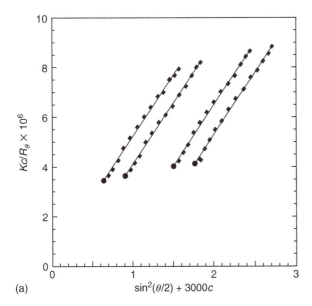

(a)

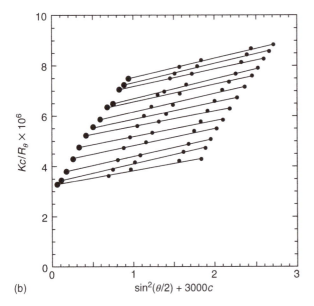

(b)

Figure 8.9 Construction of the Zimm plot of Figure 8.8 as developed in Example 8.3. (a) The extrapolation to $\theta = 0$ for each concentration. (b) The extrapolation to $c = 0$ for each angle.

(continued)

Finally, these two sets of data should be extrapolated to $c = 0$ and $\theta = 0°$, respectively. The results are shown in Figure 8.9c. From these two straight lines, the desired information can be extracted as follows:

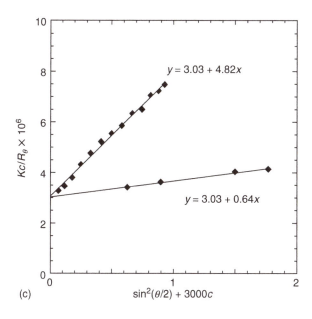

Figure 8.9 (continued) (c) The extrapolations of the $\theta = 0$ data to $c = 0$, and the $c = 0$ data to $\theta = 0$.

$\sin^2(\theta/2)$	0.0670	0.117	0.179	0.250	0.329	0.413	0.500
$Kc/R_\theta \times 10^6$ $c = 0$	3.28	3.45	3.79	4.31	4.76	5.22	5.54
$\sin^2(\theta/2)$	0.587	0.671	0.750	0.821	0.883	0.933	
$Kc/R_\theta \times 10^6$ $c = 0$	5.85	6.33	6.48	7.04	7.23	7.47	

Both fits give a common $c = 0$, $\theta = 0°$ intercept of 3.03×10^{-6}. Thus M_w is obtained as $1/(3.03 \times 10^{-6}) = 3.3 \times 10^5$ g/mol.

The concentration extrapolation is

$$\frac{Kc}{R_\theta} = \frac{1}{M_w} + 2Bc = 3.03 \times 10^{-6} + 0.64 \times 10^{-6} \,(3000c)$$

therefore

$$B = (1/2) \times 3000 \times 0.64 \times 10^{-6} = 9.6 \times 10^{-4} \text{ mol cm}^3/\text{g}^2$$

The angle extrapolation is

$$\frac{Kc}{R_\theta} = \frac{1}{M_w}\left(1 + \frac{1}{3}q^2 R_g^2\right) = \frac{1}{M_w}\left(1 + \frac{1}{3}\frac{16\pi^2 n^2}{\lambda_0^2}\sin^2(\theta/2)R_g^2\right)$$

$$= 3.03 \times 10^{-6} + 4.82 \times 10^{-6}\sin^2(\theta/2)$$

Consequently,

$$R_g^2 = \frac{4.82 \times 10^{-6}}{3.03 \times 10^{-6}} \times 3 \times \left(\frac{488}{4 \times 3.14 \times 1.33}\right)^2 = 4073 \text{ nm}^2$$

and $R_g = 64$ nm.

8.6   Scattering Regimes and Particular Form Factors

We continue this chapter with some further discussion of form factors. We return to the expansion of Equation 8.5.13, where $P(q)$ is a power series in even powers of qr_{jk}. We can delineate four general regimes of behavior, depending on the approximate magnitude of qR_g.

1. If $qR_g \ll 1$, which means either small molecules or very small scattering angles, then $P(q) \approx 1$. Thus the molecules may be considered as point scatterers, and there is no information on chain dimensions. In other words, we may call this the *Rayleigh regime*, and Equation 8.4.24a and Equation 8.4.24b apply.
2. If $qR_g \leq 1$, then only the next term in the power series matters: $P(q) \approx 1 - (q^2/3)R_g^2$. In this regime one can obtain the value of R_g without any knowledge about the shape of the molecule. A plot of $1/I_{ex}$ versus q^2 should be linear, with slope $R_g^2/3$. Note that the intensity need not be calibrated to obtain R_g; the units of the intercept cancel out when determining the slope. Alternatively, $P(q) \approx \exp(-q^2/3R_g^2)$, and a plot of $\ln I_{ex}$ versus q^2 should be linear with slope $-R_g^2/3$. This latter format is termed a Guinier plot, and this regime the *Guinier regime* [5].
3. If $1 \leq qR_g \leq 10$, more terms in the power series expansion of $P(q)$ become important, and these depend on the specific shape of the molecule. Accordingly, in this regime the mathematical form of $P(q)$ is helpful in distinguishing different molecular shapes. We will identify some specific functional forms below.
4. If $qR_g \gg 1$, the scattering is dominated by the internal structure of the molecule, and one can extract no information about R_g.

A more physical appreciation of these four regimes can be gained from the following viewpoint. Recalling the discussion in Section 8.2, the scattering vector has dimensions of inverse length, and q^{-1} is essentially a "ruler" in the sample. The choice of q dictates that the scattering experiment will explore fluctuations on the length scale q^{-1}. In Bragg diffraction, intensity is found at a particular angle when the relation $m\lambda = 2D \sin \theta/2$ is satisfied (Equation 8.2.2), where D is the spacing between planes of atoms in the crystal and m is an integer. The Bragg equation just accounts for the fact that the phase shift is an integral number of wavelengths when moving from one lattice plane to the next, and thus the interference is constructive. The Bragg equation for first order ($m = 1$) can be rewritten $D^{-1} = (2/\lambda) \sin(\theta/2) = q/2\pi$. In other words, scattering is exactly the same process as Bragg diffraction. The only difference is that in scattering, we look for spontaneous fluctuations that just happen to have the right orientation and spacing ($2\pi/q$) to scatter to the detector, whereas in diffraction there are particular lattice planes. Now we can reinterpret the four regimes above in terms of the relationship between the spacing of the lines and the size of the molecule. This is illustrated in Figure 8.10, where a set of lines spaced at $2\pi/q$ is shown with four different polymers, illustrating the four size regimes.

Example 8.4

Estimate which of the four regimes would be accessed by light scattering on solutions of polystyrene in cyclohexane at the theta temperature (34.5°C). Assume M values of 10^4, 10^5, and 10^6 g/mol, and that the angular range of the instrument is 25°–150°.

Solution

We need to calculate R_g for each polymer, and the minimum and maximum q values of the instrument. In Example 6.2, we calculated the unperturbed R_g for polystyrene with $M = 10^5$ to be 8.5 nm. As R_g varies with $\sqrt{M}$ we can estimate R_g for 10^4 to be $\sqrt{10}$ times smaller, i.e., 2.7 nm, and R_g for 10^6 to be $\sqrt{10}$ times bigger, or 27 nm. For cyclohexane, $n = 1.424$ from Table 8.1 (at a

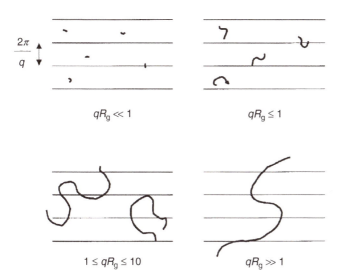

$\dfrac{2\pi}{q}$

$qR_g \ll 1$ $\qquad\qquad\qquad\qquad$ $qR_g \leq 1$

$1 \leq qR_g \leq 10$ $\qquad\qquad\qquad$ $qR_g \gg 1$

Figure 8.10 Illustration of four regimes of scattering behavior, depending on the ratio of the polymer size, R_g, to the scattering length scale, $1/q$.

slightly different wavelength and temperature, but we are only estimating here). Assuming we have an argon laser ($\lambda_0 = 488$ nm), then

$$q_{\min} = \frac{4 \times 3.14 \times 1.424}{488}\, \sin\left(\frac{25}{2}\right) = 0.0079 \text{ nm}^{-1}$$

$$q_{\max} = \frac{4 \times 3.14 \times 1.424}{488}\, \sin\left(\frac{150}{2}\right) = 0.035 \text{ nm}^{-1}$$

Therefore for $M = 10^4$, qR_g ranges from $0.0079 \times 2.7 = 0.021$ to $0.035 \times 2.7 = 0.095$. This falls entirely into the Rayleigh regime, and we would not be able to determine R_g. For $M = 10^5$, qR_g ranges from 0.067 to 0.30. This extends into the Guinier regime, and with precise measurements R_g could be determined. Finally, for $M = 10^6$, qR_g ranges from 0.21 to 0.95. Although this appears to be in the Guinier regime, in fact the next term in the expansion of $P(q)$ may contribute at larger angles, so the lower angle data should be emphasized in the determination of R_g.

There are two important conclusions to draw from these numbers. First, one can only change q by a factor of about 5 in a typical instrument. Second, for flexible polymers and typical molecular weights it is hard to get much beyond the Guinier regime. To access structural details at smaller length scales, it is necessary to increase q substantially. This is done by utilizing x-ray or neutron scattering, where the wavelength is only a few angstroms; recall $q \sim 1/\lambda$.

The form factors associated with particular distribution functions, $P(r_{jk})$, have been derived for a variety of shapes. We will just state the results for three particularly important ones: the Gaussian coil, the rigid rod, and the hard sphere.

1. For the Gaussian coil, the form factor is known as the *Debye function* after it was first developed by Debye [6]. It may be written as

$$P(q) = \frac{2}{x^2}(e^{-x} - 1 + x), \quad x \equiv q^2 R_g^2 \tag{8.6.1}$$

This function applies to chains in a theta solvent, and in the melt. It is important to realize that it no longer applies when q^{-1} becomes comparable to the persistence length or statistical segment length; in that regime the chain conformation no longer follows the Gaussian distribution.

2. For a rigid rod of length L and zero width, the form factor is [7]

$$P(q) = \frac{1}{x} \int_0^{2x} \frac{\sin z}{z} \, dz - \left(\frac{\sin x}{x}\right)^2, \quad x \equiv \frac{qL}{2} \qquad (8.6.2)$$

where the definite integral is tabulated in most mathematical handbooks.

3. For a hard sphere of radius R the result (also due to Lord Rayleigh [8]) is

$$P(q) = \left(\frac{3}{x^3}\right)^2 (\sin x - x \cos x)^2, \quad x \equiv qR \qquad (8.6.3)$$

All three $P(q)$ expressions are plotted in Figure 8.11a as a function of $(qR_g)^2$. In Figure 8.11b, $1/P(q)$ is plotted versus $q^2 R_g^2$, i.e., in the form anticipated by the Zimm plot. In this format all three have the same small q limiting slopes, as they must, but they diverge beyond $qR_g \approx 1$. Note that $R_g = L/\sqrt{12}$ for the rod, and $\sqrt{3/5}R$ for the sphere (see Table 6.3), so the different definitions of x in Equation 8.6.1 through Equation 8.6.3 need to be handled carefully. In Figure 8.11c, $P(q)$ is plotted on a logarithmic axis against qR_g; note the oscillations in $P(q)$ for a sphere. It is clear that all three $P(q)$ look similar for $qR_g < 1$, as expected; this corresponds to regions 1 ("Rayleigh") and 2 ("Guinier") above. Beyond that they begin to diverge, corresponding to regimes 3 and 4.

8.7 Experimental Aspects of Light Scattering

The technique of light scattering was first applied to polymer solutions in the 1940s. Early instruments were homemade, and limited by the quality of the available components, such as photodetectors. Two early commercial instruments, the Bryce-Phoenix and the Sofica, remained in

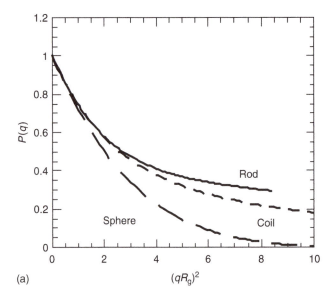

(a)

Figure 8.11 Form factors $P(q)$ for Gaussian coils, hard spheres, and very thin rods (a) as a function of $(qR_g)^2$.

(*continued*)

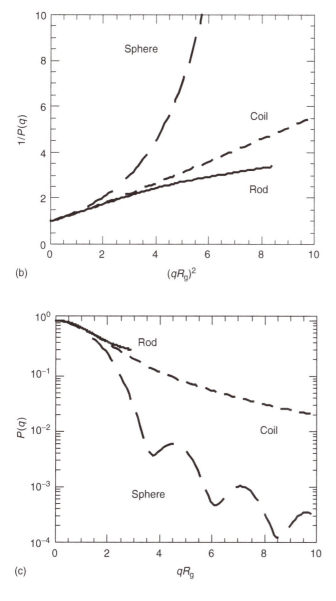

Figure 8.11 (continued) (b) Inverse form factors as in (a), showing the convergence to a common slope of 1/3 in the small qR_g limit. (c) Form factors plotted on a logarithmic scale, showing the oscillations for the hard sphere.

service in some laboratories for decades. Even with the advent of lasers, high-quality photomultipliers, and sensitive photodiodes, light scattering was largely conducted on custom instrumentation, which limited its applicability. Two trends in the last two decades have reinvigorated the field, however. One was the emergence of dynamic light scattering (DLS, see Section 9.5) as an important characterization technique for polymer, biopolymer, and colloidal solutions. A DLS instrument can be used to measure $I(q)$ as well, and therefore commercial DLS apparatuses now often serve a dual role. The second trend was the emergence of light scattering as an absolute molecular weight detector for size-exclusion chromatography (SEC, see Section 9.7). In this context light scattering is finally approaching the status of a "routine" characterization tool for polymer solutions.

8.7.1   Instrumentation

The basic components of a light scattering instrument are illustrated in Figure 8.12. The typical light source is a laser, although the unique features of a laser source (e.g., temporal and spatial coherence, collimation, high monochromaticity) are not required. The most desirable feature of the source is stability. The source may be vertically polarized or unpolarized, as discussed in Section 8.3, but it is important to establish that you have one or the other, and not a partially polarized beam. The beam should be collimated before traversing the sample cell. The solution is usually contained in a glass cell, and the glass should be carefully selected for clarity and lack of imperfections. All other things being equal, the larger the diameter of the sample cell, the better. The reason is that at each air–glass interface a significant portion of the incident light will be reflected, rather than transmitted. This reflected light may find its way to the detector as *stray light*, or back into the scattering volume at a different angle, leading to erroneous scattering signals. The reflected fraction of the intensity, R, for near-normal incidence is given roughly by

$$R \approx \frac{(n_{glass} - n_{air})^2}{(n_{glass} + n_{air})^2} \tag{8.7.1}$$

where n_{glass} is typically about 1.5 and $n_{air} = 1.0$. This gives a reflectivity of about 4% per air–glass interface. The larger the diameter of the sample cell, the further this source of stray light will be from the scattering volume, and therefore the easier it is to eliminate. Empirically, for a 1 cm sample cell it is very difficult to get to scattering angles below about 25° because of this problem. Note that there will also be two glass–sample solution interfaces, where for some solutions (aqueous ones, for example) there will also be substantial reflection.

The reflection problem is often mitigated, but not completely eliminated, by use of an index-matching fluid. In this case the sample cell is suspended in a much larger container filled with a fluid of refractive index near that of the glass (silicon oil and toluene are two common examples). By this expedient the main air–glass interfaces can be moved several centimeters from the scattering volume. Furthermore, the index-matching bath can provide an excellent heat-transfer medium for controlling the sample temperature. Good temperature stability is important for precise measurements.

The scattered light is collected by some combination of lenses and apertures, and directed onto the active surface of the detector. In routine applications it is usually assumed that the scattered

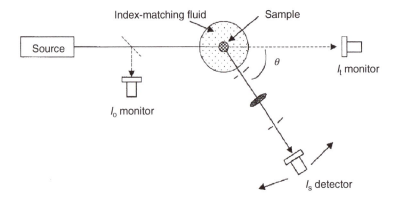

Figure 8.12   Schematic diagram of a light scattering photometer.

light is entirely polarized vertically (for vertically polarized incident light), and a polarizer is not used in front of the detector. This is a good assumption for most flexible polymers. However, in some situations, such as when examining rod-like particles, significant *depolarization* of the scattered light may occur, and a polarizer in front of the detector becomes a necessity. Three general detection schemes can be envisioned. The simplest is to have a single detector at some fixed scattering angle θ. This is of limited utility, especially in terms of gaining size or shape information, but it can be useful as an SEC detector or for monitoring the course of a chemical process whereby the scattering changes in time. The next option is to have a single detector that can rotate about a vertical axis, which runs through the center of the sample cell. This provides a continuously variable θ, which is very desirable. In recent years, it has become popular to use a movable optical fiber to collect the light, and keep the actual detector fixed in space. In this configuration, the detector is usually a photomultiplier, which can be extremely sensitive (capable of counting single photons, if need be) and which has a wide linear range (output current proportional to incident intensity over many orders of magnitude). The third approach is to arrange multiple detectors at fixed angles, which allows for simultaneous detection with the attendant increase in overall signal-to-noise (the "multiplex advantage"). This technique has only recently become practical with improvements in photodiode detectors, such that small, inexpensive units are sufficiently sensitive and linear. The multiangle light scattering technique is now the basis of a popular SEC detector, which can give real-time information about M and R_g for each slice of the chromatogram (see Section 9.7).

The final instrumental issue is the use of an incident intensity monitor, to follow fluctuations and drift in the source output. Often a small portion of the incident beam is split off, for example by inserting a piece of glass in the incident beam at $45°$ to the propagation direction, and directed to a separate detector. This permits the scattered signal to be normalized to the incident intensity in real time, which ultimately leads to a much more reliable R_θ. For example, when measuring a dilute polymer solution the scattered intensity may be less than twice that of the solvent alone. The solvent and solution would be measured at different times, so small but uncontrolled variations in source intensity can severely compromise the reliability of the excess intensity. It is also desirable to place a detector to monitor the transmitted beam. This could serve as an incident intensity monitor, but by having detectors both prior to and after the sample it is possible to detect changes in the total sample scattering, and thereby to assess whether absorption or multiple scattering is a problem.

8.7.2 Calibration

Referring back to Equation 8.4.22, we see that in order to extract a value of M_w, we will need to know the refractive index of the solvent, $\partial n/\partial c$, c, λ_0, and r. All of these are straightforward to determine, except r. In addition, and most importantly, the Rayleigh ratio involves the ratio of the scattered intensity per unit volume to the incident intensity, and as we noted in the beginning of the chapter, we rarely determine a true intensity. Furthermore, we would not know the active area of the photodetector, or its quantum efficiency (the fraction of incident photons actually detected).

The resolution of these difficulties is actually rather simple. We assume that the detector produces an output signal S (current, voltage, or "counts") proportional to the incident intensity:

$$S_s = \beta_s I_s \qquad\qquad (8.7.2)$$

where the proportionality factor β is independent of the magnitude of S over the relevant range (i.e., linear response), and the subscript s denotes "scattering." As long as we maintain the detector at a fixed distance r from the scattering volume as the scattering angle is varied, then the detector should collect light over the same range of solid angle fraction at each θ. A similar equation describes the response of the incident intensity monitor:

$$S_0 = \beta_0 I_0 \qquad (8.7.3)$$

We are now in a position to obtain the Rayleigh ratio (units of cm^{-1}) as follows:

$$
\begin{aligned}
R_\theta &= \frac{r^2 (I_s^{solution} - I_s^{solvent})}{I_0} \\
&= \frac{r^2 \beta_s (S_s^{solution} - S_s^{solvent})}{\beta_0 S_0} \\
&= \frac{\gamma (S_s^{solution} - S_s^{solvent})}{S_0} \qquad (8.7.4)
\end{aligned}
$$

where there is a single unknown proportionality factor γ. This we obtain by measuring a pure solvent for which the absolute Rayleigh ratio has been measured (by someone who took a lot of care) at the same wavelength and temperature. A second calibration step is required for instruments with multiple detectors, as each detector will have its own value of β (and also a different r and collection solid angle). This is best accomplished by using a solution with relatively high scattering, but one for which the scattering is entirely incoherent. A solution of a moderate molecular weight polymer can serve this purpose. Then the signal from each detector can be normalized to a single reference detector (usually chosen at $\theta = 90°$) by a multiplicative factor. It should be noted that Equation 8.7.2 and Equation 8.7.3 assume that there is no background signal (i.e., a finite S when the source is turned off) or stray light (i.e., contributions to S not from sample scattering at angle θ). In fact, all photodetectors have some *dark current* (output in the absence of input), but this is very small. (If it is a significant fraction of the scattering signal, then your experiment is in trouble.) Other sources of background (e.g., room light) and stray light can be minimized by appropriate instrumental design. However, the simple fact is that the detector is indiscriminate; all photons that reach it will be counted as scattering, no matter what their origins, and so great care must be taken to eliminate background and stray light.

There is an additional correction step that must be performed when using a single detector that rotates to various θ. Referring to Figure 8.13, the scattering volume is determined by the intersection of the incident beam and the collected beam (the latter being set by the geometry of the detection system). Assuming that both are square in cross section, and that the latter is comparable in diameter to the former, we can see that as θ deviates from $90°$ the volume of

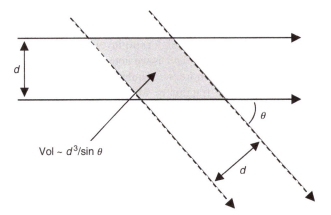

Figure 8.13 Illustration of the change in scattering volume with scattering angle, θ, for incident and scattered beams with width d.

intersection increases by a factor of $1/\sin\theta$. Thus, the signal at any angle should be multiplied by a factor of $\sin\theta$ in order to account for this variation in scattering volume. A very good indication of a functioning instrument is to plot $S_s(\theta)\sin\theta$ versus θ for an incoherent scattering solution. This product should be independent of θ. Ultimately, at high and low θ the data will deviate from the constant value due to reflections at the various interfaces discussed above. This measurement will therefore provide a guide as to the reliable range of θ for the instrument.

8.7.3 Samples and Solutions

There are two main issues here. First is the choice of solvent, and second is the preparation of "dust-free" samples. One may not have the freedom to choose the solvent, but all other things being equal it is desirable to have $|\partial n/\partial c|$ as large as possible, as the intensity is proportional to $(\partial n/\partial c)^2$. It may also be helpful to choose a solvent with a relatively small R_θ of its own, so that the polymer contribution to the excess scattering is larger. (You might be wondering "Why does the pure solvent scatter at all, given that we emphasized that homogeneous materials do not scatter?" The answer is *density fluctuations*, and the scattered intensity is determined by the isothermal compressibility of the solvent, κ, in parallel to Equation 8.4.11, the intensity is proportional to $kT\kappa$). Finally, some solvents are easier to make dust-free than others; for example, more polar solvents such as water and THF are often trickier to clean than toluene or cyclohexane. The preparation of dust-free samples takes some care and experience. It is essential to remove dust, as stray particles that are significantly larger than the polymer molecules will scatter strongly. The two standard options are filtration and centrifugation, and the former is usually preferred. Both are less than ideal, in that they may change the concentration of the solution. The use of light scattering as an SEC detector has a particular advantage here in that the column acts as an excellent filter, and a refractive index detector serves as a direct concentration monitor. There are two standard diagnostics for the presence of dust in the sample. The first is to examine the temporal fluctuations in S_s. These should be random, and have a root-mean-square amplitude close to $\sqrt{\langle S_s\rangle}$. If some dust is present, the signal is likely to increase suddenly and then decrease suddenly some seconds later, as dust particles drift in and out of the scattering volume. The scattering volume can also be examined visually; the tell-tale bright flashes of dust are often easily seen. The second diagnostic is to examine $1/I(\theta)$ versus $\sin^2(\theta/2)$ (or q^2). According to the Zimm equation this should produce a straight line. Dust will increase $I(\theta)$ selectively at low θ, resulting in a characteristic downturn in the plot. (Note that this test should only involve the range of θ deemed to be free of reflections.)

8.7.4 Refractive Index Increment

As is evident from the preceding discussion, $\partial n/\partial c$ plays a central role in the magnitude of the scattered intensity. It may be defined as follows:

$$n = n_s + \frac{\partial n}{\partial c}c + ac^2 + \cdots \tag{8.7.5}$$

where n is the refractive index of the solution, n_s the refractive index of the pure solvent, and c is the concentration in g/mL. The initial slope of $n(c)$ versus c gives $\partial n/\partial c$, but is better obtained as

$$\frac{\partial n}{\partial c} = \lim_{c\to 0}\left(\frac{n-n_s}{c}\right) = \lim_{c\to 0}\left(\frac{\Delta n}{c}\right) \tag{8.7.6}$$

Usually one is interested in concentrations on the order of $10^{-2} - 10^{-4}$ g/mL, and $\partial n/\partial c$ is of order 0.1, so it is not simply a matter of measuring n directly, because the change in n will only appear in the third or fourth place after the decimal. Rather, a *differential refractometer* should be used. A variety of designs are available, either commercially or by relatively simple assembly in the laboratory, but the principle behind the simplest version is illustrated in Figure 8.14. The light

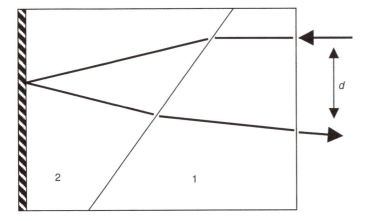

Figure 8.14   Illustration of a refractometer based on a split cell prism.

beam is incident on a divided cell that contains the solvent in one compartment and the solution in the other (or solutions of two different concentrations). The dividing glass surface is angled, so that the beam direction in compartment 2 depends on the refractive index of the fluid in compartment 2, by Snell's law. The beam is then reflected back through both compartments, with a final direction that depends on Δn. For small Δn it can be shown that the displacement of the reflected beam, d, is proportional to Δn.

The value of $\partial n/\partial c$ depends on T and λ, so it should be measured under the same conditions as the light scattering itself. It is independent of molecular weight for large M, but will develop a significant M dependence at low M, so this may need to be taken into account in, for example, SEC detection (see Section 9.7). The main source of the M dependence of $\partial n/\partial c$ lies in the M dependence of the polymer refractive index n_p, which in turn depends directly on density. The M dependence of the density reflects primarily the concentration of chain ends, and $\partial n/\partial c$ is therefore usually linear in $1/M_n$. It is tempting to try and relate $\partial n/\partial c$ to $n_p - n_s$ in some simple way, but this is risky. There is no general, rigorous expression for $n(c)$ that depends only on n_s, n_p, and c, and in fact n need not even vary monotonically with c. However, it is usually the case that $\partial n/\partial c$ will increase with the magnitude of $n_p - n_s$, so this becomes a reasonable starting point when choosing a solvent to maximize $\partial n/\partial c$. Note that $\partial n/\partial c$ can be positive, negative, or zero; as it appears squared in the scattered intensity, the sign does not matter. If one chooses an isorefractive solvent, such that $\partial n/\partial c \approx 0$, that polymer will be invisible in the scattering experiment. This can be used to advantage in examining polymer mixtures and copolymers, in order to highlight the behavior of one particular component.

8.8   Chapter Summary

In this chapter we have developed in some detail the equations governing the scattering of light from dilute polymer solutions. Light scattering is a very powerful experimental tool, providing both thermodynamic and structural information. Recent developments in commercial instrumentation have accelerated the use of light scattering in routine characterization. The main concepts in the development are the following:

1.   Completely uniform materials do not scatter. Scattering in polymer solutions arises from random fluctuations in concentration. This scattering is incoherent, which means that the intensity is independent of scattering angle.

2.   The incoherent intensity is determined by the mean-square concentration fluctuations, which in turn are set by the ratio of the thermal driving force, kT, to the free energy penalty for a given fluctuation, $\partial^2 G/\partial c^2$. It is through this relationship that the scattering experiment provides a measurement of M_w and B.

3. For polymers that are large enough, a significant phase difference can arise between portions of the incident wave that are scattered from different monomers on the same chain. This leads to angle-dependent, or coherent, scattering. The underlying process is very similar to Bragg diffraction from crystals, with the key difference being that in polymer solutions there are only average correlations in the positions of the various monomers, rather than a permanent lattice.

4. The description of coherent scattering is built around the scattering wavevector, $\vec{q}$. The magnitude of this vector depends on wavelength and scattering angle, and has units of inverse length. Depending on the magnitude of the dimensionless product qR_g, the coherent scattering can give information about the internal structure or the overall size of the polymer. It is often possible to use light scattering to measure R_g in a completely model-independent way.

Problems

1. Explain in your own words (i) why the permeation of solvent through a membrane into a polymer-rich phase and the amount of light scattered by a polymer solution are related, (ii) why even though they are closely related, the osmotic pressure and light scattering experiments measure different moments of the molecular weight distribution, and (iii) why the osmotic pressure measurement becomes increasingly difficult for degrees of polymerization $>10^3$ while light scattering becomes harder for degrees of polymerization $\leq 10^3$.

2. Estimate the largest and smallest molecular weight polystyrenes for which one could reasonably expect to measure R_g reliably in THF, using an Argon laser at 488 nm, and a usable angular range of $30° < \theta < 150°$; THF is a good solvent for polystyrene.

3. Carry though the integration of Equation 8.4.7 to obtain Equation 8.4.10.

4. An estimate of R_g can be obtained rather simply from the so-called *dissymmetry ratio*, defined as $I_{ex}(\theta = 45°)/I_{ex}(\theta = 135°)$. Explain how this works. Zimm has reported the intensity of scattered light ($\lambda = 364$ nm) at various angles of observation for polystyrene in toluene at a concentration of 2×10^{-4} g/mL[†]. The following results were obtained (values marked * were estimated and not measured):

θ (°)	I_s (a.u.)
0	4.29*
25.8	3.49
36.9	2.89
53.0	2.18
66.4	1.74
90.0	1.22
113.6	0.952
143.1	0.763
180	0.70*

Draw a plot in polar coordinates of the scattering envelope in the $x - y$ plane. How would the envelope of a Rayleigh scatterer compare with this plot? By interpolation, evaluate the dissymmetry ratio and R_g. What are some practical and theoretical objections to this procedure for estimating R_g?

5. The effect of adenosine triphosphate (ATP) on the muscle protein myosin was studied by light scattering in an attempt to resolve conflicting interpretations of viscosity and ultracentrifuge data. The controversy hinged on whether the myosin dissociated or changed molecular shape by interaction with ATP. Blum and Morales[‡] reported the following values of $(Kc/R_\theta)_{c=0}$ versus $\sin^2(\theta/2)$ for myosin in 0.6 M KCl at pH 7.0. Which of the two models for the mode of

[†] B.H. Zimm *J. Chem. Phys. 16*, 1093, 1099 (1948).
[‡] J.J. Blum and M.F. Morales *Arch. Biochem. Biophys. 43*, 208 (1953).

ATP interaction with myosin do these data support? Explain your answer by quantitative interpretation of the light-scattering data.

$\sin^2(\theta/2)$	0.15	0.21	0.29	0.37	0.50	0.85
$Kc/R_\theta \times 10^7$ (Before ATP)	0.9	1.1	1.5	1.8	2.2	2.7
$Kc/R_\theta \times 10^7$ (With ATP)	1.9	2.8	3.7	4.6	6.0	6.8

6. For poly(n-hexyl isocyanate) and poly(methyl methacrylate), estimate the range of molecular weights for which the Guinier regime ($qR_g < 1$) can be accessed. The data in Figure 6.10 and Table 6.1 may be useful.

7. Aggregation of fibrinogen molecules is involved in the clotting of blood. To learn something about the mechanism of this process, Steiner and Laki[†] used light scattering to evaluate M and the length of these rod-shaped molecules as a function of time after a change from stable conditions. The stable molecule has a molecular weight of 540,000 g mol^{-1} and a length of 840 Å. The accompanying table shows the average molecular weight and average length at several times for two different conditions of pH and ionic strength μ. Criticize or defend the following proposition: The apparent degree of aggregation x at various times can be obtained in terms of either the molecular weight or length. The ratio of the value of x based on M to that based on length equals unity exclusively for end-to-end aggregation and increases from unity as the proportion of edge-to-edge aggregation increases. In the higher pH–lower μ experiment there is considerably less end-to-end aggregation in the early stages of the process than in the lower pH–higher μ experiment.

	pH $= 8.40$ and $\mu = 0.35$ M			pH $= 6.35$ and $\mu = 0.48$ M		
t (s)	$M \times 10^{-6}$ (g mol^{-1})	Length (Å)	t (s)	$M \times 10^{-6}$ (g mol^{-1})	Length (Å)
650	1.10	1300	900	1.10	1100
1150	1.63	1600	1000	2.0	1200
1670	2.20	1900			
2350	3.30	2200			

8. Zimm plots at 546 nm were prepared for a particular polystyrene at two temperatures and in three solvents. The following summarizes the various slopes and intercepts obtained[‡]:

$T = 22°C$

Solvent	Intercept	$\left(\dfrac{\text{Slope}}{\text{Intercept}}\right)_{c=0}$	$\left(\dfrac{\text{Slope}}{\text{Intercept}}\right)_{\theta=0}$
Methyl ethyl ketone	0.896	0.608	260
Dichloroethane	1.61	1.16	900
Toluene	3.22	1.14	1060

$T = 67°C$

Solvent	Intercept	$\left(\dfrac{\text{Slope}}{\text{Intercept}}\right)_{c=0}$	$\left(\dfrac{\text{Slope}}{\text{Intercept}}\right)_{\theta=0}$
Methyl ethyl ketone	0.840	0.551	230
Dichloroethane	1.50	1.05	870
Toluene	2.80	1.09	800

[†] R.F. Steiner and K. Laki, *Arch. Biochem. Biophys.* **34**, 24 (1951).
[‡] P. Outer, C.I. Carr, and B.H. Zimm, *J. Chem. Phys. 18*, 839 (1950).

The slope–intercept ratios have units of cubic centimeters per gram, and the intercepts are $c/R_{\theta,v}$, where the subscript v indicates vertically polarized light. The following values of n and $\partial n/\partial c$ can be used to evaluate K:

	$T = 20°C$		$T = 67°C$	
	n	$\partial n/\partial c$	n	$\partial n/\partial c$
Methyl ethyl ketone	1.378	0.221	1.359	0.230
Dichloroethane	1.444	0.158	1.423	0.167
Toluene	1.496	0.108	1.472	0.118

Evaluate M_w, R_g, and B from each piece of pertinent data and comment on (a) the agreement between M_w values and (b) the correlation of R_g and B with solvent quality.

9. Plot the light scattered intensity (arbitrary units) versus scattering angle that you would expect to see for a very dilute solution of polystyrene with $M = 4,000,000$ in cyclohexane at 35°C. Assume the instrument has an angular range of $20°$–$150°$. On the same axes show the angular dependence of the intensity if the scattering object were a hard sphere with the same R_g.

10. For polystyrene in butanone at 67°C the following values of $Kc/R_\theta \times 10^6$ were measured at the indicated concentrations and angles. Construct a Zimm plot from the data below using $\gamma = 100$ mL/g for the graphing constant. Evaluate M, B, and R_g from the results. In this experiment $\lambda_o = 546$ nm and $n = 1.359$ for butanone.

	c (g/mL)	
θ (°)	1.9×10^{-3}	3.8×10^{-4}
25.8	—	1.48
36.9	1.84	1.50
53.0	1.93	1.58
66.4	1.98	1.62
90.0	2.10	1.74
113.6	2.23	1.87
143.1	2.34	1.98

11. Draw a sketch of a complete Zimm plot for a high molecular weight polymer in a theta solvent, assuming data were acquired at eight angles and at four concentrations. In a different color pen, indicate the effect on the data if a small amount of dust were present in the solution. Similarly, in a different color indicate the effect on the data of raising the solution temperature substantially. State any assumptions you make.

12. When a dilute solution of block copolymers undergoes micellization, i.e., some numbers of chains aggregate into a (usually spherical) assembly to shield one block from the solvent it does not like, the light scattering intensity increases. In fact, if micellization is induced by changing temperature at a fixed concentration, the ratio of the intensity after micellization to that before micellization is a good estimate of the average aggregation number of the micelles. Explain this observation.

13. Show that the form factors for the Gaussian coil and the hard sphere do indeed reduce to the expected $1 - q^2 R_g^2/3$ at low q, using the appropriate series expansions.

14. Imagine that you perform light scattering measurements on a diblock copolymer, and you generate a Zimm plot in the standard way. However, one of the blocks is isorefractive with the solvent ($\partial n/\partial c = 0$), so it does not contribute directly to the scattering signal. Comment on how the apparent values of M_w, R_g, and B (i.e., those obtained by assuming you are

looking at a homopolymer) might relate to the values associated with the copolymer as a whole, or those of the two blocks.

15. Imagine dissolving dilute quantities of two different polymers, A and B, in a solvent and making light scattering measurements. What is the appropriate equation to describe the results, analogous to the Zimm equation? You may assume that the chains are sufficiently small that $P(q) = 1$ for both A and B. You will need to consider two concentrations, c_A and c_B, two molecular weights M_A and M_B, two values of one important optical parameter, and more than one second virial coefficient. You do not need to do a complicated derivation to get the answer, but it is necessary to consider how the light scattering signal responds to concentration fluctuations, and how many different "kinds" of concentration fluctuations there are in a three-component mixture.

16. Suppose you had a new polymer, in which you knew the monomer structure but not the shape of the polymer in solution (e.g., rod, coil, spherical globule, etc.). A good light scattering measurement would give you M_w, R_g, and B. However, from this information alone, you could also infer the shape; explain how, perhaps with numerical estimates of the relevant parameters. Note that fitting to the form factor is not the answer; all form factors are the same in the small qR_g limit.

17. Imagine making light scattering measurements on a statistical copolymer of styrene and methyl methacrylate, in a very dilute solution, and at angles such that $qR_g \ll 1$. The sample has a mean styrene composition of f $(0 < f < 1)$. The refractive index increment of the sample in the particular solvent in question is measured, and found to be $\langle \partial n/\partial c \rangle$; as expected $\langle \partial n/\partial c \rangle = f(\partial n/\partial c)_{PS} + (1-f)(\partial n/\partial c)_{PMMA}$. Give an expression for the excess scattered intensity, for a solution of total concentration c, in terms of M_i, c_i, $(\partial n/\partial c)_i$, (where the subscript i allows for polydispersity in M), and the usual optical constants. In reality, the measured intensity may well be larger than this value; the reason lies in the fact that each chain will have a composition that may differ from f. Making reasonable assumptions about the M and f dependences of $(\partial n/\partial c)_i$, the refractive index increment of chain i, develop an expression for the observed intensity, in terms of averages involving δf_i, where $\delta f_i = f_i - f$, the difference in composition between chain i and the sample average. Your result should show the interesting (and correct) prediction that there can be excess scattered intensity from a statistical copolymer solution, even when a solvent is chosen such that $\langle \partial n/\partial c \rangle = 0$.

18. An analytical expression for the form factor of a Kratky-Porod worm-like chain (WLC, Chapter 6.4.2) is not available (although numerical approximations valid in various regimes have been developed). Nevertheless, with a little thought you should be able to make a reliable sketch of what $P(q)$ must look like. Take a WLC with $L = 100$ nm and $\ell_p = 10$ nm. Using a reasonable plotting program, generate a plot of $q^2 P(q)$ versus q (a so-called "Kratky" plot) for a Gaussian coil with the same R_g as this WLC. Run the q axis from 0 out to 0.5 nm^{-1}. Do the same thing for a rigid rid with $L = 100$ nm, and plot it on the same axes. Finally, by considering the low and high q asymptotic behavior of the various structures, draw by hand a smooth curve representing the WLC.

19. There are often differences in practice between various analysis schemes that are otherwise equivalent in principle. For example, use a computer to generate $P(q)$ "data" for a high molecular weight Gaussian coil from $qR_g = 0$ out to $qR_g = 2$. Add "noise" to the data with a reasonable amplitude using a random number generator. Then, fit the data to a straight line according to the Zimm approach ($1/P(q)$ versus $(qR_g)^2$) and to the Guinier approach (ln $P(q)$ versus $(qR_g)^2$. Vary the qR_g range of the data included in the fit at the high qR_g end (why is this appropriate?). Which fitting approach is more robust, in terms of returning the correct R_g value?

References

1. Brandrup, J. and Immergut, E.H. (Eds.), *Polymer Handbook*, 3rd ed., Wiley, New York, 1989.
2. Strutt, J.W. (Lord Rayleigh), *Phil. Mag.* 41, 177, 447 (1871).
3. Miller, T.H., *CRC Handbook of Chemistry and Physics*, 81st ed., pp. 10–160, CRC Press, Cleveland, 2001.
4. Zimm, B.H., *J. Chem. Phys.* 16, 1099 (1948).
5. Guinier, A., *Ann. Phys.* 12, 161 (1939).
6. Debye, P., *J. Appl. Phys.* 15, 338 (1944).
7. Neugebauer, T., *Ann. Physik* 42, 509 (1943).
8. Lord Rayleigh, *Proc. Roy. Soc.* A84, 25 (1910).

Further Readings

Brown, W. (Ed.), *Light Scattering: Principles and Development*, Clarendon Press, Oxford, 1996.
Huglin, M.B. (Ed.), *Light Scattering from Polymer Solutions*, Academic Press, New York, 1972.
Johnson, C.S. Jr. and Gabriel, D.A., *Laser Light Scattering*, Dover Publications, New York, 1994.
Tanford, C., *Physical Chemistry of Macromolecules*, Wiley, New York, 1961.
Yamakawa, H., *Modern Theory of Polymer Solutions*, Harper & Row, New York, 1971.
Wyatt, P.J., *Analytica Chimica Acta*, 272, 1, 1993.

9

Dynamics of Dilute Polymer Solutions

9.1 Introduction: Friction and Viscosity

This chapter contains one of the more diverse assortments of topics of any chapter in this volume. In this chapter we discuss the viscosity of polymer solutions, the diffusion of polymer molecules, the technique of dynamic light scattering, the phenomenon of hydrodynamic interaction, and the separation and analysis of polymers by size exclusion chromatography (SEC). At first glance these seem to be rather unrelated topics, but all are important to molecular weight determination in solution. Furthermore, all share a crucial dependence on the spatial extent of the molecules. In Chapter 8 we considered in detail how light scattering can provide a direct measurement of the radius of gyration. In this chapter the measure of size turns out to be roughly proportional to, but not numerically equal to, the radius of gyration. As the chapter heading suggests, we now consider for the first time in the book the time-dependent properties of polymers and particularly the rate at which polymer molecules move through a solvent. By emphasizing dilute solutions, the properties of individual polymer molecules are highlighted; in Chapter 11 we will consider the dynamic properties of more concentrated solutions and melts.

A parameter that plays an important role in unifying the concepts of viscosity and diffusion is the friction factor. We will initially define the *molecular friction factor*, *f*, by a thought experiment. Imagine a polymer molecule dissolved in a solvent. Further imagine that we have a way of pulling this molecule gently, but persistently, in one direction. After an initial start-up period, or transient response, we would find that if we apply a constant force $\vec{F}$ the molecule would move with a constant velocity $\vec{v}$ in the direction of the force. The proportionality factor between the applied force and the resulting velocity is $f = |\vec{F}|/|\vec{v}|$. The units of f are therefore (g cm/s^2)/(cm/s) = g/s in the cgs system (the SI unit is kg/s), and a typical value for a polymer dissolved in water might fall between 10^{-7} and 10^{-6} g/s. The next few sections are concerned with two general questions: how can we gain experimental access to f, and what can f tell us about the polymers themselves? Our starting point will be to consider a single hard sphere. Although this idealization might appear at first to resemble the ideal gas or the ideal solution in thermodynamics—a simple model to use in learning the ropes, but of limited practical relevance—it is not so. For a rather profound reason, which we shall explore in Section 9.6, even floppy random coils have friction factors very similar to those of hard spheres of comparable size. However, we will get to that in due course; let us start at the beginning. Before discussing hard spheres, we need to introduce the viscosity of a fluid in more concrete terms.

As a place to begin, we visualize the fluid as a set of infinitesimally thin layers moving parallel to each other, each with a characteristic velocity. In addition, we stipulate that those fluid layers that are adjacent to nonflowing surfaces have the same velocity as the rigid surface. This is another way of stating that there is no slip at the interface between the stationary and flowing phases, which is a good approximation for the slow flows of immediate interest. Now suppose we consider a sample of fluid that is maintained at constant temperature, sandwiched between two rigid parallel plates of area *A* as shown in Figure 9.1. If a force *F* in the *x*-direction is applied to the top plate that plate and the layer of fluid adjacent to it will accelerate until a steady *x*-velocity is reached (although force and velocity are vectors, we will drop the arrows for the remainder of this chapter). As long as the deforming force continues to be applied, this velocity is unchanged.

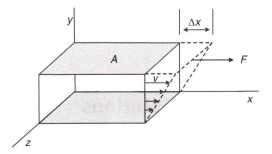

Figure 9.1 The relationship between the applied force F per unit area A and the velocity v used in the definition of viscosity.

This time-independent behavior is called steady flow and will be our primary concern. During the acceleration that precedes the stationary state, the velocity is a function of time. For our purposes, we shall simply wait until the stationary state is reached and not even question how long it will take. Force per unit area is called *stress*, and in shear flow is given the symbol σ.

In the experiment described in Figure 9.1, the bottom plate remains in place and the nonslip condition stipulated above requires that the layer of fluid adjacent to the bottom plate also has zero velocity. This situation clearly requires that the velocity of the fluid varies from layer to layer across the gap between the two rigid plates. To formulate this mathematically, we write that the top and bottom of these imaginary fluid layers are separated by a distance Δy and differ from each other in velocity by Δv. The ratio Δv$/\Delta y$ has units of reciprocal time and is called either the velocity gradient or the rate of shear. The former name is self-explanatory and the latter may be understood by considering the actual deformation the sample undergoes under the shearing force. During a short time interval Δt, the top layer moves a distance Δx relative to the bottom layer. Accordingly, Δv may be written as Δv $= \Delta x/\Delta t$ and the velocity gradient may be expressed as Δv$/\Delta y = (\Delta x/\Delta t)/\Delta y = (\Delta x/\Delta y)/\Delta t$. The shear displacement Δx divided by the distance over which it vanishes to zero, in this case Δy, is called the shear strain, which we represent by γ. These relationships show that the ratio Δv$/\Delta y$ also describes the rate at which the shear strain develops, or, more simply, the rate of shear $\Delta \gamma/\Delta t$, or $\dot{\gamma}$. In summary,

$$\frac{\Delta \text{v}}{\Delta y} = \dot{\gamma} = \frac{\Delta x/\Delta y}{\Delta t} \tag{9.1.1}$$

Now let us invoke our experience with liquids of different viscosities, say, water and molasses, and imagine the magnitude of the shearing force that would be required to induce the same velocity gradient in separate experiments involving these two liquids. Our experience suggests that more force is required for the more viscous fluid. Since the area of the solid plates in this liquid sandwich is also involved, we can summarize this argument by writing a proportionality relation between the shear force per unit area, σ, and the velocity gradient

$$\sigma = \frac{F}{A} = \eta \frac{\Delta \text{v}}{\Delta y} = \eta \dot{\gamma} \tag{9.1.2}$$

where η is called the coefficient of viscosity of the fluid or, more simply, the viscosity. Equation 9.1.2 implies that the velocity gradient is exactly the same throughout the liquid. As this may not be the case over macroscopic distances, our best assurance of generality is to consider the limiting case, in which Δy and therefore Δv approach zero. In the limit of these infinitesimal increments Δv and Δy become dv and dy, respectively, so Equation 9.1.2 becomes

$$\sigma = \eta \frac{d\text{v}}{dy} = \eta \dot{\gamma} \tag{9.1.3}$$

Equation 9.1.3 is called Newton's law of viscosity, and those fluids that follow it (with η independent of the magnitude of the shear rate) are said to be *Newtonian*.

Equation 9.1.3 describes a straight line with zero intercept if σ is plotted versus the velocity gradient; such a plot is shown in Figure 9.2. Since the coefficient of viscosity is the slope of this line, this quantity has a single value for Newtonian liquids. Liquids of low molecular weight compounds and their solutions are generally Newtonian, but quite a few different variations from this behavior are also observed. We shall not attempt to catalog all of these variations, but shall only mention the other pattern of behavior shown in Figure 9.2. This example of non-Newtonian behavior is described as *shear thinning*, and is often observed when the material under study is a polymer solution or melt. Since Equation 9.1.3 defines the coefficient of viscosity as the slope of a plot of σ versus velocity gradient, it is clear from Figure 9.2 that shear-thinning substances are not characterized by a single viscosity. The viscosity at a particular velocity gradient is given by the ratio $\sigma/(dv/dy)$; inspection of Figure 9.2 reveals that shear-thinning materials appear less viscous at high rates of shear than at low rates. For the purposes of this chapter and the next, we are only concerned with Newtonian response. For shear-thinning fluids, Newtonian response can be achieved by reducing the shear rate sufficiently to access the linear portion near the origin in Figure 9.2. This may be formalized by defining the *zero shear viscosity*, η_0:

$$\eta_0 \equiv \lim_{\dot{\gamma} \to 0} \eta \tag{9.1.4}$$

In most viscometers it is possible to vary $\dot{\gamma}$ in such a way as to ascertain whether η is constant. For the remainder of this book, the symbol η refers to the zero shear viscosity, unless explicitly stated otherwise.

To see another interpretation of viscosity, we multiply both sides of Equation 9.1.2 by $\Delta v/\Delta y$:

$$\sigma \frac{\Delta v}{\Delta y} = \frac{F}{A}\frac{\Delta v}{\Delta y} = \eta \left(\frac{\Delta v}{\Delta y}\right)^2 \tag{9.1.5}$$

To appreciate this result, we remember that Δv as introduced in Figure 9.1 is actually $\Delta x/\Delta t$. Thus the product $F(\Delta v/\Delta y)$ can be written as $F(\Delta x/\Delta t)/\Delta y$, and $F(\Delta v/\Delta y)/A$ becomes $F\Delta x/A\Delta y\Delta t$. The product of a force and the distance through which it operates equals an energy ΔE and the product of A and Δy equals the volume element of ΔV, upon which the shearing force described in

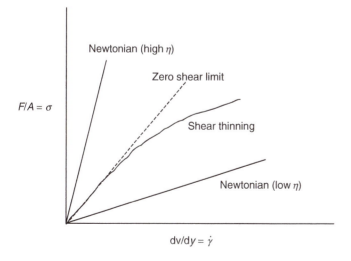

Figure 9.2   Comparison of shear stress versus shear rate for Newtonian and shear-thinning behavior.

Figure 9.1 operates. Therefore $F\Delta x/A\Delta y\Delta t$ is the same as $\Delta E/\Delta V\Delta t$. Defining the increment in shear energy dissipated per unit volume by the symbol ΔW, we obtain

$$\frac{\Delta W}{\Delta t} = \eta\left(\frac{\Delta v}{\Delta y}\right)^2 \tag{9.1.6}$$

As in the parallel case of going from Equation 9.1.2 to Equation 9.1.3, we take the limit of infinitesimal increments and write

$$\frac{dW}{dt} = \eta\left(\frac{dv}{dy}\right)^2 \tag{9.1.7}$$

The deforming forces that induce flow in fluids are not recovered when these forces are removed. These forces impart kinetic energy to the molecules, which is dissipated within the fluid as heat. Equation 9.1.7 implies that at sufficiently high shear rates the amount of heat generated could lead to measurable temperature increases; this phenomenon of *viscous heating* is indeed a concern in high shear-rate measurements and applications.

We conclude this section with a consideration of the units required for η by Equation 9.1.3. To do this, we rewrite these equations in terms of the units of all quantities except η. The units of η must make the expressions dimensionally correct. Force has units of mass times acceleration, or g cm/s^2 in cgs units. Since the viscosity gradient has the units s^{-1} and area is cm^2, the dimensional statement of Equation 9.1.3 is

$$\text{g cm s}^{-2}/\text{cm}^2 = (\eta)\text{ s}^{-1}$$

In order to satisfy this equation, η must have units g/cm/s, which is defined to be 1 poise (1 P). In the SI system the units are kg/m/s, or pascal seconds (Pa s). At room temperature, water has a viscosity of about 0.01 P or 0.001 Pa s and other low molecular weight liquids have comparable viscosities. The viscosity of a polymer liquid depends very much on the concentration and molecular weight of the polymer, as we shall see in Section 9.3 and in Chapter 11, and it can be many orders of magnitude larger than 0.01 P.

9.2 Stokes' Law and Einstein's Law

The shearing force F that is part of the definition of viscosity can also be analyzed in terms of Newton's second law and written as

$$F = ma = m\frac{dv}{dt} \tag{9.2.1}$$

When the force is divided by the area of a shearing plane, A, to obtain a stress, we would also write

$$\frac{F}{A} = \frac{m}{A}\frac{dv}{dt} = \eta\frac{dv}{dy} \tag{9.2.2}$$

If it were complete, Equation 9.2.2 would be a differential equation whose solution would give v, the velocity of the flowing liquid, as a function of time and position within the sample. Equation 9.2.2 does not tell the whole story, however. Other forces are also operative: external forces of, say, gravitational or mechanical origin are responsible for the motion in the first place, and pressure forces are associated with the velocity gradient. In general, things are not limited to one direction in space, but the forces, gradients, and velocities have x, y, and z components. It is possible to bring these considerations together in a very general form by adding together all of the forces acting on a volume element of liquid, including viscous forces, and using the net force and the mass of the volume element in Equation 9.2.2. The resulting expression is called the *equation of motion* and is the cornerstone of fluid mechanics. The equation of motion is generally written in terms of vector operators and takes on a variety of forms, depending on the system of coordinates and the vector identities that have been employed. If the equation of motion is complicated even in writing, things are even worse in the solving. Accordingly, we will not reproduce a full treatment here, but rather outline the essential points.

As with any differential equation, an important part of solving the equation of motion is defining the boundary conditions. An important boundary condition was introduced above, namely that no slip occurs at the boundary between a moving fluid and a rigid wall. The "no slip" or "stick" condition means that the fluid layer immediately adjacent to a stationary wall has a velocity of zero, with successive layers away from the wall possessing larger increments of velocity. At a sufficiently large distance from the wall, a net velocity is attained, which is unperturbed by the presence of the surface. In Section 9.4 we shall apply this idea to the flow of a liquid through a capillary tube. For now we consider two classic problems involving the effect of rigid spheres on the flow behavior of a liquid.

9.2.1   Viscous Forces on Rigid Spheres

The first of these problems involves relative motion between a rigid sphere and a liquid, as analyzed by Stokes around 1850 [1]. The results apply equally to liquid flowing past a stationary sphere with a steady-state (subscript "0") velocity v_0, or to a sphere moving through a stationary liquid with a velocity $-v_0$; the relative motion is the same in both cases. If the relative motion is in the vertical direction, we may visualize the slices of liquid described above as consisting of a bundle of layers, some of which are shown schematically in Figure 9.3.

In the horizontal plane containing the center of the sphere a limiting velocity v_0 is reached as the distance r from the center of the sphere becomes large. This would be the observable settling velocity of such a spherical particle, for example. In an infinitesimally thin layer adjacent to the surface, the tangential component of velocity would be that of the solid sphere. This is the no slip condition as it applies to this problem. This means that a velocity gradient exists that is described in terms of the distance from the center of the sphere and the component of velocity perpendicular to r. The velocity gradient dv/dr should certainly be proportional to v_0 and then we need a quantity with units of length to be dimensionally correct. The only relevant length we have is R, the radius

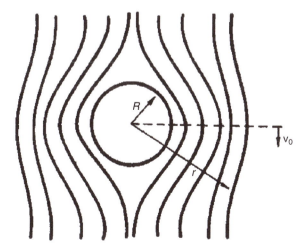

Figure 9.3 Distortion of flow streamlines around a spherical particle of radius R. The relative velocity in the plane containing the center of the sphere equals v_0 as $r \rightarrow \infty$.

of the sphere, and thus we anticipate that dv/dr is proportional to v_0/R. Formal analysis of the problem via the equation of motion verifies this argument from dimensional analysis and provides the necessary proportionality factors as well.

From Equation 9.2.2 we see that the total viscous force associated with this motion equals $\eta \times (dv/dr) \times (\text{area})$, where the pertinent area is proportional to the surface of the sphere and therefore varies as R^2. This qualitative argument suggests that the viscous force opposing the relative motion of the liquid and the sphere is proportional to $[\eta(v_0/R)] (R^2)$. The complete solution to this problem reveals that this is correct; both pressure and shear forces arising from the motion are proportional to $\eta R v_0$, and the total force of viscous resistance is given by

$$F_{\text{vis}} = 6\pi\eta R v_o = f v_o \tag{9.2.3}$$

This famous equation is *Stokes' law* for rigid spheres. We emphasize that the viscosity in Equation 9.2.3 is that of the medium surrounding the sphere by labeling it with the subscript "s" for solvent, and thus

$$f = 6\pi\eta_s R \tag{9.2.4}$$

for spherical particles of radius R.

9.2.2 Suspension of Spheres

The second classic problem arises in describing the viscosity of a suspension of spherical particles. This problem was analyzed by Einstein in 1906, with some corrections appearing in 1911 [2]. As with Stokes' law, we shall only present qualitative arguments to give plausibility to the final form. The fact that it took Einstein five years to work out the "bugs" in this theory is an indication of the complexity of the formal analysis. Derivations of both the Stokes and Einstein equations that do not require vector calculus have been presented by Lauffer [3]; the latter derivations are at about the same level of difficulty as most of the mathematics in this book, but are lengthy.

We return to Figure 9.1 with the stipulation that the volume of fluid sandwiched between the two plates is a unit of volume. This unit is defined by a unit of contact area with the walls and a unit of separation between the two walls. Next we consider a shearing force acting on this cube of fluid to induce a unit velocity gradient. According to Equation 9.1.7, the rate of energy dissipation per unit volume from viscous forces dW/dt is proportional to the square of the velocity gradient, with η_s the factor of proportionality. Thus, to maintain a unit gradient, a volume rate of energy dissipation equal to η_s is required.

Next we consider replacing the sandwiched fluid with the same liquid, but in which solid spheres are suspended at a volume fraction ϕ. Since we are examining a unit volume of liquid—a suspension of spheres in this case—the total volume of the spheres is also ϕ. We begin by considering the velocity gradient if the velocity of the top surface is to have the same value as in the case of the pure liquid. Being rigid objects, the suspended spheres contribute nothing to the velocity gradient. As far as the gradient is concerned, the spheres might as well be allowed to settle to the bottom and then be fused to the lower, stationary wall. The equivalency of the suspended spheres and a uniform layer of the same volume are illustrated schematically in Figure 9.4a and Figure 9.4b, respectively. Since the unit volume has a unit cross-sectional area, a volume ϕ fused to the base will raise the stationary surface by a distance ϕ and leave a liquid of thickness $1 - \phi$ to develop the gradient. These dimensions are also shown in Figure 9.4b. If the velocity of the top layer is required to be the same in this case as for the pure solvent, then the gradient in the liquid need only be the fraction $1/(1 - \phi)$ of that for the pure liquid. Of course, since ϕ is less than unity, this "fraction" is greater than unity.

Now we return to consider the energy that must be dissipated in a unit volume of suspension to produce a unit gradient, as we did above with the pure solvent. The same fraction applied to the shearing force will produce the unit gradient and the same fraction also describes the volume rate

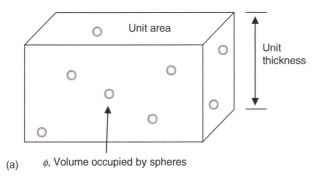

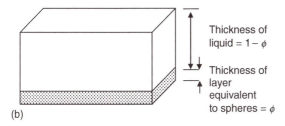

Figure 9.4 (a) Schematic representation of a unit cube containing a suspension of spherical particles at volume fraction ϕ. (b) The volume equivalent to the spheres in (a) is fused to the base, leaving $1 - \phi$ as the thickness of the liquid layer.

of energy dissipation compared to the situation described above for pure solvent. Since the latter was η, we write for the suspension, in the case of $dv/dy = 1$,

$$\frac{dW}{dt} = \eta = \frac{1}{1 - \phi}\eta_s \tag{9.2.5}$$

Again, since $\phi < 1$, $\eta > \eta_s$.

This is only one of the contributions to the total volume rate of energy dissipation; a second term that arises from explicit consideration of the individual spheres must also be included. This second effect can be shown to equal $1.5\,\phi\eta_s/(1 - \phi)^2$; therefore the full theory gives a value for η, the viscosity of the suspension:

$$\eta = \frac{\eta_s}{1 - \phi} + \frac{1.5\,\phi\eta_s}{(1 - \phi)^2} = \eta_s\frac{1 + 0.5\,\phi}{(1 - \phi)^2} \tag{9.2.6}$$

One additional assumption that underlies the derivation of the second term in Equation 9.2.6 is that ϕ is small. This being the case, $1/(1-\phi)^2$ can be replaced by the leading terms of the series expansion $(1 + \phi + \phi^2 + \cdots)^2$ to give

$$\eta = \eta_s\left(1 + \frac{1}{2}\phi\right)(1 + \phi + \cdots)^2 = \eta_s\left(1 + 2.5\,\phi + 4\,\phi^2 + \cdots\right) \tag{9.2.7}$$

This is Einstein's famous viscosity equation; the following observations are pertinent:

1. η is the viscosity of the suspension as a whole; η_s is the viscosity of the solvent; and ϕ is the volume fraction occupied by the spheres.
2. The validity of the derivation is limited to small values of ϕ, so Equation 9.2.7 is generally truncated after the first two terms on the right-hand side.
3. The viscosity does not depend on the radius of the spheres, only on their total volume fraction.

4.   By describing the concentration dependence of an observable property as a power series, Equation 9.2.7 plays a comparable role for viscosity as Equation 7.4.7 does for osmotic pressure.
5.   The volume fraction emerges from the Einstein derivation as the natural concentration unit to describe viscosity. This parallels the way volume fraction arises as a natural thermodynamic concentration unit in the Flory–Huggins theory, as seen in Section 7.3.

Both the Stokes and Einstein equations have certain features in common, which arise from the hydrodynamic origins they share:

1.   The liquid medium is assumed to be continuous. This makes the results suspect when applied to spheres, which are so small that the molecular nature of the solvent cannot be ignored.
2.   Both relationships have been repeatedly verified for a variety of systems and for spheres with a wide range of diameters. Despite item (1), both Equation 9.2.4 and Equation 9.2.7 have often been applied to individual molecules, for which they often work surprisingly well.
3.   The spherical geometry assumed in the Stokes and Einstein derivations gives the highly symmetrical boundary conditions favored by theoreticians. For ellipsoids of revolution having axial ratio a/b, friction factors have been derived by Perrin [4] and the coefficient of the first-order term in Equation 9.2.7 has been derived by Simha [5]. In both cases, the calculated quantities increase as the axial ratio increases above unity. For spheres, of course, $a/b = 1$.
4.   In the derivation of both Equation 9.2.4 and Equation 9.2.7, the disturbance of the flow streamlines is assumed to be produced by a single particle. This is the origin of the limitation to dilute solutions in the Einstein theory, where the net effect of an array of spheres is treated as the sum of the individual nonoverlapping disturbances. When more than one sphere is involved, the same limitation also applies to Stokes' law. In both cases, contributions from the walls of the container are also assumed to be absent.

We shall make further use of the Stokes equation later in this chapter; for the present, viscosity is our primary concern, and the Einstein equation is our point of departure.

9.3   Intrinsic Viscosity

9.3.1   General Considerations

The *intrinsic viscosity* $[\eta]$, a quantity that will be defined formally below, is a measure of the ability of added polymer to increase the viscosity of the solution over that of the solvent. It turns out that $[\eta]$ is directly related to the size of the polymer in solution, and before SEC (to be discussed in Section 9.8) it was the most commonly employed method for determining molecular weight. It is still useful in this regard, but it will also provide us with a basis for understanding much about the behavior of polymers in solution. We begin by proposing that the dilute solution viscosity can be written as a power series in concentration, by analogy with the osmotic pressure (Equation 7.4.7) and the Einstein relation (Equation 9.2.7):

$$\eta = \eta_s + ac + bc^2 \cdots \tag{9.3.1}$$

where a and b are unspecified coefficients and η_s is the solvent viscosity. We can rearrange Equation 9.3.1 by factoring out η_s:

$$\begin{aligned} \eta &= \eta_s \left(1 + a'c + b'c^2 \cdots \right) \\ &= \eta_s \left(1 + c[\eta] + k_h c^2 [\eta]^2 \cdots \right) \end{aligned} \tag{9.3.2}$$

where we have introduced the intrinsic viscosity, $[\eta]$, in place of a'. Note that by dimensional analysis we can see that $[\eta]$ must have units of inverse concentration, typically mL/g. Consequently it is *not* actually a viscosity; rather, it is a coefficient that quantifies the rate at which the solution viscosity increases per g/mL of added solute, when c is small. As c increases further, the next term, proportional to c^2, begins to contribute. Just as in the osmotic pressure virial expansion, the term in c^2 reflects the pairwise interactions between solute molecules; k_h is a kind of second virial coefficient for viscosity and is known as the Huggins coefficient. We can rearrange Equation 9.3.2 to obtain a formal definition of $[\eta]$:

$$[\eta] \equiv \lim_{c \to 0} \left(\frac{\eta - \eta_s}{c \eta_s} \right) \tag{9.3.3}$$

In the literature one encounters a variety of different ways of presenting the solution viscosity, and these are collected in Table 9.1 for reference purposes.

Having defined $[\eta]$, now we proceed to see what it can tell us about the polymer solute. Equation 9.3.2, which was just proposed on general principles, is analogous to Einstein's equation (Equation 9.2.7), which was derived using hydrodynamics. To compare these equations directly, we need to convert the concentration c into a volume fraction, ϕ. This can be done as follows:

$$\phi = \frac{\text{Solute volume}}{\text{Solution volume}} = \frac{c}{M}(\text{Volume of a mole of solute}) \tag{9.3.4}$$

where we have left the solute volume rather vaguely defined. For a rigid sphere, as envisioned by Einstein, this volume is simply Avogadro's number $\times (4\pi/3)R^3$. For other shapes, however, or for flexible coils, it is not so clear what volume is appropriate. The molar volume is a *thermodynamic* quantity, but the viscosity measurement is *hydrodynamic*, not thermodynamic. Accordingly, we finesse this issue by defining a *hydrodynamic volume* for the molecule, V_h:

$$\phi = \frac{c}{M} N_{av} V_h \tag{9.3.5}$$

and we insert Equation 9.3.5 into Equation 9.2.7:

$$\eta = \eta_s \left(1 + \frac{5}{2} \frac{c}{M} N_{av} V_h + \cdots \right) \tag{9.3.6}$$

which, by comparison with Equation 9.3.2 relates the intrinsic viscosity to the hydrodynamic volume:

$$[\eta] = \frac{5 N_{av}}{2} \frac{V_h}{M} \tag{9.3.7}$$

Now we make a bold step. We propose that the hydrodynamic volume is proportional to the radius of gyration cubed:

$$V_h \sim \frac{4\pi}{3} R_g^3 \tag{9.3.8}$$

Table 9.1 Summary of Names and Definitions of the Various Functions of η, η_s, and c in Which Solution Viscosities Are Frequently Discussed

Symbol	Definition	Common name	IUPAC name
η_r	$\dfrac{\eta}{\eta_s}$	Relative viscosity	Viscosity ratio
η_{sp}	$\dfrac{\eta}{\eta_s} - 1$	Specific viscosity	—
η_{red}	$\dfrac{1}{c}\left(\dfrac{\eta}{\eta_s} - 1 \right)$	Reduced viscosity	Viscosity number
$[\eta]$	$\lim\limits_{c \to 0} \left(\dfrac{\eta - \eta_s}{c \eta_s} \right)$	Intrinsic viscosity	Limiting viscosity number
η_{inh}	$\dfrac{1}{c} \ln\left(\dfrac{\eta}{\eta_s} \right)$	Inherent viscosity	Logarithmic viscosity number

In other words, the volume that matters in the viscosity experiment is not the volume actually *occupied* by the polymer segments (which would be the degree of polymerization times the volume of the monomer), but the volume *pervaded* by the entire molecule. For a random coil this means that we assume that the molecule behaves hydrodynamically like a rigid sphere of radius R_g. This might seem rather far fetched at first glance, but in fact, it is basically correct.

9.3.2 Mark–Houwink Equation

We know from Chapter 6, and Equation 6.6.1 in particular, that $R_g \sim M^\nu$, where the exponent ν takes on various characteristic values (1/3 for a solid sphere, 1/2 for a flexible coil in a theta solvent or in the melt, 3/5 for a flexible chain in a good solvent, and 1 for a rigid rod). If we insert this relation into Equation 9.3.8, and the result into Equation 9.3.7, we have

$$[\eta] \sim \frac{R_g^3}{M} \sim \frac{M^{3\nu}}{M} \sim M^{3\nu-1} \tag{9.3.9}$$

This gives what we were looking for: a direct relation between the intrinsic viscosity and the molecular weight. Using the values of ν cited above, we can see that $[\eta]$ should be independent of M for a rigid sphere, increase as $M^{1/2}$ or $M^{4/5}$ for a flexible chain in a theta solvent or good solvent, respectively, and increase as M^2 for a rod. These various possibilities are encompassed by the general relation

$$[\eta] = kM^a \tag{9.3.10}$$

This relationship with $a = 1$ was first proposed by Staudinger [6], but in this more general form it is known as the *Mark–Houwink equation* [7]. The constants k and a are called the Mark–Houwink parameters for a system. The numerical values of k and a depend on both the nature of the polymer and the nature of the solvent, as well as the temperature; extensive tabulations are available [8,9] and Table 9.2 gives a few examples. (Note, however, that the values can vary for a given system among different investigators, and that attention must be paid to details such as microstructure,

Table 9.2 Values for the Mark–Houwink Parameters for a Selection of Polymer–Solvent Systems at the Temperatures Noted

Polymer	Solvent	T (°C)	$k_s \times 10^3$ (mL/g)[a]	a
Poly(ethylene oxide)	Toluene	35	14.5	0.70
Poly(ethyleneterephthalate)	m-Cresol	25	0.77	0.95
1,4-Polybutadiene	Cyclohexane	20	36	0.70
1,4-Polyisoprene	Dioxane (θ)	34	135	0.50
Poly(hexamethylene adipamide)	m-Cresol	25	240	0.61
Poly(dimethylsiloxane)	Toluene	25	21.5	0.65
Polyethylene	p-Xylene	75	135	0.63
Polypropylene	Cyclohexane	30	20.9	0.76
Polyisobutylene	Benzene	40	43	0.60
Poly(methyl methacrylate)	Chloroform	25	4.8	0.80
Poly(vinyl chloride)	Chlorobenzene	30	71.2	0.59
Poly(vinyl acetate)	Ethanol (θ)	56.9	90	0.50
Poly(vinyl alcohol)	Water	25	20	0.76
Polystyrene	Toluene	25	17	0.69
Polystyrene	Cyclohexane (θ)	35	80	0.50
Polyacrylonitrile	dimethyl formamide	50	30	0.752
Poly(ε-caprolactam)	m-Cresol	25	320	0.62

[a] The units for k itself depend on the value of a; the indicated value gives $[\eta]$ in mL/g.

Source: From Kurata, M. and Tsunashima, Y., in *Polymer Handbook*, 3rd ed., Brandrup, J. and Immergut, E.H. (Eds.), Wiley, New York, 1989.

molecular weight range, and polydispersity of the samples.) Since viscometer drainage times (see Section 9.4) are typically on the order of a few hundred seconds, intrinsic viscosity experiments provide a rapid method for evaluating the molecular weight of a polymer. A possible drawback to the method is that the Mark–Houwink coefficients must be established for the particular system under consideration by calibration with samples of known molecular weight, but given the extensive tabulations this is often not a significant limitation.

The values of the exponent a in Table 9.2 range between 0.5 and about 1. Although in practice $a \approx 0.5$ in all theta systems, in good solvents the values vary quite a bit. Part of this is a consequence of the slow crossover to good solvent limiting behavior anticipated by the Flory–Krigbaum theory (Equation 7.7.10). In other words, even though a solvent might be "better-than-theta" for a given polymer, the experimental temperature might not be sufficiently far above $T = \theta$, or the molecular weight range might not extend to sufficiently high values, to obtain $\nu = 3/5$ and thus $a = 0.8$. In some cases $a > 0.8$, which can be attributed to a semiflexible structure (recall that the simple argument above predicts $a = 2$ for a rod). There is another, more complicated reason why the exponents can take on a range of values, the phenomenon of hydrodynamic interactions, and this will be explored in Section 9.7.

Even fractionated polymer samples are generally polydisperse, which means that the molecular weight determined from intrinsic viscosity experiments is an average value. The average obtained is called the *viscosity average molecular weight*, M_v, which can be derived as follows:

1. The experimental intrinsic viscosity is proportional to some average value, M_v, raised to the power a, according to Equation 9.3.10,

$$[\eta] = kM_v^a \tag{9.3.11}$$

2. The dilute solution viscosity for the polydisperse system can be expressed as

$$\eta = \eta_s(1 + \sum_i c_i[\eta]_i + \cdots) \tag{9.3.12}$$

where the index i refers to different molecular weights.

3. We can now obtain $[\eta]$ as

$$[\eta] = \lim_{c \to 0}\left(\frac{\eta - \eta_s}{c\eta_s}\right) = \lim_{c \to 0}\left(\frac{\sum c_i[\eta]_i}{\sum c_i}\right)$$
$$= \frac{\sum(n_iM_i/V)kM_i^a}{\sum n_iM_i/V} = k\frac{\sum n_iM_i^{1+a}}{\sum n_iM_i} \tag{9.3.13}$$

where n_i is the number of molecules with molecular weight M_i and we assume all species i have the same k and a.

4. Combining Equation 9.3.11 and Equation 9.3.13 we obtain

$$M_v \equiv \left(\frac{\sum_i n_iM_i^{1+a}}{\sum_i n_iM_i}\right)^{1/a} \tag{9.3.14}$$

For flexible polymers in general, $M_n < M_v < M_w$, and $M_v = M_w$ if $a = 1$. On the basis of this last observation, it can be argued that the Mark–Houwink coefficients should be evaluated using weight average rather than number average molecular weights as calibration standards. We saw in Chapter 8 how M_w values can be obtained from light scattering experiments.

Table 9.3 lists the intrinsic viscosity for a number of polystyrene samples of different molecular weights. The M values are weight averages based on light scattering experiments. The values of $[\eta]$ were measured in cyclohexane at the theta temperature of 35°C. In the following example we consider the evaluation of Mark–Houwink coefficients from these data.

Table 9.3 Intrinsic Viscosity as a Function of Molecular Weight for Samples of Polystyrene

(a) M_w	$[\eta]$ (mL/g)	(b) M_w	$[\eta]$ (mL/g)
266	1.49	19800	11.9
370	1.92	44000	18.0
474	2.47	50500	19.2
578	2.74	55000	20.0
680	2.93	76000	24.5
904	3.26	96200	26.0
1,480	3.70	1.25×10^5	29.0
1,780	4.08	1.60×10^5	34.0
2,270	4.50	1.80×10^5	35.4
3,480	5.43	2.47×10^5	42.0
5,380	6.59	3.94×10^5	54.5
10,100	9.00	4.06×10^5	55.0
20,500	12.3	5.07×10^5	60.0
40,000	17.2	6.22×10^5	66.0
97,300	27.3	8.62×10^5	78.0
1.91×10^5	38.0	1.05×10^6	86.0
3.59×10^5	51.2	1.56×10^6	106
7.32×10^5	73.4		
1.32×10^6	98.1		

Source: From (a) Yamakawa, H., Abe, F., and Einaga, Y., *Macromolecules*, 26, 1891, 1993; (b) Berry, G., *J. Chem. Phys.*, 46, 1338, 1967.

Example 9.1

Evaluate the Mark–Houwink coefficients for polystyrene in cyclohexane at 35°C from the data in Table 9.3. How well do the two data sets agree? What is the appropriate set of Mark–Houwink parameters for high molecular weight PS in this solvent? Does the exponent agree with expectation? At what molecular weight does this relation break down?

Solution

The two data sets are plotted in Figure 9.5 in a log–log format. This particular choice is helpful because adherence to a power law relationship, such as the Mark–Houwink relation, will result in the data falling on a straight line with the exponent as the slope. From Equation 9.3.10

$$\log [\eta] = \log k + a \log M$$

Visually it is clear that the two data sets agree extremely well over the common molecular weight range. Furthermore, for M greater than about 10,000 the data fall on a straight line. A combined least squares fit of the two sets of data in this range gives

$$\log [\eta] = -1.06 + 0.497 \log M$$

that corresponds to $k = 0.088$ and $a = 0.497$. The exponent agrees with expectation for a theta solvent; the uncertainty in a is at least ± 0.003. Below $M = 10,000$ the relationship between $[\eta]$ and M becomes more complicated. A full explanation of this dependence is not yet available, but among the contributing factors are non-Gaussian conformations for short chains, chain stiffness, chain end effects, and modification of the solvent dynamics in the vicinity of the chain.

For flexible coils the value of the Mark–Houwink exponent tells us something about the solvent quality, independent of the polymer or solvent. Is there anything similarly general to be extracted from the proportionality factor k? The answer is yes. In a theta solvent, particularly, it provides

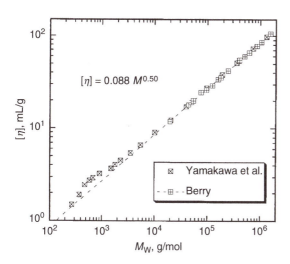

$$[\eta] = 0.088\, M^{0.50}$$

Figure 9.5 Plot of $\log [\eta]$ versus $\log M$ for the data in Table 9.3. An analysis of the Mark–Houwink parameters from these data is presented in Example 9.1.

information about the coil dimensions, as the following argument shows. For a theta solvent, we know that $R_g^2 = Nb^2/6$ (recall Equation 6.5.3). Combining this with Equation 9.3.7 and Equation 9.3.8 gives

$$[\eta] = \frac{5N_{av}}{2M} V_h \approx \frac{5N_{av}}{2M} \frac{4\pi}{3} R_g^3 = \frac{20\pi N_{av}}{6} \frac{N^{3/2}b^3}{M6^{3/2}} \tag{9.3.15}$$

We can replace N by M/M_o, where M_o is the monomer molecular weight, to obtain

$$[\eta] \approx \frac{20\pi N_{av}}{6^{5/2}} \frac{b^3}{M_o^{3/2}} M^{1/2} \tag{9.3.16}$$

or

$$k_\theta \approx \frac{20\pi N_{av}}{6^{5/2}} \frac{b^3}{M_o^{3/2}} = 4.3 \times 10^{23} \frac{b^3}{M_o^{3/2}} \tag{9.3.17}$$

Here we have used the subscript θ to remind us that we are dealing with a theta solvent. Equation 9.3.17 makes explicit the part of k that depends on the particular polymer in question. For polystyrene, $M_o = 104$ g/mol and $b = 6.7 \times 10^{-8}$ cm (see Table 6.1), and inserting these values into Equation 9.3.17 would predict that $k_\theta \approx 0.12$ (for $[\eta]$ in g/mL). From Example 9.1 above, in experiments $k_\theta = 0.088$, which is only about 30% different from the result utilizing the naive "a coil is pretty much a hard sphere with $R = R_g$" argument.

Equation 9.3.17 can be recast in another form, whereby the polymer-specific part is presented as $\langle h^2 \rangle_o/M$ (refer to Table 6.1):

$$k_\theta \approx 4.3 \times 10^{23} \frac{b^3}{M_o^{3/2}} = 4.3 \times 10^{23} \left(\frac{\langle h^2 \rangle_o}{M}\right)^{3/2}$$

$$\sim \Phi_o \left(\frac{\langle h^2 \rangle_o}{M}\right)^{3/2} \tag{9.3.18}$$

Here we have inserted Φ_o in place of the numerical prefactor of Equation 9.3.17. More detailed theories give $\Phi_o \approx 2.8 \times 10^{23}$ as a universal value for all flexible chains in a theta solvent, which is close to the polystyrene result in Example 9.1. The difference in the numerical value of the prefactor arises because in our simple model we assumed that a coil behaves as a hard sphere with $R = R_g$, whereas in fact it behaves as a hard sphere with $R \propto R_g$ and the proportionality constant is a little less than 1. Thus measurements of intrinsic viscosity in a theta solvent gives access to the coil dimensions (i.e., C_∞, b, or ℓ_p) via Equation 9.3.18.

Example 9.2

Using the arguments given above, predict the dependence of $[\eta]$ on generation number for dendrimer molecules. Recall from Chapter 1 and Chapter 4 that dendrimers have a tree-like structure, with each generation adding a new layer of material in a completely regular way.

Solution

The crucial relation we will invoke is Equation 9.3.9, that is,

$$[\eta] \sim \frac{R^3}{M}$$

and so the key step is to see how M and R grow with generation number n. The two-dimensional pictures of a dendrimer given in Figure 1.2 and Figure 4.7 suggest a roughly spherical structure, with each generation adding an approximately equal increment to R. Furthermore, as the later generations have larger numbers of units, one might suspect that they become densely packed. If this were the case, then R would grow as $M^{1/3}$, and we would recover the Einstein result for hard spheres: $[\eta]$ would be independent of M and therefore n. In fact, as the data in Figure 9.6 demonstrate, this is not true. Remarkably, $[\eta]$ goes through a maximum with increasing n, near $n = 3$; how does this arise?

The origin of this unusual behavior is rather easily understood. It turns out that, although R does increase roughly linearly with n, meaning that each generation adds a roughly constant increment to the total R, the mass added in each generation grows much more rapidly. Let us consider a

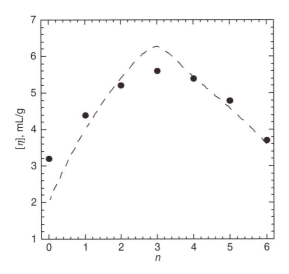

Figure 9.6 Intrinsic viscosity versus generation number for polyether dendrimers. The dashed curved represents the simple calculation described in Example 9.2, scaled by a factor of 5. (Data from Mourey, T.H., Turner, S.R., Rubinstein, M., Fréchet, J.M.J., Hawker, C.J., and Wooley, K.L., *Macromolecules*, 25, 2401, 1992.)

simple example. Assume a trifunctional reactive unit with $M = 100$ g/mol and assume that each generation adds 10 Å to the radius. We can compute a table of M, R, and R^3/M as a function of n. For $n = 0$, $M = 100$, and $R = 10$ by assumption. (Note that by convention, the initial dendrimer "core" is labeled generation 0.) For $n = 1$, $M = 100 + 3(100) = 400$, and $R = 10 + 10 = 20$. For $n = 2$, $M = 400 + 6(100) = 1000$, and $R = 20 + 10 = 30$. The multiplier 6 in the calculation of M arises because the previous generation added 3 units, each of which has two functional groups left to react further. The results through $n = 5$ computed in this way are listed in the table below:

n	0	1	2	3	4	5
R	10	20	30	40	50	60
M	100	400	1000	2200	4600	9400
R^3/M	10	20	27	29	27	23

From this simple calculation, we see that R^3/M goes through a maximum for $n = 3$, in excellent agreement with the experiments. The origin of the maximum lies in the fact that M grows geometrically with n rather than as a power law. Consequently, M in the denominator eventually increases more rapidly than R^3 in the numerator. In other words, the dendrimer density is increasing steadily, as opposed to a hard sphere, for which the density is constant with R. This density increase cannot persist indefinitely, and in fact dendrimers can typically only be grown out to $n = 6$–8 before there is no more room on the surface to complete another generation.

9.4 Measurement of Viscosity

In this section we consider two standard techniques for measuring viscosity. The first concerns the use of capillary viscometers for low-viscosity fluids, such as the dilute polymer solutions of relevance to $[\eta]$. The second describes the Couette or concentric cylinder geometry, which is more elaborate but is capable of covering a much wider range of η and $\dot{\gamma}$.

9.4.1 Poiseuille Equation and Capillary Viscometers

We defined the equation of motion as a general expression of Newton's second law applied to a volume element of fluid subject to forces arising from pressure, viscosity, and external sources. Although we shall not attempt to use this result in its most general sense, it is informative to consider the equation of motion as it applies to a specific problem: the flow of liquid through a capillary. This consideration not only provides a better appreciation of the equation of motion, but also serves as the basis for an important technique for measuring solution viscosity. We shall examine the derivation first and then discuss its application to experiment.

Figure 9.7a shows a portion of a cylindrical capillary of radius R and length L. We measure the general distance from the center axis of the liquid in the capillary in terms of the variable r and consider specifically the cylindrical shell of thickness dr designated by the broken line in Figure 9.7a. In general, gravitational, pressure, and viscous forces act on such a volume element, with the viscous forces depending on the velocity gradient in the liquid. Our first task then is to examine how the velocity of flow in a cylindrical shell such as this varies with the radius of the shell.

The net viscous force acting on this volume element is given by the difference between the frictional forces acting on the outer and inner surfaces of the shell:

$$F_{\text{vis, net}} = (F_{\text{vis}})_{\text{outer}} - (F_{\text{vis}})_{\text{inner}}$$

$$= 2\pi(r + dr) L\eta \left(\frac{dv}{dr}\right)_{r+dr} - 2\pi r L\eta \left(\frac{dv}{dr}\right)_r \qquad (9.4.1)$$

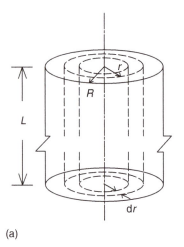
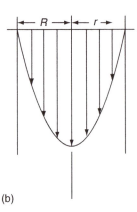

(a)   (b)

Figure 9.7   (a) Portion of a cylinder of radius R and length L showing (by broken lines) section of thickness dr. (b) Profile of flow velocity in the cylinder. (From Hiemenz, P.C., *Principles of Colloid and Surface Chemistry*, Marcel Dekker, New York, 1977.)

where the length times the circumference of the surface describes the appropriate area in Equation 9.1.2. The relationship between the velocity gradient at the two locations is given by

$$\left(\frac{dv}{dr}\right)_{r+dr} = \left(\frac{dv}{dr}\right)_r + \left(\frac{d^2v}{dr^2}\right)dr \tag{9.4.2}$$

provided that dr is small. Combining Equation 9.4.1 and Equation 9.4.2 and retaining only those terms that are first order in dr give

$$F_{vis,net} = 2\pi\eta L\left[r\left(\frac{d^2v}{dr^2}\right)dr + \left(\frac{dv}{dr}\right)dr\right] = 2\pi\eta L\frac{d}{dr}\left(r\frac{dv}{dr}\right) \tag{9.4.3}$$

Under stationary-state conditions of flow, that is, when no further acceleration occurs, this force is balanced by gravitational and pressure forces. For simplicity, we assume that the capillary is oriented vertically so that gravity operates downward and, for generality, we assume that an additional mechanical pressure Δp is imposed between the two ends of the capillary. Under these conditions, the net gravitational and mechanical forces acting on the volume element equal

$$F_{grav,mech,net} = (2\pi Lr\ dr)\rho g + (2\pi r\ dr)\Delta p \tag{9.4.4}$$

where $2\pi Lr\ dr$ is the volume of the element, $2\pi r\ dr$ is its cross-sectional area, ρ is the density of the fluid, and g is the acceleration due to gravity. Under the stationary-state conditions we seek to describe, Equation 9.4.4 and Equation 9.4.3 are equal, and the following relationship applies to the volume element:

$$\eta\frac{d}{dr}\left(r\frac{dv}{dr}\right) = \left(\rho g + \frac{\Delta p}{L}\right)r\ dr \tag{9.4.5}$$

Integration in the radial direction (along r) converts Equation 9.4.5 to

$$\eta r\frac{dv}{dr} = \frac{1}{2}\left(\rho g + \frac{\Delta p}{L}\right)r^2 \tag{9.4.6}$$

where the fact that $r(dv/dr) = 0$ at $r = 0$ is used to eliminate the integration constant. Note that the velocity gradient is directly proportional to the radial position in the fluid: it is zero at the axis and has a maximum value at $r = R$.

Equation 9.4.6 can be integrated again to give v as a function of r:

$$\int dv = \frac{\rho g + \Delta p/L}{2\eta} \int r \, dr \tag{9.4.7}$$

Because of the nonslip condition at the wall, $v = 0$ when $r = R$, and the constant of integration can be evaluated to give

$$v = \frac{\rho g + \Delta p/L}{4\eta} \left(R^2 - r^2 \right) \tag{9.4.8}$$

This result describes a parabolic velocity profile, as sketched in Figure 9.7b.

Equation 9.4.8 describes the velocity with which a cylindrical shell of liquid moves through a capillary under stationary-state conditions. This velocity times the cross-sectional area of the shell gives the incremental volume of liquid dV, which is delivered from the capillary in an interval of time Δt. The total volume delivered in this interval, ΔV, is obtained by integrating this product over all values of r:

$$\frac{\Delta V}{\Delta t} = \frac{2\pi(\rho g + \Delta p/L)}{4\eta} \int_0^R (R^2 - r^2) r \, dr$$

$$= \frac{(\rho g L + \Delta p)\pi R^4}{8\eta L} \tag{9.4.9}$$

This result is called the *Poiseuille equation*, after the researcher who discovered in 1844 this fourth-power dependence of flow rate on radius [10]; the unit of viscosity, poise, is also named after him. The following example illustrates the use of the Poiseuille equation in the area where it was first applied.

Example 9.3

Poiseuille was a physician–physiologist interested in the flow of blood through blood vessels in the body. Estimate the viscosity of blood from the fact that blood passes through the aorta of a healthy adult at rest at a rate of about 84 cm^3 s^{-1}, with a pressure drop of about 0.98 mmHg m^{-1}. Use 9 mm as the radius of the aorta for a typical human.

Solution

The pumping action of the heart rather than gravity is responsible for blood flow; hence the term $\rho g L$ can be set equal to zero in Equation 9.4.9 and the result solved for η:

$$\eta = \frac{\Delta p \pi R^4}{8L \Delta V / \Delta t}$$

The units must be expressed in a common system, with the pressure gradient requiring the most modification:

$$\frac{\Delta p}{L} = \frac{0.98 \text{ mmHg}}{m} \times \frac{133.3 \text{N m}^{-2}}{1 \text{ mmHg}} = 131 \text{ kg m}^{-2}\text{s}^{-2}$$

Therefore

$$\eta = \frac{\left(131\ \text{kg m}^{-2}\,\text{s}^{-2}\right)\pi\left(9\times 10^{-3}\ \text{m}\right)^{4}}{8\left(84\times 10^{-6}\ \text{m}^{3}\,\text{s}^{-1}\right)}$$

$$= 4.0\times 10^{-3}\ \text{kg m}^{-1}\,\text{s}^{-1} \quad (\text{or } 0.04\ \text{P})$$

At 37°C, the viscosity of pure water is about 0.69×10^{-3} kg m^{-1} s^{-1}; the difference between this figure and the viscosity of blood is due to the dissolved solutes in the serum and the suspended red cells in the blood. The latter are roughly oblate ellipsoids in shape.

The Poiseuille equation provides a method for measuring η by observing the time required for a liquid to flow through a capillary. The apparatus shown in Figure 9.8 is an example of one of many different instruments designed to use this relationship. In such an experiment, the time required for the meniscus to drop the distance between the lines etched at opposite ends of the top bulb is measured. This corresponds to the drainage of a fixed volume of liquid through a capillary of constant R and L. The weight of the liquid is the driving force for the flow in this case, so the Δp term in Equation 9.4.9 is zero and the observed flow time equals

$$\Delta t = \left(\frac{8\Delta V}{\pi g R^{4}}\right)\frac{\eta}{\rho} \tag{9.4.10}$$

or

$$\eta = A\rho\,\Delta t \tag{9.4.11}$$

where A represents a cluster of factors that are constant for a particular apparatus. The constant A need not be evaluated in terms of the geometry of the apparatus, but can be eliminated from

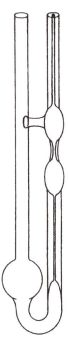

Figure 9.8 A typical capillary viscometer. (From Hiemenz, P.C., *Principles of Colloid and Surface Chemisty*, Marcel Dekker, New York, 1977.)

Equation 9.4.11 by measuring both a known (subscript 2) and an unknown (subscript 1) liquid in the same instrument:

$$\eta_1 = \frac{\rho_1}{\rho_2} \frac{\Delta t_1}{\Delta t_2} \, \eta_2 \tag{9.4.12}$$

Note that the time interval depends on both the density and the viscosity of the fluid, and the ratio η/ρ is sometimes referred to as the *kinematic viscosity*.

In more precise work an additional term, which corrects for effects arising at the ends of the tube, is added to Equation 9.4.11. This correction—which is often negligible—can be incorporated by writing

$$\eta = A\rho \, \Delta t - B\frac{\rho}{\Delta t} \tag{9.4.13}$$

where $B = \Delta V/8\pi$. As above, both A and B can be treated as instrument constants and evaluated by measuring two liquids, which are known with respect to η and ρ and then solving a pair of simultaneous equations for A and B. A better strategy might be to choose a capillary sufficiently narrow so that Δt is long enough to eliminate the second term on the right-hand side of Equation 9.4.13.

One limitation of this method is the fact that the velocity gradient is not constant in this type of instrument, but varies with r, as noted in connection with Equation 9.4.6. This would be a concern if the viscosity were shear-rate dependent over the relevant range. For dilute solutions, and the slow flows appropriate to determine the intrinsic viscosity, this usually does not matter.

9.4.2 Concentric Cylinder Viscometers

The second standard geometry for solution viscometry is based on concentric cylinders. The illustration that enabled us to define the coefficient of viscosity also suggests a modification that would be experimentally useful. Suppose the two rigid parallel plates in Figure 9.1 and the intervening layers of fluid were wrapped around the z-axis to form two concentric cylinders, with the fluid under consideration in the gap between them. The required velocity gradient is then established by causing one of these cylinders to rotate while the other remains stationary. The velocity is now in the direction described by the angle θ in Figure 9.9a, and its gradient is in the radial direction r. Thus the velocity gradient in this arrangement may be written dv_θ/dr.

Some of the reasons for our interest in this type of viscometer are the following:

1. The basic design is a direct extension of the discussion of viscosity in Section 9.1 and Section 9.2.
2. The range of applicability is very wide, extending at least from $\eta \approx 0.01$–10^{10} P.
3. The design permits different velocity gradients to be considered, so that non-Newtonian behavior (e.g., shear thinning) can be investigated, if desired.
4. The number of technically important viscosity-measuring devices may be thought of as variants of this basic apparatus.

As a practical matter, the outer cylinder is part of a cup that holds the fluid, while the inner cylinder is a coaxial bob suspended within the outer cup. Suppose the cup is centered on a turntable that rotates with an angular velocity ω, measured in radians per second. The viscous fluid now transmits a force, which can be measured in terms of the torque on a torsion wire to the suspended bob. This arrangement is sketched in Figure 9.9b. In this representation, the outer cylinder has a radius R and the inner cylinder has a radius fR, where f is some fraction. The closer this fraction is to unity, the narrower will be the gap between the cylinders and the more closely the apparatus will approximate the parallel plate model in terms of which η is defined.

A formal mathematical analysis of the flow in the concentric cylinder viscometer yields the following relationship between the experimental variables and the viscosity:

$$\text{torque} = \text{force} \times \text{radius} = 4\pi\eta LR^2 \omega \frac{f^2}{1-f^2} \tag{9.4.14}$$

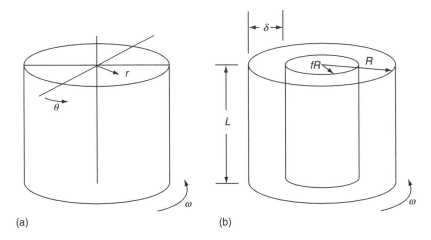

Figure 9.9 Definition of variables for concentric cylinder viscometers; (a) the rotating cylinder and (b) the coaxial cylinders.

This equation can be cast into a more recognizable form by assuming that f is very close to unity. In that case we have the following:

1. The radius R applies to the entire fluid sample. Since torque equals the product of force and R, canceling out one power of R leaves the shearing force acting on the fluid on the left-hand side of Equation 9.4.14.
2. The remaining factor R times ω on the right-hand side of Equation 9.4.14 can be replaced by the linear velocity v_θ.
3. The factor $1 - f^2$ can be replaced by $2(1 - f)$, since $1 - f^2 = (1 + f)(1 - f)$ and $(1 + f) \rightarrow 2$ as $f \rightarrow 1$.
4. The area of contact A between the cylinders and the fluid is $2\pi RL$; therefore $4\pi LR/2R(1 - f) = A/(1 - f)R$.
5. The product $(1 - f)R$ is the width of the gap, δ.

Introducing these substitutions in Equation 9.4.14 gives

$$\frac{F}{A} = \sigma = \eta\, f^2 \frac{v_\theta}{\delta} \tag{9.4.15}$$

Since v_θ is the difference in velocity between the inner and outer cylinders and δ is the difference in the radial location of the two rigid surfaces, Equation 9.4.15 becomes

$$\sigma = \eta \frac{dv_\theta}{dr} \tag{9.4.16}$$

in the limit as $f \rightarrow 1$. This is identical to Equation 9.1.3 and is the result we anticipated in rolling Figure 9.1 into a cylinder. Equation 9.4.14 is more general than Equation 9.4.16, since its applicability is not limited to vanishingly small gaps.

9.5 Diffusion Coefficient and Friction Factor

We now turn our attention to the phenomenon of diffusion. This turns out to be directly related to the viscosity through the friction factor, f. Most of us have a sense of diffusion as a randomization process: place a drop of food coloring into a glass of water, and before long the entire glass will be uniformly colored even without stirring. The time it takes for the color to spread depends on two things: how rapidly the dye molecules move, and the size of the glass. The former is quantified by a *diffusion coefficient*, D, which depends directly on the molecular friction factor. Diffusion is an important process to understand and control in many polymer applications (e.g., how long will soda in a plastic bottle hold its fizz? How long does it take a drug in a skin patch to pass through the skin?), and the diffusion coefficient itself is also a useful means of molecular characterization. In this section we will emphasize the diffusion of polymers in dilute solution, but the development has a much broader range of applicability. We begin the discussion by distinguishing two related, but conceptually distinct, diffusion coefficients, which we shall call the tracer diffusion coefficient, D_t, and the mutual diffusion coefficient D_m. Failure to distinguish clearly between these two quantities can be a major source of confusion.

9.5.1 Tracer Diffusion and Hydrodynamic Radius

Returning to our glass of water, imagine a single dye molecule somewhere in the glass and further imagine that we could actually follow its motion directly in time and space. (This is not too far fetched, in fact, with recent developments in single-molecule spectroscopy and microscopy.). What would we see? The molecule is constantly buffeted by solvent molecules, and consequently executes *Brownian motion*: it moves tiny distances between collisions, with the instantaneous direction of motion fluctuating randomly. If we watch for a relatively long time interval, how far is the dye molecule likely to move? We already have developed all the mathematics needed to answer this question, when we considered random walks in Chapter 6. Namely, if the molecule executes a total of N random steps of average length b, the mean square displacement $\langle r^2 \rangle$ will be Nb^2. We can formalize this as follows:

$$\lim_{t \to \infty} \frac{\langle r^2 \rangle}{t} = \frac{Nb^2}{t} \equiv 6D_t \tag{9.5.1}$$

The units of the diffusion coefficient are (length)^2/time, or cm^2/s in the cgs system. We take the limit of long times just to remove all memory of past collisions and directions. We may consider the factor of 6 in front of D_t to be a historical convention. The average denoted by $\langle \cdots \rangle$ is necessary because by definition we cannot make any predictions about a single random walk. Rather, if we watch a particular particle for many time intervals or watch many independent but otherwise identical particles over a given time interval, we can generate the average mean square displacement. In fact, based on our work with random walks in Chapter 6, we even know the distribution function that should describe the results of many equivalent experiments: it should be a Gaussian. Just as we did in Chapter 6, we can ask the following question: if we define the location of a particle at time $t = 0$ as the origin, what is the probability of finding it a distance r away after time t? This probability is known by a special name, the "van Hove space–time self-correlation function," but it is mathematically equivalent to Equation 6.7.12:

$$P(r, t) = 4\pi \, r^2 \left(\frac{1}{4\pi \, D_t t} \right)^{3/2} \exp\left(-\frac{r^2}{4D_t t} \right) \tag{9.5.2}$$

Equation 6.7.12 can be recovered by substituting $6D_t t = Nb^2$ into Equation 9.5.2. Also, recall that the prefactor of $4\pi r^2$ emerges because we are interested in the distance from the origin, independent of direction. Unlike the case of polymer chains, which are self-avoiding walks, a Brownian particle really does execute a random walk.

The tracer diffusion coefficient is a property of an individual molecule or particle undergoing Brownian motion. Its value, however, will generally depend on the size of the molecule and on the medium in which it is diffusing. Einstein showed that there is a beautifully simple relationship between the friction factor and D_t:

$$D_t = \frac{kT}{f} \tag{9.5.3}$$

From the definition of the friction factor, we can see that the thermal energy, kT, is playing the role of a generalized "force," and D_t is the resulting "velocity." As T increases, solvent molecules move more rapidly and are more effective at jostling the tracer molecule, so D_t should increase. At the same time, if the solvent viscosity increases, or if the particle increases in size, then f should increase and D_t will be smaller. If we now consider our tracer molecule to be a hard sphere and the solvent to be a continuum, then we can incorporate Stokes' law for f (Equation 9.2.4) to arrive at the *Stokes–Einstein relation*

$$D_t = \frac{kT}{6\pi\eta_s R} \tag{9.5.4}$$

This simple relation provides a direct connection between the tracer diffusion coefficient, the particle size, and the viscosity of the solvent. We can go an important step further, however. If our particle is not a hard sphere, we can use Equation 9.5.4 to define an equivalent radius in terms of a measured diffusivity, called the hydrodynamic radius, R_h:

$$R_h \equiv \frac{kT}{6\pi\eta_s D_t} \tag{9.5.5}$$

In other words, R_h for any polymer is the radius of a hard sphere that would have the same friction factor or diffusivity. In Section 9.3 we saw that flexible molecules behaved hydrodynamically rather like hard spheres with radius proportional to R_g. Does this hold for diffusion as well? Yes, indeed. Consequently R_h is directly proportional to R_g, with the proportionality factor depending on the particular polymer shape. Thus, R_h depends on molecular weight with the same power law exponent as does R_g, and D_t exhibits the inverse of that dependence. For flexible chains, therefore, $D_t \sim M^{-1/2}$ in a theta solvent and $D_t \sim M^{-3/5}$ in a very good solvent. Examples are shown in Figure 9.10 for polystyrene in cyclohexane at the theta temperature, and in toluene, a good solvent. There are several experimental techniques by which diffusion may be measured, and examples will be given following the next section.

9.5.2 Mutual Diffusion and Fick's Laws

Now we turn to the mutual diffusion coefficient, D_m, which describes how a collection of Brownian particles will distribute themselves in space. In the context of the food coloring analogy, D_t tells us how rapidly any individual dye molecule explores space, but D_m describes how quickly the entire droplet of food coloring disperses itself. Experience tells us that after a reasonable time interval, the glass of water will be uniformly colored. The underlying reason is that mutual diffusion acts to eliminate any gradients in concentration. Although the individual dye molecules are happily diffusing about, largely oblivious of one another, collectively they tend to spread themselves out evenly.

Fick first recognized that mass diffusion was analogous to thermal diffusion and he proposed an adaptation of Fourier's law of heat conduction for the transport of material [11]. Specifically, the flux J (in units of mass per unit area per unit time) across a plane is assumed to be proportional to the gradient in concentration (mass/volume) along the direction perpendicular to the plane. We will restrict ourselves to one-dimensional diffusion along the x-direction, and therefore

$$J = -D_m \frac{dc}{dx} \tag{9.5.6}$$

In this expression, called Fick's first law, the proportionality constant is D_m and it follows that D_m also has units of length^2/time. The minus sign in Equation 9.5.6 recognizes the fact that the direction of the flow is that of decreasing concentration.

We now consider a volume element and the flux of solute in and out of that element. Figure 9.11 schematically represents three regions of an apparatus containing a concentration gradient. The end

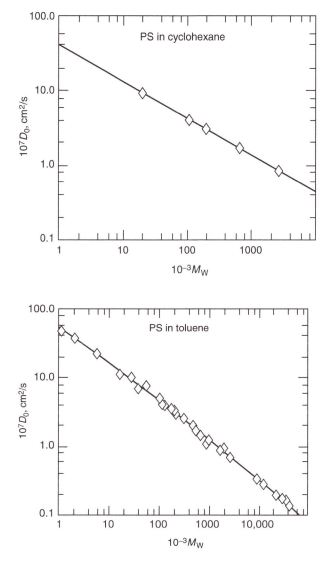

Figure 9.10 Molecular weight dependence of D_t for polystyrene in (a) a theta solvent, cyclohexane at 35°C, and (b) a good solvent, toluene. The curve in (a) has a slope of -0.50, and the curve in (b) approaches a slope of -0.60 at large M. (Reproduced from Schaefer, D.W. and Han, C.C., *Dynamic Light Scattering*, R. Pecora (Ed.), Plenum, New York, 1985. With permission.)

compartments contain the solute at two different concentrations c_1 and c_2, with $c_2 > c_1$. The center region is a volume of cross-sectional area A and thickness dx, along which the gradient exists. The arrows in Figure 9.11 represent the flux of solute from the more concentrated solutions to the less concentrated one. The incremental change in the total amount of solute dQ in the center volume element per time increment dt can be developed in two different ways. In terms of the fluxes,

$$\frac{dQ}{dt} = (J_{in} - J_{out})A \tag{9.5.7}$$

whereas in terms of a concentration change dc in the element of volume $A\,dx$,

$$dQ = dc(A\,dx) \tag{9.5.8}$$

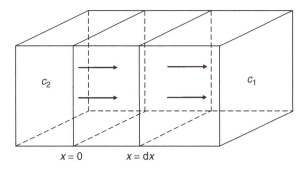

Figure 9.11 Schematic of diffusion with respect to a volume element of thickness dx located at $x = 0$ between regions with concentration c_1 and $c_2 > c_1$.

By combining these two relations, and substituting Fick's first law for J, we obtain

$$\frac{dc}{dt}(A\,dx) = (J(x) - J(x + dx))A$$

$$= -D_m \left[\left(\frac{dc}{dx}\right)_x - \left(\frac{dc}{dx}\right)_{x+dx} \right] A \qquad (9.5.9)$$

The areas cancel and the term in square brackets can be recognized as

$$\left[\left(\frac{dc}{dx}\right)_{x+dx} - \left(\frac{dc}{dx}\right)_x \right] = \frac{d^2c}{dx^2}\,dx \qquad (9.5.10)$$

that leads us directly to Fick's second law (in one dimension):

$$\frac{dc}{dt} = D_m \frac{d^2c}{dx^2} \qquad (9.5.11)$$

This differential equation can be solved (sometimes easily, sometimes not) when one specifies the appropriate initial or boundary conditions (see Example 9.4 and Example 9.5).

Fick's second law is very useful, but it leaves us with two unresolved issues. First, how is D_m related to D_t and f? Second, what is the role of thermodynamics in this process? These questions can be answered together. In general, a single phase will be at equilibrium when the temperature, pressure, and *chemical potential* of each species are everywhere the same. In other words, diffusion actually acts to remove gradients in μ rather than gradients in c. The thermodynamic "driving force" can be written (recall that the gradient of a potential is a force) as

$$F_{diff} = -\frac{1}{N_{av}}\frac{d\mu_2}{dx} \qquad (9.5.12)$$

We divide Equation 9.5.12 by Avogadro's number to convert the partial molar Gibbs free energy, μ, to a molecular quantity and the minus sign enters as in Fick's first law because the force and the gradient are in opposing directions. Recalling the definition of chemical potential from Equation 7.1.13 we write $\mu_2 = \mu_2^o + RT\ln a_2 = \mu_2^o + RT\ln\gamma_2 c$, where a_2 and γ_2 are the activity and activity coefficient, respectively, of the solute (note that when c is in g/mL, γ_2 is defined accordingly). Substituting into Equation 9.5.12 we obtain

$$F_{diff} = -\frac{1}{N_{av}}\left(\frac{d\mu_2}{dc}\right)\left(\frac{dc}{dx}\right) = -\frac{RT}{N_{av}}\left(\frac{1}{c} + \frac{d\ln\gamma_2}{dc}\right)\left(\frac{dc}{dx}\right) \qquad (9.5.13)$$

Under stationary-state flow conditions, F_{diff} equals the force of viscous resistance experienced by the particle. The latter, in turn, equals the friction factor times the stationary velocity v_0; therefore

$$-\frac{kT}{c}\left(1 + c\frac{d\ln\gamma_2}{dc}\right)\left(\frac{dc}{dx}\right) = f\,v_0 \tag{9.5.14}$$

The product cv_0 defines the flux J, and therefore Equation 9.5.14 becomes

$$J = -\frac{kT}{f}\left(1 + c\frac{d\ln\gamma_2}{dc}\right)\frac{dc}{dx} \tag{9.5.15}$$

Comparing Equation 9.5.6 and Equation 9.5.15 gives the desired result

$$D_m = \frac{kT}{f}\left(1 + c\frac{d\ln\gamma_2}{dc}\right) = D_{t,o}\left(1 + c\frac{d\ln\gamma_2}{dc}\right) \tag{9.5.16}$$

In the limit of low concentration of solute, $c \rightarrow 0$, we see that $D_m \rightarrow D_{t,o}$, where the subscript "0" refers to this "infinite dilution" limit. This is a completely general result for any two-component mixture: the mutual diffusivity approaches the tracer diffusivity of the minor component as its concentration tends to zero. At finite concentrations, however, things are not so simple. Clearly we can define tracer diffusion coefficients for both polymer and solvent and these may be very different. On the other hand, there is only one mutual diffusion coefficient, but it need not bear any simple relation to the tracer diffusivities. In Equation 9.5.16 we should also consider the possibility that the solute friction factor, f, will depend on concentration. If we propose a series expansion, that is,

$$f(c) = f_o\{1 + k_f c + \cdots\} \tag{9.5.17}$$

and insert this into Equation 9.5.16, we would obtain

$$D_m = \frac{kT}{f_o(1 + k_f c)}\left(1 + c\frac{d\ln\gamma_2}{dc}\right) = D_{t,o}\left(1 + c\left\{\frac{d\ln\gamma_2}{dc} - k_f\right\}\right) \tag{9.5.18}$$

Experimentally, a variety of tools can be used to determine either D_t or D_m. Probably the most commonly employed technique for dilute polymer solutions is dynamic light scattering, also known as quasielastic light scattering, which we will briefly describe in Section 9.6. Another common technique for solutions, which we will not discuss, is pulsed-field gradient NMR. It measures the random motions of particular nuclear spins without regard to any gradients in concentration, and thus determines D_t. A more generic approach, often applied to less fluid samples, is to prepare adjacent layers of two different compositions, and watch them interdiffuse by some depth-profiling technique. This is illustrated in the following example.

Example 9.4

In many experiments to study polymer diffusion, some fixed, small amount of the polymer is placed in contact with the medium into which it will diffuse. This medium, or matrix, could be a solvent, another polymer, or a solution. Solve Fick's second law (Equation 9.5.11) for the one-dimensional case where an extremely thin layer of polymer is placed in an "infinite" beaker of solvent.

Solution

In fact, we have already almost solved this problem in Equation 9.5.2. That case concerned three-dimensional diffusion from the origin, whereas now we have one-dimensional diffusion from a plane. Nevertheless, the solution is still a Gaussian function, just as in Section 6.7, where we showed that the end-to-end distance of a random walk was a Gaussian function in x, y, or z.

For Fick's second law in this case

$$\frac{\partial c(x,t)}{\partial t} = D_m \frac{\partial^2 c(x,t)}{\partial x^2}$$

and the solution is

$$c(x,t) = c_0 \frac{1}{\sqrt{4\pi D_m t}} \exp\left(-\frac{x^2}{4D_m t}\right)$$

In order to show that this is a solution, we can follow through with the differentiation:

$$\frac{\partial c}{\partial t} = c_0 \frac{1}{\sqrt{4\pi D_m t}} \exp\left(-\frac{x^2}{4D_m t}\right)\left(-\frac{1}{2}t^{-1} + \frac{x^2}{4D_m}t^{-2}\right)$$

which must be equal to $D_m(\partial^2 c/\partial x^2)$. Now take the derivative twice with respect to x:

$$D_m \frac{\partial c}{\partial x} = D_m c_0 \frac{1}{\sqrt{4\pi D_m t}} \exp\left(-\frac{x^2}{4D_m t}\right)\left(-\frac{2x}{4D_m t}\right)$$

and

$$D_m \frac{\partial^2 c}{\partial x^2} = D_m c_0 \frac{1}{\sqrt{4\pi D_m t}} \exp\left(-\frac{x^2}{4D_m t}\right)\left(-\frac{2}{4D_m t} + \frac{2x}{4D_m t}\frac{2x}{4D_m t}\right)$$

which is the same as the expression for $\partial c/\partial t$ above.

The character of this solution is shown in Figure 9.12a. The polymer concentration begins as a spike at $x=0$, and then as time progresses, it spreads out symmetrically to positive and negative values of x. The constant c_0 is the total amount of polymer in the initial spike, as can be seen from the following argument. If we integrate $c(x,t)$ over all x at any time, we must get the total amount of polymer:

$$\int_{-\infty}^{\infty} dx\, c(x,t) = c_0 \frac{1}{\sqrt{4\pi D_m t}} \int_{-\infty}^{\infty} dx \exp\left(-\frac{x^2}{4D_m t}\right)$$

$$= c_0 \frac{1}{\sqrt{4\pi D_m t}} \sqrt{4\pi D_m t} = c_0$$

where we use the expression for the solution to the integral of a Gaussian given in Section 6.7.

In real experiments, it is more common to have a layer of finite thickness of material in contact with the effectively infinite matrix. For simplicity, we center this layer at $x=0$ and it extends from $x=-1/2$ to $x = +1/2$, as illustrated in Figure 9.12b. This new initial condition changes the solution to Fick's law. The approach to the solution is to view the layer of thickness h as a series of infinitesimal layers and then to calculate the amount of material at each point x as a sum of the material that came from each of the infinitesimal layers. We already have the solution for one infinitesimal layer, and the result for the finite layer is

$$c(x,t) = \frac{1}{2}c_0 \left\{\text{erf}\left(\frac{x+h}{\sqrt{4D_m t}}\right) + \text{erf}\left(\frac{x-h}{\sqrt{4D_m t}}\right)\right\}$$

where erf denotes the error function

$$\text{erf}(z) = \frac{2}{\sqrt{\pi}} \int_0^z du \exp\left(-u^2\right)$$

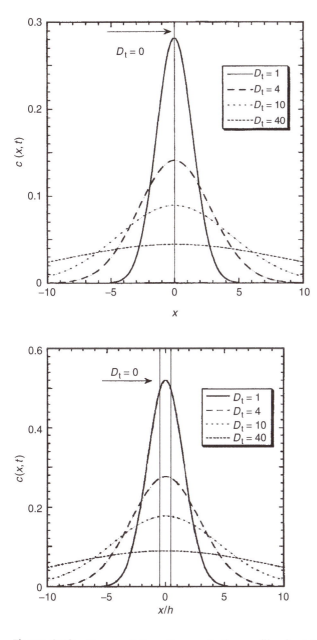

Figure 9.12 Time evolution of concentration profiles for (a) an infinitesimal layer of material at $x = 0$, $t = 0$, and (b) a layer of thickness h centered on $x = 0$ at $t = 0$. The corresponding mathematical forms for $c(x, t)$ are discussed in Example 9.4.

The error function is tabulated in many standard mathematical references. The solution for $c(x, t)$ is also shown in Figure 9.12b and to a first approximation it looks rather like the result for a single layer.

In an actual diffusion experiment, the geometry is usually slightly different, namely a single thin layer of material is placed on one thick layer of matrix. However, with a little thought it should be apparent that the answer is the same as above, by symmetry. The sum of error functions is symmetric with respect to $x > 0$ and $x < 0$ (note that $\mathrm{erf}(z) = \mathrm{erf}(-z)$), and we could imagine placing an impenetrable barrier at $x=0$. In this case no material starting between $x=-h/2$ and $x=0$ would appear at $x>0$, but an equal amount that started between $x=0$ and $x=h/2$ would not be "lost" to $x<0$. The value of D_m would be obtained by allowing diffusion to occur for some appropriate time interval, and then using some kind of depth profiling experiment capable of measuring $c(x)$. The resulting concentration profile could be fit to the solution given above, to extract D_m.

9.6 Dynamic Light Scattering

We recall from the discussion in Section 8.2 and Section 8.4 that light scattering in a dilute polymer solution arises from fluctuations in concentration (e.g., see Equation 8.4.4). However, each fluctuation must appear and disappear over some time interval and this time interval is determined by D_m. As a consequence, the total scattered intensity also fluctuates in time and these fluctuations may be analyzed to extract a characteristic relaxation time, τ. The magnitude of the scattering vector, q (defined in Equation 8.2.4), sets the relevant length scale in solution to be $1/q$ (see Section 8.6) and the relaxation time τ turns out to be equal to $1/(q^2 D_m)$.

From Section 8.2 and Section 8.6 we remember that the only fluctuations that will contribute to the scattered intensity I_s are the fluctuations that happen to have period $2\pi/q$ and that are oriented along the scattering vector, in other words the spontaneous fluctuations that satisfy the Bragg condition. Suppose that such a fluctuation occurs at some time we designate $t=0$; we could write it as $\delta c(t=0) = A\cos(qx)$, where A is its amplitude and x is the appropriate direction. This fluctuation would now relax according to Fick's second law (Equation 9.5.11):

$$\frac{\partial(\delta c)}{\partial t} = D_m \frac{\partial^2(\delta c)}{\partial x^2} \tag{9.6.1}$$

for which the solution is (check it yourself):

$$\delta c(t) = \delta c(0)\, e^{-q^2 D_m t} = A\cos(qx)\, e^{-q^2 D_m t} \tag{9.6.2}$$

From this relation it appears that, if we could measure the time decay of concentration fluctuations, we could measure D_m.

The complication arises because there is no special time $t=0$; fluctuations rise and decay all the time. However, the necessary information is there in the light scattering signal, if we measure the temporal fluctuations in I_s, rather than the time average value as we did in Chapter 8. A schematic example of $I_s(t)$ is shown in Figure 9.13a. The instantaneous value of I_s bounces around the average value, as the scattering molecules move in solution and thereby alter the exact phase relations among the waves scattered from different polymers. Although $I_s(t)$ looks like noise, it actually has in it a typical time constant, which corresponds to the correlation time over which the signal loses memory of whether it was, say, larger than average or smaller than average. (In a sense, this correlation time plays the same role as the persistence length of a random walk discussed in detail in Chapter 6.) To extract this time constant, the scattered intensity is analyzed via a time autocorrelation function, $C(t)$, defined as follows:

$$C(t) \equiv \lim_{T\to\infty} \frac{1}{T}\int_0^T I_s(t')I_s(t'+t)\, dt' \tag{9.6.3}$$

This function involves taking the intensity at some time t', multiplying it by the intensity an interval t later, and adding the results up over some long interval T. By then dividing by T, the length of time of the experiment is taken out of the answer. To obtain the full $C(t)$, this process needs to be repeated for many different values of the interval t. In the actual experiment, the scattered intensity is digitized and stored as a string of numbers, which are then manipulated by a special computer called a correlator, to generate the digital version of $C(t)$.

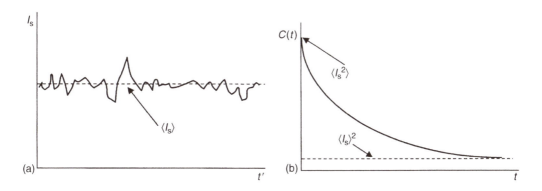

Figure 9.13 (a) Scattered intensity as a function of time, with data digitized every interval t. (b) Intensity autocorrelation function $C(t)$ as a function of t.

What does $C(t)$ look like? The instantaneous $I_s(t)$ can always be written as the sum of its average and a fluctuation: $I_s(t) = \langle I_s \rangle + \delta I_s(t)$ (recall how we used the same strategy in treating the polarizability in Chapter 8). The integrand of $C(t)$ can be written as

$$\langle I_s \rangle^2 + \delta I_s(t')\langle I_s \rangle + \delta I_s(t'+t)\langle I_s \rangle + \delta I_s(t')\delta I_s(t'+t)$$

The first term is a constant and the next two terms integrate to zero because the fluctuations are equally likely to be positive or negative. So, once again, it is the square of the fluctuations that contain all the interesting information. Now consider the two limits, of the time interval t very small and very large, compared to the timescale for molecular motion $(1/q^2 D_m)$. When t is very small, the molecules do not move between t' and $t'+t$, so I_s will not change. Thus $\delta I(t') \approx \delta I_s(t'+t)$ and thus the integrand is $(\delta I_s(t'))^2$. This is always positive and therefore $C(t)$ has its maximum $(= \langle (I_s(t))^2 \rangle)$ when $\tau = 0$. On the other hand, when t is very large, the molecules have completely randomized their positions and so $\delta I_s(t')$ and $\delta I_s(t'+t)$ have no correlation. Consequently, their product is equally likely to be positive or negative and the integral will be zero. Thus in this limit $C(t)$ reduces to the constant $\langle I_s \rangle^2$. The functional form of $C(t)$ in the intervening region turns out to be an exponential, just like the solution to Fick's law above, as shown in Figure 9.13b:

$$C(t) = \left(\langle I_s^2 \rangle - \langle I_s \rangle^2 \right) e^{-2q^2 D_m t} + \langle I_s \rangle^2 \qquad (9.6.4)$$

The additional factor of 2 in the exponential of Equation 9.6.4 arises because the motion of the molecules results in the loss of correlation of the scattered electric field, E, and the intensity is proportional to the square of the field.

In the actual experiment, the range of t can be adjusted to match the system under study over a very wide interval (about 100 ns to 100 s) (see Example 9.5). A measurement of $C(t)$ at any scattering angle is sufficient to extract a value of D_m by Equation 9.6.4, but the result will be more reliable if measurements are taken at several angles and the experimental time constants τ are plotted as $1/\tau$ versus q^2. The data should follow a straight line, with zero intercept and slope equal to D_m. The values of D_m for solutions of low concentration can be extrapolated to zero concentration and thereby the tracer diffusion coefficient D_t can be obtained (Equation 9.5.16). The data in Figure 9.10 were obtained by this method. Finally, use of the Stokes–Einstein relation (Equation 9.5.5) gives access to R_h. This is the basis of the common use of dynamic light scattering for particle sizing. It is also worth noting that this method can be used for quite small values of R_h, approximately 1 nm, in contrast to light scattering, which can only determine R_g when R_g is greater than about 10 nm.

Example 9.5

Estimate the range of time constants, τ, which would be extracted from dynamic light scattering measurements of dilute aqueous suspensions of latex particles ranging in size from 10 nm to 1 μm.

Solution

The time constants are determined by the product of q^2 and D_m, so we need to estimate both. For q, let us assume the light source is an argon ion laser operating at 488 nm and that the instrument can access scattering angles, q, from 30° to 150°. The refractive index of water is about 1.33 (see Table 8.1). From Equation 8.2.4

$$q = \frac{4\pi n}{\lambda_o} \sin\left(\frac{\theta}{2}\right) = \frac{4 \times 3.14 \times 1.33}{488}\left\{\sin\left(\frac{30}{2}\right) \text{ to } \sin\left(\frac{150}{2}\right)\right\}$$
$$= 0.0089 \text{ to } 0.033 \text{ nm}^{-1}$$

We have emphasized both here and in Section 8.6 that $1/q$ is a length, and by taking the reciprocal of these numbers we can see that the typical values for light scattering fall in the range 30–120 nm.

The values of D_m can be estimated by assuming that the latex particles are sufficiently dilute so that $D_m \approx D_t$, and using the Stokes–Einstein relation (Equation 9.5.4). For a hard sphere, remember that $R_h = R$.

$$D_t = \frac{kT}{6\pi\eta_s R_h} = \frac{1.4 \times 10^{-16} \times 300}{6 \times 3.14 \times 0.01 \times (10^{-6} \text{ to } 10^{-4})}$$
$$= 2.2 \times 10^{-7} \text{ to } 2.2 \times 10^{-9} \text{ cm}^2/\text{s}$$

In this calculation we have used cgs units, with the viscosity of water estimated to be 0.01 P. In the final calculation of τ we will need to convert these numbers to nm^2/s, which brings in a factor of 10^{14} (nm^2/cm^2). The results are

$$\tau_{max} = \frac{1}{2.2 \times 10^{-9} \times 10^{14} \times (0.0089)^2} \approx 6 \times 10^{-2} \text{ s}$$
$$\tau_{min} = \frac{1}{2.2 \times 10^{-7} \times 10^{14} \times (0.033)^2} \approx 4 \times 10^{-5} \text{ s}$$

These values are well within the range of commercial correlators.

We conclude this section on dynamic light scattering by emphasizing one further point about diffusion. Compare the three solutions to the same equation, Fick's second law, in Example 9.4 and in Equation 9.6.2. In the former example, $c(x,t)$ has in one instance a very complicated dependence on time and a Gaussian form in space, and in the other, a rather obscure answer involving error functions. In contrast, from Equation 9.6.2 $c(x,t)$ decays exponentially in time and follows an oscillatory function in space. How can the same process, diffusion, and the same equation, Fick's second law, give such different results? The answer is that the solution depends critically on the initial conditions and the boundary conditions. In Equation 9.6.2 the initial condition was a cosine wave and that particular function is preserved by Fick's law; in other words the x dependence of $c(x,t)$ is unchanged and only the amplitude of the wave decays with time. The initial conditions in Example 9.4 were different and so were the results. In fact, there are numerous practical situations with a variety of different constraints and many different functional forms for $c(x,t)$ can result. An excellent discussion of many of these cases can be found in the text by Crank [12].

9.7 Hydrodynamic Interactions and Draining

In this section we address a question that has played an important role in interpreting the intrinsic viscosity and the diffusivity: why does treating a flexible coil as a hard sphere with $R = R_h \approx R_g$ reproduce the experimental results for f (and therefore $[\eta]$ and D_t)? Before providing the qualitative answer, it is worth taking a moment to emphasize why this result is surprising. A detailed model for the dynamics of a flexible chain, called the bead–spring or Rouse–Zimm model, will be discussed in Chapter 11. Its essence is to replace the real polymer with a freely jointed sequence of N beads connected by elastic springs of average length b, where a particular bead–spring unit can be thought of as representing a persistence length or two (see Figure 9.14). The springs resist deformation of the chain and act to restore random coil conformations perturbed by the flow, but are otherwise "invisible." Each bead encounters frictional resistance as it moves relative to the solvent. The bead friction factor, ζ, is a parameter of the model but it should correspond to the net friction of a few real monomer units. In particular, the number of beads N is proportional to the degree of polymerization of the real chain; for example if $N = M/5M_0$, or five monomers per bead–spring unit, then ζ should correspond to the friction of five monomers. This model is extremely successful in many respects, as we shall see in the next chapter, but for now let us focus on the chain friction factor, f. As the chain moves through the solvent, the total friction should just be the sum of the friction experienced by each bead:

$$f = N\zeta \tag{9.7.1}$$

This simple argument leads to the conclusion that $f \sim M$, and therefore $D_t \sim 1/M$, whereas the experiments clearly show $f \sim R_g \sim M^\nu$, and $D_t \sim 1/M^\nu$ (see Figure 9.10). What is wrong with this argument? The answer is that it neglects the phenomenon known as intramolecular *hydrodynamic interactions* (HI for short).

The underlying idea is illustrated schematically in Figure 9.14. When any bead moves through the solvent, it sends out a ripple, or wave, that is felt by every other bead. The amplitude of this wave dies off only as $1/r_{ij}$, the distance between beads i and j, which makes it a rather long-ranged interaction. (Recall from Chapter 8 that the electric field around a point charge and the amplitude of the scattered electric field from a single polarizable object both fall off as $1/r$ as well, whereas the short-ranged van der Waals attractions discussed in Section 7.6 fall off as $1/r^6$.) Furthermore HI is very complicated to handle mathematically because to compute the effect on each bead at any instant, we need to sum the contributions from every other bead on the chain and therefore we need

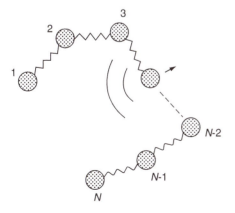

Figure 9.14 Schematic of a bead–spring chain with hydrodynamic interactions.

to know the distance and direction between every pair of beads. Clearly substantial simplifications are required.

The first thorough treatment of HI in polymers was developed by Kirkwood and Riseman in 1948 [13]. The crucial step in the development is to "preaverage" the HI, which means to replace the instantaneous positions of the beads with their average positions, in particular as given by the Gaussian distribution. This average can be evaluated for the separation of any pair of beads using the Gaussian distribution (Equation 6.7.12) because we recognize r_{ij} as the end-to-end distance of a chain from bead i to bead j, that is, with $|i-j|$ steps of length b:

$$\left\langle \frac{1}{r_{ij}} \right\rangle = \int \frac{1}{r_{ij}} 4\pi r_{ij}^2 \left(\frac{3}{2\pi|i-j|b^2} \right)^{3/2} \exp\left(-\frac{3}{2} \frac{r_{ij}^2}{|i-j|b^2} \right) dr_{ij}$$

$$= \left(\frac{6}{\pi|i-j|b^2} \right)^{1/2} \tag{9.7.2}$$

where we omit the details of how to evaluate this integral. The hydrodynamic radius, defined in Equation 9.5.5, is obtained from the average of Equation 9.7.2 over all pairs of beads on the chain, and thus

$$\frac{1}{R_h} = \frac{1}{N^2} \sum_{i=1}^{N} \sum_{j\neq i}^{N} \left\langle \frac{1}{r_{ij}} \right\rangle = \frac{\sqrt{6}}{\sqrt{\pi}} \frac{1}{N^2 b} \sum_{i=1}^{N} \sum_{j\neq i}^{N} \frac{1}{\sqrt{|i-j|}}$$

$$= \left(\frac{128}{3\pi} \right)^{1/2} \frac{1}{N^{1/2}b} = \frac{3.69}{N^{1/2}b} \tag{9.7.3}$$

where again we omit the details of the evaluation of the double sum. The Kirkwood–Riseman prediction for the hydrodynamic radius of a Gaussian chain (i.e., in a theta solvent) is finally obtained by inverting Equation 9.7.3:

$$R_h = 0.271 N^{1/2} b = 0.66 R_g \tag{9.7.4}$$

where the relation to R_g is also included. The key result is that by including HI in this way, the chain friction factor, f, is found to be given by $6\pi\eta_s R_h$, and not by $N\zeta$.

The preceding paragraphs only hint at the rather elaborate mathematical machinery required, and the extension of the Kirkwood–Riseman theory to an expression for the intrinsic viscosity is even more involved. Consequently, it can be helpful to view these results in a more qualitative, physical way by invoking the concept of *draining*. This is illustrated in Figure 9.15a and Figure 9.15b. In both panels, a flexible chain is shown, with the streamlines of the surrounding solvent following some imposed flow. We switch back to the perspective of the flowing solvent and the frictional resistance that the chain presents, rather than the friction experienced by the chain, but of course these must be equivalent, as in the discussion of Stokes' law in Section 9.2. In panel (a), the solvent streamlines pass right through the coil, a limiting behavior termed *freely draining*. In this case the total friction offered by the chain will be $N\zeta$ and thus $f\sim M$. In panel (b), the streamlines are totally diverted around the periphery of the coil, just as in the hard sphere case (see Figure 9.3). This extreme of behavior is called *nondraining*, and the friction varies as $f\sim R_h \sim M^\nu$. We can now see that the effect of HI is to make the coil essentially nondraining. The beads on the upstream edge of the coil, which the solvent encounters first, shield the other beads from the solvent and thus the net friction is reduced. If we revert to the frame of reference of the polymer, when one bead moves in a certain direction, the cumulative effect of HI is to "nudge"

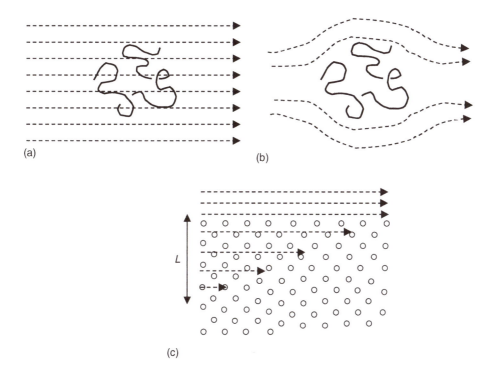

Figure 9.15 Illustration of (a) a freely draining coil, (b) a nondraining coil, and (c) the argument for the importance of (b) given in the text.

all the other beads in the same direction so that there is a concerted element to the chain motion. This cooperation among segments contributes to a reduction in f, relative to $N\zeta$.

Although the limits of freely draining and nondraining coils are easy to visualize, it may not be obvious why a random coil should tend to approximate the nondraining limit. A rather straight-forward argument shows why. The cartoon in Figure 9.15c depicts a layer of pure solvent flowing with velocity v past an infinite field of beads; the beads have a number density of n/V. Although the flow may penetrate into the field of beads by some finite distance, if we are far enough from the surface the flow velocity must tend to zero. Without worrying about the exact functional form of the decay, we will just assert that there is a characteristic penetration distance L, over which the velocity vanishes. Now we will estimate the force F required to maintain the solvent flow over an area of surface A by two separate routes; numerical prefactors will be omitted. In the first route, we add up the friction of each bead (i.e., as if the beads were freely draining within the layer of thickness L). The number of beads is given by (n/V) times the volume LA and thus the force would be

$$F_1 \approx \left(\frac{n}{V}\right) L A \zeta \, v \tag{9.7.5}$$

In the second route, we recognize that the force will be proportional to η_s, A, and the velocity gradient (recall the discussion accompanying Figure 9.1). In this case the velocity gradient should be proportional to v/L, that is, the velocity falls from v to 0 over the distance L. Therefore the force can be written as

$$F_2 \approx \left(\frac{v}{L}\right) A \eta_s \tag{9.7.6}$$

We now equate these two forces, and solve for L:

$$L \approx \sqrt{\frac{V\eta_{\mathrm{s}}}{n\zeta}} \tag{9.7.7}$$

The key question is how does this penetration distance compare to the typical coil size? If $L \gg R_{\mathrm{g}}$, then we expect freely draining behavior, but if $L \ll R_{\mathrm{g}}$, nondraining will be a better description. Note that for a coil, the number of beads per unit volume n/V will be proportional to N/R_{g}^3 or $N^{1-3\nu}$, so from Equation 9.7.7 we can see that L depends on N raised to the power of $(3\nu-1)/2$, or $1/4$ for a Gaussian coil. But, we know that R_{g} grows as $N^{1/2}$, so that for big polymers $R_{\mathrm{g}} \gg L$ and therefore the coil will be nondraining.

In real polymer solutions, the situation is more complicated and tends to fall between these two extremes, albeit closer to the nondraining limit. For example, as the solvent quality is improved beyond a theta solvent, the chain expands and thus the average distance between beads increases. Hydrodynamic interactions are therefore diminished. For chains in good solvents, part of the variation in the Mark–Houwink exponent, a, can be attributed to variable degrees of draining. The effect also depends on molecular weight. For shorter chains the degree of draining increases simply because there are not enough monomers to shield the inside of the coil from the solvent flow. In summary, the main qualitative effect of HI is to make the chain friction factor close to that of a hard sphere with radius R_{h}, but a full mathematical treatment is extremely complicated.

9.8 Size Exclusion Chromatography (SEC)

SEC is one of several modes of liquid chromatography in which a mixture of solutes is separated by passing a solution through an appropriate column. As the solution (mobile phase) passes through the column, different solutes are retained to various degrees according to their interaction with the column packing (stationary phase). Surface adsorption and ion exchange are examples of interactions that serve as the basis for other types of liquid chromatography. In SEC, the columns are filled with porous particles and the separation occurs because molecules of different sizes penetrate the pores of the stationary phase to varying degrees. The method is akin to a "reverse sieving" at the molecular level. The largest molecules are excluded from the pores to the greatest extent and, hence, are the first to emerge (elute) from the column. Progressively smaller molecules permeate the porous stationary phase to increasing extents and are eluted sequentially. The eluting liquid is monitored for the presence of solute by a suitable detector and an instrumental trace of the detector output (chromatogram) provides distinct peaks for well-resolved mixtures and broad peaks for a continuous distribution of molecular sizes. With suitable calibration or multiple detection schemes, this information can be translated into a quantitative characterization of the sample in terms of molecular weight, molecular weight distribution, chemical composition, and even architecture. Owing to this wealth of potential information, the relatively rapid sample throughput (typically ~30 min per solution), and the ease of automation, SEC is currently the single most important characterization tool in the polymer industry. In our discussion, we will first describe the basic separation process and identify associated strengths and weaknesses. Then we will explore two general strategies for column calibration, that is, how the measured quantity (the elution time or elution volume) can be related to the solute molecular weight. We conclude with a description of various detection schemes that are currently employed, and in particular how they can be used to overcome some of the difficulties in calibration.

To avoid confusion, it is helpful to realize that what is essentially the same method is known by several different names—and their acronyms—by workers in different fields. Some other terminologies are noted below:

1. *Gel permeation chromatography* (GPC) is a very common name for this method of separation; it is a less desirable term than SEC in that it emphasizes the column packing material (the gel) rather than the supposed mechanism for the separation (size exclusion).
2. *Gel filtration chromatography* (GFC) is a name often used in the biochemical literature to describe this method of separation. Under this heading, the method is primarily applied to aqueous solutions of solutes of biological origin.
3. All three of these names (SEC, GPC, and GFC) are also modified by the term *high perform-ance*, or its acronym, to give HPSEC, HPGPC, and HPGFC. The additional feature implied by this terminology is the increased speed of efficient separations due to rapid flow through the column under the influence of relatively high applied pressures. At the time of writing, this prefix is generally omitted because "low performance" instruments are little used.

9.8.1 Basic Separation Process

A cartoon of an SEC experiment is shown in Figure 9.16. The column is packed with porous particles with some characteristic pore size, which should correspond roughly to the typical sizes of the polymers to be analyzed. The first rule of SEC is that the separation is based on *size* and not on *molecular weight*; as we shall see the relevant size parameter is the hydrodynamic volume, V_h, which is roughly proportional to R_g^3 (recall Equation 9.3.8). From the radius of gyration data for polystyrenes in Figure 7.17, we can see that a typical range of pore sizes might be 10–10^4 Å. In high-resolution applications, it is often desirable to have a sequence of two to four columns with different average porosities, so that a broader range of sizes can be resolved; the price to be paid is that more time is required for each polymer to elute. The packing material can be made from a variety of substances, but the chosen substance must be compatible with the solvent, polymer, and temperature to be employed. For example, Styragel columns are made from styrene/divinylbenzene copolymers, in the form of cross-linked beads. Such columns are suitable for relatively nonpolar polymers that dissolve in good solvents for polystyrene such as toluene, THF, or chloroform.

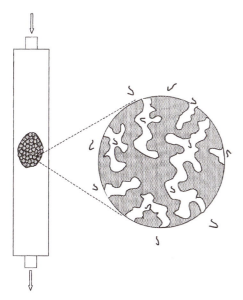

Figure 9.16 Schematic illustration of an SEC column, packed with spherical porous particles, and individual molecules either inside or outside the pores.

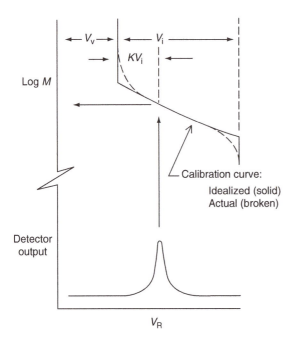

Figure 9.17 Calibration curve for SEC as log M versus the retention volume V_R, showing how the location of the detector signal can be used to evaluate M. Also shown are the void volume V_V and the internal volume V_i in relation to V_R and KV_i as a fraction of V_i.

(Consider what would happen to a Styragel column if the mobile phase were a nonsolvent for polystyrene.) A great deal of clever engineering has gone into the development of packing materials of controlled pore size that can withstand large pressure drops and remain stable for weeks of constant use. The solvent is pumped through the column at a slow but steady rate, typically on the order of 1 mL/min. The pump itself plays an important role in the experiment, in maintaining a constant pressure drop across the column; a good one is expensive. The solvent must also be good for the polymer to be analyzed so that the individual molecules are swollen and have no tendency to either precipitate or adsorb on the column. A small volume of dilute polymer solution is injected through a special port upstream of the column. The detector (or series of detectors) responds to various properties of the eluting solution. Concentration detectors, such as those based on refractive index or uv–vis absorption, have been the most commonly used, but these are now often supplemented by light scattering and/or viscometric detectors. The resulting chromatogram is a plot of detector signal versus time; the time axis is typically converted to volume by multiplying by the known flow rate.

To use SEC for molecular weight determination, we must relate the volume of solvent that passes through the column before a polymer of a particular M is eluted, to M. This quantity is called the retention volume V_R. Figure 9.17 shows schematically the relationship between M and V_R; it is an experimental fact that such calibration curves are approximately linear over about two orders of magnitude in M when plotted as log M versus V_R. In practice, the column is calibrated by constructing such a curve with standards of known molecular weight, as we will discuss later. However, for the present purposes, we will assume that this curve is known. There are three regimes of response. For all M above a certain value, the curve is vertical. This means that all these polymers elute together; there is no separation. This happens for all polymers whose size is larger than the largest pore; they are excluded from all the pores and thus travel entirely with the imposed flow through the interstices or voids between packing particles. Similarly, at low M there is a region with no separation, when all polymers that are small enough to enter all the pores will elute with the solvent. The intermediate range of M is the useful regime: a peak at a particular V_R can be related to a particular M, as shown in Figure 9.17. Note the curious fact that it is a common practice to plot the measured quantity, or dependent variable, V_R along the horizontal axis, and the independent variable, log M, along the vertical axis.

Polydisperse polymers do not yield sharp peaks in the detector output, in contrast to the one illustrated in Figure 9.17. Instead, broad bands are produced that reflect the polydispersity of synthetic polymers. Assuming that suitable calibration data are available, we can construct approximate molecular weight distributions from this kind of experimental data. An indication of how this is done is provided in the following example.

Example 9.6

A broad chromatogram, shown in Figure 9.18, is subdivided into 20 slices, each 1 mL wide, and these are indexed from $i = 1$–20. The height h of the curve above a horizontal base line is carefully measured for each slice. The molecular weight of the ith slice is assigned from independent calibration via the retention volume. Columns 2–4 in Table 9.4 list h_i, $V_{R,i}$, and M_i values, respectively, for a particular chromatogram. Explain the significance and use of the remaining columns in Table 9.4 for the determination of a molecular weight distribution from these data.

Solution

The basic premise of this method is that the magnitude of the detector output, as measured by h_i for a particular fraction, is proportional to the weight fraction of that component in the sample. (This is a reasonable premise for a detector, such as an RI detector, that has a response proportional to

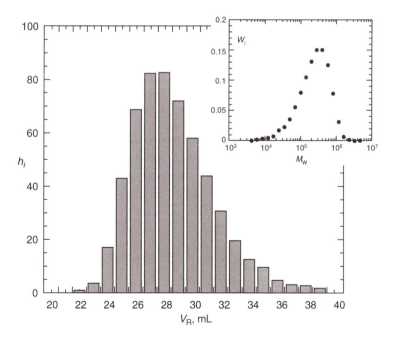

Figure 9.18 Discrete version of the SEC chromatogram for the data in Example 9.6. The inset shows the weight fraction versus molecular weight.

Table 9.4 Data for the Analysis of the Size Exclusion Chromatogram of a Polydisperse Polymer Used in Example 9.6

(1) i	(2) h_i	(3) $V_{R,i}$ mL	(4) $M_i \times 10^{-6}$ g/mol	(5) $\Sigma_i\, h_i$	(6) $h_i/M_i \times 10^6$	(7) $h_i M_i \times 10^{-6}$	(8) A_i	(9) A_i/A_{tot}
21	0.0	20	4.709	545.0	0.0	0.0	545.0	1.0
20	0.0	21	3.302	545.0	0.0	0.0	545.0	1.0
19	0.8	22	2.327	545.0	0.34	1.86	544.6	0.999
18	3.5	23	1.640	544.2	2.13	5.74	542.5	0.995
17	16.8	24	1.1555	540.7	14.54	19.40	532.3	0.977
16	42.4	25	0.8142	523.9	52.08	34.52	502.7	0.922
15	67.9	26	0.5738	481.5	118.2	38.90	447.6	0.821
14	81.5	27	0.4003	413.7	203.6	32.62	373.0	0.684
13	81.4	28	0.2821	322.2	288.6	22.96	291.5	0.535
12	71.0	29	0.1988	250.8	357.1	14.12	215.3	0.395
11	57.0	30	0.1401	179.8	406.8	7.98	151.3	0.278
10	43.0	31	0.09872	122.8	435.6	4.24	101.0	0.186
9	30.0	32	0.06887	79.8	435.6	2.07	64.8	0.119
8	19.0	33	0.04853	49.8	391.5	0.92	40.3	0.074
7	12.1	34	0.03420	30.8	356.7	0.42	24.7	0.045
6	9.0	35	0.02410	18.6	373.4	0.22	14.1	0.026
5	4.0	36	0.01698	9.6	235.6	0.07	7.6	0.014
4	2.6	37	0.01197	5.6	217.2	0.03	4.3	0.008
3	2.0	38	0.00843	3.0	237.1	0.02	2.0	0.004
2	1.0	39	0.00588	1.0	170.0	0.01	0.5	0.001
1	0.0	40	0.00414	0.0	0.0	0.0	0.0	0.0

Source: From Yau, W.W., Kirkland, J.J., and Bly, D.D., in *Modern Size Exclusion Chromatography*, Wiley, New York, 1979.

the concentration of the solute. However, the proportionality constant, related to dn/dc in this instance, must itself be independent of M. Furthermore, it assumes that all of the polymer elutes from the column, or, failing that, that any adsorption or other loss of material on the column is independent of M.) In this sense, the chromatogram itself presents a kind of picture of the molecular weight distribution. The following column entries provide additional quantifications of this distribution:

Column 5. Σh_i is proportional to the cumulative weight of all polymers in all categories up to the nth.

Column 6. The ratio h_i/M_i is proportional to the weight of materials in the ith slice, w_i, divided by M_i, that is, to the number of moles in that class n_i. Therefore M_n can be evaluated as follows:

$$M_n = \frac{\Sigma_i\, n_i M_i}{\Sigma_i\, n_i} = \frac{\Sigma_i\,(h_i/M_i)(M_i)}{\Sigma_i\,(h_i/M_i)} = \frac{\Sigma_i\, h_i}{\Sigma_i\,(h_i/M_i)} = \frac{545}{4.29 \times 10^{-3}}$$
$$= 127{,}000 \text{ g mol}^{-1}$$

Column 7. The product $h_i M_i$ is proportional to $w_i M_i$, and M_w is evaluated as follows:

$$M_w = \frac{\Sigma_i\, w_i M_i}{\Sigma_i\, w_i} = \frac{\Sigma_i\, h_i M_i}{\Sigma_i\, h_i} = \frac{1.86 \times 10^6}{545} = 341{,}000 \text{ g mol}^{-1}$$

Column 8. $A_i = \sum_{i=1}^{N}[h_i + 1/2(h_{i+1} - h_i)]$. Adding $(1/2)(h_{i+1} - h_i)$ gives h_i the height of the midpoint of each slice, and since each slice is 1 unit wide, the summation gives the area under the curve up to the nth class.

Column 9. A_i/A_{tot} gives that fraction of the area under the entire curve that has accumulated up to the *n*th class. Since the curve is a weight distribution, this is equal to the weight fraction of material in the sample having $M < M_i$.

A plot of the last entry versus M gives the integrated form of the distribution function. The more familiar distribution function in terms of weight fraction versus M is given as the inset to Figure 9.18.

9.8.2 Separation Mechanism

A more thorough examination of the correlation between V_R and M can be found in [14]. We shall only outline the problem, with particular emphasis on those aspects that overlap other topics in this book. To consider the origin of $V_R(M)$, begin by picturing a narrow band of polymer solution being introduced at the top of a solvent-filled column. The volume of this solvent can be subdivided into two categories: the stagnant solvent in the pores (subscript i for internal) and the interstitial liquid in the voids (subscript v) between the packing particles:

$$V_{solvent} = V_v + V_i \tag{9.8.1}$$

The entire interstitial volume moves through the column at the imposed flow rate and must pass through the column before any polymer emerges. Then the first polymer that does appear is the one with the highest molecular weight. This solute has spent all its time in the voids—not the pores—of the packing and passes through the column with the velocity of the solvent. Progressively smaller molecules have access to successively larger fractions of the internal volume. Therefore, as V_i emerges, consecutive fractions of the polymer come with it. Thus we can write the retention volume for a particular molecular weight fraction as

$$V_R = V_v + KV_i \tag{9.8.2}$$

where K is called the distribution coefficient. K is a function of both the pore size and the molecular size and indicates what fraction of the internal volume is accessible to the particular solute. The relationships among V_R, V_v, V_i, and KV_i are also indicated in Figure 9.17. When $K = 0$, the solute is totally excluded from the pores; when $K = 1$, it totally penetrates the pores.

It is instructive to consider a simple model for the significance of the constant K in Equation 9.8.2. For simplicity, assume a spherical solute molecule of radius R and a cylindrical pore of radius a and length L. As seen in Figure 9.19a, an excluded volume effect prevents the center of the

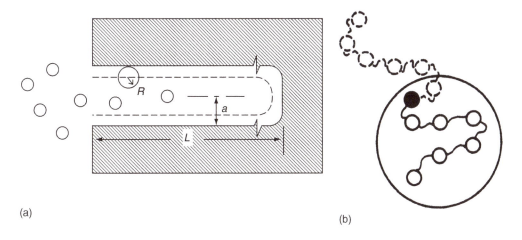

(a)

(b)

Figure 9.19 Schematic illustration of size exclusion in a cylindrical pore: (a) for spherical particles of radius R and (b) for a flexible chain, showing allowed (solid) and forbidden (dotted) conformations.

spherical solute molecule from approaching any closer than a distance R from the walls of the pore. This effectively decreases the volume accessible to the solute to a smaller cylinder of radius $(a - R)$. In this accessible cylinder, the concentration of the solute is the same as in the nearby voids, but the excluded volume near the walls of the pore is devoid of solute. Hence the average concentration of solute in the pore as a whole is less than that outside the pore. The fraction of the external concentration found in the pore is given by the ratio of the accessible volume to the actual volume of the cylindrical pore: $\pi(a - R)^2 L/\pi a^2 L$. This fraction gives K for the case of spherical solute molecules in cylindrical cavities. If we assume that the pore is long enough to neglect end effects, we have

$$K = \frac{(a - R)^2}{a^2} = \left(1 - \frac{R}{a}\right)^2 \tag{9.8.3}$$

Note that the fraction is zero when $R = a$ and unity when $R = 0$.

This simple model illustrates how the fraction K and, therefore, V_R are influenced by the dimensions of both the solute molecules and the pores. For solute particles of other shapes in pores of different geometry, theoretical expressions for K are quantitatively different, but typically involve some ratio of solute size to pore dimensions. The extension of these ideas to random coils can proceed along two lines. In one approach the coil domain is visualized as a hard sphere, as in the case above, with R_g or R_h taking the place of R; this is similar in spirit to the application of Stokes' and Einstein's laws to hydrodynamics earlier in this chapter. Alternatively, statistical methods can be employed to consider those conformations of a random chain, which are excluded for a coil confined to a pore. This latter situation is illustrated in Figure 9.19b, in which solid and broken lines represent two conformations of the same chain, with the filled-in repeat unit being held in a fixed position. If the molecule were in bulk solution, both conformations would be possible. In a pore, represented by the enclosing circle in Figure 9.19b, the broken line conformation is impossible. This is equivalent to a decrease in conformational entropy for the coil in the pore and the effect can be translated into an equilibrium constant between the solute in the pore and in the bulk solution. The factor K in Equation 9.8.2 is just such a constant—the distribution coefficient—and can be evaluated by this approach for pores of different shapes.

Figure 9.20 shows the theoretical predictions for K versus $R_g/\langle a \rangle$ compared with experimental findings. The solid line is drawn according to the statistical theory. The experimental points correspond to the same porous beads used as the stationary phase with their pore size analyzed by two different experimental procedures: mercury penetration (circles in Figure 9.20, $\langle a \rangle = 21$ nm) and gas adsorption (squares in Figure 9.20, $\langle a \rangle = 41$ nm). We can draw several conclusions from an examination of Figure 9.20:

1. Characterization of the stationary phase is also a source of discrepancy: The polymers are not the only sources of difficulty.
2. Despite item (1), the fact that one set of experimental points agrees reasonably well with the theory supports the basic soundness of this approach.
3. Since K represents the fraction of V_i, at which a particular molecular weight emerges from the column, and since $\ln M \sim \ln R_g$, we see that this model correctly accounts for the form of the calibration curve shown in Figure 9.16.

Despite the evident promise of this line of attack, quantitative predictions for $V_R(M)$ in SEC are not generally at hand. There are at least two general reasons for this. One is the complexity of the full problem. For example, the real pores are not cylindrical, but have a distribution of shapes and sizes. The analysis assumes complete equilibrium between pores and voids, but this may be hindered by diffusional limitations. Further, the separation mechanism is assumed to be entirely entropic in nature, which means that there are no interactions between polymers and pore walls other than the excluded volume. For example, any slight attraction between a monomer unit and

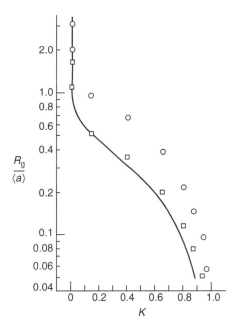

Figure 9.20 Comparison of theory with experiment for $R_g/\langle a \rangle$ versus K. The solid line is drawn according to the theory for flexible chains in a cylindrical pore. Experimental points show some data, with pore dimensions determined by mercury penetration (circles, $\langle a \rangle = 21$ nm) and gas adsorption (squares, $\langle a \rangle = 41$ nm). (From Yau, W.W. and Malone, C.P., *Polymer Preprints (Am. Chem. Soc., Div. Polym. Chem.)*, 12, 797, 1971. With permission.)

the packing material can drastically increase V_R. Furthermore, this contribution causes V_R to increase with M, in opposition to size exclusion, because the net attraction will increase with the number of monomers. In short, a quantitative theory would be very complicated. The second reason is more practical. Calibration techniques, to be described in the next section, are perfectly adequate for many applications, and modern detectors, which can even circumvent the need for column calibration, satisfy most requirements. Consequently, the need for a full theory is reduced.

9.8.3 Two Calibration Strategies

The simplest calibration strategy was alluded to above: take a series of polymers of known M, referred to as calibration standards, run them through the column, and generate an empirical plot of log M versus V_R. Some kind of polynomial function (it need not be linear) can be used to obtain a smooth function, which can be stored in a computer. That function can be used to assign a molecular weight to each value of V_R and the computer can easily calculate the molecular weight averages and other details of the distribution for each sample along the lines illustrated in Example 9.6 above.

This approach is in common practice and it is appealingly simple. The calibration should be repeated at regular intervals because column performance will drift with time, but as only minuscule amounts of the standards are required for each run, this is not particularly expensive. What are the limitations of this approach? There are several to consider, any of which may or may not be important for a particular application.

9.8.3.1 Limitations of Calibration by Standards

1. Separation is based on the hydrodynamic volume, V_h, and not on M directly. Consequently, accuracy in M requires the use of standards of the same polymer as the analyte. Although

controlled polymerization, especially anionic polymerization as discussed in Chapter 4, is used to produce standards of a variety of polymers (polystyrene, poly(methyl methacrylate), poly-butadiene, polyisoprene, poly(ethylene oxide)...), there are, of course, a much larger number of polymer structures for which standards are simply not available. One expedient often encountered in the literature is to quote "an apparent M, based on polystyrene standards," which gives some very approximate idea of the real M.

2. The absolute M of a given standard may not be known to an accuracy better than 5%–10%, which ultimately limits the accuracy achievable with this mode of SEC.

3. If one injected an absolutely monodisperse standard, an ideal instrument would give a very sharp spike at the particular value of V_R. In reality, for such a sample the instrument would show a narrow peak with a finite width. This width characterizes the *instrument response function*, which quantifies the deviation of the instrument from ideality. In typical SEC, a monodisperse sample would show a peak width corresponding to a polydispersity of about 1.01–1.05. The reason for this is the phenomenon of *band broadening*. One significant and easily visualized contribution to band broadening is termed *axial diffusion*. As molecules of identical M pass through the column, even if they were injected at precisely the same time, they will all diffuse randomly. Some will show a net displacement relative to the average that places them further down the column and others will lag behind. Indeed, based on the arguments in Section 9.5, we might anticipate a Gaussian distribution of concentration versus V_R. This has important consequences. For example, it means that no matter how narrow a "slice" of the chromatogram we select, the material eluting at that particular V_R is never a single M; each slice i has its own $M_{w,i}$ and $M_{n,i}$. Furthermore, for a given column, a particular V_R does not always correspond to one particular $M_{w,i}$ or $M_{n,i}$; the average M of the material at a particular V_R will depend on the sample.

4. In most cases calibration standards have very narrow molecular weight distributions ($M_w/M_n \leq 1.1$). However, the true polydispersity of such materials is often not known. For example, if M_w is obtained by light scattering and M_n by osmotic pressure, the combined uncertainties of, say, 5% in each number means that one cannot distinguish among polydispersities of 1.03, 1.01, or 1.001. Recall from Chapter 4 that standards prepared by living anionic polymerization should ideally follow the Poison distribution; a polystyrene with $M_n = 100,000$ would have a theoretical ideal polydispersity of 1.001. It is therefore quite probable that most calibration standards are narrower than can be determined by SEC. Recent progress in mass spectrometric methods (see Chapter 1) offers the possibility that such materials may be characterized more precisely.

5. Many polymers have nonlinear or branched architectures (recall the examples in Chapter 1). Two polymers of identical M but different amounts of branching will have different R_g and therefore different V_R. For example, in the free radical polymerization of ethylene, the product called "low density" polyethylene has a substantial degree of long-chain branching. It is a notoriously difficult problem to characterize the resulting molecular structures. In particular, it is impossible to learn anything about the degree of branching from SEC when only this simple method of calibration is employed.

6. It is often the case that polymer samples are heterogeneous not just in molecular weight or architecture, but also in composition (e.g., copolymers), microstructure (e.g., polydienes), and tacticity. All of these factors may contribute in some way to V_R and each slice of the chromatogram will also be heterogeneous in these variables.

9.8.3.2 Universal Calibration

The second mode of calibration in routine use is known by the optimistic name of *universal calibration*. The hypothesis is that in SEC, V_R depends solely on the hydrodynamic volume, V_h, which itself will be proportional to R_g^3. Now recall the discussion of intrinsic viscosity leading up to

Equation 9.3.8, where the essence of the argument is that, no matter the detailed molecular structure or shape,

$$[\eta] \sim \frac{V_h}{M} \approx \frac{R_g^3}{M} \tag{9.8.4}$$

The hydrodynamic volume, therefore, should be proportional to the product $[\eta]M$. In universal calibration, we assume that the proportionality factor between the hydrodynamic volume and $[\eta]M$ is independent of structure. Let us further suppose that $[\eta]$ for our sample follows the Mark–Houwink relation (Equation 9.3.10), with known values of k and a. We compare our sample with the standard, say polystyrene, that elutes at the same time:

$$([\eta]M)_r = k_r M_r^{1+a_r} = ([\eta]M)_s = k_s M_s^{1+a_s} \tag{9.8.5}$$

where the subscripts "r" and "s" refer to the reference (calibrant) and sample polymers, respectively. Equation 9.8.5 can be rearranged to solve for M of the unknown at any particular V_R:

$$M_s = \left(\frac{\left(k_r M_r^{1+a_r} \right)}{k_s} \right)^{1/1+a_s} \tag{9.8.6}$$

In short, if we know the appropriate Mark–Houwink parameters (and a great many have been tabulated, as noted in Section 9.3) then we can extract the absolute molecular weight of one polymer based on column calibration with another. This approach is therefore often used to overcome the first limitation listed above. The success of the underlying assumption behind universal calibration is nicely illustrated in Figure 9.21. In the first panel, plots of $\log M$ versus V_R are shown for a variety of different polymers, including some branched structures; clearly the different species are all over the map. In the second panel, the same data are shown, but with the vertical axis being $\log([\eta]M)$. In this case, there is a very satisfying collapse of the data onto one universal calibration curve.

9.8.4 Size Exclusion Chromatography Detectors

To conclude this discussion of SEC we offer a brief discussion on the four classes of detectors in common use: the refractive index (RI), absorption (uv–vis), light scattering (LS), and viscometer (V) detectors. It is increasingly common practice to utilize two or more of these in series, for reasons that should become apparent. There are different implementations of these various detectors by different companies and we shall try to present the general principles without reference to particular instrument configurations.

9.8.4.1 RI Detector

This is the most commonly employed. In Section 8.7 we discussed the central role that the refractive index increment, $\partial n/\partial c$, plays in light scattering. For a dilute polymer solution (which is almost always the case in SEC), the refractive index of the eluent may be written

$$n(c) = n_s + \left(\frac{\partial n}{\partial c} \right) c + \cdots \tag{9.8.7}$$

where c is the concentration in g/mL of the eluting polymer and n_s is the refractive index of the solvent. The refractometer is typically set up such that a transmitted light beam is deflected by an amount proportional to $n - n_s$ and thus to c. The response of the detector (e.g., in volts) can then be assumed proportional to c and the proportionality factor determined by injection of known quantities of material (assuming no adsorption on the column and that $\partial n/\partial c$ is known). However, in routine cases the proportionality factor is not necessary; the value of any M depends only on V_R, not on the height of the peak (i.e., h_i), and in the calculation of polydispersities the proportionality constant would cancel out (see Example 9.6).

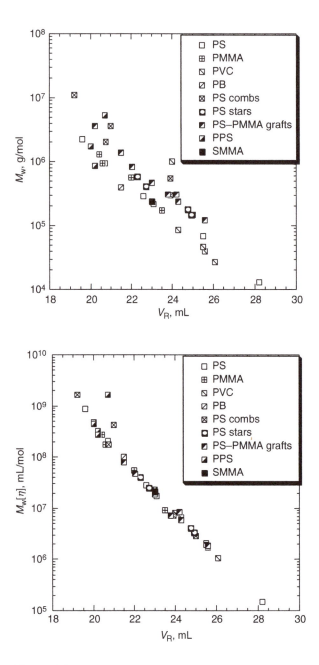

Figure 9.21 The success of universal calibration. The data in (a) as molecular weight versus retention volume collapse to a single curve when plotted in (b) as the product of molecular weight and intrinsic viscosity. (Data from Grubisic, P., Rempp, H., and Benoit, J., *J. Polym. Sci.*, B5, 753 1967.)

9.8.4.2 UV–Vis Detector

This approach takes advantage of Beer's law, whereby the transmittance of the solution (ratio of transmitted light intensity to incident light intensity, I_t/I_0) is related to the concentration of the absorbing species by

$$\varepsilon bc = A = \log \frac{I_0}{I_t} \tag{9.8.8}$$

where A is the absorbance, b is the path length through the capillary, and ε is the absorptivity of the solute at the wavelength of interest. Thus, as with the RI detector, the signal is arranged to be proportional to c. The uv–vis detector is less general than the RI detector because many polymers do not absorb at sufficiently long wavelengths to avoid solvent absorption. On the other hand, when dealing with mixtures, for example copolymers, it may be possible to choose a wavelength in the uv–vis detector that favors only one component, whereas the RI detector is likely to respond to both. In this case both detectors in series can be used to establish the average composition of the sample at each V_R.

9.8.4.3 Light Scattering Detector

The crucial advantage of the LS detector is the possibility of obtaining an absolute $M_{w,i}$ of each slice of the chromatogram without any column calibration. This makes use of the theoretical machinery developed in Chapter 8. If M is sufficiently low that the form factor, $P(q)$, is effectively 1, then a single angle detector is adequate. On the other hand, for larger polymers multiple angle detectors are desirable. Some commercial models have more than 10 photodiodes surrounding the sample volume, and both $R_{g,i}$ and $M_{w,i}$ can be determined for each slice. An illustration is provided in Figure 9.22 and discussed in the following example.

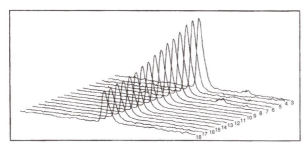

(a)

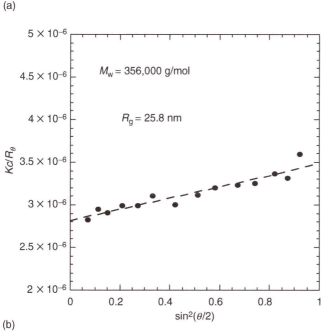

(b)

Figure 9.22 Multiangle light scattering detector signal from SEC of a polystyrene solution, as discussed in Example 9.7. (a) Intensity versus time traces for the individual detectors and (b) a plot of Kc/R_θ versus $\sin^2(\theta/2)$ for a particular slice of the chromatogram near the peak in intensity.

Example 9.7

Explain how to extract M_w and R_g from each slice of a chromatogram.

Solution

Recalling Equation 8.5.19, the scattered intensity in slice i, $I_{s,i}$, can be written as

$$\frac{I_{s,i}}{I_o} = K c_i M_{w,i}\left(1 - \frac{1}{3}q^2 R_{g,i}^2 \cdots\right)$$

where we have dropped the term containing the second virial coefficient B. (This assumption is usually safe because the concentrations coming off the column are small, but it can always be checked by reinjecting a different concentration of sample and seeing if the answer changes.) The value of q is determined by the scattering angle for each detector. The constant K contains many factors (see Equation 8.4.22), including $(\partial n/\partial c)^2$, so it must be known accurately. This can be accomplished by calibrating the detector with a modest M sample of known M_w and $\partial n/\partial c$. The intensity for each slice should be plotted against q^2 (or $\sin^2(\theta/2)$) to obtain R_g, as illustrated in Figure 9.22b using the "Zimm format" (Equation 8.5.18). Finally, to obtain M for each slice requires dividing by c_i, obtained from either the RI or uv–vis detector. In this case, it is necessary that the concentration detector be accurately calibrated as well. It should also be noted that because the two detectors are arrayed in series, there is a time delay between the arrival of slice i at each detector. This must also be determined accurately.

9.8.4.4   Viscometer

This detector takes advantage of the direct relation between the hydrodynamic volume, which determines V_R and $[\eta]$: $V_h \sim M[\eta]$. Consequently by estimating $[\eta]_i$ at each slice, M can be calculated. The detector itself is a kind of "Wheatstone bridge," or null detector, for viscosity. The eluting solution passes across one face of a sensitive pressure transducer. Pure solvent is circulated against the other face. If both liquids are flowing at the same rate, any difference in viscosity between them will be transformed into a differential shear force and thus a pressure on the transducer. This could be measured, but a more effective approach is to increase the pure solvent flow rate to null out the pressure drop. The amount the flow rate needs to be increased is directly proportional to $\eta - \eta_s$. The intrinsic viscosity is then estimated by

$$[\eta]_i = \frac{\eta - \eta_s}{\eta_s c_i} \tag{9.8.9}$$

where again we assume the concentration is small enough that extrapolation to infinite dilution is not necessary. Note that in Equation 9.8.9 an accurate measurement of concentration and the interdetector delay time is also essential.

The preceding discussion gives some insight into the rather sophisticated detection schemes now in use. It is perhaps appropriate to close the discussion of SEC with the major limitation of the technique, namely that no matter how elaborate or sensitive the detector, the separation itself has rather low resolution. Furthermore, for samples containing distributions of composition, micro-structure, and architecture, in addition to molecular weight, it is necessary to employ other separation schemes in tandem that can discriminate among these various characteristics. In normal SEC, all of these various distributions are likely to be present to some extent in each slice, and the information the detectors can provide will be limited.

9.9   Chapter Summary

In this chapter we have examined some dynamic properties of polymers in dilute solution, with a particular emphasis on the viscosity. The underlying concept is that of the molecular friction factor, f, which depends on both polymer shape (rod, coil, sphere ...) and molecular weight. The main points are the following:

1. The friction factor is proportional to the product of the solvent viscosity and a hydrodynamic radius, R_h, where R_h varies with M just as R_g does. This corresponds to an extension of Stokes' law for rigid spheres to polymers of any shape. The friction factor is experimentally accessible through either the intrinsic viscosity or the tracer diffusion coefficient.
2. The hydrodynamic volume, V_h, can also be defined, and is proportional to R_h^3 and R_g^3. The Einstein equation for the viscosity of a suspension of hard spheres can be extended to polymers of any shape, with the result that the intrinsic viscosity is proportional to the ratio of V_h/M.
3. The relation between intrinsic viscosity and molecular weight can be expressed by the Mark–Houwink equation $[\eta] = kM^a$, where the parameters k and a have been tabulated for many polymer and solvent combinations. As $[\eta]$ is relatively easy to measure, this offers a simple route to molecular weight characterization.
4. Tracer and mutual diffusion coefficients were defined and distinguished. Diffusion coefficients offer another route to molecular characterization and play a key role in many applications of polymers.
5. The technique of SEC is the most commonly applied technique for polymer characterization, as it can determine both average molecular weights and the molecular weight distribution. The separation is based on V_h and is of relatively low resolution. Application of various detection schemes can obviate the need for column calibration.
6. The reason that Stokes' law and Einstein's viscosity equation can be applied to polymers of any shape is rather subtle. It is due to the phenomenon of hydrodynamic interactions, whereby the motion of any monomer in the polymer is transmitted through the solvent to all other monomers. The net effect is that monomers tend to move collectively, like a hard sphere with radius close to R_g, rather than as N independent objects.

Problems

1. A fluid of viscosity η is confined within the gap between two concentric cylinders as shown in Figure 9.9b. Consider a cylindrical shell of radius r, length L, and thickness dr located within the gap.

   a. What is the torque acting on the shell if torque is the product of force and the distance from the axis and $F/A = \eta r \, d\omega/dr$?
   b. Under stationary-state conditions, the torques at r and at $r + dr$ must be equal, otherwise the shell would accelerate. This means that the torque must be independent of r. Show that this implies the following variation of ω with r: $\omega = -B/2r^2 + C$, where B and C are constants.
   c. Evaluate the constant B by noting that $\omega = \omega_{ex}$, the experimental velocity, at $r = R$ and $\omega = 0$ at $r = fR$.
   d. Combine the results of a, b, and c to obtain Equation 9.4.14.

2. A slightly different but useful way of defining the viscosity average molecular weight is the following:

$$M_v^a = \frac{\sum_i f_i M_i M_i^a}{\sum_i f_i M_i}$$

   where $f_i M_i$ is the weighting factor used to average M_i^a. A satisfactory way of treating many polymer distributions is to define

$$f_i = \frac{1}{M_n} e^{-M_i/M_n}$$

Then

$$M_v^a = \frac{\int\limits_0^\infty f_i M_i^{1+a}\,dM_i}{\int\limits_0^\infty f_i M_i\,dM_i}$$

Combine the last two expressions and integrate to express M_v in terms of M_n and a. The integrals are standard forms and are listed in integral tables as gamma functions.

3. A sphere of density ρ_2 and radius R falling through a medium of density ρ_1 and viscosity η experiences three kinds of forces: gravitational, buoyant, and frictional. The first is determined by the mass of the ball (and the acceleration due to gravity, g), the second by the mass of the displaced fluid and g, and the last is given by Equation 9.2.3 and Equation 9.2.4. During most of the fall (excluding the very beginning and the very end of the path), these forces balance. Use this condition to derive an equation showing how this stationary-state velocity of fall is related to R, ρ_1, ρ_2, and η. This is the basis for the so-called falling-ball viscometer.

4. Plazek, Dannhauser, and Ferry[†] measured the viscosities of a poly(dimethylsiloxane) sample of $M_w = 4.1 \times 10^5$ over a range of temperatures using the falling-ball method. Stainless steel ($\rho_2 = 7.81$ g/cm^3) balls of two different diameters, 0.1590 and 0.0966 cm, were used at 25°C, where $\rho_1 = 0.974$ g/cm^3 and $\eta = 8.64 \times 10^4$ P. Use the result derived in the last problem to calculate the ratio of the stationary-state settling velocities for the two different balls. How long would it take the smaller ball to fall a distance of 15 cm under these conditions?

5. The intrinsic viscosity of poly(γ-benzyl-L-glutamate) ($M_0 = 210$) shows such a strong molecular weight dependence in dimethyl formamide that the polymer was suspected to exist as a helix, which approximates a prolate ellipsoid of revolution in its hydrodynamic behavior.

$M \times 10^{-3}$ (g/mol)	21.4	66.5	130	208	347
$[\eta]$ (dL/g)	0.107	0.451	1.32	3.27	7.20

Using 1.32 g/cm^3 as the density of the polymer, estimate the axial ratio for these molecules, using Simha's equation (for large $p \equiv a/b$):

$$\frac{\eta_{sp}}{\phi} = \frac{p^2}{15[\ln(2p) - 3/2]} + \frac{p^2}{5[\ln(2p) - 1/2]} + \frac{14}{15} \cong 175\left(\frac{p}{50}\right)^{1.8}$$

For an α-helix, the length per residue is about 1.5 Å. Use this figure with the molecular weight to estimate the length $2a$ of the particle. Use the estimated a/b ratios to calculate the diameter $2b$ of the helix, which should be approximately constant if this interpretation is correct. Comment on the results.

6. Fox and Flory[‡] used experimental molecular weights, intrinsic viscosities, and rms end-to-end distances from light scattering to evaluate the constant Φ_0 in Equation 9.3.18. For polystyrene in the solvents and at the temperatures noted, the following results were assembled (M in kg mol^{-1}, $[\eta]$ in dL/g):

[†] D.J. Plazek, W. Dannhauser, and J.D. Ferry, *J. Colloid Sci.*, 16, 1010 (1961).
[‡] T.G Fox Jr. and P.J. Flory, *J. Am. Chem. Soc.*, 73, 1915 (1951).

Solvent	T	M	$[\eta]$	h_{rms} (Å)
Methyl ethyl ketone	22	1760	1.65	1070
	22	1620	1.61	1015
	67	1620	1.50	980
	22	1320	1.40	900
	25	980	1.21	840
	22	940	1.17	750
	22	520	0.77	545
	25	318	0.60	475
	22	230	0.53	400
Dichloroethane	22	1780	2.60	1410
	22	1620	2.78	1335
	67	1620	2.83	1295
	22	562	1.42	760
	22	520	1.38	680
Toluene	22	1620	3.45	1290
	67	1620	3.42	1280

Evaluate Φ_0 for each set of data and compare the average with the value given in the text. Does the fact that these data are not from theta solvents matter? Why, or why not?

7. Precise determination of the intrinsic viscosity, $[\eta]$, and the Huggins coefficient, k_H, is not as straightforward as one might expect, even when the instrument provides accurate and precise measurements of η_{rel} ($= \eta/\eta_s$). The primary issue becomes how many terms to include in the concentration expansion of $\eta(c)$, and what range of c is appropriate? The usual interval is $1.1 < \eta_{rel} < 2$; smaller values have too much uncertainty, and larger values require too many terms. The usual strategy is to perform two linear extrapolations, and compare the results; if they do not agree, more data or more terms are required. The first extrapolation is due to Huggins; plot η_{sp}/c versus c, where $\eta_{sp} = \eta_{rel} - 1$, and fit to a straight line:

$$\frac{\eta_{sp}}{c} = \alpha' + \beta' c \cdots$$

where α' and β' are the fit parameters. The second extrapolation is due to Kraemer: plot (ln η_{rel})/c versus c, and fit to a straight line:

$$\frac{\ln \eta_{rel}}{c} = \alpha'' + \beta'' c \cdots$$

where α'' and β'' are the fit parameters.

a. Derive the relationships between (i) α' and α'', (ii) β' and β'', and (iii) express α' and β' in terms of $[\eta]$ and k_H. It may help to recall that $\ln(1+x) = x - (1/2)x^2 + (1/3)x^3 \cdots$

b. One test of the validity of determination of $[\eta]$ and k_H is to compare the results from these two plots. The data below are for polystyrene ($M = 20,000$) in a theta solvent. Prepare the two plots, do the linear regression, and answer these questions:

   i. How well do the two fits satisfy the criteria you derived above?
   ii. What are the implied values of $[\eta]$ and k_H?
   iii. What would you estimate the uncertainties to be?

c (g/mL)	η_{sp}/c	$(\ln \eta_{rel})/c$
0.005	12.25	11.890
0.011	12.7	11.888
0.015	13.3	12.127
0.028	14.4	12.098
0.034	15.7	12.581

8. Another method of extrapolation to obtain the intrinsic viscosity is due to Schulze and Blaschke:

$$\frac{\eta_{sp}}{c} = \alpha''' + \beta''' \eta_{sp} \cdots$$

Evaluate the data in the previous problem using this approach, and relate the new parameters to $[\eta]$ and k_H. What approximation is necessary to make the Huggins extrapolation and the Schulze–Blaschke extrapolation equivalent? It is claimed in the literature that the Schulze–Blaschke extrapolation is valid over a wider range of concentration; justify or refute this claim.

9. The intrinsic viscosity of polystyrene in benzene at 25°C was measured for polymers with the following molecular weights:

M (g/mol)	$[\eta]$ (dL/g)	M (g/mol)	$[\eta]$ (dL/g)
6,970,000	11.75	277,000	1.07
4,240,000	8.15	63,000	0.358
2,530,000	5.54	63,100	0.356
838,000	2.43	43,200	0.268
784,000	2.32	16,050	0.136
676,000	2.07	10,430	0.106
335,000	1.23	8,370	0.0932
		3,990	0.0608

Estimate R_g for these polymers from these data. Use the data in Table 6.1 to compute $R_{g,0}$ and thus evaluate the coil expansion ratio α for each fraction. How does α vary with M, and how does this compare with the Flory–Krigbaum prediction (Equation 7.7.10 and Equation 7.7.12)?

10. Diblock copolymers can readily form spherical micelles in a solvent that does not dissolve one block. A typical aggregation number for such a micelle might be 100 individual polymers, with the inner "core" formed of the 100 insoluble blocks and little or no solvent. Imagine a solution of a block copolymer with $M = 100,000$ and a concentration of 0.01 g/mL, which forms micelles upon cooling below some critical micelle temperature. Estimate the ratio of viscosity of the solution before and after micellization and also the ratio of the hydrodynamic radius before and after micellization. Assume that both blocks are made of polymers with flexibilities similar to polystyrene.

11. When the mutual diffusion coefficient is measured for dilute polymer solutions, for example by dynamic light scattering, it is found that the concentration dependence of D_m is linear with c, but the slope can be either positive or negative. Furthermore, in a good solvent the slope is usually positive, but in a theta solvent it is negative. This behavior is consistent with Equation 9.5.18, but the dependence on solvent quality is not transparent. Show that, in fact,

$$\frac{d \ln \gamma_2}{dc} = 2BM - \bar{\nu}_2$$

where B is the second virial coefficient and $\bar{v}_2$ is the partial specific volume of the polymer (which we can take to be given by $\bar{V}_2/M$). To do this, start with Equation 9.5.13, and work to replace $(d\mu_2/dc)$ with $\partial\Pi/\partial c$, following the discussion surrounding Equation 8.4.16 and Equation 8.4.17.

12. The ratio of the radius of gyration, R_g (measured by light scattering as described in Chapter 8) to the hydrodynamic radius, R_h (measured by dynamic light scattering) can be a sensitive indicator of molecular conformation. Compare the value of this ratio for a high molecular weight linear chain in a theta solvent to that of a hard sphere. Now consider a sixth-generation dendrimer and a regular four-arm star polymer with long arms; where would they rank relative to each other, and to the other two shapes? Explain your reasoning.

13. Dynamic light scattering measurements were made on a very dilute aqueous suspension of latex particles at a scattering angle of 45°; the measured decay rate implied a hydrodynamic radius of 240 nm. When measurements were taken at a scattering angle of 90° to confirm this result, there was no significant scattering signal. Why? What can be inferred about the particles based on this information? (Hint: some material in Section 8.6 may be helpful.)

14. The overlap concentration c^*, which separates the dilute solution regime from the so-called semidilute regime, can be estimated by space-filling arguments, as the concentration where the individual coil-volumes begin to fill space. Derive the expression for c^* in terms of R_g and M and indicate how c^* scales with M in good and theta solvents. Alternatively, c^* can be estimated from the dilute solution viscosity expansion, such that $c^* \sim 1/[\eta]$. Use the Flory–Fox relation to relate these two definitions, that is, find the proportionality constant between c^* and $1/[\eta]$ that makes the two definitions equivalent.

15. Use the model for the size exclusion of a spherical solute molecule in a cylindrical capillary to calculate K_{GPC} for a selection of R/a values, which are compatible with Figure 9.20. Plot your values on a photocopy or tracing of Figure 9.20. On the basis of the comparison between these calculated points and the line in Figure 9.20 drawn on the basis of a statistical consideration of chain exclusion, criticize or defend the following proposition: There is not much difference between the K values calculated by the equivalent sphere and statistical models. The discrepancy between various experimental methods of evaluating $\langle a \rangle$ is much greater than the differences arising from different models. Even for random coil molecules, the simple equivalent sphere model is acceptable for qualitative discussions of V_R.

16. SEC measurements are now often made with both an RI and an LS detector. The former responds to the change in solution RI, n, as polymer elutes; n may be taken to be linear in concentration at these dilute concentrations and independent of M. The latter either measures I_s at a series of scattering angles simultaneously, or at one very low angle. For each slice, i, of the chromatogram, one now has two pieces of data: n_i and $I_{s,i}$. Show how to compute M_w and M_n from these data (i might easily run from 1 to 1000). What do you need to know in order to get absolute M averages (i.e., without calibrating the columns)? Can you suggest a way to get $\partial n/\partial c$ without making any additional measurements?

17. Both preparative and analytical GPC were employed to analyze a standard (NBS 706) polystyrene sample[†]. Fractions were collected from the preparative column, the solvent was evaporated away, and the weight of each polymer fraction was obtained. The molecular weight of each fraction was obtained using an analytical gel permeation chromatograph. The following data were obtained (mass in milligrams and $M \times 10^{-4}$ g/mol):

[†] Y. Kato, T. Kametani, K. Furukawa, and T. Hashimoto, *J. Polym. Sci. Polym. Phys. Ed.* 13, 1695 (1975).

Fraction	Mass	M_n	M_w	Fraction	Mass	M_n	M_w
6	2	109	111	19	42	9.14	9.35
7	8	90.8	92.5	20	30	7.52	7.68
8	20	76.7	78.0	21	28	6.16	6.28
9	42	62.3	63.5	22	18	5.12	5.22
10	64	51.5	52.5	23	12	4.09	4.18
11	84	41.7	42.5	24	8	3.33	3.40
12	102	34.7	35.4	25	6	2.63	2.69
13	110	28.7	29.3	26	5	2.01	2.06
14	110	23.3	23.8	27	4	1.52	1.56
15	96	18.9	19.4	28	3	1.13	1.16
16	86	15.9	16.3	29	2	0.83	0.85
17	68	13.0	13.3	30	1	0.59	0.61
18	54	11.0	11.2				

Calculate M_w and M_n and the ratio M_w/M_n for the original polymer. Also evaluate the ratio M_w/M_n for the individual fractions. Comment on the significance of M_w/M_n for both the fractionated and unfractionated polymer.

18. A polystyrene sample was prepared by living anionic polymerization (recall Chapter 4), with $M_w = 34,500$. The polydispersity was measured by four different techniques. Matrix-assisted laser desorption/ionization (MALDI) mass spectrometry (Chapter 1) gave 1.016. Temperature gradient interaction chromatography (TGIC, a higher resolution technique than SEC) gave 1.020. Standard SEC with an RI detector and calibration with PS standards gave 1.05. However, SEC on the same instrument but using an LS detector to obtain M_w of each slice gave a value of 1.005. The Poisson distribution (see Chapter 4) predicts a polydispersity of 1.003 for an ideal living polymerization, so the small value obtained by LS detection is at least conceivable. However, there is good reason to believe that both MALDI and TGIC are more accurate. Explain why using SEC with RI detection gives a polydispersity that is too large, and why the LS detection gives a value that is too small.

References

1. Stokes, G., *Trans. Cambridge Phil. Soc.*, 8, 287 (1847); 9, 8 (1851).
2. Einstein, A., *Ann. Physik*, 19, 289 (1906); 34, 591 (1911).
3. Lauffer, M.A., *J. Chem. Educ.*, 58, 250 (1981).
4. Perrin, F., *J. Phys. Radium*, 7, 1 (1936).
5. Simha, R., *J. Phys. Chem.*, 44, 25 (1940).
6. Staudinger, H. and Heuer, W., *Ber.* 63, 222 (1930).
7. Mark, H., *Der Feste Körper*, Hirzel, Leipzig, 1938; R. Houwink, *J. Prakt. Chem.*, 157, 15 (1941).
8. Brandrup, J. and Immergut, E.H. (Eds.), *Polymer Handbook*, 3rd ed., Wiley, New York, 1989.
9. Kurata, M. and Stockmayer, W.H., *Fortschr. Hochpolym.-Forsch.*, 3, 196 (1963).
10. Poiseuille, J.L., *Comptes Rendus*, 11, 961, 1041 (1840); 12, 112 (1841).
11. Fick, A., *Ann. Physik*, 170, 59 (1855).
12. Crank, J., *The Mathematics of Diffusion*, 2nd ed., Clarendon Press, Oxford, 1975.
13. Kirkwood, J.G. and Riseman, J., *J. Chem. Phys.*, 16, 565 (1948).
14. Yau, W.W., Kirkland, J.J., and Bly, D.D., *Modern Size Exclusion Chromatography*, Wiley, New York, 1979.

Further Readings

Berne, B.J. and Pecora, R., *Dynamic Light Scattering*, Wiley, New York, 1976.

Chu, B., *Laser Light Scattering*, 2nd ed., Academic Press, San Diego, 1991.

Flory, P.J., *Principles of Polymer Chemistry*, Cornell University Press, Ithaca, 1953.

Fujita, H., *Polymer Solutions*, Elsevier, Amsterdam, 1990.

Tanford, C., *Physical Chemistry of Macromolecules*, Wiley, New York, 1961.

Van Holde, K.E., *Physical Biochemistry*, Prentice-Hall, New Jersey, 1971.

Yamakawa, H., *Modern Theory of Polymer Solutions*, Harper & Row, New York, 1971.

Yau, W.W., Kirkland, J.J., and Bly, D.D., *Modern Size Exclusion Chromatography*, Wiley, New York, 1979.

10

Networks, Gels, and Rubber Elasticity

10.1 Formation of Networks by Random Cross-Linking

In this chapter we consider one of the three general classes of polymers in the solid state: infinite networks. The other two categories, glassy polymers and semicrystalline polymers, will be taken up in Chapter 12 and Chapter 13, respectively. We will shortly define the term network more precisely, but we have in mind a material in which covalent bonds (or other strong associations) link different chain molecules together to produce a single molecule of effectively infinite molecular weight. These linkages prevent flow and thus the material is a solid. There are two important subclasses of network materials: elastomers and thermosets. An *elastomer* is a cross-linked polymer that undergoes the glass transition well below room temperature; consequently, the solid is quite soft and deformable. The quintessential everyday example is a rubber band. Such materials are usually made by cross-linking after polymerization. A *thermoset* is a polymer in which multifunctional monomers are polymerized or copolymerized to form a relatively rigid solid; an epoxy adhesive is a common example. In this chapter we will consider both elastomers and thermosets, but with an emphasis on the former. The reasons for this emphasis are that the phenomenon of rubber elasticity is unique to polymers and that it is an essential ingredient in understanding both the viscoelasticity of polymer liquids (see Chapter 11) and the swelling of single chains in a good solvent (see Chapter 7). In the first two sections we examine the two general routes to chemical formation of networks: cross-linking of preformed chains and polymerization with multifunctional monomers. In Section 10.3 through Section 10.6 we describe successively elastic deformations, thermodynamics of elasticity, the "ideal" molecular description of rubber elasticity, and then extensions to the idealized theory. In Section 10.7, we consider the swelling of polymer networks with solvent.

10.1.1 Definitions

Figure 10.1 provides a pictorial representation of a network polymer. In panel (a), there is a schematic representation of a collection of polymer chains, which could be either in solution or in the melt. In panel (b), a certain number of chemical linkages have been introduced between monomers on different chains (or on the same chain). If enough such *cross-links* are created, it becomes possible to start at one surface of the material and trace a course to the far side of the material by passing only along the covalent bonds of chain backbones or cross-links. In such a case an infinite *network* is formed, and we can say that the covalent structure *percolates* through the material. The network consists of the following elements, as illustrated in Figure 10.2:

1. *Strand.* A strand is a section of polymer chain that begins at one junction and ends at another without any intervening junctions.
2. *Junction.* A junction is a cross-link from which three or more strands emanate. The *functionality* of the junction is the number of strands that are connected; in the case of the random cross-linking pictured in Figure 10.1 the functionality is usually four. Note that a cross-link might simply connect two chains, but it would not be a junction until it becomes part of an infinite network.

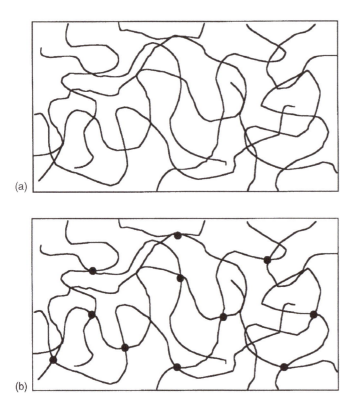

(a)

(b)

Figure 10.1   Schematic illustration of (a) an uncross-linked melt or concentrated solution of flexible chains and (b) the same material after cross-links are introduced.

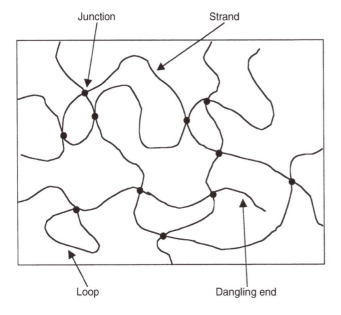

Figure 10.2   Schematic illustration of network elements defined in the text.

3. *Dangling end.* The section of the original polymer chain that begins at one chain terminus and continues to the first junction forms a dangling end. Because it is free to relax its conformation over time, it does not contribute to the equilibrium elasticity of the network, and as such it can be viewed as a defect in the structure.
4. *Loop.* Another defect is a loop, a section of chain that begins and ends at the same cross-link, with no intervening junctions. A loop might be formed by an intramolecular cross-linking reaction. Again, as with the dangling end, the loop can relax its conformation (at least in part) and is thus not fully elastically active.
5. *Sol fraction.* It is not necessary that every original polymer chain be linked into the network; a given chain may have no cross-links or it may be linked to a finite number of other chains to form a *cluster.* In either case, if the material were placed in a large reservoir of a good solvent the *sol fraction* could dissolve, whereas the network or *gel fraction* could not. Thus the sol fraction contains all the extractable material, including any solvent present.

The apparently synonymous terms network, infinite network, and gel have all appeared so far and it is time to say how we will use these terms from now on. We have used network and infinite network interchangeably; the modifier *infinite* just serves to emphasize that the structure percolates throughout a macroscopic sample and from now on we will omit it. The term *gel* is somewhat more problematic, as it is used by different workers in rather disparate ways. We will henceforth use it to refer to a material that contains a network, whereas the term network refers to the topology of the underlying molecular structure. Often, an elastomeric material containing little or no sol fraction is called a *rubber*, whereas a material containing an equivalent network structure plus a significant amount of solvent or low-molecular-weight diluent would be called a gel.

10.1.2 Gel Point

We now consider the following question: given a collection of polymer chains, how many random cross-links need to be introduced before a network will be formed? For simplicity, assume that all chains have the same degree of polymerization N, and that all monomers are equally likely to react. We will give examples of cross-linking chemistry in a moment, but for now we assume we can measure the extent of reaction, p, defined as the fraction of monomers that participate in cross-links. Suppose we start on a chain selected at random and find a cross-link; we now use it to cross over to the next chain. What is the probability that, as we move along the second chain, we will find a second cross-link? It is simply given by $(N-1)p \approx Np$. The probability of being able to hop from chain to chain x times in succession is therefore $(Np)^x$. (Recall that the probability of a series of independent events is given by the product of the individual probabilities.) For a network to be formed, we need this probability to be ≥ 1 as $x \to \infty$, and therefore we need $Np \geq 1$. Conversely, if $Np < 1$, $(Np)^x \to 0$ as $x \to \infty$. Consequently, the critical extent of reaction, p_c, at which an infinite network first appears, the *gel point*, is given by

$$p_c = \frac{1}{N-1} \approx \frac{1}{N} \tag{10.1.1}$$

This beautifully simple result indicates how effective polymers can be at forming networks; a polymer with $N \approx 1000$ only needs an average of 0.1% of the monomers to react to reach the gel point. Note that Equation 10.1.1 probably underestimates the true gel point because some fraction of cross-linking reactions will result in the formation of loops, which will not contribute to network formation.

Any real polymer will be polydisperse, so we should consider how this affects Equation 10.1.1. Let us return to our first chain, find the cross-link, and then ask, what is the average length of the next chain? As the cross-linking reaction was assumed to be random, then the chance that the next chain has degree of polymerization N_i is given by the weight fraction of N_i-mers, w_i. In other words, the probability that the neighboring monomer that forms the cross-link belongs to a chain of length N_i is proportional to N_i. (To see this argument, consider a trivial example: the sample contains 1 mole of chains of length 100 and 1 mole of chains of length 200. Any monomer selected at random has a probability of 2/3 to be in a chain of length 200, and 1/3 to be in chain of length 100; 2/3 and 1/3 correspond to the weight fractions.) The critical probability therefore becomes

$$p_c = \frac{1}{\sum\limits_{i=1}^{\infty} w_i\,(N_i - 1)} \approx \frac{1}{\sum\limits_{i=1}^{\infty} w_i N_i} = \frac{1}{N_w} \qquad (10.1.2)$$

and thus the critical extent of reaction is determined by the weight-average degree of polymerization, N_w.

Examples of postpolymerization cross-linking reactions are many. Free-radical initiators such as peroxides (see Chapter 3) can be used to cross-link polymers with saturated structures (i.e., no carbon–carbon double bonds), such as polyethylene or poly(dimethylsiloxane). Alternatively, high-energy radiation can be utilized for the same purpose. A prime example occurs in integrated circuit fabrication, where electron beam or UV radiation can be used to cross-link a particular polymer (called a negative resist) in desired spatial patterns. The uncross-linked polymer is then washed away, exposing the underlying substrate for etching or deposition. (In contrast, some polymers such as poly(methyl methacrylate) degrade rapidly on exposure to high-energy radiation, thereby forming a positive resist.) Of course, the classic example of cross-linking is that of polydienes cross-linked in the presence of sulfur. The use of sulfur dates back to 1839 and the work of Goodyear in the United States [1] and Macintosh and Hancock in the UK. The polymer of choice was natural rubber, a material extracted from the sap of rubber trees; the major ingredient is *cis*-1,4 polyisoprene. This basic process remains the primary commercial route to rubber materials, especially in the production of tires, and the cross-linking of polydienes is generically referred to as *vulcanization*. Remarkably, perhaps, the detailed chemical mechanism of the process remains elusive. For some time a free-radical mechanism was suspected, but current thinking favors an ionic route, as shown in Figure 10.3. The process is thought to proceed through formation of a sulfonium ion, whereby the naturally occurring eight-membered sulfur ring, S_8, becomes polarized or opened (Reaction A). The next stage is abstraction of an allylic hydrogen from a neighboring chain to generate a carbocation (Reaction B), which subsequently can react with sulfur and cross-link to another chain (Reaction C). A carbocation is regenerated, allowing propagation of the cross-linking process (Reaction D). Termination presumably involves sulfur anions. In practice, the rate of vulcanization is greatly enhanced by a combination of additives, called accelerators and activators. Again, the mechanisms at play are far from fully understood, although the technology for producing an array of rubber materials with tunable properties is highly developed.

Example 10.1

A sample of polyisoprene with $M_w = 150,000$ is vulcanized until 0.3% of the double bonds are consumed, as determined by spectroscopy. Do you expect this sample to have formed a network, and what is the probability of finding a polyisoprene chain that is untouched by the reaction?

Solution

The nominal monomer molecular weight for polyisoprene is 68 g/mol, so for this sample the critical extent of reaction estimated by Equation 10.1.2 is

$$p_c \approx \frac{1}{N_w} = \frac{68}{150,000} = 0.00045$$

This is a factor of 6.7 less than the stated value of $p = 0.003$, so we may be reasonably confident that the sample has passed the gel point.

For an individual chain to be untouched, every monomer must be unreacted. The probability for each monomer to be unreacted is $1 - p = 0.997$ and for a chain of N monomers

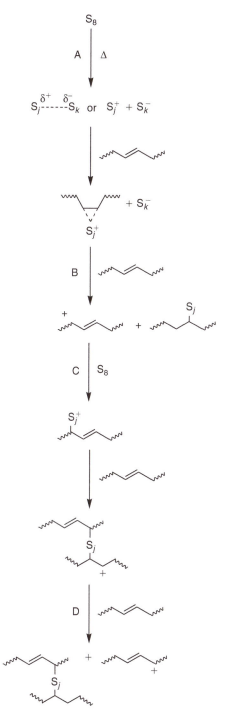

Figure 10.3 Possible mechanism for vulcanization of 1,4-polybutadiene with sulfur, following Odian. (From Odian, G., *Principles of Polymerization*, 2nd ed., Wiley, New York, 1981.)

we must raise 0.997 to the Nth power. For simplicity, we assume all chains to have the same $N = 150,000/68 = 2200$; then $(0.997)^{2200} \approx 0.0013$ or there is about 0.1% chance that a chain is untouched.

10.2   Polymerization with Multifunctional Monomers

In this section we consider the other general approach to network formation or gelation, using polymerization of multifunctional monomers. Multifunctional, as noted in Chapter 2, means functionality greater than 2. We will build on the material in that chapter by considering step-growth or condensation polymerization of monomers containing A and B reactive groups. The resulting thermosets are widely used as engineering materials because their mechanical properties are largely unaffected by temperature variation.

For simplicity, we assume that the reaction mixture contains only A and B as reactive groups, but that either one (or both) of these is present (either totally or in part) in a molecule that contains more than two of the reactive groups. We use f to represent the number of reactive groups in a molecule when this quantity exceeds 2 and represent a multifunctional molecule as A_f or B_f. For example, if A were a hydroxyl group, a triol would correspond to $f=3$. Several reaction possibilities (all written for $f=3$) come to mind in the presence of multifunctional reactants, as shown in Figure 10.4. The lower case "a" and "b" refer to the corresponding groups that have reacted.

The third reaction is interesting inasmuch as either the AA or BB monomer must be present to produce cross-linking. Polymerization of AB with only A_f (or only B_f) introduces a single branch point, but no more, since all chain ends are unsuited for further incorporation of branch points. Including the AA or BB molecule reverses this. The bb unit that accomplishes this is underlined.

What we seek next is a quantitative relationship among the extent of the polymerization reaction, the composition of the monomer mixture, and the gel point. We shall base our discussion on the system described by the first reaction in Figure 10.4; other cases are derived by similar methods (see

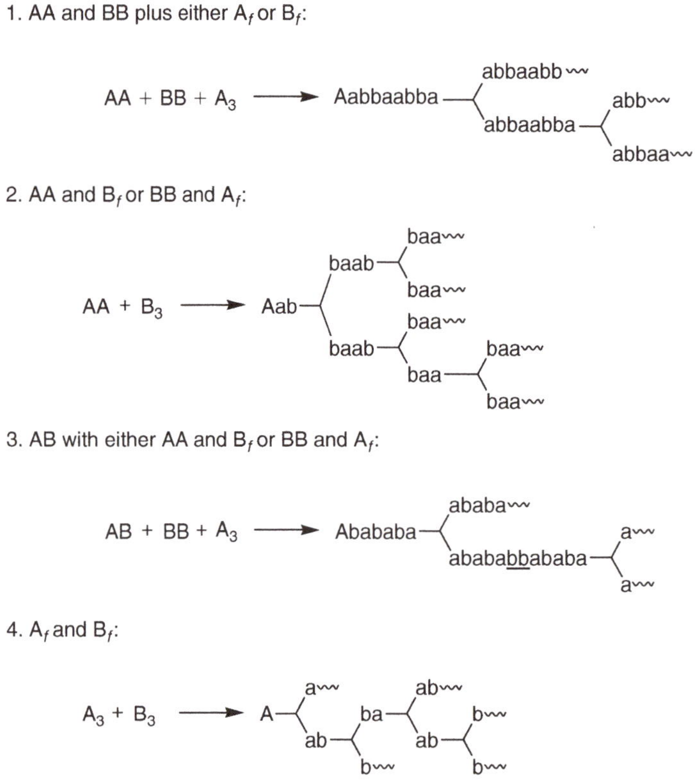

1. AA and BB plus either A_f or B_f:

$$AA + BB + A_3 \longrightarrow Aabbaabba$$

2. AA and B_f or BB and A_f:

$$AA + B_3 \longrightarrow Aab$$

3. AB with either AA and B_f or BB and A_f:

$$AB + BB + A_3 \longrightarrow Abababa$$

4. A_f and B_f:

$$A_3 + B_3 \longrightarrow A$$

Figure 10.4   Possible reaction schemes for monomer mixtures containing A and B functional groups that can lead to network formation.

Problem 3). To further specify the system, we assume that A groups limit the reaction and that B groups are present in excess. Two parameters are necessary to characterize the reaction mixture:

1. The ratio of the initial number of A to B groups, ν_A^0/ν_B^0, defines the factor r, as in Equation 2.7.1. The total number of A groups from both AA and A_f is included in this application of r.
2. The fraction of A groups present in mulifunctional molecules is defined by the ratio

$$\rho = \frac{\nu_A(\text{from } A_f)}{\nu_A(\text{total})} \tag{10.2.1}$$

There are two additional useful parameters that characterize the reaction itself:

1. The extent of reaction p is based on the group present in limiting amount. For the system under consideration, p is therefore the fraction of A groups that have reacted. (Note that this p is slightly different from p in Section 10.1.)
2. The probability that a chain segment is capped at both ends by a branch unit is described by the *branching coefficient* α. The branching coefficient is central to the discussion of network formation, as the occurrence or nonoccurrence of gelation depends on what happens after capping a section of chain with a potential branch point.

10.2.1 Calculation of the Branching Coefficient

The methods we consider were initially developed by Stockmayer [2] and Flory [3] and have been applied to a wide variety of polymer systems and phenomena. Our approach proceeds through two stages: first we consider the probability that AA and BB polymerize until all chain segments are capped by an A_f monomer, and then we consider the probability that these are connected together to form a network. The actual molecular processes occur at random and not in this sequence, but mathematical analysis is more feasible if we consider the process in stages. As long as the same sort of structure results from both the random and the subdivided processes, this analysis is valid.

The arguments we employ are statistical, so we recall that the probability of a functional group reacting is given by the fraction of groups that have reacted at any point and that the probability of a sequence of events is the product of their individual probabilities (as used in developing Equation 10.1.1). As in Chapter 2 and Chapter 3, we continue to invoke the principle of equal reactivity, that is, that functional group activity is independent of the size of the molecule to which the group is attached. One additional facet of this assumption that enters when multifunctional monomers are considered is that all A groups in A_f are of equal reactivity.

Now let us consider the probability that a section of polymer chain is capped at both ends by potential branch points:

1. The first step is the condensation of a BB monomer with one of the A groups of an A_f molecule: Since all A groups have the same reactivity by hypothesis, the probability of this occurrence is simply p.
2. The terminal B group reacts with an A group from AA rather than A_f:

   $$A_{f-1}abB + AA \rightarrow A_{f-1}abbaA$$

   The fraction of unreacted B groups is rp, so this gives the probability of reaction for B. Since ρ is the fraction of A groups on multifunctional monomers, rp must be multiplied by $1 - \rho$ to give the probability of B reacting with an AA monomer. The total probability for the chain shown is the product of the probabilities until now: $p[rp(1 - \rho)]$.
3. The terminal A groups react with another BB:

   $$A_{f-1}abbaA + BB \rightarrow A_{f-1}abbaabB$$

   The probability of this step is again p, and the total probability is $p[rp(1 - \rho)p]$.

4. Additional AA and BB molecules condense into the chain to give a sequence of i bbaa units

    $$A_{f-1}abbaabB + AA + BB \rightarrow\rightarrow\rightarrow A_{f-1}a(bbaa)_i bB$$

    We have just evaluated the probability of one such unit; the probability for a series of i units is just the product of the individual probabilities: $p[rp(1-\rho)p]^i$.
5. The terminal B groups react with an A group from a multifunctional monomer:

    $$A_{f-1}a(bbaa)_i bB + A_f \rightarrow A_{f-1}a(bbaa)_i bbaA_{f-1}$$

    The probability of B reacting is rp and the fraction of these reactions that involve A_f molecules is $rp\rho$. The probability of the entire sequence is therefore $p[rp(1-\rho)p]^i rp\rho$.
6. In the general expression above, i can have any value from 0 to ∞, so the probability for all possibilities is the sum of the individual probabilities. Note that a different procedure is used for compounding probabilities here: the sum instead of the product. This time we are interested in *either* $i=0$ *or* $i=1$ *or* $i=2$, and so forth, whereas previously we required the first A–B reaction *and* the second A–B reaction *and* the third A–B reaction, etc.

As the branching coefficient gives the probability of a chain segment being capped by potential branch points, the above development describes this situation:

$$\alpha = \sum_{i=0}^{\infty} rp^2\rho[rp^2(1-\rho)]^i \tag{10.2.2}$$

The summation applies only to the quantity in brackets, since it alone involves i. Representing the bracketed quantity by Q, we note that $\sum_{i=0}^{\infty} Q^i = 1/(1-Q)$ (see Appendix) and therefore

$$\alpha = \frac{rp^2\rho}{1 - rp^2(1-\rho)} \tag{10.2.3}$$

10.2.2 Gel Point

We have now completed the first (and harder) stage of the problem we set out to consider: we know the probability that a chain is capped at both ends by potential branch points. The second stage of the derivation considers the reaction between these chain ends via the remaining $f-1$ reactive A groups. (By hypothesis, the mixture contains an excess of B groups, so there are still unreacted BB monomers or other polymer chain segments with terminal B groups that can react with the A_{f-1} groups we have been considering.) By analogy with the discussion of the gel point in Section 10.1, we ask the question: if we choose an A_f group at random, and follow this chain to another A_f group, what is the probability that we can continue in this fashion forever? If this probability exceeds 1, we have a network, and the gel point corresponds to when it equals 1. The probability of there being a strand, that is, a chain segment between two junctions, is α. When we arrive at the next A_f, there are $f-1$ chances to connect to a new strand and the probability of there being a strand from any particular one of the $f-1$ groups is again α. Thus the total probability of keeping going from each A_f is just $(f-1)\alpha$. If we want to connect x strands in sequence, the probability that we can is $[(f-1)\alpha]^x$. Just as in the argument preceding Equation 10.1.1, therefore, the critical extent of reaction is simply given by

$$\alpha_c = \frac{1}{f-1} \tag{10.2.4}$$

which can be compared directly with Equation 10.1.1. Whenever the extent of reaction, p, is such that $\alpha > \alpha_c$, gelation is predicted to occur. Combining Equation 10.2.3 and Equation 10.2.4 and rearranging gives the critical extent of reaction for gelation, p_c as a function of the properties of the monomer mixture r, ρ, and f:

$$p_c = \frac{1}{\sqrt{r + r\rho(f - 2)}} \tag{10.2.5}$$

Corresponding equations for any of the reaction schemes depicted in Figure 10.4 can be derived in a similar fashion (see Problem 3 for an example).

Equation 10.2.5 is of considerable practical utility in view of the commercial importance of three-dimensional polymer networks. Some reactions of this sort are carried out on a very large scale: imagine the consequences of having a polymer preparation solidify in a large and expensive reaction vessel because the polymerization reaction went a little too far. Considering this kind of application, we might actually be relieved to know that Equation 10.2.5 errs in the direction of underestimating the extent of reaction at gelation. This comes about because some reactions of the multifunctional branch points result in intramolecular loops, which are wasted as far as network formation is concerned; the same comment applies to Equation 10.1.1. It is also not uncommon that the reactivity of the functional groups within one multifunctional monomer decreases with increasing p, which tends to favor the formation of linear structures over the branched ones.

As an example of the quantitative testing of Equation 10.2.5, consider the polymerization of diethylene glycol (BB) with adipic acid (AA) in the presence of 1,2,3-propane tricarboxylic acid (A$_3$). The critical value of the branching coefficient is 0.50 for this system by Equation 10.2.4. For an experiment in which $r = 0.800$ and $\rho = 0.375$, $p_c = 0.953$ by Equation 10.2.5. The critical extent of reaction was found experimentally to be 0.9907, determined in the polymerizing mixture as the point where bubbles fail to rise through it. Calculating back from Equation 10.2.3, the experimental value of p_c is consistent with the value $\alpha_c = 0.578$, instead of the theoretical value of 0.50.

10.2.3 Molecular-Weight Averages

It is apparent that numerous other special systems or effects could be considered to either broaden the range or improve the applicability of the derivation presented. Our interest, however, is in illustrating concepts rather than exhaustively exploring all possible cases, so we shall not pursue the matter of the gel point further here. Instead, we conclude this section with a brief examination of the molecular-weight averages in the system generated from AA, BB, and A$_f$. For simplicity, we restrict our attention to the case of $\nu_A^0 = \nu_B^0$. It is useful to define the average functionality of a monomer $\langle f \rangle$ as

$$\langle f \rangle \equiv \frac{\sum_i n_i f_i}{\sum_i n_i} \tag{10.2.6}$$

where n_i and f_i are the number of molecules and the functionality of the ith component in the reaction mixture, respectively. The summations are over all monomers. If n is the total number of molecules present at the extent of reaction p and n_0 is the total number of molecules present initially, then $2(n_0 - n)$ is the number of functional groups that have reacted and $\langle f \rangle n_0$ is the total number of groups initially present. Two conclusions immediately follow from these concepts:

$$N_n = \frac{n_0}{n} \tag{10.2.7}$$

where N_n is the number-average degree of polymerization, and

$$p = \frac{2(n_0 - n)}{\langle f \rangle n_0} \tag{10.2.8}$$

Elimination of n between these expressions gives

$$N_n = \frac{2}{2 - p\langle f \rangle} \qquad (10.2.9)$$

This result is known as the *Carothers equation* [4]. It is apparent that this expression reduces to Equation 2.2.4 for the case of $\langle f \rangle = 2$, that is, the result for the most probable distribution in polycondensation reactions considered in Chapter 2. Furthermore, when $\langle f \rangle$ exceeds 2, as in the AA/BB/A_f mixture under consideration, then N_n is increased over the value obtained at the same p for $\langle f \rangle = 2$. A numerical example will help clarify these relationships.

Example 10.2

An AA, BB, and A_3 polymerization mixture is prepared in which $\nu_A^0 = \nu_B^0 = 3.00$ mol, with 10% of the A groups contributed by A_3. Use Equation 10.2.9 to calculate N_n for $p = 0.970$ and p for $N_n = 200$. In each case, compare the results with what would be obtained if no multifunctional A were present.

Solution

Determine the average functionality of the mixture. The total number of functional groups is 6.00 mol, but the total number of molecules initially present must be determined. Using $3n_{AAA} + 2n_{AA} = 3.00$ and $3n_{AAA}/3 = 0.100$, we find that $n_{AA} = 1.350$ and $n_{AAA} = 0.1000$. Since $n_{BB} = 1.500$ the total number of moles initially present is $n_0 = 1.350 + 0.100 + 1.500 = 2.950$:

$$\langle f \rangle = \frac{3(0.100) + 2(1.350) + 2(1.500)}{2.950} = 2.034$$

Solving Equation 10.2.9 with $p = 0.970$ and $\langle f \rangle = 2.034$:

$$N_n = \frac{2}{2 - 0.97(2.034)} = 73.8$$

For comparison, solve Equation 10.2.9 with $p = 0.970$ and $\langle f \rangle = 2$:

$$N_n = \frac{1}{1 - p} = \frac{1}{1 - 0.97} = 33.3$$

Solve Equation 10.2.9 with $N_n = 200$ and $\langle f \rangle = 2.034$:

$$p = \frac{2(1 - 1/N_n)}{\langle f \rangle} = \frac{2(0.995)}{2.034} = 0.978$$

Solve Equation 10.2.9 with $N_n = 200$ and $\langle f \rangle = 2$:

$$p = \left(1 - \frac{1}{N_n}\right) = \left(1 - \frac{1}{200}\right) = 0.995$$

These results demonstrate how for a fixed extent of reaction, the presence of multifunctional monomers in an equimolar mixture of reactive groups increases the degree of polymerization. Conversely, for the same mixture a lesser extent of reaction is needed to reach a specific N_n with multifunctional reactants than without them. Remember that this entire approach is developed for the case of stoichiometric balance. If the numbers of functional groups are unequal, this effect works in opposition to the multifunctional groups.

The Carothers approach, as described above, is limited to the number-average degree of polymerization and gives no information concerning the breadth of the distribution. A statistical approach to the degree of polymerization yields expressions for both N_w and N_n. Ref. [4] contains a derivation of these quantities for the self-polymerization of A_f monomers. Although this specific system might appear to be very different from the one we have considered, the essential aspects of the two different averaging procedures are applicable to the system we have considered as well. The results obtained for the A_f case are

$$N_n = \frac{2}{2 - \alpha f} \tag{10.2.10}$$

and

$$N_w = \frac{1 + \alpha}{1 - \alpha(f - 1)} \tag{10.2.11}$$

from which it follows that

$$\frac{N_w}{N_n} = \frac{(1 + \alpha)(1 - \alpha f/2)}{1 - \alpha(f - 1)} \tag{10.2.12}$$

The value of α to be used in these expressions is given by Equation 10.2.3 for the specific mixture under consideration. At the gel point $\alpha_c = 1/(f - 1)$ according to Equation 10.2.4, and thus Equation 10.2.11 predicts that N_w becomes infinite, whereas N_n remains finite. This is a very important point. It emphasizes that in addition to the network molecule, or gel fraction, of essentially infinite molecular weight, there are still many other molecules present at the gel point, the sol fraction. The ratio N_w/N_n also indicates a divergence of the polydispersity as $\alpha \to \alpha_c$. Expressions have also been developed to describe the distribution of molecules in the sol fraction beyond the gel point. We conclude this discussion with an example that illustrates application of some of these concepts to a common household product.

Example 10.3

The chemistry underlying an epoxy adhesive is illustrated in Figure 10.5. An excess of epichlorohydrin is reacted with a diol to form a linear *prepolymer*, terminated at each end with epoxide

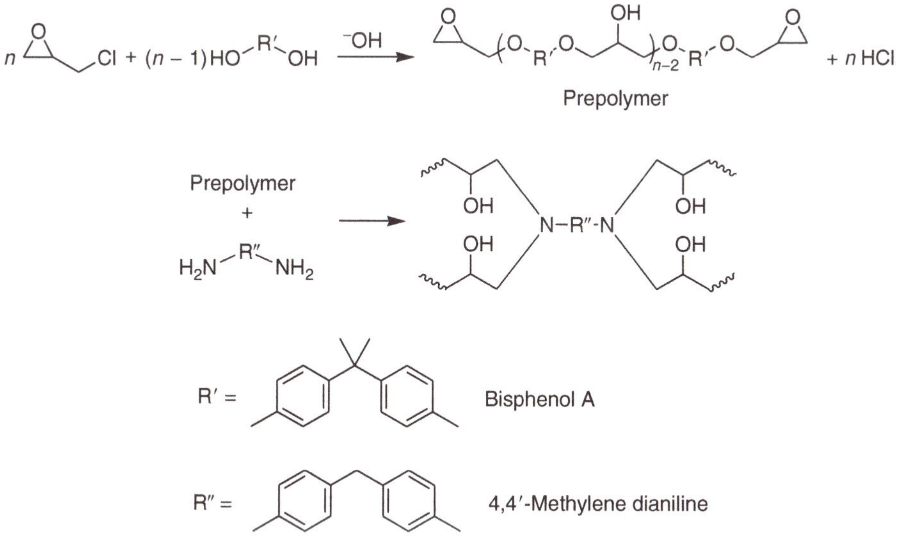

Figure 10.5 Illustration of an epoxy formulation. A prepolymer, formed by base-catalyzed condensation of an excess of epichlorohydrin with bisphenol A, is cured by cross-linking with 4,4′-methylene dianiline.

rings. For the example in Figure 10.5, the diol is based on bisphenol A. The prepolymer is then reacted (*cured*) with a multifunctional anhydride or amine (methyl dianiline in the figure) to form a highly cross-linked material. Adapt the analysis in the preceding section to find the gel point for this system, assuming that the two compounds are mixed in the weight ratio 1:10 diamine to prepolymer and that the prepolymer has $n = 4$ (see Figure 10.5). Then interpret the statement found in the instructions for a typical "two-part" epoxy that "the bond will set in 5 minutes, but that full strength will not be achieved until 6 hours."

Solution

Following the reaction scheme in Figure 10.5, the prepolymer has functionality 2 whereas the diamine has functionality $f = 4$, so we will call the epoxide group "B" and the diamine A_4. We now need to find out which group is in excess, that is, to calculate the ratio r. The molecular weight of the diamine is 198 g/mol and that of the prepolymer is 914 g/mol. If we mix 1 g of the diamine with 10 g of the prepolymer we have a molar ratio of $(1/198):(10/914)$ or 0.00505:0.0109. As there are four A groups per diamine and two B groups per prepolymer, the final ratio of A:B groups is 0.0101:0.0109 or 0.93:1. Thus the A group is limiting the reaction, albeit only just.

From Equation 10.2.1 we can see that $\rho = 1$, as all the A group are in A_4 units. This also makes the development of the branching coefficient quite simple, as every chain between two A_4 groups contains one and only one prepolymer (BB) unit. The addition of the first BB to an A_4 group takes place with probability p, and the addition of the subsequent A_4 has probability rp. Thus $\alpha = rp^2$, which we could also obtain from Equation 10.2.3 after inserting $\rho = 1$. The critical extent of reaction corresponds to $\alpha_c = 1/3$ from Equation 10.2.4, and from Equation 10.2.5 we have

$$p_c = \frac{1}{\sqrt{3r}} \approx 0.6$$

We can interpret the time for the bond to set as a time when the gel point is consistently exceeded, perhaps $p \approx 0.7$, so that the adhesive has solidified. The time to develop full mechanical strength reflects the time required for p to approach 1.

10.3   Elastic Deformation

For the remainder of this chapter we will emphasize elastomers rather than thermosets, and our primary focus will be the elasticity of such network materials. The various elastic phenomena we discuss in this chapter will be developed in stages. We begin with the simplest case: a sample that displays a purely elastic response when deformed by simple elongation. On the basis of Hooke's law, we expect that the force of deformation—related to the stress—and the distortion that results—related to the strain—will be directly proportional, at least for small deformations. In addition, the energy spent to produce the deformation is recoverable: the material snaps back when the force is released. We are interested in the molecular origin of this property for polymeric materials but, before we can get to that, we need to define the variables more precisely. One cautionary note is appropriate here. A full description of the elastic response of materials requires tensors, but we will avoid this complication by emphasizing one kind of deformation—uniaxial extension—and touching on another, shear.

A quantitative formulation of Hooke's law is facilitated by considering the rectangular sample shown in Figure 10.6a. If a force f is applied to the face of area A, the original length of the block L_0 will be increased by ΔL. Now consider the following variations:

1.   Imagine subdividing the block into two portions perpendicular to the direction of the force, as shown in Figure 10.6b. Each slice experiences the same force as before, and the same net deformation results. A deformation $\Delta L/2$ is associated with a slice of length $L_0/2$. The same

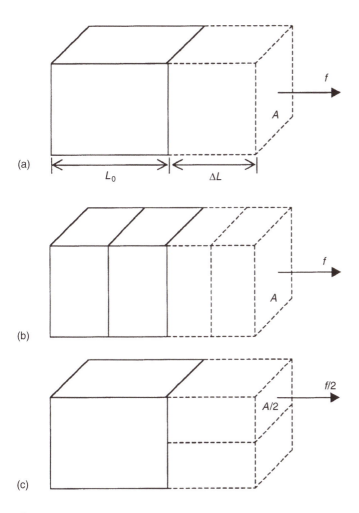

(a)

(b)

(c)

Figure 10.6 (a) A force f applied to area A extends the length of the sample from L_0 by an amount ΔL. Parts (b) and (c) illustrate the argument that $f/A \sim \Delta L/L_0$.

argument could be applied for any number of slices; hence it is the quantity $\Delta L/L_0$ that is proportional to the force.

2. Imagine subdividing the face of the block into two portions of area $A/2$. A force only half as large would be required for each face to produce the same net distortion. The same argument could be applied for any degree of subdivision; hence it is the quantity f/A that is proportional to $\Delta L/L_0$.

3. The force per unit area along the axis of the deformation is called the uniaxial tension or stress. We shall use the symbol σ as a shorthand replacement for f/A and attach the subscript t to signify tension; we will use σ for the shear stress, as in Chapter 9 and Chapter 11. The elongation expressed as a fraction of the original length, $\Delta L/L_0$, is called the strain. We shall use ε as the symbol for the resulting extensional strain to distinguish it from the shear strain (γ) also discussed in Chapter 9 and Chapter 11.

With these considerations in mind, we write

$$\sigma_t = E\varepsilon = E\left(\frac{\Delta L}{L_0}\right) \tag{10.3.1}$$

where the proportionally constant E is called the tensile modulus or *Young's modulus*. Remember, it will be different for different substances and for a given substance at different temperatures. Since ε is dimensionless, E has the same units as f/A, namely, force/length^2, or N/m^2(Pa) in the SI system. Note that for Equation 10.3.1 to be useful as a definition of E, the strain must be sufficiently small so that the stress remains proportional to the strain.

There is another aspect of tensile deformation to be considered. The application of a distorting force not only stretches a sample, but also causes the sample to contract at right angles to the stretch. If d and h represent the width and height of area A in Figure 10.6, both contract by the same fraction, a fraction that is related to the strain in the following way:

$$-\frac{\Delta d}{d} = -\frac{\Delta h}{h} = \nu\frac{\Delta L}{L_0} = \nu\varepsilon \qquad (10.3.2)$$

where the minus signs indicate that Δd and Δh are negative when ΔL is positive. The constant ν is an important property of a material called *Poisson's ratio*; it may also be written as

$$\nu = \frac{1}{2}\left(1 - \frac{1}{V}\frac{dV}{d\varepsilon}\right) \qquad (10.3.3)$$

where V is the volume of the sample (see Problem 9). Thus, if the volume does not change on elongation, the factional contraction in each of the perpendicular directions is half the fractional increase in length and $\nu = 0.5$. In general two parameters, for example E and ν, are required to describe the response of a sample to tensile force. Poisson's ratio also provides a means to relate E to the shear modulus, G, and the compressional modulus, K:

$$G2(1 + \nu) = E \qquad (10.3.4a)$$
$$K3(1 - 2\nu) = E \qquad (10.3.4b)$$

For isotropic materials such as those we are considering in this chapter, the small strain elastic response can therefore be described by any two of the parameters of E, G, K, and ν. For elastomers, where the volume change on deformation tends to be very small, $\nu \approx 0.5$ and $E \approx 3G$. For example, polyisoprene has $\nu = 0.4999$, so this approximation is excellent; in contrast, for metals, ν typically lies between 0.25 and 0.35.

10.4 Thermodynamics of Elasticity

It is not particularly difficult to introduce thermodynamic concepts into a discussion of elasticity. We shall not explore all of the implications of this development, but shall proceed only to the point of establishing the connection between elasticity and entropy. Then in the next section we shall go from macroscopic thermodynamics to statistical thermodynamics, in pursuit of a molecular model to describe the elastic response of cross-linked networks.

10.4.1 Equation of State

We begin by remembering the mechanical definition of work and apply that definition to the stretching process of Figure 10.6. Using the notation of Figure 10.6, we can write the increment of elastic work associated with an increment in elongation dL as

$$dw = f\,dL \qquad (10.4.1)$$

It is necessary to establish some conventions concerning signs before proceeding further. When the applied force is a tensile force and the distortion is one of stretching, f, dL and dw are all defined to be positive quantities. Thus dw is positive when elastic work is done on the system. The work done by the sample when the elastomer snaps back to its original size is a negative quantity.

The first law of thermodynamics defines the change dU in the internal energy of a system as the sum of the heat absorbed by the system, dq, plus the work done on the system, dw:

$$dU = dq + dw \tag{10.4.2}$$

The element of work is generally written $-p\,dV$, where p is the external pressure, but with the possibility of an elastic contribution, it is $-p\,dV + f\,dL$. With this substitution, Equation 10.4.2 becomes

$$dU = dq - p\,dV + f\,dL \tag{10.4.3}$$

A consistent sign convention has been applied to the pressure–volume work term: a positive dV corresponds to an expanded system, and work is done by the system to push back the surrounding atmosphere.

The second law of thermodynamics gives the change in entropy associated with the isothermal, reversible absorption of an element of heat dq as

$$dS = \frac{dq}{T} \tag{10.4.4}$$

This relationship can be used to replace dq by $T\,dS$ in Equation 10.4.3, since the infinitesimal increments implied by the differentials mean that the system is only slightly disturbed from equilibrium and the process is therefore reversible:

$$dU = T\,dS - p\,dV + f\,dL \tag{10.4.5}$$

We now turn to the Gibbs free energy G (recall the treatment of mixtures in Chapter 7) defined as

$$G = H - TS \tag{10.4.6}$$

where the enthalpy

$$H = U + pV \tag{10.4.7}$$

Combining the last two results and taking the derivative gives

$$dG = dU + p\,dV + V\,dp - T\,dS - S\,dT \tag{10.4.8}$$

Comparing Equation 10.4.8 with Equation 10.4.5 enables us to replace several of these terms by $f\,dL$

$$dG = V\,dp - S\,dT + f\,dL \tag{10.4.9}$$

thus establishing the desired connection between the stretching experiment and thermodynamics.

Since G is a state variable and forms exact differentials, an alternative expression for dG is

$$dG = \left(\frac{\partial G}{\partial p}\right)_{T,L} dp + \left(\frac{\partial G}{\partial T}\right)_{p,L} dT + \left(\frac{\partial G}{\partial L}\right)_{p,T} dL \tag{10.4.10}$$

Comparing Equation 10.4.10 and Equation 10.4.9 enables us to write

$$f = \left(\frac{\partial G}{\partial L}\right)_{p,T} \tag{10.4.11}$$

Note this is the same derivation that yields the important results $V = (\partial G/\partial p)_T$ and $S = -(\partial G/\partial T)_p$ when no elastic work is considered; these will arise in the discussion of the glass transition in Chapter 12.

We differentiate Equation 10.4.6 with respect to L, keeping p and T constant:

$$\left(\frac{\partial G}{\partial L}\right)_{p,T} = \left(\frac{\partial H}{\partial L}\right)_{p,T} - T\left(\frac{\partial S}{\partial L}\right)_{p,T} \tag{10.4.12}$$

The left-hand side of this equation gives f according to Equation 10.4.11; therefore

$$f = \left(\frac{\partial H}{\partial L}\right)_{p,T} - T\left(\frac{\partial S}{\partial L}\right)_{p,T} \tag{10.4.13}$$

This expression is sometimes called the *equation of state for an elastomer*, by analogy to

$$-p = \left(\frac{\partial U}{\partial V}\right)_T - T\left(\frac{\partial S}{\partial V}\right)_T \tag{10.4.14}$$

the thermodynamic equation of state for a fluid. Note the parallel roles played by length and volume in these two expressions.

10.4.2 Ideal Elastomers

Equation 10.4.12 shows that the force required to stretch a sample can be broken into two contributions: one that measures how the enthalpy of the sample changes with elongation and one that measures the same effect on entropy. The pressure of a system also reflects two parallel contributions, except that the coefficients are associated with volume changes. It will help to pursue the analogy with a gas a bit further. For an ideal gas, the molecules are noninteracting and so it makes no difference how far apart they are. Therefore, for an ideal gas $(\partial U/\partial V)_T = 0$ and the thermodynamic equation of state becomes

$$-p = -T\left(\frac{\partial S}{\partial V}\right)_T \tag{10.4.15}$$

By analogy, an *ideal elastomer* is defined as one for which $(\partial H/\partial L)_{p,T} = 0$; in this case Equation 10.4.13 reduces to

$$f = -T\left(\frac{\partial S}{\partial L}\right)_{p,T} \tag{10.4.16}$$

Although defined by analogy to an ideal gas, the justification for setting $(\partial H/\partial L)_{p,T} = 0$ cannot be the same for an elastomer as for an ideal gas. All molecules attract one another and this attraction is not negligible in condensed phases (recall the cohesive energy density in Chapter 7). What the ideality condition requires in an elastomer is that there is no change in the enthalpy of the sample as a result of the stretching process. This has two implications. On the one hand, the average energy of interaction between different molecules cannot change. For a given material this intermolecular contribution is determined primarily by the density, and therefore for a deformation that does not change the volume it may be a good approximation. The intramolecular contribution arises from the conformational energy of each chain, which is determined by the relative population of *trans* and *gauche* conformers (recall Chapter 6). In fact, moderate changes in the end-to-end distance of a chain can be accomplished with the expenditure of relatively little energy. For large deformations, or for networks with strong interactions—say, hydrogen bonds instead of dispersion forces—the approximation of an ideal elastomer may be very poor. There is certainly an enthalpy change associated with crystallization (see Chapter 13), so $(\partial H/\partial L)_{p,T}$ would not vanish if stretching induced crystal formation (which can occur, e.g., in natural rubber).

We have presented this development of the ideal elastomer in terms of the Gibbs free energy, which is generally the most appropriate for processes of importance in chemistry: p and T (and number of moles) are the natural independent variables. However, in the majority of texts the Helmholtz free energy, $A = U - TS$ is employed, so it is worthwhile to take a moment and compare the answers. For an experiment at constant temperature, we can write

$$dA = dU - T\,dS \tag{10.4.17}$$

which may then be compared to Equation 10.4.5 to yield

$$dA = f\,dL - p\,dV \tag{10.4.18}$$

At both constant temperature and constant volume, therefore,

$$f = \left(\frac{\partial A}{\partial L}\right)_{T,V} = \left(\frac{\partial U}{\partial L}\right)_{T,V} - T\left(\frac{\partial S}{\partial L}\right)_{T,V} \tag{10.4.19}$$

and the criterion for an ideal elastomer becomes $(\partial U/\partial L)_{T,V} = 0$. Because the volume changes on elastomer deformation are typically so small, a deformation carried out at constant T and p is very close to one done at constant T and V.

10.4.3 Some Experiments on Real Rubbers

Before proceeding to the statistical theory of rubber elasticity, it is instructive to examine some of the classical experiments conducted on rubbers. An example is shown in Figure 10.7, where the tensile stress (proportional to f) was measured as a function of temperature at the indicated constant

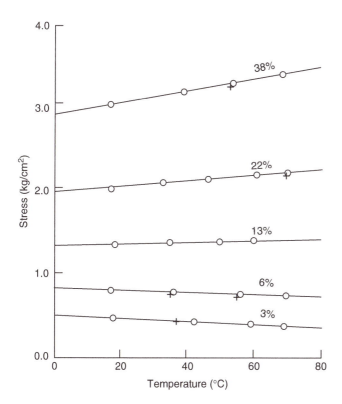

Figure 10.7 Stress at a constant length for natural rubber, at the indicated elongations, as a function of temperature. Thermoelastic inversion occurs below about 10% elongation. (Data from Anthony, R.L., Caston, R.H., and Guth, E., *J. Phys. Chem.*, 46, 826, 1942. With permission.)

length. These data show an interesting feature, known as *thermoelastic inversion*, whereby at elongations below about 10%, the stress decreases with temperature, in contrast to the larger strain behavior. As we are anticipating that the elasticity is primarily due to entropy, we expect the force to increase with temperature. The reason for the behavior at small elongation is actually quite simple; it is due to thermal expansion. The unstrained length increases with temperature due to expansion and thus the actual strain at fixed length decreases with increased temperature and consequently the force decreases. Thus the thermoelastic inversion can be eliminated by comparing the data at constant strain.

This kind of thermoelastic data can be further analyzed in terms of the thermodynamic contributions. From Equation 10.4.19 we can write

$$\left(\frac{\partial f}{\partial T}\right)_L = -\left(\frac{\partial S}{\partial L}\right)_T \tag{10.4.20}$$

and

$$\left(\frac{\partial U}{\partial L}\right)_T = f - T\left(\frac{\partial f}{\partial T}\right)_L \tag{10.4.21}$$

These expressions are useful because they permit extraction of information about S and U from the measured behavior of f. Figure 10.8a shows data for f versus elongation and the decomposition into an entropic and an internal energy contribution, following Equation 10.4.20 and Equation 10.4.21. Clearly at large elongation, the entropic part of the force dominates, but at low elongations the internal energy contribution is larger. Again, however, this effect is largely eliminated by plotting the data at constant strain, as shown in Figure 10.8b. These results and many others confirm, to a good approximation, that there is only a modest internal energy contribution to the force for a deformation at constant volume.

One further example of a "model-free" thermodynamic interpretation of rubber elasticity is given by the temperature increase observed in adiabatic extension of a rubber band. This underlies the standard classroom demonstration of the entropic origin of rubber elasticity, whereby a rubber band is rapidly extended and placed in contact with a (highly temperature-sensitive) upper lip. This kind of experiment goes back at least as far as Gough [5] and Joule [6], and some of Joule's data are shown in Figure 10.9 along with some from James and Guth [7]. At low extensions, the temperature actually decreases slightly, but then increases steadily. The interpretation of the experiment is as follows. In the adiabatic extension of an ideal elastomer, the work done on the sample is retained entirely as heat; there is a loss of entropy but no change of internal energy and $dq = -dw$. The work is given by Equation 10.4.1 and the heat by Equation 10.4.4; therefore the temperature change is

$$\Delta T = \frac{1}{C_L}\int f\, dL = -\frac{T}{C_L}\int_{L_0}^{L}\left(\frac{\partial S}{\partial L}\right)_T dL \tag{10.4.22}$$

where C_L is the appropriate heat capacity at constant length. As in the previous examples, the negative change in temperature at small extensions is due to the positive entropy of deformation, that is, it corresponds to the thermoelastic inversion.

10.5 Statistical Mechanical Theory of Rubber Elasticity: Ideal Case

We now proceed to use a molecular model to derive predictions for the stress–strain behavior of an ideal elastomer. In the subsequent section, we will consider various nonidealities that could occur in a real material, but even granted the existence of some or all of these nonidealities, the

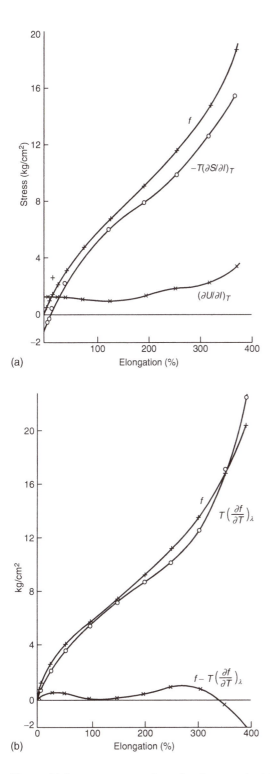

(a)

(b)

Figure 10.8 Stress versus elongation for natural rubber, resolved into internal energy and entropic contributions, at (a) constant temperature and (b) constant strain. (Data from Anthony, R.L., Caston, R.H., and Guth, E., *J. Phys. Chem.*, 46, 826, 1942. With permission.)

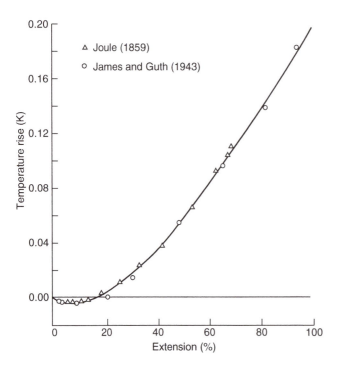

Figure 10.9 Temperature change during adiabatic extension of natural rubber. (Data from Joule, J.P., *Phil. Trans. R. Soc.*, 149, 91, 1859; James H.M. and Guth, E., *J. Chem Phys.*, 11, 455, 1943; 15, 669, 1947.) (From Treloar, L.R.G., *The Physics of Rubber Elasticity*, 3rd ed., Clarendon Press, Oxford, 1975. With permission.)

qualitative success of the ideal model is really a remarkable triumph of statistical mechanics. We have already considered the most famous equation of state, that of the ideal gas. That simple result is illuminating, but only describes the behavior of very dilute gases with any reliability and dilute gases are of limited significance from the point of view of materials science. In contrast, the ideal elastomer equations will provide a reasonable description of a practically important, but extremely complex, amorphous condensed phase, even though the derivation is not appreciably more elaborate than that for the ideal gas. We will begin by considering the force required to extend a single Gaussian chain, an example that already arose in the context of chain swelling in Section 7.7 and that will resurface in the bead–spring model of viscoelasticity in Section 11.4. Then we will apply this result to an entire ensemble of cross-linked chains.

10.5.1 Force to Extend a Gaussian Chain

Since entropy plays the determining role in the elasticity of an ideal elastomer, let us review some ideas about this important thermodynamic variable. We used a probabilistic interpretation of entropy extensively in Chapter 7 to formulate the entropy of mixing. The starting point was the Boltzmann relation:

$$S = k \ln \Omega \tag{10.5.1}$$

where k is Boltzmann's constant and Ω is the number of possible states. As then, the difference in entropy between two states of different thermodynamic probability is

$$\Delta S = S_2 - S_1 = k \ln \left(\frac{\Omega_2}{\Omega_1} \right) \tag{10.5.2}$$

so that ΔS is positive when $\Omega_2 > \Omega_1$ and negative when $\Omega_2 < \Omega_1$.

In the previous section, we identified the force of extension with the associated change in free energy (Equation 10.4.11 or Equation 10.4.19). Then, if the change in free energy is entirely due to the entropy, the material is an ideal elastomer. Figure 10.8 provides an example of how reasonable this assumption is for a material; now we apply it to one chain. Consider extending a single Gaussian chain of N units, with statistical segment length b (recall Section 6.3). The chain has one end fixed at the origin $(0,0,0)$ and the other is held in the infinitesimal cube between (x_0, y_0, z_0) and $(x_0 + dx_0, y_0 + dy_0, z_0 + dz_0)$, as shown in Figure 10.10. The imposed end-to-end distance is $h_0 = (x_0^2 + y_0^2 + z_0^2)^{1/2}$, which may be compared to the equilibrium mean square end-to-end distance $\langle h^2 \rangle = Nb^2$. The number of ways that this chain can satisfy the imposed constraint is given by the Gaussian distribution (recall Equation 6.7.1):

$$P_i(N, \vec{h}_0) = \left(\frac{3}{2\pi \langle h^2 \rangle} \right)^{3/2} \exp\left[-\frac{3h_0^2}{2\langle h^2 \rangle} \right]$$
$$= \beta^{3/2} \exp\left[-\pi \beta h_0^2 \right] \tag{10.5.3}$$

where we define the normalization factor, β as

$$\beta \equiv \frac{3}{2\pi \langle h^2 \rangle} = \frac{3}{2\pi N b^2} \tag{10.5.4}$$

and the subscript "i" on P denotes the "initial" state. We then extend the chain to a new end-to-end distance, h, with coordinates between (x, y, z) and $(x + dx, y + dy, z + dz)$. The corresponding "final" state distribution function P_f is

$$P_f(N, \vec{h}) = \beta^{3/2} \exp\left[-\pi \beta h^2 \right] \tag{10.5.5}$$

We now associate the number of possible conformations with the entropy defined by Equation 10.5.1, that is, we take $\Omega = AP$, with A as some unspecified proportionality constant. Then we can say

$$\Delta S_{\text{chain}} = k \ln A P_f - k \ln A P_i = k \ln \left(\frac{P_f}{P_i} \right)$$
$$= -k\pi\beta \left(h^2 - h_0^2 \right) \tag{10.5.6}$$

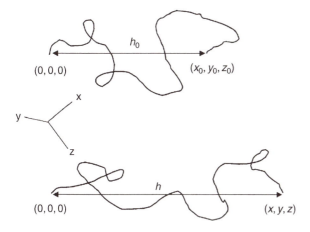

Figure 10.10 Extension of a single Gaussian chain from initial end-to-end distance h_0 to final end-to-end distance h.

where we use the subscript "chain" to emphasize that this is a single chain calculation. The unknown constant A cancels out when we calculate the change in entropy. The force to extend the chain to h is given by

$$f = -T\left(\frac{\partial \Delta S_{\text{chain}}}{\partial h}\right) = \frac{3kT}{\langle h^2 \rangle} h \tag{10.5.7}$$

This is a fundamental result, and one we will use extensively in modeling the viscoelastic properties of polymer liquids in Chapter 11. Equation 10.5.7 indicates that a single Gaussian chain behaves like a Hooke's law spring, with force constant $3kT/\langle h^2 \rangle$ and zero rest-length. Note the interesting result that this spring will stiffen as T increases, in contrast to intuitive expectation for a metal spring; this is a direct result of its entropic basis. Equation 10.5.7 contains most of the physical concepts that are required to describe rubber elasticity from a molecular viewpoint.

10.5.2 Network of Gaussian Strands

We now consider an ideal network made up of Gaussian strands. If the cross-links were introduced to a melt of Gaussian chains, for example, by vulcanization, it is plausible that the strands will be more or less Gaussian as well. For simplicity, we will assume that all strands contain an identical number of statistical segment lengths, N_x; this simplification will subsequently be removed. We now impose a macroscopic deformation on the network; for example, we might stretch it in the x direction. However, to be more general, we describe the deformation by three *extension ratios* λ_x, λ_y, and λ_z, given by $L_x/L_0, L_y/L_0$, and L_z/L_0, respectively. If we begin with a cube of material of length L_0 on each side, that cube will be deformed to a three-dimensional volume element with sides L_x, L_y, and L_z, as shown in Figure 10.11. We assume that there is no volume change on deformation, and thus

$$V = L_x L_y L_z = V_0 = L_0^3; \quad \lambda_x \lambda_y \lambda_z = 1 \tag{10.5.8}$$

This is a reasonable approximation for bulk elastomers, where Poisson's ratio is nearly 0.5, but is not appropriate, for example, when the network is swollen with solvent. The removal of this assumption will be discussed in Section 10.7.

We now make a final, very important assumption, the so-called *affine junction assumption*: each junction point moves in proportion to the macroscopic deformation. Consequently, the end-to-end vector of each strand is deformed so that the coordinates of one end transform $x_0 \rightarrow x = \lambda_x x_0$, $y_0 \rightarrow y = \lambda_y y_0$, $z_0 \rightarrow z = \lambda_z z_0$, when we take the other end as the origin. We already know the entropy change per strand associated with this process: it is simply the result for a single chain, see Equation 10.5.6, applied to a single strand. Writing it out in more detail, we have

$$\begin{aligned} \Delta S_{\text{strand}} &= k \ln AP_{\text{f}} - k \ln AP_{\text{i}} = k \ln\left(\frac{P_{\text{f}}}{P_{\text{i}}}\right) \\ &= -k\pi\beta(x^2 + y^2 + z^2) - \left(-k\pi\beta(x_0^2 + y_0^2 + z_0^2)\right) \\ &= -k\frac{3}{2N_x b^2}\left(x_0^2(\lambda_x^2 - 1) + y_0^2(\lambda_y^2 - 1) + z_0^2(\lambda_z^2 - 1)\right) \end{aligned} \tag{10.5.9}$$

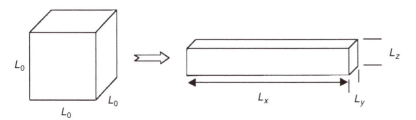

Figure 10.11   Deformation of a cube of material subjected to uniaxial elongation along x.

We now note that on average $x_0^2 = N_x b^2/3$, and the same for y_0 and z_0, so that

$$\langle \Delta S_{\text{strand}} \rangle = -\frac{k}{2} \left(\lambda_x^2 + \lambda_y^2 + \lambda_z^2 - 3 \right) \tag{10.5.10}$$

In this rather simple result, N_x does not appear, so the assumption of constant N_x was actually not necessary. To obtain the total entropy change for the material, we simply need the number of strands per unit volume. For our ideal network this is given by $\rho N_{\text{av}}/M_x$, where M_x is the (number average) molecular weight between cross-links, but in anticipation of defects such as dangling ends and loops in real networks, we will just define the total *number of elastically effective strands*, ν_e. The number of strands per unit volume is thus ν_e/V and the total entropy change becomes

$$\Delta S = -\frac{\nu_e k}{2} \left(\lambda_x^2 + \lambda_y^2 + \lambda_z^2 - 3 \right) \tag{10.5.11}$$

This equation represents the principal result of this molecular network theory. We will now consider a specific deformation to obtain expressions for the modulus, but the necessary manipulations are all results of continuum elasticity theory and require no further assumptions about what the molecules are doing.

10.5.3 Modulus of the Gaussian Network

We begin with a uniaxial extension, say along x, by a stretch ratio λ. Thus $\lambda_x = \lambda$, and by volume conservation (see Equation 10.5.8) $\lambda_y = \lambda_z = 1/\sqrt{\lambda}$. Furthermore, $\varepsilon = \lambda - 1$. In this case, then

$$\Delta S = -\frac{\nu_e k}{2} \left(\lambda^2 + \frac{2}{\lambda} - 3 \right) \tag{10.5.12}$$

and the force is given by

$$f = -T \left(\frac{\partial \Delta S}{\partial L} \right) = -\frac{T}{L_0} \left(\frac{\partial \Delta S}{\partial \lambda} \right) = \frac{\nu_e k T}{L_0} \left(\lambda - \frac{1}{\lambda^2} \right) \tag{10.5.13}$$

Note that the force changes sign, as it should, when $\lambda = 1$. If we now divide both sides by the cross-section area normal to the stretching direction, $L_y L_z = L_0^2/\lambda$, we obtain the tensile stress:

$$\sigma_t = \frac{f}{\text{area}} = \frac{\lambda f}{L_0^2} = kT \frac{\nu_e}{V} \left(\lambda^2 - \frac{1}{\lambda} \right) \tag{10.5.14}$$

Alternatively, it is often experimentally more convenient to divide by the initial cross-sectional area, L_0^2, which leads to the following result:

$$\sigma_t = kT \frac{\nu_e}{V} \left(\lambda - \frac{1}{\lambda^2} \right) \tag{10.5.15}$$

The stress given by Equation 10.5.14 is sometimes called the *true stress* to distinguish it from the quantity given Equation 10.5.15, which is known as the *engineering stress* or the *nominal stress*.

We can now obtain an expression for Young's modulus, E, recalling Equation 10.3.1 (and that $d\lambda = d\varepsilon$):

$$E = \lim_{\lambda \to 1} \frac{\partial \sigma_t}{\partial \lambda} = 3kT \frac{\nu_e}{V} \tag{10.5.16}$$

Note that the same result is obtained if we use either the true stress or the engineering stress because they coincide in the small strain limit.

404   Networks, Gels, and Rubber Elasticity

We finally obtain an expression for the shear modulus, G, using the approximate relation $G = E/3$ (Equation 10.3.4a):

$$G = kT\frac{\nu_e}{V} = \frac{\rho RT}{M_x}$$

(10.5.17)

where in the last step we have substituted the ideal value for ν_e/V in terms of the molecular weight between cross-links, M_x. From these equations (Equation 10.5.14 through Equation 10.5.17) we can extract some important conclusions:

1. The modulus increases with temperature, just as with the spring constant of a single chain, due to its entropic origin.
2. The modulus increases as a function of cross-link density, because M_x decreases; a "tighter" network is "stiffer."
3. The modulus is independent of the functionality of the cross-links.
4. The extensional stress is not a linear function of the strain, even though the individual network strands are supposed to be Hookean. (In contrast, the shear stress turns out to be linear in the strain, but we will not take the time to derive this relation.)
5. Assuming a density of 1 g/cm³ at room temperature, and $M_x = 10,000$ g/mol, Equation 10.5.17 gives a modulus of 2.5×10^6 dyn/cm², or 0.25 MPa. Typical values for elastomers fall within an order of magnitude of this number.

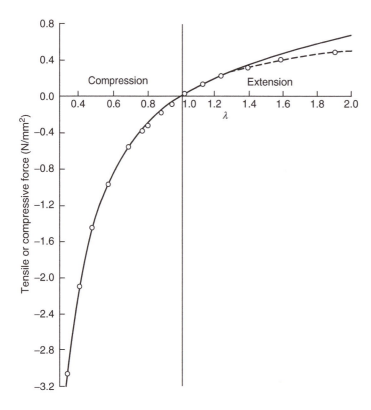

Figure 10.12 Stress for cross-linked natural rubber in compression and extension. (Data from Treloar, L.R.G., *Trans. Faraday Soc.*, 40, 59, 1944. With permission.)

An example of a test of the theory, and Equation 10.5.14 in particular, is shown in Figure 10.12. Both extensional and compressive stresses were determined as a function of λ for a piece of vulcanized rubber. The agreement between experiment and theory is impressive, particularly in compression. The same sample was subsequently extended up to its breaking point, near λ ≈ 7.5, and the results are shown in Figure 10.13. The data at low extension ratios were fit to the theory to obtain the modulus of 0.39 MPa. The theory and the data are not in perfect agreement in this case; the main difference is the sharp increase in experimental stress at high λ. This is primarily due to the failure of the Gaussian assumption for large extensions; when the end-to-end distance becomes an appreciable fraction of the contour length, the Gaussian distribution no longer applies. This point will be considered again in the next section.

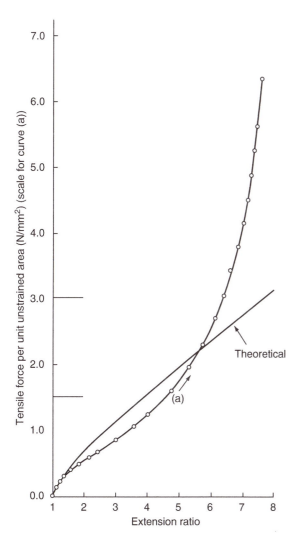

Figure 10.13 Same sample as in Figure 10.12, but now subjected to simple extension and much larger extension ratios (λ). (Data from Treloar, L.R.G., *Trans. Faraday Soc.*, 40, 59, 1944. With permission.)

10.6   Further Developments in Rubber Elasticity

The development of a statistical thermodynamic approach to rubber elasticity in the previous section involved a series of assumptions that could be questioned. In this section we touch briefly on some of these and give an indication of how they might be addressed. To begin with, we recall the central result of the theory (see Equation 10.5.11):

$$\Delta S = -\frac{\nu_e k}{2}\left(\lambda_x^2 + \lambda_y^2 + \lambda_z^2 - 3\right) \tag{10.6.1}$$

from which the stress and modulus can be computed for any deformation. The main assumptions invoked in the development and application of this equation are summarized below:

1. There is no change in internal energy, U, upon deformation at constant T and p.
2. There is no change in volume, V, upon deformation at constant T and p.
3. The number of conformations available to the strands both before and after deformation is given by a Gaussian distribution.
4. The number of conformations available to the strands before deformation is the same as for equivalent chains in the uncross-linked state.
5. The junction points deform affinely with the macroscopic deformation.
6. The number of elastically effective strands per unit volume, ν_e/V, is given by $\rho N_{av}/M_x$ for a perfect network.

It turns out that refinement or relaxation of almost any of these assumptions has generated significant amounts of controversy over the years, and to address these issues thoroughly would require an entire book. Accordingly, we will have to be content with a few examples.

10.6.1   Non-Gaussian Force Law

The approximate validity of the first two assumptions was suggested in the previous section and the effects of deviations from these assumptions should be reasonably transparent. Accordingly, except for the issue of solvent swelling, which will be taken up in the following section, we will not consider these further. The third assumption is more interesting from a molecular point of view. One violation of this assumption can be readily imagined: upon large extensions, say beyond 100% (recall Figure 10.13), a strand may no longer be Gaussian. Clearly as we approach the limit of full extension, when the end-to-end distance becomes a significant fraction of the contour length, the Gaussian force law (Equation 10.5.7) will not apply. This problem has been addressed theoretically and a reasonable solution is known. Specifically, Kuhn and Grün [8] showed that when a freely jointed chain of N_x links of length b is extended to an end-to-end distance h, the distribution function is not the familiar Gaussian, but rather

$$P(N_x, h) \propto \exp\left(-N_x\left[\frac{h}{N_x b}\beta + \ln\left(\frac{\beta}{\sinh\beta}\right)\right]\right) \tag{10.6.2}$$

where the quantity β is the so-called inverse Langevin function, $L^{-1}(x)$:

$$\beta = L^{-1}(h/N_x b) \tag{10.6.3}$$

This formulation is not particularly transparent, but it turns out that the exponential in Equation 10.6.2 can be expanded as a power series in $(h/N_x b)$ as follows:

$$P(N_x, h) \propto \exp\left(-N_x\left[\frac{3}{2}\left(\frac{h}{N_x b}\right)^2 + \frac{9}{20}\left(\frac{h}{N_x b}\right)^4 + \frac{99}{350}\left(\frac{h}{N_x b}\right)^6 \cdots\right]\right) \tag{10.6.4}$$

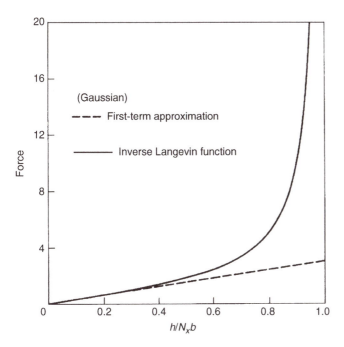

Figure 10.14 Force versus extension for a Gaussian chain (first-term approximation) and for the full inverse Langevin function.

from which we can see that the Gaussian result is just the first term in a series. When the argument $(h/N_x b)$ is small, the Gaussian function is adequate, but at large extensions Equation 10.6.4 is more accurate. This result can be carried through the analysis of the previous section (see Equation 10.5.6 and Equation 10.5.7) to obtain the corresponding force law and the result is illustrated in Figure 10.14. The main new feature is that for extensions such that $(h/N_x b) > 0.4$, the force increases rather sharply. This is in excellent qualitative agreement with the data in Figure 10.13. In fact, recent experiments have been able to measure the force of extension of single DNA molecules and the inverse Langevin function provides a good account of the results (see Problem 12), so although this approach is mathematically unwieldy, it is successful.

A second difficulty with the Gaussian assumption arises from the inevitable distribution of strand lengths. In the development leading up to Equation 10.5.10, we argued that because the quantity N_x cancels out of the final expression for ΔS, the assumption of monodisperse strands was benign. In reality the cross-linking process, however it is carried out, will leave some distribution of N_x. Consequently, the non-Gaussian character of the distribution function will become apparent at different values of the macroscopic extension; shorter strands will sooner become more fully stretched than longer ones. For example, an average strand begins with an unperturbed $h/b = \sqrt{N_x}$, which is then stretched to $\lambda \sqrt{N_x}$. The smaller the N_x, the smaller the value of λ required for $\lambda \sqrt{N_x}$ to approach N_x. This problem is much more difficult to deal with, not least because of the difficulty in characterizing the distribution of N_x.

10.6.2 Front Factor

The fourth assumption above represents a reasonable simplification, but it need not hold, even for rubbers lightly cross-linked in the melt state. Under some circumstances, such as cross-links introduced while the material is under stress, it certainly would not apply. The first step in lifting this assumption is to assume that in the undeformed state the elastomer is isotropic and the mean square strand length is $\langle h_i^2 \rangle$, where the subscript "i" denotes initial. If we return to Equation 10.5.9, we would replace $x_0^2 = y_0^2 = z_0^2$ with $x_i^2 = y_i^2 = z_i^2$, and the expression for the strand entropy (Equation 10.5.10) would be replaced by

$$\langle \Delta S_{\text{strand}} \rangle = -\frac{k}{2} \frac{\langle h_i^2 \rangle}{\langle h_0^2 \rangle} \left(\lambda_x^2 + \lambda_y^2 + \lambda_z^2 - 3 \right) \tag{10.6.5}$$

The ratio of mean square end-to-end distances in the undeformed network relative to that in a melt of the same strands is known as the *front factor*. As it is a constant, it carries through the subsequent development and therefore appears in the expressions for the stress and the modulus.

Alternatives to Assumption 5—that the junction points deform affinely with the macroscopic strain—have also been proposed. In the most common approach only the junctions on the edges of the material are so constrained; those in the bulk of the network are free to fluctuate about their mean positions [7]. This model is sometimes referred to as the *phantom network*. This modification cannot be applied to the entropy of a single strand, as in Equation 10.6.5, but must be applied to the network as a whole. The result is a reduction in the net stress, which is determined by the number of junctions, ν_x; Equation 10.5.11 so modified becomes

$$\Delta S = -(\nu_e - \nu_x) \frac{k}{2} \frac{\langle h_i^2 \rangle}{\langle h_0^2 \rangle} \left(\lambda_x^2 + \lambda_y^2 + \lambda_z^2 - 3 \right) \tag{10.6.6}$$

Again, this new front factor is a constant and would carry through to the expressions for the stress and the modulus. For the particular case of a regular network in which f strands emerge from each junction, then $\nu_e = \nu_x f/2$, and $(\nu_e - \nu_x) = \nu_e (f - 2)/f$. For a network made by vulcanization, therefore, $f = 4$ and this term is equal to $\nu_e/2$.

10.6.3 Network Defects

Any real network will contain defects in the structure, as suggested at the beginning of the chapter. These include loops, dangling ends, and the sol fraction. We have finessed this issue, in part, by using the concept of *elastically effective strands*, ν_e, which suggests that the contribution from such defects has been removed. However, it is not a straightforward matter to account for these in practice. Consider two general approaches. By NMR spectroscopy, for example, we might monitor the conversion of double bonds in a polydiene. From this information we could estimate M_x and thus ν_e by Assumption 6. But the NMR experiment could not tell us about loops or distinguish sol fraction from gel fraction (and, in fact, for modest degrees of cross-linking it is hard to even quantify that reliably). The next option is to measure the modulus itself and fit the results to the model. But, because all of these various effects contribute proportionally to the modulus, there is no easy way to resolve them from measurements on a single sample. In general, these three particular kinds of defects are treated in the following way: (a) loops are ignored, (b) dangling ends are corrected for, and (c) the conversion is high enough that the sol fraction is negligible, or it is extracted before measurement. Under (b), the number of dangling ends can be estimated from M_x, because each prenetwork chain had two ends. Therefore each prenetwork chain will contribute two dangling ends to the network, and the average length of these dangling ends will be M_x. Accordingly, the fraction of these dangling ends will be $2M_x/M$, where M is the molecular weight of the prenetwork chain and the number of effective strands may be estimated as

$$\frac{\nu_e}{V} = \frac{\rho N_{av}}{M_x} \left(1 - \frac{2M_x}{M} \right) \tag{10.6.7}$$

Another contribution to the modulus, the so-called *trapped entanglement*, although not strictly a topological defect has been considered extensively. We will see in Section 11.6 that the visco-elastic properties of high-molecular-weight polymers are dominated by the phenomenon of entanglement, which results from the intertwining of different chains. These entanglements act like temporary cross-links, imparting a rubber-like modulus to the liquid at intermediate times, before eventual relaxation and flow. By exploiting the expression for the shear modulus of an ideal elastomer, Equation 10.5.17, a molecular weight between entanglements can be defined, M_e. Now imagine we have a high-molecular-weight melt, with $M \gg M_e$, and we cross-link it enough to produce a network. (Recall from Equation 10.1.1 that we only need a few cross-links, $\sim 1/N$, to pass the gel point.) At this stage the modulus must be about the same, or perhaps slightly higher than it was before cross-linking, but $M_x \gg M_e$. In other words, we expect a modulus of about $\rho RT/M_e$, but the theory of rubber elasticity says it should be only $\rho RT/M_x$. Therefore these entanglements contribute to the modulus, but they are trapped; full stress relaxation, or flow, is eliminated by the cross-linking. Thus at low degrees of cross-linking, trapped entanglements should make the experimental modulus larger than expected by ideal elastomer theory. As the degree of cross-linking goes up, so that $M_x < M_e$, then this contribution should become progressively less important. A variety of approaches have been taken to this issue, all of which have the general effect just described; for example, the modulus could be expressed as

$$G = G_{\text{network}} + x G_{\text{entanglements}} = \rho RT \left(\frac{1}{M_x} + x \frac{1}{M_e} \right) \tag{10.6.8}$$

where x is just a parameter between 0 and 1, depending on the degree to which the entanglements are effective as cross-links. However, it is likely that the importance of this effect varies with extension as well. For example, imagine that the cross-linking process has created a "loose knot": two different strands are irreversibly intertwined, but not influencing each other very much. At low degrees of extension, this constraint on the strand conformations plays no particular role, but as the deformation increases, the "knot" might be pulled tight. This would have the effect of increasing the number of cross-links with deformation, and therefore act in the same direction as the non-Gaussian force law.

10.6.4 Mooney–Rivlin Equation

The preceding discussion of modifications to the ideal elastomer theory has emphasized additional molecular contributions and particularly ones that modify the prefactor. Following Equation 10.5.15, we may represent the tensile stress from these models as

$$\sigma_t = 2C_1 \left(\lambda - \frac{1}{\lambda^2} \right) \tag{10.6.9}$$

where the prefactor, now labeled "$2C_1$," is given by

$$2C_1 = \frac{f - 2}{f} \frac{\langle h_i^2 \rangle}{\langle h_o^2 \rangle} \frac{\rho RT}{M_x} \left(1 - \frac{2M_x}{M} \right) \tag{10.6.10}$$

where we have incorporated the modifications discussed above. Before the development of the statistical theory, however, Mooney and Rivlin [9] used continuum arguments to propose that

$$\sigma_t = 2C_1 \left(\lambda - \frac{1}{\lambda^2} \right) + 2C_2 \left(1 - \frac{1}{\lambda^3} \right) = \left(2C_1 + \frac{2C_2}{\lambda} \right) \left(\lambda - \frac{1}{\lambda^2} \right) \tag{10.6.11}$$

where C_1 and C_2 are unknown parameters of the material, but not functions of the deformation. The first term in this Mooney–Rivlin equation, therefore, has exactly the form of the statistical theory, and the second term can be viewed as a correction. The second form of Equation 10.6.11 suggests plotting the following quantity versus $1/\lambda$:

$$\frac{\sigma_t}{\left(\lambda - \frac{1}{\lambda^2} \right)} = 2C_1 + \frac{2C_2}{\lambda} \tag{10.6.12}$$

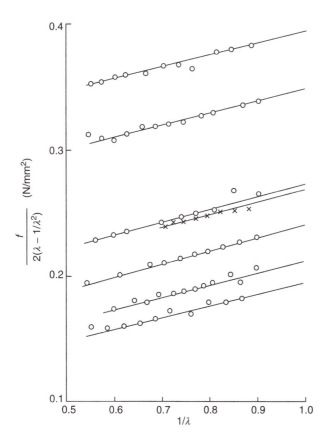

Figure 10.15 A "Mooney plot" for various rubbers in simple extension. (Data from Gumbrell, S.M., Mullins, L., and Rivlin, R.S., *Trans. Faraday Soc.*, 49, 1495, 1953. With permission.)

which should give a straight line with intercept $2C_1$ and slope $2C_2$. Adherence to the ideal theory should give a horizontal line. Examples from a variety of rubbers are shown in Figure 10.15; clearly Equation 10.6.12 gives a good description of the data, with nonzero values of C_2. There are other cases, however, such as swollen rubbers or networks prepared by cross-linking in solution, where $C_2 \approx 0$. A convincing and generally applicable molecular explanation for C_2 has proven elusive. Furthermore, for some materials the values of C_2 are even found to depend on the kind of experimental deformation employed, which suggests that Equation 10.6.11 is not the universally correct functional form.

10.7 Swelling of Gels

One of the distinguishing features of a lightly cross-linked polymer material is its ability to imbibe and retain a large volume of solvent. Examples include:

1. *Hot melt adhesives.* In this family of adhesives, the glue is applied as a liquid at high temperature and solidifies upon cooling. A typical formulation could include about 30% of a "thermoplastic elastomer," a styrene–isoprene–styrene triblock copolymer. At low temperatures, the styrene segments segregate from the isoprene blocks to form roughly spherical styrene aggregates that act as cross-links. At high temperatures, however, the segregation is disrupted and the polymer flows. The remaining 70% of the material consists of low-molecular-weight species, largely to dilute the isoprene segments and make the resulting gel

softer, but also in part to plasticize the styrene domains and lower their glass transition temperature (see Section 12.6).

2. *Soft contact lenses.* Soft contact lenses are an example of a hydrogel—a network in which the polymer is either water soluble or at least water compatible. The original soft contact lenses were made largely from cross-linked poly(hydroxyethyl methacrylate), and were developed in the late 1960s.

3. *Diapers.* Clearly, the major function of a diaper material is to imbibe and retain a large quantity of aqueous solution, without dissolving. Diaper materials can absorb up to several hundred times their own weight. A typical ingredient is poly(acrylic acid) or (sodium poly-acrylate). In this instance the network strands bear charged groups, which have the effect of greatly increasing the osmotic drive for water to enter the polymer and swell it.

4. *Biological tissue.* Much of biological tissue is essentially network material, although of course very complex. For example, in the "extra cellular matrix" collagen molecules intertwine in triple helices, which in turn aggregate to form fibrils, which in turn cross-link with the assistance of certain proteins to form three-dimensional gels.

Clearly, swollen networks are of fundamental importance in many areas of materials and biological science. In this section we will briefly address two aspects of the swelling phenomenon. First, we consider how the expression for the modulus of an ideal elastomer changes when solvent is incorporated. Then we consider swelling equilibrium; how much solvent can a network take up?

10.7.1 Modulus of a Swollen Rubber

We begin with a network formed at volume V_0 and then swollen with solvent to a new volume V. The volume fraction of polymer in the resulting gel is $\phi_2 = V_0/V$ (assuming additivity of volumes). We assume that the swelling is isotropic, so that the x, y, and z components of the end-to-end vector of each strand are increased by a factor of $(V/V_0)^{1/3} = \phi_2^{-1/3}$. We will reference the deformation of the already swollen network to the isotropically swollen dimensions so that the reference state terms (x_0^2, y_0^2, and z_0^2) in Equation 10.5.9 should each be multiplied by a factor of $\phi_2^{-2/3}$. Thus Equation 10.5.11 describing the entropy of deformation becomes

$$\Delta S = -\frac{\nu_e}{2} k \phi_2^{-2/3} \left(\lambda_x^2 + \lambda_y^2 + \lambda_z^2 - 3 \right) \tag{10.7.1}$$

(The same result could be obtained by taking the front factor of Equation 10.6.5, $\langle h_i^2 \rangle / \langle h_0^2 \rangle$, and realizing that $\langle h_i^2 \rangle$ is increased by the same factor of $\phi_2^{-2/3}$.) Now let us apply an elongation along the x direction, so $\lambda_x = \lambda$, and $\lambda_y = \lambda_z = 1/\sqrt{\lambda}$, and compute the force as in Equation 10.5.13:

$$f = -T \left(\frac{\partial \Delta S}{\partial L} \right) = -\frac{T}{L_s} \left(\frac{\partial \Delta S}{\partial \lambda} \right) = \frac{\nu_e kT}{L_s} \phi_2^{-2/3} \left(\lambda - \frac{1}{\lambda^2} \right) \tag{10.7.2}$$

where now L_s refers to the swollen but unstrained network. To compute the stress, again referenced to the swollen but otherwise undeformed network, we divide by L_s^2, recognizing that $L_s^3 = V$, not V_0.

$$\sigma_t = \frac{f}{L_s^2} = kT \frac{\nu_e}{V} \phi_2^{-2/3} \left(\lambda - \frac{1}{\lambda^2} \right)$$

$$= kT \frac{\nu_e}{V_0} \phi_2^{1/3} \left(\lambda - \frac{1}{\lambda^2} \right) \tag{10.7.3}$$

Comparing this result with Equation 10.5.15, the main result of swelling is that the stress in the swollen network is reduced by a factor of $\phi_2^{1/3}$ compared to the original network, and the modulus is reduced by the same factor when computed for constant cross-sectional area.

10.7.2 Swelling Equilibrium

Now we address the question of how much solvent a network can take up. Imagine we have a piece of lightly cross-linked polymer and we immerse it in a beaker of solvent. We have actually considered a very similar situation before. In Section 7.7 we treated a single polymer coil as an osmotic pressure experiment and saw how the good solvent exponent, $\nu = 3/5$, could be obtained from balancing the osmotic swelling of the coil (the solvent wants to dilute the monomers) and elastic resistance to swelling (loss of conformational entropy as the chain stretches out). Essentially the same balance will occur here. The chemical potential gradient will drive solvent into the piece of network, but the elasticity of the network will resist unlimited deformation. A state of *swelling equilibrium* will be reached, from which it is possible to determine both χ and the average molecular weight between cross-links.

The earliest theory of swelling equilibrium was that of Flory and Rehner [10], and we will now rederive their main result. We begin by considering the free energy of the swollen network to be composed of two parts, one due to the mixing of solvent and polymer and the other due to distortion of the network:

$$\Delta G = \Delta G_m + \Delta G_{el} \tag{10.7.4}$$

The former part can be represented by the Flory–Huggins theory expression (Equation 7.3.13), but with the network contributing no entropy of mixing ($N \to \infty$):

$$\Delta G_m = RT\{n_1 \ln \phi_1 + \chi n_1 \phi_2\} \tag{10.7.5}$$

In this case χ represents a combination of interaction energies between solvent and monomer and between solvent and cross-linking unit, but for low cross-link densities and/or systems such as styrene/divinylbenzene where the monomer and the cross-linker are chemically very similar, this complication is not important. The elastic part of ΔG is assumed to be purely entropic ($\Delta G_{el} = -T\Delta S_{el}$) and we simply invoke the rubber elasticity result (Equation 10.5.11):

$$\Delta S_{el} = \frac{-k\nu_e}{2}\left\{\lambda_x^2 + \lambda_y^2 + \lambda_z^2 - 3\right\} \tag{10.7.6}$$

where ν_e is the number of effective strands. However, it is time to address a complication. By a different analysis than that used in this chapter, Flory [4] obtained an expression for the entropy that contains an additional logarithmic term not present in Equation 10.7.6:

$$\Delta S_{el} = \frac{-k\nu_e}{2}\left\{\lambda_x^2 + \lambda_y^2 + \lambda_z^2 - 3 - \ln(\lambda_x\lambda_y\lambda_z)\right\} \tag{10.7.7}$$

A simple and very qualitative way to understand the origin of such a term is that there is an additional entropy gain for the placement of the end of each strand in space, as the volume of the network plus solvent increases. In this way the new term (which could be written as $-\ln \phi_2$) is equivalent to an ideal entropy of mixing contribution as derived in Section 7.3.1. Flory's analysis included a more explicit calculation of the entropy change associated with the cross-linking process, but in fact the appropriateness of the new logarithmic term in Equation 10.7.7 has not been without controversy [11]. Because the product of the three extension ratios is unity for a constant volume deformation, this term drops out in the unswollen network, so we have not missed anything by omitting it in the previous sections. We retain it now to be consistent with the Flory–Rehner result and we note that the factor of α^3 in Equation 7.7.10 from the Flory–Krigbaum theory for excluded volume also originates from inclusion of this term.

As in the consideration of the swollen network modulus above, we call the volume of the initial piece of polymer V_0 and the swollen volume V, but now the extension ratios are

$$\lambda_x = \lambda_y = \lambda_z = \lambda = \left(\frac{V}{V_0}\right)^{1/3} \tag{10.7.8}$$

Thus

$$\Delta S_{el} = \frac{-k\nu_e}{2}\left\{3\lambda^2 - 3 - \ln\lambda^3\right\} \tag{10.7.9}$$

At the point of swelling equilibrium, the chemical potential of the solvent inside the swollen network will equal that in the surrounding pure solvent, and so we have

$$\mu_1 - \mu_1^{\circ} = \Delta\mu_1 = \left(\frac{\partial\Delta G}{\partial n_1}\right)_{T,p}$$
$$= \left(\frac{\partial\Delta G_m}{\partial n_1}\right)_{T,p} + \left(\frac{\partial\Delta G_{el}}{\partial\lambda}\right)\left(\frac{\partial\lambda}{\partial n_1}\right)_{T,p} \tag{10.7.10}$$

If we assume no volume change on mixing as before, then the volume fraction of polymer in the swollen gel, ϕ_2, is simply given by

$$\frac{1}{\phi_2} = \frac{V}{V_0} = \lambda^3 = \frac{\left(V_0 + n_1\hat{V}_1\right)}{V_0} \tag{10.7.11}$$

where $\hat{V}_1$ is the molar volume of the solvent. If we differentiate this expression with respect to n_1 we obtain

$$3\lambda^2\left(\frac{\partial\lambda}{\partial n_1}\right) = \frac{\hat{V}_1}{V_0} \tag{10.7.12}$$

The other two derivatives on the right-hand side of Equation 10.7.10 are also straightforward. The first is the solvent chemical potential of the Flory–Huggins theory (Equation 7.4.14) in the infinite molecular weight limit:

$$\left(\frac{\partial\Delta G_m}{\partial n_1}\right)_{T,p} = RT\left\{\ln(1 - \phi_2) + \phi_2 + \chi\phi_2^2\right\} \tag{10.7.13}$$

The second is just

$$\left(\frac{\partial\Delta G_{el}}{\partial\lambda}\right)_{T,p} = \frac{RT\nu_e}{N_{av}}\left(3\lambda - \frac{3}{2\lambda}\right) \tag{10.7.14}$$

where we have multiplied and divided by Avogadro's number to put this term on a molar basis. By combining Equation 10.7.10 and Equation 10.7.12 through Equation 10.7.14 we obtain

$$\Delta\mu_1 = 0 = RT\left\{\ln(1 - \phi_2) + \phi_2 + \chi\phi_2^2 + \frac{\nu_e}{N_{av}}\frac{\hat{V}_1}{V_0}\left(\frac{1}{\lambda} - \frac{1}{2\lambda^3}\right)\right\} \tag{10.7.15}$$

or, using Equation 10.7.8 to eliminate λ, we obtain

$$\ln(1 - \phi_e) + \phi_e + \chi\phi_e^2 = \frac{\nu_e}{N_{av}}\frac{\hat{V}_1}{V_0}\left(\frac{\phi_e}{2} - \phi_e^{1/3}\right) \tag{10.7.16}$$

where $\phi_e = \phi_2$ is used to emphasize that we have swelling equilibrium. The prefactor on the right-hand side includes the number of strands per unit volume in the original network, ν_e/V_0, and from Equation 10.6.7 this can be written as

$$\frac{\nu_e}{V_0} = \frac{\rho N_{av}}{M_x}\left(1 - \frac{2M_x}{M}\right) \tag{10.7.17}$$

where M_x is the average molecular weight between cross-links. Thus if χ is known, Equation 10.7.16 and Equation 10.7.17 can be solved for M_x using the measured value of V (although in practice it is more common to actually weigh the dry and swollen polymer, and convert to volume through the known densities of polymer and solvent). The following numerical example gives an idea of the magnitude of the swelling.

Example 10.4

Calculate the predicted volumetric swelling for the polyisoprene network of Example 10.1, when exposed to a reservoir of cyclohexane at room temperature.

Solution

The polyisoprene chains had $M = 150,000$ with the extent of vulcanization $= 0.003$. The monomer molecular weight is 68 g/mol, so the average number of cross-links per chain is $(150,000 \times 0.003)/68 = 6.6$, and thus $M_x \approx 150,000/(6.6) = 23,000$. Utilizing Equation 10.7.17 and a density of 0.91 g/cm^3 for polyisoprene gives

$$\frac{\nu_e}{N_{av}V_0} = \frac{0.91}{23,000}\left(1 - \frac{46,000}{150,000}\right) = 2.7 \times 10^{-5} \text{ mol strands/cm}^3$$

The last ingredient we need in order to apply the Flory–Rehner expression is a value for χ. Back in Section 7.6.2 we considered ways to estimate χ using solubility parameters, and the necessary values were given in Table 7.1. Using Equation 7.6.6 we have

$$\chi = \frac{\hat{V}_1}{RT}(\delta_1 - \delta_2)^2 \approx \frac{108}{2 \times 300}(8.2 - 8.1)^2 \approx 0.002$$

where 108 cm^3 is the molar volume of cyclohexane.

Returning to the Flory–Rehner expression, that is, Equation 10.7.16, we have values for all the quantities except the desired ϕ_e, but the equation does not reduce to a simple algebraic expression. To proceed we anticipate that $\phi_e \ll 1$, and then the left-hand side can be simplified to $(\chi - 0.5)\phi_e^2$ using the expansion $\ln(1 - \phi_e) \approx -\phi_e - \phi_e^2/2$. Similarly, on the right-hand side we can expect $\phi_e^{1/3} \gg \phi_e/2$. This gives

$$\phi_e^{5/3} \approx \frac{108 \times 2.7 \times 10^{-5}}{0.5 - 0.002} = 0.0058$$

or $\phi_e \approx 0.046$. This gives a volumetric expansion of $1/0.046 = 22$ times. This value also justifies the small ϕ_e approximation that we have used to simplify the algebra. However, see Problem 15 for a numerically more realistic version of this calculation.

A test of the Flory–Rehner theory (Equation 10.7.16) is provided by the data in Figure 10.16. A variety of butyl rubbers with differing degrees of cross-linking were swollen to equilibrium and then extended to $\lambda = 4$. The results show a power law dependence of force on $\phi_e^{-5/3}$. This is consistent with Equation 10.7.16 in the limit of $\phi_e \to 0$, as can be seen by taking this limit on both sides of the equation as in the preceding example. (Note that force $\sim G \sim 1/M_x$.)

10.8 Chapter Summary

In this chapter we have considered two routes to network formation, cross-linking of preformed chains and direct polymerization including multifunctional monomers. The former is the preferred route to elastomers, or lightly cross-linked, low T_g materials, and the latter is commonly employed to produce thermosets for high-temperature applications. We then developed the theory of rubber elasticity in some detail, covering general thermodynamic aspects, the statistical theory in the ideal

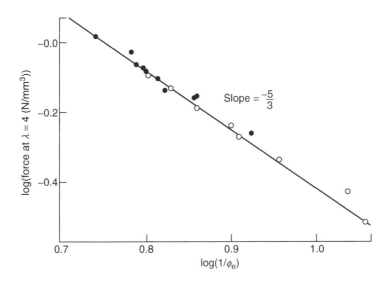

Figure 10.16 Force at constant extension ($\lambda = 4$) versus equilibrium swelling for butyl rubbers with different degrees of cross-linking, showing the predicted dependence on $\phi_e^{5/3}$. (Data from Flory, P.J., *Ind. Engng. Chem.*, 38, 417, American Chemical Society, 1948. Reproduced from Treloar, L.R.G., *The Physics of Rubber Elasticity*, 3rd ed., Clarendon Press, Oxford, 1975. With permission.)

case, some modifications to account for nonideal features of the response of real materials, and the case of networks swollen with solvent. The main points may be summarized as follows:

1. When cross-linking preformed chains with weight-average degree of polymerization N_w, the "gel point" is predicted to occur when the fraction of monomers participating in cross-links is equal to $1/N_w$. In practice, this tends to underestimate the necessary extent of reaction.
2. For multifunctional monomers, explicit predictions for the gel point can be developed using the principle of equal reactivity and probability arguments appropriate to the particular polymerization mechanism and combination of monomers. In practice, these expressions also tend to underestimate the extent of reaction needed to reach the gel point.
3. Analysis of rubber elasticity via macroscopic thermodynamics is relatively straightforward, the main new ingredient being the incorporation of the work of deformation into the free energy. An ideal elastomer is defined as one for which the force resisting deformation is entirely entropic, which is a reasonable approximation for many rubbery materials.
4. The molecular basis of rubber elasticity rests in the reduction of conformational degrees of freedom when a single Gaussian chain is extended. A single Gaussian chain acts as a Hooke's law spring, with a stiffness that is proportional to absolute temperature.
5. Straightforward expressions for the force required to deform an ideal elastomer are obtained by modeling the network as a collection of Gaussian strands and by making an assumption as to how the macroscopic deformation is transmitted to each strand. The resulting shear and extensional moduli are proportional to the number of strands per unit volume.
6. Networks or gels are often capable of absorbing more than 100 times their own weight in solvent, a phenomenon that is central to many applications, and that can be understood as a simple balance between the osmotic drive to dilute the polymer and the entropic resistance to strand extension.
7. Although the statistical theory of rubber elasticity captures the main features of a wide variety of experimental phenomenology, attempts to bring the theory into quantitative agreement with experiment have met with rather mixed success.

Problems

1. A constant force is applied to an ideal elastomer, assumed to be a perfect network. At an initial temperature T_i the length of the sample is L_i. The temperature is raised to T_f and the final length is L_f. Which is larger: L_i or L_f (remember F is a constant and $T_f > T_i$)? Suppose a wheel were constructed with spokes of this same elastomer. From the viewpoint of an observer, the spokes are heated near the 3 o'clock position—say, by exposure to sunlight—while other spokes are shaded. Assuming the torque produced can overcome any friction at the axle, would the observer see the wheel turn clockwise or counterclockwise? How would this experiment contrast, in magnitude and direction, with an experiment using metal spokes?

2. An important application of Equation 10.5.15 is the evaluation of M_x; P.J. Flory, N. Rabjohn, and M.C. Shaffer measured the tensile force required for 100% elongation of synthetic rubber with variable cross-linking at 25°C.[†] The molecular weight of the uncross-linked polymer was 225,000, its density was 0.92 g/cm^3, and the average molecular weight of a repeat unit was 68. Use Equation 10.5.15 to estimate M_x for each of the following samples and compare the calculated value with that obtained from the known fraction of repeat units cross-linked:

Fraction cross-linked	0.005	0.010	0.015	0.020	0.025
F/A (lb-force/in.^{2})	61.4	83.2	121.8	148.0	160.0

    How important for this system is the end group correction introduced in Equation 10.6.7?

3. Develop the equivalent to Equation 10.2.3 and Equation 10.2.5 for the third system in Figure 10.4, that is, $AB + BB + A_3$.

4. The Carothers equation (Equation 10.2.9) can also be used as the basis of an estimate of the extent of reaction at gelation. Consider the value implied for each of the parameters in the Carothers equation at the threshold of gelation, and derive a relationship between p_c and f on the basis of this consideration. Compare the predictions of the equation you have derived with those of Equation 10.2.5 for a mixture containing 2 mol A_3, 7 mol AA, and 10 mol BB. Criticize or defend the following proposition: the Carothers equation gives higher value for p_c than Equation 10.2.5 because the former is based on the fraction of reactive groups that have reacted and hence considers wasted loops that the latter disregards.

5. Categorize the following mixtures as to whether they can form linear, branched, or network structures:
    (a) $A_2 + B_2 + AB$ (b) $AB_2 + A_2$
    (c) $AB + AB_2$ (d) $A_3 + B_2 + A_2$
    (e) $AB + B_3$ (f) $A_2B_2 + A_2 + B_2$

6. Suppose you have a balloon made of an ideal elastomer that is inflated to a reasonable size with an ideal gas at room temperature. If the temperature of the balloon plus gas system is then increased to 100°C, will the balloon expand, contract, or stay the same size? Justify your answer.

7. Find the relation between the (true) stress σ and the strain λ for a piece of ideal rubber in biaxial extension. Assume the rubber has initial area A_0 and thickness d_0, and let the final area be $A = \lambda^2 A_0$.

8. Use the result from the previous problem to calculate the relation between the pressure of an ideal gas, p, inside a balloon made from an ideal elastomer, expanded to a radius $R = \lambda R_0$, where R_0 is the initial radius. Use a version of the Young–Laplace equation to relate the excess pressure (inside the balloon minus outside) to the stress in the rubber, $p = 2d\sigma/R$, where d is the thickness of the balloon skin. Empirically, it often seems harder to "get started" blowing up a balloon, than to blow it up further beyond a certain point. Explain this observation based on your result for p versus λ.

[†] P.J. Flory, N. Rabjohn, and M.C. Shaffer, *J. Polym. Sci.*, 4, 225 (1949).

9. Show that Equation 10.3.2 and Equation 10.3.3 are equivalent definitions of Poisson's ratio.
10. Use Equation 10.4.22 and the data in Figure 10.8a to assess whether or not the heat capacity at constant length, C_L, is comparable to typical values of C_p for rubber of 2 J/g/K. Be careful with the stated units of stress (kg/cm^2) in Figure 10.8a.
11. Estimate the temperature increase in a rubber band when extended to $\lambda = 8$ at room temperature. Assume C_L is 2 J/g/K and $\rho = 1$ g/cc.
12. The following data represent force (f, in picoNewtons) versus extension (h, in microns) for a single λ–DNA molecule, measured at room temperature in salt solutions.[†] These data therefore represent an opportunity to test basic assumptions of the theory of rubber elasticity. Try to fit these data in three ways. First, use the Gaussian expression; restrict the fit to the low extension part of the curve. Second, try the inverse Langevin function, approximated in Equation 10.6.4. Third, try the following formula derived for the worm-like chain (recall Chapter 6) in C. Bustamante et al.[‡]

h (μm)	f (pN)	h (μm)	f (pN)	h (μm)	f (pN)
10.1	0.0338	25.0	0.367	30.3	3.03
10.8	0.0335	25.6	0.364	31.2	5.57
17.6	0.102	26.4	0.486	31.3	6.82
18.2	0.102	27.1	0.604	31.7	11.9
22.4	0.218	28.5	1.08	32.0	12.2
23.0	0.217	29.0	1.53	32.1	18.3
24.3	0.382	30.3	3.39	32.2	17.0

$$f = \frac{kT}{\ell_p}\left(\frac{1}{4(1 - h/L)^2} - \frac{1}{4} + \frac{h}{L}\right)$$

Comment on the success or failure of the various expressions and provide values for the contour length, L, and the persistence length.

13. A rubber band, made of a styrene–butadiene random copolymer, is swollen to equilibrium in toluene; the volume increases by a factor of 5. Taking $\chi \approx 0.4$, estimate the number of strands per unit volume, and therefore the extent of cross-linking. Estimate Young's modulus for both the dry and swollen rubber bands.
14. According to the Flory–Rehner theory, what value of χ would be required to get absolutely no uptake of solvent for a typical rubber? What value of χ would restrict the uptake to less than 5% by volume?
15. In Example 10.4, the estimated value of χ was 0.002, which is not realistic. Revisit the discussion in Section 7.6, and use a more realistic estimate for χ. How much does the rubber swell at equilibrium in this case? Is it significantly different from the answer in Example 10.4?
16. On the following plot of a normalized equilibrium shear modulus ($G/\rho RT$) versus an inverse "effective" molecular weight between cross-links ($1/M_x$) are two curves. The first is a straight dashed line with unit slope to "guide the eye" (and the brain). The second represents the modulus of a real polymer as it undergoes progressive cross-linking (the arrow represents the course of the modulus with extent of reaction). Why does the modulus shoot up at $1/M_x \approx 3 \times 10^{-5}$ mol/g? Why does the modulus exceed the unit slope line, just after $1/M_x \approx 3 \times 10^{-5}$ mol/g? Estimate M_e for this polymer. (Note that the last two questions will be easier to answer after reading Section 11.6.)

[†] S.B. Smith, L. Finzi, and C. Bustamante, *Science*, 258, 1122 (1992).
[‡] C. Bustamante, J. Marko, E. Siggia, and S. Smith, *Science*, 265, 1599 (1994).

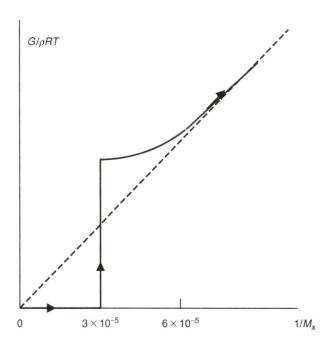

References

1. Goodyear, C., U.S. Patent 3, 633 (1844).
2. Stockmayer, W.H., *J. Chem. Phys.*, 11, 45 (1943); 12, 125 (1944).
3. Flory, P.J., *Principles of Polymer Chemistry*, Cornell University Press, Ithaca, NY, 1953.
4. Carothers, W.H., *Trans. Faraday Soc.*, 32, 39 (1936).
5. Gough, J., *Mem. Lit. Phil. Soc. Manchester*, 1, 288 (1805).
6. Joule, J.P., *Phil. Trans. R. Soc.*, 149, 91 (1859).
7. James, H.M. and Guth, E., *J. Chem. Phys.*, 11, 455 (1943); 15, 669 (1947).
8. Kuhn, W. and Grün, F., *Kolloidzschr.*, 101, 248 (1942).
9. Mooney, M., *J. Appl. Phys.*, 11, 582 (1940); Rivlin, R.S., *Phil. Trans. R. Soc.*, A241, 379 (1948).
10. Flory, P.J. and Rehner, J., *J. Chem. Phys.*, 11, 512 (1943).
11. Treloar, L.R.G., *The Physics of Rubber Elasticity*, 3rd ed., Clarendon Press, Oxford, 1975.

Further Readings

Flory, P.J., *Principles of Polymer Chemistry*, Cornell University Press, Ithaca, NY, 1953.
Graessley, W.W., *Polymeric Liquids and Networks: Structure and Properties*, Garland Science, New York, 2003.
Mark, J.E. and Erman, B., *Rubber Elasticity—A Molecular Primer*, Wiley, New York, 1988.
Odian, G., *Principles of Polymerization*, 4th ed., Wiley, New York, 2004.
Rubinstein, M. and Colby, R.H., *Polymer Physics*, Oxford University Press, New York, 2003.
Treloar, L.R.G., *The Physics of Rubber Elasticity*, 3rd ed., Clarendon Press, Oxford, 1975.

11

Linear Viscoelasticity

11.1 Basic Concepts

In this chapter we extend the study of the dynamic properties of polymer liquids in two new directions. In Chapter 9 we considered only dilute solutions, but here we will consider also very concentrated solutions and molten polymers. In Chapter 9 we also focused on the steady-flow viscosity, or the diffusion over long time intervals; now we will examine the time- or frequency-dependent response of polymer liquids to an imposed deformation or force. This response can be characterized by a variety of *material functions*, such as the viscosity, the modulus, and the compliance. In general, we will find that polymer liquids are *viscoelastic*, i.e., their behavior is intermediate between (elastic) solids and (viscous) liquids. The phenomenon of viscoelasticity is familiar to anyone who has played with Silly Putty. If you roll some into a ball, and leave it for a few hours, it flows to adopt the shape of its container. This behavior is that of a liquid; it just takes a long time because the viscosity is very high. On the other hand, if you stretch a sample very rapidly, and immediately release one end, the sample will partially recover toward its original dimensions. This recovery is an elastic response, and is more typical of solids than liquids. The previous chapter concerned the elasticity of polymer networks, and important results from that discussion will be directly incorporated into our treatment of viscoelasticity.

Viscoelasticity is one of the most distinctive features of polymers. As virtually all polymer materials are processed in the liquid state, viscoelasticity plays a central role in the optimization and control of processing. On the other hand, we will see in Chapter 12 and Chapter 13 that the concepts developed here will also have broad application to the mechanical properties of solid polymers. Furthermore, the viscoelastic properties can be readily measured and therefore provide an additional route to molecular characterization, particularly for polymers that are difficult to dissolve in convenient solvents. Finally, measurement of the viscoelastic response of a polymeric material provides direct and detailed information about how long it will take for that material to respond to any kind of perturbation.

To gain a glimpse at what lies ahead, Figure 11.1 shows the steady flow viscosity, η, of five samples of poly(α-methylstyrene) in the molten state as a function of molecular weight, M, at 186°C. The viscosities are very large; they are about 10^{10}–10^{14} times larger than the viscosity of liquid water (about 0.01 P). Furthermore, the viscosity is a very strong function of molecular weight. The plot is in a double logarithmic format and the indicated straight line has a slope of 3.24. Thus, $\eta \sim M^{3.24}$, which means that doubling the molecular weight is sufficient to increase η by about a factor of 10. In this chapter we will provide a molecular picture for the origin of this response and see that this behavior is typical of all flexible polymers. Of course, a central objective of molecular models of polymer behavior is to understand the molecular weight dependence of any experimental quantity.

Figure 11.2 shows a quantity called the stress relaxation modulus (which we will define a little later) as a function of time for the same five polymers. This quantity is also plotted in a double-logarithmic format. For now, we can just compare these modulus values with those (time-independent) values typical of other materials, including the cross-linked rubber discussed in the previous chapter. At very short times in Figure 11.2, the modulus approaches 10^9 Pa (1 GPa,

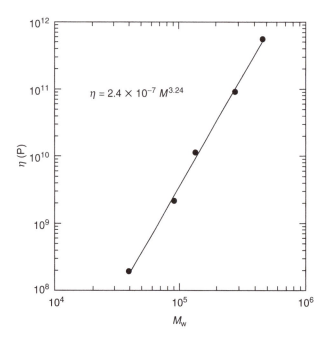

Figure 11.1 Viscosity versus molecular weight for molten poly(α-methyl styrene) at 186°C. (Data taken from Fujimoto, T., Ozaki, N., and Nagasawa, M., *J. Polym. Sci. Part A-2*, 6, 129, 1968.)

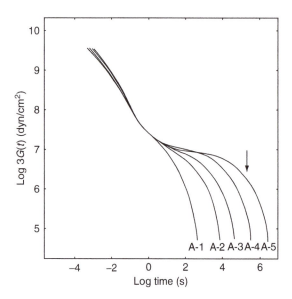

Figure 11.2 Stress relaxation modulus, $G(t)$, versus time for molten poly(α-methyl styrene) at 186°C; samples A-1 through A-5 correspond to the five samples in Figure 11.1. The arrow locates the longest relaxation time for sample A-5, as discussed in Example 11.1. (Data taken from Fujimoto, T., Ozaki, N., and Nagasawa, M., *J. Polym. Sci. Part A-2*, 6, 129, 1968. With permission.)

or 10^{10} dyn/cm^2); this can be compared with steel (80 GPa), silica (20 GPa), and granite (10 GPa). On the other hand, at intermediate times, say 1–100 s, the modulus has dropped to a value close to that of a typical rubber, approximately 1 MPa. Finally, at even longer times, the modulus tends to zero. Another interesting observation from Figure 11.2 is that in the early time, high modulus state, the modulus is independent of M, but at long times, the time decay of the modulus depends very much on M. In this chapter we will explain all of these features and even show how the viscosity values in Figure 11.1 were actually calculated from the data in Figure 11.2, rather than measured directly.

In the first three sections, we will define the basic terms and concepts, examine simple models that reveal important features of viscoelastic response, and show how the different material functions can be related to one another in a model-independent way. The subsequent four sections will examine molecular models that, collectively, are remarkably successful in capturing the viscoelastic response of polymers all the way from dilute solution to the melt. The final section covers some aspects of experimental *rheology*, the science of flow and deformation of matter.

11.1.1 Stress and Strain

In a typical experiment, a sample confined in a particular geometry is subjected to a displacement, and the resulting force is measured. An example is shown in Figure 11.3, for the deformation called *simple shear*. This is exactly the geometry used to discuss viscosity in Chapter 9.1. The material is confined between two parallel plates dy apart, and if one surface is moved a distance dx, we say that the sample has been subjected to a *strain*, γ, = dx/dy; γ is thus dimensionless. The velocity of the plate v_x = dx/dt, and d(dx/dy)/dt = $\dot{\gamma}$ is called the *strain rate* or *shear rate*. It takes the application of a force to accomplish the deformation; alternatively, we can think of the material exerting a force on the moving plate. The total force will depend on the area of the plate in contact with the material and thus we consider the force per unit area, or *stress*, σ. There are several different kinds of deformation geometries, such as uniaxial elongation, biaxial elongation, simple shear, etc. In all cases it is possible to define a stress, strain, and strain rate in an analogous fashion. In the most general case, the stress and the deformation in the material should be represented as tensor quantities. In this chapter we will restrict our attention to the simplest case of shear flow and shear stress only. This avoids the need to employ tensor algebra; the mathematics will be invigorating enough as it is. We will use the symbols σ, γ, and $\dot{\gamma}$ to denote shear stress, shear strain, and shear rate, respectively, just as in Chapter 9.

11.1.2 Viscosity, Modulus, and Compliance

In general, a *viscosity* is defined as the ratio of a stress to a strain rate, and in shear we begin with Newton's relation (Equation 9.1.3)

$$\sigma = \eta\dot{\gamma} \tag{11.1.1}$$

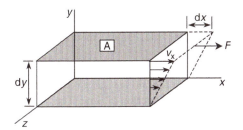

Figure 11.3 Illustration of simple shear flow between two parallel plates.

where η is the viscosity. Recall also from Section 9.1 that if η is independent of the magnitude of $\dot{\gamma}$, the fluid is said to be *Newtonian*. Many polymer fluids are non-Newtonian, but in general Newtonian behavior is recovered in the limit that $\dot{\gamma} \rightarrow 0$. Similarly, in the mechanics of solid bodies a *modulus* is defined as the ratio of a stress to a strain, and in shear we have

$$\sigma = G\gamma \tag{11.1.2}$$

where G is the (shear) modulus (recall Equation 10.5.17). Note that this relation is essentially Hooke's law for an ideal elastic spring, $F = -kx$; the modulus is a generalized spring constant, and the sign has changed because we consider the force acting on the material. When G is independent of the magnitude of γ, we say that the response is *linear*. This is generally true as $\gamma \rightarrow 0$, but in experiments we always need a finite strain to generate a measurable stress, and whether the response falls in the so-called *linear viscoelastic limit* is something that needs to be checked. For the remainder of this chapter we will assume linear response, for simplicity; treatment of the nonlinear response is substantially more complicated. However, we should note that in commercial polymer systems the nonlinear response is interesting and very important, especially for processing operations where strains and strain rates are typically high. One further useful concept is that of a *compliance*, J, which can be defined as the ratio of a strain to a stress. Therefore the compliance is the inverse of the modulus, but it will turn out that when the time dependence is involved, $J(t)$ is not simply equal to $1/G(t)$.

11.1.3 Viscous and Elastic Responses

Two crucial limiting cases in this chapter are viscous response and elastic response. Viscosity reflects the relative motion of molecules, in which energy is *dissipated* by friction. It is a primary characteristic of a liquid. A liquid will always flow until the stress has gone away and it will dissipate energy as it does so. In contrast, elasticity reflects the *storage* of energy; when a spring is stretched, the energy can be recovered by releasing the deformation. A solid subjected to a small strain is primarily elastic, in that it will remain deformed as long as the force is still applied. In flexible polymers, the elasticity arises from the many conformational degrees of freedom of each molecule and from the intertwining of different chains; it will turn out to be primarily entropic in origin, just as in Chapter 10. When the material is subject to a deformation, the individual molecules respond by adopting a nonequilibrium distribution of conformations. For example, the chains on average may be stretched and/or oriented in the direction of flow; in so doing they lose entropy. Left to themselves, the molecules will relax back to an isotropic, equilibrium distribution of conformations, just like a spring. As they relax, the relative motion of the molecules through the surrounding fluid dissipates the stored elastic energy. It is this interplay of viscous dissipation during elastic recovery that underlies the viscoelastic properties of polymer liquids. (If you are at all familiar with elementary electronic circuits, these concepts of energy dissipation and storage are well known; a resistor (resistance R) is the dissipative element, and the capacitor (capacitance C) is the storage element. Voltage (V) plays the role of force, current (I) the role of velocity, and charge (Q) the role of displacement. Thus Ohm's law ($V = IR$) is the analog of Newton's law (Equation 11.1.1), and for a capacitor $V = CQ$ in analogy to Equation 11.1.2. Furthermore, the mathematical results we will derive in the next section for simple mechanical models will be analogous to the results for V and I in simple RC circuits.)

In experimental measurements of the viscoelastic response, several different time histories are routinely employed. In a *transient* experiment, at some specific time a strain (or stress) is suddenly applied and held; the resulting stress (or strain) is then monitored as a function of time. The former mode is called *stress relaxation* and the associated modulus the *stress relaxation modulus*, $G(t) = \sigma(t)/\gamma$. The latter mode (in parentheses) is called *creep* and the associated compliance the *creep compliance*, $J(t) = \gamma(t)/\sigma$. In a steady flow experiment, the strain rate is constant; the resulting steady stress gives the *steady flow viscosity*. Finally, in what is arguably the most important mode, the sample is subjected to a sinusoidally time-varying strain at frequency ω. The resulting *dynamic modulus*, $G^*(\omega)$, is resolved into two dynamic moduli: one in-phase with the strain, called G', reflecting the elastic component of the total response and one in-phase with the strain rate, called G'', reflecting the viscous response.

11.2 Response of the Maxwell and Voigt Elements

We can learn a great deal about viscoelastic response through consideration of two simplified models, the so-called Maxwell [1] and Voigt [2] elements. We will examine the stress relaxation modulus, creep compliance, and dynamic moduli of the Maxwell element, which illustrates a viscoelastic liquid. We will also derive the creep compliance of the Voigt element, which exemplifies a viscoelastic solid. Each element will turn out to have a *characteristic time*, τ, which determines the timescale of its response.

11.2.1 Transient Response: Stress Relaxation

The Maxwell element consists of an ideal, Hookean spring with spring constant $\widehat{G}$ connected in series with an ideal, Newtonian dashpot with viscosity $\widehat{\eta}$, as shown in Figure 11.4a. Thus the stress in the two components is given by

$$\sigma = \widehat{G}\gamma \tag{11.2.1a}$$

for the spring, and

$$\sigma = \widehat{\eta}\dot{\gamma} \tag{11.2.1b}$$

for the dashpot. At time $t=0$ we apply an instantaneous strain of magnitude γ_0, and hold it indefinitely; if we follow the stress as a function of time, it is a stress relaxation experiment. Qualitatively we can anticipate what the response should be. At very short times, the dashpot will not want to move; that is the whole point of a dashpot (i.e., a shock absorber). The spring, on the other hand, only cares about how much it is stretched, not how rapidly. Thus the initial deformation will be entirely taken up by the spring. However, the stretched spring will then exert a force on the

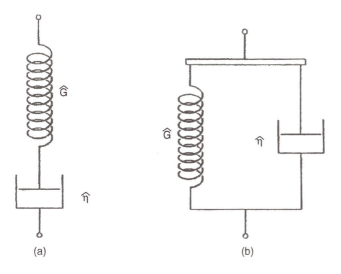

(a)   (b)

Figure 11.4   Illustration of (a) the Maxwell element and (b) the Voigt element.

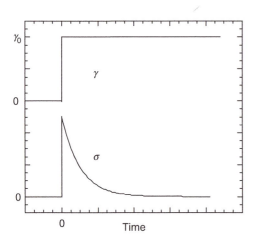

Figure 11.5 Illustration of the stress response $\sigma(t)$ after a step strain of magnitude γ_0 was applied at $t = 0$, for the Maxwell element.

dashpot, which will slowly flow in response. Ultimately, the spring will relax back to its rest length and there will be no more stress; the long time-response is that of a liquid.

To make this argument quantitative, we see that the total applied strain is distributed between the elements,

$$\gamma_0 = \gamma_{\text{spring}} + \gamma_{\text{dashpot}} \tag{11.2.2}$$

and because the strain is constant for $t > 0$,

$$\frac{d\gamma_0}{dt} = 0 = \dot{\gamma}_{\text{spring}} + \dot{\gamma}_{\text{dashpot}} = \frac{d}{dt}\frac{\sigma(t)}{\widehat{G}} + \frac{\sigma(t)}{\widehat{\eta}} \tag{11.2.3}$$

This is a linear, first-order, homogeneous differential equation for $\sigma(t)$:

$$\dot{\sigma} + \frac{1}{\tau}\sigma = 0 \tag{11.2.4}$$

where the dot denotes the time derivative. We have defined the *relaxation time* $\tau \equiv \widehat{\eta}/\widehat{G}$, and from Equation 11.2.1 it is clear that this ratio has units of time ($\widehat{\eta}/\widehat{G} = (\sigma/\dot{\gamma})/(\sigma/\gamma) = (\gamma/\dot{\gamma}) = t$). The concept of relaxation time is central to the material in this chapter. In essence, the relaxation time is a measure of the time required for a system to return to equilibrium after any kind of disturbance.

The solution to Equation 11.2.4 is an exponential decay

$$\sigma(t) = \sigma_0 \exp(-t/\tau) \tag{11.2.5}$$

as shown in Figure 11.5. The initial stress, σ_0, is obtained as suggested above: at the earliest times the deformation is all in the spring, and therefore $\sigma_0 = \widehat{G}\gamma_0$. (Note that an instantaneous deformation would make $\dot{\gamma}$ infinite and thus the stress in the dashpot would be infinite if it moved, so it does not.) The stress relaxation modulus is obtained as

$$G(t) = \frac{\sigma(t)}{\gamma_0} = \widehat{G}\exp(-t/\tau) \tag{11.2.6}$$

as shown in Figure 11.6, in both linear and logarithmic formats. The Maxwell model captures the main feature of the stress–relaxation response of any liquid; the material supports the stress for $t \ll \tau$, but flows until the stress has vanished for $t \gg \tau$. Thus the magnitude of the relaxation time

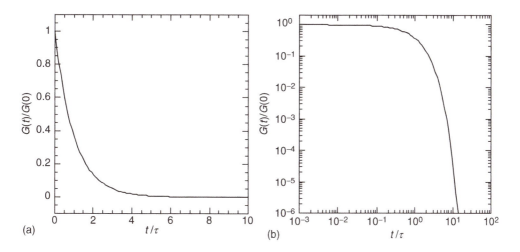

Figure 11.6 Normalized stress relaxation modulus $G(t)/G(0)$ versus reduced time t/τ for the Maxwell element, plotted in (a) linear and (b) double logarithmic formats.

is vital in determining the properties we experience. For example, a molten high M polymer not far above its glass transition temperature may not flow over a timescale of hours (see Figure 11.2 and Example 11.1); in contrast, the stress relaxation time for water is measured in picoseconds. We shall also see that polymers, with their many degrees of internal conformational freedom, show multiple relaxation times spread over many orders of magnitude, as also illustrated in Figure 11.2. However, before we get to that, we will pursue these simple models further.

Example 11.1

Use the simple Maxwell model analysis of stress relaxation to estimate the longest relaxation time for the highest molecular weight sample in Figure 11.1 and Figure 11.2.

Solution

The relaxation time is given by the ratio of a viscosity to a modulus. The highest M sample in Figure 11.1 ($M = 4.6 \times 10^5$ g/mol) has a viscosity of 5.5×10^{11} P($= $g/cm s). It is not so obvious what value to take for the modulus in Figure 11.2, however, because G is a function of t. In Figure 11.6b, the Maxwell model prediction for $G(t)$ in a log–log format looks like the long time-response in Figure 11.2, where $G(t)$ falls from an apparent plateau value of about 3×10^6 dyn/cm^2 to zero. In the Maxwell model the plateau value is $\hat{G}$. Using this result, then, we have

$$\tau = \frac{\eta}{G} \approx \frac{5.5 \times 10^{11}}{3 \times 10^6} \approx 2 \times 10^5 \text{ s} = 2.3 \text{ days}$$

This value is marked as the arrow in Figure 11.2. It has a very simple physical interpretation: if we place a chunk of this polymer on a desktop at 186°C (do not try this at home!) it will take 2–3 days before we will see it flow down into a puddle of liquid.

11.2.2 Transient Response: Creep

In a creep experiment, a sample is subjected to a constant force and the resulting deformation is monitored as a function of time. The term creep implies that the deformation will be very slow, which in turn suggests that the sample should have a rather high viscosity; think of a glacier moving under the influence of gravity. Now let us subject the Maxwell element to creep: we apply σ_0 at time $t = 0$ and watch $\gamma(t)$ evolve (using Equation 11.2.2):

$$\gamma(t) = \gamma_{\text{spring}} + \gamma_{\text{dashpot}} = \frac{\sigma_0}{\widehat{G}} + \frac{1}{\widehat{\eta}} \int_0^t \sigma_0$$

$$= \frac{\sigma_0}{\widehat{G}} + \frac{\sigma_0}{\widehat{\eta}} t \qquad\qquad (11.2.7)$$

Here the strain in the dashpot is obtained by integrating the strain rate $\dot{\gamma}$, where $\dot{\gamma} = \sigma_0 / \widehat{\eta}$ is a constant. Thus the compliance is given by

$$J(t) = \frac{\gamma(t)}{\sigma_0} = \frac{1}{\widehat{G}} + \frac{1}{\widehat{\eta}} t = J_e^0 + \frac{1}{\widehat{\eta}} t \qquad\qquad (11.2.8)$$

Figure 11.7a illustrates this behavior. At long times there is a steady-state response, with the strain increasing linearly in time; the slope is the reciprocal of the viscosity. At short times there is a transient response, reflecting the initial deformation of the spring; in this model, it is instantaneous. Consequently, if the long time linear portion is extrapolated back to $t = 0$, there is a finite intercept, J_e^0, called the *steady-state compliance*. If the stress is suddenly removed at some instant after steady flow has been achieved, then the spring will retract but the dashpot will stop moving. Consequently there will be an elastic recovery of the fluid; this is also indicated in Figure 11.7a. The amount of this recovery is called the *recoverable compliance* and if the flow achieves steady state, the recoverable compliance should be equal to J_e^0.

The Voigt element places the two components in parallel, thereby enforcing equality of strain between the spring and the dashpot (Figure 11.4b). This combination serves to illustrate the response of a *viscoelastic solid*. In a creep experiment, in which a constant stress σ_0 is applied at $t = 0$ and held indefinitely, the strain will grow slowly, as the dashpot resists rapid deformation. As the strain continues to grow, the resistance of the spring will take over, until at long times the strain will saturate. From the definitions in Equation 11.2.1a and Equation 11.2.1b, we can write

$$\sigma_0 = \widehat{G}\gamma + \widehat{\eta}\dot{\gamma} \qquad\qquad (11.2.9)$$

which we can rearrange into a linear, first-order differential equation for $\gamma(t)$:

$$\dot{\gamma} + \frac{1}{\tau}\gamma = \frac{\sigma_0}{\widehat{\eta}} \qquad\qquad (11.2.10)$$

The solution to this equation is also an exponential decay, but with a twist:

$$\gamma(t) = \frac{\sigma_0}{\widehat{G}}[1 - \exp(-t/\tau)] \qquad\qquad (11.2.11)$$

Now the strain starts from zero and increases exponentially to its infinite time value of $\sigma_0/\widehat{G}$, as shown in Figure 11.7b. The compliance, $J(t)$, can be written as

$$J(t) = \frac{\gamma(t)}{\sigma_0} = \frac{1}{\widehat{G}}[1 - \exp(-t/\tau)] = J_e[1 - \exp(-t/\tau)] \qquad\qquad (11.2.12)$$

where J_e has replaced $1/\widehat{G}$; J_e is called the *equilibrium compliance*.

11.2.3 Dynamic Response: Loss and Storage Moduli

Although the step strain and step stress experiments are both useful in characterizing the visco-elastic response of a material, the most common experimental approach is to apply a sinusoidally

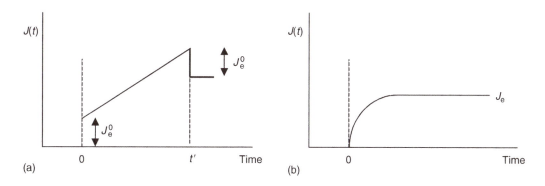

Figure 11.7   Creep compliance $J(t)$ for (a) a viscoelastic liquid (the Maxwell model) and (b) a viscoelastic solid (the Voigt model). A stress σ_0 is applied at time $t = 0$ and in (a) the stress is removed at some later time t'.

time-varying strain (or stress), e.g., $\gamma(t) = \gamma_0 \sin \omega t$ and measure the sinusoidally time-varying stress (or strain). One advantage of this approach, as we shall see, is that both the viscous and elastic character of the response can be resolved concurrently. Other advantages are technical; for example, small sinusoidal signals can be extracted reliably through lock-in amplifier detection schemes. Also, because the magnitude of the modulus can vary by several orders of magnitude (see Figure 11.2), in a stress relaxation experiment the signal will become smaller and smaller as time evolves. In contrast, in the dynamic experiment at each new driving frequency ω the strain amplitude γ_0 can be adjusted to bring the stress signal into the conveniently measurable range (of course, taking care to remain in the linear viscoelastic regime).

To see how the Maxwell element responds to a strain of $\gamma_0 \sin \omega t$, we adapt Equation 11.2.3:

$$\frac{d\gamma_{tot}}{dt} = \gamma_0 \frac{d}{dt} \sin \omega t = \gamma_0 \omega \cos \omega t$$

$$= \dot{\gamma}_{el} + \dot{\gamma}_{vis} = \frac{1}{\widehat{G}} \dot{\sigma} + \frac{1}{\eta} \sigma \tag{11.2.13}$$

Now the first-order, linear differential equation has a driving term proportional to $\cos \omega t$:

$$\dot{\sigma} + \frac{1}{\tau} \sigma = \widehat{G} \gamma_0 \omega \cos \omega t \tag{11.2.14}$$

One of the beautiful features of a linear system is that the response to a sinusoidal input is always a sinusoidal output at the same frequency; the amplitude and phase will generally differ between input and output, however. (We have already encountered this linearity in Chapter 8; in light scattering, the scattered electric field had a different amplitude and phase from the incident wave, but the frequency was the same.) It is also helpful to remember that $\sin \omega t$ and $\cos \omega t$ are the same wave, just phase-shifted by 90° (or $\pi/2$ rad): $\cos \omega t = \sin(\omega t + \pi/2)$. Thus the strain rate in this experiment, $\dot{\gamma}$, is 90° out-of-phase with the strain. We can say that the stress in the elastic element is in-phase with the strain, and the stress in the viscous element is in-phase with the strain rate and 90° out-of-phase with the strain. This is a general result: in the linear response regime, the stress can always be resolved into two components, one in-phase with the strain and one 90° out-of-phase with the strain. For a purely elastic solid the latter component would vanish, whereas for a purely viscous liquid the former component would be zero. A viscoelastic material is one for which both components are significant.

Returning now to the solution to Equation 11.2.14, the answer must be a wave with frequency ω. The most general solution is

$$\sigma(t) = A \sin \omega t + B \cos \omega t = A_0 \sin(\omega t + \phi) \tag{11.2.15}$$

where we note that any wave with frequency ω can be written as a linear combination of a sine wave and a cosine wave, with two adjustable amplitudes A and B, or as a single sine wave, with adjustable amplitude A_0 and phase ϕ. Let us insert the former version into the differential Equation 11.2.14, and work it through:

$$A\omega \cos \omega t - B\omega \sin \omega t + \frac{1}{\tau}A \sin \omega t + \frac{1}{\tau}B \cos \omega t = \widehat{G}\gamma_0 \omega \cos \omega t \tag{11.2.16}$$

recalling that $d(\sin \omega t)/dt = \omega \cos \omega t$ and $d(\cos \omega t)/dt = -\omega \sin \omega t$. The sine and cosine components are independent of each other, so we can solve for the coefficients separately. For the sine part (recall that this represents the stress component in-phase with the strain)

$$-B\omega + \frac{1}{\tau}A = 0$$
$$A = \omega\tau B \tag{11.2.17}$$

and for the cosine part (stress $90°$ out-of-phase with the strain)

$$A\omega + \frac{1}{\tau}B = \widehat{G}\gamma_0 \omega$$
$$= \omega^2 \tau B + \frac{1}{\tau}B \tag{11.2.18a}$$
$$B = \widehat{G}\gamma_0 \frac{\omega\tau}{1 + \omega^2\tau^2}$$

and thus

$$A = \widehat{G}\gamma_0 \frac{\omega^2\tau^2}{1 + \omega^2\tau^2} \tag{11.2.18b}$$

These expressions for the coefficients A and B can be substituted into the expression for the modulus:

$$\frac{\sigma(t)}{\gamma_0} = \widehat{G}\frac{\omega^2\tau^2}{1 + \omega^2\tau^2} \sin \omega t + \widehat{G}\frac{\omega\tau}{1 + \omega^2\tau^2} \cos \omega t$$
$$= G' \sin \omega t + G'' \cos \omega t \tag{11.2.19}$$

This last relation defines the elastic or *storage modulus*, G', and the viscous or *loss modulus*, G'':

$$G' = \widehat{G}\frac{\omega^2\tau^2}{1 + \omega^2\tau^2}, \quad G'' = \widehat{G}\frac{\omega\tau}{1 + \omega^2\tau^2} \tag{11.2.20}$$

The former measures the component of the stress response that is in-phase with the strain and the latter the component in-phase with the strain rate. The relationships among the applied strain, the stress response, and the two components are illustrated in Figure 11.8. We can say that a material is viscoelastic if both G' and G'' are significant and we can anticipate that when $G' \gg G''$, the material is solid like, and when $G'' \gg G'$, the material is liquid like.

The normalized dynamic moduli $G'/\widehat{G}$ and $G''/\widehat{G}$ for the Maxwell element are plotted versus reduced frequency $\omega\tau$ in Figure 11.9a in a double logarithmic format. These functions display the following features. At low frequencies, $\omega\tau \ll 1$, both G' and G'' increase with ω. The former increases as ω^2 and the latter as ω and $G'' > G'$. This scaling with frequency is characteristic of all liquids when the frequency of deformation is much lower than the inverse of the longest relaxation time of the material. Therefore, this is what one would expect to see for all polymer liquids once ω is low enough. At high frequencies, $\omega\tau \gg 1$, $G' > G''$, and G' is independent of frequency whereas G'' falls as ω^{-1}. This response is characteristic of a solid: the stress is independent of frequency (or time), and in-phase with the strain. The two functions are equal, $G' = G''$, and G'' shows a

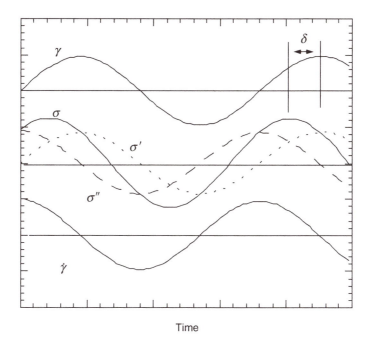

Time

Figure 11.8 Strain γ, strain rate $\dot{\gamma}$, net stress σ, and stress components in-phase with the strain σ' and in-phase with the strain rate σ'' for a sinusoidally time-varying strain. The phase angle between the stress and the strain is δ.

maximum when $\omega\tau = 1$. In other words, when the frequency of deformation is exactly the reciprocal of the relaxation time, the material is equally liquid-like and solid-like. All of these features of G' and G'' will be evident when we consider detailed molecular models for the viscoelastic response of polymer liquids.

11.2.4 Dynamic Response: Complex Modulus and Complex Viscosity

It is common to use complex notation to describe the dynamic modulus, just as we did with light waves in Chapter 8; complex numbers are reviewed in the Appendix. Thus, if we apply a sinusodial strain, $\gamma^* = \gamma_0 \exp(i\omega t)$, the response can be written as

$$G^*(\omega) = \frac{\sigma^*}{\gamma^*} = G' + iG'' = G_m \exp(i\delta) \tag{11.2.21}$$

where G^* is the complex dynamic modulus and σ^* is the complex stress ($= \sigma_0 \exp[i(\omega t + \delta)]$). The storage modulus G' is the "real" part of G^* because it is in-phase with the applied strain and G'' is the "imaginary" part because it is 90° out-of-phase with the applied strain. The complex notation is simply a way to keep track of relative phases; there is nothing imaginary about the viscosity or the modulus. One can choose to represent the modulus via its two components, G' and G'', or in terms of the magnitude, G_m, and the phase angle, δ. These are interrelated by

$$G_m = \left\{ (G')^2 + (G'')^2 \right\}^{1/2}$$

and

$$\tan\delta = \frac{G''}{G'} \tag{11.2.22}$$

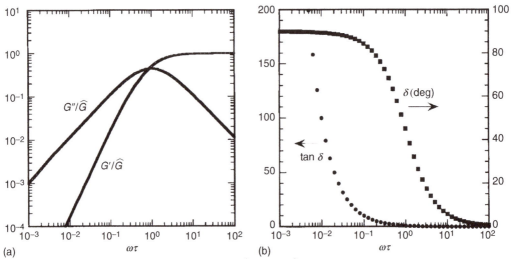

Figure 11.9 (a) Normalized dynamic moduli $G'/\widehat{G}$ and $G''/\widehat{G}$, and (b) phase angle δ and loss tangent $\tan \delta$, versus reduced frequency $\omega\tau$ for the Maxwell element.

The phase angle, δ, is often discussed in terms of the so-called *loss tangent*, $\tan \delta$, because it is the direct ratio of the viscous and elastic parts. In other words, when the material is behaving like a liquid, $\tan \delta \gg 1$, and when it is a solid, $\tan \delta \ll 1$; the crossover occurs at $\tan \delta \approx 1$. In Figure 11.9b both $\tan \delta$ and the phase angle δ itself are plotted for the Maxwell element.

Sometimes the viscoelastic response is presented in terms of a *dynamic viscosity*, η^*, rather than the dynamic modulus:

$$\eta^* = \frac{\sigma^*}{\dot{\gamma}^*} = \eta' - i\eta'' = \eta_m \exp(i\psi) \tag{11.2.23}$$

Note that $\dot{\gamma}^* = d\gamma^*/dt = i\omega\gamma^*$, so that the components of G^* and η^* are related by a factor of ω:

$$G^* = i\omega\eta^*, \quad G' = \omega\eta'', \quad G'' = \omega\eta' \tag{11.2.24}$$

Note also that there is a new phase angle, ψ, in Equation 11.2.23; the relation between ψ and δ is left to Problem 6. However, there is no new information content in η^* relative to G^*, or even in G^* relative to just discussing G' and G''. Consequently the choice of format (i.e., G' and G''; G_m and δ; η' and η'') is mostly a matter of convenience or convention.

11.3 Boltzmann Superposition Principle

The last issue we will take up before proceeding to detailed molecular models is that of the interrelationships among G', G'', $G(t)$, J_e^0, and the steady-flow viscosity, η. We begin with the *Boltzmann superposition principle* of linear viscoelasticity, which asserts that the stress in the material is the sum of the stress contributions from all strains applied in past times [3]. The linearity of this principle lies in the fact that we can simply add up all the contributions because the response of the material to any particular strain is linear and independent of whatever went on before or after. From the Maxwell model, we saw that $G(t)$ was an exponentially decaying function of time. The strain was applied at one instant and the effect of the strain was evident in the stress at all subsequent times, but with an exponentially decaying amplitude. The more detailed models we will consider subsequently all predict that $G(t)$ is sum of such exponentials, with a sequence of different relaxation times. In fact, the modulus is really a function of the *time interval*, $G(t - t')$, rather than the absolute time, t; in the previous analysis we had simply chosen $t' = 0$. Now we can express the Boltzmann superposition principle as follows. We first relate the increment in stress, $d\sigma$, to an increment in strain, $d\gamma$:

$$d\sigma = G \, d\gamma = G \frac{d\gamma}{dt} dt = G\dot{\gamma} \, dt \tag{11.3.1}$$

and thus

$$\sigma(t) = \int d\sigma = \int_{-\infty}^{t} G(t - t')\dot{\gamma}(t') \, dt' \tag{11.3.2}$$

Therefore the stress now (time t) is obtained by adding the stress increments from past strain history ($\dot{\gamma}(t')$, with t' ranging from $-\infty$ to now). The stress increments are less and less important for times further and further into the past because for a liquid $G(t - t')$ always decays; we can say that the material has a memory that fades with time. Equation 11.3.2 can be further transformed by a simple change of variable: $s = t - t'$, $ds = -dt'$, and therefore

$$\sigma(t) = \int_{0}^{\infty} G(s)\dot{\gamma}(t - s) \, ds \tag{11.3.3}$$

This equation is an example of a *constitutive equation*, an equation that allows calculation of the stress in a material based on knowledge of all past deformations and of the relevant material response function. In the general case, $\dot{\gamma}(t - s)$, $G(s)$, and $\sigma(t)$ will all be tensor quantities, but as noted at the outset, we are restricting ourselves to simple shear and have tacitly chosen the relevant tensor elements.

From Equation 11.3.2 we can develop some very useful interrelations as follows:

1. Assume we apply an instantaneous step strain at time $t' = 0$, so that $\dot{\gamma} = \gamma_0 \delta(t')$. The Dirac delta function $\delta(t')$ is infinite at $t' = 0$, zero everywhere else, and integrates to 1 over all time, so that Equation 11.3.2 becomes $\sigma(t) = G(t)\gamma_0$. In this way we recover the stress relaxation modulus as defined in Equation 11.2.6.

2. In steady flow we apply a constant shear rate, $\dot{\gamma}(t - s) = \dot{\gamma}$ and by substituting in Equation 11.3.3 we obtain:

$$\sigma(t) = \dot{\gamma} \int_{0}^{\infty} G(s) \, ds$$

$$\eta = \frac{\sigma}{\dot{\gamma}} = \int_{0}^{\infty} G(t) \, dt \tag{11.3.4}$$

Thus the steady-flow viscosity is just the integral over the entire stress relaxation modulus. The viscosity represents the superposition of all modes of relaxation in the sample (i.e., all the exponential terms in $G(t)$), but because the integral is taken over all time, it is the slowest decaying modes that will dominate the long time, steady-flow response. Equation 11.3.4 was used to obtain the viscosity values in Figure 11.1 from the $G(t)$ data in Figure 11.2.

3. To obtain the dynamic moduli, we apply a strain $\gamma = \gamma_0 \sin \omega t$, and recognize that

$$\dot{\gamma}(t - s) = \gamma_0 \omega \cos \omega(t - s)$$
$$= \gamma_0 \omega \cos \omega t \cos \omega s + \gamma_0 \omega \sin \omega t \sin \omega s \tag{11.3.5}$$

Inserting this result into Equation 11.3.3, the stress can be written as the sum of two terms:

$$\sigma(t) = \gamma_0 \omega \, \cos \omega t \int_0^\infty G(s) \cos \omega s \, ds + \gamma_0 \omega \, \sin \omega t \int_0^\infty G(s) \sin \omega s \, ds \tag{11.3.6}$$

Comparing with Equation 11.2.20, we can express G' and G'' in terms of the sine and cosine Fourier transforms of $G(t)$, respectively:

$$G'(w) = \omega \int_0^\infty G(s) \sin \omega s \, ds$$

$$G''(w) = \omega \int_0^\infty G(s) \cos \omega s \, ds \tag{11.3.7}$$

4. The steady-state recoverable compliance must also be related to $G(t)$, although it takes a few lines of manipulation to show that this is the case. Consequently we will just provide the result here:

$$J_e^0 = \frac{\int_0^\infty s G(s) \, ds}{\left[\int_0^\infty G(s) \, ds\right]^2} = \frac{1}{\eta^2} \int_0^\infty s G(s) \, ds \tag{11.3.8}$$

These various relations underscore the fact that if you have access to $G(t)$, you can calculate all the other linear viscoelastic functions.

11.4 Bead–Spring Model

The Maxwell and Voigt models provide useful insight into the nature of viscoelastic response, but are severely lacking in terms of providing a satisfying description of real polymer liquids. First, we would like to have a molecular model in which the microscopic origins of viscosity and elasticity are more apparent. Second, a key feature of polymer viscoelasticity, whether in dilute solution or in concentrated solutions and melts, is the wide range of relaxation processes that contribute to the modulus. On an isolated polymer in a solvent, for example, the individual C–C bonds along the backbone can reorient in fractions of a nanosecond, whereas the entire end-to-end vector might take microseconds or even milliseconds to undergo substantial reorientation. In a molten polymer, the difference between the timescale for monomer motion and the timescale for the entire chain motion can be much greater than in dilute solution. Consequently, it is essential that a useful molecular model be able to capture this effect. In the next two sections we will examine the bead–spring model (BSM) of Rouse [4] and Zimm [5]. The BSM is over 50 years old, but it still serves as the starting point for describing the viscoelastic character of all flexible polymers. First we will describe the BSM itself, and then make some physical arguments for the character of the main predictions. A fully detailed solution to the model is beyond the scope of this book.

11.4.1 Ingredients of the Bead–Spring Model

The BSM model represents the chain as a linear string of N Hookean springs, or harmonic oscillators, connecting $N+1$ beads or mass points, as shown in Figure 11.10. (Recall that this model was introduced briefly in Section 9.6, in the context of hydrodynamic interactions in dilute solution. The difference between the Rouse and Zimm versions of the BSM is that the latter incorporates the hydrodynamic interactions, whereas the former does not.) The assembly is

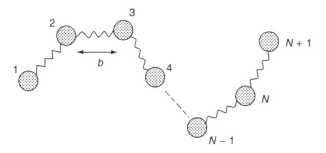

Figure 11.10 The bead–spring model of Rouse [4] and Zimm [5].

suspended in a viscous continuum, or solvent, with viscosity η_s. The actual mass of the polymer is distributed uniformly among the beads. Each bead has a friction coefficient, ζ, when it moves through its surroundings. We could imagine using Stokes' Law (Equation 9.2.4) for this friction, i.e., $\zeta = 6\pi\eta_s a$, where a is the radius of the bead, but in fact it is conventional to retain ζ as a parameter of the model rather than a. The springs have an rms end-to-end length b and the entire chain is freely jointed at each bead–spring unit. A single bead–spring unit is meant to represent a *Gaussian subchain* of the real polymer, that is, enough real backbone bonds or monomers that the end-to-end length of the subchain follows the Gaussian distribution (Equation 6.7.12). Thus this subchain corresponds to a small number of persistence lengths.

We already have enough information to make some comments about this model:

1. The viscous response in the BSM arises from the relative motion of the beads and solvent, whereas the elastic response will come from the tendency of the Gaussian subchains to resist deformation. We already discussed the entropic resistance to stretching a Gaussian chain in the context of chain swelling (see Section 7.7), and developed it in more detail in Chapter 10 (see Equation 10.5.7). The force, F, to stretch the subchain will be determined (in one dimension) from

$$F = \frac{dG}{dx} = -\frac{d}{dx}(TS) = -T\frac{d}{dx}k\ln P$$
$$= -kT\frac{d}{dx}\left(\text{constant} - \frac{3}{2b^2}x^2\right) = \frac{3kT}{b^2}x \tag{11.4.1}$$

    where in this expression G is the free energy, S is the entropy, and P is the Gaussian probability distribution applied to one subchain. The entropy $S \sim k\ln P$, where the proportionality factor is incorporated into a constant that does not matter after we take the derivative with respect to x. The last term in Equation 11.4.1 can be interpreted to mean that the subchain behaves as a Hookean spring with a spring constant of $3kT/b^2$.

2. The freely jointed nature of the chain tells us that the radius of gyration will be $Nb^2/6$ (see Equation 6.5.3) and therefore the model is designed for a theta solvent. (In principle one could insert excluded volume interactions into the detailed solution of the model, but this can only be done approximately, and at the price of great numerical complexity.)

3. The details of the actual chemical structure of the chain are subsumed into b and ζ. Thus, the BSM should make universal predictions for the character of the chain dynamics of flexible molecules, but it cannot be used to say anything about particular local motions, bond rotations, etc. of any actual polymer, which involve sections of the polymer less than a few persistence lengths.

4. The model has three parameters N, b, and ζ. N is proportional to the chain length, and therefore to M. In fact, this is a minimal set of parameters; we need to know how long the chain is, we need to know some length scale characteristic of the polymer, and we need to know the

timescale for motions on this length scale. This latter time, which we can call τ_{seg}, turns out to be proportional to $b^2\zeta/kT$.

5. It is worth pointing out that there are N springs but $N+1$ beads and, depending on the property in question, a literal solution of the model will involve either N or $N+1$. However, we expect the model to be most realistic when N is a large number, so this distinction will not matter; the important point is that $N \sim M$.

11.4.2 Predictions of the Bead–Spring Model

Now that we have defined the model, we can anticipate some of its predictions. We have a system of N coupled, identical harmonic oscillators, and by analogy with the analysis of molecular vibrations and rotations used in infrared spectroscopy, we could think in terms of degrees of freedom and normal coordinates. We have $N+1$ beads, so we need $3(N+1)$ coordinates to completely specify the instantaneous conformation of the chain. (It turns out that we do not need to worry about the $3(N+1)$ velocity or momentum values, however, because these equilibrate very rapidly due to the very high frequency of collisions between actual solvent molecules and monomers.) We need three coordinates to specify the center of mass position and the remaining $3N$ to describe the coordinates relative to the center of mass. That is, the translational diffusion of the whole chain will involve the first three coordinates and the internal motions or conformational relaxations the other $3N$. Finally, we recognize a threefold degeneracy, namely that the x, y, and z positions of any bead are uncorrelated; a force exerted on a particular spring in the x direction will not influence the y or z coordinates of a connected bead. Thus, there are really only N distinct internal degrees of freedom. If we choose the right coordinate system, the so-called *normal coordinates*, we will find N *normal modes*, or characteristic relaxations, each with its own natural frequency (or inverse relaxation time). Thus the main predictions of the BSM are the values of the N relaxation times and the character of the associated modes. The term normal here has the sense of *orthogonal* or linearly independent. Any particular motion of the chain in the laboratory coordinate system can be decomposed into a linear combination of the N normal modes and the excitation of any normal mode by some applied force will be dissipated by that mode only. (For those readers with experience in quantum chemistry, there is a strong analogy between these normal modes and the eigenvectors of the Hamiltonian in the Schrödinger equation; the frequencies of the BSM correspond to the allowed energy levels or eigenvalues of a quantum mechanical system, and the normal modes to the stationary states.)

We can illustrate the normal mode concept in Figure 11.11. We can draw a vector connecting any two beads separated by N/p springs where p is an integer between 1 and N (Figure 11.11a). We can define the relaxation time for this vector in the following way. We take all the vectors connecting the ends of (N/p)-long sections of the chain and do this for many chains. These vectors would have a Gaussian distribution in terms of their length, with an rms value of $b\sqrt{N/p}$ (see Equation 6.3.2), and all orientations would be equally likely. Now we apply a step strain. The distribution of orientations would be distorted in some way (Figure 11.11b), as would the distribution of lengths (Figure 11.11c). The overall distribution would relax back to equilibrium exponentially, with a characteristic time constant τ_p. Thus the pth relaxation time, τ_p, is the average time it takes a section of chain containing N/p springs to recover from a disturbance. There are N such relaxation times, varying from the first, τ_1, for the relaxation of the end-to-end vector of the entire chain, down to the Nth, τ_N, for the relaxation of a single bead–spring unit. Clearly $\tau_1 > \tau_2 > \tau_3 > \cdots > \tau_N$. We have already identified τ_N as being proportional to τ_{seg} above. Note that the roles played by τ_N and τ_{seg} are essentially equivalent, but τ_N is associated with a specific model, whereas τ_{seg} is a more general concept that can be discussed without reference to a particular model. The next issue is to find out how τ_p depends on p, and in particular how τ_1 depends on N.

We can obtain these dependences from a rather simple argument in the case of the Rouse model where we neglect hydrodynamic interactions. Suppose for the sake of simplicity that the chain is

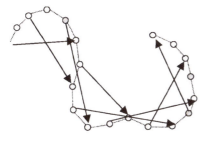

(a) Vectors connecting 5-spring units of the model chain

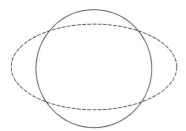

(b) Distribution of vector orientations in two dimensions immediately after deformation (dashed) and at equilibrium (solid)

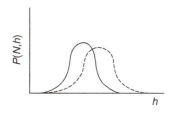

(c) Distribution of vector lengths immediately after deformation (dashed) and at equilibrium (solid)

Figure 11.11 Illustration of the relaxation process for the "end-to-end" vector of a section of chain containing N/p springs, where $1 \le p \le N$; in (a) $p = 5$. The distribution of orientations in (b) and the distribution of lengths in (c) relax together to their equilibrium states in a time τ_p.

lying along the x-axis, with all springs at the rest length of b. Suppose, we now tweak one of the beads, thereby causing a disturbance. The two neighboring springs would be distorted, thereby exerting forces on the next beads and so on down the chain. How long does it take for this disturbance to propagate down N/p springs? (Note that a bead–spring unit is not exactly a Maxwell element, because in this case no spring can be deformed without moving a bead; thus this deformation is not transmitted instantaneously along the chain.) Because this is a small disturbance and because in fact the beads are constantly undergoing Brownian motion due to collisions with the solvent, we can actually think of the disturbance diffusing down the chain, i.e., due to solvent motions, etc. it might just as well reverse direction. But we know about diffusion processes from Section 9.5. This "internal" diffusion constant, D_{int}, will be given by

$$\langle b^2 \rangle = 2D_{int}\tau_{seg} \tag{11.4.2}$$

where b represents the elementary step and τ_{seg} the associated elementary time for the diffusion of the disturbance. The factor of 2 reflects that this is a one-dimensional diffusion problem. Therefore the average time to travel N/p springs, τ_p, will be

$$\tau_p = \left(\frac{N}{p}b\right)^2 \frac{1}{2D_{int}} = \left(\frac{N}{p}\right)^2 \tau_{seg} \tag{11.4.3}$$

From Equation 11.4.3 we can see that the longest relaxation time, τ_1, will just be $N^2 \tau_{\text{seg}}$, and the pth relaxation time will be τ_1/p^2. As the local segmental time must be independent of total molecular weight (because a few monomers can rearrange themselves without disturbing the entire chain), τ_{seg} does not depend on N. Therefore the Rouse version of the BSM model predicts that the longest relaxation time of the chain will be proportional to M^2. Furthermore, the viscoelastic response will be governed by N different relaxation modes, with relaxation times spaced as $1:1/4:1/9:\cdots:1/N^2$.

This last result, which we obtained by a rather qualitative argument, is not the exact solution to this model. The correct result is

$$\tau_p = \frac{\zeta b^2}{6kT} \frac{1}{4\sin^2(p\pi/2N)} \approx \frac{\zeta b^2}{6kT} \frac{N^2}{\pi^2 p^2} \tag{11.4.4}$$

The last term in Equation 11.4.4 employs the fact that $\sin x \approx x$ for small x. Thus the simple scaling of τ_p with $(N/p)^2$ in Equation 11.4.3 is a very good approximation for small p, i.e., for the modes that involve big sections of the chain. It is these modes that we hope to describe well by the BSM; the modes for very short pieces of a real chain should depend more on local structural details and are therefore less likely to be captured by this approach. Comparing Equation 11.4.4 and Equation 11.4.3 we can see that the numerical prefactor for τ_{seg} can be specified:

$$\tau_{\text{seg}} = \frac{b^2 \zeta}{6\pi^2 kT} \tag{11.4.5}$$

Again, however, it should be emphasized that the BSM is expected to describe the longer range motions of the chain, not the local details, so the numerical prefactor in τ_{seg} should not be taken too seriously. It is also worthwhile to confirm that the collection of quantities on the right-hand side of Equation 11.4.5 gives units of time. The cgs units for ζ are g/s (recall the beginning of Chapter 9) and b is in cm, while kT has units of energy, or g cm^2/s^2. The net result is therefore (g/s)(cm^2)/(g cm^2/s^2) = s.

Now that we have a good representation of the relaxation times of the model, we need to see how they affect $G(t)$, and therefore $G^*(\omega)$ and η. First, we expect that each mode will contribute an exponential decay to $G(t)$, rather like the single mode of the Maxwell model. Because we are dealing with normal modes, they are independent of one another, so we just sum them:

$$G(t) = \sum_{p=1}^{N} G_p \exp(-t/\tau_p) \tag{11.4.6}$$

We have inserted a front factor, G_p, which gives the amplitude of each mode; G_p must have the units of a modulus. We now invoke the *equipartition theorem* of statistical mechanics: each degree of freedom, or normal mode, acquires kT of thermal energy. Furthermore, we can assume that the modulus will increase linearly with the number of chains per unit volume because each chain can store the same amount of elastic energy under deformation. The number of chains per unit volume is cN_{av}/M, where c is the concentration in g/mL. Therefore we can equate G_p with $(cN_{\text{av}}/M)kT = cRT/M$:

$$G(t) = \frac{cRT}{M} \sum_{p=1}^{N} \exp(-t/\tau_p) \tag{11.4.7}$$

You should check that cRT/M does have the units of modulus and compare this quantity to the modulus of the ideal elastomer, $\rho RT/M_x$, from Equation 10.5.17. Equation 11.4.7 is, in fact, the result that is obtained by a full solution of the model, but that would require a great deal more legwork.

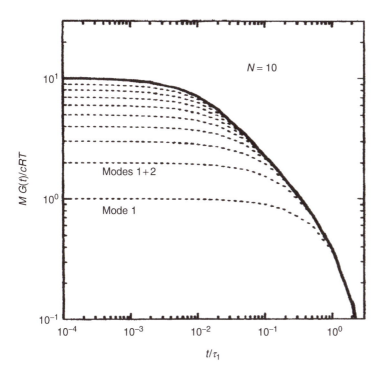

Figure 11.12 Stress relaxation modulus $G(t)$ for the Rouse model with $N = 10$ beads. The contributions of the nine individual modes as a function of reduced time t/τ_1 are indicated by dashed lines.

We plot this modulus in Figure 11.12, as $MG(t)/cRT$ versus t/τ_1. Recalling that $\tau_p/\tau_1 = 1/p^2$, we have

$$\frac{M}{cRT} G(t) = \sum_{p=1}^{N} \exp\left(-\frac{p^2}{\tau_1}t\right) \tag{11.4.8}$$

This function may be compared to $G(t)$ for the Maxwell element shown in Figure 11.6. There are now three regimes of behavior: short times, $t < \tau_N$, where little relaxation occurs; $t > \tau_1$, where the stress completely relaxes; $\tau_N < t < \tau_1$, where each mode contributes. It is this intermediate regime that is new and which is a direct consequence of the multiple relaxation times. Because $\tau_1 = N^2\tau_N$, the width of this regime on the time axis should grow as M^2. Consequently we have succeeded in achieving the two goals identified at the start of this section: we have a microscopic model that exhibits a broad range of relaxation processes.

The dynamic shear moduli can be obtained from $G(t)$ via Equation 11.3.7 (after looking up the integrals of the form $\int \cos \omega s \exp(-s/\tau_p)\, ds$):

$$G' = \frac{cRT}{M} \sum_{p=1}^{N} \frac{(\omega\tau_p)^2}{1 + (\omega\tau_p)^2}$$

$$G'' = \omega\eta_s + \frac{cRT}{M} \sum_{p=1}^{N} \frac{\omega\tau_p}{1 + (\omega\tau_p)^2} \tag{11.4.9}$$

where we note that the loss modulus includes an additive contribution from a purely viscous solvent. We could plot these functions as they stand, once values for N and τ_{seg} (or b and ζ) are specified. However, we can also form *intrinsic functions* (by analogy to the intrinsic viscosity in Section 9.3) as follows:

$$[G'] \equiv \lim_{c \to 0} \left(\frac{G'}{c} \right)$$

$$[G''] \equiv \lim_{c \to 0} \left(\frac{G'' - \omega \eta_s}{c} \right)$$

(11.4.10)

and then *reduced functions* as

$$[G']_R \equiv \frac{M}{RT}[G'] = \sum_{p=1}^{N} \frac{(\omega \tau_p)^2}{1 + (\omega \tau_p)^2}$$

$$[G'']_R = \frac{M}{RT}[G''] = \sum_{p=1}^{N} \frac{\omega \tau_p}{1 + (\omega \tau_p)^2}$$

(11.4.11)

Finally, we can normalize the frequency axis by the longest relaxation time, recalling that $(\tau_p/\tau_1) = 1/p^2$. Thus we arrive at a functional form that depends only on the dimensionless variables N and $\omega \tau_1$:

$$[G']_R = \sum_{p=1}^{N} \frac{(\omega \tau_1)^2 (\tau_p/\tau_1)^2}{1 + (\omega \tau_1)^2 (\tau_p/\tau_1)^2} = \sum_{p=1}^{N} \frac{(\omega \tau_1)^2}{p^4 + (\omega \tau_1)^2}$$

$$[G'']_R = \sum_{p=1}^{N} \frac{(\omega \tau_1)(\tau_p/\tau_1)}{1 + (\omega \tau_1)^2 (\tau_p/\tau_1)^2} = \sum_{p=1}^{N} \frac{(\omega \tau_1) p^2}{p^4 + (\omega \tau_1)^2}$$

(11.4.12)

These functions are plotted in Figure 11.13 and the results can be compared to those for the Maxwell model in Figure 11.9a. At low frequencies, $\omega \tau_1 < 1$, the liquid-like response is the same: $G' \sim \omega^2$, $G'' \sim \omega$, and $G'' > G'$. At very high frequencies, $\omega \tau_N > 1$, the material is a solid and again

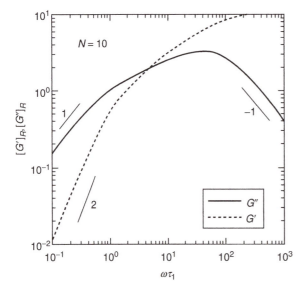

Figure 11.13   Reduced intrinsic dynamic moduli $[G']_R$ and $[G'']_R$ versus reduced frequency $\omega \tau_1$ for the Rouse model for the same 10-bead chain as in Figure 11.12. The limiting slopes are also indicated.

the response is the same: $G' \sim \omega^0$, $G'' \sim \omega^{-1}$, and $G' > G''$. As with $G(t)$, it is in the intermediate region that the results differ. Now we have N relaxations rather than the single mode of the Maxwell element and so G' and G'' evolve slowly with ω for $1/\tau_1 < \omega < 1/\tau_N$; G' and G'' both increase approximately as $\omega^{1/2}$ (although this is not evident here as N is small).

The Rouse model also makes an explicit prediction for the intrinsic viscosity, $[\eta]$ (see Section 9.3). We need to take the solution loss modulus, G'' (Equation 11.4.9), use the relation between G'' and η (Equation 11.2.24), and take the dual limits of $\omega \to 0$ and $c \to 0$.

$$\eta = \lim_{\omega \to 0} \left(\frac{G''}{\omega} \right) = \eta_s + \frac{cRT}{M} \sum_{p=1}^{N} \tau_p = \eta_s + \frac{cRT}{M} \tau_1 \sum_{p=1}^{N} \frac{\tau_p}{\tau_1} \tag{11.4.13}$$

The last transformation was taken because

$$\sum_{p=1}^{N} \frac{\tau_p}{\tau_1} = \sum \frac{1}{p^2} \approx \frac{\pi^2}{6} \quad as \ N \to \infty \tag{11.4.14}$$

where we have inserted the known result for this infinite sum. Thus

$$[\eta] = \lim_{c \to 0} \left(\frac{\eta - \eta_s}{\eta_s c} \right) = \frac{RT\tau_1 \pi^2}{M\eta_s 6} \tag{11.4.15}$$

As we demonstrated above, $\tau_1 \sim M^2$, so the Rouse model predicts that $[\eta] \sim M$. We saw in Section 9.3 that this is not experimentally correct. This is, in fact, the clear evidence that the Rouse model is missing an important ingredient in solution dynamics, namely the hydrodynamic interactions described in Section 9.7. As a final note, we can state that the tracer diffusion coefficient of the Rouse model is given by

$$D_t = \frac{kT}{(N+1)\zeta} \tag{11.4.16}$$

which is what we termed the *freely draining* result in Section 9.7. This expression has $N + 1$ in place of N because there are $N + 1$ beads, but this is of no real consequence as pointed out previously.

11.5 Zimm Model for Dilute Solutions, Rouse Model for Unentangled Melts

As we have just seen, the Rouse version of the BSM predicts that $[\eta] \sim M$ and $D_t \sim 1/M$, whereas we saw in Chapter 9 that experimentally $[\eta] \sim M^{3\nu-1}$ and $D_t \sim M^{-\nu}$, with $\nu = 0.5$ in a theta solvent and 0.6 in a good solvent. The principal reason for this discrepancy is the phenomenon of hydrodynamic interactions. In the context of the BSM, this means that the motion of any one bead through the solvent perturbs the fluid flow at the position of every other bead. This effect was incorporated into the Rouse model by Zimm, utilizing the approach of Kirkwood and Riseman introduced in Section 9.7. As described there, two simplifying assumptions were necessary to make the solution tractable. First, the description of the hydrodynamic interaction was truncated at the leading term (the so-called Oseen tensor), where the effect of bead j on bead k falls off as $1/r_{jk}$. Second, the instantaneous $1/r_{jk}$ was replaced by its equilibrium average, $\langle 1/r_{jk} \rangle$, using the Gaussian distribution; this is called the *pre-averaging approximation*. After these assumptions it proves possible to transform to a set of normal coordinates, just as in the Rouse model, and extract a set of N relaxation times, τ_p, and associated normal modes. In contrast to the Rouse model, there is no analytical solution for the normal modes and the relaxation times; the exact solution can only be obtained numerically. (However, efficient algorithms have been developed to do this [6].)

The predictions of the Zimm model are identical to those of the Rouse model in terms of the *form* of the equations (e.g., Equation 11.4.7 and Equation 11.4.9 for $G(t)$, G', and G''), but differ only in the relative *values* of the relaxation times. Qualitatively, the hydrodynamic interaction accelerates the relaxation of the chain because each bead communicates with every other directly through the solvent, rather than just by the springs. To put it another way, the relaxation times become more closely spaced than $1/p^2$. We saw in Chapter 9 that the diffusion constant of a single chain and the intrinsic viscosity were well described by picturing the coil as a sphere with a radius proportional to R_g; the same approach is successful here. A relation that plays the same role for rotational friction as Stokes' law does for translational friction (Equation 9.2.4) is called the *Stokes–Einstein–Debye* equation. It gives the rotational time for a sphere of radius R as

$$\tau_{\text{rot}} = \frac{8\pi R^3 \eta_s}{kT} \tag{11.5.1}$$

We present this result without derivation, but we can observe that it has the same structure as τ_{seg} in the BSM (Equation 11.4.5), if we recognize that $\zeta \sim b\eta_s$. On this basis we may propose that

$$\tau_1 \sim \frac{R_g^3 \eta_s}{kT} \sim \frac{M^{3\nu} \eta_s}{kT} \tag{11.5.2}$$

Thus, in a theta solvent, the longest relaxation time should scale as $M^{1.5}$, which turns out to be exactly the prediction of the Zimm model. We also can see that the longest relaxation time $(\sim R_g^3)$ and the intrinsic viscosity $(\sim R_g^3/M)$ are again intimately related:

$$\tau_1 \sim \frac{[\eta] M \eta_s}{RT} \tag{11.5.3}$$

which was also a prediction for the Rouse model (Equation 11.4.15).

We need to be a little careful about something here. The M dependences of $[\eta]$, D_t, and now τ_1 are all based on the M dependence of R_g because hydrodynamic interactions are sufficiently strong to make the dynamic behavior of the entire chain equivalent to that of a hard sphere. The effect of solvent quality is accounted for through the appropriate value of ν and thus this argument works equally well for theta solvents and good solvents. However, the Zimm model for the full viscoelastic spectrum is only strictly valid for theta chains because (a) the model is freely jointed and has no self-avoidance terms and (b) the hydrodynamic interaction is incorporated via pre-averaging over a Gaussian distribution. This may seem a little disappointing in the sense that we have simple physical arguments that give the correct M dependences of the global chain dynamics in any solvent ($[\eta]$, D_t, and τ_1), but the model for the internal chain dynamics (G', G'', and τ_p for $p > 1$) can only be applied to theta solvents. However, it turns out that there are various approximate ways that the effects of varying solvent quality can be incorporated into the Zimm model and several are sufficiently accurate to describe experimental data very well. We will briefly describe the most physically transparent approach known as *dynamic scaling*; it is in the spirit of the "a coil behaves hydrodynamically like a hard sphere" argument.

The algebra is very simple. Returning to Equation 11.5.2, we have

$$\tau_1 \sim \frac{R_g^3 \eta_s}{kT} \sim N^{3\nu} \frac{b^3 \eta_s}{kT} \sim N^{3\nu} \tau_{\text{seg}} \tag{11.5.4}$$

where again we invoke $\zeta \sim b\eta_s$. By analogy with Equation 11.4.3, we expect that the relaxation time for a subsection of the chain containing N/p units should scale as $(N/p)^{3\nu}$, and thus we assert that any relaxation time can be written

$$\tau_p \sim \left(\frac{N}{p}\right)^{3\nu} \tau_{\text{seg}} \tag{11.5.5}$$

or

$$\frac{\tau_p}{\tau_1} = p^{-3\nu} \tag{11.5.6}$$

This can be substituted directly into the expressions for the reduced intrinsic dynamic moduli, Equation 11.4.12, and thus

$$[G']_R = \sum_{p=1}^{N} \frac{(\omega\tau_1)^2 (p^{-3\nu})^2}{1 + (\omega\tau_1)^2 (p^{-3\nu})^2}$$

$$[G'']_R = \sum_{p=1}^{N} \frac{\omega\tau_1 p^{-3\nu}}{1 + (\omega\tau_1)^2 (p^{-3\nu})^2} \tag{11.5.7}$$

These expressions are straightforward to evaluate and give predictions for the moduli that are essentially indistinguishable from the full numerical evaluation of the Zimm model.

Comparisons of the Zimm model to the viscoelastic properties of flexible chains are presented in Figure 11.14. In these experiments, five high molecular weight 1,4-polybutadiene chains were dissolved in diethylhexyl phthalate, which is a theta solvent at 18.0°C. Measurements were made at a series of dilute concentrations, and extrapolated to infinite dilution to obtain the intrinsic moduli. (The actual measurements were made on a flow birefringence apparatus, which determines an optical anisotropy that is directly proportional to the stress; the optical experiment is considerably more sensitive than most rheometers, which is important for very dilute solutions where the signals are small.) The data are plotted as reduced intrinsic moduli versus reduced frequency, $\omega\tau_1$, and thus theory is compared with the data with no adjustable parameters. In Figure 11.14a, the temperature corresponds to the theta point and the fit is to the dynamic scaling result with $\nu = 0.50$, or to the full numerical solution to the Zimm theory; the two curves are identical. In Figure 11.14b, the temperature has increased to 30°C and now there is some degree of excluded volume. The data for G'' for the highest M polymer are well described by dynamic scaling with $\nu \approx 0.52$–0.53, and furthermore the longest relaxation times used to collapse the data, shown in Figure 11.15, have the expected dependence on $M^{3\nu}$. Based on many results such as these, we may conclude that the Zimm model, modified for the effects of excluded volume as needed, provides a quantitative description of the dynamics of isolated, long, flexible chains.

From the success of the Zimm model, one might be led to conclude that the Rouse model serves only as a conceptual foundation on which to build more elaborate descriptions of solution viscoelasticity and that it is not so useful in terms of actual data. However, it turns out that the Rouse model is very successful in describing the dynamics of low molecular weight polymer melts or concentrated solutions. The reason is twofold. First, at very high concentrations, the hydrodynamic interactions are effectively *screened*. Any motion of one monomer is transmitted by many other monomers before it reaches another monomer on the same chain and thus it imparts no through-space coherence to the chain motion. Second, the chains are very nearly Gaussian in the melt, as discussed in Section 7.7, so the freely jointed assumption is appropriate. In the melt there is no solvent, so the effect of frictional resistance to motion on the segmental level is subsumed into the friction factor, ζ. The Rouse model predicts that $\eta \sim M$, $\tau_1 \sim M^2$, and $D \sim M^{-1}$, all of which are at least approximately true for low molecular weight polymers in the melt. (They are not exactly true in most cases, because the subchain friction factor ζ develops an M dependence at low M, due primarily to the M dependence of the glass transition temperature. This effect will be discussed in Chapter 12, but it can be corrected for, and then the Rouse predictions hold rather well.) We have been careful to emphasize low molecular weight melts here because for higher molecular weights a new effect comes into play and the Rouse model is again inadequate. The new effect is called *entanglement* and we describe it in the next section. To conclude this section, we summarize in Table 11.1 the main predictions for the M dependence of chain dynamics from the

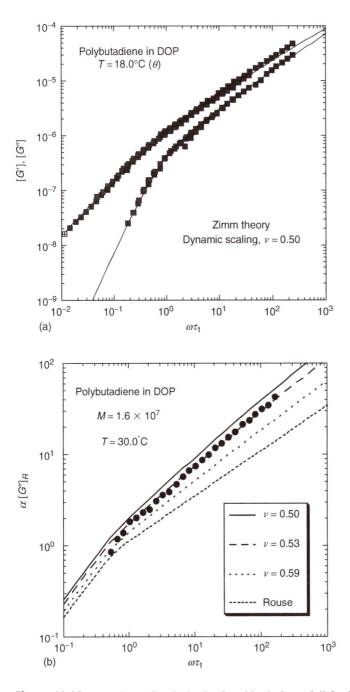

Figure 11.14 Experimentally obtained reduced intrinsic moduli for five polybutadiene samples in dioctyl phthalate, compared to the predictions of the Zimm model. (a) At the theta temperature, 18°C, with $\nu = 0.50$. (b) The data for G'' for the highest molecular weight sample only, slightly above the theta temperature, at 30°C. The curves represent the Zimm theory with three different scaling exponents, ν, and the Rouse theory. The data have been shifted vertically by a factor α, but it is clear that the experimental slope agrees with expectation for modest excluded volume ($\nu > 0.50$). (Data obtained from Sahouani, H. and Lodge, T.P., *Macromolecules*, 25, 5632, 1992. With permission.)

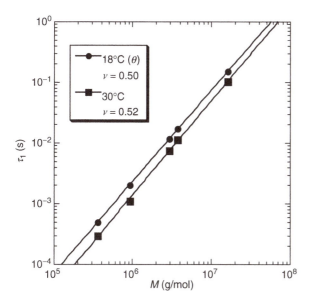

Figure 11.15 Infinite dilution longest relaxation time versus molecular weight at the theta temperature (18°C) and at 30°C, for the same solutions as in Figure 11.14. The exponents are obtained from the fitted slopes ($= 3\nu$). (Data obtained from Sahouani, H., and Lodge, T.P., *Macromolecules*, 25, 5632, 1992. With permission.)

two forms of the BSM (and also the related predictions of the reptation model to be presented in Section 11.7), and give an example of the application of the BSM.

Example 11.2

Use the BSM to estimate the longest relaxation time and the segmental relaxation time for polystyrene with $M = 10^6$ g/mol in cyclohexane at the theta temperature (35°C). Recall that values of D_t and $[\eta]$ were given as a function of M for this system in Figure 9.10a and Figure 9.5, respectively.

Solution

There are several different ways to proceed. For example, from Figure 9.10a we find that $D_t \approx 10^{-7}$ cm^2/s, and therefore we can calculate the hydrodynamic radius by the Stokes–Einstein equation (Equation 9.5.5) (the viscosity of cyclohexane is about 0.8 cP at 35°C):

$$R_h \approx \frac{1.4 \times 10^{-16} \times 308}{6 \times 3.14 \times 0.008 \times 10^{-7}} = 3 \times 10^{-6} \text{ cm} = 30 \text{ nm}$$

Table 11.1 Predictions of Three Models for the Molecular Weight Dependence of Various Chain Dynamics Quantities

Property	Rouse	Zimm	Reptation
D_t	$D_t = \frac{kT}{(N+1)\zeta} \sim M^{-1}$	$D_t = \frac{kT}{6\pi\eta_s R_h} \sim M^{-\nu}$	$D_t = \frac{kT}{N\zeta}\frac{N_e}{N} \sim M^{-2}$
τ_1	$\tau_1 \sim N^2 \tau_{seg} \sim M^2$	$\tau_1 \sim N^{3\nu}\tau_{seg} \sim M^{3\nu}$	$\tau_1 \sim \frac{N^3}{N_e}\tau_{seg} \sim M^3$
$[\eta]$	$[\eta] \sim \frac{RT\tau_1}{M\eta_s} \sim M$	$[\eta] \sim \frac{RT\tau_1}{M\eta_s} \sim M^{3\nu-1}$	—
η	—	—	$\eta = \frac{\pi^2}{12}\tau_1 G_N \sim M^3$

We can now insert this value as R in Equation 11.5.1 to obtain the longest relaxation time:

$$\tau_1 \approx \tau_{\text{rot}} = \frac{8 \times 3.14 \times (3 \times 10^{-6})^3 \times 0.008}{1.4 \times 10^{-16} \times 308} \approx 10^{-4} \text{ s}$$

Alternatively, we can use the value of $[\eta] \approx 84$ mL/g from Figure 9.5 in Equation 11.5.3:

$$\tau_1 \approx \frac{84 \times 10^6 \times 0.008}{6 \times 10^{23} \times 1.4 \times 10^{-16} \times 308} \approx 3 \times 10^{-5} \text{ s}$$

These two values are in reasonable agreement. (Compare them to the value of the longest relaxation time for molten poly(α-methylstyrene) in Figure 11.2 and Example 11.1: chains relax a lot more rapidly in dilute solution than in the melt.).

To estimate τ_{seg}, we can take Equation 11.5.4 with $\nu = 0.5$ (appropriate for a theta solvent) and $N = M/M_0 = 10^6/104 \approx 10^4$:

$$\tau_{\text{seg}} \approx \frac{10^{-4}}{(10^4)^{1.5}} \approx 10^{-10} \text{ s}$$

This suggests that motions on the length scale of a styrene monomer take place in 100 ps, which is reasonable. It might be more in the spirit of the model to take the segment as a few persistence lengths. From Table 6.1 and the discussion in Section 6.4, about 10 styrene monomers correspond to four persistence lengths and the newly computed $N = 10^6/(10 \times 104) \approx 1000$. This gives a correspondingly larger value for τ_{seg} of about 3×10^{-9} s.

11.6 Phenomenology of Entanglement

In the previous two sections we have emphasized mechanical models for the viscoelastic response of polymer chains and we have seen how well the BSM in the Zimm form is able to describe the behavior in very dilute solutions. In large measure, these models were formulated before extensive experimental tests had been performed. When we turn our attention to highly concentrated solutions and melts, however, the situation is reversed. The basic experimental phenomenology has been well known since the 1950s and 1960s, but only in the 1970s did a successful, detailed molecular model emerge. This model was initially developed by Doi and Edwards [7], based on the *reptation hypothesis* of de Gennes [8]. We will follow this historical chronology by focusing in this section on the experimental phenomena and then in the next section on the reptation model itself.

11.6.1 Rubbery Plateau

The Rouse and Zimm models for G' and G'' of isolated chains have three distinct regimes of behavior. At low frequencies, $\omega\tau_1 < 1$, the response is that of a liquid, with $G' \sim \omega^2$, $G'' \sim \omega$, and $G'' > G'$. This terminal regime is common to all liquids and is therefore common to all models, as noted in Section 11.2. At very high frequencies, $\omega\tau_N > 1$ where τ_N is the shortest relaxation time, the response is solid-like: $G' \sim \omega^0$, $G'' \sim \omega^{-1}$, and $G' \gg G''$. It is the third, intermediate zone that is characteristic of the BSM; the response depends on the detailed nature of the spectrum of relaxation times, i.e., the internal degrees of freedom of the chain.

When we turn to a molten polymer, we find that there are now *four* distinct regions of behavior. This is illustrated in Figure 11.16a for G' and G'', and in Figure 11.16b for $G(t)$. The latter figure may be compared to Figure 11.2; the results are qualitatively very similar. The long time (or low frequency) terminal regime is just as before (albeit with a considerably different value of τ_1 and η) and the short time (high frequency) regime is solid-like, with a modulus typical of a glass ($10^9 - 10^{10}$ Pa). Thus, it is in the internal dynamics that a new feature has emerged. In particular, there is a new regime of solid-like behavior, with $G' \sim \omega^0 \approx 10^5 - 10^6$ Pa. This modulus is

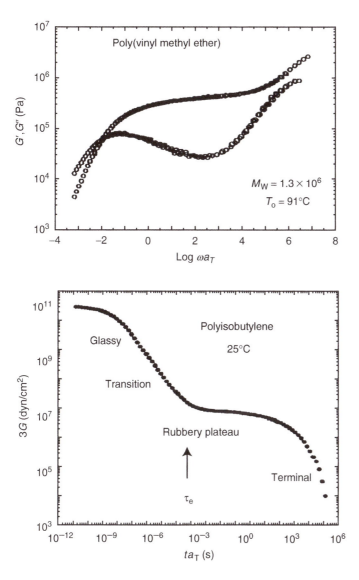

Figure 11.16 Illustration of the viscoelastic response of highly entangled polymer melts. (a) G' and G'' versus reduced frequency for poly(vinyl methyl ether). (Data from Kannan, R.M. and Lodge, T.P., *Macromolecules*, 30, 3694, 1997. With permission.) (b) $G(t)$ versus reduced time for polyisobutylene. (Data from Catsiff, E. and Tobolsky, A.V., *J. Colloid. Sci.*, 10, 375, 1955. With permission.) The significance of the shift factor a_T will be discussed in Chapter 12.

characteristic of a lightly cross-linked rubber, as we saw in Chapter 10. Qualitatively, if we follow $G(t)$ in Figure 11.16b starting from short times, the initial modulus is that of a *glass*. Rapidly, however, the stress begins to relax, in what is called the *transition zone*. This character is similar to what is seen when a glass is heated through the glass transition temperature (see Chapter 12), hence the terminology. The remarkable new feature is that at some characteristic time, τ_e in Figure 11.16b, the stress stops relaxing before it has decayed to zero. In other words, the material thinks it is a solid again, albeit with a much reduced modulus compared to the glassy state. This so-called *rubbery plateau* can persist for some decades in time before finally giving way to flow in the *terminal regime*. Correspondingly, in G'' we see two peaks, one in the transition zone and one at the onset of the terminal regime. The former corresponds to motions of a few monomers at a time and represents the transition out of the glassy state. The second corresponds to motions of entire molecules, or large subsections, and represents the final transition to liquid-like behavior. The rubbery plateau in G' and $G(t)$ then arises because of an increased separation on frequency or time between local relaxations and chain relaxations. A plateau in G' signifies a gap in the spectrum of relaxation times, which might arise for any number of reasons. In the particular case of a molten flexible polymer, the gap in the relaxation time spectrum arises for a purely topological reason, referred to as intermolecular entanglements.

For high molecular weight polymers, the individual chains become intertwined with one another. The full relaxation of one chain is thus highly dependent on its surroundings, as one chain cannot pass through another. Consequently, the longer relaxation times of the chain are severely increased by this effect, whereas relaxations of a few monomers can still proceed rapidly; this gives rise to the gap in the relaxation time spectrum. The modulus in the plateau region (either in $G(t)$ or in G') is called G_N and has a magnitude very similar to the solid modulus of the same polymer if it were lightly cross-linked like a rubber band. Thus the interchain entanglements can be thought of as *temporary cross-links*; for times shorter than the lifetime of an entanglement, the material behaves as a solid, but then at longer times the material flows. In the next section we will explore the *reptation model*, which explains how chains escape from their entanglements, but here we need to examine the nature of the entanglements themselves. As we saw in Chapter 10, to a first approximation, the modulus of a lightly cross-linked rubber is given by $G = \rho RT/M_x$ where ρ is the density and M_x is the molecular weight between cross-links. As with the prefactor for G in the BSM (see Equation 11.4.7 and associated discussion), this modulus is an example of the equipartition theorem: it is the number of network strands per unit volume, $\rho N_{av}/M_x$, multiplied by the thermal energy, kT. We now use this equation to define a *molecular weight between entanglements*, M_e:

$$M_e \equiv \frac{\rho RT}{G_N} \tag{11.6.1}$$

In other words, M_e is the average molecular weight between temporary cross-links. The parameter M_e has been tabulated for many flexible polymers; it typically corresponds to about 50–200 monomers. Examples are given in Table 11.2. We can now see that the relaxation of the polymer in the transition zone corresponds to relaxations of chain segments less than M_e long, whereas in the terminal zone we have relaxations of chains containing many (i.e., M/M_e) entanglement lengths.

Although M_e provides a convenient parameterization of entanglements, it is not a fully satisfying explanation of the effect. For example, in a real cross-linked rubber, the cross-links

Table 11.2 Plateau Modulus, Molecular Weight between Entanglements, Packing Length, and Entanglement Spacing for Some Common Polymers at 140°C

Polymer	G_N (MPa)	M_e (g/mol)	$p*$(Å)	d (Å)
Polyethylene	2.6	840	1.69	33
Poly(ethylene oxide)	1.8	1,600	1.94	37.5
1,4-Polybutadiene	1.2	1,800	2.29	44
1,4-Polyisoprene	0.42	5,400	3.20	62
Polyisobutylene	0.32	7,300	3.43	66
Poly(methyl methacrylate)	0.31	10,000	3.46	67
Polystyrene	0.20	13,000	3.95	76.5
Poly(dimethylsiloxane)	0.20	12,000	4.06	79

Source: From Fetters, L.J., Lohse, D.J., Richter, D., Witten, T.A., and Zirkel, A., *Macromolecules*, 27, 4639, 1994.

are identifiable chemical entities, different from ordinary monomers on the chain. In the entangled melt, we probably should not think of one monomer in an M_e-long section of chain as being stuck to its surroundings, while all the other ones move happily. Rather, the entanglement phenomenon represents the cumulative effect of many interactions between monomers on different chains, with no single monomer behaving differently on average from any other. Thus, the picture of a temporary network with particular points of entanglement, although physically appealing, is potentially misleading. An interesting question then arises: can we predict what M_e should be for a given polymer, based on things we already know about the chemical structure? The answer turns out to be yes, and we summarize the explanation given by Fetters and coworkers [9].

11.6.2 Dependence of M_e on Molecular Structure

From Table 11.2 we can discern a qualitative correlation between M_e and the bulkiness of the sidegroups. For example, polystyrene, poly(methyl methacrylate), and poly(dimethylsiloxane) have relatively large values of M_e and relatively large sidegroups whereas polyethylene, poly(ethylene oxide), and polybutadiene have much smaller M_e. In general, "thin" chains entangle more easily (have lower values of M_e) than "fat" chains. If we suppose that entanglement effects are due to the uncrossability and mutual intertwining of different chains, then a crucial parameter should be the amount of chain contour per unit volume. If we may be permitted a culinary analogy, a bowl of cooked cappellini has many more entanglements than the same volume of fettuccini. To be more quantitative, we introduce the packing length, $p*$, which is the ratio of the volume occupied by a chain ($M/\rho N_{av}$) to its mean square end-to-end distance; values are also listed in Table 11.2.

$$p* \equiv \frac{M}{\rho N_{av}} \frac{1}{\langle h^2 \rangle_0} = \frac{M_b}{\rho N_{av} C_\infty \ell^2} \tag{11.6.2}$$

Here M_b is the molecular weight per backbone bond, C_∞ is the characteristic ratio (recall Equation 6.3.1), and ℓ is the average backbone bond length. Therefore thinner chains will tend to have a smaller packing length because they have a smaller M_b. Note, however, that more flexible chains will have a smaller C_∞, which tends to increase $p*$.

Now consider the volume of space pervaded by a chain, $V_p = AR_g^3$, where A is a constant of order unity. The pervaded volume is considerably larger than the occupied volume for a long chain. The number of different chains, n, that pack into a volume V_p is given by

$$n = V_p \frac{\rho N_{av}}{M} \tag{11.6.3}$$

Note that because V_p increases as $R_g^3 \sim M^{3/2}$, n increases with $M^{1/2}$. Now we introduce a hypothesis: the onset of entanglement effects occurs when $n \approx 2$, i.e., when M is just big enough that a test chain and one other intertwine to pack a volume V_p. Certainly we would not expect entanglement effects if n were less than 2 and maybe the onset really occurs for some larger value, but that would just affect the proportionality constant in the following analysis.

We rewrite the radius of gyration as

$$R_g = \frac{\sqrt{C_\infty M} \ell}{\sqrt{6M_b}} \tag{11.6.4}$$

and thus

$$n = 2 = A \frac{C_\infty^{3/2} \ell^3}{(6M_b)^{3/2}} \sqrt{M} \rho N_{av} \tag{11.6.5}$$

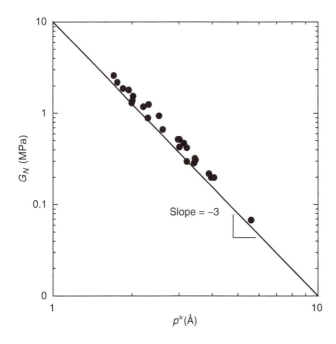

Figure 11.17 Dependence of the plateau modulus, G_N (and thus the molecular weight between entangle-ments, M_e) on the packing length, p^*, for various polymers. (From Fetters, L.J., Lohse, D.J., Richter, D., Witten, T.A., and Zirkel, A., *Macromolecules*, 27, 4639, 1994. With permission.)

We now equate M_e with the value of M for which $n = 2$:

$$M_e = \frac{4M_b^3 6^3}{A^2 C_\infty^3 \ell^6 \rho^2 N_{av}^2} = \frac{\rho N_{av}}{(A')^2}(p^*)^3 \tag{11.6.6}$$

where A' is a new constant $= A/(12\sqrt{6})$. Using the empirical definition of M_e (Equation 11.6.1), we can express the plateau modulus as

$$G_N = (A')^2 \frac{kT}{(p^*)^3} \tag{11.6.7}$$

This relation is compared to an extensive compilation of experimental data in Figure 11.17. The dependence on $(p^*)^{-3}$ is very clear and the resulting value of the unknown constant A is about 1.7. Thus we can assert that entanglement is a universal property of flexible chains and that the spacing of entanglements for a given polymer is determined entirely by the density (ρ) and the flexibility (C_∞).

Example 11.3

A well-known rule of thumb in adhesion science is the *Dahlquist criterion*, which states that in order for a pressure sensitive adhesive (PSA) to have good *tack*, it should have a plateau modulus below 3×10^5 Pa. Tack describes the ability of an adhesive to form a bond of measurable strength, quickly, under a light load (e.g., Post-it Notes). Give a qualitative explanation for this rule, and comment on whether the polymers in Table 11.2 might be useful as PSAs.

Solution

There are two aspects to tack: forming a bond quickly and developing mechanical strength. The former requires molecular motion to be rapid; higher G_N means smaller M_e (or M_x in the cross-linked case) and thus more restricted chain motions. A typical PSA is soft and squishy to the touch and distinctly softer than a rubber band ($G_N \approx 10^6$ Pa or more). At the molecular level, this means that the molecules are able to flow and conform to the surfaces to be adhered. The mechanical strength part of the problem is a little more subtle. It turns out that the strength of an adhesive is much more than the sum of the attractive forces between the glue and the substrate; most of it comes from energy dissipation in deforming the adhesive. This dissipation is achieved by molecular motion, so softer materials are favored for this reason too. However, this argument does not mean that the modulus should be made as small as possible; then there would be no strength at all. All adhesives are polymers because the many modes of viscoelastic relaxation provide many modes of dissipation, and thus good adhesive strength.

From Table 11.2 we can see that the last five polymers might be candidates for PSAs (poly-isoprene, polyisobutylene, poly(methyl methacrylate), polystyrene, and poly(dimethylsiloxane)), although their plateau moduli are very close to the limit of 3×10^5 Pa. In order to reduce the modulus of a PSA, it is common to add a lower molecular weight diluent or oil, known as a *tackifier*. Experimentally, G_N decreases at least as c^2, so adding 50% of a tackifier should drop the modulus by at least a factor of $(1/2)^2 = 1/4$. With this additional degree of freedom, we could make a PSA out of almost any polymer, except for one crucial factor: the bond also needs to form quickly. How quick is quick enough? Well, common experience tells us that we do not want to hold Scotch tape in place for 10 s while it sets, so quickly means within 1 s or less. This brings into play the time dependence of the modulus: at very short times, the material will behave as a glass, with $G \approx 10^9$ Pa or more. So, we need to make sure that timescales of say 0.1–1 s fall within the rubbery plateau of response. Figure 11.2 tells us that poly(α-methylstyrene) would not work; the rubbery plateau does not begin until 1–10 s (and then only at 186°C). Figure 11.16a, on the other hand, suggests that polyisobutylene would be a good candidate at room temperature. In fact, out of our list of five polymers from Table 11.2, polystyrene and poly(methyl methacrylate) can be eliminated; at room temperature for timescales of 1 s, they behave as hard, glassy solids (e.g., Plexiglas). This demarcation depends on a crucial parameter—the glass transition temperature—which is the subject of the next chapter. In addition to smaller values of G_N, PSAs tend to have glass transition temperatures below room temperature.

Before proceeding to the reptation model, two more phenomenological effects of entanglement deserve comment. The first is the molecular weight dependence of the viscosity. Recall from Equation 11.3.4 that the viscosity is the integral over $G(t)$. Thus, when entanglement effects set in, the rubbery plateau grows in extent and the area under $G(t)$ increases markedly. The viscosity itself, plotted for several polymers in Figure 11.18, shows a very strong but universal dependence on M, namely $\eta \sim M^{3.4 \pm 0.2}$, when $M \gg M_e$. This was illustrated earlier, in Figure 11.1, for one particular polymer. For shorter polymers, $M \leq M_e$, the dependence is much weaker because there is no entanglement and is more consistent with the Rouse model ($\eta \sim M$). The molecular weight for which the dependence changes is called M_c and generally $M_c \approx 2-3$ M_e. The exponent of 3.4 is something that any successful theory of polymer melt dynamics must explain. From the practical point of view, it means that the processing of molten polymers will be very dependent on the average molecular weight; as noted in the context of Figure 11.1, an increase by a factor of two in M results in an increase by a factor of 10 in η.

The second important phenomenological effect is that of large-scale *elastic recovery*. This is a nonlinear response and therefore not something we will cover in detail, but it serves to illustrate a crucial point. The most remarkable feature of a rubber band is that not only can it be *deformed* by 500% or more without breaking, but that it *recovers* its original shape upon letting go. The same is true of a molten polymer liquid. If a very large strain is applied quickly and then released before the stress has had a chance to relax much, the liquid will also snap back to its original shape. (How quickly is quickly? From Figure 11.16b we can see that the polymer will behave as a rubbery solid for times significantly less than τ_1.) It is this ability to show large scale elastic recovery that

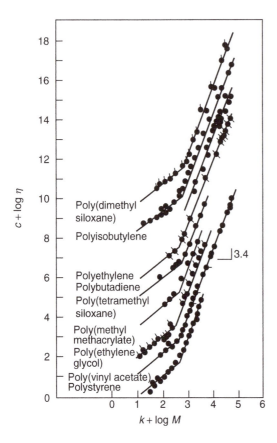

Figure 11.18 Dependence of the melt viscosity on molecular weight for a variety of polymers c and k are arbitrary constants to shift the data. (Compiled by Berry, G.C. and Fox, T.G, *Adv. Polym. Sci.*, 5, 261, 1968. With permission.)

truly is the hallmark of entanglements in polymer liquids. Many systems can show a plateau in G' or $G(t)$ that has nothing to do with entanglements; examples include dense colloidal suspensions near their glass transition, and densely packed, roughly spherical polymer objects (e.g., block copolymer micelles, hyperbranched chains, dendrimers, and microgels). One might be tempted to argue on this basis that the viscoelastic response of flexible polymers has nothing much to do with entanglements, because other nonentangling systems also show a plateau in G' with a similar magnitude to G_N. This argument is fallacious, however. A plateau in G', as noted above, only requires a wide separation between relaxation processes. Large-scale elastic recovery in a liquid, in contrast, demands some kind of interchain entanglement. Materials comprising roughly spherical objects such as those noted above can usually not be extended by even 5% or 10% without falling apart, let alone the 500% that flexible polymer liquids can sustain.

11.7   Reptation Model

The reptation model was originally developed by de Gennes for a single, flexible chain trapped in a permanent network (*reptation* was coined from the Latin *reptare*, to creep) [8]. It was then extended to a theory for linear and nonlinear rheological response of polymer liquids by Doi and Edwards [7]. Here, as throughout the chapter, we will emphasize the linear response, but it is worth noting in passing that the success of the reptation model in capturing nonlinear phenomena is at least as impressive as its description of linear viscoelasticity. We will begin with a simple

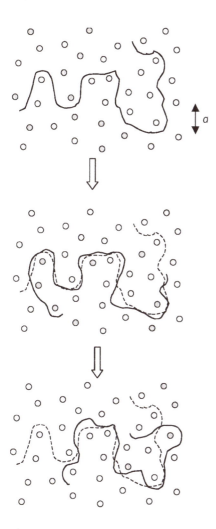

Figure 11.19 A single chain trapped in an array of obstacles, spaced an average distance d apart. By snaking backward and forward through the obstacles, i.e., by reptating, the chain eventually escapes from its original set of obstacles.

argument for the molecular weight dependence of the longest relaxation time and the diffusion coefficient, before proceeding to the stress relaxation modulus.

11.7.1 Reptation Model: Longest Relaxation Time and Diffusivity

Imagine a chain trapped in a field of obstacles, as in the two-dimensional representation in Figure 11.19. The obstacles have an average spacing d, which we will subsequently associate with the average distance between entanglements (see Table 11.2). Therefore $d \approx b\sqrt{N_e}$, where N_e is the number of monomers in M_e; d is typically 30–80 Å, i.e., longer than the persistence length but less than R_g. How do the obstacles affect the relaxation and diffusion of the chain? There are three regimes of behavior. On short times, such that individual monomers on average do not move as far as d, the obstacles have no effect. We may associate this regime of behavior with the transition zone of viscoelastic response, $t < \tau_e$; segments of the chain up to N_e units long can relax readily. On very long times, such that the chain has completely lost contact with the obstacles it was surrounded by at time $t = 0$, it has fully relaxed; this is the terminal regime $t > \tau_1$ as far as our test chain is concerned. The intermediate regime, $\tau_e < t \leq \tau_1$, is the one we must concentrate on, when the chain feels the confining effects of the obstacles. In particular we want to find the longest relaxation time, which defines when the chain has fully escaped from its initial surroundings.

The reptation hypothesis is that the chain ultimately escapes from the obstacles by snaking along its own contour. Under the influence of Brownian motion, one end of the chain comes out of the obstacles and explores space at random, while the other end penetrates further into the obstacles. However, this is a random process, so the chain is just as likely to move back in the opposite direction; the snake has two heads. The key point is that when the chain moves back, it is under no obligation to retrace its previous path through the obstacles; it will follow a random course. In this way it will escape the first set of obstacles, as illustrated in Figure 11.19. Now, as time goes on, random motion will drive the ends of the chain further and further into the initial set of obstacles and then subsequent reverse motion will erase the memory of where the chain used to be. The longest relaxation time will correspond to the moment when the middle segments of the chain finally escape from the confines of the initial set of obstacles.

We can describe this process in the following way. The set of fixed obstacles with spacing d will be replaced by a confining *tube* of diameter d, as in Figure 11.20. This tube is defined by the conformation of the test chain at time $t = 0$; it is itself a random walk in three dimensions. The real polymer segments will move freely within the tube due to the rapid segmental motions, but the whole chain will only escape the tube by reptating out of the ends. We replace the real chain (N monomers, statistical segment length b, $R_g^2 = Nb^2/6$) with an average chain that has N/N_e entanglement lengths (with $d^2 = N_e b^2$) trapped in the tube. When one end of this *coarse-grained* chain diffuses out of the end of the tube and the other end moves a distance d into the tube, that portion of the tube is erased (see Figure 11.20). The entire tube will be erased when the chain center-of-mass has diffused a distance proportional to the length of the tube, L. We can define a diffusion coefficient for the motion within the tube, D_{tube}, by

$$\langle L^2 \rangle = 2D_{\text{tube}}\tau_{\text{rep}} \tag{11.7.1}$$

where τ_{rep} is the reptation time, the time needed to completely erase the tube. This diffusion process is postulated to have a Rouse-like dependence on chain length, i.e., the friction of the chain

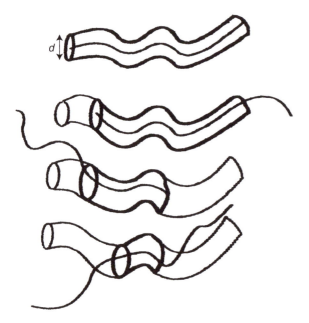

Figure 11.20 The obstacles in Figure 11.19 can be replaced by a tube with diameter d; the process of reptation gradually erases the tube from the ends inward.

moving in its tube is proportional to the number of tube segments, N/N_e, and the friction factor of each tube segment, ζ_e. The latter quantity is just the number of monomers in the segment, N_e, times the monomeric friction factor, ζ_0. (The monomeric friction factor is similar to the friction factor of a bead in the Rouse model, but is calculated to correspond to the friction per repeat unit.) Using this argument we find

$$D_{\text{tube}} = \frac{kT}{\text{friction}} = \frac{kT}{(N/N_e)\zeta_e} = \frac{kT}{(N/N_e)(N_e\zeta_0)} = \frac{kT}{N\zeta_0} \quad (11.7.2)$$

Of course, the contour length of the tube is directly related to the length of the chain, namely

$$L = \frac{N}{N_e}d = \frac{N}{N_e}\sqrt{N_e}b \quad (11.7.3)$$

so that we can solve Equation 11.7.1 for τ_{rep}

$$\tau_{\text{rep}} = \frac{L^2}{2D_{\text{tube}}} = \frac{N^2}{N_e}b^2\frac{N\zeta_0}{2kT} = \frac{N^3b^2\zeta_0}{2kTN_e} \quad (11.7.4)$$

This equation can be recast in terms of molecular weight by recognizing that $N = M/M_0$, where M_0 is the monomer molecular weight, $N/N_e = M/M_e$, and $\tau_{\text{seg}} \sim b^2\zeta_0/kT$:

$$\tau_{\text{rep}} \approx \frac{M^3}{M_0^2 M_e}\tau_{\text{seg}} \quad (11.7.5)$$

The crucial feature of this result is that the longest relaxation time of the chain is predicted to vary with M^3. This dependence is much stronger than the Zimm ($\tau_1 \sim M^{3/2}$) or Rouse ($\tau_1 \sim M^2$) results and arises because of the fact that the chain conformation (or stress) is relaxed only by the motion of the chain ends, rather than by the concurrent relaxation of all segments of the chain. As noted at the beginning of the chapter, describing the molecular weight dependences of the various dynamic properties is the first goal of molecular models.

In addition, we can consider the long-time translational diffusion of the chain. At the instant the chain finally escapes the initial tube, one of its ends must just be touching the initial tube at some point. The initial tube and the new tube are uncorrelated random walks, except that they touch in this manner. Therefore in the time τ_{rep} the center of mass of the chain moves a mean-square distance of approximately $\langle h^2 \rangle$, and therefore we can write

$$\langle h^2 \rangle = Nb^2 = 6D_t\tau_{\text{rep}}$$
$$D_t = \frac{1}{3}\frac{N_e}{N}\frac{kT}{N\zeta_0} \quad (11.7.6)$$

Thus D_t decreases as M^{-2}, which is also a much stronger dependence than the Zimm ($D_t \sim M^{-1/2}$) and Rouse ($D_t \sim M^{-1}$) models. This particular prediction of the reptation model inspired a great deal of experimental activity, including the development and application of several new approaches to measuring D_t. The results overall are in reasonable agreement with Equation 11.7.6, except that the M exponent (about -2.3) is slightly stronger than anticipated. An example is shown in Figure 11.21 for hydrogenated 1,4-polybutadiene (essentially polyethylene) measured by different researchers using several different techniques. Two possible reasons for the difference between the experimental and theoretical M dependences will be discussed in Section 11.7.3.

11.7.2 Reptation Model: Viscoelastic Properties

Now that we have the reptation prediction for the longest relaxation time, we can also develop predictions for $G(t)$ and therefore G', G'', and η. At short times (in the transition zone, $t < \tau_e$) the chain sections between entanglements relax in a Rouse-like fashion before the relaxation effectively ceases with $G(t > \tau_e) = G_N$. The reptation model does not really address this regime.

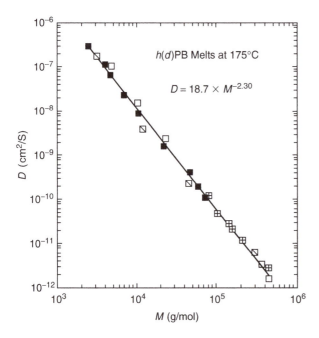

Figure 11.21 Experimental data for the diffusion of hydrogenated (or deuterated) polybutadienes from various sources. (Compiled by Lodge, T.P., *Phys. Rev. Lett.*, 83, 3218, 1999. With permission.)

At longer times, in the rubbery plateau and in the terminal regime, the escape from the tube also involves a spectrum of Rouse-like modes, except that the chain is confined to the tube. The argument above only addressed the longest relaxation time, but to avoid a good deal more mathematics we will just state the result:

$$G(t) = G_N \sum_{\text{odd } p} \frac{8}{\pi^2} \frac{1}{p^2} \exp\left(-\frac{t}{\tau_p}\right) = G_N \sum_{\text{odd } p} \frac{8}{\pi^2} \frac{1}{p^2} \exp\left(-\frac{p^2 t}{\tau_{\text{rep}}}\right) \tag{11.7.7}$$

The time constants for the modes have a p^2 dependence, which echoes the Rouse model, but the amplitude of each mode is further attenuated by a factor of $1/p^2$. Furthermore, only odd numbered modes contribute due to the symmetry of the reptation process; the center of mass does not move for even numbered modes. Consequently, Equation 11.7.7 is not much different from a single exponential decay, dominated by the longest relaxation time. The dynamic moduli, G' and G'', can be obtained from Equation 11.7.7 through the sine and cosine Fourier transforms (Equation 11.3.7):

$$G' = G_N \frac{8}{\pi^2} \sum_{\text{odd } p} \frac{1}{p^2} \frac{(\omega \tau_p)^2}{1 + (\omega \tau_p)^2}$$

$$G'' = G_N \frac{8}{\pi^2} \sum_{\text{odd } p} \frac{1}{p^2} \frac{\omega \tau_p}{1 + (\omega \tau_p)^2} \tag{11.7.8}$$

These predictions are compared with experimental data for a hydrogenated polybutadiene in Figure 11.22. The main discrepancy occurs in $G''(\omega)$ for frequencies in the plateau zone where experimentally the decrease in G'' with ω is rather less rapid than predicted. It will turn out that this difference can be accounted for in much the same way as the difference in M exponents for D_t identified above, but again we will defer this discussion to the next section.

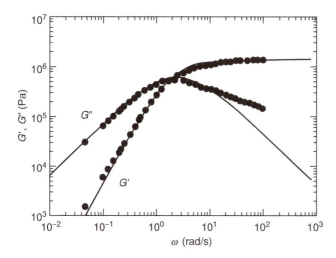

Figure 11.22 Comparison of the basic reptation theory with experiment for the dynamic moduli of hydrogenated polybutadiene melts. The main discrepancy is in the slope of G'' at higher frequencies. (Data from Tao, H., Huang, C.-I., and Lodge, T.P., *Macromolecules*, 32, 1212, 1999.)

Lastly we turn to the steady flow viscosity. This can be obtained by integrating over the modulus (Equation 11.3.4):

$$\eta = \frac{\pi^2}{12} \tau_{rep} G_N \tag{11.7.9}$$

Note that in performing this integration we ignore the modulus for times shorter than τ_e, or in other words, $t = 0$ for the stress relaxation process corresponds to the onset of the rubbery plateau. This approximation is fine, because the time ranges are so different that the early time portion contributes essentially nothing to the integral for a high M chain. The main result of Equation 11.7.9, and arguably the main result of the basic reptation model, is that the viscosity is predicted to scale as M^3. This should be compared with the universal experimental result of $M^{3.4}$, illustrated in Figure 11.1 and Figure 11.18. On the one hand, the fact that such a simple idea can get a nontrivial answer ($\eta \sim M^3$), which is not too different from reality, is very encouraging. On the other hand, the difference is significant. Accordingly, after the following example we briefly consider two omissions of the reptation model that are thought to account for the discrepancies with the experimental results for D_t, G'', and η. The main predictions of reptation were also summarized in Table 11.1.

Example 11.4

Consider the following polymer processing problem. Imagine you are making a plastic hoop by injecting molten polymer into a mold. The molten polymer travels in two directions from the injection point and meets on the opposite side of the hoop. How long do you need to wait for the two liquid streams to merge after they first meet and develop full mechanical strength at the junction? Assume the polymer is polystyrene, with $M \gg M_e$ and a viscosity of 10^5 P at the processing temperature, and use the reptation model. (The answer is of practical importance, in the sense that the faster the part sets, the faster it can be ejected from the mold and the more parts can be made per unit time. And, although the proposed geometry is rather artificial, the issue of healing a polymer–polymer interface is a very general one.)

Solution

The mechanical strength of a polymer part will depend on achieving full interpenetration or entanglement of the various molecules. Assume that we start with two flat polymer surfaces that are brought into contact at time $t = 0$. To achieve full interpenetration, we need to wait until the average chain has diffused a distance of about R_g. We have not been given direct information about M for our polymer, so estimating R_g and D_t (from η) could be tedious. However, in the reptation model τ_{rep} is nothing more than the time to diffuse R_g, and τ_{rep} itself is simply related to η and the plateau modulus, G_N, through Equation 11.7.9. We can take the value of G_N from Table 11.2; it is about 2×10^5 Pa. This gives

$$\tau_{rep} \approx \frac{10^5 \text{ P}}{2 \times 10^5 \text{ Pa}} \frac{1 \text{ Pa s}}{10 \text{ P}} \approx 0.05 \text{ s}$$

where we ignore the numerical prefactor of $\pi^2/12$ in this estimate, and are careful to make the units of viscosity and modulus match.

The answer suggests that interfacial healing will not be a problem in this processing operation, as it will presumably take several seconds to fill the mold, and several more to cool the part sufficiently to remove it from the mold. Note also that the form of this solution is exactly the same as in Example 11.1, where we estimated the flow time from the modulus and the viscosity using the Maxwell model. As a last comment, the issue of interfacial healing is much richer than this analysis reveals. For example, significant mechanical strength at an interface will actually develop after the chains have interpenetrated a distance of only about the tube diameter or entanglement spacing and then it will continue to grow slowly as the chains become completely intermixed. A full description of the evolution of interfacial strength with time therefore requires the complete spectrum of relaxation times.

11.7.3 Reptation Model: Additional Relaxation Processes

The experimental scaling laws $(D_t \sim M^{-2.3}, \eta \sim M^{3.4})$ have stronger M dependences than predicted by the reptation model. Numerically, the experimental results indicate more rapid chain motion than by reptation alone. This is illustrated in Figure 11.23 for the viscosity. For chains containing only a few M_e the viscosity is significantly lower than expected by the model. However, as M increases, the steeper experimental dependence suggests that the experiments might converge on the reptation prediction at very high M. One obvious candidate for this more rapid relaxation is the fact that the reptation model was developed for a chain moving in an array of fixed obstacles. In reality, of course, the obstacles are a way of accounting for the entanglements with other chains, and because all chains are moving, some entanglements should disappear while a test chain is trying to reptate. In the tube language, we could imagine that the tube develops occasional leaks, thereby allowing the chain to escape through the leaks rather than the ends (see Figure 11.24a). This process is known as *constraint release*, and because it accelerates the chain relaxation, it could explain why η is lower and D_t is higher compared to pure reptation. However, a quantitative development of constraint release is complicated, and it is not yet clear how much of the discrepancy between experiment and theory can be accounted for in this manner.

A second correction, termed *contour length fluctuations*, can actually reproduce the experimental M dependences of η and D_t, plus the frequency dependence of $G''(\omega)$, rather well. This process applies even to a chain in an array of fixed obstacles. In the discussion preceding Equation 11.7.8, we argued that relaxation modes for which the center of mass does not move down the tube do not contribute to $G(t)$. In fact, this is not quite true. Imagine that we pin the center of the chain, but let the ends wiggle around. There are "accordion" modes, in which the two ends of the chain penetrate into (and out of) the average tube by a certain amount. If they do so, they can select a different path on their way out, and thus they accelerate the relaxation of the chain ends. This is illustrated in Figure 11.24b. (Of course, the ends must also occasionally fluctuate out of the tube,

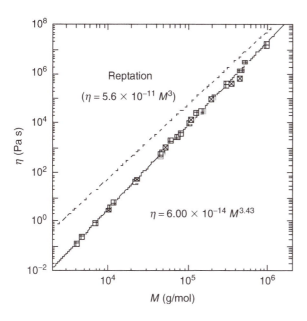

Figure 11.23 Discrepancy between experimental results and the reptation prediction for the viscosity. The data are for hydrogenated polybutadiene at 140°C. (Reported in Tao, H., Lodge, T.P., and von Meerwall, E.D., *Macromolecules*, 33, 1747, 2000. With permission.)

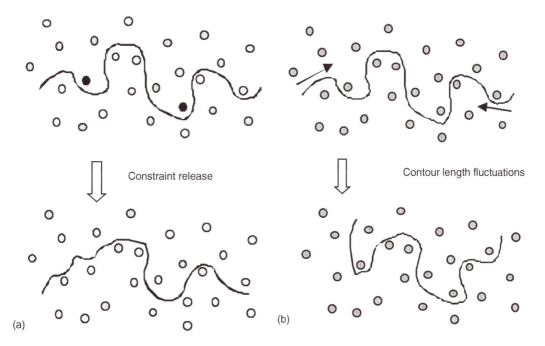

Figure 11.24 Illustration of additional relaxation processes that can bring reptation predictions into closer agreement with experiment. (a) The two entangling chains indicated by solid circles diffuse away, allowing the test chain to move sideways by the process of constraint release. (b) The ends of the chain can move some distance into the tube by Brownian motion, without moving the center of mass, and therefore relax some stress without reptation; these movements are called contour length fluctuations.

but that has no effect on the escape from the tube if we consider the center to be pinned.) As this process is a random fluctuation, we expect that the characteristic distance the ends can penetrate should be roughly the square root of the total, i.e., a distance of about $(dN/N_e)^{1/2}$. This is a negligible fraction of the tube for very long chains, but turns out to be significant for most experimentally accessible chain lengths.

11.8 Aspects of Experimental Rheometry

In Chapter 9 we described two flow geometries in common use for the measurement of the viscosity in steady flow, namely the capillary (Poiseuille flow) and the concentric cylinder (Couette flow). Although both of these can serve for transient and dynamic measurements, at least two other geometries are also commonly employed, and particularly for the higher viscosities associated with molten polymers. These are represented by the parallel plate (shear sandwich) rheometer for dynamic measurements and the cone-and-plate rheometer capable of dynamic, transient, and steady shear measurements. Both of these will be described briefly below and then we conclude by identifying several important general issues that arise in rheometry.

11.8.1 Shear Sandwich and Cone and Plate Rheometers

The shear sandwich geometry is illustrated in Figure 11.25a. A central flat plate is driven up and down between two fixed, parallel plates and the sample is contained within the two narrow gaps on each side of the moving plate. This arrangement is nothing more than an experimental realization of the parallel surface configuration used in Figure 9.1 and Figure 11.1. It is made possible by the use of an oscillatory strain; clearly, steady flow cannot be achieved with this design. As drawn in Figure 11.25a, the plates have an area $A = Lh$, and the total volume of sample contained in the two gaps is $2dLh$. If the moving plate has a displacement along the x axis of $x_0 \sin \omega t$, its velocity v_x will be $x_0 \omega \cos \omega t$. Furthermore, the fluid velocity at each fixed plate will be zero (under the no-slip assumption). If the velocity profile across each gap is linear, as is the case for sufficient small gap widths d, then the velocity at any point across the gap is $(v_x/d)y$, where y is the distance from the fixed surface. Recalling the discussion in Section 9.1 and Section 11.1, the shear rate in the gap is therefore given by

$$\dot{\gamma} = \frac{d\gamma}{dt} = \frac{dv_x}{dy} = \frac{v_x}{d} = \frac{x_0}{d}\omega \cos \omega t \tag{11.8.1}$$

So in this case, both the shear rate and the shear strain are oscillatory functions of time. The shear rate is not, however, a function of position in the gap, and therefore by Equation 9.1.3 or Equation 11.1.1 the stress is the same everywhere in the sample. This geometry is less useful for lower viscosity fluids because they tend to leak out of the gap.

The primary limitation of the shear sandwich is the exclusion of steady flow experiments. To overcome this, rotational rheometers are the common solution. The Couette geometry described in Section 9.4.2 is one approach; another is the cone and plate, illustrated in Figure 11.25b. In this apparatus, the sample is confined in the narrow gap between a flat, fixed plate, and a conical piece, which makes a small angle θ (typically 1°–3°) with the flat plate. The cone rotates around the vertical axis with rotation rate Ω (rad/s), sweeping out an angle ϕ with time. The key feature of this design is that the shear rate is homogeneous throughout the sample. To see how this comes about, consider the strain rate at any radial distance r. The width of the gap d increases linearly with r, namely $d = r \tan \theta \approx r \sin \theta \approx r\theta$ for small θ (also in radians). The instantaneous linear velocity of the moving cone in the tangential direction, v_ϕ, also increases linearly with r: $v_\phi = r\Omega$. The net result is that for any value of r the instantaneous shear rate is $v_\phi/d = \Omega/\theta$.

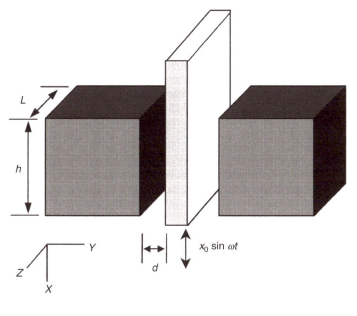

(a)

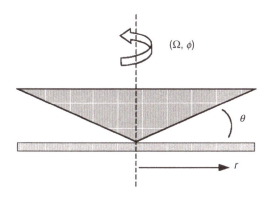

(b)

Figure 11.25 Illustration of the (a) shear sandwich and (b) cone and plate geometries.

11.8.2 Further Comments about Rheometry

There exist several texts in the field of rheometry that interested readers may consult for much fuller treatments of this important topic. There are also several commercial ventures that specialize in rheological equipment, so that a wide variety of instruments are readily accessible. We conclude this section with a brief listing of some of the important issues that can arise in choosing and using a rheometer:

1. All rheometers measure a force (or torque) related to the stress and a displacement (or velocity) related to the strain (or strain rate). Some rheometers are strain-controlled, meaning that the operator inputs a desired strain amplitude and the instrument measures the stress required to achieve that strain; others are stress-controlled and impose a stress and monitor the strain. Strain control is necessary to measure the stress relaxation

modulus, whereas stress control is required for a creep experiment. In the linear viscoelastic regime, either approach should be suitable, given the interrelationships among the various viscoelastic functions outlined in Section 11.3. However, in many cases it is the nonlinear properties that are of interest and then the choice of control mode becomes more important.

2. In the measurement of force and displacement, *transducers* are required to convert the measured quantity into an electrical current or voltage. The force transducer is particularly important, in terms of sensitivity, range, speed of response, linearity, and reproducibility. The data in Figure 11.1, Figure 11.2, Figure 11.16, and Figure 11.18, for example, illustrate how the modulus and viscosity can vary over many orders of magnitude and no single transducer can cover this entire range. The measured force for a low viscosity fluid can be increased somewhat by increasing the area of the moving surface and by increasing the strain rate (and vice versa for a force that is too high to measure reliably); but in the end, a range of transducers or even a range of instruments are required to cover the full spectrum of materials from dilute solutions to highly entangled melts.

3. In the dynamic mode, the accessible frequency range is also important. Most commercial instruments can manage approximately 0.001–10 Hz, and some custom instrumentation can achieve kHz or even MHz frequencies. However, the data in Figure 11.14 extend over 10–15 orders of magnitude along the time or frequency axis, well beyond the capability of any single instrument. How are such measurements accomplished? The answer will be given in the next chapter, but it involves the use of variable temperature, and the *principle of time–temperature superposition*.

4. We have not discussed the temperature dependence of the viscoelastic properties in this chapter; that subject, too, is deferred to Chapter 12. However, the temperature dependence is generally very strong, and so accurate rheometry requires good temperature control. Unfortunately, this is not so easy to attain in many cases, particularly far above or far below room temperature. Part of the difficulty arises from having the sample and moving parts surrounded by air or inert gas, neither of which have good heat transfer characteristics. A related problem is viscous heating, which arises from the substantial dissipation of energy at high viscosities or high flow rates (see Equation 9.1.7).

5. Rheometric experiments integrate the response over the entire sample, in the sense that only a single measurement of force is used to determine the stress. For example, if the strain field is inhomogeneous for any reason, the stress will be different at different locations in the sample, but the measurement will not take that into account. The inhomogeneity might arise from temperature gradients, sample nonuniformity, or secondary flows induced at high flow rates, but in addition it is always present under the category of edge effects. In every geometry there is a sample surface that is not in contact with either the fixed or moving surfaces of the apparatus, and the flow profile in the vicinity of this extra surface must differ from that assumed in the calculation of the stress. Schemes have been developed for partially correcting edge effects in all of the common rheometer geometries.

11.9   Chapter Summary

In this chapter we have examined the linear viscoelastic properties of flexible polymers, both in dilute solution and in the molten state. We have defined the basic concepts and experimental approaches, emphasizing shear flow. Molecular models have been presented for both dilute solutions and melts and they are quite successful in describing the experimental phenomena. The main points are as follows:

1. Polymer liquids generally show viscoelastic behavior, i.e., a response to an imposed deformation that is intermediate between the viscous flow of liquids and the elastic deformation of solids. This response can be characterized by a variety of material functions, such as the steady flow viscosity, stress relaxation modulus, creep compliance, and dynamic modulus or dynamic viscosity.

2. The basic character of viscoelastic response is revealed by the simple Maxwell and Voigt mechanical models. In general, a polymer liquid behaves more like an elastic solid at short times or high frequencies and more like a viscous liquid at long times or low frequencies. The demarcation between these limits is determined by the relaxation times of the material.

3. In the limit of linear response, i.e., sufficiently small strain amplitudes and strain rates such that the material functions do not depend on the strain amplitude or rate, the Boltzmann superposition principle provides a direct route to calculate the viscosity, dynamic moduli, and recoverable compliance from the stress relaxation modulus.

4. The BSM of Rouse and Zimm provides a molecular explanation for the viscoelastic response of polymers. The Zimm version, which includes intramolecular hydrodynamic interactions, is very successful in dilute theta solutions, and variable solvent quality can be incorporated by a simple dynamic scaling argument. The Rouse version, without hydro-dynamic interactions, applies very well to low molecular weight molten polymers.

5. For high molecular weight polymers in concentrated solutions or melts, the phenomenon of entanglement dominates the viscoelastic properties. The moduli exhibit four regimes of behavior denoted glassy, transition zone, rubbery plateau, and terminal, as a function of increasing time or decreasing frequency. In the rubbery plateau the liquid behaves as a soft solid, with a modulus similar to a lightly cross-linked rubber (Chapter 10). A characteristic molecular weight between entanglements is inferred, and may be predicted based solely on knowledge of the chain flexibility and density.

6. The reptation model provides a physically appealing description of chain motion and stress relaxation in entangled polymers. The theoretical predictions for diffusion and viscosity do not quite match the experimental results, but good agreement can be obtained when the additional processes of constraint release and contour length fluctuations are included.

Problems

1. The following are approximate σ (in dyn/cm^2) versus $\dot{\gamma}$ data for three different samples of polyisoprene in tetradecane solutions of approximately the same concentration:[†]

M_w (g/mol)	1.61×10^6	1.95×10^6	1.45×10^6
	Linear	Four-armed star	Six-armed star
c (g/cm^3)	0.0742	0.0773	0.0078
$\dot{\gamma}$ (s^{-1})	$\sigma \times 10^{-3}$	$\sigma \times 10^{-3}$	$\sigma \times 10^{-3}$
0.6	0.7	—	—
0.8	0.9	—	—
1	1	0.3	—
2	2	0.6	—
4	4	1	0.15
8	5	2	0.3
10	6	3	0.4
20	7	4	0.8
60		7	2
100		8	4

From plots of these data, estimate the Newtonian viscosity of each of the solutions and the approximate rate of shear at which non-Newtonian behavior sets in. Are these two quantities better correlated with the molecular weight of the polymer or the molecular weight of the arms?

[†]W.W. Graessley, T. Masuda, J.E.L. Roovers, and N. Hadjichristidis, *Macromolecules*, 9, 127 (1976).

2. Newtonian viscosities of polystyrene samples of different molecular weights were determined at 200°C by Spencer and Dillon.[†] Use these data to determine the exponent of M in the relationship between η and M.

$M \times 10^{-3}$	η (P)
86	3.50×10^3
162	4.00×10^4
196	6.25×10^4
360	4.81×10^5
490	1.89×10^6
508	1.00×10^6
510	1.64×10^6
560	3.33×10^6
710	6.58×10^6

3. In a dynamic experiment with $\gamma(t) = \gamma_0 \sin(\omega t)$ the power loss per cycle of oscillation is given by

$$\int_{\omega t=0}^{\omega t=2\pi} \sigma \, d\gamma$$

    (a) Evaluate the power loss per cycle if the material is a Hookean solid: $\sigma = G\gamma$.
    (b) Evaluate the power loss per cycle if the material is a Newtonian liquid: $\sigma = \eta \, (d\gamma/dt)$.
    (c) Briefly comment on the significance of these results.

4. Using complex notation, derive the Maxwell model predictions for the dynamic shear modulus, $G^*(\omega)$, and the dynamic viscosity, $\eta^*(\omega)$, and the relations between the elastic and viscous components of each; assume a dynamic strain, $\gamma^* = \gamma_0 \exp(i\omega t)$.

5. For polystyrene ($M = 600,000$) at 100°C, the following values describe the creep compliance, $J(t)$, at long times:[‡]

log $J(t)$ (m²/N)	−1.8	−1.4	−1.0	−0.6	−0.2	+0.1
log t (s)	12.6	13.0	13.4	13.8	14.2	14.6

Use Equation 11.2.8 to evaluate the viscosity of the polymer at this temperature. Then use Equation 11.7.5 and Equation 11.7.9 and Table 11.2 to estimate the segmental relaxation time for polystyrene at this temperature.

6. Find the relation between the phase angles δ and ψ given in Equation 11.2.21 and Equation 11.2.23.

7. Estimate the high molecular weight values of the dimensionless group $D\tau_1/R_g^2$ for dilute flexible chains in a theta solvent, and for the same polymer in the melt, using appropriate theoretical models.

8. Sketch a careful log–log plot of the Zimm theory prediction for $[\eta']_R$ and $[\eta'']_R$ versus $\omega\tau_N$, where τ_N is the shortest relaxation time of the model. Mark the decades on the axes to make the curves realistic. Perform two pairs of curves on the same axes, one for a large N, i.e., a high molecular weight polymer and one for a small N polymer to indicate the effects of molecular weight on the response.

[†]R.S. Spencer and R.E. Dillon, *J. Colloid Sci.*, 4, 241 (1949).
[‡]D.J. Plazek and V.M. O'Rourke, reported in J.D. Ferry, *Viscoelastic Properties of Polymers*, 3rd ed., Wiley, New York, 1980.

9. Propose two possible reasons why the Zimm model fails to exhibit shear-thinning behavior at high shear rates.

10. Estimate the longest relaxation time, τ_1, for polystyrenes with $M = 10^5$ and 10^6, in cyclohexane at 35°C and toluene at 25°C. There is no single correct way to do this, so be sure to identify any assumptions you make.

11. Pearson, et al.[†] reported measurements of the viscosity and diffusivity of narrow distribution polyethylenes at 175°C. The data are given below. Prepare log–log plots of η versus M and D versus M and compare with expectations based on the Rouse and reptation models. How well do the data agree with these theories? What do you propose for the origin of any discrepancies? What is M_c for PE? Prepare a plot of the product $D\eta$ versus M. Does this agree better with theory in the putative Rouse regime? Why is this the case?

M (g/mol)	$10^6\, D$ (cm^2/s)	η (P)
506	6.6	0.0185
590	5.4	0.0232
618	4.8	0.0248
695	3.5	0.035
1,280	1.4	0.092
2,390	0.35	0.338
3,310	0.15	0.873
4,100	0.093	1.63
13,600	0.012	37.7
32,100	0.0020	1,100
119,600	0.00013	125,000

12. Consider the tracer diffusion coefficient of polystyrenes in the melt, at 176°C. At this temperature, $M_e = 13,000$ g/mol, $M_0 = 104$ g/mol, and $kT/\zeta = 10^{-9}$ cm^2/s. What is D_t for $M = 13,000$ under these conditions? What is D_t for $M = 65,000$ dissolved as a tracer in a melt with $M = 13,000$? Suppose the $M = 13,000$ matrix polymer was end-functionalized at both ends, so that it could be cross-linked. Suppose that a small amount of bifunctional agent was added and reacted to completion; the $M = 13,000$ polymers were therefore all end-linked to form very long linear chains. What would D_t be for the $M = 65,000$ tracer then? Suppose that the cross-linking agent was tetrafunctional, so that a complete network was formed. What would D_t be for the $M = 65,000$ tracer in this case?

13. Imagine you have a narrow distribution sample of 1,4-polybutadiene with $M_w = 54,000$ g/mol. On one set of logarithmic axes, sketch the shape of the stress relaxation modulus $G(t)$ versus time that you would expect to see for this polymer in the melt. Extend your curve to cover the full range of viscoelastic response, and estimate any numerical values that you can. (For polybutadiene, $\rho = 0.9$ g/mL, $b = 6.9$ Å, $M_e = 1800$ g/mol.) Then add two more curves, corresponding to the expected response after cross-linking: (i) 0.05% of the monomers and (ii) 1% of the monomers. Indicate clearly which curve is which, and briefly explain why the three curves differ from each other (if they do), and why in some respects they are the same. (It may be helpful to recall Section 10.1.)

14. Use the correlation of plateau modulus and packing length developed in Section 11.6.2 to predict the molecular weight between entanglements, M_e, for poly(vinyl acetate). Table 6.1 provides useful data for the chain dimensions, and the density is approximately 1.08 g/cm^3. How does your value compare with the experimentally reported value of 7000 g/mol?

[†]D.S. Pearson, et al., *Macromolecules*, 20, 1133 (1987).

15. Use the reptation model to estimate the longest relaxation time for the linear polyisoprene in Problem 1. How does the inverse relaxation time compare with the onset of non-Newtonian response? Explain.

References

1. Maxwell, J.C., *Phil. Trans.* 157, 49 (1867).
2. Voigt, W., *Ann. Phys.* 47, 671 (1892).
3. Boltzmann, L., *Ann. Phys. Chem.* 7, 624 (1876).
4. Rouse, P.E., Jr., *J. Chem. Phys.* 21, 1872 (1953).
5. Zimm, B.H., *J. Chem. Phys.* 24, 269 (1956).
6. Sammler, R.L. and Schrag, J.L., *Macromolecules* 21, 1132 (1988); Lodge, A.S., and Wu, Y.-J., *Rheol. Acta* 10, 539 (1971).
7. Doi, M. and Edwards, S.F., *J. Chem. Soc., Faraday Trans.* 2, 74, 1789, 1802, 1818 (1978).
8. de Gennes, P.G., *J. Chem. Phys.* 55, 572 (1971).
9. Fetters, L.J., Lohse, D.J., Richter, D., Witten, T.A., and Zirkel, A., *Macromolecules* 27, 4639 (1994).

Further Readings

Bird, R.B., Armstrong, R.C., and Hassager, O., *Dynamics of Polymer Liquids*, Volume 1, 2nd ed.; Bird, R.B., Curtiss, C.F., Armstrong, R.C., and Hassager, O., *Dynamics of Polymer Liquids*, Volume 2, 2nd ed., Wiley, New York, 1987.
Doi, M. and Edwards, S.F., *The Theory of Polymer Dynamics*, Clarendon Press, Oxford, 1986.
Ferry, J.D., *Viscoelastic Properties of Polymers*, 3rd ed., Wiley, New York, 1980.
Graessley, W.W., *Polymeric Liquids and Networks: Structure and Properties*, Garland Science, New York, 2003.
Macosko, C.W., *Rheology: Principles, Measurements, and Applications*, VCH, New York, 1994.
Morrison, F.A., *Understanding Rheology*, Oxford University Press, New York, 2001.
Rubinstein, M. and Colby, R.H., *Polymer Physics*, Oxford University Press, New York, 2003.
Shaw, M.T. and MacKnight, W.J., *Introduction to Polymer Viscoelasticity*, 3rd ed., Wiley, Hoboken, NJ, 2005.
Walters, K., *Rheometry*, Chapman and Hall, London, 1975.

12

Glass Transition

12.1 Introduction

Here we resume a sequence of three chapters that treat polymers in the solid state. In Chapter 10 we examined the formation of polymer networks and rubber elasticity. In Chapter 13 we will consider crystallinity in polymers and the associated crystallization transition from the high temperature, liquid state. In this chapter we take up the subject of the glass transition, whereby a polymer liquid is cooled in such a way as to solidify without adopting a crystalline packing. Among the three classes of polymer solid—network, crystal, and glass—the glassy state is the most universal; relatively few polymers are used to form networks; a significant fraction can never crystallize, but all can form glasses. Furthermore, all three topics are central to understanding the utility of polymer materials. In Chapter 9 and Chapter 11 we covered some properties of polymer liquids, especially those pertaining to flow. In almost all cases, polymers are synthesized, characterized, and processed in the liquid state, and consequently the material in Chapter 9 and Chapter 11 represents a foundation for many diverse areas of polymer science. However, most polymer applications rely on the properties in the solid state; consequently, Chapter 10, Chapter 12, and Chapter 13 provide the background for understanding how polymers are chosen or developed for one application or another.

12.1.1 Definition of a Glass

To begin, we need a working definition of a glass. A reasonable one may be simply stated: a glass is an *amorphous solid*. By amorphous, we mean that there is no long-range order or symmetry in the packing of the molecules. In this sense, the structure of a glass looks very much like the structure of a liquid. However, a glass is a solid: it does not flow over relevant timescales. Although the preceding statement is apparently innocent, the phrase "relevant timescales" hints at a fundamental complexity: if we are talking about an equilibrium state, time should play no role. What we will see is that glasses are generally nonequilibrium states, metastable in a sense, and kinetic issues will be of central importance. When a polymer liquid is cooled, the density increases and the molecular relaxation times increase. Over some range of temperature, the molecular motion will become so slow that an equilibrium packing of the molecules cannot be attained during the experiment. When this happens we say that the sample has undergone the glass transition, or has *vitrified*, and we associate with each polymer a *glass transition temperature*, T_g. The value of T_g is the single most important characteristic in choosing a polymer for a given application. It must lie significantly above any temperature at which we intend to use the polymer as a solid, but below any temperature at which we intend to process the polymer as a liquid. As we will see, in practice T_g is not a thermodynamic property of the polymer, and it can adopt a range of values for a given polymer, but nevertheless it is an extremely useful parameter.

Although we will cover crystallization in polymers in the next chapter, a few pertinent points are appropriate here. Upon cooling from the liquid state, a polymer might either crystallize or turn into a glass; thus these two transitions are, in a sense, in competition. Polymers with irregular microstructure, such as atactic vinyl polymers or mixed-microstructure polydienes, cannot crystallize, because the lack of a regular structure at the monomer length scale prevents the formation of a unit cell. Therefore crystallization is not an option for many polymers. For polymers that do crystallize, the kinetics associated with finding the correct packing are sufficiently slow that only a fraction of the molecules succeed; consequently the material is termed semicrystalline. The remaining fraction is amorphous, and can undergo a glass transition. Thus the glass transition temperature plays an important role even for crystallizable polymers.

The remainder of the chapter is organized as follows. We conclude this introductory section with a more detailed comparison of the glass–liquid and crystal–liquid transitions, focussing on the temperature dependence of the specific volume (or density). In the second section, we consider the glass transition from a thermodynamic point of view, especially emphasizing the fundamental question of whether there is actually a thermodynamic transition hidden beneath the kinetically dominated T_g. The third section describes the most common experimental routes to characterizing T_g, which leads into the fourth section concerning kinetic models of the transition. The most popular of these, the free volume model, provides a simple basis for understanding one of the most dramatic consequences of the glass transition, namely the very strong temperature dependence of molecular relaxation above T_g. This temperature dependence underlies the utility of the principle of time–temperature superposition, which is of central importance to the experimental characterization of viscoelasticity. This topic is covered in Section 12.5. The chapter concludes with a discussion of how T_g can be modified (Section 12.6), and an introduction to the properties of glassy polymers (Section 12.7).

12.1.2 Glass and Melting Transitions

In the preceding chapter, time (or frequency) was the primary independent variable under consideration. We saw that at short times of observation, polymers exhibit high values of the modulus, roughly three or four orders of magnitude higher than those shown in the rubbery state of these materials. The transition between the two values of the modulus occurs over a range of time at fixed temperature in the so-called transition zone of viscoelasticity. It turns out that temperature variation at fixed time can produce changes in mechanical properties that parallel those resulting from shifts of timescale. This change in mechanical behavior signals the glass transition, and when monitored during temperature variation it occurs near the glass transition temperature. We shall return to an examination of the equivalency of time and temperature with respect to effects on mechanical properties later in this chapter. For now, however, it is desirable to consider some other properties of matter that change near T_g; although a variety of observables are available, we shall emphasize volume.

Figure 12.1 illustrates schematically the range of possibilities for the variation in specific volume, V_{sp}, with temperature. Remember that V_{sp} is the reciprocal of the density; V_{sp}, rather than density, is chosen to describe these changes in anticipation of the "free volume" interpretation to be presented in Section 12.4. Path ABDG in Figure 12.1 shows how V_{sp} changes upon freezing a low molecular weight compound. (A few substances—water is the best known example—occupy a larger volume per unit mass in the solid state than in the liquid, and for these the transition at the melting point T_m would correspond to a jump rather than a drop in volume.) What is significant is that this transition occurs at a single temperature, the melting point T_m. The slopes of AB and DG reflect the coefficients of thermal expansion of the liquid, α_l, and the crystalline solid, α_c, respectively. These coefficients are approximately independent of temperature (i.e., AB and DG are nearly straight lines) but the main point is that V_{sp} is different for solids and liquids; it shows a discontinuity at the melting point.

An entirely different pattern of behavior is shown along lines ABHI. In this case there is no discontinuity at T_m. The line AB, which characterizes the liquid, changes slope at T_g to become HI. Actually, the change in slope occurs over a range of temperatures (about 20°C), as suggested by Figure 12.1, but extrapolation of the two linear portions permits a single T_g to be defined. The region HI characterizes the glassy state, and the threshold for its appearance is the glass transition temperature. In the region BH, the liquid is said to be *supercooled*.

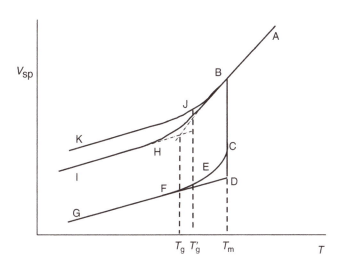

Figure 12.1 Schematic illustration of possible changes in the specific volume of a polymer with temperature. See text for a description of the significance of the various lettered features.

Each of the two paths we have discussed could describe the behavior of either high or low molecular weight compounds. This is not to say, however, that each is equally probable for the two classes of compounds. For most low molecular weight materials, special effort must be made to suppress crystallization and achieve glass formation. With polymers, on the other hand, the glassy state is always obtained, whether a particular polymer is crystallizable or not. The mere fact that molecular structure allows the possibility of crystal formation does not mean that the latter occurs rapidly or completely. The line ABCEFG in Figure 12.1 describes the situation of a partly crystalline, partly amorphous polymer. At T_m crystallization begins and the characteristic discontinuity in specific volume occurs. The sharpness of T_m is not as pronounced for polymers as for low molecular weight compounds, as evidenced by the trailing off between C and E. In the region EF the volume contraction reflects the supercooling of the amorphous portion of the polymer. The change in slope between EF and FG occurs at T_g, just as it would in the absence of crystallization. If partial crystallization occurs, the amount of amorphous material is decreased and the change in slope at T_g may be harder to detect in this case.

The line ABJK in Figure 12.1 is a displaced variation of ABHI in which AB is a liquid, BJ is a supercooled liquid, and JK is a glass. The experimental variable that causes region JK to be offset from HI is the cooling rate, ABJK being the route for the more rapidly cooled polymer. Since T_g is identified from the change in slope, it is apparent that T_g is also displaced, appearing at a higher temperature (T_g' in the figure) for higher rates of cooling. The change in slope that defines T_g may also be viewed as the first departure from the behavior extrapolated from the liquid state. In other words, although the supercooled liquid is not at thermodynamic equilibrium, its specific volume follows the same T dependence as the equilibrium liquid. Thus the glass transition on cooling represents the first obvious departure from equilibrium, and it is intuitively reasonable that the higher the rate of cooling, the sooner this departure will become apparent.

To summarize some basic observations for polymers:

1. Above T_m the material is liquid. The zero-shear viscosity depends strongly on the molecular weight of the polymer (see Chapter 11) and temperature (see Section 12.4) but it would be considered high by all standards.
2. Between T_m and T_g, depending on the regularity of the polymer and on the experimental conditions, this domain may be anything from almost 100% crystalline to 100% amorphous. The amorphous fraction, whatever its abundance, behaves like a supercooled liquid in this region.

Table 12.1   Representative Values of the Glass Transition Temperature and the Melting Temperature (for Stereoregular Forms, where Applicable) for Some Common Polymers

Polymer	T_g (°C)	T_m (°C)
Poly(dimethylsiloxane)	−123	− 40
Polyethylene	−120	135
1,4-Polybutadiene (*cis*)	−112	12
Polyisobutylene	−75	44
1,4-Polyisoprene (*cis*)	−70	28
Poly(ethylene oxide)	−70	65
Polypropylene	−10	188
Poly(vinyl acetate)	30	—
Poly(hexamethylene adipamide)	50	265
Poly(ethylene terephthalate)	70	265
Poly(vinyl alcohol)	90	240
Poly(vinyl chloride)	90	270
Polystyrene	100	240
Poly(methyl methacrylate)	110	183
Poly(tetrafluoroethylene)	130	330
Polycarbonate of bisphenol A	150	330
Poly(oxy-2,6-dimethyl-1,4-phenylene)	210	310
Poly(*p*-phenylene terephthalamide)	240	325

3. Below T_g the material is hard and rigid with a coefficient of thermal expansion equal to roughly half that of the liquid. With respect to mechanical properties, the glass is closer in behavior to a crystalline solid than to a liquid. In terms of molecular order, however, the glass more closely resembles the liquid. In this temperature region, the noncrystalline fraction acquires the same glassy properties it would have if the crystallization had been suppressed completely.

4. The location of T_g depends on the rate of cooling. In principle the location of T_m is not subject to this variability, but in fact, the degree of crystallinity does depend on the conditions of the experiment, as well as on the nature of the polymer. For example, if the rate of cooling exceeds the rate of crystallization, there may be no observable change at T_m, even for a crystallizable polymer (see Chapter 13).

The foregoing description introduces the phenomena with which we shall be dealing in this chapter. As noted above, both high and low molecular weight compounds are capable of displaying these effects, but the chain structure of the polymer molecules is responsible for the reversed probabilities. The specific identity of the polymer anchors these transitions to some particular region of the temperature scale; a list of representative values of T_g and T_m is provided in Table 12.1. The regularity of the microstructure of the polymer molecule, along with experimental conditions, determines the extent of crystallization. The glassy state is thus seen as a lowest common denominator shared by all polymers, because 100% crystallinity is virtually impossible. This promotes T_g to the position of importance assumed by T_m for low molecular weight compounds. The fact that the mechanical properties undergo such profound change at T_g also contributes to the significance of this parameter.

12.2   Thermodynamic Aspects of the Glass Transition

The kinetic nature of the experimental glass transition was noted in the previous section, but it is nevertheless instructive to consider the possibility of a thermodynamic description of the transition that occurs near T_g. Most phase equilibria in common experience, such as boiling and melting, are examples of what are called *first-order transitions*. There are other, less familiar but also well-known transitions in nature that are not first order. The disappearance of ferromagnetism at a particular temperature (called the Curie point) is an example of such a transition. Rather than the discontinuities in S, V, and H characteristic of first-order transitions, these variables merely exhibit a change in slope with increasing temperature. Since this is similar to the behavior near T_g, it is important to consider whether the glass transition is actually a *second-order phase transition*, or, perhaps, whether the kinetically affected experimental transition actually masks an underlying thermodynamic transition.

12.2.1 First-Order and Second-Order Phase Transitions

There is no discontinuity in volume at the Curie point, but there is a change in the temperature coefficient of V, as evidenced by a change in slope. To understand why this is called a second-order transition, we begin by recalling the definition of some relevant physical quantities:

1. The coefficient of thermal expansion α:

$$\alpha \equiv \frac{1}{V}\left(\frac{\partial V}{\partial T}\right)_p \tag{12.2.1}$$

2. The isothermal compressibility κ:

$$\kappa \equiv -\frac{1}{V}\left(\frac{\partial V}{\partial p}\right)_T \tag{12.2.2}$$

Since V experiences a change of slope at the second-order transition, i.e., $(\partial V/\partial T)_p$ and $(\partial V/\partial p)_T$ have different values on each side of the transition, it is α and κ that show the discontinuities at the second-order transition rather than V itself. The term second order comes about because the quantities may be written as second derivatives of the free energy, G, as follows. Recalling Equation 7.1.3 and Equation 7.1.4, here applied to a one-component system

$$V = \left(\frac{\partial G}{\partial p}\right)_T, \quad S = -\left(\frac{\partial G}{\partial T}\right)_p \tag{12.2.3}$$

we can expand Equation 12.2.1 and Equation 12.2.2 to obtain

$$\alpha = \frac{1}{V}\left(\frac{\partial V}{\partial T}\right)_p = \frac{1}{V}\frac{\partial}{\partial T}\left[\left(\frac{\partial G}{\partial p}\right)_T\right]_p \tag{12.2.4}$$

and

$$\kappa = -\frac{1}{V}\left(\frac{\partial V}{\partial p}\right)_T = -\frac{1}{V}\left(\frac{\partial^2 G}{\partial p^2}\right)_T \tag{12.2.5}$$

Figure 12.2a and Figure 12.2b describe a second-order transition schematically, in terms of V, S, α, and κ. By extension, an nth-order phase transition is associated with discontinuities in nth-order derivatives of the free energy.

Another useful quantity in this context is the heat capacity at constant pressure, C_p:

$$C_p = \left(\frac{\partial H}{\partial T}\right)_p = T\left(\frac{\partial S}{\partial T}\right)_p = -T\left(\frac{\partial^2 G}{\partial T^2}\right)_p \tag{12.2.6}$$

where the last form exploits Equation 12.2.3. From this relation it is apparent that the heat capacity should also be discontinuous at a second-order transition, whereas the enthalpy (H) should behave

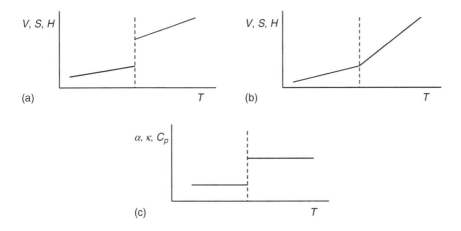

Figure 12.2 Schematic illustration of the behavior of V, S, and H at a (a) first-order and (b) second-order phase transition, and of (c) α, κ, and C_p at a second-order transition.

like V and S and show only a change in slope. In the following section the importance of the heat capacity in the calorimetric determination of T_g will become apparent. The entropy term in Equation 12.2.6 can be inverted to provide another useful relation that shows how measurements of heat capacity versus temperature can be used to determine the entropy:

$$S(T_2) - S(T_1) = \int_{T_1}^{T_2} \frac{C_p}{T}\, dT = \int_{T_1}^{T_2} C_p\, d\ln T \tag{12.2.7}$$

The behavior of these various thermodynamic functions at first-order and second-order transitions is compared schematically in Figure 12.2.

The Clapeyron equation is a well-known thermodynamic relation that applies to first-order transitions:

$$\left(\frac{dp}{dT}\right)_{1st} = \left(\frac{\Delta S}{\Delta V}\right)_{1st} \tag{12.2.8}$$

This expression relates the variation of the pressure–temperature coordinates of a first-order transition (i.e., the phase boundary) in terms of the changes in S and V that occur there. The Clapeyron equation cannot be applied to a second-order transition because ΔS and ΔV would be zero and their ratio undefined. However, we may apply L'Hôpital's rule to both the numerator and denominator of the right-hand side of Equation 12.2.8 to establish a limiting value of dp/dT. In this procedure we may differentiate either with respect to p,

$$\left(\frac{dp}{dT}\right)_{2nd} = \frac{(\partial\Delta S/\partial p)_{T,2nd}}{(\partial\Delta V/\partial p)_{T,2nd}} = \left(\frac{\Delta\alpha}{\Delta\kappa}\right)_{2nd} \tag{12.2.9}$$

or T

$$\left(\frac{dp}{dT}\right)_{2nd} = \frac{(\partial\Delta S/\partial T)_{p,2nd}}{(\partial\Delta V/\partial T)_{p,2nd}} = \left(\frac{\Delta C_p}{TV\Delta\alpha}\right)_{2nd} \tag{12.2.10}$$

to generate some additional useful expressions. All of the Δ's in these equations refer to the difference in the value of the variable from one side (prime) of the transition temperature to the other (double prime):

$$\Delta\alpha = \alpha' - \alpha''; \quad \Delta\kappa = \kappa' - \kappa''; \quad \Delta C_p = C_p' - C_p'' \tag{12.2.11}$$

The following example illustrates how results like these can be applied.

Example 12.1

On the assumption that the glass transition is a second-order thermodynamic transition, estimate the pressure dependence dT_g/dp of T_g using the following data for poly(vinyl chloride): $T_g = 347$ K, $V_{sp} = 0.75$ cm^3/g, $\Delta\alpha = 3.1 \times 10^{-4}$ K^{-1} and $\Delta C_p = 0.068$ cal/ K/g.[†]

Solution

Invert Equation 12.2.10 and substitute. The ratio of gas constants is convenient for unit conversions:

$$\frac{dT_g}{dp} = \frac{T_g V \Delta\alpha}{\Delta C_p} = \frac{(347 \text{ K})(0.75 \text{ cm}^3 \text{ g}^{-1})(3.1 \times 10^{-4} \text{ K}^{-1})}{(0.068 \text{ cal g}^{-1} \text{ K}^{-1})} \times \frac{1.99 \text{ cal}}{82 \text{ atm cm}^3} = 0.029 \text{ K atm}^{-1}$$

This quantity has been measured directly to be 0.016 K/atm. Note that a pressure change of about 60 atm is required to change T_g by 1 K. Note also that the stated value of T_g is somewhat different from that given in Table 12.1, underscoring the variability from sample to sample or from technique to technique.

Despite these useful thermodynamic relationships, it is clear that the glass transition in practice is not truly a second-order phase transition. A moment's reflection reveals that the source of this reservation is the doubt about the state of equilibrium for the glass transition. Implicit throughout the thermodynamic arguments above has been the notion that the phases on either side of a transition are in thermodynamic equilibrium. This enters the mathematical formalism from the start: Equation 12.2.3 assumes that the free energy of a phase is described by only two variables, in this case p and T. Although the glass transition is certainly affected by p and T, it is also dependent on the time of observation, as indicated in Figure 12.1. Because of this time dependence, the experimental glass transition must involve more than a simple second-order transition. In terms of stability (recall Section 7.5) the equilibrium liquid (above T_m) and the crystalline state (below T_m) are stable, meaning that the free energy is at a minimum. The glassy state is unstable, i.e., the system is constantly evolving into a state of lower free energy (although this evolution may proceed at a glacial pace). The supercooled liquid can be viewed as metastable, because the free energy is in a local minimum but not the global minimum (i.e., the crystal); but for a noncrystallizable polymer, the supercooled state is effectively at equilibrium.

12.2.2 Kauzmann Temperature

The preceding discussion indicates what should be expected for a true thermodynamic second-order transition, and therefore underlines how the glass transition both resembles and is distinct from such a transition. One way to reconcile these characteristics of T_g is to invoke an underlying thermodynamic transition that in practice is masked by kinetic effects. For example, if we were to perform a cooling experiment at progressively slower rates, would the obtained values of T_g continue to decrease steadily, or would they converge to a limiting value? In the latter case, we

---

[†] Data from J.M. O'Reilly, *J. Polym. Sci.*, 57, 429 (1962).

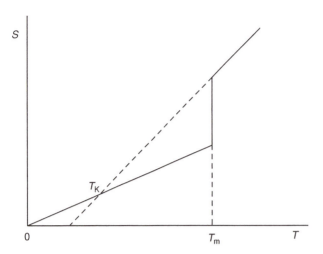

Figure 12.3 Illustration of the "Kauzmann paradox": the specific entropy of the glass (dashed line) would be less than that of the crystal (solid line) below the Kauzmann temperature, T_K.

would have located an apparently thermodynamic transition. It turns out that experimentally it has not proven possible to answer this question definitively, but an argument can be made that some kind of a transition must intervene even for infinitely slow cooling. This argument was originally made by Kauzmann [1], and has since become known as the "Kauzmann paradox" or the "entropy crisis." Kauzmann showed that if the entropy of the supercooled liquid were to continue to follow the temperature dependence seen just above T_g, then eventually the entropy of the glass would be less than the entropy of the crystal. This in itself violates no laws of thermodynamics, but does seem highly counterintuitive. However, if this state of affairs persisted, the entropy of the glass would go to zero at a temperature above 0 K, which would violate the third law of thermodynamics. On the other hand, if a transition intervened, at which the heat capacity dropped sufficiently, then neither of these difficulties would arise.

Kauzmann's argument can be seen from the schematic diagram in Figure 12.3. The specific entropy (entropy per gram of material) drops with decreasing T in the liquid state, down to T_m. Along the crystal branch, there is a discontinuous drop in entropy at the first-order transition. Below T_m the entropy goes smoothly to zero at 0 K (by the third law of thermodynamics), and may be computed from the experimental heat capacity by Equation 12.2.7. Along the supercooled liquid branch, the heat capacity is larger than in the crystal, and thus the entropy drops more rapidly with decreasing T than for the crystal. By extrapolation, therefore, the supercooled liquid entropy will equal the crystal entropy at a finite temperature, the Kauzmann temperature, T_K. Experimentally, such extrapolations give values of T_K that are about 50° below the measured T_g. On the other hand, if a thermodynamic transition intervenes, then the heat capacity of the glass would be lower than that of the supercooled liquid, perhaps very close to that of the crystal, and the problem would be averted. Note, however, that a smooth reduction in heat capacity with decreasing temperature could also avoid the Kauzmann paradox without the necessity for an intervening phase transition.

12.2.3   Theory of Gibbs and DiMarzio [2]

We will not develop this theory in detail, but its physical content and basic conclusions can be readily appreciated. The treatment begins with a lattice, similar in spirit to that employed in the Flory–Huggins theory of mixing in Chapter 7. In this case, however, the solvent is replaced by voids, or vacancies on the lattice, which will play a role qualitatively similar to that of the "free volume" to be discussed in Section 12.4. There is an energy associated with the voids, because a polymer segment adjacent to a void will have lost the interaction energy that it might have had with another segment. In the notation of Section 7.2 and Section 7.3, the total energy associated with the voids E_v will be given as

$$E_v \approx \frac{z|w_{22}|}{2} m_1 \phi_2 \qquad (12.2.12)$$

where z is the number of nearest neighbors on the lattice, w_{22} is the energy of interaction between two polymer segments (species 2), m_1 is the total number of voids (species 1), and ϕ_2 is the polymer volume fraction. This result may be compared to Equation 7.3.12a for the enthalpy of mixing in the Flory–Huggins theory, as $z|w_{22}|/2$ plays the role of χkT. A key feature of m_1 is that it can vary with temperature; as T decreases, the energy penalty for having empty space plays an increasingly important role, and so the material will contract. As m_1 decreases, so will the entropy of placement of the chains on the shrinking lattice.

The second, crucial modification to the Flory–Huggins approach is to assign different energies to various nearest neighbor conformations of the chain on the lattice. In the simplest case, an energy ε_1 is assigned to one, lowest energy conformation, and ε_2 is assigned to other possibilities. This is analogous to having one energy for a trans conformer, and a higher energy for either gauche plus or gauche minus states in polyethylene, as discussed in Chapter 6. As temperature decreases, the chains will tend to adopt more and more ε_1 conformations, which also serve to reduce the entropy of placement of the chains on the lattice.

The main calculation in this theory involves enumerating the number of ways the chains may occupy the lattice. This is similar in spirit to calculation of the entropy of the Flory–Huggins theory in Section 7.3.2, but is complicated by including the different energies of each state (via both E_v and $\Delta\varepsilon = \varepsilon_1 - \varepsilon_2$) and by a more accurate accounting of the effects of chain ends. The inclusion of the energy terms means that the result (called the partition function in statistical mechanics) can be used to find the free energy, and not just the entropy. The central consequence is the emergence of a temperature, T_2, at which the number of possible states shrinks to 1, and thus where the entropy vanishes. The value of T_2 depends only on N, E_v, and $\Delta\varepsilon$. The calculation shows the free energy, entropy, and volume to all be continuous at T_2, whereas the thermal expansivity and heat capacity are discontinuous, so this represents a second-order transition.

The results of this calculation exhibit several features that are in general agreement with the characteristics of the experimental glass transition (the corresponding experimental dependences will be discussed in Section 12.6):

1. T_2 increases with E_v, just as the experimental T_g tends to increase with cohesive energy density (recall Equation 7.6.5 to relate the cohesive energy density to w_{22}).
2. T_2 increases with $\Delta\varepsilon$, just as T_g tends to increase with chain stiffness.
3. T_2 increases with N at low N, but approaches a limiting value as $N \to \infty$.
4. T_2 depends on the number average molecular weight for polydisperse samples.
5. T_2 decreases with added solvent.
6. T_2 increases with degree of crosslinking.

Given all of these characteristics, it is tempting to interpret T_2 as corresponding to the T_g that would be obtained in the limit of infinitely slow cooling. In so doing, the Kauzmann paradox would also be resolved.

Despite these successes, it is probably fair to say that the Gibbs–DiMarzio theory has not been widely adopted as a general description of the glass transition. One reason for this lies in the fact that many different classes of materials can exhibit a glass transition, such as inorganic networks (e.g., $-SiO_2-$, our everyday "glass"), ionic liquids, small organic molecules (e.g., glycerol, o-terphenyl), and even colloidal particles. Consequently, a description based on chain flexibility is unlikely to provide a universal description of the glass transition. However, a successful, universal description of the glass transition has not yet been achieved even 40 years after the Gibbs–DiMarzio theory, so the successes of this theory should not be taken lightly.

12.3   Locating the Glass Transition Temperature

In this section we briefly describe three approaches to characterizing the glass transition in polymers. There are, in fact, many other possible experimental probes, but these three represent the most commonly employed and also serve to illustrate several of the important aspects of the glass transition. The first is to measure the density or volume directly as a function of temperature, by *dilatometry*; this ties directly to the introduction to the glass transition in Section 12.1. The second is to use *differential scanning calorimetry* (DSC) to determine the heat capacity versus temperature. This is probably the most commonly employed method, as it combines speed, ease of use, and potentially quantitative thermodynamic information. The third method is mechanical analysis, which is nothing more than a particular application of the viscoelastic properties described in Chapter 11.

12.3.1   Dilatometry

This is about as unglamorous an experiment as one can imagine. As a property of matter, we take density very much for granted. The fact that it is conceptually simple, readily accessible in handbooks for many materials, and relatively monotonous in its variations all contribute to this attitude. Yet the phenomena represented schematically in Figure 12.1 require careful experimentation on well-defined samples to yield reproducible results. The device that is used to follow volume changes with temperature is called a *dilatometer*; an example is shown in Figure 12.4. The

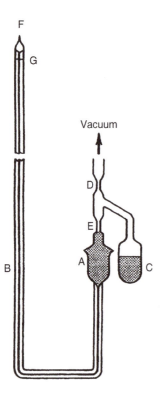

Figure 12.4   Schematic illustration of a dilatometer. The polymer is in bulb A, the height of the polymer plus Hg is determined in capillary B, and C contains extra Hg. The cell can be sealed at D and E. G is a calibration capillary sealed at F. (Reproduced from Sperling, L.H., *Physical Polymer Science*, Wiley, New York, 1986. With permission.)

sample is placed in a bulb that is then filled with an inert liquid, generally mercury. Mercury is suitable because it has negligible solubility in most polymers, and because it does not undergo any transitions of its own the relevant temperature range. The bulb is connected to a capillary so that changes in volume register as variations in the height of the mercury column, just as in a thermometer. For a constant temperature experiment, say, monitoring crystallization at T_m, the volume changes in the capillary correspond identically to changes occurring in the sample. When temperature variation is involved, the expansion of the mercury due to the temperature change is superimposed on the expansion of the specimen and must be taken into account. To obtain meaningful results it is necessary to standardize the rate at which temperature changes are made and, of course, to have an accurately measured and uniform temperature in the bath surrounding the dilatometer. The sample must be degassed to prevent entrapment of air; a gas bubble can really raise havoc in this kind of experiment. This experimental protocol favors slow rates of heating and cooling, to allow for temperature stabilization throughout the bath. For this reason, measurements during cooling are relatively straightforward to make. This turns out not to be the usual case in other common methods where measurements during heating are the norm. Measurements on cooling have one fundamental advantage, namely, the sample is initially at equilibrium, and T_g represents the first observable departure from equilibrium. Experiments conducted during heating begin with a nonequilibrium sample, and therefore the results generally depend on sample history.

An example of dilatometric data is provided in Figure 12.5, taken from the classic study of Kovacs [3]. Two sets of specific volume versus temperature data for poly(vinyl acetate) are shown.

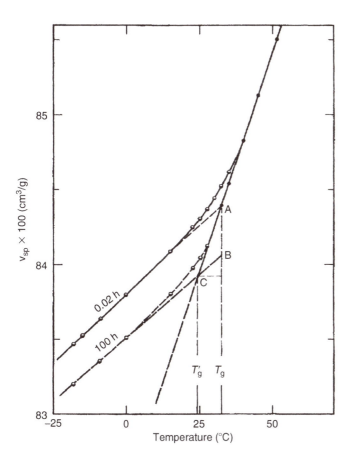

Figure 12.5 Specific volume versus temperature for poly(vinyl acetate), measured at 0.2 and 100 h after cooling rapidly from well above the glass transition temperature. (Reproduced from Kovacs, A.J., *J. Polym. Sci.*, 30, 131, 1958. With permission.)

In each case the measurements were taken after a direct *temperature quench* from an initial $T \gg T_g$. The upper curve corresponds to measurements taken 0.02 h after the quench, and the second set 100 h later. The two traces closely resemble the schematic diagram in Figure 12.1 and in particular the longer time data lead to a lower value of T_g, as expected. However, there is an important distinction to be noted. The 100 h data were acquired not upon cooling more slowly, but rather after waiting for a longer time after cooling below T_g. The process of isothermal volumetric contraction with time after cooling below T_g is called *aging*. It can play an important role in the longtime stability of the mechanical properties of glassy polymers, because densification can lead to undesirable changes in dimensions, increases in brittleness, and even failure.

12.3.2 Calorimetry

Differential Scanning Calorimetry (DSC) is a common example of a *thermal analysis* method. A small quantity of the sample is confined within an aluminum pan and subjected to controlled temperature variation. A reference material is placed in an equivalent pan and the two pans are heated simultaneously. The temperatures of the two pans are monitored continuously and the rate of heat flowing to the sample is adjusted to keep the temperatures of the two pans equal. The heat flow is proportional to an electrical current in a resistive heating element, so it is straightforward both to control and to monitor. Whenever the sample undergoes a thermal transition, so that there is a change in heat capacity, the DSC registers both the amount and the direction of the additional heat flow. A schematic example is shown in Figure 12.6. The first feature upon heating is the increase in heat flow required by the increase in heat capacity at T_g. This would be a step function for a genuine second-order transition examined at very slow rates of heating. The second feature indicated is an exothermic peak, indicating that some of the material has crystallized on heating, at a crystallization temperature T_c. This often occurs in crystallizable polymers because insufficient time was allowed on cooling for extensive crystallization; but once the glass transition is traversed, chain segments acquire enough mobility to crystallize. As will be discussed in Chapter 13, the area under this peak is proportional to the amount of crystallized material if the enthalpy of fusion is known and the DSC has been calibrated with a standard. The final feature in the DSC trace is an endothermic peak, corresponding to the melting of all crystalline material in the sample near the melting temperature, T_m.

Figure 12.7 shows an experimental DSC trace for a polystyrene sample with molecular weight 13,000. In this case there are no peaks associated with crystallization, because the polymer is

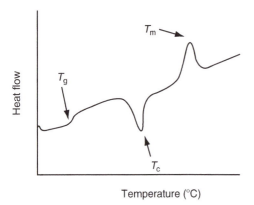

Figure 12.6 Schematic illustration of heat flow into the material versus temperature, showing the glass transition, crystallization, and melting.

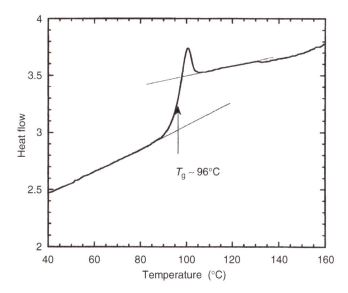

Figure 12.7 Calorimetric determination of T_g for a polystyrene with $M = 13,000$ at a heating rate of 10°C/ min. (Data described in Milhaupt, J.M., Lodge, T.P., Smith, S.D., and Hamersky, M.W., *Macromolecules*, 34, 5561, 2001.)

atactic. There is a peak associated with T_g, which, however, is often observed. It is sometimes called an "enthalpy overshoot," and is attributed to a "superheating" of the glassy state, i.e., as temperature increases and the enthalpy increases along the glassy branch, it crosses the equilibrium line at T_g but does not recover immediately. Consequently there is an extra increment of enthalpy required, beyond that dictated by T_g; this leads to a more rapid change with time in the enthalpy, and thus an overshoot in the DSC trace. Another notable feature of the curve in Figure 12.7 is that the transition itself extends over an interval of 15°C–20°C, which is quite typical. There are various conventions for extracting a single value of T_g from such a trace, but the most common is to take the midpoint, as indicated in the figure.

As alluded to above, the DSC measurement is generally made on heating, and therefore the sample begins in a nonequilibrium state. If we return for a moment to Figure 12.1, and suppose that the sample has just been cooled along path ABJK. If immediately reheated, this path would be retraced, giving T_g' as the result. Suppose, however, the sample sat overnight at the temperature corresponding to point K. The volume would actually contract towards the extrapolation of the equilibrium curve AB, as illustrated by the volumetric data in Figure 12.5. We can see that if the sample had contracted as far as point I in Figure 12.1, then on reheating it would follow the path IHBA, and give a different (lower) value of T_g. In general, then, the observed value of T_g in a heating experiment will depend not only on the heating rate, but also on how long the sample was held below T_g, and also at what temperature it was held. In this way one could obtain many different values of T_g from a single sample, even without varying the heating rate. To avoid this complication, the standard measurement protocol is as follows. The sample is loaded, usually at room temperature, and is then heated above any suspected transition. A few minutes at elevated temperature are sufficient to *anneal* the sample, i.e., erase all memory of past thermal history and stress. The sample is then rapidly cooled or quenched to the beginning temperature, and a second heating scan begun immediately. This second scan is taken as the measurement; a typical heating rate is 10°C/min. Examination of Figure 12.1 and Figure 12.5 suggests that this protocol will tend to give relatively high values of T_g because of the rapid cooling, but often the more important issue is to make the result reproducible.

12.3.3 Dynamic Mechanical Analysis

As a polymer sample is cooled through T_g, the motions of individual segments undergo a dramatic slowing down. Consequently, any experimental measurement of such local relaxation times should be sensitive to T_g. As commercial rheometers are widely available, measurements of G' and G'' versus temperature are commonly employed to locate T_g. An example of the dynamic moduli measured versus temperature at a fixed frequency of 1 Hz is shown in Figure 12.8. Note the similarity in shape between this G' curve and those in Figure 11.2 and Figure 11.16b for $G(t)$. The former reflects the modulus at fixed frequency (or time) as a function of temperature, whereas the latter shows the variation with time at fixed temperature. The origin of this time–temperature equivalence will be explored in Section 12.5, but for now we can consider it intuitively. Below T_g, all relaxation modes are frozen, so the material behaves as a solid with a very high modulus. Upon heating, all relaxation processes accelerate, and when a process can occur in 1 s or less, it will show up as a relaxation if the modulus measurement is made at about 1 Hz. The first processes to become fast enough upon heating a glass correspond to motions of small pieces of chain, in what we called the transition zone of viscoelastic response in Chapter 11. Gradually, on further heating we enter the rubbery plateau, and, eventually, at temperatures far above T_g the molecules can fully relax and the material can flow within 1 s, and the moduli enter the terminal regime. During the transition zone G'' exhibits a peak at a particular temperature (as does the loss tangent, tan δ, given

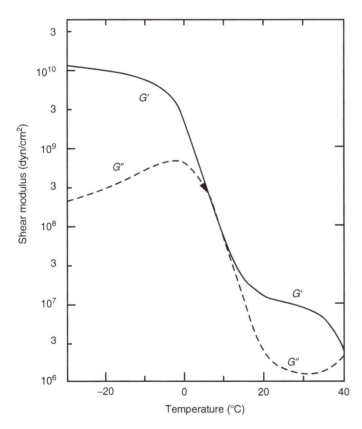

Figure 12.8 The dynamic moduli G' and G'' at 1 Hz, measured as a function of increasing temperature, for a poly(styrene-*ran*-butadiene) copolymer. (Reproduced from Nielsen, L.E., *Mechanical Properties of Polymers*, Reinhold, New York, 1962. With permission.)

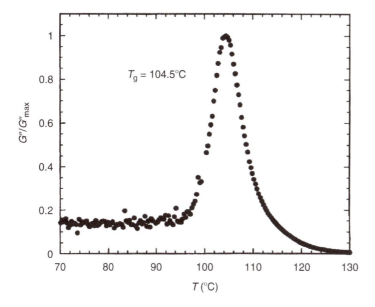

Figure 12.9 Determination of T_g for polystyrene with $M = 13,000$ from the maximum in G'' obtained at 1 rad/s and 2°C/min. (Data described in Milhaupt, J.M., Lodge, T.P., Smith, S.D., and Hamersky, M.W., *Macromolecules*, 34, 5561, 2001.)

by G''/G'). By analogy with the Maxwell model in Section 11.2, this means that the frequency equals an inverse relaxation time, so the location of this peak corresponds to the temperature at which segmental motion occurs in 1 s. This can be taken as an alternative, empirical definition of T_g. Figure 12.9 shows an example for the same polystyrene sample as in Figure 12.6, with the data obtained while heating at 2°C/min. The values of T_g obtained by the two methods are quite comparable (although there is no fundamental reason why they should agree to better than about 10°C).

12.4 Free Volume Description of the Glass Transition

In this section we set aside the issue of thermodynamic equilibrium and simply consider why the viscosity, or other dynamic and mechanical properties of polymers, should undergo a dramatic change over a relatively narrow range of temperatures, even though the structure of the material remains liquid-like. The essence of the argument is based on the concept of *free volume*, which is intuitively very appealing although difficult to pin down precisely. The actual volume of a sample can be written as the sum of the volume "occupied" by the molecules (subscript occ) and the free volume (subscript f). Acknowledging that each of these is a function of temperature, we write

$$V(T) = V_{\text{occ}}(T) + V_{\text{f}}(T) \tag{12.4.1}$$

The variation in V_{occ} with T arises from changes in the amplitude of molecular vibrations with changing T, a variation that affects the excluded volume of the molecules. The free volume, on the other hand, may be viewed as the "elbow room" between molecules, and is required for molecules to undergo rotation and translational motion. As the average kinetic energy increases with increasing temperature, so the associated free volume is also expected to increase with T. Free volume plays a similar role to the "voids" in the Gibbs–DiMarzio theory.

12.4.1 Temperature Dependence of the Free Volume

These concepts may be represented schematically as shown in Figure 12.10 where V is plotted against T. The solid line represents the actual volume, which shows a change in slope at T_g. The dashed line indicates the increase in V_{occ} with T. In all cases the lines are straight, i.e., a linear dependence of V on T, which is a very reasonable approximation for our purposes. The shaded area represents the free volume, V_f. The key feature of V_f is that it decreases upon cooling from the liquid state, until at T_g it reaches some critical, small value. Physically the idea is that once V_f becomes too small, molecular rearrangements are effectively frozen out and the system can no longer continue contracting. Below T_g, V and V_{occ} have approximately the same T dependence and so V_f becomes roughly independent of T.

On the basis of these ideas, we can write the following expression for the volume of the sample:

1. Below T_g

$$V\left(T < T_g\right) = V_{occ}(T = 0) + V_f(T = 0) + \left(\frac{dV_{occ}}{dT}\right)_g T \qquad (12.4.2)$$

2. At T_g

$$V\left(T_g\right) = V_{occ}(T = 0) + V_f(T = 0) + \left(\frac{dV_{occ}}{dT}\right)_g T_g \qquad (12.4.3)$$

3. Above T_g

$$V\left(T > T_g\right) = V\left(T_g\right) + \left(\frac{d(V_{occ} + V_f)}{dT}\right)_1 \left(T - T_g\right) \qquad (12.4.4)$$

The subscripts g and 1 on the coefficient of volume variation with T indicate that these are determined for the glass and liquid states, respectively. Each of the differences used to describe the second-order phase transition as given by Equation 12.2.9 or Equation 12.2.10 can be

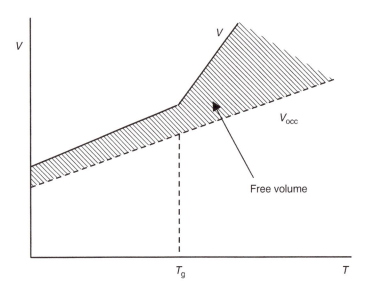

Figure 12.10 Schematic representation of the actual volume (solid line) and occupied volume (dashed line) versus temperature. The hatched area represents the free volume.

qualitatively traced to the sort of change implied by Figure 12.10. Around T_g, for example, the coefficients of expansion of the liquid and glassy states are, respectively,

$$\alpha_1 = \frac{1}{V_g} \left(\frac{d(V_{occ} + V_f)}{dT} \right)_1 \tag{12.4.5}$$

$$\alpha_g = \frac{1}{V_g} \left(\frac{dV_{occ}}{dT} \right)_g \tag{12.4.6}$$

and therefore $\Delta\alpha = \alpha_1 - \alpha_g$ is a measure of the "opening up" of V_f at T_g. The additional volume above T_g also accounts for the increase in compressibility $\Delta\kappa$, and the emergence of the associated modes of energy storage, i.e., additional degrees of motional freedom, accounts for ΔC_p.

The preceding introduction to the free volume interpretation of the glass transition provides a qualitative picture, but is not immediately of much utility because, as discussed in the preceding section, volume is not the generally measured quantity. Furthermore, even if it were we would be faced with the tricky problem of resolving the measured V into the contributions from V_f and V_{occ}. Instead, we shall turn our attention to the steady flow viscosity, η, which will serve many purposes. First, it will illustrate the dramatic effect approaching T_g can have on the flow properties. Second, it will show how powerful the free-volume concept can be in describing experiments. Third, it will lead us naturally to the principle of time–temperature superposition, which forms the topic of Section 12.5.

12.4.2 Free Volume Changes Inferred from the Viscosity

We begin with some experimental results. Figure 12.11a and Figure 12.11b show the viscosity of an oligomeric polystyrene as a function of temperature, from about 180°C down to 37.5°C (the nominal T_g for this sample). The most remarkable feature of the data is the smooth, 12 orders of magnitude variation in η over an otherwise unremarkable temperature interval. This behavior is actually typical of all polymers as T_g is approached from above, although η is rarely measured over such a wide range. For most small molecule fluids we would expect an Arrhenius temperature dependence:

$$\eta(T) = A \exp\left(\frac{E_a}{RT}\right) \tag{12.4.7}$$

where A is a prefactor with units of viscosity, and E_a is the activation energy. In this view, the limiting process to flow is the energetic barrier to molecules sliding past one another. For toluene at room temperature, for example, E_a is approximately 9 kJ/mol. (This may be compared with the energy of a hydrogen bond, approximately 15–20 kJ/mol, or a carbon–carbon bond in polyethylene, about 350 kJ/mol. These numbers indicate that there is little energetic resistance to flow in a small molecule fluid.) If the viscosity of polystyrene followed Equation 12.4.7, then the data would fall on a straight line when plotted as log η versus $1/T$, as in Figure 12.11b; clearly they do not. One can take the slope of a small portion of the curve, and extract an apparent E_a. This gives a value almost 10 times greater than toluene in the high temperature range, and 60 times higher near T_g (see Problem 12.8). The local energy of interaction between styrene monomers cannot be too different from that between toluene molecules, so this factor of 60 must have a different origin. It is not a simple result of molecular weight, because the T dependence of η is found to be more or less independent of chain length (see Section 12.5). The reason is that flow in glass-forming liquids is impeded primarily by a lack of free volume, rather than by an energy barrier (although such barriers should still contribute).

Doolittle [4] studied the viscosity of n-alkanes in detail and found that the following equation was able to describe the data:

$$\eta = A' \exp\left(\frac{B'V_{occ}}{V_f}\right) \tag{12.4.8}$$

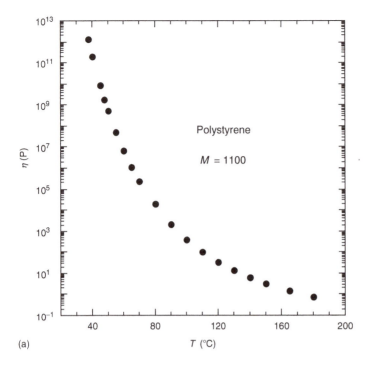

(a)

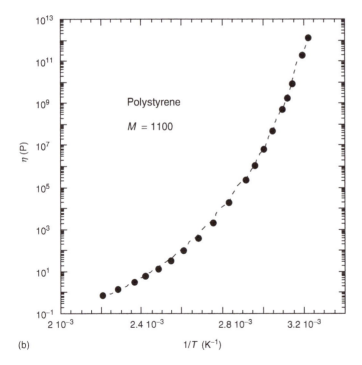

(b)

Figure 12.11   Viscosity of oligomeric polystyrene ($M = 1100$) plotted (a) versus temperature and (b) as log viscosity versus inverse temperature. The smooth curve in (b) represents the fit to the VFTH or WLF equations. (Data described in Plazek, D.J. and O'Rourke, V.M., *J. Polym. Sci. A2*, 9, 209, 1971.)

where A' and B' are empirical constants. Although they relate η to different variables, both the Doolittle and Arrhenius equations have the same functional form. Just as the activation energy E_a represents the height of a barrier relative to thermal energy, RT, the Doolittle equation compares the space needed for a molecule, V_{occ}, to the space available, V_f. In a sense Equation 12.4.8 implies that the primary impediment to molecular motion is entropic in nature (is there space available to move?) rather than energetic as in Equation 12.4.7 (is there enough thermal energy to overcome the activation barrier?). As we shall see shortly, the Doolittle equation describes the remarkable T dependence of η shown in Figure 12.11 very well.

To proceed, we define the *fractional free volume*, $f = V_f/V$, and then Equation 12.4.8 can be written as

$$\eta = A' \exp\left[B'\left(\frac{1}{f} - 1\right)\right] \tag{12.4.9}$$

We can simplify this expression as follows:

$$\eta = A' \exp\left(\frac{B'}{f} - B'\right) = A \exp\left(\frac{B'}{f}\right) \tag{12.4.10}$$

simply by incorporating a factor of $\exp(-B')$ into A. To proceed further, we need to insert an expression for f, particularly in terms of its T dependence. On the basis of Figure 12.10, we propose a linear dependence for $f(T)$, for temperatures at and above T_g:

$$f = f_g + \alpha_f\left(T - T_g\right) \tag{12.4.11}$$

In this expression α_f is the coefficient of expansion of the free volume only, but it should be close to $\Delta\alpha$. Equation 12.4.11 can be inserted into Equation 12.4.10 as follows:

$$\begin{aligned}
\eta &= A \exp\left(\frac{B'}{f_g + \alpha_f(T - T_g)}\right) \\
&= A \exp\left(\frac{B'/\alpha_f}{f_g/\alpha_f + T - T_g}\right) = A \exp\left(\frac{B}{T - T_0}\right)
\end{aligned} \tag{12.4.12}$$

where the new parameters are $B = B'/\alpha_f$ and $T_0 = T_g - f_g/\alpha_f$. The last form of Equation 12.4.12 is often known as the Vogel–Fulcher–Tammann–Hesse (VFTH) equation and the parameter T_0 is referred to as the *Vogel temperature* [5]. By comparison with Equation 12.4.7, we can see that the VFTH equation reduces to the Arrhenius equation when $T_0 = 0$ K. However, experimentally one finds that $T_0 > 0$, and therefore the viscosity is predicted to become infinite at T_0, or some f_g/α_f degrees below T_g.

The data in Figure 12.11 have been fit to the VFTH equation, resulting in the smooth curve shown in Figure 12.11b. Clearly the data follow this form very well. The resulting parameter values are $A = 5.1 \times 10^{-5}$ P, $B = 1743$ K, and $T_0 = 265$ K. Thus in this case T_0 is $45°$ below T_g, which turns out to be quite typical. This is intriguing, because the Vogel temperature is therefore rather close to the Kauzmann temperature discussed in Section 12.2; T_0, based purely on a semiempirical fit to a dynamic property, lends some support to the concept of T_K, a quantity anticipated on thermodynamic grounds. Furthermore, recalling the derivation above, we can equate $45°$ with the ratio f_g/α_f. Given that α_f should be close to $\Delta\alpha$, and that this quantity is on the order of $5 \times 10^{-4}\,°C^{-1}$, the implication is that $f_g \approx 0.023$. It is sometimes suggested that the glass transition corresponds to a particular value of the fractional free volume. We will consider this issue further later in this section.

12.4.3 Williams–Landel–Ferry Equation

The VFTH equation is capable of describing rather well the temperature dependence of η or, in fact, of any relaxation time of a polymer chain, based on three parameters. It is often the case that one has measurements over a range of temperatures, and it is convenient to choose some temperature as a reference. The value of the measured quantity at the reference temperature, T_r, can then be used to eliminate the front factor A from the VFTH equation, thereby reducing the number of parameters to two. This can be seen by taking the ratio of two values of η with two different values of f (subscript r denotes the property at T_r) from Equation 12.4.10, remembering that B' is a constant:

$$\frac{\eta}{\eta_r} = \exp\left[B'\left(\frac{1}{f} - \frac{1}{f_r}\right)\right] \tag{12.4.13}$$

or

$$\ln\left(\frac{\eta}{\eta_r}\right) = B'\left(\frac{1}{f} - \frac{1}{f_r}\right) \tag{12.4.14}$$

Now we reintroduce the linear temperature dependence of f as in Equation 12.4.11, except that we choose T_r as the reference instead of T_g: $f = f_r + \alpha_f(T - T_r)$. This is legitimate as long as $T_r > T_g$.

$$\begin{aligned} \ln\left(\frac{\eta}{\eta_r}\right) &= B'\left(\frac{1}{f_r + \alpha_f(T - T_r)} - \frac{1}{f_r}\right) \\ &= B'\left(\frac{f_r - f_r - \alpha_f(T - T_r)}{f_r^2 + f_r\alpha_f(T - T_r)}\right) = -B'\left(\frac{(1/f_r)(T - T_r)}{(f_r/\alpha_f) + (T - T_r)}\right) \end{aligned} \tag{12.4.15}$$

We now convert to base 10 logarithms, and collect the various constants and rename them:

$$\log\left(\frac{\eta}{\eta_r}\right) = -\frac{C_1(T - T_r)}{C_2 + (T - T_r)} \tag{12.4.16}$$

This relation is known as the *Williams–Landel–Ferry* (WLF) *equation* [6], where the two new parameters are $C_1 = B'/2.303 f_r$ and $C_2 = f_r/\alpha_f$. The utility of the WLF equation will become apparent in the next section, but for now it is important to realize that it has the same physical content as the VFTH equation. The two main assumptions required to derive both relations were that the viscosity followed the Doolittle equation, with an exponential dependence on $1/f$, and that f has a linear dependence on T above T_g. The WLF equation tends to be more familiar to polymer scientists, whereas the VFTH equation is used more by scientists studying the glass transition in low molecular weight materials.

As noted above, one approach to the glass transition is to assign it to a certain, critical value of fractional free volume. This is an appealingly simple picture, but it is not quantitatively reliable. One difficulty is that there are several different ways to estimate f, but no direct and unambiguous way to measure it. Table 12.2 provides values of the various free volume parameters for common

Table 12.2   Representative Values of VFTH/WLF Equation Parameters for Various Polymers

Polymer	T_g (°C)	α_l^a	α_g^a	f_g	C_1^g	C_2^g (°C)
Poly(dimethylsiloxane)	−130	8.5	4.5	0.071	6.1	69
1,4-Polybutadiene	−95	7.8	2.0	0.039	11.2	60.5
Polyisobutylene	−75	6.2	1.5	0.026	16.6	104
1,4-Polyisoprene	−70	6.2	2.1	0.021	10.8	51.1
Poly(vinyl acetate)	30	6.0	2.1	0.028	15.6	46.8
Polystyrene	100	5.5	1.8	0.032	13.7	50
Poly(methyl methacrylate)	110	4.6	2.2	0.013	34	80

[a] in units of 10^{-4} °C^{-1}.

Source: Data collected in Ferry, J.D., *Viscoelastic Properties of Polymers*, 3rd ed., Wiley, New York, 1980; Sperling, L.H., *Physical Polymer Science*, Wiley, New York, 1986.

polymers, and the WLF coefficients C_1 and C_2 when T_g is taken as the reference temperature. The extracted value of f_g varies significantly, and the sensitivity of the VFTH and WLF equations to the choice of parameters is rather strong. Conversely, when fitting experimental data to extract these parameters, the exact location of T_g plays an important role. And, although α_l and α_g can be measured, the expansion of free volume α_f is not necessarily exactly equal to $\alpha_l - \alpha_g$, and one is still left with B as an adjustable parameter in extracting a value for f_g. Some further observations about the free volume approach are listed below:

1. Although Equation 12.4.12 or Equation 12.4.16 describes the variation of viscosity over a wide range of conditions quite well, they tend to break down both very far above T_g, where free volume is not so important, and very close to T_g. In the latter range some workers advocate a power law relationship in $(T - T_0)$, where T_0 is an adjustable critical temperature analogous to the Vogel temperature; such power laws are the rule for the divergence of experimental quantities approaching a phase transition. This idea will be explored further in Problem 12.11.
2. Although it is easy to discuss free volume, it is necessary to come up with a numerical value for this quantity in order to test these concepts. There is considerable disagreement as to which of several different methods of computation gives the best value for V_f and V_{occ}, and its extraction from experimental data requires assuming a particular model.
3. It is possible to derive an expression equivalent to Equation 12.4.12 or Equation 12.4.16 starting from entropy rather than free-volume concepts. We have emphasized the latter approach, since it is easier to visualize and hence to use for qualitative predictions about T_g.
4. Completely aside from the theories that attempt to explain it, the empirical usefulness of the VFTH and WLF equations is beyond doubt. We shall examine this in detail in the next section.

Example 12.2

A process for molding transparent plastic cups from polystyrene has been optimized to run at 150°C. When the supplier of the raw material introduces a new resin with a 20% higher M_w, the increased viscosity slows the process down. At what temperature should the process be run in order to recover the viscosity of the original raw material?

Solution

From Chapter 11 we recall that $\eta \sim M^{3.4}$, so a 20% increase in M_w increases η by about $(1.2)^{3.4} = 1.86$. We can write two WLF equations (Equation 12.4.16) for the two temperatures $T_1 = 150°C$ and T_2, the unknown, using the parameters from Table 12.2:

$$\log\left(\frac{\eta_1}{\eta_g}\right) = -\frac{C_1^g(T_1 - T_g)}{C_2^g + (T_1 - T_g)} = -\frac{13.7 \times 50}{100} = -6.85$$

$$\log\left(\frac{\eta_1}{1.86\eta_g}\right) = -\frac{C_1^g(T_2 - T_g)}{C_2^g + (T_2 - T_g)} = -\frac{13.7(T_2 - 100)}{T_2 - 50}$$

where the factor of 1.86 is inserted in the second equation in order to find the temperature T_2 at which the viscosity has been lowered by that amount. We now take the difference between these two equations to obtain

$$\log(1.86) = 0.269 = -6.85 + \frac{13.7(T_2 - 100)}{T_2 - 50}$$

which can be solved to give $T_2 = 154°C$. This calculation illustrates the remarkable influence of the glass transition on dynamics. In this instance, although the experimental temperature is about 50° above T_g, only a 4° increase in temperature is sufficient to cut the viscosity almost in half.

12.5 Time–Temperature Superposition

One of the main points of the previous section was that the viscosity of a polymer can change by many orders of magnitude over only a few tens of degrees in temperature. This feature is routinely exploited to extend the range of measurement in any viscoelastic experiment. As an example, consider Figure 11.16b, where the stress relaxation modulus of polyisobutylene at 25°C was plotted against time. Note that the time axis was actually t/a_T, where the significance of the factor a_T will emerge shortly. The range of the time axis extended from a few picoseconds to megaseconds—nearly 18 orders of magnitude. There are no rheometers capable of measuring the picosecond response of materials, or even the nanosecond response. Some custom instrumentation has been developed that can extend down to microseconds, but that was not the case for these measurements. Similarly, at the long time end, the measurements extend to almost two weeks—hardly experimentally convenient. What was done in fact was to make measurements over a range of temperatures, and to reduce these to one master curve using the *time–temperature superposition* (TTS) *shift factor*, a_T. The underlying principle is one of *corresponding states*; a measurement at a certain temperature and time (or frequency) is equivalent to a measurement at a lower temperature and longer time.

This correspondence can be understood quite simply. We begin with a generic expression for the stress relaxation modulus, $G(t,T)$, which also happens to be consistent with the Rouse, Zimm, and reptation models (see Chapter 11):

$$G(t,T) = \frac{\rho(T)RT}{M} \sum \exp\left(-\frac{t}{\tau_p(T)}\right) \tag{12.5.1}$$

where we have explicitly indicated the quantities that depend on temperature: density, ρ, and the relaxation times, τ_p. Of these, the dependence of ρ is rather weak; recall that the thermal expansion factor is usually less than one tenth of 1% per degree. The relaxation times, however, follow the VFTH or WLF dependence derived in the previous section, and might change by an order of magnitude over a few degrees (see Figure 12.11). Now we define the TTS shift factor, a_T, as the ratio of any relaxation time at one temperature to its value at the chosen reference temperature, T_r

$$a_T \equiv \frac{\tau_p(T)}{\tau_p(T_r)} = \frac{\eta(T)}{\eta(T_r)} \tag{12.5.2}$$

The crucial assumption here is that *all relaxation times have the same temperature dependence* and thus that one value of a_T applies to any dynamic property including the viscosity. Now we insert this definition into Equation 12.5.1 and rearrange:

$$G(t,T) = \frac{\rho(T)RT}{M} \sum \exp\left(-\frac{t}{\tau_p(T)}\right) = \frac{\rho(T)RT}{M} \sum \exp\left(-\frac{t}{a_T\tau_p(T_r)}\right)$$
$$= \frac{\rho(T)T}{\rho(T_r)T_r} G(t/a_T, T_r) \tag{12.5.3}$$

The significance of Equation 12.5.3 is this: a measurement of G at a particular combination of (t,T) is exactly equivalent to a measurement at a new time, t/a_T, and new temperature, T_r. (The front factor involving ρ and T contributes a small correction that is often ignored in practice; note that ρ decreases as T increases, so the net effect is even smaller than that due to ρ alone.)

Now we can see how TTS is used in practice. The actual measurements of Catsiff and Tobolsky [7] are shown in Figure 12.12. They employed an instrument for which they could resolve measurements from a few seconds up to a few hours, and they varied the temperature from $-80.8°C$ to $50°C$. At the lowest temperatures the modulus is very high, characteristic of a glass. As temperature increases, the modulus decreases at fixed time. Near $-20°C$ the modulus seems to be independent of time and the value is characteristic of a lightly crosslinked rubber. Finally, by $50°C$ the material flows on the measurement timescale. The data may be shifted

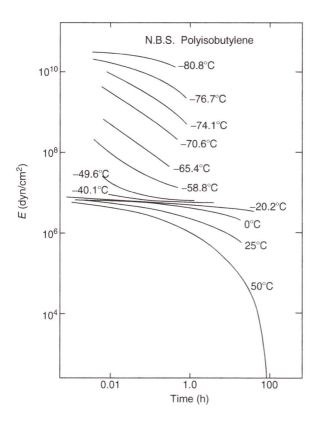

Figure 12.12 Stress relaxation modulus for polyisobutylene at the indicated temperatures. Data are actually Young's modulus, which is approximately equal to $3G(t)$ (see Section 10.3). (Reproduced from Catsiff, E. and Tobolsky, A.V., *J. Colloid Sci.*, 10, 375, 1955. With permission.)

horizontally by an arbitrary shift factor, until they overlap to produce a master curve; a reference temperature of 25°C was selected. (A small vertical shift corresponding to the ρT front factor was also applied.) The result is shown in Figure 12.13 (and previously as Figure 11.16b). This empirical generation of a master curve produces a set of shift factors, a_T, and we can ask how a_T depends on T. The answer should come as no surprise: it follows the WLF equation, as in Equation 12.4.16:

$$\log a_T = -\frac{C_1(T - T_r)}{C_2 + (T - T_r)} \tag{12.5.4}$$

The shift factors are used to generate the master curve in Figure 12.13 and plotted against temperature in Figure 12.14, along with a fit to Equation 12.5.4. They follow the WLF equation very well except for the lowest temperatures. The reason for this discrepancy is simple; some of the data were obtained below $T_g \approx -68°C$. The WLF equation relies on a linear variation of free volume with temperature, which should not hold when traversing T_g, as Figure 12.10 indicates.

The principle of TTS applies to any dynamic property, a generality that can be understood from the discussion in Section 11.3, where we showed that any linear viscoelastic function is derivable from $G(t)$. In particular, TTS is often used in dynamic experiments, and G' and G'' may be superposed to form master curves versus reduced frequency ωa_T. This approach was used, for example, to produce the dynamic moduli for poly(vinyl methyl ether) shown in Figure 11.16a. In this case the superposed data extend over almost 10 orders of magnitude in reduced frequency,

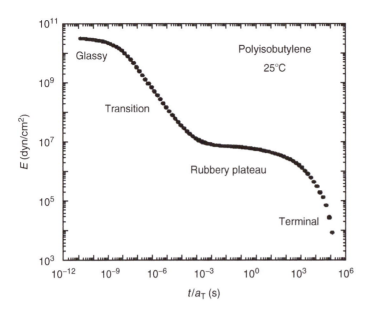

Figure 12.13   Master curve of modulus versus reduced time for the polyisobutylene data [7] in Figure 12.12.

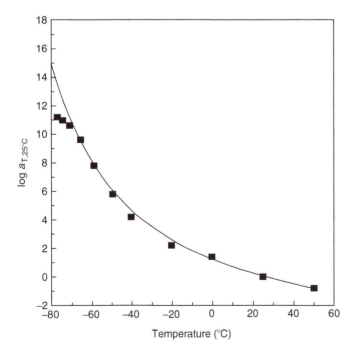

Figure 12.14   Temperature dependence of the shift factors used in Figure 12.13, and the corresponding fit to the WLF equation.

even though the rheometer employed could only be used from about 0.001–10 Hz. Another illustration of how TTS is applied in practice is provided by the following example.

Example 12.3

The dynamic moduli for a polyisoprene sample ($M = 80,000$) were measured at five temperatures: $-40°C$, $-20°C$, $0°C$, $20°C$, and $50°C$.[†] The unshifted data (G' and G'' versus ω) are shown in Figure 12.15a and Figure 12.15b, respectively. Generate the master curve of G' and G'' versus ωa_T, using $20°C$ as the reference temperature.

Solution

To obtain the shift factors, one can simply play with the values of a_T in a spreadsheet until the data superpose nicely, or each set of data can be plotted on separate sheets of paper (keeping the axes scales the same) and superposed in front of a bright light. However, because small vertical shifts may be permitted due to the ρT term in the front factor, a more rigorous approach to determine a_T is actually to shift $\tan \delta$ (recall Equation 11.2.22). As $\tan \delta$ involves the ratio of G'' and G', the front factor cancels, and the data can be shifted exclusively along the horizontal axis to obtain the best superposition. Finally, small vertical shifts can be applied based on knowledge of $\rho(T)$. Figure 12.15c shows the unshifted values of $\tan \delta$, and Figure 12.15d the shifted values, using $20°C$ as the reference temperature and values of a_T of 12000, 130, 7, and 0.11 for $-40°C$, $-20°C$, $0°C$ and $50°C$, respectively. The final master curves for G' and G'' are shown in Figure 12.15e and the superposition can be seen to be excellent. Vertical shifts have not been applied in this instance.

The preceding discussion illustrates how TTS can be used to expand the accessible range of time (or frequency) scales dramatically. Furthermore, once the temperature dependence of a_T has been determined for a particular property and a particular polymer, it should apply to all dynamic properties and all reasonably high molecular weights of the same polymer. In general, this is the case, but we still have one complication if we hope to just look up the WLF parameters C_1 and C_2 for a given polymer in some handbook. The problem is that the values of C_1 and C_2 depend on the choice of reference temperature and there are obviously many possible choices of T_r. Suppose, however, we make the particular selection $T_r = T_g$. We may find it experimentally inconvenient to make the measurement at T_g, but that does not matter; we can extrapolate our data using the WLF function. When this choice of reference is made, the corresponding parameters are designated C_1^g and C_2^g, as in Example 12.2. It turns out that the values of these parameters are very approximately universal, as illustrated in Table 12.2. For many systems C_1^g is about 10–15 and C_2^g is about $50°C$–$60°C$, for example. This approximate universality permits reasonable estimation of the T dependence for any polymer, once T_g is known. Also, if we compare the VFTH and WLF parameters directly (Equation 12.4.12 and Equation 12.4.16), we see that $C_2^g = T_g - T_0$, i.e., C_2^g corresponds to the interval between the glass transition temperature and the Vogel temperature.

We conclude this section with a brief discussion of when TTS is not, or may not, be applicable. The single, crucial assumption is that all the relevant relaxation times have the same temperature dependence over the measured temperature range. Note that adherence to the WLF dependence is not a requirement; the remarkably strong T dependence as T_g is approached is what makes TTS

[†] Data from J.C. Haley, Ph.D. Thesis, University of Minnesota, 2005.

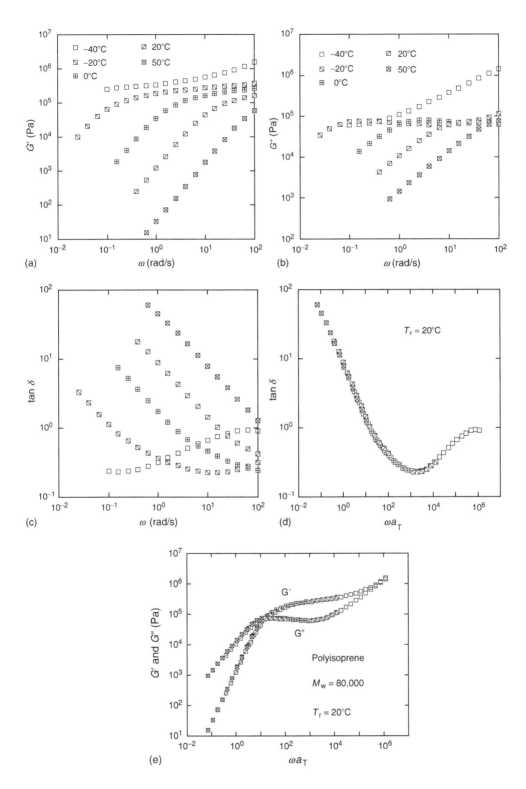

Figure 12.15   Time–temperature superposition of dynamic moduli for polyisoprene. (a) G', (b) G'', and (c) tan δ versus frequency at five temperatures, (d) tan δ shifted horizontally to obtain a master curve and determine the shift factors, (e) G' and G'' master curves. (Data from Haley, J.C., Ph.D. Thesis, University of Minnesota, 2005.)

useful, but it is not necessary for its *validity*. One way to violate the necessary assumption is for the sample to undergo some kind of transition with temperature, such as crystallization (Chapter 13) or crosslinking (Chapter 10). A second situation arises in polymer mixtures or blends, where it is sometimes observed that the temperature dependences of the relaxation times of the two components differ. A more subtle failure can arise at high effective frequencies, or short effective times, in the transition zone of viscoelastic response. The origin of the problem can be understood as follows. In the bead–spring formulation of polymer dynamics (see Section 10.4 through Section 10.6), all of the relaxation times are proportional to an underlying segmental relaxation time, $\tau_{seg} \sim \zeta b^2/kT$, where ζ is the bead friction factor. It is this friction factor (divided by T) that follows the WLF form, and therefore all the relaxation times do too. However, this model does not describe the relaxation of very short pieces of chain, on the scale of a few monomers and below. On this very local scale, it is reasonable to anticipate that some relaxations may have a different T dependence. For example, the rotation of a particular functional group might be limited by a particular conformational barrier rather than free volume, and therefore have an approximately Arrhenius dependence. Consequently, it is to be expected, and some careful measurements have shown, that somewhere in the transition zone of viscoelastic response TTS will break down.

12.6 Factors That Affect the Glass Transition Temperature

In this section we consider the major factors that affect the value of T_g in a given polymer material, independent of the role of kinetics and measurement technique. These experimental observations should also be compared with expectations based on the Gibbs–DiMarzio theory outlined in Section 12.2.3.

12.6.1 Dependence on Chemical Structure

The primary factor, of course, is monomer structure. A list of representative T_g values for common polymers was provided in Table 12.1. They range from $-130°C$ for poly(dimethylsiloxane) to $240°C$ for poly(*p*-phenylene terephthalimide) (Kevlar®). As indicated at the beginning of the chapter, the value of T_g is the single most important parameter in selecting a polymer for a given application. Those polymers with T_g below room temperature are sometimes called *elastomers*; when crosslinked into a permanent network structure (see Chapter 10) they exhibit tremendous elasticity, as in rubber bands, o-rings, gaskets, and tires. Polymers with T_gs near or above $100°C$ are called *thermoplastics*; they are processed above T_g, but then are solidified into plastic parts by cooling. Polymers with particularly high T_gs, approaching or above $200°C$, are termed *engineering thermoplastics*. They are in high demand for more strenuous applications and as such tend to be rather expensive. Polymers with T_gs between room temperature and $100°C$ that do not crystallize are rather less widely applicable as bulk materials, but are useful as adhesives.

It is natural to seek a general correlation between the value of T_g and some other, familiar property of the polymer, but in fact no robust correlation exists. Some broad generalizations may be made, however.

1. Backbone flexibility increases as T_g decreases, in general. This may be simply understood from a conformational barrier argument. Flexible polymers tend to have smaller potential barriers between conformations and thus at a given temperature conformational rearrangements should be more rapid (see Section 6.1). The correlation with T_g follows because the glass transition corresponds to the freezing out of long-range backbone rearrangements. However, the correlation is not strict, as can be seen by comparing the T_g values in Table 12.1 with the persistence lengths listed in Table 6.1. Based on flexibility alone, poly(ethylene oxide) has an anomalously high and poly(dimethylsiloxane) an anomalously low T_g.
2. The larger the rigid sidegroup, the larger the T_g. This correlation follows the previous one, in that bulky sidegroups impede backbone rearrangements. Exceptions here include polyisobutylene,

which has a significantly lower T_g than polypropylene, even though it has double the number of sidegroups. Also, adding flexible side chains to relatively stiff backbones will have the effect of lowering T_g.

3. Polymers that have weak interactions, such as the purely dispersive interactions of the polyolefins, have lower T_gs than more strongly interacting materials, such as the more polar poly(vinyl chloride) or poly(vinyl alcohol). This arises from the retarding effect of stronger intermolecular constraints on chain relaxation.

4. To obtain the highest T_gs, high backbone stiffness is essential, which is most easily conferred by incorporating aromatic groups within the backbone. Poly(tetrafluoroethylene) (Teflon®) is another example in this class; the bulky fluorine atoms render the all-carbon backbone rather stiff.

12.6.2 Dependence on Molecular Weight

Once a polymer has been selected, how may its T_g be modified? There are two simple routes; one is to vary molecular weight and the other is to add some amount of a low molecular weight diluent, called a plasticizer. The molecular weight dependence follows from the molecular weight dependence of density, and is thus easily understood via the free-volume concept. For a homologous series of compounds, such as the linear alkanes C_nH_{2n+2}, the density increases with n. This arises because of the chain ends; they are less dense essentially because covalent bonds are shorter than intermolecular nearest neighbor distances. Therefore the lower the n, the higher the fraction of the material that is made up of chain ends, and the lower the density. The chain ends may be viewed as a kind of impurity, and as with colligative properties such as boiling point elevation, freezing point depression, or osmotic pressure (see Section 7.4), the change in transition temperature should be linear in the mole fraction of impurity. In this case, the concentration of chain ends varies as $1/M_n$, and therefore we anticipate:

$$T_g(M_n) = T_g(M \to \infty) - \frac{A}{M_n} \tag{12.6.1}$$

where A is an empirical parameter. This expression turns out to be a very reasonable description for many polymers. For example, for polystyrenes shown in Figure 12.16, the value of A is about 10^5 g/mol, which implies that T_g becomes effectively independent of M when M exceeds that value. A $10°$ depression of T_g below the high molecular weight limit would be seen for a polymer with $M = 10,000$. (Recall that the polystyrene in Figure 12.11 with $M = 1,100$ had a T_g of 37.5°C, which is not in good agreement with this relation; Equation 12.6.1 is more reliable for larger values of M.) Problem 12.12 provides another example of the M dependence of T_g. This M dependence of T_g is important to appreciate, but it is not a particularly useful design parameter because the range over which T_g may be varied while retaining other desirable properties, such as mechanical strength, turns out to be rather small.

12.6.3 Dependence on Composition

A general expression for the glass transition temperature of a mixture of two components was developed by Couchman [8], beginning with a thermodynamic relation for the transition similar in spirit to those discussed in Section 12.2. Furthermore, the same relation can be applied to a polymer mixed with a low molecular weight compound, a blend of two polymers, or a statistical copolymer. We begin by equating the (specific) entropies of the mixture in the glass (superscript g) and liquid (superscript l) states at the "second-order transition" temperature, T_g:

$$S^g(T_g) = S^l(T_g) \tag{12.6.2}$$

For each phase, we construct the entropy of the mixture as follows:

$$S = w_1 S_1 + w_2 S_2 + \Delta S_m \tag{12.6.3}$$

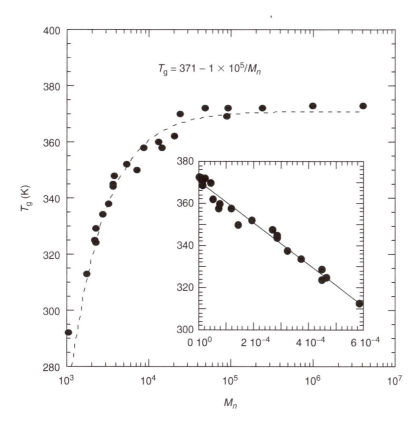

Figure 12.16 Dependence of the glass transition temperature on molecular weight for polystyrene. The inset shows a plot of T_g versus $1/M_n$ and a linear regression fit. (Data from Fox, T.G and Flory, P.J., *J. Polym. Sci.* 14, 315, 1954; Santangelo, P.G. and Roland, C.M., *Macromolecules*, 31, 4581, 1998.)

where w_1 and w_2 are the weight fractions of components 1 and 2, respectively (the appropriate composition variable because we are dealing with specific entropies), S_1 and S_2 are the component entropies, and ΔS_m is the entropy of mixing. This last quantity might be given by the Flory–Huggins theory discussed in Chapter 7, for example, but if we assume that it is independent of state at a given temperature, it will cancel out from both sides of Equation 12.6.2. The component entropies can be constructed via Equation 12.2.7 using the pure component T_g as the lower bound and the mixture T_g as the upper bound:

$$S_1(T_g) = S_1(T_{g,1}) + \int_{T_{g,1}}^{T_g} C_{p,1} \, d \ln T$$

$$S_2(T_g) = S_2(T_{g,2}) + \int_{T_{g,2}}^{T_g} C_{p,2} \, d \ln T \qquad (12.6.4)$$

In these expressions the only quantities that depend on the state of the mixture are the heat capacities (specific heats); from Equation 12.6.2 we can see that the reference state entropy of each component in the liquid and glassy states is the same at the pure component T_g. Armed with this information we can insert Equation 12.6.4 and Equation 12.6.3 into Equation 12.6.2 to obtain

$$w_1 \int_{T_{g,1}}^{T_g} C_{p,1}^g \, \mathrm{d}\ln T + w_2 \int_{T_{g,2}}^{T_g} C_{p,2}^g \, \mathrm{d}\ln T = w_1 \int_{T_{g,1}}^{T_g} C_{p,1}^l \, \mathrm{d}\ln T + w_2 \int_{T_{g,2}}^{T_g} C_{p,2}^l \, \mathrm{d}\ln T \qquad (12.6.5)$$

which can be rewritten

$$w_1 \int_{T_{g,1}}^{T_g} \Delta C_{p,1} \, \mathrm{d}\ln T = -w_2 \int_{T_{g,2}}^{T_g} \Delta C_{p,2} \, \mathrm{d}\ln T \qquad (12.6.6)$$

Assuming that the heat capacities do not change with temperature, the integration gives

$$w_1 \Delta C_{p,1} \ln\left(\frac{T_g}{T_{g,1}}\right) + w_2 \Delta C_{p,2} \ln\left(\frac{T_g}{T_{g,2}}\right) = 0 \qquad (12.6.7)$$

This expression can be rearranged to give the general expression

$$\ln T_g = \frac{w_1 \Delta C_{p,1} \ln T_{g,1} + w_2 \Delta C_{p,2} \ln T_{g,2}}{w_1 \Delta C_{p,1} + w_2 \Delta C_{p,2}} \qquad (12.6.8)$$

Simpler versions of this expression can be obtained with additional assumptions. For example, if the condition $\Delta C_{p,1} T_{g,1} \approx \Delta C_{p,2} T_{g,2}$ holds, and the T_gs are not too different, Equation 12.6.8 reduces to the more commonly applied *Fox equation*:

$$\frac{1}{T_g} = \frac{w_1}{T_{g,1}} + \frac{w_2}{T_{g,2}} \qquad (12.6.9)$$

These relations are compared with data for polymer blends and statistical copolymers in Figure 12.17 and Figure 12.18, respectively. The agreement is generally very good. The use of a plasticizer is illustrated in Example 12.4 below.

Example 12.4

Di-*n*-ethylhexyl phthalate (DEHP) and related compounds are commonly used to plasticize poly(vinyl chloride) (PVC) to produce the pliable material generically referred to as "vinyl." A good plasticizer is miscible with the polymer in question, does not crystallize itself, and has a very low vapor pressure. What fraction of DEHP should be added to PVC to bring T_g down below room temperature, given that T_g for DEHP is about $-86°C$?

Solution

This is a straightforward application of the Fox equation. If we take 300 K as our target T_g, and 360 K as the T_g of pure PVC (see Table 12.1):

$$\frac{1}{300} = \frac{w_1}{187} + \frac{1 - w_1}{360} = w_1\left(\frac{1}{187} - \frac{1}{360}\right) + \frac{1}{360}$$

then we need a weight fraction of DEHP $w_1 = 0.22$. Of course, in an actual application the desired T_g would be confirmed by measurements on compositions in the neighborhood of the estimated value.

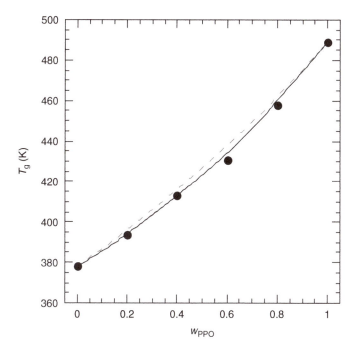

Figure 12.17 Glass transition temperature versus composition for miscible polystyrene/poly(oxy-2,6-dimethyl-1,4-phenylene) (PPO) blends, with fits to Equation 12.6.8 (solid curve) and Equation 12.6.9 (dashed line). (Data described in Couchman, P.R., *Macromolecules*, 11, 1156, 1978.)

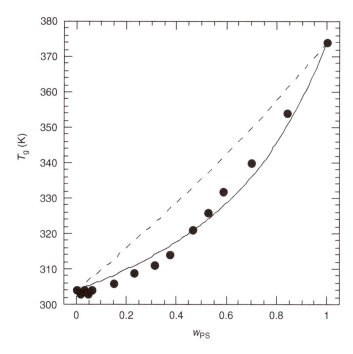

Figure 12.18 Glass transition temperatures for poly(styrene-stat-*n*-butylmethacrylate) copolymers, with fits to Equation 12.6.8 (solid curve) and Equation 12.6.9 (dashed line). (Data described in Kahle, S., Jorus, J., Hempel, E., Unger, R., Höring, S., Schröter, K., and Donth, E., *Macromolecules*, 30, 7214, 1997.)

12.7   Mechanical Properties of Glassy Polymers

Although all polymers can exhibit a glass transition, we are concerned here with thermoplastics: materials that may be processed into desired shapes or forms above T_g, and then cooled below T_g for use. For semicrystalline polymers to be discussed in Chapter 13, T_m will play the same role as T_g in setting the boundary between liquid state processing and solid state application. As most common applications occur in the vicinity of room temperature, say from $-25°C$ to $50°C$, thermoplastics must have T_g well above that. Three prevalent thermoplastics are polystyrene ($T_g \approx 100°C$), poly(methyl methacrylate) ($T_g \approx 110°C$), and polycarbonate ($T_g \approx 150°C$), and this subset will be sufficient to illustrate the main aspects of mechanical response.

Clearly, there are many different physical attributes beyond the value of T_g (or T_m) that could either favor or disfavor a particular polymer for a given application, and we will make no attempt to treat these comprehensively. We will emphasize mechanical strength in this section, but even in that context we will only be able to cover a few aspects of a very rich subject. In general semicrystalline polymers have superior mechanical strength compared to amorphous polymers, but in many cases the latter class is perfectly satisfactory. Amorphous polymers do offer some processing advantages, because vitrification is essentially instantaneous upon cooling, whereas the kinetics of crystallization can be highly dependent on the polymer, the flow profile, and the presence of additives. On the other hand, processing is usually carried out tens of degrees above T_g, because of the strong temperature dependence of the viscosity (recall Sections 12.4 and 12.5), whereas semicrystalline polymers can be processed just a few degrees above T_m. One further attribute of amorphous polymers that makes them the materials of choice in numerous applications is optical clarity. As the molecular packing in the glassy state is that of a liquid, the refractive index is spatially homogeneous and isotropic; as discussed at length in Section 8.2, this minimizes the scattering of visible light. Accordingly, the three polymers identified above are familiar in everyday use: polystyrene in clear plastic cups; poly(methyl methacrylate) as Plexiglas®; polycarbonate in compact discs. In contrast, semicrystalline polymers contain more dense crystalline regions within an amorphous matrix; the resulting fluctuations in refractive index often render the material opaque, or at best, hazy.

12.7.1   Basic Concepts

The mechanical properties of polymer solids represent an extremely rich but complicated field of study, and we must necessarily limit our focus. For those students with a background in materials science and engineering, the topic of this section should be familiar, but for those trained in the chemical sciences an introduction may be necessary. It could also be helpful to review some of the material in Section 10.3 and Section 10.5 on elastic deformation, elastic modulus, and the stress–strain behavior of elastomers. We begin with the following observations.

1. Every day adjectives such as *strong*, *hard*, and *tough*, which might be considered to be roughly synonymous, now acquire specific and distinct meanings. Table 12.3 provides a glossary of pertinent terminology.

2. Throughout our consideration of viscoelasticity in Chapter 11, we emphasized the limit of small strains and strain rates: the material response was linear. Now we are concerned with large strain, nonlinear response, and in particular we ask the question, how will a given material break?

3. There are actually many questions that one might ask about the nonlinear response of a material, such as: how large a strain (or stress) can the material withstand before failure? How large a strain (or stress) is required to achieve a nonlinear response? How large a strain (or stress) is required to undergo a nonrecoverable (plastic) deformation? How many times can a material undergo a relatively small deformation without deteriorating? The answers to these questions provide "figures of merit," but they need not be simply related to one another for a given material, and a different figure of merit may be most crucial for a particular application.

Table 12.3   Terminology Relevant to the Mechanical Strength of Materials

Term	Significance
Brittle fracture	Failure by rapid crack propagation without much deformation
Craze	Localized yielding consisting of microvoids interspersed with fibrils
Ductility	Ability to undergo substantial plastic deformation before failure
Elastic deformation	Completely recoverable deformation (note: not synonymous with Hookean)
Engineering stress	Force divided by initial cross-sectional area
Hardness	Ability to withstand surface abrasion or indentation
Plastic deformation	Nonrecoverable deformation
Strain hardening	Stress increasing with strain during plastic deformation
Modulus	Stress divided by strain during small elastic deformation
Tensile strength	Tensile stress at point of fracture
Tensile stress	Maximum engineering stress without fracture
Toughness	Amount of energy absorbed during fracture (proportional to the area under the stress–strain curve)
True stress	Force divided by instantaneous cross-sectional area
Yield point	Onset of plastic deformation
Yield strength	Magnitude of stress at the yield point

Source: Adapted from Callister, W.D., *Materials Science and Engineering: An Introduction*, 5th ed., Wiley, New York, 2000.

4.   Just as there are many questions to ask, there are also many different testing protocols. We will emphasize tensile testing (i.e., uniaxial extension as in Section 10.3 and Section 10.5), but response to torsion, compression, shear, bending, and sudden impact are other common modes. Molecular level information about the origins of strength and mechanisms of fracture is often obtained from controlled crack propagation experiments. The response of a given material may vary significantly from one mode of deformation to another.

Figure 12.19 illustrates three schematic stress–strain curves for polymers in uniaxial extension. Curves A and B represent glassy materials, whereas curve C is a rubber such as that discussed in

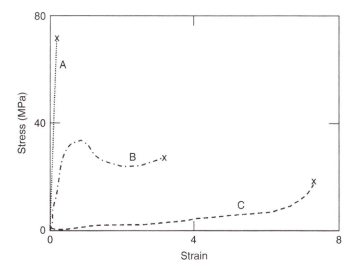

Figure 12.19   Schematic illustration of the stress–strain curves for polymers in uniaxial extension, drawn approximately to scale. Curve A: brittle, curve B: ductile (plastic), curve C: elastomeric.

Table 12.4   Representative Values of Mechanical Properties for Common Thermoplastics
at Room Temperature

Polymer	E (GPa)	Tensile strength (MPa)	Yield strength (MPa)	Elongation at break (%)
Polycarbonate	2.4	60–70	62	110–150
Poly(methyl methacrylate)	2.2–3.2	48–72	54–73	2–6
Polystyrene	2.3–3.3	36–52	—	1.2–2.5

Source: Data compiled in Callister, W.D., *Materials Science and Engineering: An Introduction*, 5th ed., Wiley,

Chapter 10 (see Figure 10.14 for an example). Curve A illustrates a material that undergoes *brittle fracture*. The modulus E, which corresponds to the slope of the curve in the small strain limit, is relatively high, and there is not much deviation from linearity in response up to the point of failure. Failure typically occurs at strains of less than 10%. Curve B illustrates a much richer response, with the stress first rising steeply, then exhibiting a maximum followed by a decrease, which evolves into a broad interval of nearly constant or slightly increasing values before fracture ultimately intervenes. The stress maximum corresponds to the *yield point*, the strain at which the material begins to undergo *plastic deformation*. The ensuing large range of strains, during which the material may exhibit some degree of both strain softening and strain hardening, is characteristic of a *ductile* material. The strain at break may correspond to extensions of up to a factor of 100%. As indicated in Table 12.3, the *toughness* of the material is given by the area under the stress–strain curve, so materials that undergo ductile deformation can be extremely tough. At room temperature, polystyrene and poly(methyl methacrylate) show behavior closer to curve A, whereas polycarbonate follows curve B. These differences may be attributed to differences in chain stiffness and entanglement density, as will be discussed in Section 12.7.3. Representative values of modulus, tensile strength, yield strength, and elongation at break are given in Table 12.4. Curve C for the rubber is markedly different. First, the modulus is lower by 2–3 orders of magnitude; for most polymer glasses E is in the range 2–4 GPa, whereas for rubbers the values have a wider range (depending on crosslink density), but 0.1–1 MPa is typical. Second, there is no yielding; the deformation is largely recoverable throughout the deformation. Third, the stress increases monotonically up to the point of failure. Fourth, the strain at break may correspond to extensions approaching 1000%, and generally exceeds 500%.

12.7.2   Crazing, Yielding, and the Brittle-to-Ductile Transition

Figure 12.20 shows stress–strain curves for poly(methyl methacrylate) at various temperatures below T_g (approximately 110°C). The most notable feature of these curves is the brittle-to-ductile transition that occurs between 40°C and 50°C. For lower temperatures the behavior follows Curve A of Figure 12.19. The stress rises almost linearly with strain up to a maximum strain of a few percent, at which point the specimen breaks. In contrast, for temperatures between 50°C and T_g the response follows Curve B of Figure 12.19. The yield point corresponds to a strain amplitude similar to that at the point of brittle fracture at slightly lower temperatures. As temperature increases within the ductile regime, the strain-to-break increases substantially. This transition from brittle failure to a ductile response as the glass transition is approached from below is common, but the temperature interval over which the (usually more desirable) ductile behavior is seen depends on the polymer. The phenomenon of increasing brittleness with decreasing temperature is a familiar one; plastic toys, automobile parts, etc. are often noticeably more brittle, or at least stiffer, when left outside on a cold winter day. Similarly most of us have seen a demonstration of brittle fracture for a rubber ball, flower, or similar soft material after immersion in liquid nitrogen. Macroscopic failure of a polymeric material corresponds to rupture of covalent

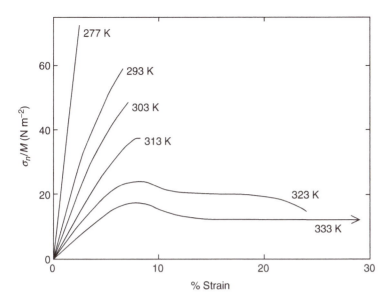

Figure 12.20 Stress–strain response for poly(methyl methacrylate) at various temperatures, showing the brittle-to-ductile transition. (Reproduced from Young, R.J., and Lovell, P.A., *Introduction to Polymers*, 2nd ed., Chapman & Hall, London, 1991. With permission.)

bonds at the molecular level. Brittle fracture implies that there is little deformation of the material prior to bond rupture and as reduced temperature leads to reduced molecular mobility, it is reasonable that the material undergoes less deformation before fracture at lower temperatures. In contrast, ductility requires substantial molecular mobility, and is thus favored at higher temperatures. In a qualitative sense the process of yielding is analogous to melting; the material is able to "flow" under the externally imposed deformation. In support of this notion is the fact that the brittle-to-ductile transition occurs at progressively higher temperatures as the rate of deformation is increased. At higher rates there is less time for molecular rearrangements to relieve the stress, and thus bond rupture is more likely.

Brittle fracture in polymers occurs by the formation of cracks, as in other materials such as metals and ceramics. However, there are major differences between the way cracks propagate in glassy polymers as compared to ceramics. In the latter, cracks tend to follow grain boundaries and other defects, which often leads to significant variation in performance depending on the grain size, defect density, etc. of the material. In glassy polymers the lack of structural regularity, combined with the substantial spatial extent and interpenetration of the covalently bonded chains, makes the mechanical response more uniform. Cracks in polymers, therefore, are usually initiated by defects on the surface, such as scratches. In polymers, the mode of crack propagation after initiation is also qualitatively different than in metals and ceramics; the new phenomenon is called crazing. A *craze* is formed by a localized cavitation process, in which microvoids are created in the polymer to accommodate the applied strain. The microvoids are surrounded by a fibrillar structure where the fibrils contain extended polymer chains. An electron micrograph of a craze in polystyrene is shown in Figure 12.21, along with a cartoon version of the craze structure. The thickness of a typical craze is on the order of 100 nm and the fibrils are typically about 10–20 nm thick. As the macroscopic strain is increased, the crazes propagate roughly perpendicular to the strain direction and then give way to cracks when the individual fibrils break. The formation of crazes has two major implications for the fracture process. First, a significant amount of energy is dissipated in the drawing of polymer chains into fibrils; mechanical energy expended in this manner is no longer available for bond rupture, so the ability to craze greatly enhances the mechanical strength of the

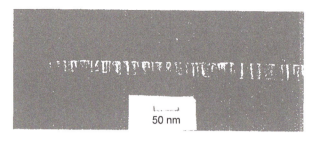

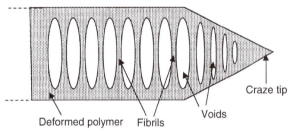

Figure 12.21   (a) Transmission electron micrograph of a craze in polystyrene. (Reproduced from Donald, A.M., *The Physics of Glassy Polymers*, Haward, R.N., and Young, R.J. (Eds.), 2nd ed., Chapter 6, Chapman & Hall, London, 1997. With permission.) (b) Cartoon of a craze, illustrating voids, fibrils, and the deformation zone.

material. Second, the significant deformation that takes place during crazing means that typically the strain at break is several percent. This should be contrasted with ceramic materials, such as everyday silicon dioxide–based glasses, where brittle fracture occurs at strains well below one tenth of 1%.

Yielding occurs when the material is able to undergo significant deformation without crazing. The crucial difference between crazing and yielding is that the former is a highly localized deformation, whereas the latter involves macroscopic deformation. Crazing leads to brittle failure because the imposed strain must be accommodated locally, whereas yielding allows the strain to be distributed over larger volumes of material. In such a case, there is less chance of a localized stress buildup that is sufficiently large to break bonds. It is often the case that the macroscopic sample under tension displays a shape transformation just after yielding known as *necking* (see Figure 12.22), whereby the specimen becomes visibly thinner at some point along its length. In this region the individual molecules have been extended to some significant degree. By analogy to rubber elasticity (Chapter 10) we know that the force to extend a sample increases with extension. Therefore it is easier to stretch the unnecked portion than to continue to extend the necked region. The consequence of this is that once necking has occurred in one location, the size of the necked region tends to grow while the original neck thickness is more or less preserved. During this region the engineering stress will remain roughly constant, as individual portions of the sample deform. Eventually the neck encompasses the entire specimen, and further extension leads to a more uniform deformation along the sample, often accompanied by some strain hardening. Detailed analysis of the material response throughout the postyield regime is complicated, especially because the strain is not homogeneous throughout the material; the distinctions between the engineering stress (strain) and the true stress (strain) (see Table 12.2) become very important.

In contrast with polystyrene and poly(methyl methacrylate), which tend to craze at any temperature 50° or more below T_g, polycarbonate shows yielding behavior over 200° below T_g, as shown in Figure 12.23. Clearly there must be some important aspects of the molecular structure

Figure 12.22   Schematic illustration of necking in a ductile polymer undergoing uniaxial extension.

that accounts for this difference. The dependence of the mechanical response of glassy polymers on molecular variables will be taken up in the next section.

12.7.3   Role of Chain Stiffness and Entanglements

The molecular mechanisms of plastic deformation and failure in glassy polymers are far from completely understood, but some important basic principles have been elucidated. A crucial parameter turns out to be the molecular weight between entanglements, M_e (recall Section 11.6) or equivalently the number of entanglement strands per unit volume, ρ_e:

$$\rho_e = \frac{\rho N_{av}}{M_e} \tag{12.7.1}$$

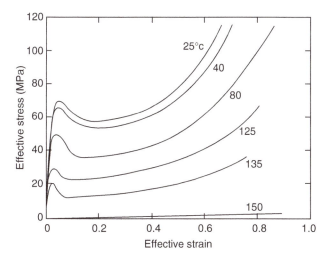

Figure 12.23   Stress–strain behavior of polycarbonate in uniaxial extension at various temperatures. (Reproduced from G'Sell, C., Hiver, J.M., Dahoun, A., and Souahi, A., *J. Mater. Sci.*, 27, 5031, 1992. With permission.)

where ρ is the mass density. In Chapter 11 we discussed at some length how entanglements play a central role in the dynamic properties of a polymer melt, such as the viscosity and the stress relaxation modulus. This analysis, in turn, was developed by analogy to rubber elasticity, where the fundamental parameter was the density of elastically active strands. It was argued that the entanglement strand in a melt acts like an elastically active strand in a network until the reptation process finally allowed each chain to escape its entanglements and permit flow. The value of M_e was also shown to be well correlated with the chain stiffness (i.e., the characteristic ratio, C_∞) and the average thickness of the chain, through the packing length. Now we are asserting that the same concepts have relevance to deformation below the glass transition. This might seem surprising at first because after all, in both melts and rubbers local relaxations (i.e., on a length scale smaller than M_e) are facile, whereas the main consequence of undergoing the glass transition is to freeze out such motions. However, we are now dealing with large imposed strains and as a consequence the molecules must do something; one possibility is that chain segments less than M_e long can slip past one another.

What are the possible responses of a polymer glass to increasing strain? Here are some to consider:

1. *Bond rupture*: The material could fracture by scission of enough backbone bonds, where "enough" means all the bonds traversing some fracture surface. In the case of an "ideal" chain of links under tension, the tension is equal in all links, and they would all fracture simultaneously when the tension in each link reached a critical value. Of course, the tension will not be uniformly distributed at the level of individual carbon–carbon bonds in a real polymer material, so some would fracture first, thereby relieving the tension on the others. A simple calculation to estimate the stress required to fracture all the bonds in a polymer (Problem 12.15) shows that this value is significantly larger than both the yield and tensile strengths (see Table 12.4).
2. *Molecular separation*: In this limit the molecules would simply move apart without bond rupture. The energy required to do this would be related to the intermolecular interactions, quantified by the cohesive energy density (see Section 7.6). The dispersive interactions that hold nonpolar molecules together in liquid or solid phases are orders of magnitude weaker than covalent bonds, so simple estimates of the required stresses are much lower than observed values (see Problem 12.16).
3. *Long chain pullout*: It should be apparent that as high molecular weight polymers are thoroughly intertwined with one another, and because mobility is very low below T_g, the simple molecular separation mode is not feasible. Another possibility is that under tension the individual molecules disentangle by a kind of forced reptation, and are pulled apart without bond rupture. The friction associated with disentanglement would certainly dissipate energy, giving rise to a larger tensile strength than case 2. However, as we saw in Chapter 11, it takes many orders of magnitude longer for chains to disentangle than to undergo local rearrangements; and as the local rearrangements themselves are already very slow, such a process seems unlikely.
4. *End strand or short chain pullout*: Chain end segments that are shorter than M_e, or chains with $M < M_e$, could pull out from one another without the need for forced reptation.
5. *Chain extension*: Perhaps polymers that are sufficiently flexible can be induced to undergo local conformation rearrangements (such as *trans* to *gauche* for a carbon–carbon bond), and thereby the material can extend without either rupture or pullout. In essence, this response would be equivalent to the extensional flow of an entangled melt, but of course, requiring substantially higher stresses to bring it about.

This set of possibilities is not exhaustive, and none of these possibilities are mutually exclusive. At a superficial level, however, we can match these possibilities to the observed phenomena. First of all, responses 2 and 3 are not important to polymers. The fact that response 3 is not relevant is indicated by the molecular weight independence of tensile strength and yield strength for high molecular weight materials. Fracture ultimately involves some degree of response 1. Yielding and ductility are largely encompassed by response 5. The fact that ductility involves substantial chain extension is demonstrated by the fact that once the plastically deformed specimen is heated above T_g, it undergoes large scale elastic recovery just as a deformed rubber or entangled melt would. Crazing involves a combination of responses 1, 4, and 5, but the chain extension part is restricted both in extent and localized in space.

If we consider first the propagation of a craze, the material responds both by forming extended fibrils and by cavitation. The latter process requires the formation of free surface and therefore costs an amount of energy proportional to the product of the void surface area and the surface tension of the material. The formation of fibrils requires local extension of the polymer chains. The telling feature of these fibrils is that the local extension ratio λ (recall Section 10.5) within the fibril remains constant, even as the craze widens. Thus the fibrils grow in length by drawing in more material, not by thinning the existing fiber. Furthermore, the extension ratio varies with the material, but is well approximated by the ratio of the contour length of an entanglement strand to the root mean square end-to-end distance of the same strand. In other words, the material within the fibrils corresponds to nearly fully extended entanglement strands. The material ultimately fails when chemical bonds within the fibrils undergo rupture, presumably when fresh material cannot be drawn into the fibrils at the necessary rate.

The preceding discussion indicates that an entanglement network is a necessary precondition for crazing to occur. This is also nicely illustrated in Figure 12.24, where the elongation at break is plotted as a function of total molecular weight for polystyrene. For high molecular weights the value is constant, but at low molecular weights it tends to vanish. The curve extrapolates to a value of about 35,000, indicating that there is essentially no tensile strength for chains shorter than this. Recalling from Table 11.2 that M_e for polystyrene is about 13,000, we can see that 2–3 entanglement strands per chain are necessary to develop appreciable tensile strength and beyond about 10 strands per chain the strength becomes constant. These values are easily appreciated.

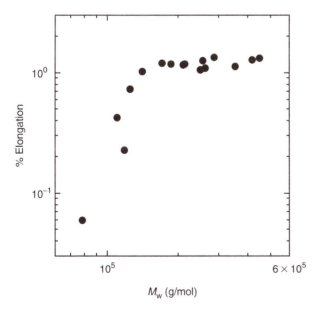

Figure 12.24 Elongation at break in polystyrene as a function of weight average molecular weight. (Data reported in McCormick, H.W., Brower, F.M., and Kim, L., *J. Polym. Sci.*, 39, 87, 1959.)

Imagine a plane through the material along which we wish to drive a crack. For any chain that is shorter than 2–3 M_e long that traverses this plane, at least one end will be less than M_e long and will therefore be able to pull out readily (case 4 above). On the other hand, for chains longer than about 10 M_e, almost all chains that traverse the plane will be well anchored by several entanglement strands on each side of the interface.

We now turn to one last question, namely, which molecular characteristics determine whether the material responds by crazing, leading to brittle fracture, or by yielding? The accumulated evidence indicates that increasing chain flexibility (i.e., smaller C_∞) and higher entanglement densities, ρ_e, favor yielding [10]. For example, polystyrene tends to craze ($C_\infty = 9.5$, $\rho_e = 4.8 \times 10^{19}$ cm^{-3}), poly(methyl methacrylate) tends to craze but yields at higher temperature ($C_\infty = 9.0$, $\rho_e = 6.8 \times 10^{19}$ cm^{-3}), whereas polycarbonate yields ($C_\infty = 2.4$ [10], $\rho_e = 5 \times 10^{20}$ cm^{-3}). Of course, external variables such as temperature, deformation type, and deformation rate can affect the answer in a particular case, but all other things being equal, this demarcation is generally correct. The question can be reformulated slightly when we recognize that for any material there must be critical stress values for both crazing and yielding to occur and the observed response will correspond to the process with the lower critical stress value. Smaller values of C_∞ are generally associated with chain structures that have smaller energy differences between *trans* and *gauche* conformers and thus chain extension should be more facile in such cases; this would lower the critical yield stress. On the other hand, crazing requires cavitation. In addition to the surface energy penalty noted above, the formation of a cavity requires either chain scission or pullout across the interface where the cavity forms. The higher the entanglement density, the shorter will be the dangling entanglement strands at the ends of chains and the smaller the extension ratio of one entanglement strand. Both factors increase the critical stress for crazing and thus more highly entangled chains favor yielding. Simply put, a higher entanglement density corresponds to a material that is more tightly stitched together, and therefore more resistant to localized yielding (i.e., crazing).

12.8 Chapter Summary

In this chapter, we have examined the transition between the liquid state and the glassy state, which takes place over a range of temperatures near a characteristic glass transition temperature, T_g. The principal points are the following:

1. The glass transition is a kinetic transition, but it approximates a second-order thermodynamic transition. A completely satisfactory theory of the glass transition is not yet available.
2. The glass transition temperature may be located in a variety of ways, but the most common tools are DSC and rheology.
3. The glass transition temperature is the single most important parameter in determining whether a given polymer may be suitable for a certain application. There is no simple way to correlate T_g with a particular chemical structure, although some general rules of thumb exist.
4. The value of T_g may be modified by changing molecular weight or by blending; the molecular weight and composition dependences of T_g are generally straightforward.
5. The concept of free volume is a particularly useful and physically intuitive way to understand the glass transition, and the profound effect that proximity to T_g has on the temperature dependence of any viscoelastic or transport property. The free volume approach provides a natural explanation for the widely used Vogel–Fulcher–Tammann and Williams–Landel–Ferry equations, which describe the temperature dependence of viscoelastic properties above T_g.
6. The principle of time–temperature superposition is an essential ingredient in the study of polymer viscoelasticity because small changes in temperature produce large changes in the polymer relaxation times. Consequently measurements over a finite range of time or frequency at one temperature can be superposed with measurements at other temperatures to generate

master curves of dynamic response, which can extend over as many as 20 orders of magnitude in reduced time or frequency.

7. Noncrystallizable thermoplastics are in common use, for their ease of processing, optical clarity, and mechanical strength. Glassy polymers under large deformation may undergo either brittle failure, through a distinctive localized yielding process known as crazing, or yield macroscopically, leading to very large elongations before failure. Both processes involve extension of individual chains; the operative response mode is strongly influenced by the flexibility and entanglement density of the material.

Problems

1. Hirai and Eyring[†] assembled the following data from diverse sources (scarcely any two pieces of data were measured in the same laboratory, much less on the same sample):

	V_{sp} (cm^3/g)	ΔC_p (erg/K/g)	$\Delta\alpha$ (K^{-1})	$\Delta\kappa$ (cm^2/dyn)
Rubber	1.1	5×10^6	4.0×10^{-4}	1×10^{-11}
Polystyrene	1.0	7.7×10^6	1.75×10^{-4}	3×10^{-12}
Polyisobutylene	1.1	4.0×10^6	4.5×10^{-4}	3×10^{-11}

Use these data to evaluate T_g, assuming that the latter is a true second-order transition. Compare your results with the values in Table 12.1 and comment on the agreement or lack thereof.

2. Time–temperature superposition was applied to the maximum in dielectric loss factors measured on poly(vinyl acetate).[‡] Data collected at different temperatures were shifted to match at $T_g = 28°C$. The shift factors for the frequency (in hertz) at the maximum were found to obey the WLF equation in the following form: $\log \omega + 6.9 = [19.6(T - 28)]/[42 + (T - 28)]$. Estimate the fractional free volume at T_g and α for the free volume from these data. Recalling from Chapter 11 that the loss factor for the mechanical properties occurs at $\omega\tau = 1$, estimate the relaxation time for poly(vinyl acetate) at $40°C$ and $28.5°C$.

3. Imagine a high molecular weight polymer that is a cycle. How would its T_g differ from a linear polymer of the same molecular weight? Make an argument based on the spirit of the Gibbs–DiMarzio theory, and one based on free volume ideas. Do they make the same qualitative prediction? Incidentally, the experimental evidence suggests that at infinite molecular weight linear and cyclic polymers have the same T_g, but as the chains get shorter, T_g for cycles actually increases.

4. A polystyrene sample is split into two, and the T_g for each sample is measured by DSC in two different laborataries. The reported results are $98°C$ and $106°C$. Propose three possible simple explanations for this difference.

5. Suppose now that the two technicians in the previous problem consult one another on their measurement techniques. After repeating the measurements, the two laboratories agree as to the value of T_g. However, in one lab the increase in heat flow associated with the transition is about 30% larger than in the other. What could cause this discrepancy?

6. The figure illustrates plots of heat capacity for a particular polymer sample.[§] The scans were obtained at a constant heating rate of $5.4°C/min$, but different cooling rates were used to bring the sample down to the starting temperature (350 K). Explain why the shape of the curve

---

[†] N. Hirai and H. Eyring, *J. Polym. Sci.* 37, 51 (1959).
[‡] S. Matsuoka, G.E. Johnson, H.E. Bair, and E.W. Anderson, *Polym. Prepr.* 22(2), 280 (1981).
[§] B. Wunderlich, D.M. Bodily, and M.H. Kaplam, *J. Appl. Phys.* 35, 95 (1964).

depends on the prior cooling rate, and identify which curve corresponds to the smallest cooling rate. Speculate on the identity of the polymer.

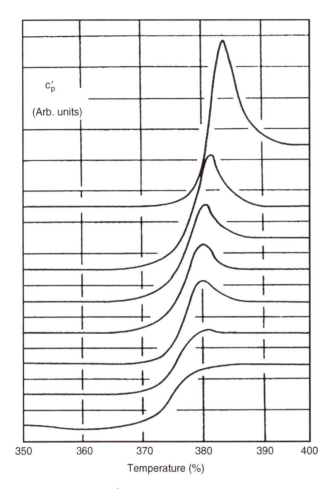

7. Williams and Ferry[†] measured the dynamic compliance of poly(methyl acrylate) at a number of temperatures. Curves measured at various temperatures were shifted to construct a master curve at 25°C, and the following shift factors were obtained. Assess whether these data obey the WLF equation; if so, evaluate the constants C_1 and C_2, and also C_1^g and C_2^g. Note that $T_0 \neq T_g = 3°C$ for these data.

T (°C)	$\log a_T$	T (°C)	$\log a_T$
25.00	0	54.90	−3.88
29.75	−0.98	59.95	−4.26
34.85	−1.80	64.70	−4.58
39.70	−2.42	69.50	−4.88
44.90	−3.00	80.35	−5.42
49.95	−3.47	89.15	−5.72

[†] M.L. Williams and J.D. Ferry, *J. Colloid Sci.* 10, 474 (1955).

8. Plazek and O'Rourke[†] measured the viscosity of polystyrene ($M = 3400$) as a function of temperature. Plot these data in the Arrhenius format (ln η versus $1/T$) and find apparent activation energies for the lowest and highest temperatures measured. Compare these values to those obtained from the data in Figure 12.11b, and comment on their physical significance.

T (°C)	log η	T (°C)	log η
70.0	12.967	94.3	6.888
75.0	11.152	100.6	5.995
79.8	9.749	109.4	5.047
84.3	8.619	130.3	3.418
89.9	7.609	144.6	2.627

9. A certain extruder has optimum performance when $\eta = 2 \times 10^4$ P. The polymer of choice had this viscosity at 150°C, and its M_w was 80,000. Its T_g was about 75°C. An error in polymerization control led to a batch of the same polymer with $M_w = 60,000$. At what temperature should the extruder be run?

10. In small molecule glass formers, a common empirical definition of T_g is the temperature for which the viscosity equals 10^{13} P. At first glance, this definition seems wholly inappropriate for polymers, because the viscosity at fixed temperature varies so strongly with M (i.e., $\eta \sim M^{3.4}$), while T_g varies weakly with M. To assess the worth of this definition, consider polystyrene with $M = 10^5$ and $M = 10^6$. What would be the difference in the two T_gs based on $\eta = 10^{13}$, and how does this compare with what you would expect from calorimetry?

11. It is sometimes suggested that the temperature dependence of the viscosity should follow a power law, with an exponent ν and a divergence (viscosity becomes infinite) at some $T_0 < T_g$:

$$\eta \sim (T - T_0)^{-\nu}$$

Such a dependence is common for many experimental quantities in the vicinity of a true phase transition (so-called critical phenomena). Assess whether the following a_T data for polyisoprene (in part from Example 12.3) can be modeled in this way, and compare the quality of the fit to that with the WLF or VFTH function. How do the values of the Vogel temperature and the putative critical temperature compare?

T (°C)	−40	−30	−20	−10	0	10	20	30	50
a_T	12,000	950	130	28	7	2.5	1	0.43	0.11

12. The following dependence of T_g on M for poly(dimethylsiloxane) was determined by DSC.[‡] Prepare a plot to compare with Figure 12.16, and determine whether the functional relationship of Equation 12.6.1 is followed in this case. Then select some measure of the crossover to the high molecular weight asymptotic value of T_g (e.g., perhaps M where T_g is 5° less than the infinite M limit), and compare these values of M for poly(dimethylsiloxane) and polystyrene. Criticize or defend the proposition that this crossover corresponds to a certain number of persistence lengths (see Table 6.1) independent of chain structure.

[†] D.J. Plazek and V.M. O'Rourke, *J. Polym. Sci. Part A-2*, 9, 209 (1971).
[‡] S.J. Clarson, K. Dodgson and J.A. Semlyen, *Polymer*, 26, 930 (1985).

M_n	T_g (K)	M_n	T_g (K)	M_n	T_g (K)
240	123.4	1630	146.1	7,720	149.5
310	129.1	2260	147.5	10,060	149.5
530	137.4	2460	148.0	12,290	149.4
630	139.4	2920	148.3	14,750	149.7
810	141.2	4080	149.2	18,250	149.8
990	143.3	4880	148.8	21,390	149.8
1290	144.8	6330	149.3	25,460	149.7

13. Sketch what you would expect to see for the dynamic modulus, G', measured at 1 rad/s as a function of temperature, for polyisoprene ($M = 50,000$), polystyrene ($M = 100,000$), and polystyrene ($M = 1,000$). Make the vertical axis logarithmic, and recall that the low temperature modulus for all three polymers will be roughly 10^{10} dyn/cm^2.

14. Show that the Fox equation, Equation 12.6.9, can be obtained from the Couchman equation, Equation 12.6.8. Hint: use the proximity of the component T_gs to eliminate the logarithms.

15. Estimate the tensile strength of polystyrene by assuming that the required stress corresponds to breaking backbone carbon–carbon bonds. The bond energy is about 80 kcal/mol, and the density of polystyrene is about 1.05 g/cm^3. How does this compare to the values in Table 12.4 (be careful to match units?) How do you account for the difference?

16. It is known that in fracturing polystyrene, by driving a crack through the material, the fracture energy released is on the order of 1 kJ/m^2. From Chapter 7 we recall that the solubility parameter of polystyrene is 9.1 (cal/cm^3)$^{1/2}$. Use this value to estimate the surface energy of polystyrene, for example by estimating the "lost" monomer–monomer interactions per unit area by creating the new surface. How do these two values compare? What is the main origin of the difference?

References

1. Kauzmann, W., *Chem. Rev.* 43, 219 (1948).
2. Gibbs, J.H. and DiMarzio, E.A., *J. Chem. Phys.* 28, 373 (1958).
3. Kovacs, A.J., *J. Polym. Sci.* 30, 131 (1958).
4. Doolittle, A.K., *J. Appl. Phys.* 22, 1031; 1471 (1951); 23, 236 (1952).
5. Vogel, H., *Physik. Z.,* 22, 645 (1921); Fulcher, G.S., *J. Am. Chem. Soc.* 8, 339; 789 (1925); Tammann, G. and Hesse, G.Z., *Anorg. Allg. Chem.* 156, 245 (1926).
6. Williams, M.L., Landel, R.F., and Ferry, J.D., *J. Am. Chem. Soc.* 77, 3701 (1955).
7. Catsiff, E. and Tobolsky, A.V., *J. Colloid Sci.* 10, 375 (1955).
8. Couchman, P.R., *Macromolecules* 11, 1156 (1978).
9. Callister, W.D., *Materials Science and Engineering: An Introduction*, 5th ed., Wiley, New York (2000).
10. Wu, S., *Polym. Eng. Sci.*, 30, 753 (1990).

Further Readings

Bower, D.I., *An Introduction to Polymer Physics*, Cambridge University Press, Cambridge, UK, 2002.
Donth, E., *The Glass Transition*, Springer, Berlin, 2001.
Ediger, M.D., Angell, C.A., and Nagel, S.R., *Supercooled Liquids and Glasses, J. Phys. Chem.* 100, 13200, 1996.
Ferry, J.D., *Viscoelastic Properties of Polymers*, 3rd ed., Wiley, New York, 1980.

Haward, R.N. and Young, R.J., *The Physics of Glassy Polymers*, 2nd ed., Chapman & Hall, London, 1997.

Sperling, L.H., *Physical Polymer Science*, Wiley, New York, 1986.

Strobl, G., *The Physics of Polymers*, 2nd ed., Springer, Berlin, 1997.

Ward, I.M., *Mechanical Properties of Solid Polymers*, 2nd ed., Wiley, New York, 1983.

Wunderlich, B., *Thermal Analysis*, Academic Press, New York, 1990.

Young, R.J. and Lovell, P.A., *Introduction to Polymers*, 2nd ed., Chapman & Hall, London, 1991.

13

Crystalline Polymers

13.1 Introduction and Overview

In this chapter we consider several important aspects of crystallinity in polymers. It is a remarkably rich field, and our coverage necessarily limited, but we will raise most of the central issues and show how many of them have been addressed. In the previous chapter we examined the glass transition, and introduced it by contrast to crystallization and melting. The first section of Chapter 12, therefore, serves as part of the introduction to the current topic as well. To continue this introduction, we pose a series of basic questions, and the answers form the outline for the rest of the chapter.

Why is crystallization in polymers important?

The world's most popular synthetic polymer, in terms of volume produced per year, is polyethylene; polyethylene can crystallize. Other high-volume polymers such as isotactic polypropylene, poly(hexamethylene adipamide) (Nylon 6,6), and poly(ethylene terephthalate) crystallize, as do many specialty materials, such as poly(tetrafluoroethylene) (Teflon) and poly(p-phenylene terephthalamide) (Kevlar). In general, crystallinity conveys enhanced mechanical strength, greater resistance to degradation, and better barrier properties.

Which polymers crystallize and which do not?

The simple answer is: stereoregular polymers (polyethylene, isotactic or syndiotactic polypropylene, poly(ethylene oxide), etc.) crystallize, stereoirregular polymers (atactic polystyrene and poly(methyl methacrylate)) or polymers of mixed microstructure (mixed *cis* and *trans* polydienes) do not. In order to crystallize, it is necessary for a few monomers to pack into a regular unit cell, which can then be stacked on a lattice to fill space. It is almost impossible for an atactic polymer such as polystyrene to form a regular unit cell, because the side groups are placed on one side of the backbone or the other at random. However, even this rule is not always obeyed. For example, a vinyl polymer with the formula $-(CH_2-C(AB))_n-$, in which the groups A and B have similar sizes, may be able to pack into a regular array. Or, an atactic polymer in which the side chain is strongly polar may sometimes crystallize; poly(vinyl alcohol) and poly(vinyl fluoride) are examples.

What is the structure of a polymer crystal and how do we characterize it experimentally?

Polymers crystallize with three levels of structure, as illustrated in Figure 13.1. On the first level, individual chain backbones form helices (of which an all-*trans* conformation is a special case), and pack with neighboring chains to form *unit cells*. The typical unit cell contains only a few monomers, and has dimensions of 2–20 Å on a side. The structure of unit cells will be considered in Section 13.2. On the second level, unit cells pack into thin sheets, called *lamellae*, which are typically 100–500 Å thick and several microns wide in the other two dimensions. The chain backbones lie at some fixed angle relative to the thin direction, and often fold by 180° at the lamella surface in order to reenter the crystal. Chain-folded lamellae, illustrated in Figure 13.1, are a unique morphological feature of polymers. These structural details are explored primarily through x-ray diffraction and electron microscopy, as will be discussed in Section 13.4. In

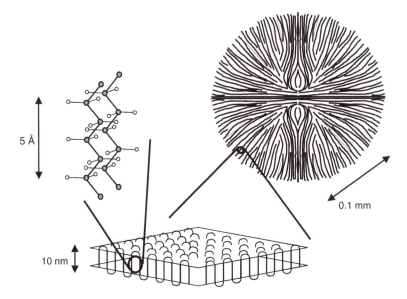

Figure 13.1 Three levels of structure in a crystalline polymer. The packing of individual helical chains gives unit cells with dimensions of a few angstroms. The unit cells are packed into chain-folded lamellae, with characteristic thicknesses on the order of 10 nm. The lamellae splay, bend, and branch to form spherulites, which can exceed millimeters in size. The space between the individual lamellae is filled with amorphous material.

favorable cases, polymer single crystals can be grown from solution, which greatly facilitates detailed structural analysis of the unit cell and the lamellae. Finally, in a bulk sample the lamellae grow to fill space, often producing a three-dimensional structure called a *spherulite*, which can be tens or hundreds of microns across. These can be observed in an optical microscope. The structure of spherulites, and some other bulk morphologies, will be taken up in Section 13.6.

Compared to the glass transition, crystallization and melting are thermodynamic transitions; does this mean that a thermodynamic analysis will be more successful?

Would that it were so, but it is not. The process of crystallization is dominated by kinetics. The underlying reason is simple. A molten polymer is a random jumble of intertwined chains, whereas the crystal has long sections of chains fully extended and closely packed in parallel with one another. Once crystallization starts, different sections of one chain may be in several different crystallites, which prevents the full extension of the chain. This is illustrated schematically in Figure 13.2. Consequently, and recalling the characteristically sluggish dynamics of polymers (see Chapter 11), it is not at all surprising that crystallization of bulk polymers may never achieve the state of minimum free energy. Furthermore, for the same kinetic reasons it is never possible to achieve 100% crystallinity in a bulk polymer, and thus there is always a nonequilibrium mixture of amorphous and crystalline regions within the material. Such polymers are more correctly termed *semicrystalline*. Nevertheless, thermodynamics still provides a good deal of insight, particularly when the process of melting is considered; in Section 13.3 and Section 13.4 in particular we will pursue a thermodynamic analysis.

How can we understand the kinetics of crystallization in general?

Crystallization is a first-order phase transition (recall Section 12.2), and it proceeds by the process of nucleation and growth, that is, stable nuclei of the new (crystal) phase appear, and then

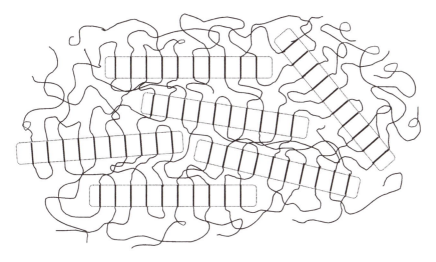

Figure 13.2 Illustration of a semicrystalline polymer melt. Individual crystal lamellae are indicated by dashed lines.

grow in size by incorporation of more chains or chain segments from the amorphous (liquid) phase. We will examine growth kinetics on two different levels, that of individual crystals in Section 13.5, and that of the crystalline fraction of the material in Section 13.7.

13.2 Structure and Characterization of Unit Cells

In this section we shall examine the smallest level of structure displayed by polymer crystals, the unit cell. These unit cells typically have dimensions between 2 and 20 Å, which lie in the same range as atomic and small-molecule crystals. Accordingly, the same experimental technique is employed to determine the spacings and the symmetries of the unit cell, namely x-ray diffraction (XRD), or, as it is more commonly known in polymer science, wide-angle x-ray scattering (WAXS). We will not describe WAXS in detail, but the basic process can be understood through comparison with the determination of molecular structure and size by light scattering, which was covered in Chapter 8. Similarly, we will only review the basics of crystallography; more details are available in a number of monographs and texts.

13.2.1 Classes of Crystals

The unit cell contains the smallest number of atoms, in the appropriate spatial relationships, necessary to enable prediction of the full structure of a macroscopic single crystal by repetitive close stacking of unit cells. A schematic unit cell is illustrated in Figure 13.3a. The lengths of the three sides are designated a, b, and c, and the corresponding angles are α, β, and γ. A single crystal can thus be generated by filling space with unit cells, so that the structure repeats exactly every a Å as one moves along the a direction, etc. However, the precise location of the various atoms within the unit cell requires further information than is contained in the three lengths and three angles. There are seven *crystal classes* (cubic, trigonal, hexagonal, tetragonal, orthorhombic, monoclinic, and triclinic), which are defined by different constraints on the values (a, b, c) and (α, β, γ). These are given in Table 13.1. Within these classes there are further subdivisions, for example a cubic unit cell could be face-centered, body-centered, or primitive. When these possibilities are included it turns out that there are 14 distinct structures, called *Bravais lattices*, within these seven classes. Finally, within a certain Bravais lattice there can be many different ways in which the atoms are

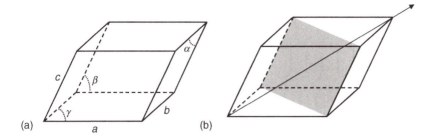

Figure 13.3 Schematic of a unit cell, showing (a) axes a, b, and c, and associated angles α, β, and γ and (b) the (110) plane (shaded) and the [111] direction (arrow).

arranged in detail. These arrangements are described by sets of symmetry operations that leave the structure unchanged, such as rotation about an axis by an angle of 60°, 90°, or 180°, or reflection through a plane, etc. The set of symmetry operations that applies to a particular crystal specifies a *space group*; in total there are 230 different space groups. The determination of the space group and the full unit cell structure of a polymer crystal including bond angles, bond lengths, and interchain distances is the first goal of polymer crystallography.

Polymers are extended one-dimensional objects in the crystalline state, and the overall direction of the backbone corresponds to one axis of the unit cell. For reasons that will become apparent this axis is termed the *fiber axis*. By convention, it is assigned to the c axis in the unit cell (except for monoclinic crystals, where it is b). All of the seven crystal classes in Table 13.1 are found in polymers, except one: cubic. This exclusion can be easily understood. In a cubic unit cell all three axes must be equivalent, but because of the covalent backbone it is virtually impossible for monomers to pack equivalently in three orthogonal directions.

Particular crystallographic planes or directions are commonly labeled with the assistance of Miller indices hkl. One vertex of the unit cell is chosen as the origin. The plane of interest intersects the three unit cell axes at particular coordinates x, y, and z. The Miller indices are obtained as $h = a/x$, $k = b/y$, and $l = c/z$. By choosing an equivalent plane so that it intersects all three axes within a single unit cell, the Miller indices are always integers. Note that if a plane is parallel to a particular axis the intersection occurs at infinity, and so the corresponding Miller index is 0. The rules for specifying directions are analogous. In the case of a negative Miller index, the value is indicated with an overbar. By convention, a particular plane is referred to by the Miller indices in parentheses, that is (hkl), the family of equivalent planes by $\{hkl\}$, a particular direction as $[hkl]$, and the family of equivalent directions by $\langle hkl \rangle$. By these conventions, the ab face of the unit cell is (001), and the c axis is [001]; the (110) plane and [111] axis are illustrated in Figure 13.3b.

Table 13.1 Constraints on Unit Cells

Crystal class	Unit-cell sides	Unit-cell angles
Cubic	$a = b = c$	$\alpha = \beta = \gamma = 90°$
Trigonal	$a = b = c$	$\alpha = \beta = \gamma \neq 90°$
Hexagonal	$a = b \neq c$	$\alpha = \beta = 90°,\ \gamma = 120°$
Tetragonal	$a = b \neq c$	$\alpha = \beta = \gamma = 90°$
Orthorhombic	$a \neq b \neq c$	$\alpha = \beta = \gamma = 90°$
Monoclinic	$a \neq b \neq c$	$\alpha = \gamma = 90°,\ \beta \neq 90°$
Triclinic	$a \neq b \neq c$	$\alpha \neq \beta \neq \gamma$

13.2.2 X-Ray Diffraction

As noted above, XRD or WAXS is the standard tool for crystal structure determination. In Section 8.2 we derived Bragg's law, and noted that the criterion for observing a diffraction peak is that the scattering vector, q, match a reciprocal lattice vector of the crystal (see Figure 8.3). Bragg's law may be written as

$$m\lambda = 2D \sin\left(\frac{\theta}{2}\right) \qquad (13.2.1)$$

where λ is the wavelength, D is the spacing between lattice planes, θ is the angle between the incident and diffracted radiation, and m is a positive integer. Recall that a *reciprocal lattice vector* points in the direction normal to a series of lattice planes, with a magnitude given by $2\pi/D$. The magnitude of the *scattering vector* is defined as

$$q = \frac{4\pi}{\lambda} \sin\left(\frac{\theta}{2}\right) \qquad (13.2.2)$$

and is illustrated schematically in Figure 8.4. Each of the 230 space groups has its own set of "allowed reflections," namely a set of values of q (or θ) that will satisfy Bragg's law. In principle, therefore, an experimental scattering pattern could determine the space group uniquely. In practice there is a lot more to it, first because many space groups have several reflections in common, and second because whether or not particular reflections are actually seen will depend on a host of factors, most important of which is the orientation of the incident x-ray beam relative to the crystal.

In an x-ray diffraction experiment, a collimated beam of monochromatic x-rays is directed on to the sample and the diffracted radiation is monitored as a function of θ. The detector itself can be one-dimensional or two-dimensional, with the latter becoming increasingly common. The condition of the specimen itself plays a crucial role in the experiment; the limiting cases are those of a single crystal and a polycrystalline sample. In a single crystal, the entire portion of the specimen that is illuminated by the x-rays has the same orientations of a, b, and c in the laboratory coordinate system, and thus the beam is incident along a single direction through the unit cell. The result is that Bragg's law is satisfied not only for particular values of q but also for particular directions in space and the scattering pattern on an area detector will be a series of particular spots. This is illustrated in Figure 13.4a for the particular cases of a layered sample and a hexagonal crystal of rods. The pattern that is observed will depend critically on the particular angle between the incident beam and the unit cell. This is illustrated in Figure 13.4b for the same two crystals, now rotated by 90° relative to Figure 13.4a. In the case of the layered sample, the beam is now incident normal to

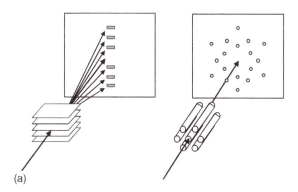

(a)

Figure 13.4 Diffraction patterns on an area detector for (a) a set of parallel sheets viewed edge-on, and a hexagonal array of rods viewed end-on.

(continued)

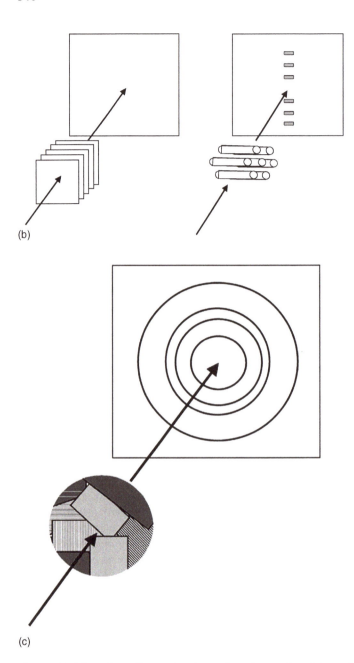

(b)

(c)

Figure 13.4 (continued)   (b) the sheets viewed through and the rods viewed edge-on; and (c) a polycrystalline sample.

the layers, and no diffraction is seen, whereas when viewed side-on, the hexagonal sample appears to be layers. This cartoon is overly simplified, in that we have tacitly assumed that the layers and rods are smooth and structureless. The arrangement of atoms within the layers could give rise to additional reflections beyond those indicated. The important conclusion, however, is that for a full crystallographic analysis of a single crystal sample, the sample must be rotated systematically relative to the incident beam, and the resulting scattering patterns collected and interpreted as a whole. Furthermore, depending on the sample and the size of the detector, it may be necessary to move the detector in space to collect a different angular range.

A polycrystalline sample comprises many little crystals perhaps microns in size, with approximately random relative orientations. Consequently, the incident beam simultaneously samples almost all possible incident angles relative to the unit cell. The result is a so-called *powder pattern*, concentric rings of scattered intensity at radial positions on the detector corresponding to particular values of q, as illustrated in Figure 13.4c. Such a pattern may be sufficient to narrow down the possible crystal structure to a few candidates, but it is unlikely to determine a space group uniquely. Consequently, single crystals are greatly preferred when determination of the unit cell structure is the goal. However, production of a macroscopic single crystal in polymers is no easy feat. It was first shown in the 1950s that single polymer crystals could be grown by careful crystallization from solution, but this strategy is not always convenient or practical. A more common approach is to draw a fiber of the polymer during or prior to crystallization, by applying a uniaxial extensional deformation (see Section 10.5). In this case the individual chains can become highly extended along the draw direction, and the crystals develop with a strong preference for the c axis to lie along this direction (hence the term fiber axis). Although the resulting morphology is not exactly a single crystal (see Section 13.6), it is much more highly organized than a polycrystalline material, and the analysis correspondingly more definitive.

We will not explore this analysis in any further detail, but conclude this overview with some additional comments, especially in comparison to light scattering discussed in Chapter 8:

1. Diffraction is the same phenomenon as coherent scattering; it is merely a convention to use the term *diffraction* when discussing crystalline samples, and *scattering* when amorphous materials are studied. The former gives rise to sharp peaks at particular angles, whereas the latter leads to more smoothly varying intensity versus angle curves.

2. The x-ray diffraction community often employs slightly different terminology, such as k or s in place of q, and θ may be defined as one half of the θ we have employed.

3. The scattering of x-rays is caused by differences in electron density from atom to atom, whereas light scattering arises from fluctuations in refractive index (i.e., polarizability). The atoms of primary interest in polymers, C, H, O, and N, are not particularly strong scatterers of x-rays, but this limitation is overcome by the long-range order in the sample. (Recall that in light scattering from solutions, we have to wait for a spontaneous concentration fluctuation to have the correct spacing and orientation to satisfy the Bragg condition; in a crystalline sample, these "fluctuations" are much larger, and locked in place.)

4. The main difference between light and x-rays as a structural tool lies in the value of λ. We saw in Chapter 8 that R_g needed to be at least 100 Å or larger in order to be determined by light scattering ($\lambda \approx 4000$–7000 Å). In contrast, a common x-ray source involves an inner electronic transition in copper, which gives off a photon at 1.542 Å. For a 10 Å unit cell size (i.e., $D = 10$ Å) inserted into Bragg's law (Equation 13.6.1), we would need a diffraction angle of 8.8°, a very reasonable value.

5. XRD or WAXS measurements can readily be made on laboratory-scale instruments, but the advantages of utilizing synchrotron radiation (e.g., at a National Laboratory such as Argonne or Brookhaven) should be noted. Synchrotron sources provide an incident flux of x-rays that is several orders of magnitude larger than a laboratory source. Furthermore it can be much better collimated, and the wavelength may be tuned to a convenient value. These features combine to provide scattering patterns with much better resolution in much shorter time intervals.

6. Electron diffraction measurements can also be very useful in determining the unit cell structure. The experimental concept is identical to XRD, except that the incident beam consists of electrons. Electrons have a de Broglie wavelength of ~ 0.03 Å, and so are well suited to structures on the 1–10 Å scale. The measurement can be made on an electron microscope, to be discussed in Section 13.4.4.

13.2.3 Examples of Unit Cells

What determines the unit cell that a particular polymer will adopt? This is sometimes difficult to predict a priori, but can usually be rationalized after the fact. From a thermodynamic point of view, the crystal is a low temperature state and is therefore dominated by enthalpic considerations. In particular, the monomers want to maximize their favorable energetic interactions, which generally means to pack as closely as possible; remember from Chapter 7 that the van der Waals energy of attraction between molecules falls off approximately as $1/(distance)^6$. (Note that certain interactions such as hydrogen bonding have a preferred distance.) However, there are both intramolecular and intermolecular interactions to consider. In most instances, polyethylene for example, each chain first adopts its lowest energy conformation (all-*trans* in this case), and then packs as closely as it can to its neighbors. This prioritization corresponds to the first two of *Natta and Corradini's Rules* [1] for polymer crystallization. On the other hand, there may be situations for which the chain conformation in the crystal is not the lowest energy conformation of the isolated chain. Furthermore, the optimum packing of chains is often quite subtle, as we shall see when we consider some examples. In general, polymers adopt one conformational motif for the backbone: a helix. The helix is described by three numbers, such as 2*1/1 for polyethylene. This terminology means 2 backbone atoms constitute a basic repeat unit, and there is 1 repeat unit in each full turn of the helix. In fact, such a 1/1 helix is identical to the all-*trans* conformation. As indicated in Table 13.2, many more interesting helices are found.

Figure 13.5 shows the arrangement of molecules in the polyethylene unit cell. X-ray measurements show that the dimensions are $a = 7.4$ Å, $b = 4.9$ Å, and $c = 2.5$ Å, and that it is orthorhombic. In regard to this unit cell, we observe the following:

1. The c-axis corresponds to both the short axis of the unit cell and the axis along the molecular chain. The observed repeat distance in the c direction is what would be expected between successive substituents on a fully extended hydrocarbon chain with normal bond lengths and angles (see Section 6.1).
2. The distances between all hydrogen atoms are approximately the same in this structure, so there is no problem with overcrowding.
3. While not overcrowded, the polyethylene structure uses space with admirable efficiency, the atoms filling the available space to about 73%. For comparison, recall that close-packed spheres fill space with 74% efficiency, so polyethylene does about as well as is possible.

Table 13.2 Unit Cell Parameters for Several Polymers

Macromolecule	Crystal class	Helix	a, b, c (Å)	α, β, γ (°)	# of units
Polyethylene I	Orthorhombic	1*2/1	7.42, 4.95, 2.55*	90, 90, 90	4
Polyethylene II	Monoclinic	1*2/1	8.09, 2.53*, 4.79	90, 107.9, 90	4
Poly(tetrafluoroethylene) I	Triclinic	1*13/6	5.59, 5.59, 16.88*	90, 90, 119.3	13
Poly(tetrafluoroethylene) II	Trigonal	1*15/7	5.66, 5.66, 19.50*	90, 90, 120	15
Polypropylene (iso)	Monoclinic	2*3/1	66.6, 20.78, 6.495*	90, 99.62, 90	12
Polypropylene (syndio)	Orthorhombic	4*2/1	14.50, 5.60, 7.40*	90, 90, 90	8
Polystyrene (iso)	Trigonal	2*3/1	21.9, 21.9, 6.65*	90, 90, 120	18
Poly(vinyl alcohol) (atac)	Monoclinic	2*1/1	7.81, 2.51*, 5.51	90, 91.7, 90	2
Poly(vinyl fluoride) (atac)	Orthorhombic	2*1/1	8.57, 4.95, 2.52*	90, 90, 90	2
1,4-Polyisoprene (*cis*)	Orthorhombic	8*1/1	13.46, 8.86, 8.1*	90, 90, 90	8
1,4-Polyisoprene (*trans*)	Orthorhombic	4*1/1	7.83, 11.87, 4.75*	90, 90, 90	4
Poly(ethylene oxide)	Monoclinic	3*7/2	8.02, 13.1, 19.3	90, 126, 90	28
Poly(hexamethylene adipamide), α	Triclinic	14*1/1	4.9, 5.4, 17.2	48.5, 77, 63.5	1
Poly(hexamethylene adipamide), β	Triclinic	14*1/1	4.9, 8.0, 17.2	90, 77, 67	2

Note: Asterisks in column 4 denote the chain axis.

Source: From Wunderlich, B., in *Macromolecular Physics, Vol. I: Crystal Structure, Morphology, Defects*, Academic Press, New York, 1973.

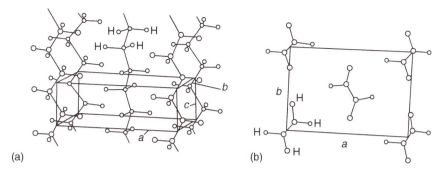

Figure 13.5 Crystal structure of polyethylene: (a) unit cell shown in relationship to chains and (b) view of unit cell perpendicular to the chain axis. (Reprinted from Bunn, C.W., *Fibers from Synthetic Polymers*, R. Hill (Ed.), Elsevier, Amsterdam, 1953. With permission.)

One of the things that can be done with a knowledge of the unit-cell dimensions is to calculate the crystal density. This is examined in the following example.

Example 13.1

Use the unit cell dimensions cited above to determine the crystal density of polyethylene. Examine Figure 13.5 to determine the number of repeat units per unit cell.

Solution

Figure 13.5 shows that the equivalent of two ethylene units are present in each unit cell. Accordingly, the mass per unit cell is

$$\frac{2 \text{ repeat units}}{\text{unit cell}} \times \frac{1 \text{ mol repeat units}}{6.02 \times 10^{23} \text{ repeat units}} \times \frac{28.0 \text{ g}}{1 \text{ mol repeat units}} = 9.30 \times 10^{-23} \text{ g (unit cell)}^{-1}$$

Since all angles in the cells are 90°, the volume of the unit cells is

$$\frac{7.4 \text{ Å} \times 4.9 \text{ Å} \times 2.5 \text{ Å}}{\text{unit cell}} \times \left(\frac{1 \text{ cm}}{10^8 \text{ Å}}\right)^3 = 9.07 \times 10^{-23} \text{ cm}^3 (\text{unit cell})^{-1}$$

The density of the crystal is obtained from the ratio of these two quantities:

$$\rho = \frac{9.30 \times 10^{-23}}{9.07 \times 10^{-23}} = 1.025 \text{ g cm}^{-3}$$

(In fact, x-ray diffraction can usually determine unit-cell dimensions to three or even four significant figures, but we have rounded off in this calculation to avoid specifying more sample information or experimental conditions.) This density may be compared with a typical value of 0.94 g cm^{-3} for polyethylene in the molten state.

Table 13.2 provides a list of representative unit cell parameters. Several interesting observations may be made on the basis of these data:

1. Polyethylene, along with poly(tetrafluoroethylene) (Teflon) and poly(hexamethylene adipamide) (Nylon 6,6) exhibits two different crystal forms. This *polymorphism* is not uncommon, and it indicates that there is a subtle balance of terms in the free energy, so that each structure

has a temperature range where it is in the equilibrium state. These issues are often challenging to sort out experimentally, because of the preponderance of kinetic influences. For instance, the observed form may simply be the one that crystallizes more rapidly, rather than the state of lowest free energy.

2. Although poly(tetrafluoroethylene) is structurally very similar to polyethylene, the increase in size and interactions in going from H to F causes a significant change in the conformation. Both polymorphs are almost all-*trans*, but not quite; form I (which is reported to be stable below 19°C) takes 13 backbone bonds to undergo six turns of the helix, and form II takes 15 bonds to make seven turns.

3. Poly(vinyl alcohol) and poly(vinyl fluoride) provide two examples of atactic polymers that can crystallize, both in the all-*trans* conformation.

4. The structure of poly(hexamethylene adipamide) provides an excellent illustration of how specific, strong interactions can dictate the structure. In this instance it is hydrogen bonding between the carbonyl group and the amide hydrogens that determine the packing. Figure 13.6a provides a schematic illustration of the hydrogen bonding pattern in sheets, and Figure 13.6b shows the difference between the α and β forms in terms of the registry of the hydrogen bonds.

5. Isotactic polypropylene forms a 3/1 helix, which corresponds to a *trans–gauche* plus–*trans–gauche* plus ... sequence of backbone bond conformations. The chain packing forms a mono-clinic unit cell, as shown in Figure 13.7.

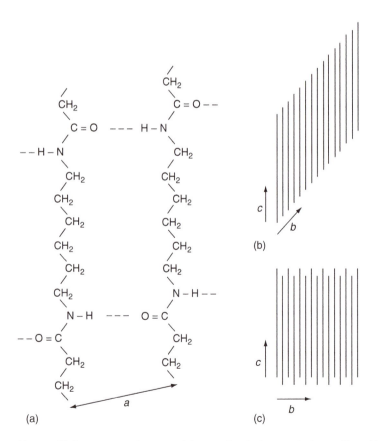

Figure 13.6 The α and β crystal forms of poly(hexamethylene adipamide) (Nylon 6,6), showing (a) the hydrogen bonding pattern and (b) the molecular registry in the two forms. (Reprinted from Young, R.J. and Lovell, P.A., *Introduction to Polymers*, 2nd ed., Chapman and Hall, London, 1991. With permission.)

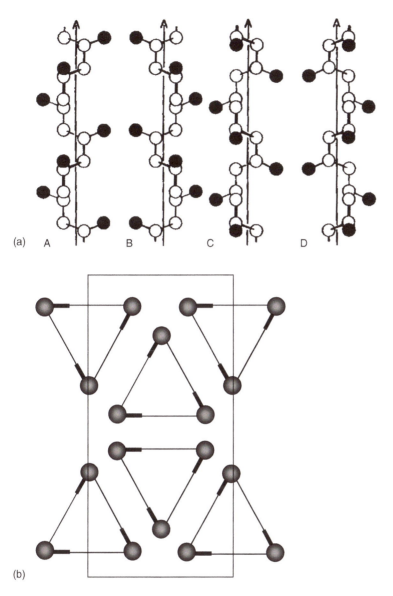

Figure 13.7 (a) Illustration of four possible 3/1 helices for an isotactic vinyl polymer such as polypropylene. A and D are right-handed, B and C are left-handed, A and D are equivalent if inverted, as are B and C. (Reprinted from Wunderlich, B., *Macromolecular Physics, Vol. I: Crystal Structure, Morphology, Defects*, Academic Press, New York, 1973. With permission.) (b) The polypropylene unit cell looking down the chain axis. Each apex of a triangle indicates a methyl group. The thick line segments suggest the projection of the bond to the methyl group; there are three left-handed and three right-handed helices shown.

13.3 Thermodynamics of Crystallization: Relation of Melting Temperature to Molecular Structure

In dealing with experimental thermodynamics, one of the criteria for a true equilibrium to have been established is to achieve the state of interest from opposite directions. Accordingly, we are in the habit of thinking of the equilibrium melting point of a crystal, or the equilibrium freezing point of the corresponding liquid, as occurring at the same temperature. In dealing with polymer crystals, we are not so fortunate as to observe this simple behavior. The transition from liquid $\rightarrow$ crystal is so overshadowed by kinetic factors that some even question the value of any thermodynamic discussion of the transition in this direction. Furthermore, because of the kinetic complications occurring during the formation of the crystal, the resulting transition from crystal $\rightarrow$ liquid also becomes more involved.

If polymers had infinite molecular weight and formed infinitely large crystals, the thermodynamics of the transition would be simpler, although the kinetics might very well be worse. Assuming, temporarily, that any kinetic complications can be overlooked, we will define the temperature of equilibrium (subscript e) between crystal and liquid for a polymer meeting the infinity criteria (superscript ∞) stated above as T_e^∞. Subsequently, we will use the superscript ∞ to indicate either infinite crystal dimension, infinite molecular weight, or both; it will be clear from the context which is meant. In such a case, the melting point of the crystal would be T_e^∞ and the freezing point of the liquid would also be T_e^∞. The facts that actual molecular weights are less than infinite and that crystals have finite dimensions both tend to drive the equilibrium transition temperature below T_e^∞. The fact that kinetic complications also interfere means, in addition, that the experimental temperature of crystallization T_c does not equal the temperature of melting T_m, and that neither equals T_e^∞.

Figure 13.8 illustrates some of these points for *cis* 1,4-polyisoprene. The temperature at which the crystals are formed is shown along the abscissa, and the temperature at which they melt, along the ordinate. Note the following observations:

1. The lower the crystallization temperature, the lower the melting point. This correlation will be understood in the next section through consideration of crystal dimensions, and particularly the lamellar thickness, ℓ.
2. Melting occurs over a range of temperatures, as shown previously in Figure 12.1. The range narrows as the crystallization temperature increases. This is probably due to a wider range of crystal dimensions, and less perfect crystals, for lower temperatures of formation.
3. There is a suggestion of convergence of these lines in the upper right-hand portion of Figure 13.8. For this polymer T_m^∞ is estimated to be 28°C—not an unreasonable point of convergence in the lines in Figure 13.8.
4. The value of T_m^∞ is often estimated by extrapolation of the experimental T_m versus T_c curve until it intersects the $T_m = T_c$ line, as illustrated for poly(p-phenylene sulfide) in Figure 13.9.

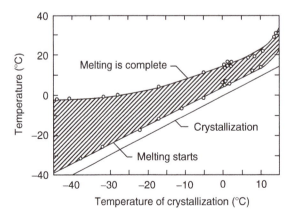

Figure 13.8 Melting temperature of crystals versus temperature of crystallization for *cis* 1,4-polyisoprene. Note the temperature range over which melting occurs. (Reprinted from Wood, L.A. and Bekkedahl, N., *J. Appl. Phys.*, 17, 362, 1946. With permission.)

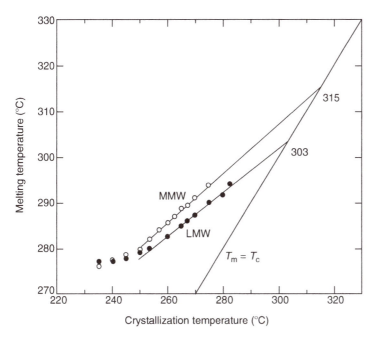

Figure 13.9 Hoffmann–Weeks plot for low molecular weight (LMW) and medium molecular weight (MMW) poly(p-phenylene sulfide). (Reproduced from Lovinger, A.J., Davis, D.D., and Padden, F.J. Jr., *Polymer*, 26, 1595, 1985. With permission.)

Such an extrapolation is known as a *Hoffmann–Weeks plot*. However, note that the data presented here, and many other data sets, suggest that a linear extrapolation may not be completely accurate.

We shall take up the kinetics of crystallization in more detail in Section 13.5 and Section 13.7. For the present, our only interest is in examining what role kinetic factors play in complicating the crystal–liquid transition. The main issue is that the lamellar thickness, ℓ, depends on the crystallization temperature, as a result of kinetic considerations. Accordingly, ℓ is related to T_c, but may not have much to do with T_e^∞. The melting point T_m of the resulting crystal is less than it would be if the crystal had infinite dimensions (T_m^∞). This latter temperature approaches T_e^∞ as $M \to \infty$. In the end, T_m gives a better approximation to a valid equilibrium parameter, although it will still be less than T_m^∞ owing to the finite dimensions of the crystal and the finite molecular weight of the polymer. We shall deal with these considerations in the next section. For now we assume that a value of T_e^∞ has been obtained and consider the thermodynamics of this phase transition.

We begin our application of thermodynamics to polymer phase transitions by considering the fusion (subscript f) process: crystal $\to$ liquid. Figure 13.10 shows schematically how the Gibbs free energy of liquid (subscript 1) and crystalline (subscript c) samples of the same material vary with temperature. For constant temperature–constant pressure processes the criterion for spontaneity is a negative value for ΔG (just as in our consideration of phase equilibria in Chapter 7), where the Δ signifies the difference between the final and initial states for the property under consideration. Applying this criterion to Figure 13.10, we conclude immediately that above T_m^∞, $\Delta G_f = G_1 - G_c$ is negative and melting is spontaneous, whereas below T_m^∞, $\Delta G_f > 0$ and it is fusion that is spontaneous. At T_m^∞ both phases have the same value of G; at that temperature, therefore, $\Delta G_f = 0$ and a condition of equilibrium exists between the phases.

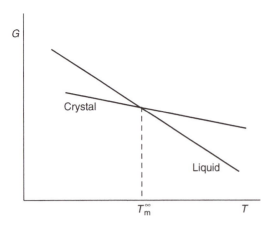

Figure 13.10 Behavior of Gibbs free energy near T_m^∞ for an idealized crystal to liquid phase transition.

At T_m^∞, $\Delta G_f = 0$, but ΔS_f, ΔV_f, and ΔH_f have nonzero values. Figure 12.2a showed how V, S, and H (as first derivatives of G) undergo a discontinuous change at a first-order transition, such as fusion. For any constant-temperature process,

$$\Delta G = \Delta H - T\Delta S \tag{13.3.1}$$

and therefore at equilibrium

$$T_m^\infty = \frac{\Delta H_f}{\Delta S_f} \tag{13.3.2}$$

This fundamental relationship points out that the temperature at which crystal and liquid are in equilibrium is determined by the balancing of entropy and enthalpy effects. Remember, it is the *difference* between the crystal and liquid free energies that is pertinent; sometimes these differences are not what we might expect.

Table 13.3 lists representative values of T_m, as well as ΔH_f and ΔS_f per mole of repeat units, for several polymers. A variety of experiments and methods of analysis have been used to evaluate these data, and because of an assortment of experimental and theoretical limitations the values should be regarded as approximate. We assume $T_m \cong T_m^\infty$. In general, both ΔH_f and ΔS_f may be broken into contributions of H_0 and S_0, which are independent of molecular weight, and increments $\Delta H_{f,1}$ and $\Delta S_{f,1}$ for each repeat unit in the chain. Therefore, $\Delta H_f = H_0 + N\Delta H_{f,1}$, where N is the degree of polymerization. In the limit of large N, $\Delta H_f \cong N\Delta H_{f,1}$ and $\Delta S_f = N\Delta S_{f,1}$, so $T_m^\infty = \Delta H_{f,1}/\Delta S_{f,1}$. Some observations concerning these data sets are listed here:

Table 13.3 Values of T_m, $\Delta H_{f,1}$, and $\Delta S_{f,1}$ for Several Polymers

Polymer	T_m (°C)	$\Delta H_{f,1}$ (J mol^{-1})	$\Delta S_{f,1}$ (J K^{-1} mol^{-1})
Polyethylene	137.5	8,220	19.8
cis 1,4-Polyisoprene	28	8,700	28.9
Poly(ethylene oxide)	66	8,700	25.1
Poly(decamethylene sebacate)	80	50,200	142.3
Poly(decamethylene azelate)	69	41,840	121.3
Poly(decamethylene sebacamide)	216	34,700	71.1
Poly(decamethylene azelamide)	214	36,800	75.3

Source: From Mandelkern, L., in *Crystallization of Polymers*, McGraw-Hill, New York, 1964.

1. Polyethylene. The crystal structure of this polymer is essentially the same as those of linear alkanes containing 20–40 carbon atoms, and the values of T_m and $\Delta H_{f,1}$ are what would be expected on the basis of an extrapolation from data on the alkanes. Since there are no chain substituents or intermolecular forces other than dispersion (London) forces in polyethylene, we shall compare other polymers to it as a reference substance.

2. *cis* 1,4-Polyisoprene. Although $\Delta H_{f,1}$ is slightly higher than that of polyethylene (on a per repeat unit basis, not per gram), it is still completely reasonable for a hydrocarbon. The lower T_m is the result of the value of $\Delta S_{f,1}$, which is 50% higher than that of polyethylene. The low melting point of this polymer makes natural rubber a useful elastomer at ordinary temperatures.

3. Poly(ethylene oxide). Although $\Delta H_{f,1}$ is similar to that of polyethylene, the effect is offset by an increase in $\Delta S_{f,1}$ similar to polyisoprene. The latter may be due to increased chain flexibility in the liquid caused by the regular insertion of ether oxygens along the chain backbone.

4. Polyesters. The next two polyesters have $\Delta H_{f,1}$ values significantly higher than polyethylene. Our first thought might be to attribute this to a strong interaction between the polar ester groups. The repeat units of these compounds are considerably larger than in the reference compound, so the $\Delta H_{f,1}$ values should be compared on a per gram basis. When this is done, $\Delta H_{f,1}$ is actually less than for polyethylene. This suggests that the large value for $\Delta H_{f,1}$ is the result of a greater number of methylene groups contributing London attraction for the polyesters, with the dipole–dipole interaction of ester groups about the same in both liquid and crystal and therefore contributing little to $\Delta H_{f,1}$. When compared on the basis of the number of bonds along the backbone, $\Delta S_{f,1}$ is not exceptional either. Accordingly, T_m is less for these esters than for polyethylene.

5. Polyamides. The next two compounds are the amide counterparts of the esters listed under item (4). Although the values of $\Delta H_{f,1}$ are less for the amides than for the esters, the values of T_m are considerably higher. This is a consequence of the very much lower values of $\Delta S_{f,1}$ for the amides. These, in turn, are attributed to the low entropies of the amide in the liquid state, owing to the combined effects of hydrogen bonding and chain stiffness from the contribution of the following resonance form: $-(HN^+=CO^-)-$.

These examples show that it is often easier to rationalize an observation or trend with respect to T_m than to predict it a priori. This state of affairs is not unique to polymers, however. The following example provides another illustration of this type of reasoning.

Example 13.2

The melting points of a series of poly(α-olefin) crystals were studied. All of the polymers were isotactic and had chain substituents of different bulkiness. Table 13.4 lists some results. Use Equation 13.3.2 as the basis for interpreting the trends in these data.

Table 13.4   Values of T_m for Poly(α-olefin) Crystals in Which the Polymer Has the Indicated Substituent

Substituent	T_m (°C)
$-CH_3$	165
$-CH_2CH_3$	125
$-CH_2CH_2CH_3$	75
$-CH_2CH_2CH_2CH_3$	−55
$-CH_2CH(CH_3)CH_2CH_3$	196
$-CH_2C(CH_3)_2CH_2CH_3$	350

Source: From Billmeyer, F.W., in *Textbook of Polymer Science*, 2nd ed., Wiley-Interscience, New York, 1971.

Solution

The bulkiness of the substituent groups increases moving down Table 13.4. Also moving down the table, the melting points decrease, pass through a minimum, and then increase again. As is often the case with reversals of trends such as this, there are (at least) two different effects working in opposition in these data:

1. As the bulkiness of the substituents increases, the chains are prevented from coming into intimate contact in the crystal. The intermolecular forces that hold these crystals together are all London forces, and these become weaker as the crystals loosen up owing to substituent bulkiness. Accordingly, the value for the heat of fusion decreases moving down Table 13.4.
2. As the bulkiness of the chain substituents increases, the energy barriers to rotation along the chain backbone increase. As seen in Chapter 6, this decreases chain flexibility in the liquid state. It is this flexibility that permits the molecules to experience a large number of conformations and therefore have high entropies. If the flexibility is reduced, the entropy change on melting is less than it would otherwise be. Accordingly, the entropy of fusion decreases moving down the table.
3. Since $T_m = \Delta H_f / \Delta S_f$, the observed behavior of this series of polymers may be understood as a competition between these effects. For the smaller substituents, the effect on ΔH_f dominates and T_m decreases with bulk. For larger substituents, the effect on ΔS_f dominates and T_m increases with bulk.

All the polymers compared have similar crystal structures, but are different from polyethylene, which excludes the possibility for also including the latter in this series. Also note that the isotactic structure of these molecules permits crystallinity in the first place. With less regular microstructure, crystallization would not occur at all.

In the discussion of Table 13.3, we acknowledged that there might be some uncertainty in the values of the quantities tabulated, but we sidestepped the origin of the uncertainty. In the next section we shall consider the most important of these areas: the effect of crystal dimensions on the value of T_m.

13.4   Structure and Melting of Lamellae

13.4.1   Surface Contributions to Phase Transitions

Whenever a phase is characterized by at least one linear dimension that is small (a few microns or less), the properties of the surface begin to make significant contributions to the observed behavior. In contrast, most thermodynamic analyses are conducted on the assumption of bulk (i.e., effectively infinite) phases. As illustrated in Figure 13.1, lamellae tend to have thicknesses on the order of 100 Å, and thus surface effects can be substantial. The following summary of generalizations about these crystals will be helpful:

1. The dimensions of the lamellae perpendicular to the smallest dimension depend on the conditions of the crystallization, but are many times larger than the thickness of a well-developed crystal.
2. The chain direction within the crystal tends to be along the short dimension of the crystal, indicating that the molecule folds back and forth, fire-hose fashion, with successive layers of folded molecules accounting for the lateral growth of the platelets. The section of chain that traverses a lamella once is called a *stem*.
3. A crystal lamella does not consist of a single molecule, nor does a molecule need to reside exclusively in a single lamella.
4. The loop formed by the chain as it emerges from the crystal, turns around, and reenters the crystal may be regarded approximately as amorphous polymer, but is insufficient to account for the total amorphous content of most crystalline polymers.

5. Polymer chain ends disrupt the orderly fold pattern of the crystal, and tend to be excluded from the crystal and relegated to the amorphous portion of the sample. The same is true of stereochemical or microstructural defects, or comonomers.

As noted above, since the polymer crystal habit is characterized by plates whose thickness is small, surface phenomena are important. During the early development of the crystal, the lateral dimensions are also small and this effect is even more pronounced. The key to understanding this fact lies in the realization that all phase boundaries possess *surface tension*, and that this surface tension reflects the Gibbs free energy stored per unit area of the phase boundary. As a qualitative illustration, consider cutting a piece of polymer into two along a selected plane. Before cleavage, there were cohesive (attractive) interactions across the plane, which are now lost. This energy per unit area becomes the surface energy of the newly exposed material. Now place two different materials in contact across a plane; unless their surface energies happen to be identical, there would be an interfacial energy, or surface tension, γ. Now suppose we consider a spherical phase of radius r, density ρ, and surface tension γ. The total surface free energy associated with such a particle is given by the product of γ and the area of the sphere, or $\gamma(4\pi r^2)$. The total mass of material in the sphere is given by the product of the density and the volume of the sphere, or $\rho(4\pi r^3/3)$. The ratio of the former to the latter gives the Gibbs free energy arising from surface considerations, expressed per unit mass; that is, the surface Gibbs free energy per unit mass is $3\gamma/\rho r$. Since γ is small compared to most other chemical and physical contributions to the free energy, surface effects are not generally considered when, say, the ΔG° of formation is quoted for a substance. The above argument shows that this becomes progressively harder to justify as the particle size decreases. The emergence of a new phase implies starting from an r value of zero in the argument above, and the surface contribution to the energy becomes important indeed. (Since two phases with their separating surface must already exist for γ to have any meaning, we are spared the embarrassment of the surface freeenergy becoming infinite at $r = 0$.) Nevertheless, it is apparent that the effect of the surface free-energy contribution is to increase the total G. Inspection of Figure 13.10 shows that an increase in the G value for the crystalline phase arising from its small particle size has the effect of shifting T_m to lower temperatures. The smaller the particle size, the bigger the effect. This is the origin of all superheating, supercooling, and supersaturation phenomena: an equilibrium transition is sometimes overshot because of the kinetic difficulty associated with the initiation (nucleation) of a new phase. Likewise, all nucleation practices—cloud seeding, bubble chambers, and the use of boiling chips—are based on providing a site on which the emerging phase can grow readily.

13.4.2 Dependence of T_m on Lamellar Thickness

To develop a more quantitative relationship between particle size and T_m, suppose we consider the melting behavior of the cylindrical crystal sketched in Figure 13.11. Of particular interest in this model is the role played by surface effects. The illustration is used to define a model and should not be taken too literally, especially with respect to the following points:

1. The geometry of the cylinder is a matter of convenience. Except for numerical coefficients, the results we shall obtain will apply to plates of any cross-sectional shape.
2. The thickness of the plate, although small, is greater than the few repeat units shown.
3. The specific nature of the reentry loops is not the point of this illustration. The sketch shows both hairpin turns and longer loops. Problem 6 at the end of the chapter examines the actual nature of the reentry loop.

To develop this model into a quantitative relationship between T_m and the thickness of the crystal, we begin by realizing that for the transition crystal $\rightarrow$ liquid, ΔG_f is the sum of two contributions. One of these is ΔG^∞, which applies to the case of a crystal of infinite size

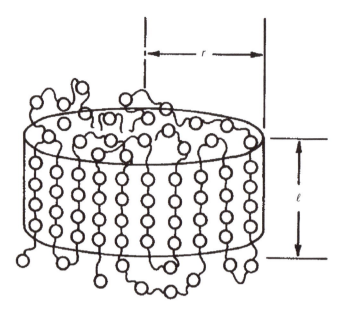

Figure 13.11 Idealized representation of a polymer crystal as a cylinder of radius r and thickness ℓ.

(superscript ∞); the other, ΔG^{s}, arises specifically from surface effects (superscript s), which reflect the finite size of the crystal:

$$\Delta G_{\mathrm{f}} = \Delta G^{\infty} + \Delta G^{\mathrm{s}} \tag{13.4.1}$$

Now each of these can be developed independently.

As in the qualitative discussion above, let γ be the Gibbs free energy per unit area of the interface between the crystal and the surrounding liquid. This is undoubtedly different for the edges of the plate than for its faces, but we shall not worry about this distinction. The area of each of the circular faces of the cylinder is πr^2, and the area of the edge is $2\pi r \ell$, where r is the radius of the face and ℓ is the length of the side as shown Figure 13.11. Since surface is destroyed by the melting process, the net contribution to ΔG_{f} is

$$\Delta G^{\mathrm{s}} = -\left[2\pi r^2 + 2\pi r \ell\right]\gamma = -2\pi r^2 \gamma \left(1 + \frac{\ell}{r}\right) \tag{13.4.2}$$

For the bulk effect, we proceed on the basis of a unit volume (subscript V) and immediately write

$$\Delta G^{\infty} = \pi r^2 \ell \Delta G_{\mathrm{V}}^{\infty} \tag{13.4.3}$$

and

$$\Delta G_{\mathrm{V}}^{\infty} = \Delta H_{\mathrm{V}}^{\infty} - T \Delta S_{\mathrm{V}}^{\infty} \tag{13.4.4}$$

When this infinite phase is in equilibrium with the melt, $\Delta G_{\mathrm{V}}^{\infty} \to 0$ and $T \to T_{\mathrm{m}}^{\infty}$. Accordingly, we can solve Equation 13.4.4 for $\Delta S_{\mathrm{V}}^{\infty} = \Delta H_{\mathrm{V}}^{\infty}/T_{\mathrm{m}}^{\infty}$ and substitute this back into Equation 13.4.4:

$$\Delta G_{\mathrm{V}}^{\infty} = \Delta H_{\mathrm{V}}^{\infty}\left(1 - \frac{T}{T_{\mathrm{m}}^{\infty}}\right) \tag{13.4.5}$$

This gives the value of ΔG_V at any temperature in terms of the two parameters ΔH_V^∞ and T_m^∞. Combining Equation 13.4.1 through Equation 13.4.3 and Equation 13.4.5 enables us to write

$$\Delta G_f = \left(\pi r^2 \ell\right)\Delta H_V^\infty \left(1 - \frac{T}{T_m^\infty}\right) - 2\pi r^2 \gamma \left(1 + \frac{\ell}{r}\right) \tag{13.4.6}$$

When the value of this ΔG is zero, we have the actual melting point of the crystal of finite dimension T_m. That is,

$$\left(\pi r^2 \ell\right)\Delta H_V^\infty \left(\frac{T_m^\infty - T_m}{T_m^\infty}\right) = 2\pi r^2 \gamma \left(1 + \frac{\ell}{r}\right) \tag{13.4.7}$$

or

$$\Delta T = T_m^\infty - T_m = \frac{2}{\ell}\frac{\gamma}{\Delta H_V^\infty}\left(1 + \frac{\ell}{r}\right)T_m^\infty \tag{13.4.8}$$

Note that this equation is dimensionally correct, as γ has units of energy area^{-1} and ΔH_V^∞ has units energy volume^{-1}. Therefore the units of $\ell\Delta H_V^\infty$ and γ cancel, as do the units of ℓ and r, leaving only temperature units on both sides of the equation. All of the quantities on the right-hand side of the equation are positive (ΔH_V^∞ is the heat of fusion), which means that $T_m^\infty > T_m$, as anticipated. The difference ΔT between a thermodynamic boundary and the temperature of interest is often referred to as the *undercooling*; as derived in Equation 13.4.8 it reflects the melting point depression due to particle size. Several limiting cases of this equation are of note:

1.   If $\gamma = 0$, $\Delta T = 0$, regardless of particle size. This is not likely to apply, however, since chains emerging from a crystal face either make a highly constrained about-face and reenter the crystal or meander off into the liquid from a highly constrained attachment to the solid. In either case, a surface free-energy contribution is inescapable.
2.   As $r \to 0$, $\Delta T \to \infty$, showing that the lateral dimensions of the plate are critical for very small crystals. This makes the crystal nucleation event especially crucial.
3.   As $r \to \infty$, which describes well-developed crystals, Equation 13.4.8 becomes

$$\Delta T = 2T_m^\infty \frac{\gamma}{\Delta H_V^\infty}\frac{1}{\ell} \tag{13.4.9}$$

which shows that an undercooling is still important because of the platelike crystal habit of polymers with limited crystal dimensions along the chain direction. Equation 13.4.9 is often referred to as the *Thompson–Gibbs equation*.

Equation 13.4.9 shows that a direct proportionality relationship should exist between crystal thickness ℓ and the ratio $T_m^\infty/\Delta T$; a plot of ℓ versus $T_m^\infty/\Delta T$ should result in a straight line of zero intercept with a slope proportional to $\gamma/\Delta H_V^\infty$. Figure 13.12 shows such a plot for polyethylene in which T_m^∞ was taken to be 137.5°C and the ℓ values were determined by x-ray scattering. While there is considerable and systematic divergence from the predicted form at large undercoolings, the data show a linear relationship for the higher-temperature region. In the following example we analyze the linear portion of Figure 13.12 in terms of Equation 13.4.9.

Example 13.3

Use Equation 13.4.9, the results in Figure 13.12, and the data in Table 13.3 to estimate a value for γ for polyethylene. Figure 13.5 shows the unit cell of polyethylene; the equivalent to two chains emerges from an area 0.740 by 0.493 nm^2. On the basis of the calculated value of γ and the characteristics of the unit cell, estimate the free energy of the fold surface per mole of repeat units.

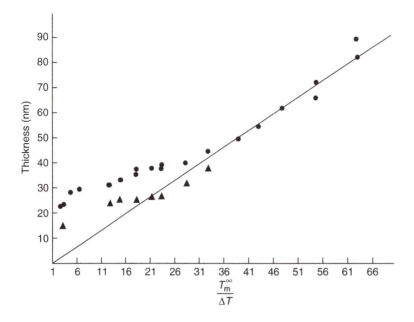

Figure 13.12 Crystal thickness versus $T_m/\Delta T$ for polyethylene. (Reprinted from Mandelkern, L., *Crystallization of Polymers*, McGraw-Hill, New York, 1964. With permission.)

Solution

Equation 13.4.9 predicts a straight line of zero intercept and slope of $2\gamma/\Delta H_V^\infty$ when ℓ is plotted versus $T_m^\infty/\Delta T$. The solid line in Figure 13.12 has a slope of 650 Å/51 = 13.7 Å. Therefore $2\gamma/\Delta H_V^\infty = 13.7 \times 10^{-10}$ m. The value of $\Delta H_{f,1}$ given in Table 13.3 is used for ΔH_V^∞ after the following change of units:

$$\Delta H_V^\infty = \frac{8220\ \text{J}}{\text{mole}} \times \frac{1\ \text{mole}}{28\ \text{g}} \times \frac{1\ \text{g}}{\text{cm}^3} \times \frac{10^6\ \text{cm}^3}{1\ \text{m}^3} = 2.8 \times 10^8\ \text{J m}^{-3}$$

Therefore $\gamma = 1/2(13.7 \times 10^{-10}\ \text{m})\,(2.8 \times 10^8\ \text{J m}^{-3}) = 0.192\ \text{J m}^{-2}$. From the data on the unit cell

$$\frac{0.192\ \text{J}}{\text{m}^2} \times \left(\frac{0.074 \times 0.493\ \text{nm}^2}{\text{unit cell}} \times \frac{1\ \text{m}^2}{10^{18}\ \text{nm}^2}\right) \times \frac{1\ \text{unit cell}}{2\ \text{molecules}}$$

$$\times \frac{6.02 \times 10^{23}\ \text{molecules}}{\text{mol}} = 2.1\ \text{kJ/mol}$$

Although it applies to a totally different kind of interface, the value of γ calculated in the example is on the same order of magnitude as the γ value for the surface between air and liquids of low molecular weight.

13.4.3 Dependence of T_m on Molecular Weight

Before concluding this section, there is one additional thermodynamic factor to be mentioned, which also has the effect of lowering T_m. The specific effect we consider is that of chain ends (and therefore the number-average molecular weight), but the role they play is that of an "impurity" from the viewpoint of crystallization. As such the treatment is similar to the effect of a solute on any colligative property, such as the osmotic pressure considered in Section 7.4. Furthermore, other "impurities" such as comonomers or low molecular-weight species can be treated in a similar fashion. In this context the repeat units in a polymer may be divided into two classes: those at the ends of the chain (subscript e) and the others, which we view as being in the middle (subscript m) of the chain. The mole fraction of each category in a sample is x_e and x_m, respectively. Since all segments are of one type or the other,

$$x_m = 1 - x_e \qquad (13.4.10)$$

The proportion of chain ends increases with decreasing molecular weight, and hence for a linear chain (two ends) $x_e = 2M_o/M_n$, where M_o is the molecular weight of the repeat unit. Now we take the molar Gibbs free energy of a finite chain crystal compared to that of an infinite chain crystal ($x_m = 1$):

$$
\begin{aligned}
G(x_m) - G(x_m = 1) = -RT \ln a_m &= -RT \ln x_m \\
&= -RT \ln (1 - x_e) \approx RT\, x_e
\end{aligned}
\qquad (13.4.11)
$$

This equation follows from the definition of the "activity" of middle segments, a_m (see Equation 7.1.13) and the approximation of ideality ($a_m = x_m$) as the impurity concentration vanishes ($x_e \ll 1$). Equation 13.4.11 is the analog of Equation 7.4.4 for the osmotic pressure. We can now insert this expression into Equation 13.4.5, but now applied to melting a crystal of chains of molecular weight M at temperature T_m compared to a crystal of infinite chains:

$$\Delta H_f \left(1 - \frac{T_m}{T_m^\infty} \right) = RT_m\, x_e \qquad (13.4.12)$$

or, after some rearranging,

$$T_m^\infty - T_m = \Delta T_m = \frac{RT_m T_m^\infty}{\Delta H_f} \frac{2M_o}{M_n} \qquad (13.4.13)$$

where T_m is the melting point of the polymer under consideration. Equation 13.4.13 indicates that a freezing point depression is to be expected from an increased concentration of chain ends. Qualitatively, at least, the presence of other types of defects is also expected to lower T_m. Remember that in the present discussion T_m^∞ is the melting point of a polymer of infinite molecular weight without regard to the crystal size, whereas in Equation 13.4.8 it was the melting point for a crystal of infinite dimension without regard to molecular weight. The two effects are therefore complementary, and both are operative if both particle dimension and molecular weight are small enough to lower the freezing point appreciably.

Throughout this section we have focused attention on thermodynamic melting points. The same thermodynamic arguments can be applied to the raising and sharpening of this transition temperature through annealing. When a crystal is maintained at a temperature between the crystallization temperature and the equilibrium melting point, an increase in T_m is observed. This may be understood in terms of the melting of smaller, less perfect crystals and the redisposition of the polymers into larger, more stable crystals. This is analogous to the procedure of digesting a precipitate before filtration. There is more to the story than this, however. The digestion analogy would suggest that those crystals that are enlarged simply add more folded chains around their perimeter. In fact, x-ray diffraction studies reveal progressive *thickening* of lamellae with annealing, that is, as T_m increases. This requires large-scale molecular reorganization throughout the crystal. Such rearrangements apparently require the molecule to snake along the chain axis, with segments being reeled in and out across the crystal surface. The process of annealing, therefore, not only involves crystal thickening, but also provides the opportunity to work out kinks and defects.

13.4.4   Experimental Characterization of Lamellar Structure

Current understanding of the structure of chain-folded lamellae has been greatly facilitated by study of solution-grown single crystals. There are many parameters of a single crystal that are of interest, in addition to the structure of the unit cell itself and the lamella thickness. These include the shape and size of the crystal in the other two dimensions, the orientation of the unit cell relative to the thickness direction, and the nature of the fold surface. Valuable information about all of these features has been obtained with electron microscopy (EM) techniques, and so we will provide a brief introduction to this important characterization tool.

The electron microscope uses the de Broglie waves associated with accelerated electrons to produce an image in much the same way as visible light produces an image in an optical microscope. Electromagnets and imposed electrical fields function as lenses for the electron beam and the image is formed on a phosphorescent screen, photographic plate, or charge-coupled device (CCD) camera. The wavelength of an electron under typical operating conditions in an electron microscope is on the order of $\lambda \approx 0.03$ Å; it depends primarily on the energy of the electrons, which in turn is dictated by the accelerating voltage. In all types of microscopy it is the resolving power rather than the magnification per se, which is the limiting factor. Waves are diffracted from the edges of illuminated bodies and this diffraction blurs their boundaries. The resolving power measures the minimum separation between objects that will produce discernibly different images in a microscope. In a well-designed instrument this separation is on the order of the wavelength of the illuminating radiation. Therefore, the resolving power of an electron microscope is potentially smaller by some five orders of magnitude than that achieved by optical microscopes. In reality, imperfections in electron optics limit the actual resolution to something closer to 100 λ, but nevertheless this resolution matches atomic dimensions, and so EM is extremely useful.

Although the concepts of wave optics apply equally well to visible light, x-rays, and electrons, there are important differences that affect the information that can be obtained, and the experimental design. First, the incident electrons carry charge, and so interact relatively strongly with matter. Consequently, the scattering cross-section is high, which means that an electron does not have to go very far through a material before being scattered. The primary consequence of this fact is that samples must be very thin (perhaps 100 nm) in order for much of the electron beam to be transmitted. Second, in one way or another the image obtained is based on the ability of the material to scatter or diffract the electron beam. In Chapter 8 we discussed how a perfectly uniform material would not scatter light. However, in that case a typical wavelength is 5000 Å, and a small-molecule liquid or glass could be almost completely transparent; in contrast, the electron wavelength is so small that no material made up of nuclei and electrons can appear transparent. Third, the short wavelength of electrons is determined by their high energy. One consequence is that the electron beam is likely to damage the sample. Part of the art of EM is to limit the sample exposure while maintaining sufficient flux to obtain a good image.

In microscopy, one obtains some kind of direct picture of the structure in question, a *real-space image*. This should be contrasted with scattering or diffraction, in which the structural information is contained in the angle-dependent intensity, providing a so-called *reciprocal-space image*. Diffraction experiments suffer from the *inversion problem*, namely that there is no unique way to take the diffraction pattern and invert it into the actual structure. This problem becomes more acute as the structure under examination becomes more complicated. All other things being equal, therefore, a real-space image is preferable. Of course, all other things are often not equal. There are several potential limitations to EM in its application to polymers. Among these are:

1. The need for very thin samples, with its attendant challenges in sample preparation.
2. The generally low contrast between different monomers (in a mixture or copolymers) or different structures in a single polymer. This arises because the chemical compositions (C, H, O, maybe N ...) and material densities (typically 0.9–1.1 g cc^{-1}) of most organic polymers are similar.

3. The need for selective staining techniques (using heavy atoms such as Os and Ru) to generate contrast.
4. The inability to view the samples directly under different conditions of interest, such as temperature, pressure, or flow.
5. The need for sample fixation to prevent changes in structure occurring between the original sample and the sample that is imaged in the microscope.
6. The likelihood of electron-beam damage.
7. An EM image represents a projection completely through the sample, so care must be taken in interpreting features that may represent objects that are separated along the beam direction.
8. A tightly focussed electron beam will only interrogate a small portion of the sample (less than $1 \ \mu m^3$), so care must be taken to establish how representative a given image is.

Fortunately, most of these issues are of less importance when examining polymer single crystals collected from a solution crystallization process; the crystals are inherently thin, fixed in structure, and small in lateral dimensions.

There are a variety of different ways in which an image can be generated in EM, and we briefly identify some of the terminology that is encountered. Two procedures of electron microscopy find particular applicability in the study of polymer crystals: *shadow casting* and *dark-field operation*. Shadow casting is used to improve the contrast between a sample and its background and between various details of the sample surface. Because polymer crystals are so thin and mostly consist of atoms of low atomic number, some sort of contrast enhancement is important. In the shadowing method the sample is placed in an evacuated chamber and a heavy metal is allowed to evaporate in the same chamber. The position of the metal source is such that the metal vapor strikes the sample at an oblique angle and condenses on the cool surface. The thin metal film thus formed literally casts shadows, which enhance the image of the sample. If the angle of incidence of the heavy metal beam is known, the thickness of a crystal or the height of surface protuberances can be determined from the length of the shadow by simple trigonometry.

In dark-field electron microscopy it is not the transmitted beam that is used to construct an image but, rather, a beam diffracted from one facet of the object under investigation. One method for doing this is to shift the aperture of the microscope so that most of the beam is blocked and only those electrons scattered into the chosen portion of the diffraction pattern contribute to the image. This decreases the intensity of the illumination used to produce the dark-field image and therefore requires longer exposure times, with the attendant modification or even degradation of the polymer. Nevertheless, dark-field operation distinguishes between portions of the sample with different orientations, as the diffracted beams will appear in different directions, and therefore produces a more three-dimensional representation of the sample. *Bright-field operation*, on the other hand, utilizes the transmitted main beam to form the image.

Figure 13.13a through Figure 13.13c provide examples of shadow-contrast, bright-field, and dark-field transmission electron micrographs of solution-grown polyethylene crystals, respectively. The effect of shadowing is evident in Figure 13.13a, where those edges of the crystal that cast the shadows display sharper contrast. The roughly diamond shape of the crystal lamella is also clearly evident. The cracks evident in the larger crystals result from the fact that in solution the crystals actually form hollow pyramids, whereas they are flattened onto a viewing grid for the EM. There is also evidence of multilamellar growth in the lower-right corner of the image. Figure 13.3b shows the clear outlines of a "truncated" diamond, and the same kind of crack as in Figure 13.3a. The dark-field image of the same crystal in Figure 13.3c is particularly revealing, because now the single crystal is clearly shown to be made up of four sectors, two of which are dark because the unit-cell orientation is such as to diffract the electrons away from the detection aperture. This sectorization and the hollow pyramid form are illustrated schematically in Figure 13.14.

The electron micrographs of Figure 13.13 are more than mere examples of EM technique. They are the first occasion we have had to actually look at single crystals of polymers. Although there is

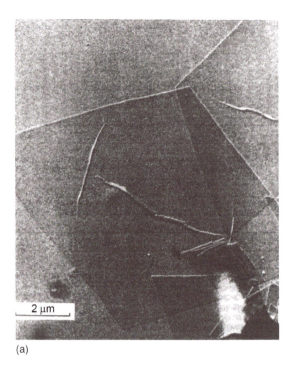

(a)

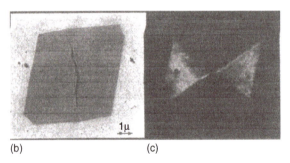

(b)   (c)

Figure 13.13 Electron micrographs of polyethylene single crystals, illustrating the use of (a), shadow casting. (Reprinted from Reneker, D.H. and Geil, P.H., *J. Appl. Phys.*, 31, 1916, 1960. With permission.) (b) Bright-field imaging. (Reprinted from Niegisch, W.D. and Swan, P.R., *J. Appl. Phys.*, 31, 1906, 1960. With permission.) (c) Dark-field imaging. (Reprinted from Niegisch, W.D. and Swan, P.R., *J. Appl. Phys.*, 31, 1906, 1960. With permission.)

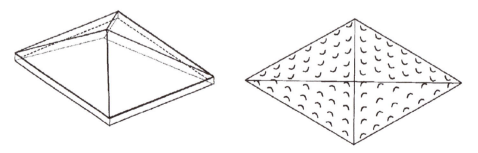

Figure 13.14 Schematic side and top views of a hollow pyramid crystal, with four sectors distinguished by different fold directions.

a great deal to be learned from studies of single crystals by EM, we shall limit ourselves to just a few observations:

1. Single crystals such as those shown in Figure 13.13 are not observed in crystallization from the bulk. Crystallization from dilute solutions is required to produce single crystals with this kind of macroscopic perfection. Polymers are not intrinsically different from low molecular-weight compounds in this regard.

2. Crystallization conditions such as temperature, solvent, and concentration can influence crystal form. One such modification is the truncation of the points at either end of the long diagonal of the diamond-shaped crystals seen in Figure 13.13b and Figure 13.13c. The facets of the diamond-shaped crystal correspond to {110} planes of the polyethylene unit cell (see Figure 13.5), whereas the truncated sections in Figure 13.13b and Figure 13.13c are {100} planes. If the rate of crystallization onto {100} planes is much greater than onto {110} planes, then the {100} facets will disappear and a diamond will result. Twinning and dendritic growth are other examples of such changes of habit, and these features can usually be attributed to the relative growth rates for different crystallographic faces.

3. Hollow pyramids are thoroughly documented and fairly well understood. The underlying factor is the nature of the fold as a chain exits and reenters the lamella. If we consider two stems plus one fold to form a "U," defining a fold plane, it turns out that the fold does not remain in this plane, but is tilted. This can be understood from Figure 13.5. The fold plane is a (110) plane, and consequently the all-*trans* orientation for each adjacent stem is rotated by 90°. Each successive fold plane is displaced vertically by one CH_2 unit, which leads to the pyramidal form; the chain axis is perpendicular to the base of the pyramid, not to the fold surface.

4. The nature of the chain folding at the fold surface has been a subject of great interest (and no little controversy). The implication of the cartoon in Figure 13.14 is that the folding is very regular, and immediate in the sense that an emerging chain folds back directly into the crystal as the adjacent stem in the fold plane. This limiting behavior is known as *adjacent reentry*, and may be contrasted with the opposite extreme of *random reentry*, or the switchboard model, as shown in Figure 13.15. For solution-grown single crystals adjacent reentry is certainly prevalent. However, for crystals grown from the melt, the evidence favors a much more random folding process; in particular, the radius of gyration of a single chain in the melt (measured by neutron scattering) changes little on crystallization, which suggests that it enters and departs from several lamellae (see Figure 13.2).

The foregoing is by no means a comprehensive list of the remarkable structures formed by the crystallization of polymers from solution. The primary objective of this brief summary is the

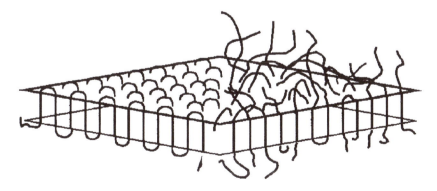

Figure 13.15 Schematic illustration of adjacent reentry (left side of crystal) and random reentry (right side).

verification that single crystals can be formed, and characterized in detail, not only by x-ray diffraction, but also by electron microscopy.

13.5 Kinetics of Nucleation and Growth

In this section our objective is to introduce the basic factors that govern the rate of growth of polymer crystals, once a sample has been cooled to a temperature below T_m^∞. Some of these factors pertain to many phase transitions, whereas others are particular to polymer crystallization. We will emphasize the rate of growth of individual lamellae, whether in bulk or in solution, as the fundamental process of importance. In Section 13.7 we will return to crystallization kinetics, but from a broader perspective; there we will consider the rate at which a macroscopic sample becomes (semi)crystalline.

Before we begin a more systematic treatment, consider the following illustrative example. If we take a polymer sample that has been allowed to crystallize at some temperature T_c, we will find a predominant average lamellar thickness. What determines this thickness? The answer is a combination of thermodynamics and kinetics, but mostly the latter. Thermodynamics tells us that only lamellae that have a lower free energy than the liquid state can grow spontaneously. In the previous section, we saw that the melting temperature of lamellae increased with thickness, so we may conclude that as T_c decreases below T_m^∞, we will progressively increase the range of smaller lamellar thicknesses that may grow. The role of kinetics is to dictate which of the thermodynamically allowed values of ℓ is observed, namely, the one that grows most rapidly. As a rule, thinner lamellae tend to grow more rapidly, so as T_c is decreased, the observed ℓ will decrease. The resulting lamellae must therefore be viewed as metastable, because a crystal with larger ℓ should have a lower free energy.

This general principle is beautifully illustrated by the isothermal crystallization data for modest molecular-weight poly(ethylene oxide) shown in Figure 13.16. These single crystals were grown in solution, and the dominant morphology is indicated on the plot. As T_c decreases, the rate of crystallization increases, but with abrupt jumps. Each jump in rate is associated with a discrete change in ℓ, from straight-chain lamellae to once-folded, then from once-folded lamellae to twice-folded, etc. The particular temperatures at which the jumps in crystallization rate occur can be

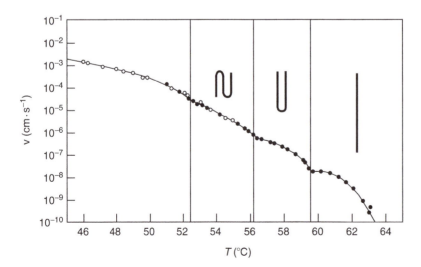

Figure 13.16 Crystal growth rate for low molecular-weight poly(ethylene oxide) crystallized from solution. (Reproduced from Strobl, G., *The Physics of Polymers*, Springer, Berlin, 1996. With permission.)

easily interpreted based on the foregoing discussion. Thus 63.5°C is the temperature below which straight-chain lamellar crystals have a lower free energy than the liquid, and so they grow. Then 59.5°C is the point at which once-folded lamellae become possible, and they grow much more rapidly than the straight-chain lamellae. Similarly, below about 56.5°C twice-folded lamellae become possible, and because they grow significantly more rapidly than straight-chain or once-folded lamellae, they become the predominant form.

13.5.1 Primary Nucleation

Crystallization, like many other first-order phase transitions, proceeds by the process known as nucleation and growth. *Nucleation* refers to the appearance of domains of the new phase that are sufficiently large to become stable; recalling the discussion in Section 13.4.1, this amounts to the domains becoming large enough for the bulk energy gain of the new phase to outweigh the unfavorable surface energy. In many transitions, nucleation is the rate-limiting step. The subsequent growth of the particles will be considered in the following section. Nucleation is commonly classified as either *heterogeneous* or *homogeneous*. The former denotes the situation where a foreign particle, impurity, or surface provides a site for facile nucleation, whereas the latter indicates the spontaneous formation of nuclei by random fluctuations. In practical situations heterogeneous nucleation is almost always more important, and often dominant; indeed, in many polymer processes nucleating agents such as talc powder are added to accelerate crystallization. However, in the laboratory the homogenous case is also important and furthermore a simple treatment of homogenous nucleation will allow us to understand a good deal about the subsequent growth processes as well.

We begin by adapting our treatment of surface effects in Section 13.4.2 to nucleation. The free energy of a spherical droplet with radius R of the new phase, ΔG, can be written as the sum of a surface term and a volume term (see Equation 13.4.1):

$$\Delta G(R) = 4\pi R^2 \gamma + \frac{4\pi}{3} R^3 \Delta G_V \qquad (13.5.1)$$

where γ is the surface energy and ΔG_V is the free energy change per unit volume; note that when $T < T_m$, ΔG_V is negative. Figure 13.17 illustrates the functional form of Equation 13.5.1, with the

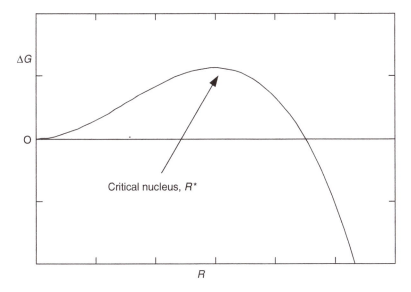

Figure 13.17 Dependence of free energy on drop size according to Equation 13.5.1, illustrating the radius of the critical nucleus.

essential feature that there is a special size, R^*, where ΔG has a positive maximum. For the new phase to form, droplets must somehow grow larger than R^*, so a droplet of this size is called a *critical nucleus*. Once a drop exceeds this size, addition of further molecules will only decrease ΔG, and therefore growth is spontaneous. In contrast, it is thermodynamically "uphill" for a droplet smaller than R^* to grow. By differentiating Equation 13.5.1 with respect to R, and setting the result equal to zero, we arrive at an expression for R^*:

$$R^* = -\frac{2\gamma}{\Delta G_V} \tag{13.5.2}$$

This, in turn, can be substituted into Equation 13.5.1 to find the free-energy barrier associated with achieving the critical nucleus:

$$\Delta G^* = 4\pi R^{*2}\gamma + \frac{4\pi}{3}R^{*3}\Delta G_V = \frac{16\pi}{3}\frac{\gamma^3}{\Delta G_V{}^2} \tag{13.5.3}$$

We can also use Equation 13.4.5 to replace ΔG_V with a term involving the enthalpy of fusion and the undercooling:

$$\Delta G^* = \frac{16\pi}{3}\frac{\gamma^3}{\Delta G_V^2} = \frac{16\pi}{3}\frac{\gamma^3}{\Delta H_V^2}\frac{(T_m^\infty)^2}{(\Delta T)^2} \tag{13.5.4}$$

It is difficult to predict with certainty the rate at which critical nuclei form, but the two most important factors can be identified. The probability of a critical nucleus should be proportional to a Boltzmann factor, $\exp(-\Delta G^*/kT)$, which would determine the "equilibrium" concentration of such nuclei. Secondly, the rate of formation will be proportional to the rate of arrival of new molecules at the droplet surface, or, in the case of polymers in the bulk, the rate of segmental rearrangements at the surface in order to fit into the lattice. In either case we know from the discussion in Section 12.4, and Equation 12.4.12 in particular, that the temperature dependence of polymer transport follows the Vogel–Fulcher or Williams–Landel–Ferry form:

$$\text{Polymer or segment mobility} \sim A\exp\left(-\frac{B}{T-T_0}\right) \tag{13.5.5}$$

where A, B, and T_0 are parameters that can be related to free volume. (However, in practice the values of A, B, and T_0 will be different for crystallization rate than for, say, the macroscopic viscosity.) Combining Equation 13.5.4 and Equation 13.5.5, we can extract the temperature dependence of the nucleation rate:

$$\begin{aligned}\ln(\text{rate}) &\propto \ln\left\{\exp\left(-\frac{B}{T-T_0}\right)\exp\left(-\frac{\Delta G^*}{kT}\right)\right\} + \text{constant}\\ &= -\frac{B}{T-T_0} - \frac{K}{T(\Delta T)^2} + \text{constant}\end{aligned} \tag{13.5.6}$$

where K denotes a collection of temperature-insensitive quantities. The temperature dependence is complicated in detail, but the qualitative consequences of the two terms are clear. The transport term indicates that at low temperatures ($T \to T_0$) the rate will go to zero, as nothing can move. However, we recall from Chapter 12 that the glass-transition temperature is usually significantly below the melting temperature, so this effect can be avoided. The barrier term indicates that as temperature decreases, the rate will increase, because $T(\Delta T)^2$ will increase. This is a simple consequence of an increased thermodynamic driving force to crystallize as the undercooling increases. From Equation 13.5.2 we can see that the critical nucleus shrinks as the crystallization temperature decreases, so nuclei are easier to form.

We now take a closer look at the nucleation of a polymer crystal. Experimental evidence, much of it indirect, suggests that the critical nucleus size is on the order of 10 nm. Furthermore, even in the early stages the nuclei can be faceted to reflect the underlying unit cell. The important new issue is: what sets the thickness of the crystal nucleus, ℓ? We already know that lamellae tend to grow with a (nearly) single value of ℓ that depends primary on ΔT, so presumably the critical nucleus has to set the stage for this choice of ℓ. A straightforward extension of the critical nucleus analysis can give some insight into how this might come about. Assume a cylindrical nucleus just as in Figure 13.11, with height ℓ and radius r. Assume the nucleus contains p stems (which might come from several chains, but not necessarily p different chains), and that each stem occupies a cross-sectional area d^2. The volume of this nucleus can be written as $\pi r^2 \ell$, or as $pd^2\ell$, so $r = d\sqrt{p/\pi}$. We can now write the analogous expression to Equation 13.5.1, Equation 13.4.2, and Equation 13.4.3 for the free energy of the nucleus, but now in terms of p and ℓ:

$$\Delta G(p, \ell) = pd^2\ell\Delta G_V + 2pd^2\gamma + 2d\ell\gamma\sqrt{\pi p} \tag{13.5.7}$$

Note that we are making the simplifying assumption that both top and side faces have the same surface energy. The critical nucleus size in terms of the number of stems can be found as before, by differentiating Equation 13.5.7 with respect to p and setting the result equal to zero. The answer is (see Problem 8)

$$p^* = \frac{\pi\ell^2\gamma^2}{d^2(\ell\Delta G_V + 2\gamma)^2} \tag{13.5.8}$$

The next step is to repeat this process, but differentiating Equation 13.5.7 (with p set equal to p^*) with respect to ℓ and setting the result equal to zero. This process locates that value of ℓ, ℓ^*, which minimizes the nucleation barrier $\Delta G^*(p^*, \ell)$. In other words, $\Delta G(p, \ell)$ is a surface with respect to p and ℓ; we have found the curve that represents the maximum with respect to p ($p = p^*$), and now we seek the minimum along that curve in terms of ℓ ($\ell = \ell^*$). The resulting point is called a *saddle point*; just as a hiker seeks a low altitude pass through a mountain range, the crystallization process will seek the easiest route across the nucleation barrier. When this is done (after some algebra, Problem 8), the resulting value of the critical stem length ℓ^* turns out to be simply

$$\ell^* = -\frac{4\gamma}{\Delta G_V} \tag{13.5.9}$$

Substituting Equation 13.5.9 and Equation 13.5.8 back into Equation 13.5.7 gives the critical barrier height:

$$\Delta G(p^*, \ell^*) = \frac{8\pi\gamma^3}{\Delta G_V^2} \tag{13.5.10}$$

This relation has exactly the same form as Equation 13.5.4; all that has changed is the numerical prefactor. In particular, the nucleation barrier height still depends on the inverse square of the undercooling, and so nucleation should be more facile at lower temperatures. What is new from this analysis is the existence of a preferred lamellar thickness for nucleation, ℓ^*. It depends linearly on the inverse undercooling, so we expect thicker nuclei (and crystals) as T_c is lowered. The dependence of ΔG from Equation 13.5.7 on p and ℓ is illustrated in Figure 13.18, for $d^2 = 20$ Å^2 per chain, and $\ell^* = 100$ Å. For each value of ℓ, ΔG shows a maximum in p, and the lowest value of this maximum occurs for $\ell^* = 100$ Å and $p^* = 400$. These curves can be used to estimate the relative rates of nucleation of different sized nuclei, as illustrated in Problem 9.

13.5.2 Crystal Growth

Although the processes of lamellar growth are still far from fully understood on the molecular scale, a reasonable understanding of the principal factors can be extracted by extension of the

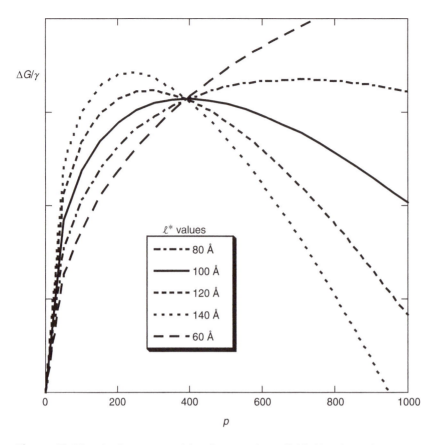

Figure 13.18 The free energy of forming a nucleus, divided by the surface energy, as a function of the number of stems, p, for different values of the lamella thickness, according to Equation 13.5.7.

concepts developed in the previous section. It is well established that under conditions of isothermal crystallization the growth velocity v of a lamellar face is a constant, and that the lamellar thickness also remains constant. The temperature dependence of v can be quite interesting, however, as we shall see.

We begin by assuming we have a perfect crystal face of height ℓ and width W, as shown in Figure 13.19. To start a new layer of chains (whether folded immediately or not), a single stem must attach to the crystal. There is a barrier to this process, because the new stem has increased the surface area of the crystal. Indeed, this process is termed *secondary nucleation*, because it nucleates the growth of a single new layer, in contrast to the primary nucleation process considered

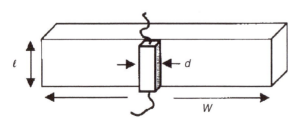

Figure 13.19 Schematic of the secondary nucleation process, whereby a single complete new stem adds to a perfectly flat crystal face.

in the previous section. For simplicity we assume that the stem has a square cross-section with side d. The barrier to the addition of a single stem, ΔG_s, contains three terms just as in Equation 13.5.7. There is a favorable contribution from the increase of the crystal bulk, given by $\ell d^2 \Delta G_V$ (recall that ΔG_V is negative). There are two unfavorable terms for the added surface, one from the top and bottom of the lamella, and the other from the new faces of the stem. The former is given by $2d^2 \gamma_f$ and the latter by $2d\ell\gamma_g$, where we distinguish the two surface energies with subscripts f for "fold surface" and g for "growth surface," respectively. Thus the free-energy change associated with secondary nucleation can be written as

$$\Delta G_s = d^2 \ell \Delta G_V + 2d^2 \gamma_f + 2d\ell\gamma_g \qquad (13.5.11)$$

The overall rate of secondary nucleation per unit width of the growth surface, which we will call r_s, should be proportional to the product of a dynamics term and the appropriate Boltzmann factor, just as in Equation 13.5.6:

$$r_s \propto \exp\left(-\frac{B}{T - T_0}\right) \exp\left(-\frac{\Delta G_s^*}{kT}\right) \qquad (13.5.12)$$

where ΔG_s^* is the barrier to secondary nucleation. Once again the relevant process is governed by the competition between the gain in bulk free energy, proportional to ΔG_V, and the surface energy penalty, involving γ_f and γ_g. Note that we cannot differentiate Equation 13.5.11 to find a "critical" nucleus, because we have already assumed that it is a single stem of length ℓ. A much more detailed treatment, due to Hoffmann and Lauritzen [2], gives $\Delta G_s^* \sim -d\gamma_f\gamma_g/\Delta G_V$, and therefore $\ln r_s$ will be proportional to a term in $1/T\Delta T$ (see Equation 13.5.6).

Once the secondary nucleus is in place, the rest of the layer could fill in by adding adjacent stems. Each new stem increases the crystal volume by the same amount, $d^2\ell$, but also increases the fold surface area by $2d^2$; it does not increase the exposed growth-surface area. Thus for the growth process we can write

$$\Delta G_g = d^2 \ell \Delta G_V + 2d^2 \gamma_f \qquad (13.5.13)$$

The rate of growth of a single face, g, should follow an expression analogous to Equation 13.5.12:

$$g \propto \exp\left(-\frac{B}{T - T_0}\right) \exp\left(-\frac{\Delta G_g}{kT}\right) \qquad (13.5.14)$$

However, under normal conditions we expect ΔG_g to be negative, or crystallization is not really going to proceed at all, and so there is no barrier and therefore g does not depend on the undercooling, ΔT.

In the above scenario, it is not unreasonable to expect that the rate of growth, g, will exceed the rate of secondary nucleation, r_s, so much so that each new layer will fill in completely before the next layer starts. If true, the overall growth velocity (units of length time^{-1}) will simply be given by the product of the layer thickness, d, the width, W and r_s (units of length^{-1} time^{-1}):

$$v = dWr_s \qquad (13.5.15)$$

and the temperature dependence of v, once corrected for the transport term, will be linear in $1/T\Delta T$. In fact this is often observed; an example from the solution crystallization of polyethylene is shown in Figure 13.20a (where the transport term is not so important).

A strong indication that the real process is much more complicated than the above description can be seen in Figure 13.20b. Here, logarithmic growth velocities, corrected for the transport term, are plotted as a function of $1/T\Delta T$, for four different polyethylenes. Each curve exhibits two different linear portions, separated by rather sharp changes; for each curve the higher-temperature regime has the larger slope. Several polymers display this general behavior, whereas in others the

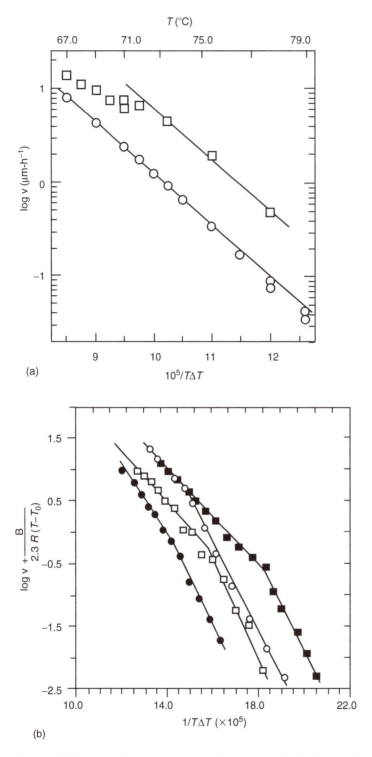

(a)

(b)

Figure 13.20 Crystallization rates for polyethylenes. (a) Single crystals grown from solution. (Reprinted from Toda, A. and Kiho, H., *J. Polym. Sci., Polym. Phys. Ed.*, 27, 53, 1989. With permission.) (b) Spherulites grown from the melt. (Reprinted from Lambert, W.S. and Phillips, P.J., *Macromolecules*, 27, 3537, 1994. With permission.)

lower-temperature regime has the larger slope. Still other polymers show three or even four regimes, but in most cases the slopes differ by factors of about 2 (or 1/2). These abrupt changes suggest that the growth mechanism is changing in some distinct way. A good possibility to consider is that the assumption $g \gg r_s$ may not always apply. In particular, the Boltzmann factor for r_s depends on $1/T\Delta T$, whereas for g it depends only on $1/T$. A small change in T will have no appreciable effect on g, but it can have a profound effect on r_s. Consequently as the undercooling is increased, perhaps r_s and g become competitive, or even $r_s \gg g$.

These possibilities are also considered in the theory of Hoffmann and Lauritzen. The main conclusion of interest to us can be anticipated in a rather straightforward way. Suppose that the undercooling has increased to the point where r_s and g are similar. In that case a new layer will fill in by both random attachment and adjacent attachment of stems. The growth velocity v will depend on both r_s and g. As r_s has units of length^{-1} time^{-1}, and g has units of length time^{-1}, dimensional analysis suggests that

$$\mathrm{v} \propto d\sqrt{r_s g} \qquad\qquad\qquad (13.5.16)$$

If this assumption is correct, the dependence of v on temperature will still come largely from r_s, as noted above. Because of the square root, the argument of the Boltzmann factor will now be multiplied by 1/2, and thus the slope will decrease by a factor of 2, as in Figure 13.20b. In the context of the Hoffmann and Lauritzen theory, this transition in mechanism corresponds to passage from so-called *Regime I* crystallization, where $r_s \ll g$, to *Regime II*, where $r_s \approx g$. With still deeper undercooling, the system can undergo another transition into *Regime III*, where $r_s \gg g$. In this case, v $\sim r_s$ again because there is essentially no lateral growth of a crystal layer; as a consequence the slope of ln v versus $1/T\Delta T$ will increase by a factor of 2 from that in Regime II.

We have deliberately avoided any more detailed examination of the secondary nucleation and layer growth processes, in part because a full molecular level description would be very compli-cated, and in part because these issues are far from fully resolved. Details we have not considered include the following: Does a secondary nucleus have a stem of length ℓ, where ℓ is the thickness of the primary nucleus, or is it different? Does growth proceed by one chain at a time folding regularly, like a fire hose (as the Hoffmann–Lauritzen theory proposes), or do new stems from other chains participate? Does a new layer immediately grow with thickness ℓ, or does it anneal to full thickness after first attaching to the surface? Does a new stem begin by one repeat unit sticking to the face of a unit cell, or do longer helical sections form first? Is the melt in the immediate vicinity of the growth surface completely disordered, or is there some intermediate level of order that precedes attachment of new stems? Is the lamella growth surface actually flat, or is it rough, thereby removing the need for secondary nucleation?

These considerations aside, there is one other general feature of the growth process outlined above that should be brought out. In Regime I, the thermodynamic drive to grow is relatively small, and the system can be thought of as being close to a local equilibrium between stems attaching and detaching. In other words, a new stem might be formed and then melt off the surface several times before actually being locked in place. Under such quasi-equilibrium growth condi-tions, very smooth and regular crystals can be grown. This is consistent with the familiar experience of growing small-molecule crystals, where modest undercoolings and long times are necessary to obtain large single crystals. In contrast, for deeper undercoolings in Regime III, stems can be envisioned as sticking virtually irreversibly on the first attempt. This kind of a disorganized process leads to rapid growth, but more defect-laden, irregular structures.

One final issue to consider: what is the role of molecular weight in all of this? We have so far completely ignored the dependence of nucleation or growth on this most important polymer variable. Two figures serve to illustrate the main points. Figure 13.21 shows the temperature dependence of crystal growth for various molecular weights of poly(tetramethyl-*p*-phenylene siloxane). There is a strong peak in growth velocity near 65°C, with the rate tending to zero

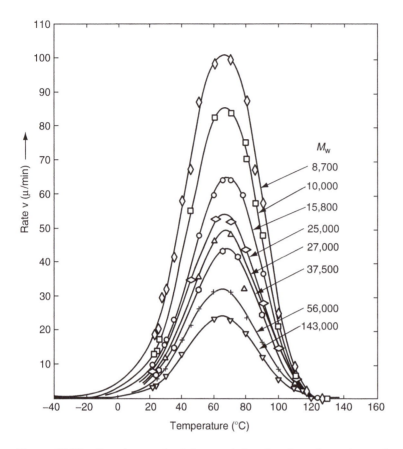

Figure 13.21 Growth rates of poly(tetramethyl-*p*-phenylene siloxane) crystals as a function of temperature, for the indicated molecular weights. (Reprinted from Magill, J.H., *J. Appl. Phys.*, 35, 3249, 1964. With permission.)

near 120°C and 0°C. This feature is exactly what we would anticipate based on the discussion above, namely that at relatively small undercoolings the rate increases with lower temperatures, but eventually the transport term takes over and the rate goes to zero. For comparison purposes, the glass-transition temperature for this polymer is about −20°C, and so the rate becomes negligible a few degrees above T_g. The important new information in this figure is that the peak position is independent of molecular weight, but the peak growth velocity decreases significantly with increasing molecular weight. The peak position is controlled by a balance between the thermo-dynamics associated with adding stems (whether by secondary nucleation or by layer growth), and the dynamics of molecular rearrangements. Neither ΔG_V nor the various surface energies depend appreciably on molecular weight, and neither do the Vogel–Fulcher or WLF temperature dependence of chain dynamics (at least for reasonably long chains, see Section 12.4); thus the position of the peak is also insensitive to chain length.

The molecular weight dependence of the growth velocity can be better seen when plotted directly against inverse molecular weight, as shown in Figure 13.22. At high molecular weights, when the chains are well entangled (see Chapter 11) the rate is lowest, but independent of molecular weight. At lower molecular weights the rate increases as the chains become shorter. The molecular weight independence for long chains argues for a rate-determining step that involves rearrangements of a portion of the chain, perhaps a few times the length of a stem, rather than diffusion of the whole polymer. Lower molecular weight chains can accommodate the necessary conformational rearrangements more rapidly, as entire molecules.

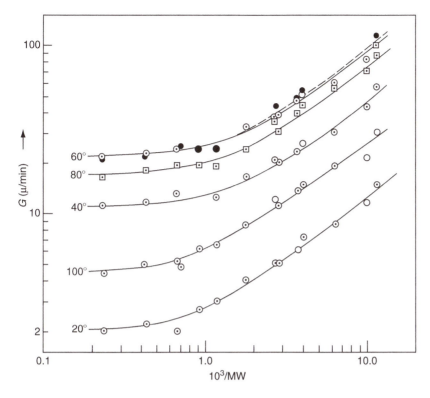

Figure 13.22 Same data as in Figure 13.21, now plotted versus inverse molecular weight at the indicated temperatures.

13.6 Morphology of Semicrystalline Polymers

At this point we have a good picture of the organization of polymer molecules at the unit cell level (~1 nm) and within a lamella (~10 nm). To complete the picture we need to consider how the lamellae and the intervening regions of amorphous material arrange themselves to fill up the bulk of the material. By far the most commonly observed morphology is that of the spherulite, to be considered first, but other structures such as hedrites, dendrites, and shish kebabs are also found under certain crystallization conditions.

13.6.1 Spherulites

Suppose a bulk-crystallized polymer sample is observed in a polarizing optical microscope, with the sample placed between two polarizers oriented at right angles to each other. In the absence of any sample, no light would be transmitted owing to the 90° angle between the vectors describing the light transmitted by the two polarizers (see Section 8.1 for a discussion of polarized light). With a crystalline sample of polymer in place, however, an image such as that shown in Figure 13.23 is generally observed. The field of view becomes at least partially filled with domains called *spherulites*, which grow in time, impinge upon one another, and eventually fill space. They generally exhibit the following features:

1. They possess spherical symmetry around a single center of nucleation. This symmetry projects a perfectly circular cross-section if the development of the spherulite is not stopped by contact with another expanding spherulite. The spherical structure indicates a single growth rate in

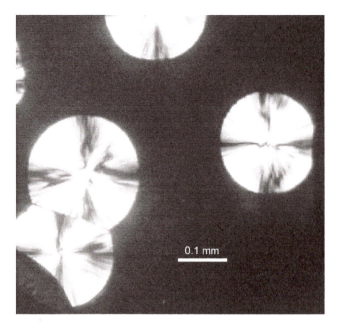

0.1 mm

Figure 13.23 Spherulites of poly(L-lactide) growing from the melt. Courtesy of R. Taribagil.

three dimensions, which we will need to reconcile with the one- or two-dimensional growth mode of individual lamellae.

2. Each spherulite is revealed by the characteristic Maltese cross optical pattern under crossed polarizers, although the Maltese cross is truncated in the event of impinging spherulites.

3. Superimposed on the Maltese cross may be such additional optical features as *banding*, illustrated in Figure 13.26.

4. A system of mutually impinging spherulites ultimately develop into an array of irregular polyhedra, the dimensions of which can be as large as a millimeter or more. The size of the domains will obviously increase as the number of nuclei decreases, and so information about nucleation density may be inferred even after crystallization is complete.

5. A larger number of smaller spherulites are produced at larger undercoolings, as the barrier to nucleation is reduced. Various details of the Maltese cross pattern, such as the presence or absence of banding, may also depend on the temperature of crystallization.

6. Spherulites have been commonly observed in organic and inorganic systems of synthetic, biological, and geological origin, including moon rocks, and are therefore not unique to polymers.

On the basis of a variety of experimental observations, including an analysis of the ubiquitous Maltese cross, a number of aspects of the structure of spherulites have been elucidated. The spherulites are aggregates of lamellar crystals radiating from a single nucleation site. The latter can be either a spontaneously formed single crystal or a foreign body. The spherical symmetry is not present at the outset, but develops with time. Fibrous or lathlike lamellar crystals grow away from the nucleus, and begin branching and fanning out. As the lamellae spread out radially and three dimensionally, branching of the crystallites continues to sustain the spherical morphology. Figure 13.24 represents schematically the leading edge of some of these fibrils, one of which has just split.

The molecular alignment within these radiating fibers is tangential, that is, perpendicular to the radius of the individual spherulite. The individual lamellae are similar in organization to single crystals: they consist of ribbons on the order of 10–100 nm in thickness, built up from successive

Figure 13.24 Schematic illustration of the leading edge of a lathlike crystal within a spherulite.

layers of folded chains. Growth is accomplished primarily by the addition of successive layers of chains to the ends of the radiating laths. This feature is also indicated schematically in Figure 13.24. Portions of an individual long polymer chain can be incorporated into different lamellae, and thus link them in a three-dimensional network. These interlamellar links are not possible in spherulites of low molecular-weight compounds, which show much poorer mechanical strength as a consequence.

The molecular chain folding is the origin of the Maltese cross. The Maltese cross pattern arises from a spherical array of birefringent particles through the following considerations (see also Figure 13.25):

1. The ordered polymer chains, that is the stems, are consistently oriented perpendicular to the radius of the spherulite.
2. The index of refraction of all polymers differs for light polarized parallel to the chain axis versus normal to it. Recalling the discussion in Section 8.1, the refractive index of a material reflects the polarizability of the constituent molecules. Individual polymer molecules have a polarizability anisotropy, which leads to anisotropy in refractive index when the molecules are aligned. Substances showing this anisotropy of refractive index are said to be *birefringent*.
3. Items (1) and (2) indicate that the refractive index in the tangential direction of the spherulite differs from that along the radius. It actually does not matter in this context which refractive index is greater; for polyethylene and poly(ethylene oxide), it is greater along the chain backbone, but for polystyrene and poly(vinyl chloride) it is greater normal to it.
4. The electric vector of the polarized light emerging from the first polarizer (let us say the beam is traveling along lab direction z and is polarized along y) may be resolved into components along the radial and tangential directions of the spherulite. These two components propagate at different speeds (recall Equation 8.1.6), and thus will not recombine to recover perfectly y-polarized light. The resulting x-component of light is transmitted by the second polarizer, leading to a bright image.
5. As we proceed around the spherulite, at 90° intervals the polarization axis of the light will coincide with either the radial or tangential direction in the crystal. At that point, the electric vector has no component along the orthogonal axis, and so there is no component to sense the different refractive index. Consequently the light emerges with polarization preserved, and is extinguished by the second polarizer. Thus there are four dark sectors in the image, producing the Maltese cross.

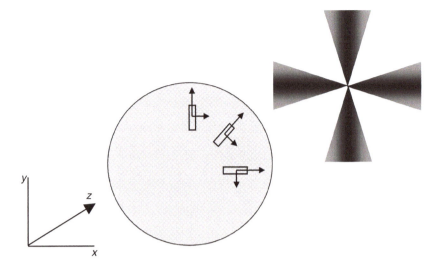

Figure 13.25 *y*-Polarized light is incident on a spherulite. The small rectangles and arrows illustrate the local orientation of the refractive indices of the crystal. Light is transmitted through an *x*-polarizer only when the local orientation is not parallel to *x* or *y*.

5. You should convince yourself that if the polarizers are held fixed and the sample rotated between them, the Maltese cross remains fixed because of the symmetry of the spherulite.

In many cases, the microscope image reveals another feature, namely banded spherulites, as illustrated in Figure 13.26. In addition to the Maltese cross, there are concentric dark rings at regular intervals moving out from the center. This is due to twisting of the individual lamellar ribbons along the radial direction; from the spacing of the bands, the period of the twist can be calculated and is found to depend on crystallization conditions. The fact that the dark rings are more or less continuous around a circle implies that the material within the ring is optically isotropic, rather than oriented with its optical axes parallel to the polarizers. As an individual lamella twists, the fold plane containing the chain stems rotates about the radial direction. The refractive index is also different parallel and perpendicular to the stem within the fold plane, and so the value of the mean refractive index along the tangential direction varies continuously between these limits with the twist angle. At certain twist angles, this projected tangential value of refractive index can be identical to the radial value, leading to optical isotropy and light extinction. The band spacing reflects the distance over which a lamella completes a helical twist; the origin of the twist is thought to lie in the particular chain conformations of the fold surface.

13.6.2 Nonspherulitic Morphologies

Spherulitic growth is the natural consequence of a crystallization process that proceeds steadily from a single nucleus, with a constant growth rate (although that is not strictly required), and is allowed to proceed with equal probability in three dimensions. A more subtle consideration is that the growth rate has to be sufficiently slow so that the relevant lamellar facets extend outward with no significant change in structure. Such a growth process is a natural consequence of a situation where secondary nucleation is the rate-limiting step, as for example in the theory of Hoffmann and Lauritzen described in Section 13.5.2. In this process, it takes a while for a growth face to add the first stem of the next layer, but once it does, the layer fills in completely and relatively rapidly.

Figure 13.26 Spherulites of poly(1-propylene oxide) observed through crossed polarizers by optical microscopy. (Reproduced from MaGill, J.H., *Treatise on Materials Science and Technology*, Vol. 10A, Schultz, J.M. (Ed.), Academic, New York, 1977. With permission.)

In this manner the structure of the lamellae is preserved. In another limit, addition of a new stem to a smooth facet could be more rapid than filling out the new layer. This is reminiscent of a general growth process known as *diffusion limited aggregation*, in which new particles stick to the first part of a growing cluster that they encounter. Under these conditions, a crystal or aggregate would grow much more haphazardly, leading to *dendritic structures* (which are often akin to the form of a fir tree). An example of dendritic growth of polyethylene crystals in solution is shown in Figure 13.27. This growth mode was accessed by the simple expedient of a deeper undercooling; the increased thermodynamic drive to form crystals reduces the barrier to secondary nucleation.

Hedrites, or *axialites* as they are sometimes termed, represent an alternative morphology that is sometimes encountered. A hedrite may be defined as a crystallite that looks like a polygon when viewed from at least one direction [3]; an example is shown in Figure 13.28a. A cartoon version of one possible hedrite structure is shown in Figure 13.28b. In this case several lamellae are stacked on top of one another, but then splay out when growing further from the center. This is somewhat analogous to splaying the pages in a book or sheets in a stack of paper. Qualitatively, one can imagine such a structure emerging when the individual lamellae grow at comparable rates in two dimensions; this growth mode inhibits structures with spherical symmetry. In contrast, the spherulite results from primarily one-dimensional growth of individual lamellae, that fan out in three dimensions over time.

We conclude this section on crystal morphology by briefly describing crystallization under applied stress. This is a fascinating topic in its own right, and of great importance in processing of some semicrystalline polymers. We noted this possibility in Chapter 10 when discussing deviations

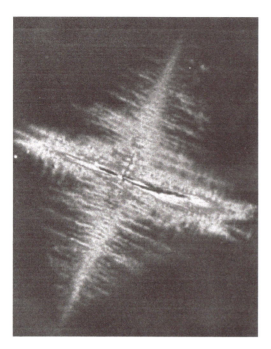

Figure 13.27   Polyethylene dendrite grown from dilute toluene solution and observed by interference (optical) microscopy; the long dimension is approximately 100 μm. (Reproduced from Wunderlich, B. and Sullivan, P., *J. Polym. Sci.*, 61, 195, 1962. With permission.)

from the simple model for rubber elasticity under large deformations. Stress-induced crystallinity is important in film and fiber technology. For example, when dilute solutions of polymers are stirred rapidly, or when fibers are spun from relatively dilute solutions, characteristic structures develop, which are described as having a *shish-kebab* morphology. A beautiful example is shown in Figure 13.29a, and a cartoon of the underlying structure is provided in Figure 13.29b.

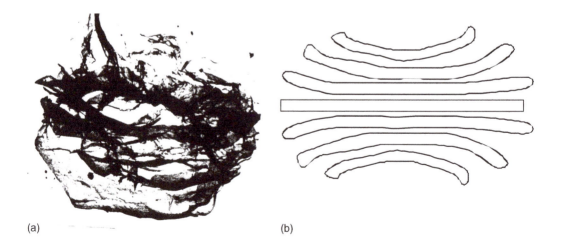

(a)   (b)

Figure 13.28   (a) A polyethylene hedrite. (b) Cartoon of a hedrite viewed end-on. (Reproduced from Bassett, D.C., Keller, A., and Mitsuhashi, S., *J. Polym. Sci.*, 1A, 763, 1963. With permission.)

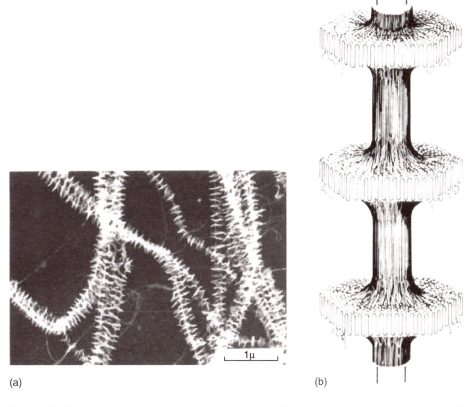

(a)   (b)

Figure 13.29 Transmission electron micrograph of polyethylene shish kebabs crystallized from xylene solution during flow (a), and (b) schematic of the underlying chain structure. (Reproduced from Pennings, A.J., van der Mark, J.M.A.A., and Kiel, A.M., *Kolloid Z.Z. Polym.*, 237, 336, 1970. With permission.)

These consist of chunks of folded chain crystals (kebabs) strung out along a fibrous central column (shish). In both portions the polymer chain axes are parallel to the overall axis of the structure. The essence of this process is to extend the individual chains to a substantial degree, before crystallization. In this way the extended chain shish provides nucleation sites for the chain-folded kebabs. Relatively dilute solutions are favored, because in a highly entangled state it is very difficult to extend individual chains before crystallization. Similarly, relatively high molecular weight chains are favored because it is difficult to apply enough stress in solution to extend a short chain. Extremely strong fibers can be fabricated from such shish kebabs, as the high degree of crystallinity combined with the uniform orientation of the chain axis imparts remarkable tensile strength. An example of such a material is *gel-spun polyethylene*, which is prepared in two stages. In the first stage, a hot solution is extruded to partly align the chains and cooled into a gel (crystallinity provides the cross-link sites). In the second stage, the material is drawn into fibers while the remaining solvent is removed. The resulting fibers have far superior mechanical properties to standard fibers, and are used in demanding specialty applications such as bullet-resistant garments and in racing yachts.

13.7 Kinetics of Bulk Crystallization

In the previous sections we have discussed the thermodynamic factors that influence crystallization, and we have considered the role of kinetics in the growth of individual crystals or lamellae. In the preceding section we have examined the diverse morphologies that can result from polymer crystallization under various situations. In this section we turn our attention back to kinetics, and in particular we consider the following questions: How long will it take a macroscopic sample to crystallize? If we have information about the crystalline fraction as a function of time, what, if anything, can we infer about the crystal growth mechanism?

13.7.1 Avrami Equation

The following derivation will illustrate how the rates of nucleation and growth combine to give the net rate of crystallization [4,5]. The theory we shall develop at first assumes a specific picture of the crystallization process, but then we can generalize the result. The assumptions of the model and some comments on their applicability follow:

1. The crystals are initially assumed to be circular disks. This geometry is consistent with previous thermodynamic derivations. It has the advantage of easy mathematical description.
2. The disks are assumed to lie in the same plane. Although this picture is implausible for bulk crystallization, it makes sense for crystals grown in ultrathin films, adjacent to surfaces, and in stretched samples. A similar mathematical formalism will be developed for spherical growth and the disk can be regarded as a cross-section of this.
3. Nucleation is assumed to begin simultaneously from centers positioned at random throughout the liquid. This is more descriptive of heterogeneous nucleation by foreign bodies introduced at a given moment than of random nucleation. We shall subsequently dispense with the requirement of simultaneity.
4. Growth in the radial direction is assumed to occur at a constant velocity. There is ample experimental justification for this in the case of three-dimensional spherical growth.

Figure 13.30a represents the top view of an array of these disks after the crystals have been allowed to grow for a time t after nucleation. The three disks on the left are separated widely enough to still have room for further growth; the three disks on the right have impinged upon one another and can grow no more. We saw in the previous section that this latter situation can be observed microscopically.

Suppose we define the rate of radial growth of the crystalline disks as $\dot{r}$. Then disks originating from all nuclei within a distance $\dot{r}t$ of an arbitrary point, say, point x in Figure 13.30a, will reach that point in an elapsed time t. If the average concentration of nuclei in the plane is N (per unit area), then the average number of fronts $\overline{F}$, which converge on x in this time interval is

$$\overline{F} = \pi(\dot{r}t)^2 N \tag{13.7.1}$$

If a second growth front were to impinge on a point like this, its growth would terminate at x. Suppose we imagine point x to be "charmed" in some way such that any number of growth fronts

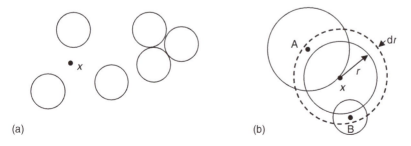

(a)   (b)

Figure 13.30 The growth of disk-shaped crystals. (a) All crystals have been nucleated simultaneously. All crystals have the same radius $\dot{r}t$ after an elapsed time t. (b) Nucleation is sporadic. Crystal A has had enough time to reach point x, while B has not, although both originate in the same ring a distance r from x.

can pass through it without interference. If we were to monitor the number of (noninterfering) fronts that cross x in a series of observations, we would expect a distribution of values because of the random placement of the nuclei. Furthermore, the distribution of F values is expected to pass through a maximum. Fronts arising from nuclei very close to x can easily cross x in the allotted time, but the area of melt under consideration in this case is small, so the number of fronts is small. As the area around the charmed point x is enlarged, a larger number of nuclei will be encompassed so the number of fronts crossing x will increase. This increase is offset by the fact that fronts originating from more distant nuclei will require more time to reach x. Therefore the number of fronts that cross x (remember that these are free from interference by hypothesis) will increase, pass through a maximum, and decrease as we allow them to originate from all parts of the sample. This distribution of values of F is our next interest.

We propose to describe the distribution of the number of fronts crossing x by the Poisson distribution function, discussed in the context of living polymerization in Chapter 4. This probability distribution function describes the random partitioning of a set of objects into a fixed number of boxes. In this case, the probability $P(F)$ describes the likelihood of a specific number of fronts, F, arriving per unit time in terms of F and the average number $\bar{F}$, as follows (see Equation 4.2.19):

$$P(F) = \frac{e^{-\bar{F}} \bar{F}^F}{F!} \tag{13.7.2}$$

Next we apply this distribution to the case where $F = 0$, that is, to the case where no fronts have crossed point x. There are several aspects to note about this situation:

1. Since $\bar{F}^0 = 1$ and $0! = 1$, Equation 13.7.2 becomes

    $$P(0) = e^{-\bar{F}} \tag{13.7.3}$$

    for $F = 0$.
2. The condition of no fronts crossing x is automatically a condition of noninterference, so the special magic previously postulated for point x poses no problem.
3. Since point x is nonspecific, Equation 13.7.3 describes the fraction of observations in which no fronts cross any arbitrary point, or the fraction of the area in any one experiment that is crossed by no fronts.
4. This last interpretation makes $P(0)$ the same as the fraction of a sample in the amorphous state. It is conventional to focus on the fraction crystallized, ϕ_c; therefore the amorphous fraction is $1 - \phi_c$ and

    $$1 - \phi_c = P(0) = e^{-\bar{F}} \tag{13.7.4}$$

5. Inverting and taking the logarithm of both sides of Equation 13.7.4, we obtain

    $$\ln\left(\frac{1}{1 - \phi_c}\right) = \bar{F} \tag{13.7.5}$$

Equation 13.7.1 and Equation 13.7.5 both describe the same situation, and can be equated to give

$$\ln\left(\frac{1}{1 - \phi_c}\right) = \pi \dot{r}^2 N t^2 \tag{13.7.6}$$

or

$$\phi_c = 1 - \exp\left(-\pi \dot{r}^2 N t^2\right) \tag{13.7.7}$$

Remember the units involved here: for $\dot{r}$ they are length time^{-1}; for N, length^{-2}; and for t, time. Therefore the exponent is dimensionless, as required. The form of Equation 13.7.7 is such that at small times the exponential equals unity and $\phi_c = 0$; at long times the exponential approaches zero and $\phi_c = 1$. In between, an S-shaped curve is predicted for the development of crystallinity with time. Experimentally, curves of this shape are indeed observed. However, we shall see presently that this shape is also consistent with other mechanisms in addition to the one considered so far.

Equation 13.7.7 may be written in the following general form, known as the *Avrami equation*

$$\phi_c = 1 - \exp(-Kt^m) \tag{13.7.8}$$

where in the previous case the so-called *Avrami exponent* $m = 2$, and the associated rate constant, K, is $\pi N \dot{r}^2$. Suppose rather than growing in two dimensions, the crystal fronts grew uniformly in three dimensions. An analysis similar to the one we just conducted would give $m = 3$ and $K = (4/3)\pi N \dot{r}^3$ in Equation 13.7.8. In this case N would be the number of nuclei per unit volume at $t = 0$, and the volume swept out per nucleus in time t would be $(4/3)\pi(\dot{r}t)^3$. Similarly, if the crystals tends to grow in one dimension, as in a growing rod or fibril, we would find $m = 1$.

In terms of spontaneous crystallization, the assumption that N nuclei begin to grow simultaneously at $t = 0$ is unrealistic. It corresponds most closely to the case of heterogeneous nucleation, where a fixed number of nucleation sites are in place at $t = 0$, but that by itself does not guarantee simultaneous onset of growth. We can modify the model to allow for random, spontaneous nucleation, a description more appropriate for homogenous nucleation, by the following argument. We draw a set of concentric rings in the plane of the disks around point x as shown in Figure 13.30b. If the radii are r and $r + dr$ for the rings, then the area enclosed between them is $2\pi r\, dr$. We postulate that spontaneous random nucleation occurs with a frequency of $\dot{N}$, having units area^{-1} time^{-1}. The rate of formation of nuclei within the ring is therefore $\dot{N} 2\pi r\, dr$.

We continue to assume that the crystals so nucleated display a constant rate of radial growth $\dot{r}$. This means that it takes a crystal originating in a ring of radius r around point x a time given by $r/\dot{r}$ to cross x. The crystal labeled A in Figure 13.30b has had just enough growth time to reach x. On the other hand, a crystal nucleated in this ring after $t - r/\dot{r}$ will not have had time to grow to x. The crystal labeled B in Figure 13.30b is an example of the latter case. It is only nucleation events that occur up to $t - r/\dot{r}$, which have time to grow from the ring of radius r and cross point x by their growth front. The increment in this number of fronts for the ring of radial thickness dr is

$$dF = \left(\dot{N} 2\pi r\, dr\right)\left(t - \frac{r}{\dot{r}}\right) \tag{13.7.9}$$

The average number of fronts crossing point x at a time of observation t is the sum of contributions from all rings, which are within reach of x in time t. The most distant ring included by this criterion is a distance $\dot{r}t$ from x. The average number of fronts, therefore, is given by integrating Equation 13.7.9 for all rings between $r = 0$ and $r = \dot{r}t$:

$$\overline{F} = 2\pi \dot{N} \int_0^{\dot{r}t} r\left(t - \frac{r}{\dot{r}}\right) dr \tag{13.7.10}$$

As far as this integration is concerned, $\dot{r}$ and t are constants, so Equation 13.7.10 is readily evaluated to give

$$\overline{F} = \frac{1}{3}\pi \dot{N} \dot{r}^2 t^3 \tag{13.7.11}$$

As before, this quantity in relation to the degree of crystallinity is given by Equation 13.7.5, so equating the latter to Equation 13.7.11 gives

$$\ln\left(\frac{1}{1-\phi_c}\right) = \frac{\pi}{3}\dot{N}\dot{r}^2 t^3 \tag{13.7.12}$$

or

$$\phi_c = 1 - \exp\left(-\frac{\pi}{3}\dot{N}\dot{r}^2 t^3\right) \tag{13.7.13}$$

Equation 13.7.7 and Equation 13.7.13 are analogous, except that the former assumes instantaneous nucleation at N sites per unit area while the latter assumes a nucleation rate of $\dot{N}$ per unit area per unit time. It is the presence of this latter rate that requires the power of t to be increased from 2 to 3 in this case. Again, Equation 13.7.13 is a particular case of the Avrami equation, Equation 13.7.8; the effect of switching from instantaneous nucleation to a constant nucleation rate is to increase the value of the Avrami exponent, m, by 1. For instantaneous nucleation, $m = 1$, 2, or 3, and for a constant nucleation rate, $m = 2$, 3, or 4, depending on the dimensionality of the growth process.

To acquire some numerical familiarity with the Avrami function, consider the following example.

Example 13.4

Three different crystallization systems show m values of 2, 3, or 4. Calculate the value required for K in each of these systems so that all will show $\phi_c = 0.5$ after 10^3 s. Use these m and K values to compare the development of crystallinity with time for these three systems.

	t in seconds		
ϕ_c	$m=2$ ($K=6.93 \times 10^{-7}$)	$m=3$ ($K=6.93 \times 10^{-10}$)	$m=4$ ($K=6.93 \times 10^{-13}$)
0.1	3.89×10^2	5.33×10^2	6.24×10^2
0.2	5.67×10^2	6.85×10^2	7.53×10^2
0.3	7.18×10^2	8.02×10^2	8.47×10^2
0.4	8.59×10^2	9.03×10^2	$9.27 - 10^2$
0.5	$1 - 10^3$	$1 - 10^3$	1×10^3
0.6	1.15×10^3	1.10×10^3	1.07×10^3
0.7	1.32×10^3	1.20×10^3	1.15×10^3
0.8	1.52×10^3	1.32×10^3	1.23×10^3
0.9	1.82×10^3	1.49×10^3	1.35×10^3

Solution

Solve Equation 13.7.8 for K and evaluate at $t = 10^3$ s for each of the m values: $K = [-\ln(1 - \phi_c)]/t^m$. For $m = 2$, $K = (\ln 0.5)/(10^3)^2 = 6.93 \times 10^{-7}$ s^{-2}; for $m = 3$, $K = 6.93 \times 10^{-10}$ s^{-3}; for $m = 4$, $K = 6.93 \times 10^{-13}$ s^{-4}. Note that the units of K depend on the value of m. Solve Equation 13.7.8 for t and evaluate at different ϕ_c's for the m and K values involved.

These three systems describe a set of crystallization curves that cross at $\phi_c = 0.5$ and $t = 10^3$ s, as shown in Figure 13.31. For the case where $m = 2$, the time interval over which the change occurs is widest (1430 s from $\phi_c = 0.1$–0.9) and the maximum slope is smallest (7.8×10^{-4} s^{-1} between $\phi_c = 0.4$ and 0.6). For $m = 4$, the range is narrowest (726 s) and the maximum slope is steepest (1.4×10^{-3}).

A further extension to the Avrami equation concerns the rate-determining step of the crystallization process. Equation 13.7.1 and those following it imply that contact between the growing disk and the surrounding melt for time t is sufficient for crystallization. Another possibility is that allowance must be made for the diffusion of crystallizable molecules to (or noncrystallizable

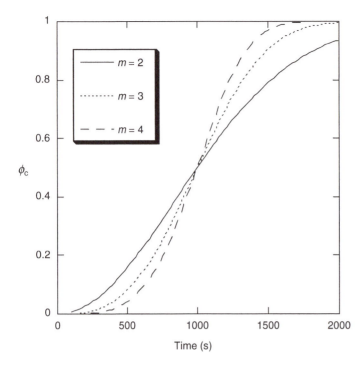

Figure 13.31 Crystallized fraction versus time for the indicated Avrami parameters, as discussed in Example 13.4.

molecules away from) the growth site. For example, it may be that amorphous molecules must diffuse out of the crystal domain to allow space for the crystallizing molecules. For a crystal of radius r, the time required for molecules to diffuse out of this domain can be determined from Equation 9.5.1 as $r = (6Dt)^{1/2}$. In Equation 13.7.1 this radius is written $r = \dot{r}t$. Thus, if the growth rate is diffusion controlled, these two expressions for r can be equated and solved for $\dot{r}$:

$$\dot{r} = \left(\frac{6D}{t}\right)^{1/2} \tag{13.7.14}$$

If this result is substituted into the previous expressions containing $\dot{r}$, the effect is to replace $\dot{r}$ with $(6D)^{1/2}$ and to multiply those t's that accompany $\dot{r}$ by $t^{-1/2}$.

This rather complex array of possibilities is summarized in Table 13.5. Table 13.5 lists the predicted values for the Avrami exponent for the following cases:

1. Growth geometry: 1D (e.g., fibrillar rod), 2D (e.g., disk or sheet), and 3D (e.g., sphere)
2. Nucleation mode: simultaneous (heterogeneous) and sporadic (homogeneous)
3. Rate determination: contact and diffusion

While there are several instances of redundancy among the Avrami exponents arising from different pictures of the crystallization process, there is also enough variety to make the experimental value of this exponent a valuable way of characterizing the crystallization process. In the next section we shall examine the experimental side of crystallization kinetics.

13.7.2 Kinetics of Crystallization: Experimental Aspects

In order to carry out an experimental study of the kinetics of crystallization, it is first necessary to be able to measure the fraction of polymer crystallized, ϕ_c. While this is necessary, it is not

Table 13.5 Summary of Exponents in the Avrami Equation (Equation 13.7.8) for Different Crystallization Mechanisms

Avrami exponent	Crystal geometry	Nucleation mode	Rate determination
0.5	Rod	Simultaneous	Diffusion
1	Rod	Simultaneous	Contact
1	Disk	Simultaneous	Diffusion
1.5	Sphere	Simultaneous	Diffusion
1.5	Rod	Sporadic	Diffusion
2	Disk	Simultaneous	Contact
2	Disk	Sporadic	Diffusion
2	Rod	Sporadic	Contact
2.5	Sphere	Sporadic	Diffusion
3	Sphere	Simultaneous	Contact
3	Disk	Sporadic	Contact
4	Sphere	Sporadic	Contact

sufficient; we must also be able to follow changes in the fraction of crystallinity with time. So far in this chapter we have said nothing about the experimental aspects of determining ϕ_c. We shall now briefly rectify this situation by citing some of the methods for determining ϕ_c. It must be remembered that not all of these techniques will be suitable for kinetic studies.

Since the fractions of crystalline (subscript c) and amorphous (subscript a) polymer account for the entire sample, it follows that we may measure whichever of the two is easiest to determine, and obtain the other by difference. Generally, it is some property P_c of the crystalline phase that we are able to monitor. If this property can be measured for a sample that is 100% crystalline (superscript °), we can compare the value of P_c measured on an actual sample (no superscript) to evaluate ϕ_c:

$$\phi_c = \frac{P_c}{P_c^\circ} \tag{13.7.15}$$

This relationship is sketched in Figure 13.32a, which emphasizes that P_c must vary linearly with ϕ_c, and that P_c° must be available, at least by extrapolation. The heat of fusion is an example of a property of the crystalline phase that could be used this way. However, it might be difficult to show that the value of ΔH_f° is constant per unit mass at all percentages of crystallinity, and to obtain a value for ΔH_f° for a crystal free from defects. Therefore, while conceptually simple, the actual utilization of Equation 13.7.15 in precise work may not be straightforward.

Figure 13.32b shows a variation in which a property of the sample (no subscript) is found to vary linearly with ϕ_c, having a value P_a when $\phi_c = 0$ and a value P_c when $\phi_c = 1$. The slope of this line is simply $P_c - P_a$, since the difference of ϕ_c values is unity for this difference in P. The equation for the line in Figure 13.32b is

$$P = P_a + \phi_c(P_c - P_a) \tag{13.7.16}$$

which can easily be solved for ϕ_c as a function of P, P_a, and P_c:

$$\phi_c = \frac{P - P_a}{P_c - P_a} \tag{13.7.17}$$

Specific volume (or density) is an example of a property that has been extensively used in this way to evaluate ϕ_c. If the amorphous component contributes nothing to the measured property (as with the heat of fusion), then Equation 13.7.17 reduces to Equation 13.7.15.

Figure 13.32c illustrates how x-ray diffraction techniques can be applied to the problem of evaluating ϕ_c. If the intensity of scattered x-rays is monitored as a function of the angle of diffraction, a result like that shown in Figure 13.32c is obtained. The sharp peak is associated with

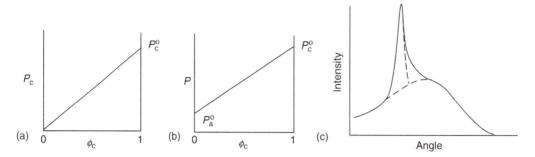

Figure 13.32   Various representations of the properties of mixture of crystalline and amorphous polymer. (a) The monitored property is characteristic of the crystal and varies linearly with ϕ_c. (b) The monitored property is characteristic of the mixture and varies linearly with ϕ_c between P_a and P_c. (c) X-ray intensity is measured with the sharp and broad peaks being P_c and P_a, respectively.

the crystalline diffraction, and the broad peak with the amorphous contribution. If the area A under each of the peaks is measured, then

$$\phi_c = \frac{A_c}{A_c + A_a} \tag{13.7.18}$$

An obvious difficulty here arises in deciding the location of the broken-line portions of the peaks in the region of overlap. Some features of the infrared absorption spectrum may also be analyzed by the same procedure to yield values for ϕ_c.

As noted above, not all techniques that provide information regarding crystallinity are useful to follow the rate of crystallization. In addition to possessing sufficient sensitivity to monitor small changes, the method must be rapid and suitable for isothermal regulation, quite possibly over a range of different temperatures. The spectroscopic techniques of infrared absorption, Raman scattering, and NMR have all been used successfully for this purpose, as has WAXS with a synchrotron source. Specific volume measurements are also convenient, and we shall continue this discussion using specific volume as the experimental method.

Although the extent of crystallinity is the desired quantity, time is the experimental variable. Accordingly, what is done is to identify the specific volume of a sample at $t = 0$ (subscript 0) with V_a, the volume at $t = \infty$ (subscript ∞) with V_c, and the volume at any intermediate time (subscript t) with the composite volume. On this basis, Equation 13.7.17 becomes

$$\phi_c = \frac{V_t - V_0}{V_\infty - V_0} \tag{13.7.19}$$

and the amorphous fraction becomes

$$1 - \phi_c = \frac{V_\infty - V_t}{V_\infty - V_0} \tag{13.7.20}$$

Figure 13.33a shows how this quantity varies with time for polyethylene crystallized at a series of different temperatures. Several aspects of these curves are typical of all polymer crystallizations and deserve comment:

1. The decrease in amorphous content follows an S-shaped curve. The corresponding curve for the growth of crystallinity would show a complementary but increasing plot. This aspect of the Avrami equation was noted in connection with the discussion of Equation 13.7.8.
2. The greater the undercooling, the more rapidly the polymer crystallizes, as discussed in Section 13.5. Although the data in Figure 13.33 are not extensive enough to show it, this trend does not continue without limit. As the crystallization temperature is lowered still

further, the rate of crystallization passes through a maximum and then drops off as T_g is approached, as illustrated in Figure 13.21.

3. Because bulk samples never become 100% crystalline, there is a potential ambiguity to quantities like V_∞; does it refer to 100% crystallinity, a state that is not achievable, or to the final value that is actually attained? Clearly the value of ϕ_c determined via Equation 13.7.19 will depend on the meaning employed. Depending on circumstances, either interpretation can be useful, but it is important to be clear as to which one is being used.

4. Replotting the data on a logarithmic timescale has an interesting effect: Figure 13.33b shows that this modification produces a far more uniform set of S curves. As a matter of fact, if the various curves are shifted along the horizontal axis, they may be superimposed to produce a reasonable master curve. By comparing the times corresponding to 50% crystallinity at 120°C and 130°C, there is a shift of over a factor of 1000. In other words, increasing the undercooling by 10 degrees increases the rate of crystallization by more than three orders of magnitude.

The preceding example of superpositioning is an illustration of the principle of time–temperature equivalence, as was discussed extensively in Section 12.5 in connection with the viscoelastic behavior of polymer samples. The current application differs in that the reason for the strong change in rate with temperature is the thermodynamic driving force, as well as partly to the reduction in free volume as T_g is approached, but the basic idea of time–temperature equivalence is the same.

Now let us examine an experimental test of the Avrami equation and the assortment of predictions from its various forms as summarized in Table 13.5. Figure 13.34 is a plot of $\ln [\ln(1 - \phi_c)^{-1}]$ versus $\ln t$ for poly(ethylene terephthalate) at three different temperatures. This format is suggested by rearranging Equation 13.7.8, and then taking the natural logarithm twice:

$$1 - \phi_c = \exp(-Kt^m)$$
$$\ln(1 - \phi_c) = -Kt^m$$
$$\ln\left[\ln\left(\frac{1}{1 - \phi_c}\right)\right] = m \ln t + \ln K \tag{13.7.21}$$

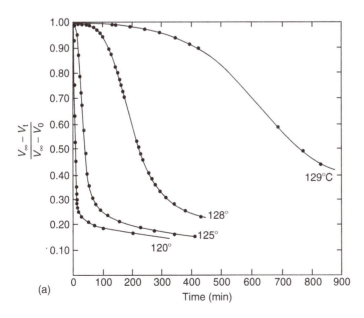

(a)

Figure 13.33 Fractions of amorphous polyethylene as a function of time for crystallization at the indicated temperatures, plotted on a (a) linear scale

(continued)

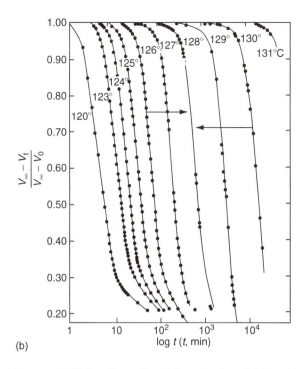

(b)

Figure 13.33 (continued)   (b) logarithmic scale. (Reprinted from Mandelkern, L., *Growth and Perfection of Crystals*, Doremus, R.H., Roberts, B.W., and Turnbull, D. (Eds.), Wiley, New York, 1958. With permission.)

According to Equation 13.7.21, this representation should yield a straight line, the slope of which corresponds to the Avrami exponent m.

The data in Figure 13.34 show that linearity is indeed obtained and that the slope equals 2 when the crystallization is carried out at 110°C and changes to 4 at higher temperatures. The melting point of poly(ethylene terephthalate) is 267°C, so that when the undercooling is about 25°C, three-dimensional growth with sporadic nucleation is suggested. With an undercooling of 150°C, the mechanism of crystallization is clearly different, although it is not possible to identify the specific combination of factors responsible for the exponent 2. The values of K in Equation 13.7.21 are best obtained analytically, once the exponent has been determined graphically. The two K values for the case where $m = 4$ in Figure 13.34 are 2.94×10^{-7} min^{-4} at 236°C and 3.13×10^{-8} min^{-4} at 240°C. The mechanism is apparently the same in these two cases, but the rate is more than nine times faster when the temperature is lowered by only 4°C. Note that it is not possible to resolve K, which is a cluster of nucleation and growth parameters, into its constituent factors, even when the value of the exponent identifies the mechanism unambiguously. At both 110°C and 120°C (not shown), $m = 2$ and the values of K are 7.93×10^{-4} and 7.45×10^{-3} min^{-2}, respectively. In this region the rate is about 10 times slower when the temperature is lowered by 10°C. Thus, both the value of m and the effect on K of changing temperature are different for these two regimes of behavior.

The testing of the Avrami equation reveals several additional considerations of note:

1. The multiple use of logarithms in the analysis presented by Figure 13.34 can obscure much of the deviation between theory and experiment. More stringent tests can be performed by other numerical methods.
2. Deviations from the Avrami equation are frequently encountered in the long time limit of the data.

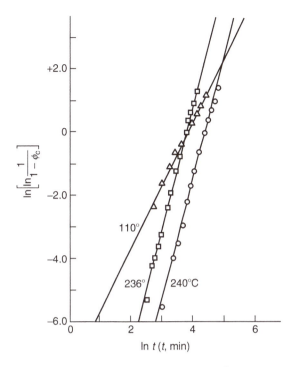

Figure 13.34 Log–log plot of $\ln(1-\phi_c)^{-1}$ versus time for poly(ethylene terephthalate) at three different temperatures. (Reprinted from Morgan, L.B., *Philos. Trans. R. Soc. London*, 247A, 13, 1954. With permission.)

3. Exponents other than integral multiples of one-half are observed. In fact, a method for determining the Avrami exponent, which is based on graphical differentiation rather than logarithmic analysis yields instantaneous m values at particular values of ϕ_c rather than a single value averaged over the entire transition. When this method is used, it is found that m increases initially, before eventually leveling off.

4. These unpredicted Avrami exponents may be indications that multiple mechanisms are operative or that $\dot{r}$ or $\dot{N}$ is itself a function of ϕ_c.

5. In general, one must exercise caution in inferring too much about the crystallization process from the Avrami analysis alone. This situation is analogous to that touched on in considering polymerization kinetics in Chapter 2 through Chapter 4, namely that it is dangerous to infer a polymerization mechanism from the resulting molecular weight distribution alone. Among the difficulties to bear in mind are the following: different mechanisms can lead to the same exponent; the nucleation may be due to a combination of heterogeneous and homogenous processes; each spherulite contains both crystalline and amorphous material; and the relative proportion may change with time.

13.8 Chapter Summary

In this chapter we have examined many aspects of the fascinating field of polymer crystallization. The main topics emphasized were the complex structural features of crystalline polymers, the interplay of thermodynamic and kinetic factors that dictate the structural details, and an introduction to techniques for the experimental characterization of both structure and crystallization kinetics:

1. At the smallest structural length scale, the unit cell, individual chains form helices to minimize intramolecular energetic constraints, and the helices pack together to maximize intermolecular

interactions. Polymer unit cells represent many of the 230 possible space groups, except those with cubic symmetry, and it is not unusual for a given polymer to exhibit two or more different polymorphs under different conditions.

2. On intermediate length scales the unit cells are organized into chain-folded lamellae, such that portions of the individual chain backbones, or stems, lie parallel to each other and approximately parallel to the thin axis of the lamella. Within an isothermally crystallized sample, the lamellae are of roughly constant thickness, but the thickness varies from a few tens to a few hundreds of angstroms, depending on crystallization conditions. Upon exiting the lamellar surface, a particular chain may execute a tight fold back into the crystal to become an adjacent stem, or it may wander off to reenter the same crystal at a different site, or even a different lamella. The prevalence of adjacent reentry is much greater in solution-grown single crystals than in melt-crystallized materials, and in crystals grown at smaller undercoolings.

3. On larger length scales the lamellae grow into spherulites, which may be viewed and characterized with a polarizing optical microscope. Other morphologies such as dendrites, hedrites, and shish kebabs can also be observed under particular conditions. A bulk sample never becomes 100% crystalline, and the lamellae are interspersed with amorphous regions that can often comprise the majority of the material.

4. Polymer crystals emerge by a process of nucleation and growth. The nucleation may be heterogeneous, homogeneous, or a combination of both. The barrier to homogeneous nucleation is dependent on the competition between bulk and surface contributions to the free energy; in general nucleation is more rapid, and the critical nuclei size smaller, the greater the undercooling. The growth process at the level of an individual lamella is still not fully understood, but in many cases the temperature dependence of the growth rate can be interpreted via the competition between the rate of addition of a single stem to a growth face and the rate of adding stems at neighboring sites.

5. The overall evolution of crystallinity often follows the Avrami equation, in which the type of nucleation, the spatial dimensionality of growth, and the presence or absence of diffusion limitations interact to yield a particular Avrami exponent. At relatively small undercoolings the rate of crystallization increases as temperature decreases, but eventually the rate decreases and vanishes as the approach to the glass transition inhibits any kind of chain or segmental motion. The overall rate of crystallization is typically independent of molecular weight for high molecular-weight polymers, but increases with decreasing molecular weight for shorter chains.

6. A wide variety of experimental tools are useful in the study of polymer crystallization. We have highlighted the use of electron microscopy to visualize single crystals in exquisite detail, and the use of x-ray and electron diffraction methods to characterize the unit cell structure.

Problems

1. Illers and Hendus measured the melting points of polyethylene crystals whose thickness was varied by controlling the conditions of crystallization, and which was measured by x-ray diffraction. The following results were obtained:[†]

T_m (°C)	139.4	137.5	136.0	134.9	131.9	127.9	117.9
ℓ (Å)	1750	758	481	392	258	177	100

Prepare a plot of T_m versus ℓ^{-1}, and from the slope and intercept, respectively, evaluate T_m^∞ and γ from these data. Compare the values obtained with quantities given in Table 13.3 and Example 13.3.

[†] K.H. Illers and H. Hendus, *Makromol. Chem.*, 113, 1 (1968).

2. There is a general correlation between the nature of the helix and the symmetry of the unit cell. For example, 3/1 and 6/1 helices often lead to hexagonal or trigonal unit cells. Why is this? What would you predict for 4/1 helices?
3. The polymers listed below are all known to form unit cells in which all of the angles are 90°. Use this fact plus the data given to complete the following table:

Polymer	M_o	a	b	c	Number of repeat units per cell	Density (g cm^{-3})
Polystyrene	104.1	—	—	6.63	18	1.126
Polyisobutene	56.1	6.94	11.96	—	16	0.937
Poly(vinyl chloride)	62.5	10.11	5.27	5.12	4	—
Nylon 8	—	4.9	4.9	22	2	1.038
Poly(methyl methacrylate)	100.1	21.08	12.17	10.55	—	1.23

4. For polyethylene, make an estimate of how much the molecular weight would need to change to bring about a 1° change in T_m, all other things being equal.
5. For n-alkanes, there is little doubt that the lowest free energy form of the crystal is the straight-chain lamella. For high molecular-weight polyethylenes, one could make the same argument, if the sample were perfectly monodisperse. Criticize or defend the following proposition: for a typical polydisperse sample of linear polyethylene, a chain-folded crystal could have a lower free energy than a straight-chain crystal, due to the greater entropy of chain end placement and to the avoidance of the need for the lamella thickness to vary with degree of polymerization.
6. Chemical evidence for chain folding in polyethylene crystals is obtained by etching polymer crystals with fuming nitric acid, which cleaves the chain at the fold surface. The resulting chain fragments are separated chromatograhically and their molecular weights determined by osmometry. The folded chain is pictured as crossing through the crystal, emerging and folding back, then reentering and recrossing the crystal, and so on. According to this picture, the shortest chain showing up in the chromatograms should equal the crystal thickness in length. The second shortest chain exceeds twice this value by some amount, which measures the length of the loop made by the chain outside the crystal. Molecular weights for the two shortest chains observed in an experiment of this sort were 1260 and 2530. Since the cleaved chains end in nitro and carboxyl groups, 60 should be subtracted from each of these molecular weights to give the polyethylene chain weight. Calculate the degree of polymerization of each molecule and the chain length (use the length of the unit cell along the chain axis, 2.53 Å, as the distance per repeat unit). Compare the latter with the crystal thickness determined by x-ray diffraction, 105 Å. What does the ratio of chain lengths for the first and second peaks suggest about the tightness of folding?
7. In discussing Figure 13.8 and Figure 13.9, the issue was raised about the accuracy of a linear Hoffmann–Weeks extrapolation. What conditions should be met for a linear extrapolation to be valid, and why might they not apply?
8. Follow through the analysis of Equation 13.5.7 to obtain Equation 13.5.8 and Equation 13.5.9.
9. In the context of Figure 13.18 and associated discussion, estimate by how much the nucleation rate would change for 20% change in ℓ away from ℓ^*.
10. Isotactic polypropylene crystallizes in 3/1 helices in a monoclinic unit cell as illustrated in Figure 13.7. Individual helices can either be right-handed or left-handed; they are energetically degenerate. In this particular polymorph, the helices are paired off. The same polymer can also be induced to crystallize in a hexagonal unit cell. In this case, how do

the helices pack? Some possibilities to consider are (a) Three right-handed and three left-handed helices occupy particular relative positions within the unit cell. (b) The crystalline packing does not care; right and left are mixed randomly. (c) Only right-handed or left-handed forms crystallize together; each lamella is pure right or left. Explain your reasoning.

11. Consider a polymer that tends to crystallize in needle-shaped crystals, rather than in lamellae. Assuming the needles have length L and radius R, with $R/L \ll 1$, derive the dependence of the melting temperature T_m on R. Why is the answer the same/different from the dependence of T_m on the lamella thickness?

12. The crystallization of poly(ethylene terephthalate) at different temperatures after prior fusion at 294°C had been observed to follow the Avrami equation with the following parameters applying at the indicated temperatures:[†]

T (°C)	m	K (min)
110	2	3.49×10^{-4}
180	3	1.35
240	4	5.05×10^{-8}

Calculate the time required for ϕ_c to reach values of 0.1, 0.2, ..., 0.9 for each of these situations. Graph ϕ_c versus t using the results calculated at 110°C and 240°C, plotting both in the same figure. Because of the much larger K at 180°C, the crystallization occurs much more rapidly at this temperature than at either 110°C or 240°C. Multiply each of the times calculated at 180°C by the arbitrary constant 60 and plot the data thus shifted on the same coordinates as the other curves. What generalization appears concerning the relative slopes at $\phi_c = 0.5$?

13. Poly(ethylene terephthalate) was crystallized at 110°C and the densities were measured after the indicated time of crystallization.[‡] Using density as the property measured to determine crystallinity, evaluate ϕ_c as a function of time for these data. By an appropriate graphical analysis, determine the Avrami exponent (in doing this, ignore values of $\phi_c < 0.15$, since errors get out of hand in this region). Calculate (rather than graphically evaluate) the value of K consistent with your analysis.

t (min)	ρ (g cm^{-3})	t (min)	ρ (g cm^{-3})
0	1.3395	35	1.3578
5	1.3400	40	1.3608
10	1.3428	45	1.3625
15	1.3438	50	1.3655
20	1.3443	60	1.3675
25	1.3489	70	1.3685
30	1.3548	80	1.3693

14. The crystallization rate of isotactic polypropylene ($M_w = 181,000$, $T_m = 172$°C) was studied under various patterns of temperature change.[§] Solids were melted at T_f, held at T_f for 1 h, and then crystallized at T_c. The following Avrami exponents were observed:

[†] F.D. Hartley, F.W. Lord, and L.B. Morgan, *Phil. Trans. R. Soc. London*, 247A, 23 (1954).
[‡] A. Keller, G.R. Lester, and L.B. Morgan, *Phil. Trans. R. Soc. London*, 247A, 1 (1954).
[§] P. Parrini and G. Corrieri, *Makromol. Chem.*, 62, 83 (1963).

Avrami exponent			
T_f (°C)	$T_c = 150°C$	$T_c = 155°C$	$T_c = 160°C$
190	—	3.1	3.5
210	2.9	3.3	4.1
220	3.1	3.8	—
230	3.1	4.0	—

On the basis of these observations, criticize or defend the following propositions:

1. When both T_f and T_c were low, the Avrami exponents are consistent with three-dimensional growth on contact with sporadic nucleation.
2. The change in m can be interpreted as arising from a change in either the growth geometry or nucleation situation. That is, the change in m for [T_f and T_c low]→ [T_f and T_c high] could arise from either the change spherical → disk geometry or the change sporadic → simultaneous nucleation.
3. Changes in m are consistent with the idea that under some conditions, nuclei from the original solid survive the period in the melt and nucleate the recrystallization.

15. Suppose a polymer spherulite grew by wrapping each chain tightly around and around the surface like a ball of string. What would you expect to see in a polarizing microscope? What would you expect to see if the chains stretched straight out along the radial direction from the center?
16. In understanding the mechanical properties of metals and alloys, crystal defects such as dislocations play a key role. Although polymer crystals certainly exhibit many analogous structural defects, such defects play almost no role in discussions of the mechanical properties of polymer materials. Why is this?
17. For a crystal that grows in an n-dimensional space, with dimensions $R_1, \ldots, R_n$ along its various facets, show that the melting temperature T_m always varies as $1/R_j$, where R_j is the smallest dimension.

References

1. Natta, G. and Corradini, P. *J. Polym. Sci.*, 39, 29 (1959).
2. Lauritzen, J.I. and Hoffmann, J.D., *J. Res. Natl. Bur. Stds.*, A64, 73 (1960); Hoffmann, J.D. and Lauritzen, J.I., *J. Res. Natl. Bur. Stds.*, A65, 297 (1961).
3. Geil, P.H., *Polymer Single Crystals*, Wiley, New York, 1963.
4. Avrami, M. *J. Chem. Phys.*, 7, 1103, (1939); 8, 212, (1940); 9, 177 (1941).
5. Evans, U.R. *Trans. Faraday Soc.*, 41, 365 (1945).

Further Readings

Bassett, D.C., *Principles of Polymer Morphology*, Cambridge University Press, Cambridge, UK, 1981.
Geil, P.H., *Polymer Single Crystals*, Wiley, New York, 1963.
Mandelkern, L., *Crystallization of Polymers*, McGraw-Hill, New York, 1964.
Schultz, J.M., *Polymer Crystallization*, Oxford University Press, New York, 2001.
Tadokoro, H., *Structure of Crystalline Polymers*, Wiley-Interscience, New York, 1979.
Wunderlich, B., *Macromolecular Physics, Vol I: Crystal Structure, Morphology, Defects*, Academic Press, New York, 1973.

Appendix

A.1 Series Expansions

Many common functions (such as $\sin x, \cos x,$ $e^x, \ln(1+x) \cdots$) can be represented by power series, i.e., a sum of terms with increasing powers of the relevant argument x. Such series are useful in allowing the function itself to be replaced by an algebraically simple approximation appropriate in some limit (e.g., $x \to 0$, $x \to 1$, $x \to \infty$). These series approximations can be looked up in many handbooks, but they can also often be derived from the *McLaurin series*. A function $f(x)$ is said to be *analytic* if all derivatives (first, second, third, ...) exist over the relevant range of x. The McLaurin series representation of an analytic function $f(x)$ is given by

$$f(x) = \sum_{i=1}^{\infty} \frac{1}{i!} \left(\frac{d^i f}{dx^i} \right)_{i=0} x^i \tag{A.1.1}$$

where the ith derivative of $f(x)$ is to be evaluated at $i=0$, and where i factorial is $i! = i \times (i-1) \times (i-2) \times \cdots \times 1$. By definition, $0! = 1$.

As an example, consider e^x, and recall that $d(e^x)/dx = e^x$. Therefore from Equation A.1.1

$$e^x = \frac{1}{0!} e^{(0)} x^0 + \frac{1}{1!} e^{(0)} x^1 + \frac{1}{2!} e^{(0)} x^2 + \cdots$$
$$= 1 + x + \frac{1}{2!} x^2 + \frac{1}{3!} x^3 + \cdots \tag{A.1.2}$$

Series expansions for trigonometric functions can also be readily obtained, recalling that $d(\sin x)/dx = \cos x$, $d(\cos x)/dx = -\sin x$, $\sin 0 = 0$, and $\cos 0 = 1$:

$$\sin x = \frac{1}{0!} \sin(0) x^0 + \frac{1}{1!} \cos(0) x^1 - \frac{1}{2!} \sin(0) x^2 + \cdots$$
$$= x - \frac{1}{3!} x^3 + \frac{1}{5!} x^5 + \cdots \tag{A.1.3}$$

$$\cos x = \frac{1}{0!} \cos(0) x^0 - \frac{1}{1!} \sin(0) x^1 - \frac{1}{2!} \cos(0) x^2 + \cdots$$
$$= 1 - \frac{1}{2!} x^2 + \frac{1}{4!} x^4 + \cdots \tag{A.1.4}$$

The natural logarithm of $(1+x)$ where $|x| < 1$ also arises often. Recall that $d(\ln x)/dx = 1/x$, and that $d(x^{-i})/dx = -ix^{-(i+1)}$:

$$\ln(1+x) = \frac{1}{0!} \ln(1+0) x^0 + \frac{1}{1!} \frac{1}{(1+0)} x^1 - \frac{1}{2!} \frac{1}{(1+0)^2} x^2 + \cdots$$
$$= x - \frac{1}{2} x^2 + \frac{1}{3} x^3 + \cdots \tag{A.1.5}$$

Then, when $f(x) = (1+x)^n$ we have

$$(1+x)^n = \frac{1}{0!}(1+0)^n x^0 + \frac{1}{1!}n(1+0)^{n-1}x^1 + \frac{1}{2!}n(n-1)(1+0)^{n-2}x^2 + \cdots$$

$$= 1 + nx + \frac{1}{2!}n(n-1)x^2 + \frac{1}{3!}n(n-1)(n-2)x^3 + \cdots \tag{A.1.6}$$

These results can be readily extended to related functions, for example by replacing x with $-x$, ax, or a complex number z.

A.2 Summation Formulae

These arise in several contexts, especially molecular weight distributions. For example, let x_i be the mole fraction of i-mer in a polycondensation that follows the most probable distribution (Equation 2.4.1),

$$x_i = (1-p)p^{i-1} \tag{A.2.1}$$

where p is the probability that a monomer has reacted. Are we sure that this distribution is normalized, that is does

$$\sum_{i=1}^{\infty} x_i = 1 = (1-p)\sum_{i=1}^{\infty} p^{i-1} ?$$

Comparison with the distribution expression therefore requires that

$$\sum_{i=0}^{\infty} p^i = \frac{1}{1-p} \tag{A.2.2}$$

(Note an important but subtle point: the mole fraction of i-mer only makes sense for $i \geq 1$, but the summation above runs from $i=0$. This is because the sum of p^{i-1} starting from $i=1$ is the same as the sum of p^i starting from $i=0$, and the solution is easier to obtain in the latter case.) To show that this is, in fact, correct, consider a slightly different, *finite* sum:

$$S_1 = \sum_{i=0}^{n} p^i = 1 + p + p^2 + p^3 + \cdots + p^n \tag{A.2.3}$$

If we multiply S_1 by p and subtract it from S_1, we have a term-by-term cancellation:

$$S_1 - pS_1 = (1 + p + p^2 + \cdots + p^n) - (p + p^2 + p^3 + \cdots + p^{n+1})$$

$$= 1 - p^{n+1} \tag{A.2.4}$$

and therefore

$$S_1 = \frac{1 - p^{n+1}}{1-p}$$

$$= \frac{1}{1-p} \quad \text{as } n \to \infty \text{ (and assuming } p < 1) \tag{A.2.5}$$

Of course, for the polymerization case p will always be <1.

To obtain the number average degree of polymerization, we require the related summation (Equation 2.4.4)

$$S_2 = \sum_{i=1}^{\infty} ip^{i-1} \tag{A.2.6}$$

The trick here is to recognize ip^{i-1} as the derivative of p^i with respect to p, and that the derivative with respect to p can be taken outside the summation:

$$S_2 = \sum_{i=1}^{\infty} \frac{\mathrm{d}p^i}{\mathrm{d}p} = \frac{\mathrm{d}}{\mathrm{d}p}\left(\sum_{i=1}^{\infty} p^i\right) = \frac{\mathrm{d}}{\mathrm{d}p}\left(\left(\sum_{i=0}^{\infty} p^i\right) - 1\right)$$

$$= \frac{\mathrm{d}}{\mathrm{d}p}(S_1 - 1) = \frac{\mathrm{d}}{\mathrm{d}p}\left(\frac{1}{1-p}\right) = \frac{1}{(1-p)^2} \tag{A.2.7}$$

Similarly, on the way to obtaining the weight average degree of polymerization we encountered the following sum:

$$S_3 = \sum_{i=1}^{\infty} i^2 p^{i-1} \tag{A.2.8}$$

and this can be evaluated using the same "derivative trick":

$$\sum_{i=1}^{\infty} i^2 p^{i-1} = \frac{\mathrm{d}}{\mathrm{d}p}\left(\sum_{i=1}^{\infty} ip^i\right) = \frac{\mathrm{d}}{\mathrm{d}p}\left(p\sum_{i=1}^{\infty} ip^{i-1}\right)$$

$$= \frac{\mathrm{d}}{\mathrm{d}p}(pS_2) = \frac{\mathrm{d}}{\mathrm{d}p}\left(\frac{p}{(1-p)^2}\right) = \frac{(1-p)^2 + 2(1-p)p}{(1-p)^4}$$

$$= \frac{1+p}{(1-p)^3} \tag{A.2.9}$$

A.3   Transformation to Spherical Coordinates

In situations where we need to integrate something over all space, and there is no preferred direction, a transformation to spherical coordinates can be extremely useful. A prime example occurred in Chapter 6, where we convert the Gaussian distribution function for the end-to-end vector into the distribution function for the end-to-end distance. Another instance arose in Chapter 8, in considering the form factor for an arbitrary particle.

Suppose we wish to find the integral over all space of some function of $f(x,y,z)$:

$$\int_{-\infty}^{\infty}\int_{-\infty}^{\infty}\int_{-\infty}^{\infty} f(x, y, z)\,\mathrm{d}x\,\mathrm{d}y\,\mathrm{d}z$$

There are two steps required to transform this integral into spherical coordinates: transform $f(x,y,z)$ itself, and transform the volume element $\mathrm{d}x\,\mathrm{d}y\,\mathrm{d}z$. These steps are facilitated by the coordinate axes below.

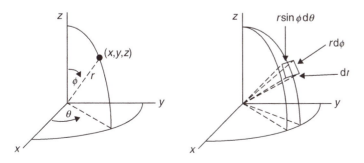

Figure A.1   Illustration of the transformation from Cartesian to spherical coordinates.

An arbitrary point (x, y, z) is represented by a distance from the origin, r, an angle away from the x-axis in the $x - y$ plane, θ, and an angle away from the z-axis, ϕ: (r, θ, ϕ). From the figure it can be seen that

$$x = r \sin \phi \cos \theta, \quad y = r \sin \phi \sin \theta, \quad z = r \cos \phi \tag{A.3.1}$$

These expressions can be substituted directly into $f(x, y, z)$ to obtain $f(r, \theta, \phi)$. Note also that

$$\begin{aligned} x^2 + y^2 + z^2 &= r^2 \left(\sin^2 \phi [\cos^2 \theta + \sin^2 \theta] + \cos^2 \phi \right) \\ &= r^2 \left(\sin^2 \phi + \cos^2 \phi \right) = r^2 \end{aligned} \tag{A.3.2}$$

Thus, in the case where $f(x, y, z)$ can be written as $f(x^2 + y^2 + z^2)$, as with the Gaussian distribution, then $f(r, \theta, \phi)$ becomes simply $f(r)$.

The volume element $dx\,dy\,dz$ is now replaced by a volume element with sides dr, $r\,d\phi$, and $r \sin \phi\,d\theta$, as shown in the figure. For a function such as the Gaussian which is only a function of r, the integral over all space can be reduced to a single integral:

$$\begin{aligned} \int_{-\infty}^{\infty} \int_{-\infty}^{\infty} \int_{-\infty}^{\infty} f(x,y,z)\, dx\, dy\, dz &= \int_0^{\infty} \int_0^{2\pi} \int_0^{\pi} f(r) r^2 \sin \phi\, dr\, d\theta\, d\phi \\ &= \int_0^{\infty} f(r) r^2\, dr \int_0^{2\pi} \int_0^{\pi} \sin \phi\, d\theta\, d\phi = \int_0^{\infty} f(r) r^2\, dr \left(2\pi (-\cos\phi)|_0^{\pi} \right) \\ &= 4\pi \int_0^{\infty} f(r) r^2\, dr \end{aligned} \tag{A.3.3}$$

A.4   Some Integrals of Gaussian Functions

A common class of integrals that arose for example in Chapter 6 are these:

$$I_n = \int_0^{\infty} x^n \exp(-ax^2)\, dx \tag{A.4.1}$$

where n is an integer and a is a positive number. The results are quite simple, and can of course be looked up in any table of integrals, but it is actually instructive to work out the answers. In so doing, we will utilize the transformation to spherical coordinates just described, as well as use the two most common methods for simplifying integrals: change of variable and integration by parts.

The hardest one to do is actually the first, namely I_0. All of the higher powers can be reduced back to this one, as we shall see. We begin by taking I_0^3, and recognizing it can be written as the product of the same integrals along x, y, and z:

$$\begin{aligned} I_0^3 &= \left(\int_0^{\infty} \exp(-ax^2)\, dx \right)^3 = \left(\int_0^{\infty} \exp(-ax^2)\, dx \right) \left(\int_0^{\infty} \exp(-ay^2)\, dy \right) \left(\int_0^{\infty} \exp(-az^2)\, dz \right) \\ &= \int_0^{\infty} \int_0^{\infty} \int_0^{\infty} \exp(-a[x^2 + y^2 + z^2])\, dx\, dy\, dz \\ &= \frac{1}{8} \int_{-\infty}^{\infty} \int_{-\infty}^{\infty} \int_{-\infty}^{\infty} \exp(-a[x^2 + y^2 + z^2])\, dx\, dy\, dz \end{aligned} \tag{A.4.2}$$

The last step was allowed because the argument of I_0 (and I_n for all even values of n) is an *even function*, that is one for which $f(x) = f(-x)$. The integral of an even function from 0 to ∞ is just half the integral from $-\infty$ to ∞. Now the integrals extend over all of space, and we make the transformation to spherical coordinates r, θ, ϕ. This is particularly simple in this case, because the argument of the integral only depends on $r^2 = x^2 + y^2 + z^2$:

$$\frac{1}{8} \int_{-\infty}^{\infty} \int_{-\infty}^{\infty} \int_{-\infty}^{\infty} \exp(-a[x^2 + y^2 + z^2]) \, dx \, dy \, dz = \frac{1}{8} 4\pi \int_{0}^{\infty} r^2 \exp(-ar^2) \, dr = \frac{\pi}{2} I_2 \qquad \text{(A.4.3)}$$

So far, this is not looking promising; we only have a simple relation between I_0^3 and I_2. However, let us attack I_0 directly by integration by parts:

$$\int_{a}^{b} u \, dv = uv \Big|_{a}^{b} - \int_{a}^{b} v \, du \qquad \text{(A.4.4)}$$

where we make the substitutions $u = \exp(-ax^2)$, $v = x$, so $du = -2ax \exp(-ax^2) \, dx$ and $dv = dx$:

$$I_0 = \int_{0}^{\infty} \exp(-ax^2) \, dx = \exp(-ax^2)x \Big|_{0}^{\infty} - \int_{0}^{\infty} (-2a)x^2 \exp(-ax^2) \, dx$$
$$= 0 + 2aI_2 \qquad \text{(A.4.5)}$$

Thus there is another simple relation between I_0 and I_2. Combining these, we see

$$I_0^3 = \frac{\pi}{2} I_2 = \frac{\pi}{4a} I_0 \qquad \text{(A.4.6)}$$

or

$$I_0 = \frac{\sqrt{\pi}}{2\sqrt{a}}; \quad I_2 = \frac{\sqrt{\pi}}{4a\sqrt{a}} \qquad \text{(A.4.7)}$$

Continuing along this simple line, we apply integration by parts to I_2, with $u = \exp(-ax^2)$ again but now $v = x^3/3$ (so $dv = x^2 \, dx$):

$$I_2 = \int_{0}^{\infty} x^2 \exp(-ax^2) \, dx = \exp(-ax^2)\frac{x^3}{3} \Big|_{0}^{\infty} - \int_{0}^{\infty} \left(-\frac{2a}{3}\right) x^4 \exp(-ax^2) \, dx$$
$$= 0 + \frac{2a}{3} I_4 \qquad \text{(A.4.8)}$$

In this way, one can arrive at the general formula for even n:

$$I_n = \frac{(n-1)(n-3)\cdots(1)}{2(2a)^{n/2}} \sqrt{\frac{\pi}{a}}, \text{ even } n \qquad \text{(A.4.9)}$$

The situation for odd n can be approached by a change of variable, e.g., $u = x^2$, $du = 2x \, dx$:

$$I_1 = \int_{0}^{\infty} x \exp(-ax^2) \, dx = \frac{1}{2} \int_{0}^{\infty} \exp(-au) \, du$$
$$= \frac{1}{2}\left(\frac{-1}{a}\right) \exp(-au) \Big|_{0}^{\infty} = \frac{1}{2a} \qquad \text{(A.4.10)}$$

and so forth. The general result for odd n becomes:

$$I_n = \frac{1}{2}((n-1)/2)! \frac{1}{a^{(n+1)/2}}, \text{ odd } n \qquad \text{(A.4.11)}$$

A.5   Complex Numbers

A complex number z can always be written as the sum of two parts, referred to as the *real part, a*, and the *imaginary part, ib*:

$$z = a + ib \tag{A.5.1}$$

where a and b are real numbers and $i = \sqrt{-1}$. The rules for addition and subtraction of two complex numbers are straightforward:

$$z_1 \pm z_2 = (a_1 + ib_1) \pm (a_2 + ib_2) = (a_1 \pm a_2) + i(b_1 \pm b_2) \tag{A.5.2}$$

Multiplication also follows directly, recalling that $i^2 = -1$:

$$\begin{aligned} z_1 z_2 &= (a_1 + ib_1)(a_2 + ib_2) = (a_1 a_2) + i(b_1 a_2) + i(a_1 b_2) - (b_1 b_2) \\ &= (a_1 a_2 - b_1 b_2) + i(a_1 b_2 + b_1 a_2) \end{aligned} \tag{A.5.3}$$

Division is a little more complicated, and is helped by introduction of the *complex conjugate* of a complex number, z^*, which is obtained by replacing i with $-i$:

$$z = a_1 + ib_1, \quad z^* = a_1 - ib_1 \tag{A.5.4}$$

The product of a complex number and its complex conjugate is always purely real:

$$zz^* = a_1 a_1 + b_1 b_1 \tag{A.5.5}$$

To divide by a complex number, it is helpful to multiply numerator and denominator by the complex conjugate of the denominator, thereby restricting complex numbers to the numerator alone:

$$\begin{aligned} \frac{z_1}{z_2} &= \frac{a_1 + ib_1}{a_2 + ib_2} = \frac{a_1 + ib_1}{a_2 + ib_2} \frac{a_2 - ib_2}{a_2 - ib_2} \\ &= \frac{a_1 a_2 + b_1 b_2 + i(b_1 a_2 - a_1 b_2)}{a_2^2 + b_2^2} = \frac{a_1 a_2 + b_1 b_2}{a_2^2 + b_2^2} + i \frac{b_1 a_2 - a_1 b_2}{a_2^2 + b_2^2} \end{aligned} \tag{A.5.6}$$

As a complex number is represented by a pair of numbers (a,b) it can also be mapped uniquely onto a point on a Cartesian coordinate system, with horizontal axis reflecting the real part and the vertical axis representing the imaginary part. Similarly, as the following figure illustrates, a complex number can be viewed as a vector from the origin, with a length given by A and a direction specified by the angle θ:

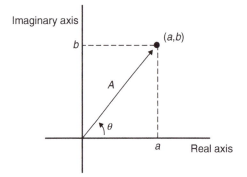

Figure A.2   Illustration of the mapping of complex numbers onto a Cartesian axis system.

$$A = \sqrt{a^2 + b^2}$$

$$\tan\theta = \frac{b}{a} \tag{A.5.7}$$

The standard trigonometric relations apply, such that

$$a = A\cos\theta$$

$$b = A\sin\theta \tag{A.5.8}$$

and therefore any complex number can be written as

$$z = A\cos\theta + iA\sin\theta \tag{A.5.9}$$

Recall the series expansions of e^x, $\cos x$, and $\sin x$ given above, and consider the complex number e^{ix}:

$$\begin{aligned}
e^{ix} &= 1 + ix + \frac{1}{2!}(ix)^2 + \frac{1}{3!}(ix)^3 + \cdots \\
&= 1 - \frac{1}{2!}x^2 + \frac{1}{4!}x^4 \cdots + i\left(x - \frac{1}{3!}x^3 + \frac{1}{5!}x^5 \cdots\right) \\
&= \cos x + i\sin x
\end{aligned} \tag{A.5.10}$$

Thus any complex number can also be written $z = Ae^{i\theta}$. This particular form is extremely useful in various mathematical operations, for example taking powers and roots:

$$z^n = \left(Ae^{i\theta}\right)^n = A^n e^{in\theta} \tag{A.5.11}$$

The product of z and its complex conjugate $z*$ is easily seen to be A^2

$$zz* = \left(Ae^{i\theta}\right)\left(Ae^{-i\theta}\right) = A^2 e^{(i\theta - i\theta)} = A^2 \tag{A.5.12}$$

In this way the product of a complex number and its conjugate is analogous to taking the dot product of a vector with itself; the result is a real number (scalar), equal to the length squared.

Index